Interfacial Aspects of Phase Transformations

NATO ADVANCED STUDY INSTITUTES SERIES

Proceedings of the Advanced Study Institute Programme, which aims at the dissemination of advanced knowledge and the formation of contacts among scientists from different countries

The series is published by an international board of publishers in conjunction with NATO Scientific Affairs Division

A	Life Sciences	Plenum Publishing Corporation
B	Physics	London and New York
C	Mathematical and Physical Sciences	D. Reidel Publishing Company Dordrecht, Boston and London
D	Behavioural and Social Sciences	Martinus Nijhoff Publishers
E	Engineering and Materials Sciences	The Hague, London and Boston
F	Computer and Systems Sciences	Springer Verlag
G	Ecological Sciences	Heidelberg

Series C – Mathematical and Physical Sciences

Volume 87 – Interfacial Aspects of Phase Transformations

Interfacial Aspects of Phase Transformations

Proceedings of the NATO Advanced Study Institute held at Erice, Silicy, August 29 - September 9, 1981

edited by

BOYAN MUTAFTSCHIEV
Center for Crystal Growth - CNRS, Marseille, France

D. Reidel Publishing Company
Dordrecht : Holland / Boston : U.S.A. / London : England

Published in cooperation with NATO Scientific Affairs Division

Library of Congress Cataloging in Publication Data

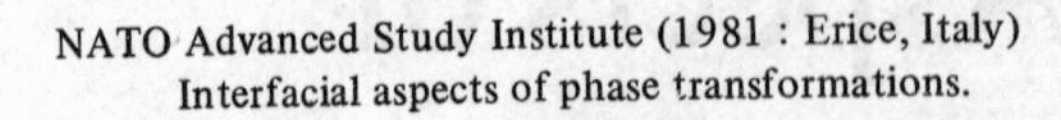

NATO Advanced Study Institute (1981 : Erice, Italy)
Interfacial aspects of phase transformations.

(NATO advanced study institutes series. Series C, Mathematical and physical sciences ; v. 87)
"Published in cooperation with NATO Scientific Affairs Division."
Includes index.
1. Phase transformations (Statistical physics)–Congresses.
2. Crystals–Growth–Congresses. 3. Surfaces (Physics)–Congresses.
I. Mutaftschiev, Boyan, 1932– . II. North Atlantic Treaty Organization. Division of Scientific Affairs. III. Title. IV. Series.
QC175.16.P55N37 1981 530.4 82-9074
ISBN 90-277-1440-1 AACR2

Published by D. Reidel Publishing Company
P.O. Box 17, 3300 AA Dordrecht, Holland

Sold and distributed in the U.S.A. and Canada
by Kluwer Boston Inc.,
190 Old Derby Street, Hingham, MA 02043, U.S.A.

In all other countries, sold and distributed
by Kluwer Academic Publishers Group,
P.O. Box 322, 3300 AH Dordrecht, Holland

D. Reidel Publishing Company is a member of the Kluwer Group

Printed in The Netherlands

TABLE OF CONTENTS

PREFACE

This volume is a collection of the lectures presented at the NATO Advanced Study Institute "Interfacial Aspects of Phase Transformations", held in Erice (Sicily) in 1981.The Institute was the seventh course of the International School of Crystallography, established in the Center of Scientific Culture "Ettore Majorana" in 1974, with the guidance and inspiration of L. Riva di Sanseverino and A. Zichichi. The course organizers, R. Kern and myself, were advised and helped in developing the program, and in the choice of lecturers and participants, by R.F. Sekerka, F. Bedarida, J.L. Katz and J.G. Dash.

Although the scope of the Institute (as reflected in the content of this book) might appear too wide, we believe that it responds to a real necessity. Both Surface Science and Crystal Growth (including problems of nucleation, thin films, recrystallization etc.) have developed to an unusual extent during the last few decades. Each of these two fields have benefited from the knowledge of the other; good examples are the progress in evaporation/condensation thermodynamics and kinetics of organized (sub-)monolayers, the developement of new methods of preparing and characterizing atomically smooth or controlled vicinal (stepped) surfaces, etc.

On closer examination, however, one finds that the interpenetration of surface science and crystal growth is still limited to "one molecular surface layer - ultra high vacuum on the one side and smooth perfect foreign substrate on the other". Besides the fact that most existing experimental surface characterization methods are applicable to this system, it is also far more amenable to various theoretical formulations. Monomolecular condensed layers appear in the very special case of strong adsorption forces decaying rapidly with increasing distance from the surface. For the crystal growth theoretician, this special case encompasses relatively well known thermodynamics (free ledge energies, nucleation barriers)and transport kinetics (surface diffusion). For the surface science theoretician, however, this special case has opened a new world of two-dimensional phases, with order-disorder and commensurate-incommensurate transitions, critical phenomena, etc.

When the forces between a surface layer and a substrate are equal to those of lateral interaction within the layer, and if we no longer consider the substrate as a "dead body" but instead allow its molecules to participate, along with the layer, towards thermodynamic equilibrium with the vapor, the substrate can no longer be considered as "foreign". The two-dimensional physics, as pursued by theoreticians, thus becomes "less two-dimensional". One enters the realm of thick adsorption layers, the structure and properties of wich continuously change in the direction normal to the surface.

B. Mutaftschiev (ed.), Interfacial Aspects of Phase Transformations, vii–x.

Crystal growers are familiar with these types of self-adsorbed layer, although present theoretical models still do not correctly describe their properties; one calls them "rough crystal faces" and the critical phenomena they undergo, "roughening transition". The interest of surface scientists in surface roughening is very recent.

If the self-adsorbed layer on a crystal surface undergoes an order-disorder transformation below the two-dimensional critical temperature, one speaks about "surface melting". In this case the interface between the molten "layer" and the solid "substrate" can be quite sharp. Computer simulation studies favor the existence of surface melting at temperatures below the bulk melting point. However, to our knowledge, there is no convincing experimental evidence for it. For this reason, the concept of surface melting is rarely evoked in crystal growth studies (mainly as a possibility of interpreting experimental results). Surface physicists have yet to show interest in the problem.

At the melting temperature, the thickness of the molten layer on a crystal face, by definition, tends to infinity. The crystal is then in equilibrium with its bulk melt. When approaching the melting point from below, however, three types of behavior might be expected: (i) the thickness of the molten layer increases monotonically to infinity, or (ii) below the melting point the crystal surface is covered by a limited number of molten layers, or (iii) there are no molten layers up to the melting point. Case (i) might occur if a smooth transition between crystal and melt is structurally possible; in case (ii), the crystal face might impose a particular intermediate structure to several monomolecular liquid-like layers in its vicinity. If this structure cannot fit with the bulk liquid structure, a division between them is inevitable; in case (iii), this division is at the crystal face itself because of structural incompatibility and loose bonding between solid and melt. Case (i) can be considered as that of a kinked crystal-melt interface while case (iii) as that of of a smooth interface. To which type belongs case (ii) is not yet clear, particulary since a roughening transition has so far been observed in the crystal-melt system only for the case of Helium.

The depicted three possibilities of pre-melting are deduced in analogy with experimentally known cases of thickening of two-dimensional condensed layers, adsorbed from the vacuum on foreign substrate, when the saturation pressure is approached. In both cases, the problem, of whether or not equilibrium layers can exist at undersaturation (undercooling), and how their properties evolue when three-dimensional equilibrium is approached, depends on the structural and energetic relations of the substrate-layer interface. Conversely, it seems that the key to understanding the structure and the equilibrium behavior of the crystal-melt interface is to study properties of disordered self-adsorbed layers with increasing thickness. Adding a second component to such layers results in a system analogous to the interface between a crystal and its saturated solution.

At the present time, surface physics is still far from playing the major role in phase transformation problems, as imagined above. On the other hand, when faced with the difficulties of understanding the properties of more complex interfaces, there is a tendency in some crystal growth studies to reduce the role of the interface to a simple parameter in the transport boundary conditions.

To confront such problems and bring together researchers involved in their different aspects was the principal goal of this Advanced Study Institute. The reader will find in the first three chapters (Bauer, Gaspard, Mutaftschiev) structural, energetic and thermodynamic descriptions of clean crystal surfaces, followed by similar treatment of two-dimensional phases on foreign substrates (Gaspard, Domany). The consideration of the structure and behavior of more complicated interfaces, contained in three more chapters(Bonissent, Franks, Gleiter) is clearly less rigorous and demonstrates the extent of the difficulties encountered. The next part is devoted to problems of interfacial kinetics. It begins with a general overview of elementary processes (Cole-Toigo) and covers their principal applications in phase transformations : homogeneous nucleation (Katz), substrate nucleation and coalescence of small particles (Kern), and crystal growth (Rosenberger).

At this point the reader is assumed to have acquired some basic knowledge in surface physics and phase transformations. The confrontation of ideas from both branches is induced through examples of real systems in the next 12 chapters. The cases treated are : physisorption (Webb-Bruch), chemisorption (Bauer, Oudar), and crystal growth in the several practically important systems, vapor (Cadoret), melt (Sekerka), aqueous solutions (Simon), biological systems (Franks), non-aqueous solutions (Boistelle), electrocrystallization (Budevski) and recrystallization (Gleiter), together with two special topics, the action of impurities (Boistelle) and dissolution (Simon).

Considerations of the relation between interfacial kinetics and morphology would have been incomplete without an overview of some experimental methods for studying morphology. The last two chapters (Bedarida, Bethge) attempt to satisfy this need.

Melting together surface physics and crystal growth problems also resulted in the great diversity of lecturers and participants of the school. A wide range of disciplines, from solid state physics to biology, were represented and discussion often continued well into the night. This productive atmosphere was further enhanced by the pleasant manifestations of the italian *savoir vivre*, led by the master hand of Lodovico Riva di Sanseverino.

The merit of publishing this book falls to all lecturers who focused their efforts towards presenting their specialties in a manner accessible to a pluridisciplinary audience, thus sometimes sacrificing personal tastes.

In editing the manuscripts I was greatly helped by J. Abernathey, whose critical sense and fine knowledge of the language prevented many of the chapters from being in a "kind of English". This is the place to express to him my gratitude. I am also indebted to Mrs C. Sekerka who helped us in Erice and to numerous friends and colleagues from the Centre des Mécanismes de la Croissance Cristalline, Marseille, for their support at different stages of the School organization.

It would be remiss not to mention that both the Advanced Study Institute and the present book would have never gotten off the ground without the help of the Scientific Division of the North Atlantic Treaty Organization which was crucial in allowing all lecturers and many students to attend. The Italian Ministry of Foreign Affairs provided four travel grants and the European Physical Society, the Istituto Italo-Latino-Americano and the Istituto Italo-Africano provided nine scholarships. The Consiglio Nazionale delle Ricerche (Italy) generously helped the publication of this book. We are also grateful to the Sicilian Regional Governement and to IBM-Italy for financial assistance.

Finally, there are things this book cannot reflect, but which will remain engraved in the memories and hearts of all participants. These are the sober beauty of Erice, the majestic landscape of its coast, and Sicily itself, still full of life after three millenaries of civilization.

B. Mutaftschiev.

THE STRUCTURE OF CLEAN SURFACES

E. BAUER

Physikalisches Institut
Technische Universität Clausthal
3392 Clausthal-Zellerfeld
Federal Republic of Germany

I. EXPERIMENTAL METHODS

I.1. Diffraction Methods

Diffraction has been and still is the major means for the determination of the structure of a surface. Waves which can be used for this purpose must fulfill three requirements : (i) their wavelength must be smaller than the distance d between neighboring atoms; (ii) they must be sufficiently coherent and (iii) they must strongly emphasize the surface relative to the bulk. The first two conditions must also be fulfilled for the structure analysis of the bulk. The condition $\lambda_{max} \leqslant d \simeq 2.5 - 5Å$ means for X-rays energies from 4keV on upwards, for particles energies from $E = (h/\lambda)^2/2m$ on upwards. For electrons $E \gtrsim 15eV$ (minus mean inner potential), for neutrons and H atoms $E \gtrsim 10meV$, for He atoms $E > 2meV$. These conditions can be easily fulfilled as well as the coherency condition, which requires a sufficiently small monochromatic source.

The third condition is a more difficult one. It can be realized in the following cases, depending upon the type of wave : (i) strong elastic (atoms) or inelastic scattering (slow electrons); (ii) grazing incidence (X-rays, fast electrons); (iii) wave source located on the surface or in the surface region (electrons). Neutrons interact so weakly with matter that discrimination between surface and bulk is very difficult, unless the surface consists of a different species (suitable adsorption layer on suitable substrate). Therefore, they will not be discussed further, irrespective of the merits they have achieved in physisorption studies. X-ray diffraction at grazing incidence has recently been applied to the

B. Mutaftschiev (ed.), Interfacial Aspects of Phase Transformations, 1–32.

study of clean surfaces with the hope that a simple structure analysis due to the weak interaction between wave and scatterer should be possible.[1] It is still too early to estimate the potential of the method. Thus only electrons, both slow (10-1000eV) and fast (10-100keV), and atoms will be discussed below.

The surface must also fulfill certain requirements : (i) it may not be irreversibly modified by the wave during the measurement time, e.g. due to dissociation or contamination, (ii) it must have sufficient long range order (if an external wave source is used) or at least sufficient short range order (if an internal source is used). External sources, such as the electron gun or the skimmer in an atomic beam source produce waves which may be considered to be plane waves. Internal sources, which are practical only for electrons, are produced by ionization of an inner shell of an atom in the surface region; in this case a spherical photoelectron wave or Auger electron wave is emitted by the ionized atom and diffracted by the surrounding atoms.

Finally the diffraction pattern can be observed in two modes: (i) in the Fraunhofer (far field) mode the observation point is at infinity or focussed to a finite distance from the surface by a lens; (ii) in the Fresnel (near field) mode the observation point is close to the surface or to the image of the surface. The former is standard diffraction, the latter defocussed imaging.

Irrespective of the observation mode and of the type of source, the amplitude of the wave at any point $\vec{r}$ is given by the integral form of the Schrödinger equation

$$\psi(\vec{r}) = \psi_o(\vec{r}) + \int_{V_c} G(\vec{r},\vec{r}')V(\vec{r}')\psi(\vec{r}')d\vec{r}'. \qquad (1)$$

The integration is over the coherently scattering region and the quantities in the equation have the following significance : $V(\vec{r}')$ is the effective interaction potential between the wave and the scatterer, $V(\vec{r}')\psi(\vec{r}')$ is the amplitude of the wave scattered from a volume element $d\vec{r}'$ about $\vec{r}'$ upon incidence of the wave $\psi(\vec{r}')$, and $G(\vec{r},\vec{r}')$ is the outgoing Green's function which describes the propagation of the scattered wave from $\vec{r}'$ to $\vec{r}$. $\psi_o(\vec{r})$ is the amplitude of the wave in the absence of the scatterer (when an external source is present) (see Fig. 1). Although the physical significance of this equation is simple, its solution in the general case in which each wave in the crystal produces secondary waves which are again scattered, i.e. when multiple scattering occurs, is quite difficult. A simple solution can be found only in two limiting cases : (i) when the incident wave interacts so weakly with the scatterer that the secondary waves in the scatterer are weak compared to the incident wave; (ii) when the incident wave interacts so strongly with the scatterer that only the topmost layer diffracts. The first case occurs in x-ray diffraction, the second case is typical of atomic beam diffraction. The simplifications which result in these two cases can be seen easily.

Fig. 1 : The physical meaning of eq. (1)

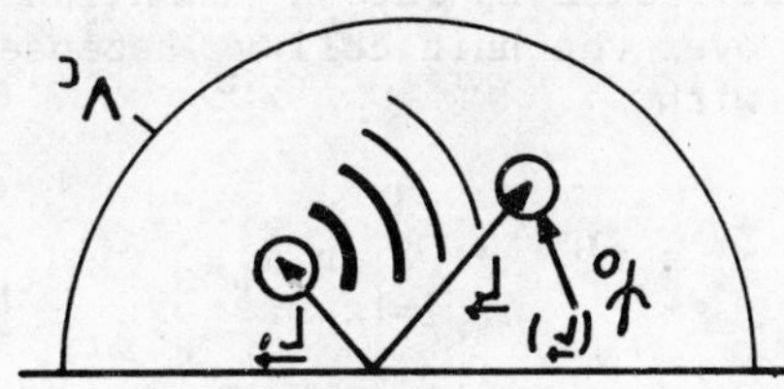

In the first case $\psi(\vec{r}')$ in the integral may be replaced by $\psi_o(\vec{r})$, $V(\vec{r}')$ by the potential of the undisturbed crystal and $G(\vec{r},\vec{r}')$ may be approximated by the spherical wave free space Green's function, i.e. a spherical wave originating at $\vec{r}'$,

$$G(\vec{r},\vec{r}') = -\frac{1}{4\pi}\ \frac{\exp(ik|\vec{r}-\vec{r}'|)}{|\vec{r}-\vec{r}'|},$$

because rescattering of the wave scattered in $\vec{r}'$ may be neglected. In Fraunhofer (far field) diffraction the observation point $\vec{r}$ is far away from all scattering points $\vec{r}'$ ($|\vec{r}| >> |\vec{r}'|$) in which case $G(\vec{r},\vec{r}')$ may be approximated by

$$G(\vec{r},\vec{r}') \simeq -\frac{1}{4\pi}\ \frac{e^{ikr}}{r}\ e^{-i\vec{k}.\vec{r}'},$$

where $\vec{k}$ is the wave vector of the scattered wave. If the incident wave is a plane wave with wave vector $\vec{k}_o$,

$$\psi_o(\vec{r}') = e^{i\vec{k}_o.\vec{r}}.$$

The total scattered wave, i.e. the integral in eq. (1) may be then written as

$$\psi_s(\vec{r}) = -\frac{1}{4\pi}\ \frac{e^{ikr}}{r}\int_{V_C} e^{-i\vec{K}.\vec{r}'}\ V(\vec{r}')d\vec{r}' = -\frac{1}{4\pi}\ \frac{e^{ikr}}{r}\ F(\vec{K}) \quad (2)$$

with $\vec{K} = \vec{k}-\vec{k}_o$. Thus the scattered wave is essentially the Fourier transform of the scattering potential. This is the basis of conventional x-ray diffraction but it is not true for electron diffraction.

If the scatterer is periodic with unit cell vectors $\vec{a}_i$ and has $\prod_{i=1}^{3} N_i$ coherently scattering unit cells, the integral over v_c may be broken up into a summation over all unit cells of the integral over the unit cell v_u because all unit cells are equivalent. Then with

$$\vec{r}' = \vec{r}_\circ' + \sum_{i=1}^{3} n_i \vec{a}_i ,$$

and because of $V(\vec{r}') = V(\vec{r}_\circ')$ (periodicity!), the amplitude $F(\vec{K})$ of the spherical wave from the scatterer becomes

$$F(\vec{K}) = \int_{v_u} e^{-i\vec{K}\cdot\vec{r}_\circ'} V(\vec{r}_\circ') d\vec{r}_\circ' \cdot \prod_{i=1}^{3} \sum_{n_i=0}^{N_i-1} e^{-in_i\vec{K}\cdot\vec{a}_i} = A(\vec{K})\cdot B(\vec{K}). \tag{3}$$

The integral over the unit cell, i.e. the structure amplitude A, describes the interference of the waves by the atoms within the unit cell. The lattice amplitude B describes the interference of the waves scattered from the different unit cells and depends upon their size and shape as well as on the size and shape of the coherently scattering region. If this is large enough $|B|^2$ is nonzero essentially only when

$$\vec{K} = \vec{h} = \sum_{i=1}^{3} h_i \vec{b}_i ,$$

i.e. when $\vec{K}$ is a reciprocal lattice vector of the crystal. The spherical scattered wave may then be described by a sum of plane waves

$$\psi_s(\vec{r}) = \sum_{\vec{h}} A_{\vec{h}}\, e^{i(\vec{k}_\circ + \vec{k})\cdot\vec{r}} . \tag{4}$$

The structure factors

$$|A_{\vec{h}}| = |\int_{v_u} e^{-i\vec{h}\cdot\vec{r}'} V(\vec{r}')d\vec{r}'|^2 \tag{5}$$

determine the relative intensities of the various diffracted beams.

In atomic beam diffraction, for typical beam energies (10-100meV), the atoms are already reflected at the very edge of the repulsive potential. This occurs typically at about 3.5Å and an electron density of 10^{-4} - 10^{-3} electrons/Å . Thus the atomic beam does not probe the location of the atom cores but rather the profile of the electron distribution parallel (x,y) to the surface. Because of the steep rise of the repulsive part of the potential V(z) normal to the surface, this profile may be approximated by a two-dimensionally corrugated hard wall

$$V(z) = \{{0 \atop \infty}\} \text{ for } z \{{> \atop \leq}\} \xi(\vec{R}) = \xi(x,y) \qquad (6)$$

with the two-dimensional unit mesh dimensions $\vec{a}_1$, $\vec{a}_2$.

Fraunhofer diffraction from such a two-dimensional lattice can be described in complete analogy with the three-dimensional case above, with the only exception that now the lattice periodicity determines not $\vec{K} = \vec{k} - \vec{k}_\circ$ but rather

$$\vec{K}_{||} = \vec{k}_{||} - \vec{k}_{\circ\,||} = \vec{h} = \sum_{i=1}^{2} h_i \vec{b}_i \ ,$$

where $\vec{h}$ is a lattice vector of the two-dimensional reciprocal lattice. $k_{\vec{h}\perp}$ is obtained from the condition

$$k_{\vec{h}}^2 = k_{h||}^2 + k_{h\perp}^2 = k_\circ^2$$

for elastic scattering. Thus the directions of the scattered waves are completely determined. Because all unit meshes are equivalent, the integral in eq. (1) again may be broken up into a summation over all unit meshes as before (Eq. (3)), with one important difference which is obvious from Fig. 2.

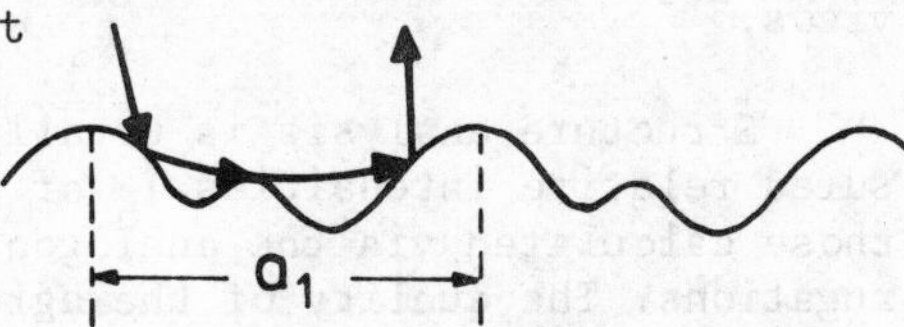

Fig. 2 : Profile of a one-dimensional corrugation function and multiple scattering within a unit mesh.

It shows the profile of a one-dimensional corrugation and some waves indicated by their propagation directions. It is apparent that multiple scattering can occur and may not be neglected. Thus the replacement of $\psi(\vec{r}')$ by $\psi_\circ(\vec{r}')$ which led to eq. (2) is no longer permissible and a different approach is needed. The basically planar nature of $V(\vec{r}')$ suggests furthermore to use the plane wave form of the outgoing free space Green's function

$$G(\vec{r},\vec{r}') = \sum_{\vec{h}} \frac{1}{k_{\vec{h}\perp}} \exp\{i[(\vec{K}_{||} + \vec{h})\cdot(\vec{R} - \vec{R}') + k_{\vec{h}\perp} |z - z'|]\}, \tag{7}$$

For the particular form of the potential (eq. 6) the scattered wave in eq. (1) originates only at the corrugated wall and can, therefore, be written in the form

$$V(\vec{r}')\psi(\vec{r}') = f(\vec{R}')\delta(z' - \xi(\vec{R}')), \tag{8}$$

where $f(\vec{R}')$ is the amplitude of the wave scattered from the surface element centered at $\vec{R}'$ ("source function"). Inserting (7) and (8) into (1) gives, in complete analogy with the total scattered wave solution above,

$$\psi_s(\vec{r}) = \sum_{\vec{h}} A_{\vec{h}} \exp\{i[(\vec{k}_{\circ ||} + \vec{h})\cdot\vec{R} + k_{\vec{h}\perp} z]\} \tag{9}$$

with the structure factor

$$|A_{\vec{h}}|^2 = \left|\frac{1}{k_{\vec{h}z}} \int \exp\{-i[(\vec{k}_{\circ ||} + \vec{h})\cdot\vec{R}' + k_{\vec{h}\perp}\xi(\vec{R}')]\} f(\vec{R}')d\vec{R}'\right|. \tag{10}$$

The relative intensities $|A_{\vec{h}}|^2$ of the diffracted beams can be calculated by determining the source function with the condition that the total wave field must be zero for $z \leqslant \xi(\vec{R}) : \psi(\vec{R}, z = \xi(\vec{R})) \equiv 0$. The details of the theoretical treatment and of the computational aspects can be found in recent reviews.[2,3]

Structure analysis is usually performed by comparing the measured relative intensities $I_{\vec{h}}$ of several diffracted beams $\vec{h}$ with those calculated via the analogon to eq.(5) for various model corrugations. The quality of the agreement is judged by reliability factors such as

$$R = \frac{1}{N}\left[\sum_{\vec{h}} (I^{calc}_{\vec{h}} - I^{exp}_{\vec{h}})^2\right]^{1/2} \text{(for N } \vec{h} \text{ values)}.$$

In electron diffraction, the situation is more complicated. In low energy (LEED) and refection high energy (RHEED) diffraction, several atomic layers contribute to the diffraction pattern. Their number is determined by the mean free path for inelastic scattering λ_{ee} and by the angles of incidence and observation. The electrons used in LEED (≃ 30-600eV) have λ_{ee} values between 3Å and 10 Å, those used in RHEED (some 10 to some 100keV) have λ_{ee}'s of the order 100 Å. Thus surface sensitivity in LEED is given even at normal incidence, while in RHEED grazing incidence is necessary. In transmission high energy electron diffraction (THEED), surface sensitivity is achieved indirectly via the strain fields which reconstructed surface layers produce in the bulk. Not only is inelastic scattering strong, which reduces the amplitude of the coherent wave (electrons with incidence energy), but elastic scattering is also, so that strong multiple scattering occurs and the approximation $\psi(r') \simeq \psi_o(r')$ in eq. (1) is no longer valid. Furthermore, the electron modifies the potential of the scatterer by exchange and correlation interactions with the electrons in the scatterer which can be taken into account by an (energy dependent) exchange-correlation contribution to the potential. Many approaches have been taken during the past 15 years to solve this diffraction problem, in particular for LEED. Most of them have in common that the crystal is divided into layers parallel to the surface and that the calculation is divided up into intralayer and interlayer scattering. The way in which this is done and the computational procedures and computer programs are described in detail in several books.[4-6] It is important to note that in spite of strong multiple scattering , all unit meshes parallel to the surface are equivalent. Therefore, it is possible to write $F(\vec{K})$ in eq. (3) in the form

$F(\vec{k},\vec{k}_o) = A(\vec{k},\vec{k}_o)\cdot B(\vec{K})$. The maxima of $|B|^2$ which occur for

$$\vec{K}_{||} = \vec{h} = \sum_{i=1}^{2} h_i \vec{b}_i$$

together with the condition for elastic scattering $k_h^2 = k_o^2$ determine the direction of the diffracted waves with intensity $|A|^2$ which must be calculated as described before. Partial structure analysis, i.e. determination of the unit mesh dimensions and the size and shape of the coherently scattering region, is therefore possible from the geometry of the diffraction pattern alone. This is of great importance for the analysis of surface imperfections which is one of the major present and future applications of LEED (see reviews 6a-d).

Complete structure analysis, i.e. determination of the atomic positions within the unit mesh, consists of the comparison of the measured intensities of as many as possible diffracted beams as a function of electron energy (I(V) curves) with intensities calculated for several structural models and selecting the model which gives the best agreement with experiment. Agreement is judged by reliability factors R for which several definitions (R_{ZJ} [7,8], R_{ANvH} [9], R_P [10], etc.) are in use and whose sensitivity to structural parameters has been checked recently.[11] The most frequently used R factor is the Zanazzi-Jona R factor R_{ZJ} which takes not only the agreement of calculated and measured intensities into account but also that of their first and second derivatives. A value $R_{ZJ} < 0.25$ is considered to indicate a reliable structure.

The calculations have not only structural parameters as inputs (up to 8 atoms in the surface layer and 4 atoms in the bulk per unit mesh in the most flexible program) but also nonstructural parameters such as mean inner potential V_o, absorption potential V_{im} and surface and volume Debye temperatures θ_s, θ_b. All of them are basically energy dependent but this dependence is generally neglected. Another basically energy dependent input is the atomic scattering potential because of the energy dependence of the exchange-correlation potential. Experience has shown however that neglect of this dependence is not too critical.[12] Calculated I(V) curves have recently become reliable enough so that the differences between curves calculated by different authors with different approximations have in some cases become smaller than those between curves measured by different authors. The experimental aspects of the LEED structure analysis have been reviewed recently.[13,14] Therefore, it should suffice to mention that surface imperfections and impurities seem to have little influence on the reliability of the intensity data in the case of nonreconstructing metal surfaces[15] but are important on reconstructing surfaces, where they may suppress reconstruction. A more critical source of discrepancies are probably differences in the precise direction of incidence onto the crystal. The quality of agreement between different experiments can be judged by a comparison of Si(100) data.[16]

All diffraction techniques discussed so far make use of an external source which produces a plane wave incident on the surface. In recent years diffraction techniques with internal sources have rapidly gained popularity, mainly due to the availability of tuneable high intensity x-ray sources (synchrotron radiation). The internal source is "turned on" by bombarding the surface with soft x-rays or electrons up to several keV energy which causes the emission of spherical photoelectron or Auger electron waves from the atoms ionized by the incident radiation. Two detection modes of the diffraction of the spherical wave by the atoms in its environment are in use : angle-integrated measurements ("total yield spectroscopies"; "extended x-ray absorption fine structure" (EXAFS) and "extended appearance potential fine structure" (EAPFS) analysis) and angle resolved measurements, both at fixed angle and variable x-ray energy ("normal photoelectron diffraction", NPD) and at variable azimuthal angle and fixed x-ray energy ("azimuthal phootelectron diffraction", APD). Both detection modes have been reviewed recently[17,18] so that only the basic principles and the limitations of the methods will be discussed.

In EXAFS the energy of the photoelectron varies with the energy of the incident x-ray energy. The photoelectron wave backscattered by the atoms surrounding the source will, therefore, as a function of energy, interfere constructively and destructively at the source. Constructive interference means strong photoelectron creation and therefore strong x-ray absorption and as secondary processes strong x-ray fluorescence and/or Auger electron emission. Monitoring the photoelectron, fluorescence or Auger electron yield, or the yield of energy selected secondary electrons created by the Auger electrons,[19] as a function of x-ray energy therefore gives information on the diffraction of the photoelectron wave by the surrounding of the source, from which via eq. (1), with $\psi_o(r)$ missing, the radial distribution of the atoms around the source atom may be obtained. A multiple scattering theory is available for this analysis[20,21] but most work to date is based on a single scattering analysis. This is done by first subtracting from the total signal $\mu(E)$ (Fig. 3a) the smoothly varying background so that only the oscillatory part $\mu_{osc}(E)$ remains (Fig. 3b), then converting the E axis (with absorption edge as zero) via $E \simeq k^2$ into a momentum axis (Fig. 3b top) and finally Fourier transforming this curve :

$$F(r) = \frac{1}{2\pi}\int\mu_{osc}(k)\exp(-2ikr)dk.$$

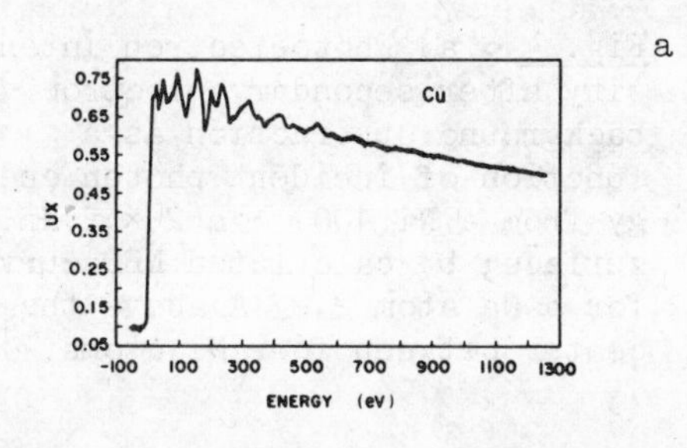

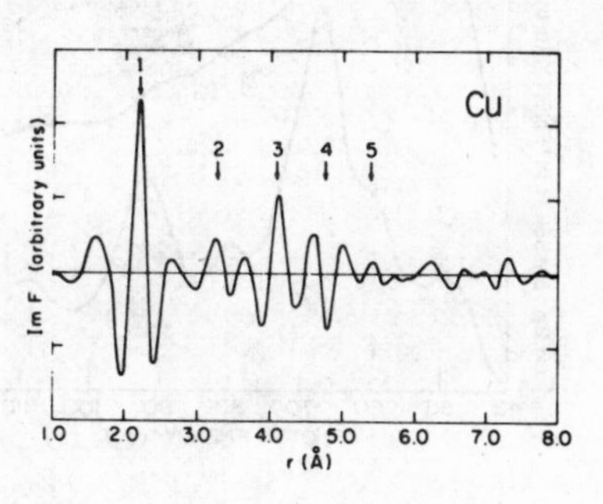

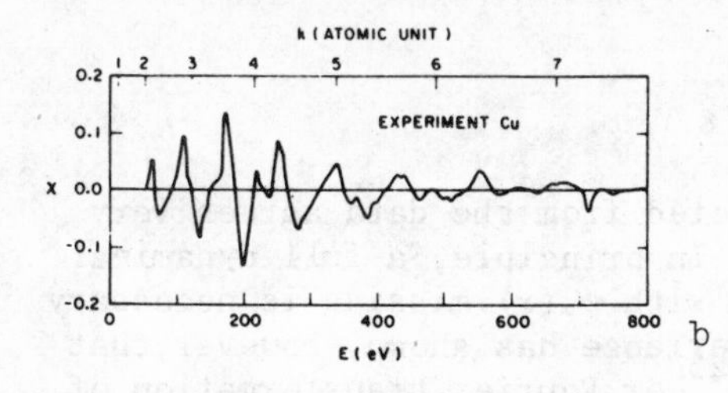

Fig.3: EXAFS proceduce a)measured spectrum; b)spectrum after subtraction of smooth background; c)Fourier transform of b).[21]

The maxima of F(r) (Fig. 3c) determine the distance to the surrounding neighbors. Obviously, $\mu_{osc}(k)$ must be measured over a sufficiently wide k-range to obtain a meaningful Fourier transform. This was not the case in the first and apparently only attempt to study a clean surface (Al(111), (100)) with EXAFS[19] which produced data in disagreement with reliable LEED results.[22]

EAPFS works in principle in a very similar manner. The electron wave from the atomic source is here created by electron bombardment. The detection of the oscillations of its excitation probability due to interference with the waves backscattered from the surrounding atoms is carried out via the Auger electron yield or via the elastic backscattering yield. The experiments performed to date (V, Ti and Fe[17]) have not produced any surface structure information competitive with LEED.

In NPD the energy of the incident synchrotron radiation is varied over such an energy range that the energy of the emitted core photoelectrons varies over the range which is generally used in LEED, i.e. between about 30eV and 200eV. For example, ionization of the Na 2p level requires 47 eV so that the incident radiation has to be varied from about 80eV on upwards. The intensity of the 2p photoelectrons emitted and scattered normal to the surface is measured with an energy analyzer whose pass energy is synchronized with the x-ray energy. The resulting intensity modulation is shown in Fig. 4.[24]

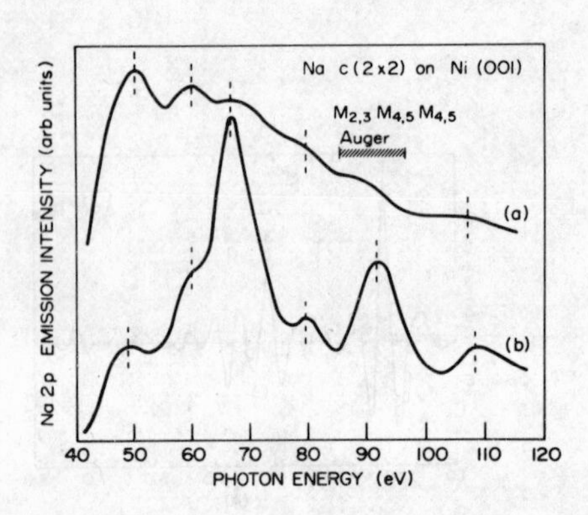

Fig. 4 : a) Photoelectron intensity after secondary electron background subtraction as a function of incident photon energy from a Ni(100) - c(2 × 2)Na surface; b) calculated NPD curve for a Na atom 2.23 Å above the center between four Ni atoms.[24]

The location of the Na atom extracted from the data agrees very well with that from LEED studies. In principle, a full dynamical LEED calculation based on eq. (1) with $\psi_o(r)$ missing is necessary for the analysis of the data. Experience has shown, however, that simplified dynamical calculations[25] or Fourier transformation of the oscillations after proper background correction analogous to EXAFS analysis[26] give surprisingly good agreement with full dynamical calculations.

In APD the surface is irradiated with monochromatic x-ray radiation, e.g. Al K_α with 1487 eV energy, the crystal is rotated about its normal and the photoelectrons from the atoms of interest are measured as a function of azimuth at small grazing angles of emission, typically 10°. This small angle is necessary for sufficient scattering of the photoelectron from a surface atom by its neighboring atoms. Fig. 5 illustrates this for oxygen 1s photoemission on Cu(100).[27] Because of the high photoelectron energy (950eV for oxygen 1s electrons (binding energy ≈ 530eV)) a single scattering calculation is sufficient to analyze the data[28,29] which agree in general well with LEED data.

Finally, angle resolved Auger electron spectroscopy (ARAES) still has to be mentioned. Here, the angular distribution of the Auger electrons, usually produced by electron bombardment of the atom of interest is measured. It is determined mainly by the diffraction of the Auger electron wave by the surrounding atoms and requires in principle a full dynamic calculation for a reliable structure analysis. Recently[30] it was shown, however, that considerably simplified calculations give about the same results as a full dynamical treatment. Nevertheless the analysis is more complicated because of the two-electron nature of the process as compared to the one-electron process basic to the methods discussed before.

All internal source diffraction techniques have in common that they identify the atom whose location is to be determined via the energy level involved in the emission process. This is a great advantage over external source diffraction techniques. Furthermore, no long range order is required because the amplitude of the spherical wave decreases rapidly with distance. This statement may not be misunderstood : it is strictly valid only in the angle integrated techniques, in the angle resolved techniques it is still necessary that the same local configuration is repeated over the whole area contributing to the signal, without rotations. Only random lateral displacements of these configurations are allowed. The advantage of atomic identification is lost in the study of the structure of clean surfaces. Here only grazing incidence or the monitoring of low energy electrons can emphasize the surface contribution to the signal but an analysis on the same confidence level as in LEED is more complicated.

Surface structure analysis by near field diffraction (defocussed imaging) or focussed imaging and by THEED is still in its infancy. Although it is possible to determine monolayer heights, i.e. the distance of the topmost layer from the second layer, via the monolayer contrast[31] in transmission electron microscopy (TEM) the measurement is presently too inaccurate for this purpose. In conjunction with THEED it is, however, useful now for the (partial) structure analysis of surfaces with superstructures, such as the Au(111) surface, which has a surface layer contracted by 4.2% in the [110] direction with[32] (without[33]) misfit dislocations. In reflection electron microscopy (REM), atomic step contrast is much stronger[34] but the analysis is not sufficiently developed to extract reliable step heights. Recently, RHEED has been developed to a level at which useful contributions to complete structure analysis can be expected; an example is the (7 × 7) superstructure of the clean Si(111) surface[35].

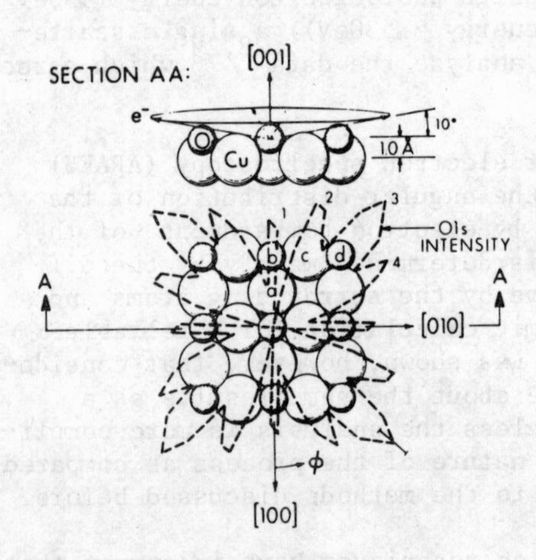

Fig. 5 : APD of 1s photoelectrons from oxygen atoms in the c(2×2)O structure on a Cu(100) surface. The O 1s intensity (averaged over all 4 quadrants) obtained by rotating the crystal about the [001] axis (top) is shown as a polar plot (bottom).[27]

I.2. Scattering Methods

These methods are based on the elastic scattering of ions with energies of the order $10^2 - 10^6$eV by the ion cores of the atoms of the scatterer. The surface is bombarded with an ion beam with well defined energy E_o and direction of incidence (θ_o, ϕ_o) and the quantity measured is the energy distribution of the scattered ions at well-defined scattering angles $I(E,\theta,\phi)$. This quantity gives not only information on the structure of the surface but also on the mass of the scattering atom, due to the fact that the energy transfer from the ion with mass m_o to an ion core with mass m upon scattering by an angle Θ is given to a good approximation by classical momentum ($\vec{p} = m\vec{v}$) and energy $E = mv^2/2$ conservation which leads to

$$\frac{E}{E_o} = \left\{\frac{\sqrt{1 - \left(\frac{m_o}{m}\right)^2 \sin^2\Theta} \pm \frac{m_o}{m} \cos\Theta}{1 + \frac{m_o}{m}}\right\}^2. \qquad (11)$$

If double or multiple scattering occurs this expression can be applied to the individual collisions which allows determination of the type of neighboring atoms. Typical ion scattering spectra and the calculated origin[36] of the peaks are shown in Fig. 6 (600eV K^+ ions from W(110)).

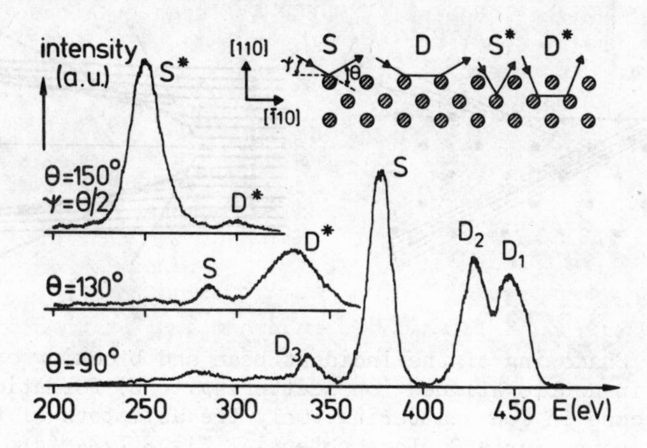

Fig. 6 : LEIS spectra of 600eV K^+ ions from a W(110) surface in the [1̄10] azimuth for specular scattering into various scattering angles. The scattering processes are indicated in the insert.[36]

The peaks D , D^* arise from linear, D_2 and D_3 from zigzag double scattering.

Structural information via double scattering can be obtained usually only in the low energy range (10^2 - 10^4eV, LEIS) and with sufficiently heavy ions (e.g. Ne^+, Na^+, Ar^+, K^+; high scattering cross section!) with low neutralization probability (unless the neutralized particles can be detected). In medium energy (10^4 - 10^5eV) and high energy (10^5 - 10^6eV) ion scattering (MEIS, HEIS) and with light ions (e.g. H^+, He^+) the process which allows structure analysis is the shadowing of the incident beam and the blocking of the scattered beam. These two processes are indicated in Fig. 7a[37] : each atom seen by the incident beam produces a shadow cone in which atoms at lower levels are hidden,each atom above an unhidden atom produces a blocking cone for the scattered ions. If the incident beam is aligned ("single alignment") with a densely packed direction, only few atoms are seen by the incident beam, and if the detector is aligned too (double alignment) with a densely packed direction, a minimum in the scattered intensity occurs ("blocking dip"). Deviations from the expected number of atoms ("atoms per row") indicate atomic displacements in the topmost layers as do angular shifts of the blocking dips. The directions of the displacements may be obtained by varying incidence and scattering directions, the depth distribution by variing the energy which changes the radius R of the shadow cone. Fig. 7b[38] indicates how the shadow cone is produced by the trajectories of particles passing the scattering center at various distances.

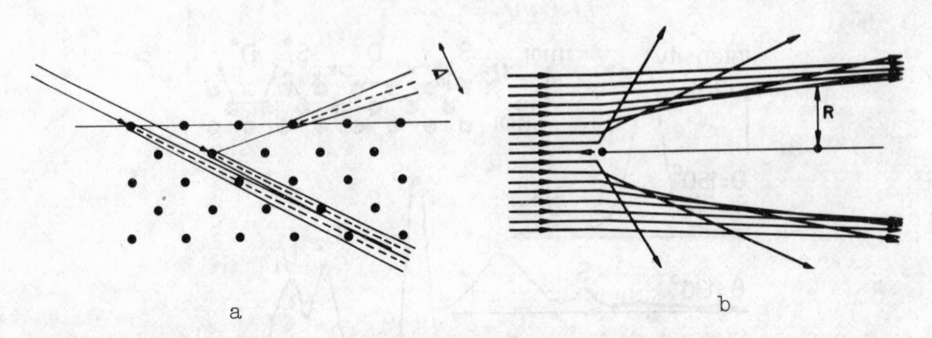

Fig. 7 : a) Shadowing of the incident beam and blocking of the scattered beam in double-aligned ion scattering.[37] b) Formation of the shadow cone in ion scattering. Only the asymptots of the trajectories are shown. For a Coulomb potential $R = 2(Z_1Z_2e^2d/E)^{1/2}$ (Z_1, Z_2 nuclear charges, d atomic spacing along row).[38]

These trajectories can be easily calculated by considering the ion as a point charge in the effective interaction potential between ion and atom. For MeV H^+ and He^+ ions this is a very good approximation so that the absolute values of the scattering cross-sections can be calculated. The area of the "surface peak" in the IS spectrum can thus be obtained for various structure models and compared with the experimental value (usually expressed in atoms per row). In this calculation the thermal vibrations of the atoms, which make a considerable contribution to the width of the shadow cone, must be taken into account including the correlation of the vibrations.[39] Fig. 8[40] shows a typical HEIS (or RBS = Rutherford backscattering) spectrum of 1.0 MeV He^+ from a Si(100) crystal bombarded in the [100] direction and measured in single alignment. In Fig. 9[40] the number of atoms per row extracted from the surface peak is compared with the expected value for a Si crystal truncated at a (100) surface without atomic displacements as a function of energy of the incident ions. From this comparison it is concluded that at least 3 monolayers are laterally displaced more than ≃ 0.15 Å and two of them have lateral displacements of more than 0.2 Å. A more detailed discussion of structure analysis by ion scattering can be found in recent reviews.[41,42]

The brevity of this section as compared to that of the diffraction section does not indicate that scattering is less useful in surface structure analysis but only that is younger and much less developed than LEED. In particular, the use of double scattering in

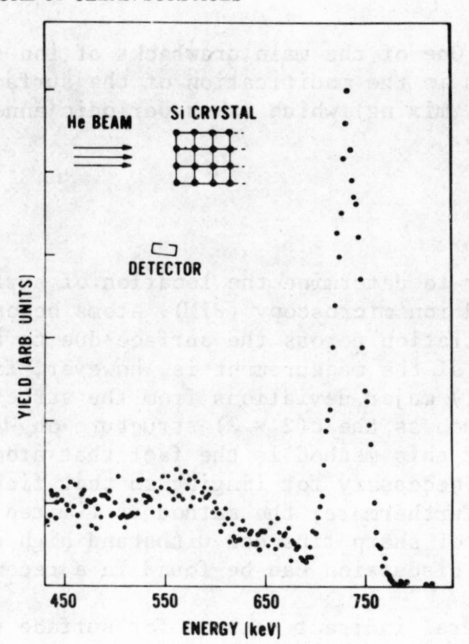

Fig. 8 : HEIS spectrum of 1.0 MeV He^+ ions scattered ≈ 95° from a [100] ligned Si(100) surface.[40]

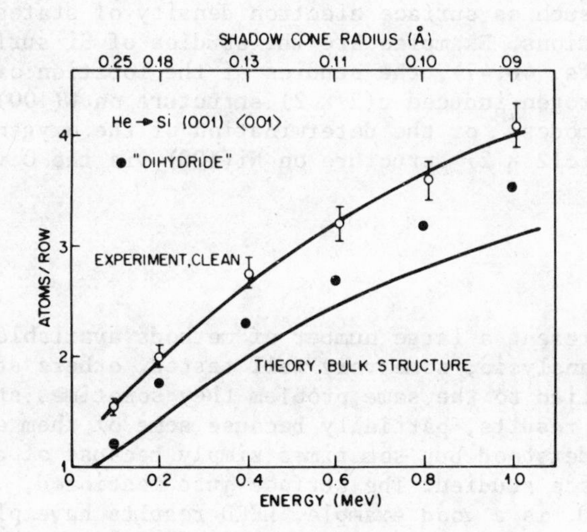

Fig. 9 : Number of atoms per row as a function of the He^+ ion energy. Open circles : Si(100) - (2 × 1) (clean), solid circles : Si(100) - (1 × 1)H (hydrogen covered),[40] lower curve : calculated with bulk Debye temperature (Θ_D = 543K). The solid circles can be fit with a surface Debye temperature of Θ_D = 230K.[78]

LEIS appears promising. One of the main drawbacks of ion scattering, especially of heavy ions is the modification of the surface by ion bombardment (sputtering, mixing) which makes periodic annealing of the surface necessary.

I.3. Other Methods

The most direct way to determine the location of surface atoms is by imaging with field ion microscopy (FIM). Atoms become visible due to the field variation across the surface due to its atomic roughness. The accuracy of the measurement is, however, far below that of LEED so that only major deviations from the structure of the bulk can be seen, such as the c(2 × 2) structure on W(100).[43,44] A fundamental problem of this method is the fact that atoms can move in the high fields necessary for imaging so that field-induced structures may appear. Furthermore, the method is limited to materials which in the form of sharp tips can withstand high electric fields. A more detailed discussion can be found in a recent review.[45]

There are also several indirect methods for surface structure analysis. Only two will be mentioned which are based on the correlation between atomic structure and electronic and vibronic structure. For a given atomic structure model the electronic and vibronic structure of the surface may be calculated and compared with experimental data such as surface electron density of states or vibrational excitations. Examples are the studies of Si surfaces (for review see refs. 46,47), the studies of the location of the W atoms in the hydrogen induced c(2 × 2) structure on W(100) via the H vibrational modes[48] or the determination of the oxygen atom locations in the O c(2 × 2) structure on Ni(100) via the O vibrational modes.[49]

I.4. Conclusion

There is at present a large number of methods available for surface structure analysis, some very well tested, others still very immature. Applied to the same problem they sometimes still give contradictory results, partially because some of them are not well enough understood but sometimes simply because of differences in the surface studied. The surface just mentioned, Ni(100) - c(2 × 2)O, is a good example. LEED results have placed the O atom in the center between four Ni atoms 0.9 Å above the plane through the center of the Ni atoms,[50-52] LEIS[53] and NPD[26] give the same result while APD[29] and vibrational loss spectroscopy[49] locate the O atom considerably lower, nearly coplanar with the Ni atoms. This discrepancy may not be due to inappropriate data analysis but rather may be caused by experimental differences : the c(2 × 2)O structure and NiO nuclei co-exist over a wide oxygen exposure range so that depending upon the reliability of the O_2 pressure measurements some authors may have looked mainly at the c(2 × 2) structure, others more at the NiO structure.

In another recently reviewed case,[54] Pt(111), the discrepancies were clearly due to incomplete understanding of the methods in the past. Most of them are eliminated now. In Section II, some of the still existing discrepancies will be mentioned.

II. RESULTS

II.1. Unreconstructed Surfaces

In this section examples will be given for structures which have the same lateral periodicity ("(1 × 1)") as the bulk. The problem of interest here is the location of the atoms in the (1 × 1) unit mesh. These examples are representative for the many structures analyzed which, for the case of LEED studies, are compiled critically in several recent reviews and books.[5,13,54a,b,55] Complete LEED structure analysis has been done up to now only for (1 × 1), (2 × 1), p(2 × 2) and c(2 × 2) structures because of the limitations of LEED programs and computer time. The other techniques, applied to clean surfaces, have only given partial information on atomic locations on more complex surfaces which will be dealt with in sect. II.2.

fcc Metals. The most densly packed planes ((111) and (100)) are, within the limits of error of the LEED analysis, simple terminations of the bulk structures without significant lateral or normal displacements of the atoms from their normal lattice sites, Au(111) and (100), Pt(100) and Ir(100) excepted. Conclusions to the contrary from LEED and other studies, e.g. for Al(111)[19] or for Pt(111) have been disproven.[22,54] Some structure determinations have been done with very high precision, e.g. for Cu(100) with R_{ZJ} = 0.068[56] or for Cu(111) with R_{ZJ} = 0.055.[11] An even smaller R_{ZJ} factor, 0.035, is obtained for Cu(100), if two structural parameters, the first two layer spacings d_{12}, d_{23} are optimized. The best agreement with experiment is obtained for a 1 % contraction of d_{12} and a 2 % expansion of d_{23}.[57]

The structure of the next low index plane, (110), is less well established. In all cases studied (Au, Pt and Ir again excepted) no lateral displacements are found but strong normal displacements are. For Cu(110) a contraction of the topmost layer spacing d_{12} by about 10 % is obtained with an R factor R_{ZJ} = 0.12,[58] for Ag(110) a 7 % contraction of d_{12} gave the best, though only good to poor agreement with experiment,[59] while for Rh(110) the smallest R_{ZJ} factor (0.11) was obtained for a 2.5 % contraction.[60] For Ni(110) LEED and MEIS[61] agree with 5 % and 4 % contraction, respectively.

The simple picture of the (110) surface has been reputed recently for Ni(110), however;[62] the understanding of the (110) surface must, therefore, be considered insufficient at present. This is illustrated by a recent re-examination of Ag(110) which for a 6 % contraction of d_{12} and a 3 % expansion of d_{23} gave $R_{ZJ} = 0.10$.

hcp Metals. The basal planes (0001) of all metals studied, e.g. of Be, Co, Cd, Zn, Ti, Zr, have their normal stacking sequence and show negligable d_{12} contraction ($\leq$ 2 %), Re(0001) excepted for which a 5 % contraction has been deduced with $\bar{R}_{ZJ} = 0.14$.[64] The only prism plane studied up to now, the Re(10$\bar{1}$0) surface, is reported to have a 17 % d_{12} contraction and a negligable d_{23} contraction (1 - 2 %). The type of (10$\bar{1}$0) termination of the crystal could also be determined.[65]

bcc Metals. Within the limits of error, the most densely packed surfaces (110) have the same d_{12} values as in the bulk. This has been established with high reliability, e.g. with $R_{ZJ} = 0.10$ for Fe(110)[66] or with $R_{ANvH} = 0.12$ for V(110).[66a] The (100) surface, which is much less densely packed than the fcc(100) surface, does not have a simple (1 × 1) structure at or below room-temperature for W, Mo and V (see sect. II.2.). At higher temperatures, e.g. above 300K and 620K for W and V, respectively, a (1 × 1) structure is seen which in spite of many studies in the case of W(100) has defied general agreement. The values of the d_{12} contraction reported vary from 4.4 % to 11 %, the most recent values being 7.5%±1.5% (from spin-polarized LEED, "SPLEED")[67] and 6.7%±2% (from LEED)[68] in good agreement with each other. The reasons for the discrepancies were recently found to be mainly in the quality of the experimental data.[69] In the case of Fe, in which the (111) and (211) surfaces were also studied, the results are less disputed : in Fe(100) d_{12} is only 1.4%±3% contracted,[70] in Fe(111) 15.4 %[71] and in Fe(211) only a slight contraction is deduced.[72]

Concluding the section on metal surfaces with (1 × 1) structures ("unreconstructed" surfaces) it must be stated that atomic beam diffraction (see review 3) and ion scattering up to now have contributed little to the analysis of their structure.

Semiconductors with Diamond Structure (Si, Ge). All Si and Ge surfaces are reconstructed or facetted over a wide temperature range, unless stabilized by impurities or quenched from the high temperature (1 × 1) structure. The only surface for which these procedures have been successful up to now is the Si(111) surface while the Si(100) - (1 × 1) structure could not be obtained by quenching (laser annealing) but only by adsorption of a considerable amount of hydrogen (0.5 monolayers). The Si(111) - (1 × 1) structure was briefly studied above the transition temperature from the reconstructed (7 × 7) structure (see sect. II.2) to the (1 × 1) structure at about 1140K with the result that a disordered surface layer rests on top of a crystal with bulk structure.[73] This conclusion was drawn from the temperature dependence of the background although the I(V) curve of the specular reflected beam is very similar to that of the (1 × 1) structures stabilized by small amounts of Cl or Te. For the Te stabilized (1 × 1) structure a d_{12} contraction of 21 % without significant (⩽ 5 %) change of d_{23} and d_{34} was deduced with $R_{ZJ.}$ = 0.21.[74] The d_{12} contraction of the (1 × 1) structure obtained by laser annealing (quenching) is quite similar (25.5%±2.5%), but here also a d_{23} expansion by 3.2%±1.5% was derived with $R_{ZJ.}$ = 0.115.[75] Recently, however, it was reported on the basis of indirect (photoemission spectroscopy) and LEED evidence that the laser annealed (1 × 1) structure has no long range order - similar to the high temperature (1 × 1) structure[73] - but a short range (2 × 1) reconstruction.[76] It is apparent that the Si(111) - (1 × 1) structure is still a matter of discussion. The Si(100) - (1 × 1)H structure (with adsorbed hydrogen) is bulk-like,[77] according to LEED work, while according to MEIS studies d_{12} is contracted by 6%±3%.[37,78]

Semiconductors with Zincblende and Wurtzite Structure. The (110) surface which has the lowest surface energy in zincblende structure has been studied for GaAs, InSb, InP, ZnTe and ZnS. The evalution of the understanding of GaAs(110) has been reviewed recently[54] as well as the general trends of the structures.[79] Although the surfaces have (1 × 1) structures, large atomic displacements occur in the topmost layer and smaller ones in the second layer as indicated in Fig. 10.

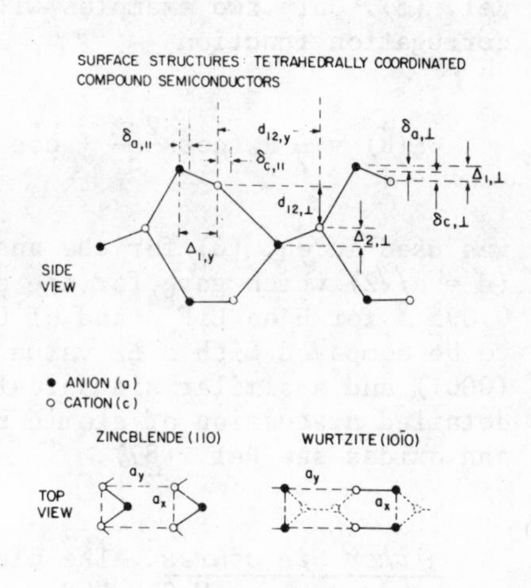

Fig. 10 : Side and top view of the surface geometries of the (110) zincblende and the related (10$\bar{1}$0) wurtzite structures.[79] 1 : first layer, 2 : second layer, δ displacement from bulk lattice position.

The cation 1 always makes the largest displacement (up to 0.6 Å downward and towards the anion) while the anion 1 moves upward up to 0.2 Å and up to 0.4 Å sideways. This results in height differences up to 0.8 Å. All bond lengths remain essentially as in the bulk except that of bond $a_1 - c_2$. This bond is usually contracted about 5 %; in ZnTe it is dilated by about 3 %. The distortions decrease with increasing ionicity so that the LEED data of ZnS are fit best by a surface with approximately bulk atomic positions (R_{ZJ} = 0.23).[80] The surface corrugation of the GaAs(110) surface for He atomic beams has recently been studied;[81] for its discussion the reader is referred to a recent review.[3]

Of the wurtzite structure surfaces, the (0001), (10$\bar{1}$0) and (11$\bar{2}$0) surfaces of ZnO have been studied by LEED with the result that all faces represent, within 6 % distortion, simple terminations of the bulk structure.[54B]

Crystals with NaCl Structure (Semiconductors and Insulators). Many of these materials are problematic for LEED studies because of charging and dissociation. Nevertheless, the (100) surfaces of several oxides (MgO, CaO, CoO, MnO, NiO and EuO) have been studied with the result the $d_{12} = 1/2(d_{12}^a + d_{12}^c)$ is generally less than 3 % contracted and that the "rumple" $d_{12}^a - d_{12}^c$, i.e. the relative normal displacement of first layer anions and cations, is less than 5 %.[82] The most recent values for NiO are : 2 % d_{12} contraction, 0 % rumple.[83] Because of their strong dissociation by ionizing particles (electrons, ions) alkali halides have been the domain of atomic beam diffraction. The relevant work is reviewed in Ref. (3). Only two examples will be quoted : LiF and NaCl. The corrugation function

$$\xi(\vec{R}) = \frac{1}{2}\,\xi_o(\cos\frac{2\pi x}{d} + \cos\frac{2\pi y}{d})$$

was used in eq. (6) for the analysis of the intensity data ($d = a/\sqrt{2}$) which gave for the corrugation parameter ξ_o values of 0.095 Å for H on LiF[84] and of 0.20±0.02 Å for H on NaCl;[85] this is to be compared with a ξ_o value of 0.010±0.001 for H on graphite (0001) and a similar sinusoidal corrugation function.[86] For a more detailed discussion of atomic beam diffraction from alkali halides and oxides see Ref. (87).

Other Structures. The cleavage surface of a number of layer compounds such as MoS_2, $NbSe_2$ has been studied by LEED with the result that there is generally a slight d_{12} contraction up to 5 %,

$TiSe_2$ excepted (5 % expansion).[5] The most complex surface studied up to now is Te(10$\bar{1}$0) : here the top layer atoms move inward 0.21Å the second layer atoms move outward 0.46 Å in such a manner that no nearest-neighbor bonds lengths are changed.[88]

II.2. Reconstructed Surfaces

In this section those surfaces whose periodicity parallel to the surface differs from that of the bulk will be discussed.

fcc Metals. The following surfaces have been found to be reconstructed : Au(111), Au, Pt, Ir(100) and (110). The (111) and (100) surfaces have large surface unit meshes and are, therefore, at present not amenable to LEED structure analysis. The (110) surfaces have a (1 × 2) structure, i.e. double periodicity in the [001] direction and have been the subject of LEED (Au,[89,91] Pt,[92] Ir[93]), LEIS[94] and atomic beam diffraction studies.[95] Some of the models proposed on the basis of LEED studies are shown in Fig. 11.

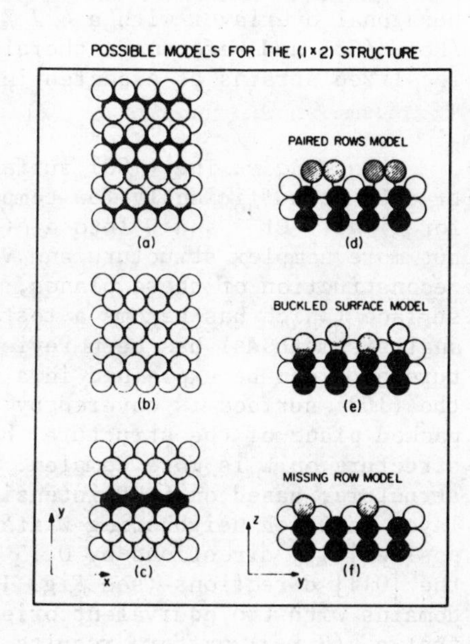

Fig. 11 : Some surface structure models proposed for the (110) - (1 × 2) surfaces of Au,Pt and Ir.[93]

In addition, a distorted hexagonal overlayer similar to that on the (100) and on the Au(111) surfaces has been considered. Although the missing row model has been favored by some authors[89,92,93] the R factors either do not differ very much for the various models (R_{ZJ} = 0.24 - 0.32[93]) or are unacceptably large (R_{ZJ} = 0.6[90], R_{ANvH} = 0.42[92]). Further doubts in the models come from 600 eV K^+ LEIS scattering experiments which cannot be reconciled with either the distorted hexagonal overlayer or the missing row models.[94] Atomic beam diffraction leads to an estimated corrugation amplitude of ξ_o = 1.65 Å which is significantly higher than those of the Ni(110) and the similar W(112) surfaces.[95] This favors a missing row model (rough) over a hexagonal overlayer (smooth) model. In summary the (110) - (1 × 2) surface is still not understood.

The same statement is valid for the (100) surfaces and the Au(111) surface. The present state of understanding has recently beem summarized.[96] A (distorted) hexagonal overlayer is the preferred model (see, however, ref. (97)). It causes, on Ir(100), a simple (1 × 5) superstructure, on Au(100), a (5 × 20) structure, recently more accurately described as c(26 × 68)[96] and on Pt(100) several structures with similar complicated periodicities. Recent combined HEIS and LEED studies of the transition from the reconstructed to the unreconstructed Pt(100) surface due to CO adsorption indicate, however, that the picture is not as simple as just mentioned but that also atoms in the second and possibly deeper layers are displaced.[98] The reconstruction of the Au(111) surface described as (23 × 1) superstructure can be explained by a distorted hexagonal overlayer with a 4.2 % contraction in the [1$\bar{1}$0] direction (Ref. (33) and references therein). This layer is reported to have localized strains as expected in monolayer misfit dislocations.[32]

bcc Metals. The (100) surfaces of W, Mo, Cr and V are reconstructed at sufficiently low temperature (e.g. <300K for W, <620K for V) as : Cr[99] and W into a c(2 × 2) structure, Mo into a similar but more complex structure and V into a (5 × 1) structure.[100] The reconstruction of these planes, in particular that of the W(100) surface (which has become a testing ground for surface structure analysis methods) has been reviewed recently.[45] The (5 × 1) structure on V can be explained in a manner very similar to that on Ir; the (100) surface is covered by a monolayer of the most closely packed plane of the structure, here a (110) plane.[100] The c(2 × 2) structure on W is more complex. The widely accepted model of this structure, based on LEED intensity analysis[101,102] is a surface layer in which neighboring W atoms are laterally displaced in opposite [011] directions by 0.15 - 0.3 Å forming zig-zag rows along the [01$\bar{1}$] directions (see Fig. 12). On a well oriented surface, domains with two equivalent orientations of these rows exist so that a c(2 × 2) pattern results. An alternating vertical displacement (see Fig. 12a) as proposed on the basis of FIM work[44] is not compatible with the LEED evidence. HEIS also indicates lateral displacements by ≈ 0.23 Å but only one-half of the surface atoms are displaced.[103] For a further discussion of the discrepancies see Ref. (45).

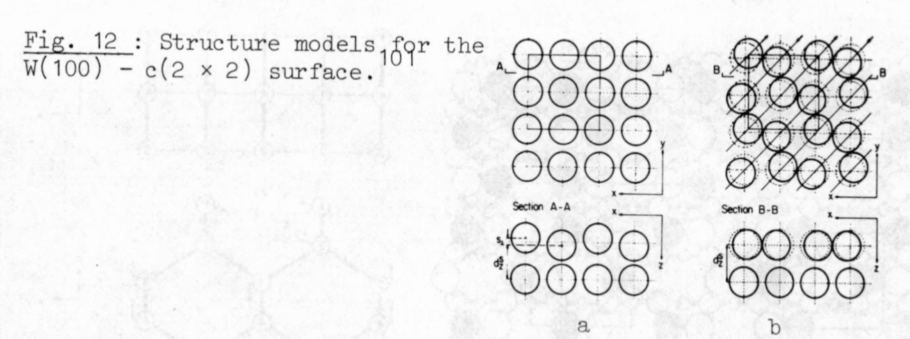

Fig. 12 : Structure models for the W(100) - c(2 × 2) surface.[101]

Semiconductors with Diamond Structure (Si, Ge). All surfaces studied to date are reconstructed, but only (111) and (100) surfaces are considered here. The structure of the (100) surface is usually described as a (2 × 1) structure but the LEED patterns shows strong background and/or half-order streaking parallel to the <10> directions of the surface unit mesh. Most authors, however, have seen sharp quarter order diffraction beams of a c(4 × 2) pattern with weak background on Ge[104] but some also on Si,[105] including some of the authors coworkers.[106] Apparently, the well-ordered surface has a c(4 × 2) structure, but it is difficult to achieve the necessary degree of order. It is, therefore, not surprising that attempts[104,107] to analyze LEED intensity data on the basis of a (2 × 1) unit mesh have met with little success, even taking into account atomic displacements down to the third layer.[107] All that can be said is that an asymmetric dimer model (see below) and subsurface layer distortions improve the agreement between model calculation and experiment. An interesting model proposed on the basis of a quasi-kinematical LEED analysis of the c(4 × 2) structure[105] is compatible with HEIS[40] and atomic beam diffraction[108] but not with MEIS[78] results. Apparently the problem is beyond the capabilities of today's LEED structure analysis. The evidence from the non-LEED techniques is as follows. (i) At least 3 monolayers have lateral displacements $\gtrsim$ 0.15 Å, two of them $\gtrsim$ 0.20 Å;[40] (ii) The surface atoms are shifted more than 0.45 Å in the dimerization direction ([001], [01̄1], see Fig. 13), the atoms in lower layers less than ≈ 0.2 Å;[78] (iii) The surface has a very strong corrugation for He atomic beam scattering.[108] All data are compatible with variants of the asymmetric dimer model (including subsurface displacements)[109] shown in Fig. 13. The displacements of the topmost atoms obtained by an energy-minimization procedure are (in Å) : $\Delta X_1 = +0.46$, $\Delta X_{1'} = -1.08$, $\Delta Z_1 = +0.04$, $\Delta Z_{1'} = -0.435$, $\Delta X_2 = -\Delta X_{2'} = 0.115$ and $\Delta Z_2 = \Delta Z_{2'} = 0.014$.[109] Indirect evidence (electronic structure studies with photoelectron spectroscopy (PS)) is also in agreement with an asymmetric dimer model,[110] as well as recent x-ray diffraction studies.[111]

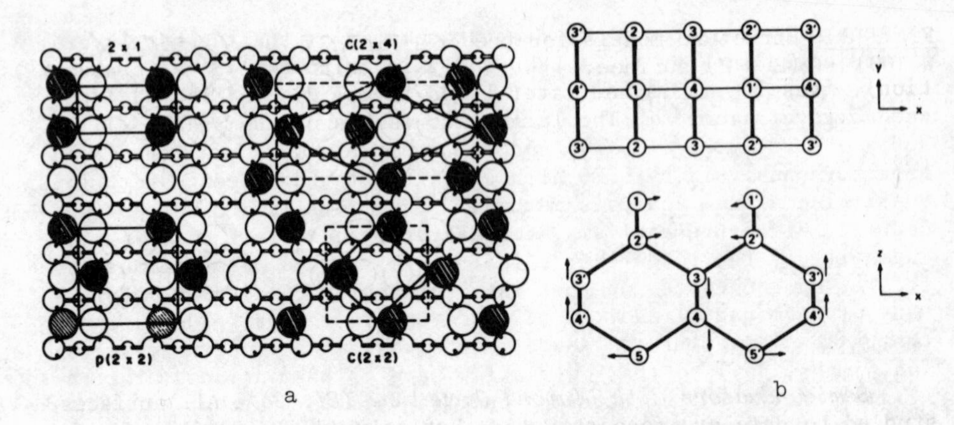

Fig. 13 : a) Top views of the asymmetric dimer models of Si (and Ge) (100) surfaces. The smallest circles represent atoms in the third layer, the largest indicate atoms in the first layer, the shaded ones being raised with respect to its dimer partner. Depending upon what atom is raised in neighboring dimers a variety of superstructers is formed.[108] b) Top and side view of ideal Si(100) surface with arrows indicating the relaxations in the (011) plane.[109]

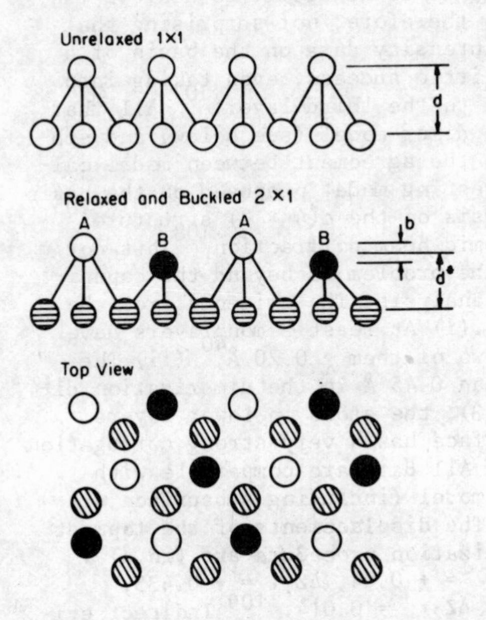

Fig. 14 : Schematic views of the (1 × 1) and buckled (2 × 1) Si(111) surface.[111]

On the (111) surface two structures occur : a (2 × 1) structure which is obtained upon cleaving at low temperature, and a stable (7 × 7) and c(2 × 8) structure on Si and Ge respectively, which is formed irreversibly by heating the (2 × 1) structure or by standard surface preparation procedures. The (2 × 1) structure is generally accepted to be due to buckling of the topmost layer as indicated in Fig. 14. A LEED analysis (witout R factor evaluation) gives b = 0.3 Å and lateral and normal displacements of the second layer atoms.[112] The large unit meshes of the stable (111) surface structures ((7 × 7) and c(2 × 8)) preclude presently a structure analysis by LEED although attempts have been made using quasi-kinematical analysis methods.[113,114] For lack of proper procedures, structure analysis, therefore, is here left mostly to the ingenuity of the researcher to patch the various observations and theoretical concepts together into internally consistent models. Many of them had already to be discarded in the past 20 years on the basis of accumulating experimental evidence (see the reviews (46,47,54b,115)) so that only a few have to be mentioned. For the Ge(111) - c(2 × 8) structure a buckling similar to that of the (2 × 1) structure but with different lateral periodicity has been suggested.[115] For the Si(111) - (7 × 7) structure two principally different competing models are presently left : the ring-buckling (RB) model[115,117] and the double-layer island (DI) model.[118] A third model, in which atoms pair up with lateral displacements of ± 0.4 Å in hexagonal regions[113] is excluded by HEIS results which set an upper limit of ≃ 0.15 Å for the lateral displacements, in strong contrast to the results for the Si(100) - (2 × 1) surface.[119] Furthermore, the reason for excluding the RB mode, the strong intensity of the 3/7 and 4/7 LEED beams expected for this model, are just in favor of it, because this is the most striking LEED feature observed. Both the RB and the DI model were conceived on the basis of energy minimization considerations, one (RB) starting from displacements of individual atoms the other one (DI) starting from epitaxial misfit strain. In the RB model the surface atoms are in hexagonal (one of them in triangular) rings raised and lowered by about the same amount as in the case of the (2 × 1) surface (Fig. 15 a).

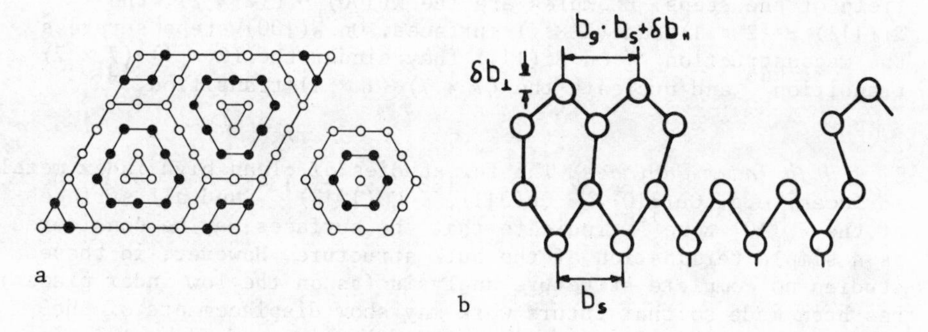

Fig. 15 : Si(111) - (7 × 7) structure models. a) Ring-buckling model,[115,117] top view; raised atoms : open circles, lowered atoms : full circles; insert shows the surrounding of the corner atoms of the unit mesh. b) Double-layer island model,[118] side view; only part of the unit mesh section is shown.

In the DI model the enhanced backbonding leads to a compression $\delta b_{\perp}$ of the outermost double layer. This is associated with a lateral expansion $\delta b_{||}$ which leads to a misfit with the substrate (Fig. 15b). The DI model and variants of it[120] are supported by gross features of the LEED patterns (3/7, 4/7 spot intensities), by HEIS results which indicate large normal displacements (< 0.4 Å) of surface atoms but small (< 0.15 Å) if no lateral displacements[119] and by a significant amount of indirect evidence.[117] It also gives a simple explanation of the Ni impurity stabilized Si(111)-($\sqrt{19}\times\sqrt{19}$) R 23.5° structure.[120,121] Recent HEIS and UPS studies show, however, that (7 × 7) and ($\sqrt{19}\times\sqrt{19}$)R23.5° structure differ considerably in atomic displacements and electronic structure[121a] so that this explanation seems questionable. The DI model is supported by He atomic beam scattering experiments[122] and by LEED studies of hydrogen adsorption on the (7 × 7) surface.[123] It is difficult to guess which of the two models will survive. There is no doubt that surface strain plays a major role[124] but it is not clear how this strain is accommodated. For more details see the reviews.[46,47,54b,115]

II.3. Vicinal, High Index and Imperfect Surfaces

Vicinal Surfaces (surfaces with less than about 10° deviation from a low index plane). Extensive LEED work (see reviews (6a, 125, 126)), some MEIS studies[127] and atomic beam diffraction studies (see review (3)) have shown that these surfaces may be well described by the terrace-ledge model in which monatomic steps are in general separated by terraces of the low index plane. The step atoms are sometimes somewhat depressed, e.g. on Ge and Si vicinals[127,128] or on Cu(410) by 5.0±1.5 %.[129] Sometimes diatomic steps occur[127,128] but the terraces have the same structure as the low index plane unless this has a reconstruction which is sensitive to the strain field of the steps. Examples are the W(100) - c(2 × 2), the Si(111) - (2 × 1) and (7 × 7) surfaces. On W(100) steps surpress the reconstruction[130] on Si(111) they hinder the (2 × 1)→(7 × 7) transition[46] and nucleate the (7 × 7)→(1 × 1) transition.[131]

High Index Surfaces. The few studies of clean high index metal surfaces, e.g. Cu(210),[132] Cu(311),[133] Co(10$\bar{1}$2)[134] and all surfaces of the W [110] zone[126] indicate that the surfaces may be considered as a simple termination of the bulk structure. However, in these studies no complete structure analysis (as on the low index planes) has been made so that future work may show displacements of the surface layers similar to those on (110) planes of fcc metals (see sect. II.1). Semiconductor surfaces (Si,Ge) tend to facet (as do vicinals in certain temperature ranges[127,128]) but can also be obtained as flat surfaces with specific superstructures (see the references in Ref. (128,135)).

Imperfect Surfaces. Most surfaces have many imperfections of which steps can be particularily easily detected by LEED. Numerous studies in this field are reviewed in Ref. (6a-6d).

III. CONCLUDING REMARKS

The analysis of the structure of surfaces is much more difficult than that of the bulk because (i) the number of surface atoms is small compared to that available in the bulk and (ii) the signal from the surface atoms is in general superimposed on a strong background signal from the atoms in the bulk; atomic beam diffraction, LEIS and to a certain extent MEIS and HEIS excepted. It is therefore not surprising that in spite of a great amount of effort no more substantial information on surface structure is available. With increased understanding and application of the non-LEED techniques this situation should, however, change rapidly in the next few years.

ACKNOWLEDGEMENTS

The author wishes to thank H.-D. Shih for stimulating discussions and his help with LEED references and several colleagues for sending (p)reprints of their recent work which helped considerably to ensure that this review was reasonably up-to-date at the time of writing (July/August 1981).

The figures 3, 5, 6, 9, 13, 14, and 15 are reproduced with permission from the American Physical Society; the figures 4 and 12 are reproduced with permission from Pergamon Press Ltd.; the figures 7, 8, and 11 are reproduced with permission from the North-Holland Publ. Co.; figure 10 is reproduced with permission from the American Institute of Physics.

REFERENCES

1 . EISENBERGER P and MARRA W.C., Phys. Rev. Letters 46(1981)1081
2 . GOODMAN F.O., CRC Crit. Rev. Solid State Sci. 7(1977)33
3 . ENGEL T. and RIEDER K.H., preprint
4 . PENDRY J.B., *Low Energy Electron Diffraction*, Academic Press, London, 1974
5 . Van HOVE M.A. and TONG S.Y., *Surface Crystallography by LEED*, Springer, Berlin, 1979
6 . *Proc. IBM LEED Structure Determination Conference*, Plenum Press, Marcus P.M. ed., 1981
a. HENZLER M., in *Electron Spectroscopy for Surface Analysis*, Ibach H. ed., Springer, Berlin, 1977
b. HENZLER M., Surf. Sci. 73(1978)240
c. HENZLER M., in *Festkörperprobleme*, Vol. XIX, Treusch J. ed., Vieweg, Braunschweig, 1979, p. 193
d. HENZLER M., 2nd Intern. Conf. on Thin Films and Solid Surfaces; 1981, to be published in Appl. Surface Sci.
7 . ZANAZZI E. and JONA F., Surf. Sci. 62(1977)61
8 . JONA F. and SHIH H.D., J. Vac. Sci. Technol. 16(1979)1248-1251
9 . ADAMS D.L., NIELSEN H.B. and van HOVE M.A., Phys. Rev. B20 (1979)4789
10 . PENDRY J., J. Phys. C13(1980)937
11 . TEAR S.P., ROLL K. and PRUTTON M., J. Phys. C,(1981) to be published
12 . MEYER R.J., DUKE C.B. and PATON A., Surf. Sci. 97(1980)512
13 . JONA F., J. Phys. C11(1978)4271
14 . HEINZ K. and MULLER K., Springer Tracts in Modern Physics, 1981, to be published
15 . ZANAZZI E., BARDI U. and MAGLIETLA M., J. Phys. C13(1980) 4001
16 . IGNATIEV A., JOVA F., DEBE M., JOHNSON D.E., WHITE S.J. and WOODRUFF D.P., J. Phys. C10(1977)1109
17 . EINSTEIN T.L., 2nd Intern. Conf. on Thin Films and Solid Surfaces, 1981, to be published in Appl. Surface Sci.
18 . McFEELY F.R., 1981, ibid.
19 . BIANCONI A. and BACHRACH R.Z., Phys. Rev. Letters 42(1979) 104
20 . LEE P.A. and PENDRY J.B., Phys. Rev. B11(1975)2795
21 . LEE P.A., Phys. Rev. B13(1976)5261
22 . JONA F., SONDERICKER D. and MARCUS P.M., J. Phys. C13(1980) L155
23 . PARK R.L., Surf. Sci. 86(1979)504
24 . WILLIAMS G.P., CERRINA F., McGOVERN I.T. and LAPEYRE G.J., Solid State Commun. 31(1979)15
25 . LI C.H. and TONG S.Y., Phys. Rev. Letters 43(1979)526
26 . ROSENBLATT D.H., TOBIN J.G., MASON M.G., DAVIS R.F., KEVAN S.D. SHIRLEY D.A., LI C.H. and TONG S.Y., Phys.Rev.B23(1981)3828

27 . KONO S., FADLEY C.S., HALL N.F.T. and HUSSAIN Z., Phys. Rev. Letters 41(1978)117
28 . KONO S., GOLDBERG S.M., HALL N.F.T. and FADLEY C.S., Phys. Rev. Letters. 41(1978)1831, Phys. Rev. B22(1980)6085
29 . PETERSSON L.G., KONO S., HALL N.F.T., GOLDBERG S., LLOYD J.T., FADLEY C.S., and PENDRY J.B., Mater. Sci. Eng. 42(1980)111
30 . PLOCIENNIK J.M., BARBET A., and MATHEY L., Surface Sci. 102 (1981)282
31 . LEHMPFUHL G. and TAKAYANAGI K., Ultramicroscopy 6(1981)195
32 . TANISHIRO Y., KANAMORI H., TAKAYANAGI K., KOBAYASHI K., YAGI K. and HONJO G., in Proc. 4th Intern. Conf. Solid Surfaces, Cannes, pp. 683, 1980; Surface Sci., to be published.
33 . HEYRAND J.C. and METOIS J.J., Surface Sci. 100(1980)519
34 . OSAKABE N., TANISHIRO Y., YAGI K and HONJO G., Surf. Sci. 102 (1981)424
35 . INO S., Jap. J. Appl. Phys. 19(1980)1277
36 . v.d. HAGEN T. and BAUER E., Phys. Rev. Letters 47 (1981)579; ECOSS 4, Münster 1981
37 . TROMP R.M., SMEENK R.G. and SARIS F.W., Surf. Sci. 104(1981) 13
38 . STENSGAARD I., FELDMAN L.C. and SILVERMAN P.J., Surf. Sci. 77 (1978)513
39 . JACKSON D.P. and BARRETT J.H., Phys. Letters 71A(1979)359
40 . STENSGAARD I., FELDMAN L.C. and SILVERMAN P.J., Surf. Sci. 102 (1981)1
41 . FELDMAN L.C., Surface Science : Recent Progress and Perspectives (edt. Vanselow R.), CRC Press, Cleveland, 1981, to be published
42 . FELDMAN L.C. and STENSGAARD I., Progress in Surface Science, 1981, to be published
43 . NISHIKAWA O., WADA M. and KONISHI M., Surf. Sci. 97(1980)16-24
44 . MELMED A.J., TUNG R.T., GRAHAM W.R. and SMITH G.D.W., Phys. Rev. Letters 43(1979)1521
45 . MELMED A.J. and GRAHAM W.R., 2nd Intern. Conf. on Thin Films and Solid Surfaces; 1981; Appl. Surface Sci., to be published
46 . MÖNCH W., Surf. Sci. 86(1979)672-699
47 . EASTMAN D.E., J. Vac. Sci. Technol. 17(1980)492-500
48 . WILLIS R.F., Surf. Sci. 89(1979)457; and references therein
49 . RAHMAN R.S., BLACK J.E. and MILLS D.L., Phys. Rev. Letters 46 (1981)1469-1472
50 . van HOVE M. and TONG S.Y., J. Vac. Sci. Technol. 12(1975)230
51 . MARCUS P.M., DEMUTH J.E. and JEPSEN D.W., Surf. Sci. 53(1975) 501
52 . HAUKE G., LANG E., HEINZ K. and MULLER K., Surf. Sci. 91(1980) 551
53 . BRONGERSMA H.H. and THEETEN J.B., Surf. Sci. 54(1976)519
54 . DUKE C.B., 2nd Intern. Conf. on Thin Films and Solid Surfaces, 1981; Appl. Surface Sci., to be published
a. MARCUS P.M., 1981, ibid.

b. DUKE C.B., CRC Crit. Rev. Solid State Mater. Sci., December, 1978, pp.69-91
55 . SOMORJAI G.A., in *Chemistry in Two Dimensions : Surfaces*, Cornell University Press, Ithaca, 1981,pp. 136-142
56 . NOONAN J.R. and DAVIS H.L., J. Vac. Sci. Technol. 17(1980)194-197
57 . DAVIS H.L., Bull. Amer. Phys. Soc. 25(1980)327
1981, private communication
58 . DAVIS H.L., NOONAN F.R. and JENKINS L.H., Surf. Sci. 83(1979) 559-571
59 . ZANAZZI E., JONA F., JEPSEN D.W. and MARCUS P.M., J. Phys.C10 (1977)375-381
60 . FROST D.C., HENGRASMEE S., MITCHELL K.A.R., SHEPHERD F.R. and WATSON P.R., Surf. Sci. 76(1978)L585-L589
61 . SMEENK R.G., TROMP R.M. and SARIS F.W., Surf. Sci. 107(1981) 429-438
62 . ONUFERKO J.H. and WOODRUFF D.P., Surf. Sci. 91(1980)400
63 . DAVIS H.L. and NOONAN J.R., 1981, to be published
64 . DAVIS H.L.and ZEHNER D.M., Bull. Am. Phys. Soc. 24(1979)468
65 . DAVIS H.L. and ZEHNER D.M., J. Vac. Sci. Technol. 17(1980)190-193
66 . SHIH H.D., JONA F., BARDI U. and MARCUS P.M., J. Phys. C13 (1980)3801-3808
a. ADAMS D.L. and NIELSEN H.B., Surf. Sci. 107(1981)305-320
67 . FEDER R. and KIRSCHNER J., Surf. Sci. 103(1981)75-102
68 . CLARKE L.J. and MORALES de la GARZA L., Surf. Sci. 99(1980) 419-439
69 . STEVENS M.A. and RUSSELL G.J, Surf. Sci. 104(1981)354-364
70 . LEGG K.O., JONA F., JEPSEN D.W. and MARCUS P.M., J. Phys. C10 (1977)937-946
71 . SHIH H.D., JONA F., JEPSEN D.W. and MARCUS P.M., Surf. Sci. 104(1981)39-46
72 . SOKOLOV J., SHIH H.D., BARDI U., JONA F., JEPSEN D.W. and MARCUS P.M., Bull. Am. Phys. Soc. 25(1980)327
73 . BENNETT P.A. and WEBB M.B., Surf. Sci. 104(1981)74-104; J. Vac Sci. Technol. 18, pp. 847-851
74 . JEPSEN D.W., SHIH H.D., JONA F. and MARCUS P.M., Phys. Rev. B22(1980)814-824
75 . ZEHNER D.M., NOONAN J.R., DAVIS H.L. and WHITE C.W., J. Vac. Sci. Technol. 18(1981)852-855
76 . CHABAL Y.J., ROWE J.E. and ZWEMER D.A., Phys. Rev. Lett. 46 (1981)600-603
77 . WHITE S.J., WOODRUFF D.P., HOLLAND B.W. and ZIMMER R.S., Surf. Sci. 74(1978)34-46
78 . TROMP. R.M., SMEENK R.G. and SARIS F.W., Phys. Rev. Letters 46(1981)939-942
79 . DUKE C.B., MEYER R.J. and MARK P., J. Vac. Sci. Technol. 17 (1980)971-977
80 . DUKE C.B., MEYER R.J.,PATON A., KAHN A., CARELLI J. and YEH J.L., J. Vac. Sci. Technol. 18(1981)866-870

81 . CARDILLO M.J., BECKER G.E., SIBENER S.J. and MILLER D.R., Surf. Sci. 107(1981)469-493
82 . PRUTTON M., WALKER J.A., WELTON-COOK M.R., FELTON R.C. and RAMSEY J.A., Surf. Sci. 89(1979)95-101
83 . WELTON-COOK M.R. and PRUTTON M., J. Phys. C13(1980)3993-4000
84 . CARACCIOLO G., IANNOTTA S., SCOLES G. and VALBUSA U., J. Chem. Phys. 72(1980)4491-4499
85 . IANNOTTA S. and VALBUSA U., Surf. Sci. 100(1980)28-34
86 . ELLIS T.H., IANNOTTA S., SCOLES G. and VALBUSA U., J. Vac. Sci. Technol. 18(1981)488-489
87 . HOINKES H., Rev. Mod. Phys. 52(1980)933-970
88 . MEYER R.J., SALANECK W.R., DUKE C.B., PATON A., GRIFFITHS C.H., KORNAT L. and MEYER L.E., Phys. Rev. B21(1980)4542-4551
89 . MORITZ W. and WOLF D., Surface Sci. 88(1979)L29-L34 and references therein
90 . NOONAN J.R. and DAVIS H.L., J. Vac. Sci. Technol. 16(1979) 587-589
91 . REIHL B., Z. Phys. B41(1981)21-34
92 . ADAMS D.L., NIELSEN H.B., van HOVE M.A. and IGNATIEV A., Surf. Sci. 104(1981)47-62
93 . CHAN C.M., van HOVE M.A., WEINBERG W.H. and WILLIAMS E.D., Surf. Sci. 91(9180)440-448
94 . OBERBURY S.H., HEILAND W., ZEHNER D.M., DATZ S. and THOE R.S. 1981, to be published
95 . RIEDER K.H. and ENGEL T., Proc. 4th Intern. Conf. Solid Surfaces and 3rd Europ. Conf. Surface Sci., 1980, pp. 861-864
96 . van HOVE M.A., KOESTNER R.J., STAIR P.C., BIBERIAN J.P., KESMODEL L.L., BARTOS I. and SOMORJAI G.A., Surf. Sci. 103 (1981)I : 189-217, II : 218-238
97 . BIBERIAN J.P., Surf. Sci. 97(1980)257-263
98 . NORTON P.R., DAVIS J.A, CREBER D.K., SITTER C.W. and JACKMAN T.E., Surf. Sci. 108(1981)205-224
99 . GEWINNER G., PERUCHETTI J.C., JAEGLE A. and RIEDINGER R., Phys. Rev. Lett. 43 (1979)935-938
100. DAVIS P.W. and LAMBERT R.M., Surf. Sci. 107(1981)391-404
101. BARKER R.A., ESTRUP P.J., JONA F. and MARCUS P.M., Solid State Commun. 25(1978)375-379
102. WALKER J.A., DEBE M.K. and KING D.A., Surf. Sci. 104(1981) 405-418
103. FELDMAN L.C., SILVERMAN P.J. and STENSGAARD I., Surf. Sci. 87 (1979)410-414
104. JONA F., SHIH H.D., JEPSEN D.W. and MARCUS P.M., J. Phys. C12 (1979)L455-L461 and references therein
105. POPPENDIECK T.D., NGOC T.C. and WEBB M.B., Surf. Sci. 75(1978) 287-315
106. WARSOW H., M.S. thesis, Clausthal, 1975; GREEN A.K, 1978, unpublished.
107. FERNANDEZ J.C., YANG W.S., SHIH H.D., JONA F., JEPSEN D.W. and MARCUS P.M., J. Phys. C14(1981)L55-L60

108. CARDILLO M.J. and BECKER G.E., Phys. Rev. B21(1980)1497-1510
109. CHADI D.J., Phys. Rev. Letters 43(1979)43-47
110. HIMPSEL F.J., HEIMANN P., CHIANG T.C. and EASTMAN D.E., Phys. Rev. Letters 45(1980)1112-1115
111. BRENNAN S., STÖHR J., JAEGER R. and ROWE J.E., Phys. Rev. Letters 45(1980)1414-1418
112. FEDER R., MÖNCH W. and AUER P.P., J. Phys. C 1979 (1979) L179-L184
113. MILLER D.J. and HANEMAN D., J. Vac. Sci. Technol. 16(1979)1270 1285; Surf. Sci. 104, pp. L237-L244
114. MILLER D.J., HANEMANN D. and WALKER L.W., Surf. Sci. 94(1980) 555-563
115. CHADI D.J., Surf. Sci. 99(1980)1-12
116. CHADI D.J. and CHIANG C., Phys. Rev. B23(1981)1843-1846
117. CHADI D.J., BAUER R.S., WILLIAMS R.H., HANSSON G.V., BACHRACH R.Z., MIKKELSEN J.C.Jr., HOUZAY F., GUICHAR G.M., PINCHAUX R. and PETROFF Y., Phys. Rev. Letters 44(1980)799-802
118. PHILLIPS J.C., Phys. Rev. Letters. 45(1980)905-908
119. CULBERTSON R.J., FELDMAN L.C. and SILVERMAN P.J.,Phys. Rev. Letters 45(1980)2043-2046; J. Vac. Sci. Technol. 18(1981)871
120. LELAY G., Surf. Sci. 108(1981)L429-L433
121. HANSSEN G.V., BACHRACH R.Z., BAUER R.S. and CHIARADIA P.,Phys. Rev. Letters 46(1981)1033-1037
a. CHABAL Y.J., CULBERTSON R.J., FELDMAN L.C. and ROWE J.E., J. Vac. Sci. Technol. 18(1981)880-882
122. CARDILLO M.J., Phys. Rev. B23(1981)4279-4282
123. McRAE E.G. and CALDWELL C.W., Phys. Rev. Letters. 46(1981)1632-1635
124. HANEMAN D., Phys. Rev. Letters(1981) to be published
125. WAGNER H., Springer Tracts Mod. Phys. 85(1979)151-221
126. GARDINER T.M., KRAMER H.M.and BAUER E., Surf. Sci. (1981) to be published
127. OLSHANETSKY B.Z., REPINSKI S.M. and SHKLYAEV A.A., Surf. Sci. 69(1977)205-217
128. OLSHANETSKY B.Z. and SHKLYAEV A.A., Surf. Sci. 82(1979)445-452
129. ALGRA A.J., LUITJENS S.B., SUURMEIJER E.P.Th.M. and BOERS A.L., Phys. Letters 75 A (1980)496-498; Surf. Sci. 100, pp. 329-341
130. DEBE M.K. and KING D.A., Surf. Sci. 81(1979)193-237
131. OSAKABE N., YAGI K. and HONJO G., Jap. J. Appl. Phys. 19(1980) L309-L312
132. McKEE C.S., RENNY L.V. and ROBERTS M.W., Surf. Sci. 75(1978)92-108
133. STREATER R.W., MOORE W.T., WATSON P.R., FROST D.C. and MITCHELL K.A.R., Solid State Commun. 24(1977)139
134. PRIOR K.A, SCHWAHA K., BRIDGE M.E. and LAMBERT R.M., Chem. Phys. Letters 65(1979)472
135. GREEN A.K. and BAUER E., Surf. Sci. 103(1981)L127-L133

ELECTRONIC STRUCTURE AND COHESION

J.P. GASPARD

Institut de Physique
Université de Liège
B-4000 Sart-Tilman
Belgium.

ABSTRACT

The different types of cohesion of condensed matter (rare gases, ionic solids, normal and transition metals, semiconductors) are analyzed in relation to their electronic structures.

The electronic properties of surfaces are reviewed and the various theories of surface tension are discussed.

I. INTRODUCTION

One of the main goals of solid state physics is to explain the stability of condensed matter and to analyze the origin and the nature of the bonds between the atoms in the solid or in the liquid.

Let us state from the beginning that the treatment of the above problems are very schematic and that solid state theory rarely makes quantitative predictions in this field.

However, it is possible to

(i) propose models for cohesion in different cases as shown in this chapter;
(ii) find the origin and trends in the properties of solid state matter, e.g. variation of the cohesive energy, elastic moduli, etc., with the number of valence electrons (see Fig. 1). Let us call this a vertical analysis : from the basic principles to the physical properties.

B. Mutaftschiev (ed.), Interfacial Aspects of Phase Transformations, 33–61.

(iii) discover correlations between different properties of solid state matter, e.g. the relation between the bulk cohesive energy* and the surface tension. This analysis could be called a horizontal analysis.

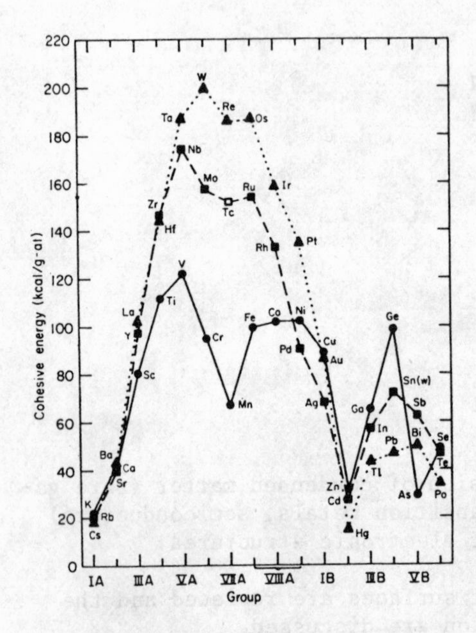

Fig. 1 : Variation of the cohesive energy along the periodic table, after GSCHNEIDER.[1]

The study of the properties of condensed matter at the atomic level is based entirely on the quantum theory. The electrons (and the nuclei) are responsible for the cohesion of molecules, polymers and condensed phases. All the ingredients of the theory are known : the energy operator (the hamiltonian H), containing essentially the kinetic energy and the electrostatic interactions of electrons and nuclei, and the Schrödinger equation

$$-\frac{\hbar}{i}\frac{\partial \psi}{\partial t} = H\,\psi \qquad (1)$$

where t is the time and ψ the wave function of the system.

* We call here *cohesive energy* the binding energy (per atom), also called "lattice energy" in the case of crystals.
It is different from the *energy of cohesion* defined mechanically as the work required to separate a condensed phase along a plane of unit area.

However, even in the case of a periodic system, the Schrödinger equation cannot be solved without drastic assumptions. Indeed, it involves $\simeq 10^{23}$ variables for a macroscopic piece of matter. Let us recall for comparison that the classical problem of three bodies interacting via gravitational interaction has no exact solution. The problem of electrons and nuclei densely packed and interacting quantum mechanically is far more difficult; in practice the number of severe assumptions that must be made is important. The physical sense comes into play. It is why different models of cohesion are suggested, depending on the nature of the atoms and the crystallographic structure of the matter.

In the following, we consider schematically the different types of cohesion that occur in condensed matter : we analyse their origin and their nature and we discuss the validity of empirical descriptions such as the pair interaction approximation widely used in theories such as crystal growth, calculation of phase diagrams, etc.

Part II is devoted to the study of cohesion in solids. Part III develops the properties of surfaces (electronic, surface tension, etc.) of pure elements and alloys.

II. Cohesion in Solids

In any material cohesion results from the quantum mechanical interaction of electrons (and nuclei). However, the quantum mechanical nature of the interaction is more apparent in some systems (covalent materials, metals) than in others (rare gases or ionic salts). In the latter systems the atoms behave classically to a good approximation, i.e. according to the laws of classical mechanics and electrostatics.

Whenever possible, we discuss the relevance of a (classical) isotropic pair potential approximation; the effect of many body interaction is estimated.

We consider successively different types of cohesion in solids :

a) Rare gases : the van der Waals forces insure the (weak) cohesion of atoms with closed electronic shells. Many body effects slightly alter the simple scheme.
b) Ionic solids (salts : e.g. NaCl) : their structure is determined by classical electrostatics.
c) Normal metals : the conduction band is wide and the s,p electrons are largely delocalized in the bulk : they can be described as being weakly scattered by the ionic pseudopotentials. The weak interaction treated to second order by perturbation theory gives an effective pair potential.
d) Transition metals : in addition to a wide conduction band, a

narrower valence band of d symmetry gives the large cohesive energy and explains the characteristic properties of the transition metals (e.g. magnetism).

e) Covalent structures : the valence electrons form strongly interacting s,p hybrids and the valence bond has an important directional character.

Rare earth and actinides are briefly discussed in the transition metal section.

In the course of this chapter, it is interesting to relate the electronic properties (responsible for the cohesion) to other physical properties like conductivity, optical properties, etc. We do not want to give a detailed theoretical analysis of the different types of cohesion in solids, rather we stress the principles, simple models and results. More detailed theoretical developments can be found in various works.[1-4]

II.1. Rare Gases (group VIIIA)

Rare gas atoms are characterized by closed electronic shells (e.g. Ar : $3s^2 3p^6$). Solid rare gases are transparent and insulating. Boiling point and cohesive energy are very low. The latter can be described by a weak pairwise (van der Waals) potential with an attractive component due to fluctuating dipole-dipole interactions and a repulsive part originating from the difficulty for two closed shells to penetrate.

Two neutral atoms with closed shells (or molecules with saturated bonds, e.g. CH_4) have neither Coulomb nor permanent dipole-dipole interactions. The dominant (however weak) interaction is due to fluctuations of the dipole moment of an atom that produces on an induced dipole on the other atom. Let us call $\vec{p}_1$ the instantaneous dipole on the atom 1 (at a given time t); it produces on atom 2 an electric field $\vec{E}_1$. Its modulus is given by :

$$E_1 \simeq \frac{1}{4\pi\varepsilon_o} \frac{P_1}{r^3}, \tag{1}$$

where r is the separation of atoms 1 and 2, and ε_o is the static dielectric constant. This field gives rise to an induced dipole $\vec{p}_2$,

$$\vec{p}_2 = \alpha\, \varepsilon_o\, \vec{E}_1, \tag{2}$$

where α is the polarizability of atom 2.

The interaction energy is then

$$V_a(r) = - < \vec{p}_2 \cdot \vec{E}_1 > = - \frac{C}{r^6}, \tag{3}$$

which clearly shows the r^{-6} dependence of the interaction potential. The classical theory explains neither the origin of the fluctuating dipole nor the way to calculate its average. The parameter C is of the order of 10^{-58} erg/cm^2 so that $V_a(r)$ at the equilibrium separation is of the order of 10^{-2} eV ≈ 100 K.

Quantum mechanically the van der Waals attraction (also called dispersion force) can be described by a simple model. An atom is characterized by its electronic polarizability α and an oscillator frequency ω_o. The frequency ω_o corresponds to the mean energy of the transition from the ground state to electronic excited states, usually approximated by the ionization energy.

When two atoms interact, the oscillator frequency changes according to the following scheme. Before they interact, two atoms have a threefold degenerate oscillator frequency ω_o. After the interaction, four coupled oscillator frequencies are present.

$$\omega_1 < \omega_2 = \omega_3 < \omega_4 = \omega_5 < \omega_6$$

They are given by the following expressions.

$$\omega_{\substack{1\\6}} = \omega_o \sqrt{1 \pm 2\frac{\alpha}{r^3}};$$

$$\omega_{\substack{2,3\\4,5}} = \omega_o \sqrt{1 \pm \frac{\alpha}{r^3}}; \tag{4}$$

The interaction energy is equal to the shift of the oscillator frequencies

$$V_a(r) = \frac{h}{2} \sum_{i=1}^{6} (\omega_i - \omega_o) \simeq - \frac{3}{4} h \omega_o \frac{\alpha^2}{r^6} \tag{5}$$

when developed to the second order in perturbation theory.

In this case, despite the quantum mechanical nature of the interaction, it is possible to approximate correctly the interaction with a classical isotropic pair interaction (centred forces).

The repulsive energy originates in the Pauli exclusion

principle. When atoms are brought close together, the electron clouds interpenetrate but since two electron cannot be in the same quantum state, they must occupy excited states and their kinetic energy increases. The variation with distance of the repulsive potential cannot be estimated simply on theoretical grounds. It is usually admitted that a r^{-12} potential correctly reproduces the experimental results; an exponential (Born-Mayer) potential could be adapted as well. Let us remark that the elements with open shells have a smoother repulsive potential, r^{-m} $(m < 12)$. In summary, the potential due to dispersion forces can be written in the form

$$V(r) = 4\, V_o \left(\frac{\sigma^{12}}{r^{12}} - \frac{\sigma^6}{r^6} \right), \tag{6}$$

called Lennard-Jones potential, where V_o is the depth of the potential well at the equilibrium distance $r_o = (2)^{1/6}\sigma$. Note that, for the stability of condensed matter, the repulsive potential must decay more strongly with distance than the attractive part of the potential.

When several atoms are packed together (crystal, liquid) the total potential energy is the sum of the central potentials (neglecting many-body interactions)

$$E_c = \frac{1}{2} \sum_{i<j} V(r_{ij}), \tag{7}$$

where the coefficient 1/2 is put to avoid double counting.

The most stable configuration of atoms linked by a pair potential (at 0°K) is the compact structure : face centred cubic (fcc) or hexagonal close packed (hcp), essentially for geometrical reasons and due to the "hard" repulsive potential. In fact, around a hard sphere it is just possible to put twelve spheres in contact. The rare gases crystallize in the fcc structure except the lightest He which is hcp. The equilibrium distance is $R_o = 1.09\ \sigma$ and the cohesive energy per atom is $E_c = 8.6\ V_o$. Notice that the relative energy difference between fcc and hcp is $\simeq 10^{-5}$, even after including the many body corrections.

Let us notice finally that liquid (or amorphous) structures are also dominated by the repulsive part of the potential as shown first by BERNAL and FINNEY and subsequently by molecular dynamics or Monte Carlo computer "experiments"[6]. The nearest neighbor distance is slightly dispersed around its value in the crystal lattice and the average coordination is between 11.5 and 10.5, depending of the temperature, as shown by diffraction experiments (X-rays or neutrons). More details are given in the chapter "Crystal-Melt,

Crystal-Amorphous and Liquid-Vapor Interfaces".

II.2. Ionic Solids

Let us consider a compound in which the two components have a very different electronegativity, e.g., NaCl. A charge q "flows" from the less to the more electronegative element. The Coulomb interaction dominates the cohesion. Charge alternation gives rise to a net attractive potential. The so-called Madelung energy,

$$V_a(r) = -\alpha\frac{q^2}{2\pi\varepsilon_o r} \tag{8}$$

(where r is the distance between nearest neighbors and α is a geometrical parameter that varies slightly with the crystalline structure, see table 1) gives the balance between attraction of charges of opposite sign and repulsion of charges of the same sign.

Table 1 : Values of parameter α and coordination number for some ionic structures.

Structure	α	Coordination number
NaCl	1.747	6
CsCl	1.762	8
ZnS	1.638	4

From the point of view only of electrostatic contributions, the CsCl structure is slightly favored among the alternant structures listed in table 1. In order to prevent collapse of the structure, an empirical repulsive pair potential must be assumed similar to that of rare gases (Na^+ and Cl^- have close shells) in which the Pauli principle does not allow the electronic clouds to interpenetrate. One usually assumes a Born-Mayer repulsive potential

$$V_r(r) = Ae^{-pr}, \tag{9}$$

where p is a constant. $V_r(r)$ is of the order of 10 % of the total energy E_c at the equilibrium separation.

The cohesive energies of ionic solids are usually high ($E_c = E_a + E_r \simeq 160$ kcal/mole) and the NaCl structures are usually encountered (Fig. 2).

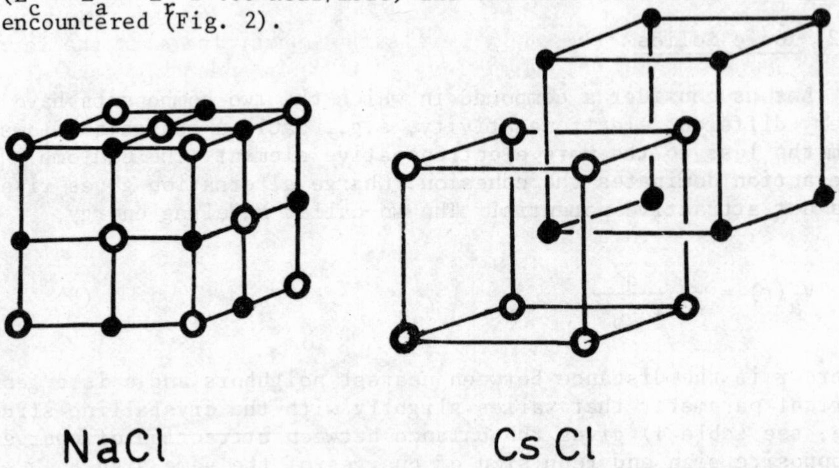

NaCl Cs Cl

Fig. 2 : Rocksalt and Cesium chloride structures.

Let us add that in the liquid phase (molten salts) a[7] coordination number between unlike ions $z_{+-} \simeq z_{-+} \simeq 7$ is measured and the coordination number between like atoms is $z_{++} \simeq z_{--} \simeq 12$-13. This is quite close to the values of the NaCl crystalline structure ($z_{+-} = 6$ and $z_{++} = 12$) showing again the similarity of the local environment in crystalline and liquid phases.

II.3. Metals

Metals occupy the largest area in the periodic table. Pure metals crystallize in compact structures (hcp, fcc, double hcp) or nearly compact structures (bcc); the same remark can be drawn for liquid metals (average coordination number z smaller but still close to 12).

Metals have in common many physical properties related to their low energy excitations : electrical conductivity, magnetic susceptibility, brillance, etc.

However, the metals have different types of behaviours and can be classified according to the electronic structure of their atoms into normal metals (monovalent, divalent, trivalent, resp. groups IA, IIA, IIIA), transition metals (IIIB to VIIIB), noble metals (IB) rare earths, and actinides. In the following sections, we study the electronic and cohesive properties of normal and transition metals. Their different behaviours (large conductivity and weak cohesion in normal and noble metals, smaller conductivity and large cohesion in transition metals, cf. Fig. 1 can be understood qualitatively as follows.

Let us consider the enegy levels that participate to the formation of bands around the Fermi level : e.g. 4s and 3d for the elements ranging from K to Zn.

Fig. 3 : 4s and 3d wave functions (schematic); r is the distance from the nucleus.

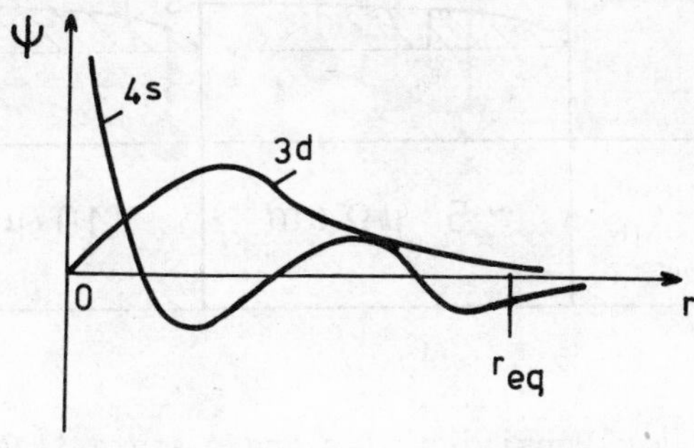

Near the equilibrium separation r_{eq}, the 4s wavefunction has a larger amplitude than the 3d wave function which is more localized around the ion core. (Fig. 3). When atoms form a condensed phase, the overlap ($S_{ij} = \langle\psi_i, \psi_j\rangle$) and the resonance interaction ($\beta_{ij} = \langle\psi_i|V_i|\psi_j\rangle$, V_i is the ionic potential) between the s wave functions is larger than between the d wave functions. The s band width ($W_s \simeq 15eV$) is consequently larger than the d band width ($W_d \simeq 5eV$). In other words, the residence time around the ion core $\tau \simeq h/W$ is larger for the d electrons compared to the more delocalized s electrons.[8] The latter are in fact highly delocalized (nearly free electrons); their wave function extends uniformly over all the crystal. When going from the left to the right of the periodic table, the d band is filled with electrons. The normal metals have their Fermi level in the s band while the transition metals have a partially filled d band. In noble metals, the d band is completely filled so that the Fermi level again falls in the s-p band (Fig. 4). Notice that the conduction electrons are called s and p although their properties and wave functions in the solid are very different from the s and p orbitals of the atom.

One observes a larger electrical resistivity in transition metals due to the scattering of[9] the conduction s electrons by the (more) localized d electrons.

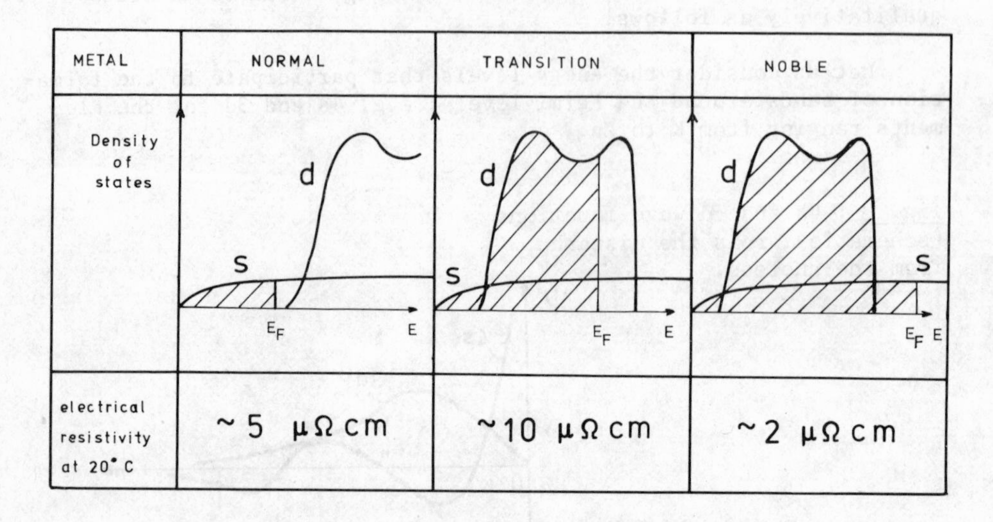

Fig. 4 : Densities of s and d states (schematic) and electrical resistivity of normal, transition and noble metals.

II.4. Normal Metals

The cohesive properties of normal metals (monovalent alcaline, divalent and some trivalent IIIA) are dominated by the s (and p) delocalized electrons. They behave as nearly free electrons i.e. electrons weakly scattered by the *pseudopotentials* of the atoms. The weak pseudopotential is the result of the (nearly compensated) bare electrostatic potential of the ion and the orthogonality condition to the core electrons. The pseudopotential is a Coulomb potential outside the ionic cores but remains finite and small in the cores. A detailed description of the important concept of pseudopotential is outside the scope of this lecture and can be found in text books on Solid State Physics[4,5] and reviews[11].

In the description of the electronic properties of a normal metals, one starts with the uniform electron gas; the density of states (i.e. the number of electronic states per energy unit) is then proportional to the square root of their energy. The positive ion cores are replaced by a uniform positive background of the same density in order to preserve the charge neutrality. This is called the jellium model. The pseudopotentials are then added and treated to second order in perturbation theory. As a result of a rather technical calculation, one finds that the total energy is the sum of a volume dependent term E(V), which does not depend on the crystallographic structure, and a sum of pair interactions (the volume

V bieng fixed) :

$$E_c = E(V) + \frac{1}{2}\sum_{i,j} w(r_{ij}). \qquad (10)$$

E(V) is the leading term containing the kinetic energy of the electrons 3/5 E_F (E_F is the energy of the Fermi level) the various electrostatic interactions, the exchange (due to the Pauli principle) and correlation (Coulomb interactions) terms. The second term is plotted in Fig. 5. It has an oscillatory behavior

$$w(r) = C\,\frac{osc(2k_F r)}{r^3}, \qquad (11)$$

where k_F is the Fermi wavevector and *osc* is an oscillating function.

Fig. 5 : Pair interaction potential deduced from the pseudopotential theory.

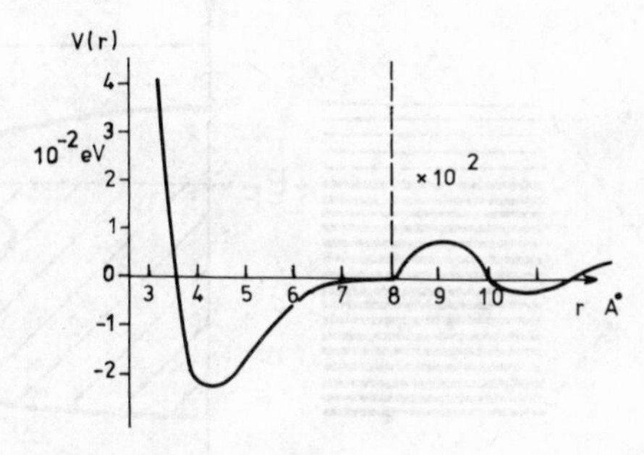

Let us remark that in the theory of normal metals, the attractive and repulsive terms are treated together. The most stable crystalline structures are correclty predicted by the pseudopotential theory : hcp for the mono and divalent metals (at 0°K), fcc for Al, as well as the atomic radii (within ≃ 10 %). The pseudopotential theory can be applied without additional complications to the liquid normal metals since the structural parameter required g(r), that is the pair correlation function, is just the quantity given by diffraction techniques.

One finds liquids of average coordination number Z ≃ 11 at the melting temperature. Note that the normal liquid metals are closer to the free electron gas than the crystal, because of the reduced coherent scattering of electrons (van Hove singularities in the case of crystals).

II.5. Transition Metals

The d electrons are largely responsible for the high cohesion of transition metals.[3] Figure 1 shows the typical variations of the cohesive energy in the 3d, 4d and 5d series. In the last two series, one finds a maximum of cohesion for an approximately half filled d band.

A complete description of the electronic properties of transition metals is difficult because s-p and d electrons must be considered simultaneously. However, a simple treatment proposed by FRIEDEL (tight binding scheme) accounts for the general trends and the order of magnitude of E_c. The bonding in transition metals is of the same nature as the covalent bond in the hydrogen molecules in which energy is gained by filling the bonding states of the molecule.

When transition metal atoms are packed, the atomic energy level E_d broadens into a d-band that is partially filled (Fig. 6).

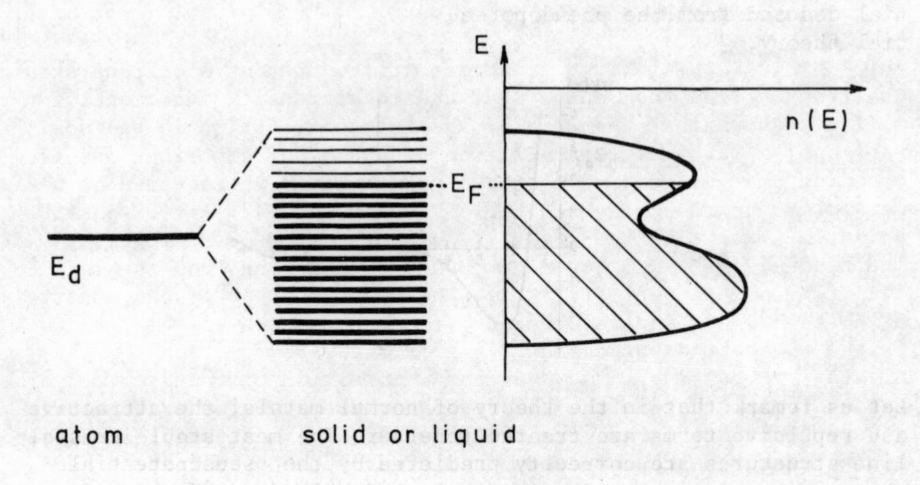

Fig. 6 : Formation of the d band in transition metals.

The cohesive energy (per atom) is given by the difference between the average energy of the occupied levels and the atomic d level E_d, i.e.

$$E_c = \int^{E_F} E\, n(E)dE - E_d n_d \tag{12}$$

where n_d is the number of d electrons.

A simple calculation assuming a constant density of states between $E_d - W/2$ and $E_d + W/2$ (W is the bandwidth) shows that the cohesive energy varies parabolically with the number of d electrons (band broadening contribution)

$$E_c = E_o(n_d) + 5\,\frac{n_d}{10}\left(1 - \frac{n_d}{10}\right) W \qquad (13)$$

where $E_o(n_d)$ is an extra contribution arising from the atom.

This formula shows that the maximum of cohesion $E_c = 1.25$ W is realized for the half filled band ($n_d = 5$) as shown experimentally for the 4d and 5d series (cf. Fig. 1). The 3d series has a strong departure from the simple parabolic law, presumably due to correlation effects[17] (these elements are magnetic). Since the bandwidth does not vary much along a given series but increases from 3d to 4d and 5d series (4 to 10 eV), one finds a correct order of magnitude for E_c and the ordering $E_c^{3d} < E_c^{4d} < E_c^{5d}$.

A more sophisticated treatment of the d and s electrons such as the renormalized atom model[13] or the local density functional formalism[14] shows that the band broadening contribution is the dominant one; a fortunate cancellation of the other contributions is observed. Let us mention that the relatively high cohesion of noble metals is due to s-d hybridization. Using eq. (12), it is possible to explain the relative stability of crystalline structures at 0°K.[5] The sequence is fcc - hcp - bcc - hcp - fcc going from the left to the right of the periodic table. The precise shape of the density of d states has been calculated[15] for the purpose in the framework of the tight binding scheme (Fig. 7). The wave functions of the crystal are considered as a linear combination of the atomic d orbitals (fivefold degenerate). A resonance energy between two atomic d wave functions ψ_i and ψ_j, $\beta_{ij} = \langle\psi_i|V|\,\psi_j\rangle$ is assumed, i and j being nearest neighbor sites. The larger β, the wider the d band. A second moments argument shows that the effective bandwidth W_{eff} varies like

$$W_{eff} \propto \sqrt{z}\ \beta \propto E_c \qquad (14)$$

as does E_c according to eq. (13). This shows that the cohesive energy does not vary like the number of nearest neighbors but rather like $\sqrt{z}$. Thus, the cohesive energy is *not* a sum of pair energies.

The bcc density of states shows a typical splitting of the band into bonding and antibonding states with a dip in the middle of the band. Hence, this structure is favored for a half filled band.

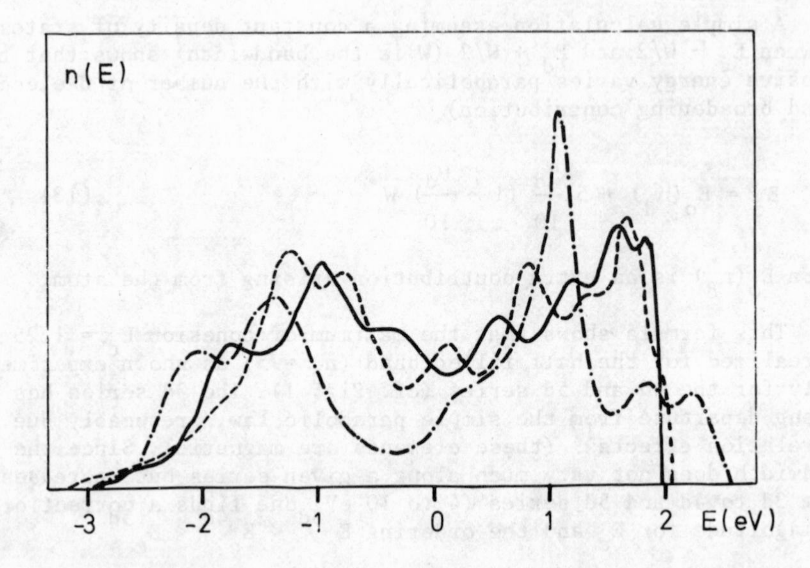

Fig. 7 : Densities of states of various structures --- fcc; —— hcp; -·-·- bcc [15].

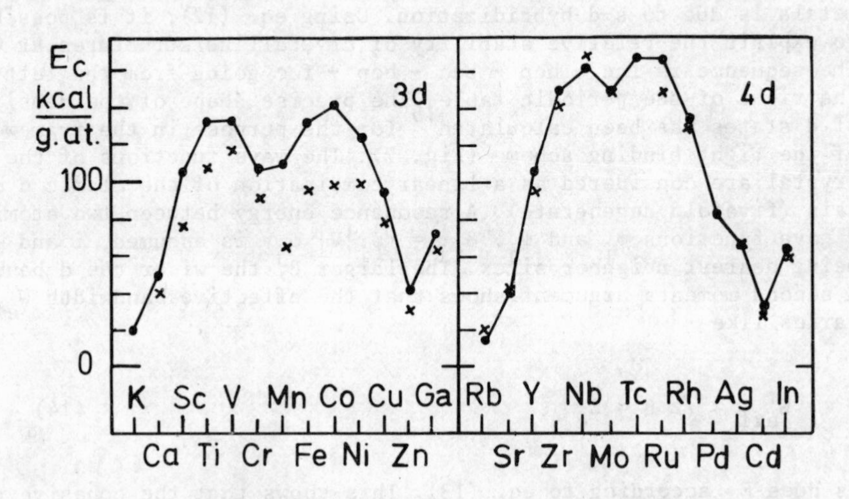

Fig. 8 : Calculated [14] (—) and experimental (×) values [1] of the cohesive energies in the 3d and 4d series.

This can be thought as a weak covalency effect. In order to calculate the elastic moduli or the phonon spectrum, an empirical repulsive pair potential must be added; it is of the order of 30 % of the total cohesive energy and is due essentially to the s electrons.[5] More sophisticated numerical calculations have been performed on crystalline transition metals. Fig. 8 shows the accuracy (≃ 10 %) of the local density functional method applied by WILLIAMS et al. to transition metals;[14] no adjustable parameter are involved. These results require a long computation time and the method is restricted to simple crystalline structures.

Finally, the rare earth elements also have a cohesion dominated by s-p-d electrons; the very localized f electrons does not play a significant role. The actinides should have a cohesion dominated by their f-d electrons. All these metals are also compact (hcp, double hcp, Sm, fcc).

We will not discuss here the cohesion (also due to d electrons of transition metal alloys. A charge transfer from one component to the other arises relative to the pure elements. In disordered alloys, the configurational average greatly complicates the calculations and restricts them to simple descriptions of the electronic structure. However in completely disordered alloys, mean field approximations have been shown useful, such as the coherent potential approximation (CPA).[16] The general trends of the variation of the formation energy are obtained and significant deviations from the regular solution behaviour is observed.[17] However, one is still far from predicting the phase diagram on the basis of electronic properties of the components.

II. 6. Covalent Structures

For the sake of simplicity, we restrict ourselves to group IV A elements (C, Si, Ge, Grey Sn) that show the most typical covalent bond; extensions to elements containing 3 to 6 sp valence electrons can be made. The covalent bond can be defined as the sharing of electrons in bonding orbitals similar to the formation of H_2 molecule (Fig. 9). We have learned a great deal about the mechanism of covalent bonding from organic chemistry.

In the bonding orbital, the density of charge is increased between the two nuclei so that electrostatic energy is gained. The situation is quite similar to the formation of covalent bonds in diamond; we discuss it in a rough but physically transparent model.

We start with isolated atoms of carbon : having the electron configuration $1s^2\ 2s^2\ 2p^2$. In order to produce a three dimensional crystal, one of the two 2s electrons must be promoted to a 2p state. Hybrid orbitals directed towards the four neighbouring atoms can then be formed at the expense of a promotion energy $\varepsilon_p - \varepsilon_s$, where ε_p and ε_s are the atomic 2p and 2s levels.

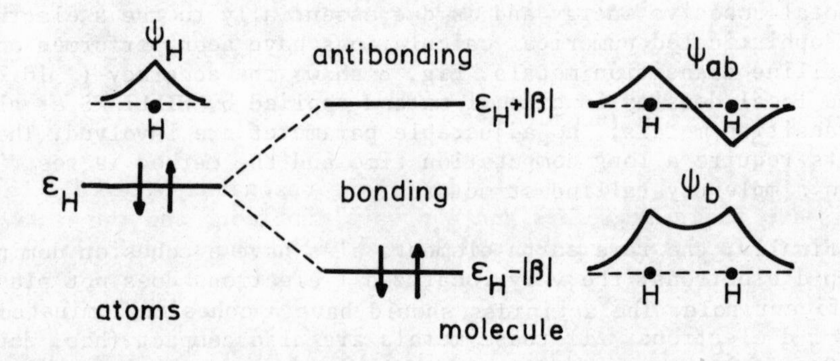

Fig. 9 : Bonding and antibonding levels of the H_2 molecule and their wave functions.

One then forms four sp^3 hybrid orbitals, labeled $|1\rangle$, $|2\rangle$, etc. :

$$
\begin{aligned}
|1\rangle &= \frac{1}{2}(s + p_x + p_y + p_z), \\
|2\rangle &= \frac{1}{2}(s - p_x - p_y + p_z), \qquad (15) \\
|3\rangle &= \frac{1}{2}(s - p_x + p_y - p_z), \\
|4\rangle &= \frac{1}{2}(s + p_x - p_y - p_z),
\end{aligned}
$$

pointing in the $[111]$, $[\bar{1}\bar{1}1]$, $[\bar{1}1\bar{1}]$ and $[1\bar{1}\bar{1}]$ directions respectively at a bond angle $\Theta = 109°$ (Fig. 10). s, p_x, p_y, p_z are the wave functions.

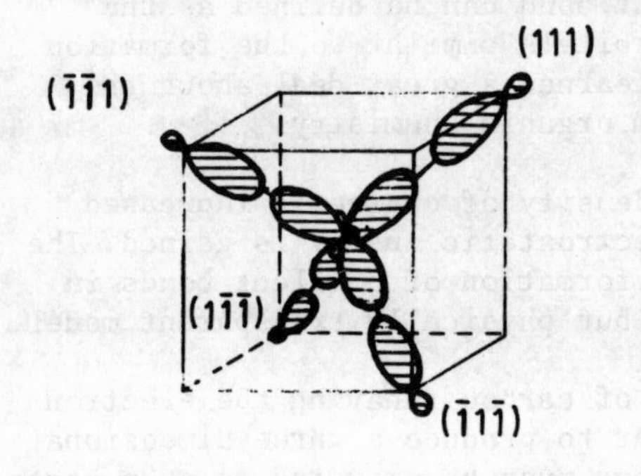

Fig. 10 : sp^3 orbitals in the diamond structure.

Each of these orbitals interacts strongly with another orbital located on a neighboring site that points in its direction thus forming bonding and antibonding orbitals separated by an energy $2|\beta|$. β is the resonance integral defined by

$$\beta = \langle sp^3|V|sp^{3\prime}\rangle = \frac{1}{4}(\beta_{ss} + 2\sqrt{3}\,\beta_{sp} + 3\beta_{pp}), \tag{16}$$

where β_{ss}, β_{sp}, β_{pp} are respectively the resonance integrals between two s wave functions, an s and a p wave function, and two p wave functions. In the molecular model (Fig. 11), the cohesive energy is

$$E_c = -(\varepsilon_p - \varepsilon_s) + 4|\beta|. \tag{17}$$

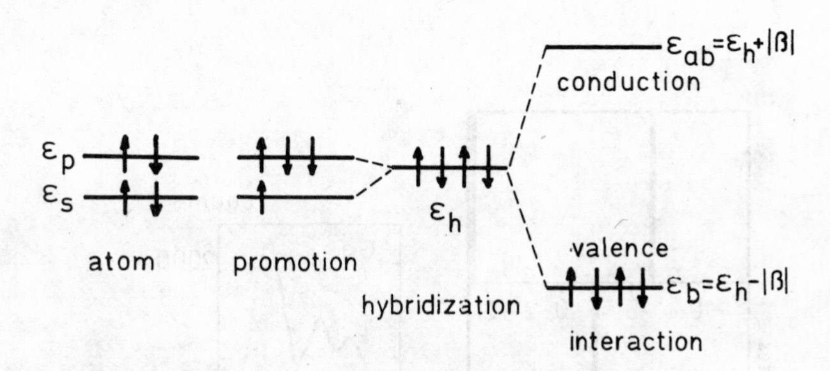

Fig. 11 : Bonding in the molecular model interactions.

Equation (17) expresses the balance between the energy gained by the filling of bonding states and the energy spent for the promotion of one s electron. Some typical values are given in Table 2. It is observed that the theoretical values obtained from this rough model (called the molecular model) are exagereted by a factor or two. Better values are obtained when corrected for the repulsion and correlations.[18]

The inclusion in the model of additional interaction parameters (e.g. interaction between sp^3 orbitals not pointing towards each other) and a proper treatment of the promotion energy broadens the bonding and antibonding levels into valence and conduction bands respectively separated by a gap (Fig. 12). In the valence band, one observes three distinct peaks, the two extremes are the s and p type respectively and the central peak has a mixed nature, in good agreement with the photoelectron spectrum.

Table 2 : Values of the different energies entering eq. (17) and of the experimental cohesive energy E_c^{exp} in a eV for same elements of the IV group.

	$\varepsilon_p - \varepsilon_s$	$-\beta$	$-E_c$	$-E_c^{exp}$
C	8	6.8	19	7.4
Si	6.5	3.8	8.7	4.6
Ge	7	3.9	8.6	3.9

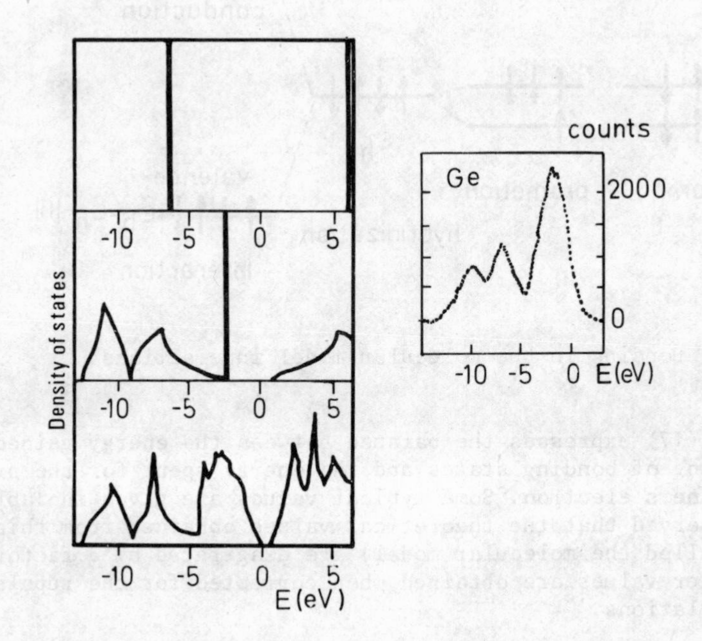

Fig. 12 : Densities of states in various approximations. Top : molecular model; middle : tight binding; bottom : full calculation; insert : photoelectron spectrum.

Let us remark that the trends of the cohesive energy curve (Fig. 1, right part) of group II to VI can be explained using similar arguments to those for transition metals. The maximum energy is obtained for a half filled band (4 sp electrons).[18]

In covalent systems, the structure is dominated by the directional character of the bond. Contrary to the metallic structures, C, Si, Ge are observed in the diamond structure with a relative bond angle of 109°. Si and Ge are also found in amorphous structures containing five fold and seven fold rings in addition to the six fold rings. The diffraction experiments show relative variations $\Delta r/r \simeq 1$ % and $\Delta\Theta/\Theta \simeq 7$ % and weak variations of the photoelectron spectrum relatively to the crystal. The gap still exists even if influenced by dangling bond states.

The case of molten Si and Ge is very special : the average coordination number equals ≃ 6.5 at the melting temperature,[19] a quite unusual value. It is conjectured that, similar to AuSi alloys,[20] the s states are completely filled, leaving the p band with 4 holes. The cohesion could then be of p type with bond angles aroung 90° and 6 nearest neighbors. Notice that white Sn is metallic with 6 nearest neighbors.

Using the molecular model just described and an empirical repulsive pair potential

$$V_r(r) = C\, e^{-p(r-r_o)} \qquad (18)$$

with $pr_o \simeq 4$ (where r_o is the equilibrium distance), it is possible to calculate deformation energies, force constants, or phonon dispersion curves. One obtains to second order[21]

$$E = \frac{pq}{1 - q/p}\, \beta(r_o) \sum_{ij}{}' d_{ij}^2 + \beta(r_o) \sum (\delta\Theta)^2 \qquad (19)$$

where d_{ij} is the length variation of nearest neighbors and d the angular variation; $\beta(r)$ has an exponential variation $\beta(r) \simeq \beta(r_o) \exp|-q(r - r_o)|$. One then finds good agreement with experimental data using realistic values for $\beta(r_o)$, p and q as shown on table 3, except for the transverse acoustic modes of phonons in Si and Ge, presumably because of the screening effect not included in the formalism.[21]

Finally, iono-covalent systems can be treated in the same way, with inclusion of charge transfer effects. III-V and II-VI semiconductors, perovskites spinels are discussed extensively in ref. 4.

Table 3 : Elastic constants $C_{11} - C_{12}$ and transverse phonon frequencies ω_{TA} for some elements of the IV group.[21]

	$C_{11} - C_{12}(10^{12}dyn/cm^2)$		$\omega_{TA}(10^{12}Hz)$	
	calc.	exp.	calc.	exp.
C	7.86	9.5	27	24.1
Si	0.96	1.01	7.5	4.49
Ge	0.78	0.8	4.3	2.39

III. SURFACES

A piece of matter is necessarily limited by a surface through which it interacts with the outside world. The understanding of the mechanisms of adsorption, crystal growth, heterogeneous catalysis, etc., is based on the understanding of the properties of clean surfaces.

In this section, we analyze the relations between the geometry the electronic properties and some thermodynamic properties of surfaces. In particular, we consider the existence of surface electronic states, the work function, the surface tension and the surface composition of alloys. The surface is the region where the atomic and electronic density goes from the bulk value to zero. This quasidiscontinuity produces various effects on the electronic states. Near the surface, one can have either a vanishing bulk wave function (Fig. 13a) or a resonant surface state (its amplitude is increased near the surface) (Fig. 13b) or a surface state (its amplitude decays exponentially in the bulk) (Fig. 13c).

In the following we consider the surfaces of normal metals, transition metals and semiconductors.

III. 1. Normal Metals

In the first attempt to calculate the surface energy of a normal metal the crystal[22] was assumed to be limited by a barrier of infinite height so that the wave function and the electronic density are zero at the surface (Fig. 14). This increases the kinetic energy of the electrons : or surface tension appears with a value several times higher than the experimental results (Fig. 15).

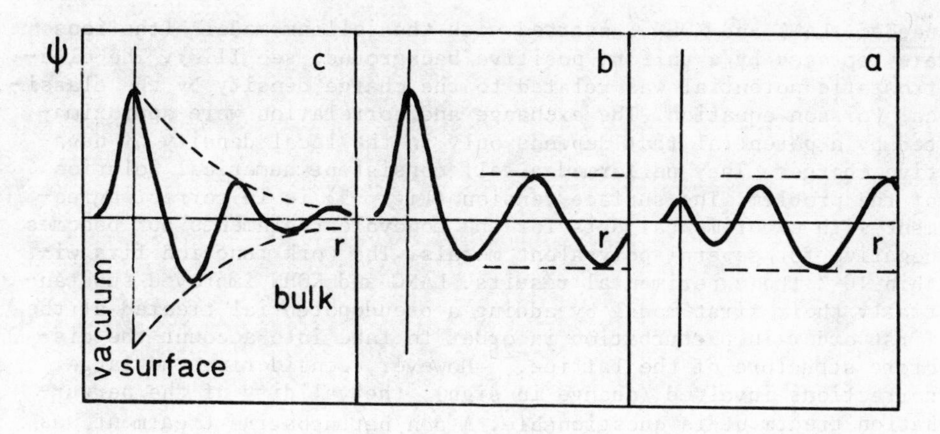

Fig. 13 : Wave function near the surface. r is the distance from the surface.

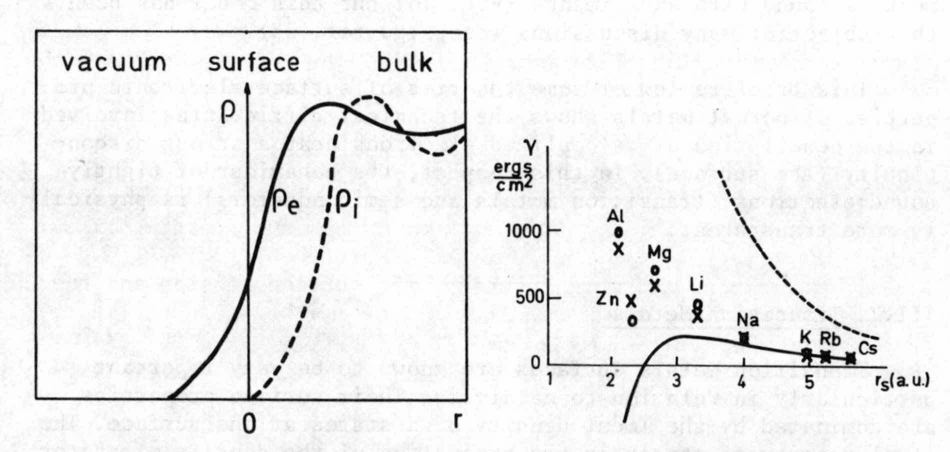

Fig. 14 : Density of charge near the surface
ρ_e : LANG and KOHN[24]
ρ_i : impenetrable barrier[22]

Fig. 15 : Surface tension
0 experimental
--- impenetrable barrier[22]
—— LANG and KOHN[24]
× id.+ pseudopotential corrections[25]

This crude model also does not predict correctly the work function, as the electrons cannot flow outside the surface and produce a surface dipole. Moving the barrier[23] outside the crystal or reducing it has not been very successful.

A more proper account of the potential energy is obviously

needed. LANG and KOHN[24] started with the jellium model (the ions are replaced by a uniform positive background, see II.3). The electrostatic potential was related to the charge density by the classical Poisson equation. The exchange and correlation were approximated by a potential that depends only on the local density of negative charges. They performed a self consistent numerical solution of the problem. The surface tension (Fig. 15) is in correct agreement with experimental data for the monovalent elements but becomes negative for several polyvalent metals. The work function fits within 10 % the experimental results. LANG and KOHN improved substantially their first model by adding a pseudopotential treated at the first order in perturbation in order to take into account the discrete structure of the lattice.[25] However, considering the large corrections involved (change in sign), the validity of the perturbation treatment is questionable. A non perturbative treatment was successfully proposed by APPELBAUM et al.[26] Finally, let us mention the surface plasmon model of surface tension where the surface tension is related to the zero point energy $h\omega_s$ of the surface plasmons created at the expense of volume plasmons.[27] A remarkable agreement is found with experiments (Fig. 16) but this model has been the subject of many discussions (see,e.g.,ref. 23).

This brief review of some theories of surface electronic properties of normal metals shows the technical difficulties involved in the description of delocalized electrons near a strong discontinuity (the surface). In this respect, the behaviour of tightly bound electrons (transition metals and semiconductors) is physically more transparent.

III.2. Transition Metals

Transition metals surfaces are known to be very important particularly in relation to catalysis. Their surface properties are dominated by the local density of d states at the surface. The local density of states is the projection of the density of states for a given atom (or a given orbital). If in a perfect crystalline solid all the local densities of states are identical, they differ, however, in a disordered system or near a surface.In the latter case, the surface density of states can be analyzed by photoemission experiments at low penetration depth ($h\nu \simeq 80$eV).

Pure metals. Due to the variation of the local environment at the surface (mainly the reduction of the number of nearest neighbors), the surface density of states shows a reduction of its effective width compared to the bulk (see eq. (14) and Fig. 16). This has two consequences. First, the cohesive energy of a surface atom is decreased (quantitatively it is different from the broken bond model); second, a charge transfer may occur from the bulk atom to the surface. Indeed, the number of d electrons, equal to the integral of n(E) (Fig. 16) up to the Fermi level, is different in the bulk and at the surface. Consequently, a self-consistency of the charge at the surface must be considered taking the Coulomb energy into account. Roughly speaking, when there is an excess charge at the surface, self-consistency moves the band towards high energies.

Extensive calculations have been performed for various surfaces of transition metals,[15] including stepped surfaces[28] using a parametrized tight binding model. It is observed that the dense planes (111) in fcc, (110) in bcc, behave not too differently from the bulk while the less dense planes show a central peak in the density of states[23] (Fig. 13).

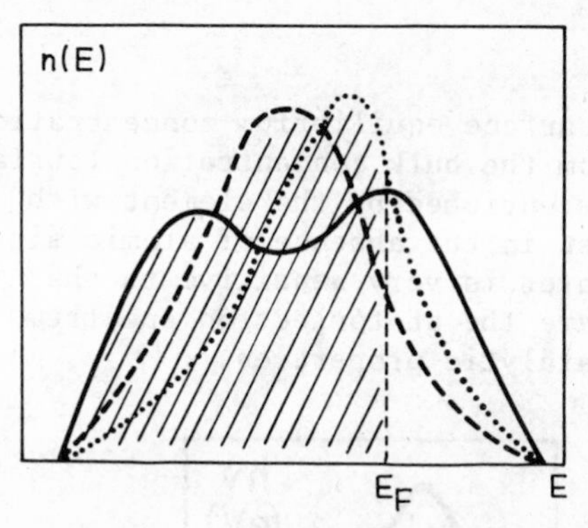

Fig. 16 :
—— bulk density of states
--- surface density of states
... id. with charge self consistency (schematic)

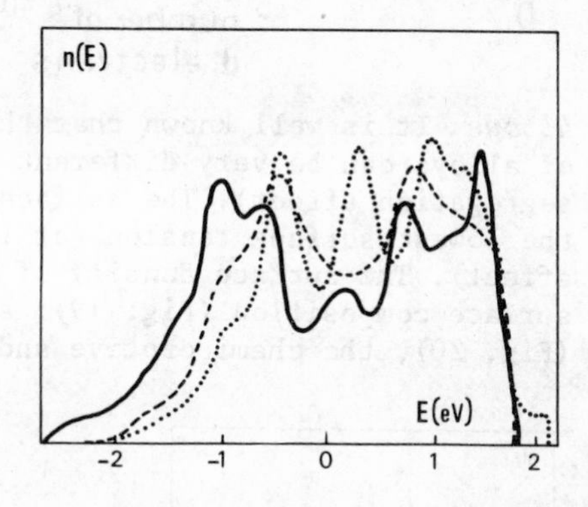

Fig. 17 : fcc transition metal
—— bulk density of states
--- (111) density of states
... (110) density of states (ref. 15).

The local density of states on the different planes below the surface converges rapidly towards the bulk density of states as expected in a system of electrons tightly bound to the ionic cores. The local density of states depends mainly on the local atomic environment in the first two or three shells of neighbours.

The surface tension and its anisotropy calculated on the basis of this model is show in Fig. 18. In general, there is agreement with a simple model of broken bonds ($\gamma_{110}<\gamma_{111}<\gamma_{100}$ in bcc) but one observes a different behavior for a small number or a large number of d electrons ($\gamma_{100}\leqslant \gamma_{111}$).

For more details, references 15, 23 should be consulted.

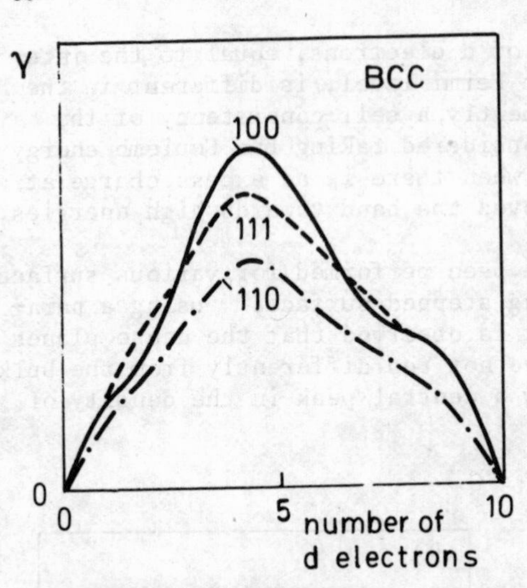

Fig. 18 : Variation of the surface tension with the number of d electrons.

Alloys. It is well known that the surface equilibrium concentration of alloys can be very different from the bulk concentration (surface segregation effect). The surface is enriched by the element with the lowest surface tension (at least in the absence of atomic size effect). The surface density of states is very sensitive to the surface composition (Fig. 19); so are the photoelectron spectrum (Fig. 20), the chemisorptive and catalytic properties.

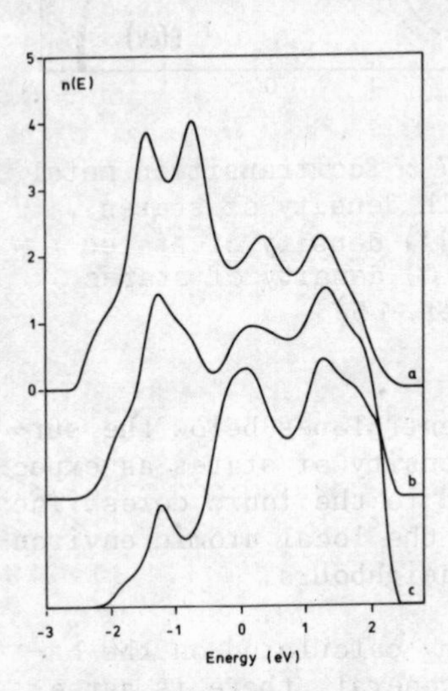

Fig. 19 : Densities of states of Cu_1N_9 with various surface compositions.[29]

a) $Cu_{0.7}\ Ni_{0.3}$
b) $Cu_{0.4}\ Ni_{0.6}$
c) $Cu_{0.1}\ Ni_{0.9}$

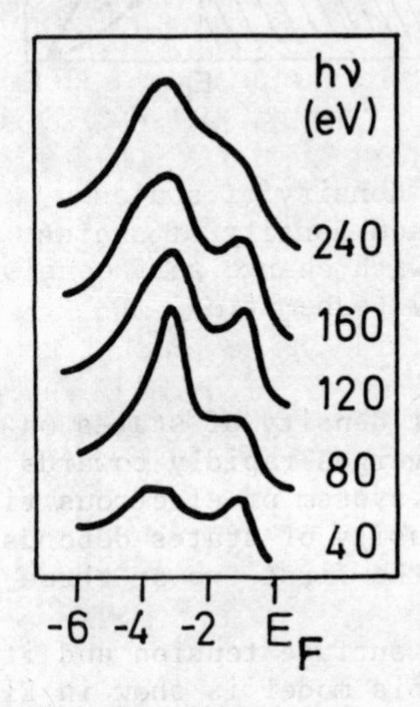

Fig. 20 : Photoelectron spectrum for h = 40-240 e V for Cu/Ni (110). Bulk atomic concentration $Cu_{0.1}\ Ni_{0.9}$
Surface composition $Cu_{0.65}Ni_{0.35}$

Using a simple tight binding scheme, it is possible to compute the energy gained when transferring the less cohesive atoms to the surface. The entropy variations are estimated by standart techniques. The minimum of the free energy F gives the surface concentration (Fig. 21).

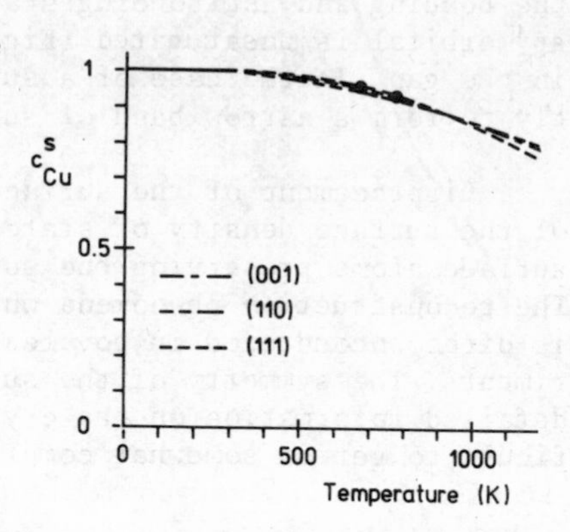

Fig. 21 : Variation of the surface composition with temperature. The dots are the experimental values.

This calculation does not assume any thermodynamic parameter : it is entirely based on the electronic properties. Qualitative differences from a broken bond model are observed regarding the anisotropy.

A systematic study of transition metal alloys gives good agreement with the experimental results except for some elements in the middle of the 3d series. When correlation effects are included with a value U/W = 0.5, much better agreement is obtained.[29]

III. 3. COVALENT STRUCTURES

Again for the sake of simplicity we restrict ourselves to group IV A elements and particularly to the semiconducting structures of Si and Ge, that are the most widely studied. Semiconductor surfaces raise important questions in solid state physics and technology. They have a fundamental interest and plays an important role in the behaviour of electronic components. Indeed large scale integration increases dramatically the contact surfaces : the (still not well understood) properties of semiconductor-metal junctions requires a sound knowledge of the properties of the isolated simple surfaces of semiconductors.

The keywords in the study of semiconductor surfaces are : dangling bond, relaxation and reconstruction. The experimental techniques and results are discussed in chapter "Structure of Clean Surfaces".

The surface electronic states are related to the presence of unsaturated (dangling) bonds at the surface which create a surface electrostatic charge responsible for the surface band bending. In the molecular model presented in II.6, it is easy to explain the occurence of dangling bond states in the gap of the semiconductor. The valence band and conduction band are formed by broadening of the bonding and antibonding states of the molecular model. If an sp^3 orbital is unsaturated (Fig. 22), it gives an electronic state in the gap. In the case of a surface, the gap states broaden slightly to form a narrow band of surface gap states (Fig. 23)

Displacement of the surface atoms changes the characteristics of the surface density of states. A uniform displacement of the surface atoms preserving the surface unit mesh is called relaxation. The reconstruction phenomena which changes the surface unit mesh is often encountered in covalent structures, as shown by LEED experiments. The symmetry of the surface plane is determined but a more detailed information on the crystallography of the surface is difficult to get in somewhat complicated systems.

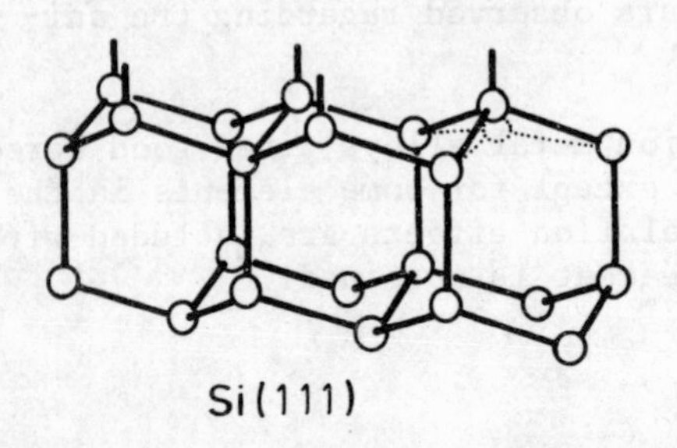

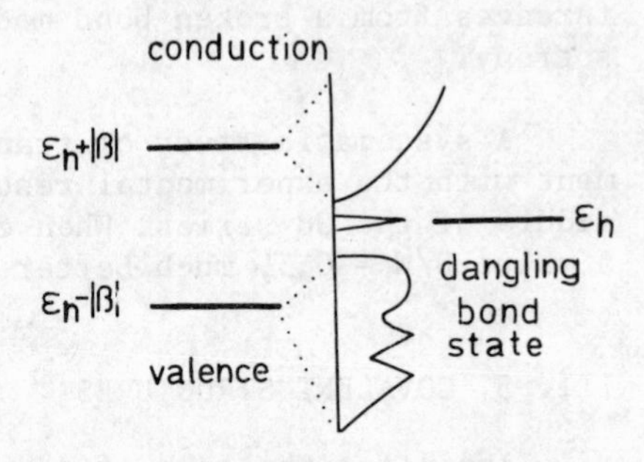

Fig. 22 : Dangling bond orbitals of the (111) diamond structure without (——) and with (...) inwards relaxation.

Fig. 23 : Dangling bond states in the gap of sp^3 bonded structures.

In the sp^3 bonded systems, an inwards displacement of the surface atoms (see Fig. 22) highers the dangling bond level : it becomes more p-like and the back bonds become stiffer as observed by angularly resolved photoemission.[30] LEED experiments demonstrate that the cleaved (111) surface of Si and Ge show a metastable (2 × 1) structure. Si anneals at 350°C to form a stable 7 × 7 structure (Ge shows a 2 × 8 superstructure). If it is easy to explain the occurence of a 2 × 1 reconstruction, the 7 × 7 reconstruction is still an open question, despite the amount of work and models devoted to it.[31] The (111) surface is represented schematically in Fig. 24, in the unreconstructed and 2 × 1 reconstructed model (the so-called buckling model). It is shown, that the surface Si atoms are moved by 0.29 Å inwards and 0.34 Å outwards. The 2 × 1 reconstruction can be explained in the following way. The surface gap states have one electron per dangling orbital (they may accomodate two with opposite spins). By doubling the surface unit mesh, the band of surface gap states is split into two parts; the lowest part is completely filled and the higher part is empty, so that electronic energy is gained relatively to the unreconstructed case (Fig. 25).

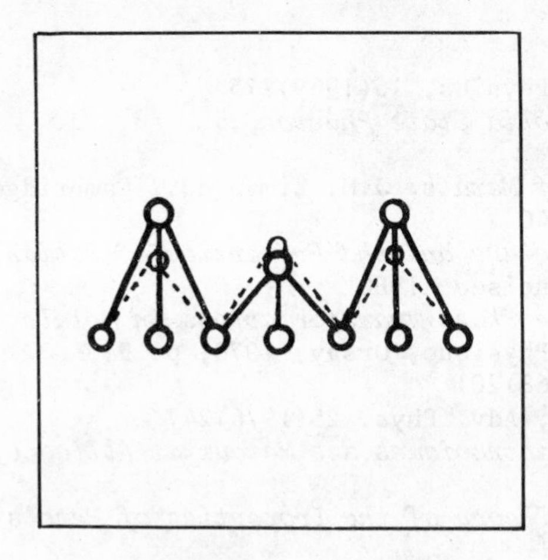

Fig. 24 : Unreconstructed (...) and reconstructed (——) surface (side view)

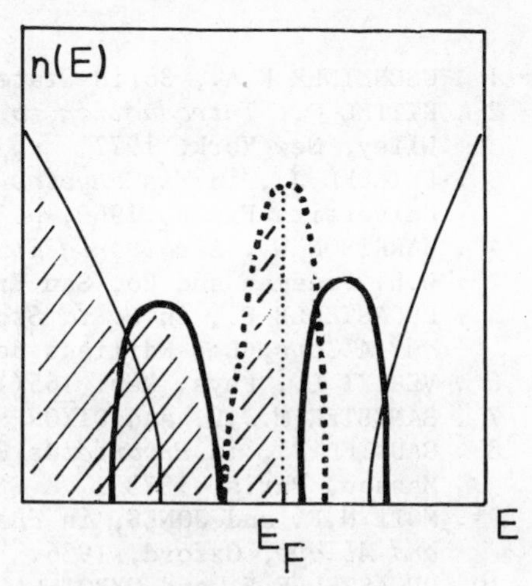

Fig. 25 : Densities of states in the gap region; unreconstructed (...) and reconstructed (——) surface.

The 7 × 7 stable reconstructed surface of Si received much attention. Different models based on vacancies or different rippled surfaces have been proposed.[31] There is no definite evidence for any of these models. The (100) surface of Si is characterized by two dangling bonds per surface atom : it is 2 × 1 reconstructed. The dimer model, in which atoms lying in parallel rows[32] are paired, explains all the experimental data (LEED, photoemission, etc.)

Beyond the simple descriptions of the semiconductor surfaces that allow discussions of the underlying physics, even when the system is not periodic, sophisticated calculations have been performed on simple crystalline surfaces. We do not give details here on these methods, the reader should consult ref. 33. Both type of approaches have their own usefulness and are complementary. We do not consider here the case of polar surfaces of binary compounds (e.g. GaAs) which is of increasing interest.[34]

Despite the inherent complexity of solid state matter at the atomic level, a schematic understanding of the physics is reached. We insisted here on the simplest description of the cohesion, electronic properties, etc. that allow general discussions on more complicated systems.

REFERENCES

1 . GSCHEINER K.A., Solid State Physics, 16(1969)275
2 . KITTEL C., *Introduction to Solid State Physics*, 5th ed., J. Wiley, New York, 1977
3 . FRIEDEL J., in *The Physics of Metals*, J.M. Ziman ed., Cambridge University Press, 1969, p. 340
4 . HARRISON W., *Electronic Structure and the Properties of Solids*, W.H. Freeman and Co, San Francisco, 1980
5 . DUCASTELLE F., in *Solid State Phase Transformations in Metals and Alloys*, Les Editions de Physique, Orsay, 1978, p. 51
6 . VERLET L., Phys. Rev. 165(1968)201
7 . SANGSTER M.J.L. and DIXON M., Adv. Phys. 25(1976)247
8 . GAUTIER F., in *Propriétés Electroniques des Metaux et Alliages*, Masson, Paris, 1973
9 . MOTT N.F. and JONES, in *The Theory of the Properties of Metals and Alloys*, Oxford, 1936.
10. DUCASTELLE F. and CYROT-LACKMANN F., J.P.C.S., 31(1970)1295 and 32(1970)285
11. HEINE V. and WEAIRE D., Solid State Physics, 24(1970)250
12. FRIEDEL J. and SAYERS C.M., J. de Phys. 38(1977)697
13. GELATT D.D. Jr, EHRENREICH H. and WATSON R.E., Phys. Rev. B15 (1977)1613
14. MORUZZI V.L., WILLIAMS A.R. and JANAK J.F., Phys. Rev. B15(1977) 2854
15. DESJONQUERES M.C. and CYROT-LACKMANN F., J. Phys. F, 5(1975) 1368
16. ELLIOTT R.J., KRUMHANSL J.A. and LEATH P.L., Rev. Mod. Phys. 46 (1974)465
17. van der REST J., GAUTIER F. and BROUERS F., J. Phys. F, 5(1975) 1884 and 2283
18. FRIEDEL J., J. de Phys. 39(1978)651
19. GABATHULER J.P. and STEEB S., Zeitschr. für Naturforsch., 34A (1979)1314
20. MOUTTET C., GASPARD J.P. and LAMBIN P., Surf. Sci. (to appear)

21. LANNOO M., J. de Phys. 40(1979)30 and in *Electronic Structure of Crystal Defects and Disordered Systems*, Les Editions de Physique, Orsay, 1981, p. 45.
22. BREGER A. and SCHUCHOWITZKI A., Acta Physicochimic, U.R.S.S. 21(1946)13
23. FRIEDEL J., Ann. de Phys. 1(1976)257
24. LANG N.D. and KOHN W., Phys. Rev. B1(1970)4555
25. LANG N.D. and KOHN W., Phys. Rev. B3(1971)1215
26. APPELBAUM J.A. and HAMANN D.R., Rev. Mod. Phys. 48(1976)479
27. SCHMIT J. and LUCAS A.A., Solid State Comm. 11(1972)415
28. DESJONQUERES M.C. and CYROT-LACKMANN R., Solid State Comm. 18 (1976)1127
29. LAMBIN P. and GASPARD J.P., J. Phys. F, 10(1980)2413; LAMBIN P., Thesis, Liège, 1981.
30. MOUZAY F., GUICHAR G.M., PINCHAUX R., THIRY P., PETROFF Y. and DAGNEAUX D., Surf. Sci. 99(1980)28
31. CHADI D.J., Surf. Sci. 99(1980)1
32. PANDEY K.C. and PHILLIPS J.C., Phys. Rev. B13(1976)740
33. APPELBAUM J.A. and HAMANN D.R., in *Theory of Chemisorption*, J.R. Smith ed., Springer Berlin, 1980, p. 43
34. POLLMANN in *Festkörperprobleme (Advances in Solid State Physics)*, XX, J. Trench ed., Vieweg, Braunschweig, 1980, p. 117.

SURFACE THERMODYNAMICS

Boyan MUTAFTSCHIEV
Centre de Recherche des Mécanismes de la Croissance Cristalline
Campus de Luminy, 13288 Marseille Cedex 9
France

I. INTRODUCTION

The purpose of this chapter is to make an inventory of the surface problems arising in the various treatments of phase transitions. As we shall see, both the number and the diversity of the problems are so large that we can be neither exhaustive nor rigorous within the space alotted. We will attempt just to interest readers already involved in different domains of surface science (structuralists, adsorption-, nucleation-, and crystal growth-men) in problems and results from other domains of the same discipline. Such a task requires a more or less unified approach to the topics considered which, if successfully applied, might also eliminate some intransigency of the different schools.

The chapter deals mainly with the principal thermodynamic characteristics of surfaces, the surface free energy. Its classical thermodynamic definition and properties, not attached to any particular system or model are treated in section II which is short since it essentially resumes what the reader can find in advanced courses of physical chemistry,[1] thermodynamics[2,3a] or surface chemistry.[4] In section III the notions of surface tension and surface work, edge and corner tension, etc., are discussed and the principles of their estimation for crystal surfaces are given. Section IV deals with the characteristic parameters of the interfaces between condensed phases. Some of these parameters are transposed to the treatment of two-dimensional layers (Section V), where the notions of spreading pressure and ledge free energy are stressed. Section VI treats the problems of equilibrium shape (in homogeneous phase and on substrate). The statistical-mechanical aspects are limited to two topics. In section VII we consider the behavior of the different faces of a

B. Mutaftschiev (ed.), Interfacial Aspects of Phase Transformations, 63–102.

crystal in contact with its vapor as a special case of adsorption. The problems of surface roughness and melting then follow as a natural sequel. Finally, in section VIII we discuss some doubts concerning the capability of macroscopic parameters to correctly express the free energy of small phases. However, we show that even for cluster sizes of only several atoms, classical thermodynamics can give astonishingly precise results.

II. SURFACE THERMODYNAMICS

Perhaps the most general classical definition of the surface tension originates from the basic thermodynamic differential equation which relates the energy E of the system to its parameters of state[3b] :

$$dE = \frac{\partial E}{\partial S}\, dS + \Sigma \frac{\partial E}{\partial x}\, dx + \Sigma \frac{\partial E}{\partial N}\, dN. \qquad (1)$$

Here S is the (total) entropy of the system, x are generalized geometric parameters and N are the numbers of molecules of the different chemical species in the different phases. It is well known that

$$\left(\frac{\partial E}{\partial S}\right)_{x,N} = T \qquad (2a)$$

is the thermodynamic temperature of the system, while

$$\left(\frac{\partial E}{\partial N}\right)_{T,x} = \mu \qquad (2b)$$

are the partial potentials introduced by GIBBS.[5] The quantity

$$-\left(\frac{\partial E}{\partial x}\right)_{T,N} = X \qquad (2c)$$

can be considered as a generalized mechanical force exerted by a phase tending to increase x. This force can be intrinsic to the phase or due to an external field. We shall limit our considerations to intrinsic forces only.

A moment's reflection leads to the following set of definitions :

(i) $x = V$ (volume) $|cm^3|$;
$X = p$ (pressure) $|dyn/cm^2|$;
$-Xx = -pV$ (work) $|erg|$,

(ii) $x = A$ (surface/interface/area) $|cm^2|$;
$X = -\gamma$ (surface/interface/tension) $|dyn/cm|$;
$-Xx = \gamma A$ (work) $|erg|$;

(iii) $x = L$ (edge/coexistence line/length) $|cm|$;
$X = -\rho$ (edge/line/tension) $|dyn|$;
$-Xx = \rho L$ (work) $|erg|$;

(iv) $x = n_c$ (number of corners) dimensionless;

$X = -\varepsilon_c$ (corner free energy) erg;
$-Xx = \varepsilon_c n_c$ (work) $|erg|$.

First, two remarks : it is clear why the generalized forces ρ and γ must have opposite signs. The pressure of a phase is assumed positive if the phase tends to expand spontaneously, while its surface tension is considered as positive if the area of its surface tends of decrease spontaneously. The product ρL and $\varepsilon_c n_c$ commonly appear as positive terms in equation (1); however, their sign depends on the type of intermolecular forces and the crystallographic orientation. Furthermore, in the above classification, the words between slashes regard systems where surfaces are formed by contact between two condensed phases. We consider this case in section IV, while the basic definitions and estimates are given here and in the next section (III) for a system constituted by one condensed phase (crystal, liquid) and one diluted phase (vapor).

II.1. The "Surface Phase"

The most convenient way of doing surface thermodynamics is to assume, as did GIBBS,[5] that the interface between two bulk phases α and β is a mathematical dividing plane (Gibbs dividing surface) with the interfacial region extending to certain distances from this plane into the respective phases, beyond which the phases have their bulk structure and thermodynamic properties. The values of these distances are of no importance since all extensive parameters and functions of state of the "surface phase" are defined as excess quantities, as can be seen from the following consideration.

No model of the surface phase is required to write a formal expression for the total energy of a system containing the two phases and the interface as

$$E = E^{\alpha} + E^{\beta} + E^{\sigma} \qquad (3)$$

(the superscript σ meaning "surface phase"). The Gibbs convention specifies on the one hand the values of the energies entering eq. (3), namely, E is the total energy of the system when the surface is present, while E^{α} and E^{β} are the energies of the bulk phases α and β with the same extensions as in the real system. On the other hand, it assumes that all extensive parameters of the surface are defined by differences like equation(3). Note that according to the Gibbs convention $V^{\sigma} = 0$.

Equation (1), written for the surface phase of a system containing i chemical species, yields :

$$dE^{\sigma} = TdS^{\sigma} - pdV^{\sigma} + \gamma dA + \sum_i \mu_i^{\sigma} dN_i^{\sigma}. \qquad (4)$$

This relationship, as well as the equations resulting from the Legendre transformations [6] of E as a function of T and V,

$$dF^{\sigma} = -S^{\sigma}dT - pdV^{\sigma} + \gamma dA + \sum_i \mu_i^{\sigma} dN_i^{\sigma}, \qquad (5)$$

or as a function of T and p,

$$dG^{\sigma} = -S^{\sigma}dT + V^{\sigma}dp + \gamma dA + \sum_i \mu_i^{\sigma} dN_i^{\sigma}, \qquad (6)$$

are fundamental formulas for the surface phase for each set of experimental variables. F^{σ} is the Helmholz free energy and G^{σ} the Gibbs free energy of the surface. In the Gibbs surface model ($V^{\sigma} = 0$), $F^{\sigma} = G^{\sigma}$.

II.2. Surface Tension and Surface Free Energy

Equations (5) and (6) can be integrated, holding constant the intensive parameters T, p, γ, μ_i^{σ}. Keeping in mind $V^{\sigma} = 0$, one has :

$$G^{\sigma} = F^{\sigma} = \gamma A + \sum_i \mu_i^{\sigma} N_i^{\sigma}, \qquad (7a)$$

or per unit surface area :

$$f^\sigma = \gamma + \sum_i \mu_i^\sigma \Gamma_i, \quad (7b)$$

where $\Gamma_i = N_i^\alpha/A$ is sometimes called "surface density of the i-th species".

The specific surface free energy f^σ is clearly different from the surface tension γ. The two quantities are equal only in the case of a monocomponent system (no foreign adsorption), when the two phases from both sides of the interface keep their structures and densities unchanged up to the Gibbs dividing surface. This is rather unrealistic, since structural and concentration variations on a pure surface can result from lattice relaxation and/or surface roughness. However, besides the thermal roughness, which will be treated in section VII, the structural changes accompagnying the creation of a new surface of a pure cristalline substance have a small constant contribution to the energy ballance (7b) and can be included in the γ-term. In this case, and for temperature much lower than the melting point, one may consider the surface free energy and the surface tension as identical.

Equations (7) raise an ambiguity regarding the location of the Gibbs dividing surface, from which depend the values of $\mu_i^\sigma \Gamma_i$. The problem is beyond the scope of this chapter. The interested reader is advised to consult more specialized literature.[5,7]

II.3. The Gibbs Adsorption Isotherm

The so called Gibbs adsorption isotherm,

$$-d\gamma = \sum_i \Gamma_i d\mu_i \quad (T = \text{const}) \quad (8)$$

is easily obtained from differentiation of (7a) and comparison with (6). It is a two-dimensional version of the Gibbs-Duhem equation and shows that the surface tension of a substrate follows variations of the chemical potential and composition of the adsorbed layer just as the pressure does in a three-dimensional bulk phase. The analogy between bulk pressure and surface tension has already been made. One usually calls "spreading pressure" Φ of an adsorbed layer, the negative variation of the surface tension of the substrate,

$$d\Phi = - d\gamma, \quad (9a)$$

$$\Phi = \gamma_o - \gamma = \Delta\gamma, \quad (9b)$$

where γ_o is the surface tension of the pure surface (no adsorption).

III. ESTIMATES OF THE SURFACE TENSION OF CRYSTALS

III.1. Surface Tension and Surface Work

Direct measurements of surface tension, that is a force per unit length, are practically limited to the liquids, and sometimes to solid metal wires or foils in a temperature range of high plasticity.[8] Since the increase of the surface area of solids occurs mostly by cleavage, the measurement of a work rather than that of a force seems to be the better choice. The so called "specific surface work" has the same dimensions and numerical value as surface tension. It is equal to one half the cohesion energy of the crystal along the given crystallographic plane, plus the work of surface relaxation or reconstruction, plus the work due to variation of the vibrational modes of the lattice after cleavage.(The coefficient of one half comes from the fact that the specific cohesion energy is the work spent to separate two crystal blocks *along a unit area*, while the specific surface work is the work spent *to create a unit area*. When one cleaves along a unit area, one creates two units of surface area.)

Calculation of the energy balance resulting from the "mental" cleavage of crystals along different crystallographic planes was the first and is still the most used method for estimating surface tension of solids.

III.2. The Surface Energy

Let us suppose that we know the structure and the potential laws of the forces in the crystal lattice, and that in a direction normal to the plane (hkl) whose surface energy (surface tension, surface work) we wish to calculate, we can distinguish Z types of molecules having different distances to this plane. The choice of Z is dictated by the range of action of the intermolecular forces; one assumes that the Z + 1 - th type of molecule does not "feel" the existence of the molecules of the plane. Moreover, Z should be a multiple of the periodicity z of the lattice in this direction ($Z = i_{max} z$). From the latter condition it follows that the binding energy (separation work) between the slice number j parallel to the plane, with a thickness of z molecules, and an identical slice, number j' ($i = j + j' < i_{max}$), situated on the opposite side of the plane, is equal to the binding energy between the slice number j - 1 and the slice j' + 1 on the opposite side of the plane (Fig. 1). Let now decrease the dimensions of the slices j in the directions parallel to the plane, leaving the slices j' with infinite extension in the same directions, until we reach the smallest repetitive mesh of the two-dimensional structure. Suppose that the so formed "elementary blocks" have a base parallel to (hkl) with area a, and call ψ_i the binding energy of the molecules contained in the block number j to all molecules contained in the slice number j' on the other side of the plane. The surface energy is then clearly

$$\gamma_{(hkl)} = \frac{1}{2a}\sum_{1}^{\infty} i\psi_i \simeq \frac{1}{2a}\sum_{1}^{imax} i\psi_i. \qquad (10)$$

Note that the approximate calculation of γ described aan be corrected to the precision one needs, through computer simulation, by estimating the energy gain due to surface relaxation. The method is worked out for potential energy variations only and thus remains valid for low temperatures.

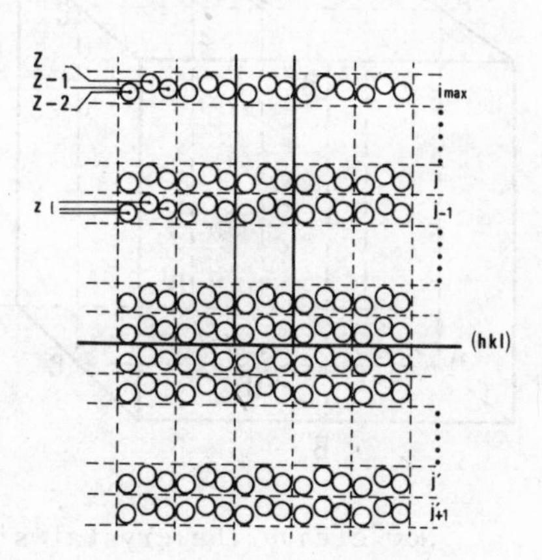

Fig. 1 : Two-dimensional lattice showing schematically the way of calculating surface energy by cleavage along the plane (hkl). For details see text.

III.3. Edge Energy and Corner Energy

BORN and STERN[9] proposed the following way of calculating edge and corner energies. Suppose that we have a cubic block of crystal with edge length L ($L \to \infty$) and an unspecified (continuum) structure. If the work W spent to cleave the crystal along a plane parallel to the cubic faces is known, the surface energy of these faces is :

$$\gamma_{(100)} = W/2L^2. \qquad (11)$$

Suppose that the initial crystal has been mentally divided into four quadratic prisms with a height L and a base (L/2) , labeled 1 to 4 (Fig. 2). Equation (11) can then be written in the form

$$2L^2\gamma_{(100)} = W_{13} + W_{14} + W_{23} + W_{24} \quad (12a)$$

for a cleavage along the plane AA'C'C (the meaning of the subscripts can be deduced from the figure), and

$$2L^2\gamma_{(100)} = W_{13} + W_{14} + W_{23} + W_{24} \quad (12b)$$

for a cleavage along the plane BB'D'D.

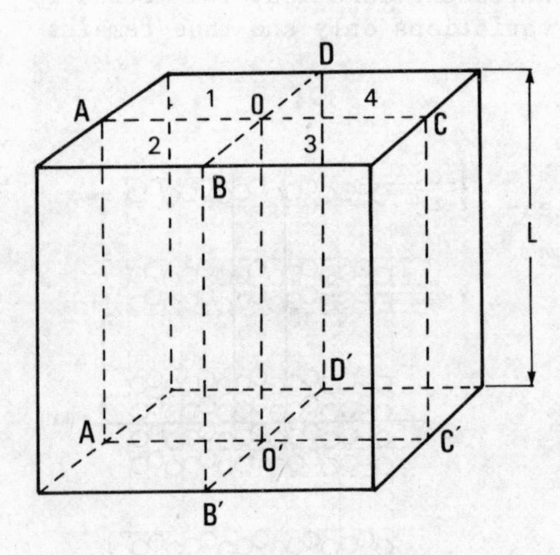

Fig. 2 : A cube of continuum matter used in the calculation of the specific edge energy according to BORN and STERN.[9] For details see text.

Now cleave the crystal simultaneously along the two {100} planes, AA'C'C and BB'D'D. The only difference in respect with the preceding cases is the appearence of four cubic edges OO' with lengths L and edge energy ρ. The energy balance is

$$4L^2\gamma_{(100)} + 4L\rho = W_{12} + W_{13} + W_{14} + W_{23} + W_{24} + W_{34}\,. \quad (13)$$

Taking into account (12a) and (12b), one has :

$$\rho = -\frac{1}{4L}(W_{13} + W_{24}) = -\frac{W_{13}}{2L}\,; \quad (W_{13} \equiv W_{24}). \quad (14)$$

The specific edge energy of the cubic edge is thus equal to minus half the work spent to separate two cubic blocks that touch each other along a unit length of this edge.

By a similar procedure, but dividing the cubic crystal into eight blocks and subsequently creating only faces, then faces and edges, and finally faces, edges and corners,one can calculate the corner energy. For the corners of the cube it is equal to minus half the work spent to separate two cubic blocks with corners touching along the cube diagonals.

The fact that it is easier to cut a polyhedron having edges and corners out of a crystal block than to create the same surface area by cutting (infinite) planes only, is not astonishing in itself. Edges and corners weaken the crystal lattice in their vicinity. However, one can wonder how increasing the length of the edges or the number of corners, at constant volume and surface area, can decrease the free energy of the crystal. The answer to this apparent contradiction is the following. The assertion in both procedures, described by the equations (12) and (13), that the surface area remains constant, is true only for a continuum. When one increases the length of the edges of a real (atomistic) crystal, one increases the number of edges atoms (molecules) which have, according to the continuum approximation, "two molecular areas", one on the side of each face. Physically this is an absurdity. In the simple cubic case, the atomistic expression that should replace (13) is

$$4(L - \delta)^2\gamma_{(100)} + 4L\,\rho_a = W_{12} + W_{13} + W_{14} + W_{23} + W_{24} + W_{34} \tag{13a}$$

where δ is of the order of the linear dimension of a molecule, and the edge free energy ρ_a now includes all interactions regarding the chain of molecules building the cubic edge. Comparing eq. (13a) to the eqs. (12), one obtains instead of (14) the expression

$$\rho_a = 2\delta\,\gamma_{(100)} - \frac{W_{13}}{2L} = 2\delta\,\gamma_{(100)} - \rho, \tag{14a}$$

in which ρ_a is clearly positive.

In any case, the effects due to edge and corner free energies are of second order with respect to those due to surface tension and become noticeable for condensed phases of only few molecules.

III.4. An Alternative Treatment

The so called "capillarity approximation", that is the use of the surface free energy, is only one possible way of expressing the excess thermodynamic properties of small condensed phases. When temperature T, pressure p and number of molecules N are used as independent variables of state, the natural excess thermodynamic quantity (per molecule) is the difference between the size dependent chemical potential μ_N of the small phase and the chemical potential μ_o of the infinite phase at T and p constant.

One can imagine the following procedure to build a (non compressible) condensed phase of any finite size. Molecules are taken from an infinite condensed phase, brought to the saturated vapor with the pressure p_o, then compressed to a pressure p so that their chemical potential becomes equal to that of the phase to be built. Since both evaporation and condensation are performed at equilibrium, the only work spent for the transfer of a molecule from the infinite to the finite phase is the volume work

$$\mu - \mu_o = \int_{p_o}^{p} v dp.$$

The sum of these elementary works is exactly the excess free energy of the molecules in the finite phase with respect to their free energy in the infinite phase. Recalling the capillarity approximation, we can write

$$\gamma A_N = \sum_{i=1}^{N} (\mu_i - \mu_o) \tag{15a}$$

(where A_N is the surface area of the N-molecular cluster), or in continuum form,

$$\gamma A_N = \int_0^N \mu(i) di - \mu_o N. \tag{15b}$$

The described procedure should be terminated by bringing the cluster back from the vapor with pressure p to the initial vapor with saturation pressure p_o. Since we assumed the condensed phase to be noncompressible, no work is done, and equations (15a) and (15b) remain valid. The fact that the cluster will evaporate again under these conditions has no relation to the present problem but will be discussed in the chapter "Three Dimensional Nucleation".

To illustrate the method, we shall consider the case of a size dependent surface energy.

Start with an infinite crystal and form by successive evaporations and condensations of molecules a monomolecular layer with orientation (hkl) of unit surface area (Fig. 3). The elementary work per molecule for these processes will be constant if one evaporates/condensates the molecules from/into repeatable step sites (kinks). Boundary effects can be neglected when the sizes of the initial crystal and the layer under formation are enough large. The chemical potential per molecule of the infinite crystal is

$$\mu_o = \phi_o + Ts_o, \qquad (16)$$

where ϕ_o is the binding energy in a kink position and s_o is the differential vibrational entropy per molecule. We make the common assumption that this entropy depends very little on the surface site and on the size of the crystal, and drops from the differences between chemical potentials. The binding energy ϕ_o can be written as

$$\phi_o = \frac{\omega}{2} + \sum_{1}^{i_{max}} \psi_j, \qquad (16a)$$

where ω is the binding energy of the molecule with its neighbors in the plane of the surface and ψ_j is the binding energy of the molecule with the j-th monomolecular slice of infinite size parallel to the surface below the top lattice plane (we use the same model as in III.2, but simplify it to a structure where the "elementary block" contains one molecule (see Fig. 3).

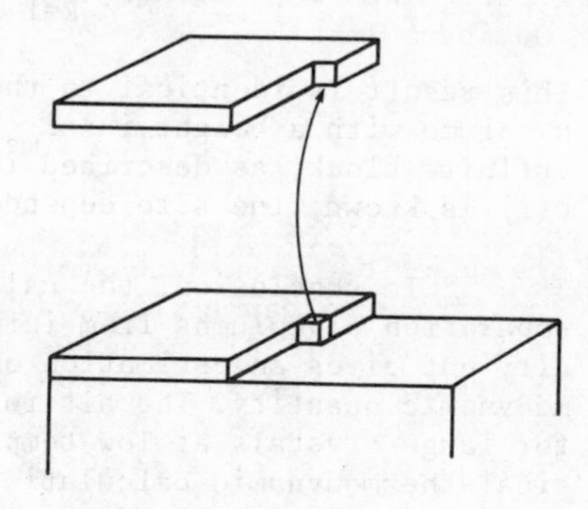

Fig. 3 : Elementary evaporation-condensation process used in the calculation of the surface free energy by the alternative method. For details see text.

The chemical potential in a kink position of the free monomolecular layer depends on the energy of lateral bonds only ($\omega/2$). Thus the work of formation of the layer containing $N_s = 1/a$ molecules is :

$$W_1 = \sum_{N_s} (\mu_i - \mu_o) = \frac{1}{a} \sum_{1}^{i_{max}} \psi_j.$$

The condensation of a second layer on the top of the monomolecular one should be done at chemical potential μ_2. Since the potential energy of a molecule in a kink position of this layer is $\phi_2 = \omega/2 + \psi_1$, the work spent for the formation of the second layer is equal to

$$W_2 = \frac{1}{a}\left(\sum_{1}^{i_{max}} \psi_j - \psi_1\right).$$

Similarly, the work of condensation of the i-th layer ($i < i_{max}$) is :

$$W_i = \frac{1}{a}\left(\sum_{j=1}^{i_{max}} \psi_j - \sum_{k=1}^{i-1} \psi_k\right)$$

The total work of formation of an i-molecular layer with unit surface area, equal to twice its surface energy, is

$$\Sigma W_i = \frac{1}{a}\left[i \sum_{1}^{i_{max}} \psi_i - \sum_{1}^{i-1} (i-k)\psi_k\right].$$

Accordingly,

$$\gamma_i = \frac{1}{2}\sum_{1}^{i} W_i \left[\frac{1}{2a} \sum_{k=1}^{i-1} k\psi_k + i \sum_{j=1}^{i_{max}} \psi_j\right]$$

This result is identical to the result obtained by separation of a column with a height $i < i_{max}$ and a base of unit area from an infinite block, as described in section III.2. If the potential law $\psi(i)$ is known, the size dependence of can be determined.

In conclusion, the calculation of surface energy based on separation of columns from infinite blocks has the merit of simplicity but gives an estimation only of the potential part of this thermodynamic quantity. The alternative method gives identical results for large crystals at low temperature. When combined with a statistical thermodynamic calculation of the (size dependent) chemical potentials μ_i, this method results in the precise computation of the excess free energy of clusters of every shape and (also non-periodic) structure. However, for practical reasons, this computation is still limited to rather small clusters.

III.5. The Stefan Rule

We saw that the surface energy of the (hkl)-face of a crystal at low temperature is very nearly equal to half the cohesion energy along a plane with the same orientation. Basically, a parameter characteristic for the boundary between two phases can not depend on a bulk property of one of the phases only, but in the case when the second phase is the vacuum. Thus, the process described in Sec. III.2, strictly speaking, gives values for the surface tension of the crystal-vacuum interface, the surface tension of the crystal-vapor interface being assimilated to that quantity.

An old empirical rule, proposed by STEFAN,[13] can be then easily understood. STEFAN stated that the molar surface energy (the surface energy per area occupied by one mole) is equal, for a large number of liquids, to one half the molar enthalpy of vaporization. A qualitative explanation of the Stefan rule is that the work spent to create the area occupied by a single molecule, when transporting it from the bulk to the surface, is half the work necessary to transport the molecule from the bulk to the vapor. This argument, valid more or less for a liquid, is clearly quantitatively wrong when different faces of a crystal are considered. However, one can retain from the Stefan rule that for both liquid-vapor and solid-vapor interfaces the value of the surface energy is proportional to the value of an intrinsic bulk quantity, that is the enthalpy of phase transformation (vaporization, sublimation).

IV. THE INTERFACE BETWEEN TWO CONDENSED PHASES

IV.1. Interface Free Energy and Interface Tension

In order to calculate the free energy of the boundary between two condensed phases α and β, we can follow the general procedure of section II.1, defining the thermodynamic functions of the surface according to the Gibbs convention. For this purpose we should be able to determine the potential energy and the entropy of the molecules in a slice with a thickness at least equal to that of the interfacial region, and to substract from the so-calculated free energy the free energy of two blocks of the bulk phases α and β, the total volume being constant. Because one cannot expect that the structures of both phases remain identical to their bulk structures up to the interface, the Gibbs convention ($V^{\sigma} = 0$) implies a variation of the number of molecules ($N^{\sigma} \neq 0$). The calculated parameter is, therefore, the interface free energy f^{σ} and not the surface tension γ(see eq. (7b)); the two quantities can in this case be quite different. The choice of the Gibbs dividing surface should not pose particular problems if the phases α and β are in thermodynamic equilibrium. Then $\mu^{\alpha} = \mu^{\beta} = \mu^{\sigma}$ and the reference free energy of the bulk phases can be obtained by multiplying the total number of molecules in the two bulk blocks with any of those potentials.

If the difference between the free energy of the system containing the interface and that of the reference system of the two bulk phases is calculated not at constant volume but at a constant number of molecules, its value is very nearly equal to the interface tension $\gamma^{\alpha\beta}$ (the two quantities differ by a negligible volume work pV^{σ}). This procedure has been applied [14,15] to model calculations of crystal-melt interface tension, as shown in chapter "Crystal-melt, Crystal-Amorphous and Liquid-Vapor Interfaces".

IV.2. The Adhesion Energy

Even if practically useful, the method described in the previous section does not stress the main characteristics of the interaction between the condensed phases meeting at the interface. Therefore, it is interesting to recall an old equation, due to DUPRE, which defines the interface energy between two condensed phases α and β :

$$\gamma^{\alpha\beta} = \gamma^{\alpha} + \gamma^{\beta} - \beta. \qquad (17)$$

The physical meaning of this expression is that, if one wants to create a unit interface $\alpha\beta$, one should first create separately a unit area of each of those phases and then join them together. The work β won in the latter stage is called *adhesion work* or *adhesion free* energy. The adhesion energy is the heterophase analogy of the cohesion energy. It is equal to the work to divide the system *along a unit interface area*.

The dissection of the interface energy into surface energies of the two phases and adhesion energy, demonstrates the fundamental difference between the interface separating two condensed phases and the surface between one such phase and the vacuum (or a diluted vapor). In the second case (condensed phase-vacuum) the surface energy can be expressed through an intrinsic parameter of the condensed phase alone. The interface energy $\gamma^{\alpha\beta}$ cannot be calculated either from a bulk quantity characteristic for one of the phases, or from a combination of two such quantities. The importance of the typical surface parameter, that is the adhesion energy, is evident. In the rest of this section we show how different the properties of the interface can be, when keeping the value of the surface energies γ^{α} and γ^{β} constant, varying only the values of β.

If the adhesion energy is zero (no interaction between the two phases), eq. (17) yields

$$\gamma^{\alpha\beta} = \gamma^{\alpha} + \gamma^{\beta} \ . \qquad (18)$$

The creation of the interface by joining the two free surfaces of the phases α and β does not gain any free energy. The interface has no reason to form. One says that the phase α does not "wet" the phase β(or vice versa).

If the adhesion energy is larger than zero but smaller than the smallest cohesion energy, e.g., that of the phase α,

$$0 < \beta < 2\gamma^{\alpha},$$

the Dupré equation gives

$$\gamma^{\alpha\beta} > \gamma^{\beta} - \gamma^{\alpha}; \qquad (19)$$

the "wetting" of β by α is imperfect.

When β is higher than the cohesion energy $2\gamma^{\alpha}$,

$$\gamma^{\alpha\beta} < \gamma^{\beta} - \gamma^{\alpha}; \qquad (20)$$

the "wetting" of β by α is "better than perfect".

Finally, if the adhesion energy is an arithmetic average of the cohesion energies of α and β,

$$\beta = \gamma^{\alpha} + \gamma^{\beta},$$

or larger, the interface energy is zero; the two phases mix together.

When the adhesion energy is *exactly equal* to the cohesion energy of the phase α,

$$\beta = 2\gamma^{\alpha},$$

the molecules of the latter "feel" the presence of phase β as if it were their own phase(α). This case of "perfect wetting" should be extremely rare, to not say inexistent, when α and β have different chemical composition, different structure (e.g., crystal and melt) or different orientation (e.g., a grain boundary). On the other hand, this is the only case in which the value of $\gamma^{\alpha\beta}$ depends formally only on the surface energies of the two phases,

$$\gamma^{\alpha\beta} = \gamma^{\beta} - \gamma^{\alpha}, \tag{21}$$

i.e., through the Stefan rule, on some combination of their bulk properties.

We emphasized the importance of adhesion energy for the properties of the interface, since widespread opinion exists still in the literature[16] that the free energy of the interface between two condensed phases can be calculated from a combination of heats of transformation of the two phases. In fact, in all papers supporting this opinion, the explicit[16] or the implicit[17] hypothesis of perfect wetting is made; thus the results obtained are predetermined.

V. SURFACE PROPERTIES OF TWO-DIMENSIONAL PHASES

V.1. Two-Dimensional Gases

The problem of the thermodynamic properties of two-dimensional phases will be largely treated in other chapters. Here we would like to discuss some of their surface properties only.

As with three-dimensional (3D) phases, two-dimensional (2D) phases or adsorbed layers on foreing substrates, exist in two main states : the diluted state (2D-gas) and the condensed state (2D-liquid or solid). Various phase equilibria between them can be forseen. Moreover, except for some very strongly bound chemisorbed layers, their continuous molecular exchange with the 3D-phase(s) should be taken into account.

A moment's reflection shows that a system including 2D-phases has one degree of freedom more per interface compared to a system containing 3D-phases only. The extra independent variable of state can be,for example, the spreading pressure Φ. For a 2D-gas the spreading pressure is always positive. Its value can be calculated by integration of the Gibbs isotherm (8), taking into account (9a), and using the relationship between the adsorbed quantity Γ and the chemical potential μ_g^{σ} of the 2D-gas (equal to the chemical potential of the 3D-gas μ^g) given by the adsorption isotherm. This is shown graphically in Figure 4a on a Langmuir type isotherm.

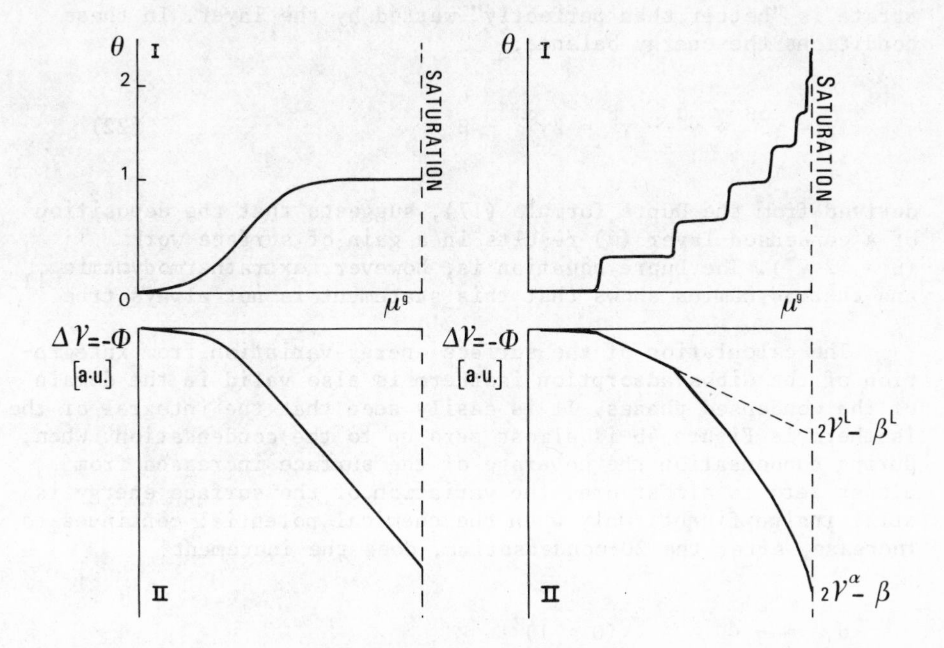

Fig. 4 : Adsorption isotherms (I) and variation of the surface energy of the substrate (II) for Langmuir-type (a) and multi-layer (b) adsorption.

V.2. Two-Dimensional Condensed Phases

When the forces among the adsorbed molecules in a 2D-gas are attractive and when the substrate temperature is below the 2D-critical point, the 2D-gas can condense to give 2D-liquid or solid. A typical low temperature adsorption isotherm, such as the isotherm of Xenon on graphite at 97 K, is shown in Figure 4b. One sees that for a wide range of increasing chemical potentials of the 3D-gas μ^g, the degree of coverage of the surface is nearly zero and then suddenly increases to an almost complete coverage, $\theta = 1$ (the degree of coverage is defined as $\theta = \Gamma/N_s = \Gamma_a$). The vertical line corresponds to the 2D-phase equilibrium, where the diluted and the condensed adsorbed layers coexist. At higher chemical potentials the density of the 2D-condensed phase slightly increases (vacancies disappear). This phase can either be subject to another phase transition (e.g., 2D-liquid to 2D-solid) or serve as a substrate for the adsorption of another layer (Figure 4b).

Condensed 2D-phases are stable only at undersaturation (at

the saturation a condensation of the 3D-phase occurs) when the substrate is "better than perfectly" wetted by the layer. In these conditions the energy balance,

$$\Delta\gamma = \gamma^{\alpha\beta} + \gamma^{\alpha} - \gamma^{\beta} = 2\gamma^{\alpha} - \beta \tag{22}$$

derived from the Dupré formula (17), suggests that the deposition of a condensed layer (α) results in a gain of surface work ($\beta > 2\gamma^{\alpha}$). The Dupré equation is, however, extrathermodynamic and thermodynamics shows that this statement is not always true.[11]

The calculation of the surface energy variation from integration of the Gibbs adsorption isotherm is also valid in the domain of the condensed phases. It is easily seen that the integral of the isotherm is Figure 4b is almost zero up to the condensation. When, during condensation the coverage of the surface increases from almost zero to almost one, the variation of the surface energy is still insignificant. Only when the chemical potential continues to increase, after the 2D-condensation, does the increment

$$d\gamma = -\,d\mu^{\sigma} \qquad (\theta \simeq 1)$$

become significant. The surface energy then decreases proportionally to the increasing chemical potential of the vapor. When saturation is reached, the variation of the surface energy due to the deposition of the single condensed layer (if no other layers are formed on its top) becomes equal to that expected from the equation (22) :

$$\Delta\gamma = 2\gamma^{1} - \beta^{1} \text{ (per unit surface area)}, \tag{22a}$$

where β^1 and γ^1 are the adhesion energy and the surface energy of the monomolecular layer (different from those of the bulk phase) calculated by some mechanical procedure, e.g., the one of Sec. III.2.

If a large number of condensed 2D-phases are formed on the top of each other until saturation is reached(see the isotherm of Fig. 4b), the sum of terms like (22a) for the n_ℓ successive layers ($n_\ell \to \infty$) is equal (only at saturation) to the macroscopic work

$$2\gamma^{\alpha} - \beta$$

expended when a three-dimensional phase α is deposited on unit area of the substrate.

V.3. The Two-Dimensional Space; the Ledge Free Energy

In the preceding two sections we still were in the three-dimensional space of the bulk phases and deduced some special properties of the two-dimensional phases. All the exotic features of the adsorbed layers disappear if we place ourselves in a 2D-space and notice that all deduced surface properties, such a spreading pressure, its invariability (and hence invariability of the surface energy) during 2D-condensation at constant temperature, etc. are exact 2D-analogies of well-known properties and phenomena in 3D-systems.

A further important thermodynamic parameter in the 2D-space is the excess free energy of a small two-dimensional condensed phase, that is the ledge free energy or ledge tension. Its definition and calculation brings nothing new compared to those of its three-dimensional analogue, the surface tension. For example, the Born-Stern procedure (cf. sec. III.3) can be applied to the calculation of the ledge and corner tensions of a layer, the two half-layers being slipped away on the substrate, so that no surface work is expended.

The ledge tension should not be confused with the edge tension. The former is one of the main variables of state of a 2D-phase, the latter a quite negligible parameter of a 3D-phase.

VI. EQUILIBRIUM SHAPE AND RELATED TOPICS

VI.1. The Wulff Theorem

The problem of the equilibrium shape, i.e. of the shape for which the total surface energy is minimum, all other parameters of state being constant, is not exactly a thermodynamic problem. A low viscosity liquid can reach its equilibrium shape long before it reaches thermodynamic equilibrium and a large crystal can be in thermodynamic equilibrium but maintain a non-equilibrium shape indefinitely. For this reason, the search for the form with minimum surface energy at a constant volume can be regarded as a purely mechanical problem, as it was originally treated by CURIE[18] and WULFF.[19]. We will consider that this part of the problem is already known and review only the principle results.

(i) The equilibrium shape of a liquid (with isotropic surface tension)is a sphere.

(ii) The equilibrium shape of an anisotropic crystal can be

obtained by constructing from a given point (Wulff point) radius vectors in all directions of the space, each vector $h_{(hkl)}$ having a modulus $\underline{h}_{(hkl)}$ proportional to the surface tension of the face (hkl). The polar diagram joining the tips of the radius vectors as a function of their orientation is the so-called γ-plot.[20] The equilibrium shape is the most interior of the set of polyhedra, obtained when all (hkl)-planes are constructed, each containing the tip of the corresponding vector [hkl]. The simple expression of the Wulff theorem is :

$$\frac{\gamma_1}{\underline{h}_1} = \frac{\gamma_2}{\underline{h}_2} = \ldots = \frac{\gamma_i}{\underline{h}_i} \tag{23}$$

where γ_i and h_i are the surface tension and the modulus of the radius-vector (the "Wulff distance") of the i-th face.

(iii) When a crystal is in contact with a foreign substrate in such a manner that j of its i faces are replaced by crystal-substrate interfaces, characterized by adhesion energies β_j, the Wulff theorem yields[21]:

$$\frac{\gamma_j - \beta_j}{\underline{h}'_j} = \frac{\gamma_j}{\underline{h}_j} = \frac{\gamma_i}{\underline{h}_i}\ , \tag{24}$$

where $\underline{h}'_j$ and $\underline{h}_j$ are the Wulff distances in the j-th direction in the presence and in the absence of substrate respectively. Figure 5 illustrates the Wulff construction for crystal in the homogeneous phase (a) and on substrate (b). In this case of centrosymmetric crystal one sees, taking into account (24), that the Wulff distance to the substrate $\underline{h}'_j$ varies from $\underline{h}_j$, when $\beta_j = 0$, to $-\underline{h}_j$, when $\beta_j = 2\gamma_j$.

(iv) The well known Young equation, relating the equilibrium contact angle Θ of a liquid on a plane solid substrate to the surface tension of the liquid (γ^L), of the solid (γ^S) and of the solid-liquid interface (γ^{SL}),

$$\gamma^S = \gamma^{SL} + \gamma^L \cos\Theta, \tag{25}$$

can be considered just as particular case of the Wulff theorem for a polyhedron with infinite number of faces with equal surface tensions. Combining (25) with (17) and assuming $\gamma^L \equiv \gamma^\alpha$; $\gamma^S \equiv \gamma^\beta$; $\gamma^{SL} \equiv \gamma^{\alpha\beta}$, we obtain

$$\cos \Theta = - \frac{\gamma^L - \beta}{\gamma^L} \tag{25a}$$

which is the analogue, for a liquid drop, of equation (24). Similarly, $\cos \Theta = -1$ ($\Theta = 2\pi$) for $\beta = 0$ and $\cos \Theta = 1$ ($\Theta = 0$) for $\beta = 2\gamma^L$.

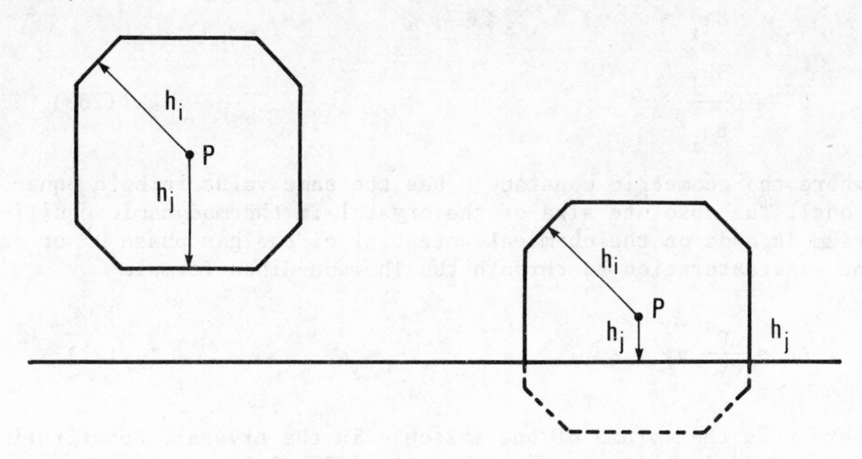

Fig. 5 : Equilibrium shape of crystals of a given substance in the homogeneous phase (a) and on substrate (b), at the same temperature and supersaturation; P ≡ Wulff point.

Note : All above considerations are valid if the condensed phases are not subject to any exterior force field. The effect of gravitation on the equilibrium shape of liquid drops on substrates is well known. As to the effect of the force field induced by a substrate on the surface tension of solid clusters[22] or on the line tension of liquids,[23] it seems to be significant for sizes of few atoms only and is generally neglected.

VI.2. Equilibrium Shape and Nucleation

The reason why equilibrium shape plays a major role in the thermodinamics of nucleation is clear. The probability for formation of a small phase, that is the crystal embryo, is larger, the smaller its excess free energy. The absolute minimum of the excess energy for a given number of molecules is reached for the equilibrium shape. Therefore, only clusters with equilibrium shapes are of interest for the thermodynamics of nucleation. We would like to draw some phenomenological inferences from this statement.

A first consequence is that one can express the volume V and the total surface energy F^σ of an equilibrium shaped crystal by mean of only one ratio γ_i/h_i or $(\gamma_j - \beta_j)/h'_j$.[24] A simple geometric reflection leads to the formulas

$$V = C \frac{h_i^3}{8\gamma_i^3}, \quad (26a)$$

$$F^\sigma = C \frac{3h_i^2}{8\gamma_i^2} \quad (26b)$$

(where the geometric constant C has the same value in both equations). The absolute size of the crystal in thermodynamic equilibrium depends on the chemical potential of the gas phase μ^g or on the supersaturation $\Delta\mu$ through the Thomson-Gibbs formula

$$\Delta\mu = \frac{dF^\sigma}{dV} v, \quad (27)$$

where v is the volume of one molecule in the crystal. Substitution of V and F^σ from (26a) and (26b) into (27) yields

$$\Delta\mu = \frac{2\gamma_i v}{h_i}. \quad (28)$$

Equation (28) makes the link between Wulff theorem and thermodynamics. It shows that the chemical potential, or the equilibrium pressure can be expressed by any ratio γ_i/h_i, provided that the crystal is simultaneously in thermodynamic and capillary equilibrium (has an equilibrium shape). The considerations leading to (28) can be repeated for the same crystal in equilibrium on a foreigh substrate. The only difference with the preceding case is in the value of the geometric factor C in eqs. (26a) and (26b). Equation (28) is still valid and can be extended as follows :

$$\Delta\mu = \frac{2\gamma_i v}{h_i} = \frac{2\gamma_j v}{h_j} = \frac{2(\gamma_j - \beta)v}{h_j'}. \quad (28a)$$

The generalised Thomson-Gibbs-Wulff equation (28a) is fundamental since it shows that the vapor pressure of a three dimensional phase in equilibrium on a substrate is independent of the adhesion energy. The crystal, in contact with the substrate, apparently "loses" part of its volume (see Fig. 5). However the general shape, linear dimensions and vapor pressure remain those of the crystal in the homogeneous vapor at the given supersaturation, as determined by eq. (28).

It has been known since GIBBS[5] and VOLMER[25] that the reversible work of formation of a small condensed phase from its supersaturated vapor (that is the activation energy of nucleation) is equal to one third of its total surface energy

$$\Delta G^* = \frac{1}{3} F^* . \qquad (29)$$

According to (26a), (26b) and (28), eq. (29) can be written as :

$$\Delta G^* = \frac{\gamma_i}{h_i^*} V^* = \frac{1}{2} \frac{V^*}{v} \Delta\mu . \qquad (30a)$$

If at the same supersaturation $\Delta\mu$ the nucleus is formed on a substrate, its total surface energy F^σ and volume V change through the geometric factor C in (26a) and (26b). Since equation (28) is still valid, the reversible work of nucleation is :

$$\Delta G^*_{(s)} = \frac{1}{3} F^{*\sigma}_{(s)} = \frac{\gamma_i}{h_i^*} V^*_{(s)} = \frac{1}{2} \frac{V^*_{(s)}}{v} \Delta\mu . \qquad (30b)$$

Comparison of (30a) and (30b) at the same supersaturation $\Delta\mu$ yields :

$$\frac{\Delta G^*_{(s)}}{\Delta G^*} = \frac{V^*_{(s)}}{V^*} . \qquad (31)$$

This equation, proposed by KAISCHEW[24], is the most general expression relating the nucleation work on a substrate to that in the homogeneous phase. Whatever the profile of the substrate (concave edge of corner, various shaped holes, etc.), a sufficient condition for the calculation of $\Delta G^*_{(s)}$ is to determine the equilibrium shape of the nucleus on the substrate and to compare its volume to the corresponding volume in the homogeneous phase. It is clear from the same equation that if a nucleus which forms on a substrate has the choice between many different orientations (contact planes or epitaxial positions), the orientation requiring the smallest nucleation work will be the one for which (to use a figurative expression illustrated by Fig. 5) "the nucleus sinks most deeply in the substrate". One can state that for three-dimensional nucleation on a substrate, the preferential contact plane or epitaxial orientation are invariant parameters given by the couple nucleus-substrate, independent of the external conditions (supersaturation, temperature).

The situation is very different for two-dimensional nucleation. The chemical potential of an infinite 2D-phase depends on its adhesion energy with the substrate. For this reason the reversible work of nucleation (which depends on supersaturation and ledge free energy) of a given 2D-phase with a given epitaxial orientation can be lower or higher, as a function of the supersaturation, than the nucleation work of another 2D-phase with another orientation. The epitaxy of the 2D-phases is thus supersaturation dependent.

VI.3. Equilibrium Shape and Some Problems of Crystal Growth

One might wonder whether effects due to the equilibrium shape could be of any importance when, after nucleation, a condensed phase increases its dimensions far above the critical nucleus size. It is evident that the surface/volume ratio decreases so rapidly with increasing linear dimension that the excess free energy per molecule of the cluster becomes vanishingly small. However, in the absence of other driving force, deviations from the capillary equilibrium are sometimes considered as responsible for growth, e.g., during recrystallization.[26]

The equilibrium shape is more strongly related to the crystal growth kinetics via the microscopic structure of the different faces, which we shall briefly stress here from rather a formal point of view. The physics of the phenomena will be discussed in the next section.

Formally, terraces or other closed steps on surface of a given face can be treated as two-dimensional condensed layers perfectly wetting their substrate. In this particular case, the interface energy between the layer and the substrate is zero and the energetic barrier for two dimensional nucleation depends only on the ledge free energy. The ledge free energy varies with orientation just as the surface tension does. In some sense the values of these two parameters are complementary, as can be seen from the following consideration. The two terms composing the binding energy of a molecule in a kink position in equation (16a) are complementary, since they change from one face to another, while ϕ_o itself is invariant. They are also closely related to the ledge and surface energies. The stronger the lateral bonds, the higher the ledge energies; the larger the desorption work $\Sigma\psi_i$, the higher the surface energy. Therefore, one can assert that for a given substance and crystal structure, the lower the surface free energy of a face, the higher will be the ledge free energies of the two-dimensional terraces or islands and vice versa.

If we now turn to the γ-plot of a given crystal, we notice that at the absolute zero point many crystallographic directions (theoretically all directions with rational indices[27] are represented by discontinuities (cusps) in γ. This is understandable if one assumes that two-dimensional layers on faces with those orientations have non-zero ledge feee energies. Small angle deviations which lead to the creation of vicinal faces,result in a linear increase of the density of steps, i.e. in a linear increase of the effective surface free energy. The tendency of such faces, called "singular" of F-faces, to keep their profile is explained by the sharp increase of their surface energy for any arbitrary variation of orientation. If only one particular variation of orientation induces a sharp surface energy increase, only one cross-section of the (three-dimensional) γ-plot has a cusp. Such faces keep their profile in only one direction; they are called "stepped" or S-faces. Finally, faces which are either not represented by a minimum on the γ-plot, or show a flat minimum instead of a cusp, do not resist to a profile modification; they are called "kinked" or K-faces. Note that in the literature the same names are sometimes given to the faces depending on their morphology.[28] This kind of definition has no thermodynamic justification, especially since it cannot forsee any modification of the crystal habit with the temperature.

When the temperature is raised, the two dimensional condensed phases with the lowest ledge free energies are the first to attain their critical points. Beyond this point an F-face transforms into an S- or K-face, depending on the values of the different ledge energies. We have already seen, however, that faces with the lowest ledge energy are the ones with the highest surface energy. An increase of temperature, therefore, induces a flattening of the cusps farthest from the Wulff point and a decrease in the number of singular faces. At still higher temperatures, the only singular faces are the ones situated very near the Wulff point. Meanwhile, the surface energy of the nonsingular faces decreases sensitively with the temperature, because of their large configurational entropy.[29] The equilibrium shape at high temperature is thus limited by a small number of singular faces and important rounded regions.

There is a tremendous difference in the behavior of different types of faces during growth. The adsorption energy $\Sigma\psi_i$ of the K-faces is nearly equal to the binding energy in a kink position ϕ_o ($\omega \simeq 0$, cf. eq. (16a)). Accordingly, molecules arriving from the vapor have more chance to be built in the lattice than to desorb, even at the lowest supersaturation. Interfacial kinetics of K-faces is thus very fast and their growth is limited by transport phenomena.

The F-faces have relatively low adsorption energies and the only kinks on their surfaces are situated in the ledges of terraces or other steps. Imperfections in the crystal lattice can provide some inexhaustible sources of steps;[30] However, when the crystal is perfect, the only way for an F-face to grow is two-dimensional condensation of molecules arriving from the vapor onto its surface. Condensation starts by nucleation at high supersaturation and continues by spreading of layers. This type of growth is usually called "lateral" or "layer by layer" growth. By raising the temperature to the 2D-critical point, the surface condensation is not more a first order transition. 2D-nucleation barriers vanish and the face starts to grow by a "diffuse" or "continuous" mechanism, just as a K-face. At the same temperature, the cusp of the face in the γ -plot disappears.

VII. STATISTICAL VIEW OF A CRYSTAL SURFACE

VII.1. Models of Rough Surfaces

In the preceding section we showed the possibility of considering the growth of a crystal surface as a succession of 2D-condensations in layers of self-adsorbed molecules. Here we would like to study the behaviors of these layers in relation to other surface properties.

Independent of the particular crystal lattice and orientation of the face under consideration, one realizes that the self-adsorbed layer is of *at least* bimolecular thickness.[31,32] The top monomolecular layer is a 2D-gas of admolecules, the next lower layer is an *almost* compact 2D-solid containing a 2D-*gas of surface vancancies*. The potential energy increase, when a molecule is taken from a kink position (the reference thermodynamic state at saturation) and brought to a position of an admolecule, is equal to the potential energy increase when a molecule from the top lattice layer is removed to a kink, leaving a monomolecular hole. The concentration of admolecules and monomolecular holes is thus of the same order of magnitude, as is their contribution to the free energy of the surface.

When the thickness of the surface increases under the effect of high temperature, the symmetric treatment of admolecules and holes is no longer useful. A common feature of the more complicated n-level models[33] and the simple 2-level model is that the existence of only open holes is assumed. "Closed holes", due to vacancies in lower levels covered by admolecules of higher levels, and "overhangs" are highly improbable, whence the name "solid on solid" (SOS !)" models.[34]

The main features of surface roughening are fairly well represented by the two-level model, although its quantitative validity suffers when the surface approaches the 2D-critical point. As to the theoretical approach, the early Ising-type calculations[35] give good quantitative results, but are, in our eyes, less suitable for a discussion. For this reason we limit ourself to strictly two-level models and the mean field approximation.[36,37]

VII.2. Adsorption Isotherms of Admolecules and Surface Vacancies

The mean field approximation results in FRUMKIN[36]-FOWLER[37] type isotherms. In the case of the two level model, the degrees of coverage of the admolecules θ^+ and of the holes θ^- are given by the equations[32]

$$k_BT \ln \frac{\theta^+}{1 - \theta^+ - \theta^-} = -\frac{\omega}{2}(1 - 2\theta^+) + \Delta\mu,$$

$$k_BT \ln \frac{\theta^-}{1 - \theta^+ - \theta^-} = -\frac{\omega}{2}(1 - 2\theta^-) - \Delta\mu. \tag{32}$$

The coupling in the configurational energy terms on the left hand sides is typical for the SOS model.

A visible result from equation (32) is that at saturation ($\Delta\mu = 0$) $\theta^+ = \theta^-$ as expected from pure energetic arguments. Furthermore, when $\Delta\mu > 0$, $\theta^+ > \theta^-$; the surface becomes "asymmetric". Whether the increase of coverage with admolecules can continue smoothly to $\theta \simeq 1$ or passes through an instability region, depends essentially on the energy of the lateral interaction ω and on the temperature. If $\omega/k_BT < 4$, the condensation of the 2D-gas of admolecules is a first order phase transition. The critical temperature is given by the condition $\omega/k_BT_c = 4$, which is only a rough estimate, since both two-level model and mean field theory collapse near the critical point. Above the critical point, the surface can be considered as "rough". From the standpoint of growth kinetics this means that the overcritical surface layers can be completed without any energetic barrier. Below the critical point, the completion of a new layer of admolecules can take place only at considerable supersaturation through 2D-nucleation. Under these conditions, the concentration of holes in the underlying layer is negligible.

The isotherms (32) reveal another interesting feature of the layer of admolecules at high supersaturations. At temperatures lower than T_c, when the supersaturation exceeds the critical value

$$\Delta\mu = \frac{\omega}{2}\xi + k_BT \ln \frac{1 - \xi}{1 + \xi},$$

$\xi = (1 - 4\,k_BT/\omega)^{1/2}$, the energetic barrier for 2D-nucleation of the 2D-gas vanishes and condensation of the 2D-gas becomes continuous. The growth of the face is typically spinodal. This type of transition, obtained also by computer simulation[38] is sometimes called "dynamic roughening", although it has been forseen following thermodynamic arguments.[32]

VII.3. Surface Roughness and Surface Energy

The effect of surface roughening on the surface energy of a crystale face is analogous to the lowering of the surface tension of a substrate by a foreing adsorbed layer. Its treatment by the two-level mean-field model is instructive. Quantitative values should be obtained by more precise models or computer simulation.

The Gibbs adsorption isotherm (8) can be integrated using the relationship between θ^+, θ^- and $\Delta\mu$ of equations (32). The result,

$$\gamma = \gamma_{sm} + \frac{k_BT}{a} \ln (1 - \theta^+ - \theta^-) + \frac{\omega}{2a} (\theta^{+2} - \theta^{-2}) \tag{33}$$

(γ_{sm} is the surface tension of the smooth surface at the same temperature), is completly analogous to the result for simultaneous foreign adsorption of two species θ^+ and θ^- following the Frumkin-Fowler isotherm.

Two effects on the surface energy are of interest : the effect of the temperature at saturation (infinite crystals), and the effect of supersaturation (small phases) at constant temperature. Consider, for simplicity, the case of small coverages (θ^+, $\theta^- \ll 1$), when the isotherms (32) can be used in the form

$$\theta^+ = \exp\left(-\frac{\omega}{2k_BT}\right) \exp\left(\frac{\Delta\mu}{k_BT}\right) = \theta_o^+ \exp\left(\frac{\Delta\mu}{k_BT}\right),$$
$$\theta^- = \exp\left(-\frac{\omega}{2k_BT}\right) \exp\left(-\frac{\Delta\mu}{k_BT}\right) = \theta_o^- \exp\left(-\frac{\Delta\mu}{k_BT}\right) \tag{32a}$$

(θ_o^+ and θ_o^- are the coverages of admolecules and holes at saturation). Neglecting the quadratic terms in eq. (33) and developing the logarithm, we have :

$$\gamma = \gamma_{sm} - \frac{k_BT}{a}(\theta^+ + \theta^-)$$
$$= \gamma_{sm} - \frac{k_BT}{a}\left[\theta_o^+ \exp\left(\frac{\Delta\mu}{k_BT}\right) + \theta_o^- \exp\left(-\frac{\Delta\mu}{k_BT}\right)\right]. \tag{34}$$

The coverages θ_o^+ and θ_o^- at saturation are equal for the very simple model that neglects the difference between vibrational entropies of admolecules and holes. In general, this equality does not exist[39] although the two coverages have the same order of magnitude. Consider both cases.

(i) $\theta_o^+ = \theta_o^-$: Equation (34) becomes :

$$\gamma = \gamma_{sm} - (2k_B T\theta_o/a)\cosh(\Delta\mu/k_B T). \qquad (34a)$$

At saturation, the dependence of γ_o on the temperature is[40]:

$$\gamma_o = \gamma_{sm} - 2k_B T\theta_o/a = \gamma_{sm} - (2k_B T/a)\exp(-\omega/2k_B T) \qquad (34b)$$

The dependence of γ on supersaturation is non significant at low supersaturation. This can be understood, since, on the one hand the effects of the admolecules and of the holes on the surface tension are additive, and on the other hand, at low supersaturation, the increase in the number of admolecules is compensated by the decrease in the number of holes, as seen from the developement of (34) to the linear term :

$$\gamma = \gamma_{sm} - (\theta_o^+ + \theta_o^-)\,kT/a - (\theta_o^+ - \theta_o^-)\Delta\mu/a. \qquad (34c)$$

(ii) $\theta_o^+ \neq \theta_o^-$: The inequality in the coverages of admolecules and holes at saturation accentuates the influence of supersaturation. It is interesting to relate the supersaturation to the radius r of the small phase in (unstable) equilibrium with the supersaturated vapor, through the Thomson-Gibbs formula (28), $\Delta\mu = 2\gamma v/r$. If one takes into account that v/a is of the order of the linear dimension δ of a molecule, eq. (34c) can be written in the form :

$$\gamma = \frac{\gamma^o}{1 + 2(\theta_o^+ - \theta_o^-)\delta/r} = \frac{\gamma^o}{1 + 2\delta_{eff}/r} \qquad (35)$$

This is the well known TOLMAN[41] equation relating the surface energy of a small phase to its linear dimension r. TOLMAN obtained the equation following classical thermodynamic arguments and thus did not precisely determine the value of the effective length δ_{eff}, which was supposed to be approximately the molecular dimension. Eq. (35) shows that δ_{eff} is not only much smaller than δ, but can be positive or negative as well, depending on the "asymmetry" of the surface density profile at saturation. Note that the size variation of the surface tension, described by eq. (35) is not due to the long range intermolecular forces, as the one considered in sec. III.4, but is induced exclusively by the surface roughness.

VII.3. Surface melting

The idea that crystal surfaces start to melt below the bulk melting point is rather old.[42] Since, on the one hand, it has not been yet checked experimentally in a unambiguous way, and on the other hand, it has not been subject to detailed theoretical treatments, defenders or opponents of this idea have mostly used general arguments that we shall try to summarize here.

2D-melting of adsorbed layers on foreign substrate is known both experimentally and theoretically.[43-46] In general layers melt when the substrate does not have a very pronounced energetic structure (shallow potential wells) and when the layer is not too compact.

In the case of the faces of a crystal, the two conditions above are somewhat contradictory. Faces with shallow wells are the ones with the highest two-dimensional density (these are mostly F-faces) and vice-versa. High surface roughness seems to be the necessary condition to make surface melting possible, at least for the dense faces. Liquid-like behavior of the top surface layer, due to high configurational entropy, can be favored by some particular two dimensional structures of the next lower layers. For example, a hexagonal close packed lattice plane of (spherical) atoms offers to the self-adsorbed atoms a honeycomb array of adsorption sites, resulting from the superposition of two triangular lattices. The occupation of one site from one of the lattices prohibits the occupation of three next nearest neighbor sites of the other lattice.[47] As long as the surface roughness is negligible ($\theta < 0.25$), the top lattice plane occupies only one of the sublattices (depending on whether the crystal has a face centered cubic or a hexagonal close packed structure). With increasing roughness, however, the simultaneous random occupation of sites of both sublattices becomes possible and the top lattice plane undergoes an order-disorder phase transition.[48]

This very simple case, realized in adsorbed layers on foreign substrate (cf. chapter "Phase Changes in Two-Dimensions"), shows only one possible way of surface disordering. Many questions remain open, e.g. whether or not the disordering occurs below the bulk melting point, if the disordered layer remains commensurate, as above, or if it acquires a true liquid character (adatoms leaving the potential wells of the substrate), if the melting is limited to one monolayer and if not, how does the thickness of the molten layers vary with the temperature, etc.

VIII. EXCESS FREE ENERGY OF SMALL CONDENSED PHASES AND THE CAPILLARITY APPROXIMATION.

The question of how far the capillarity approximation can be useful in describing the excess free energy of a small phase, when this phase approaches molecular dimensions, is fundamental for the theoretical studies of nucleation. The nucleation, e.g., of a condensed phase from the vapor, is very similar to a chemical polymerization reaction; therefore, its rate equations could be developed on the basis of molecular parameters only. It is quite evident that such an approach encounters formidable computational difficulties as soon as the embryos become larger than several molecules. The chance of nucleation theory has been the use of the surface free energy[5] which, through the Thomson-Gibbs equation and the Wulff theorem, is able to account for the main thermodinamic characteristics of the critical nulcei, and leads to general rate equations. In the same time, the use of surface energy was a handicap. The exact relationship between microscopic (molecular) and macroscopic (capilarity) parameters gave rise to numerous controversies,[49-53] and even if now one believes that they are eliminated, one is never sure, for a particular substance or model, what is the minimum size for which the capillarity approximation still works. We shall try in the following sections, first to stress the thermodynamic relations between micro- and macroscopic parameters, and then, through simple models of monatomic solids, to show the limits of the capillarity approach.

VIII.1. The Free Energy of a Small Phase

Consider a small N-molecular monocomponent phase moving freely in a vapor of volume V. When expressing by means of the classical thermodynamics the total free energy of the small phase, we have the choice between size dependent surface free energy and size dependent chemical potential. If we opt for the first, the variation of the Gibbs free energy of the small phase at constant p and T is :

$$(dG)_{p,T} \simeq (dF)_{V,T} = \gamma dA + \mu_o dN, \tag{36}$$

where μ_o is the (size independent) chemical potential of the infinite phase at the temperature T and saturation pressure p_o. The fact that the small phase is in equilibrium at $p > p_o$ is not disturbing as long as the phase is incompressible, and can be neglected in the general case since p and p_o are not very different. For the same reason, $dG \simeq dF$.

Equation (36) can be integrated in the limits i_o to N. The number i_o is the smallest size for which an "area" can still be attributed to the cluster. As we shall see later, this parameter cannot be adjusted but by numerical calculations on particular models. The free energy of the cluster is then

$$F_N = F_{i_o} + \int_{i_o}^{N} \gamma dA + \mu_o(N - i_o) = \mu_o N + \gamma_o A + \int_{i_o}^{N} (\gamma - \gamma_o) dA + F_{i_o} , \tag{37}$$

where γ_o is the surface free energy of the infinite phase, A is the area of the N-molecular cluster, and F_{i_o} is the free energy of the i_o-molecular cluster. The excess free energy of the small phase can be written in the form :

$$\Phi_N = F_N - \mu_o N = \left[\gamma_o + \frac{\int_{i_o}^{N} (\gamma - \gamma_o) dA}{A}\right] A + F_{i_o} = \tilde{\gamma} A + F_{i_o} . \tag{38}$$

Classical thermodynamics assumes : (i) that the second term in the brackets of (38) is nil, and (ii) that the constant F_{i_o} is also equal to zero. Both assertions are obviously unrealistic. First, one knows that γ *cannot* be equal to γ_o for small cluster sizes but *must* tend asymptotically to this limit with increasing N (and A). Thus, only above a given cluster size the second term in the brackets of (38) is negligible and $\tilde{\gamma}$ reaches the constant macroscopic value γ_o. Furthermore, even if formally i_o can be taken equal to unity, still $F_{i_o} \neq 0$. We shall consider, however equation (38) as an expression of the capillarity approximation which includes a non-constant $\gamma \neq \gamma_o$ for small phases and an arbitrary value of the constant F_{i_o}. The objection that in these conditions the extensive character of the free energy F_N is lost, has already been discussed by other authors.[10] Only model calculations can show how rapidly the second term in the brackets of (38) converges to zero. The same equation becomming then linear in respect to A, the constant F_{i_o} can be determined by extrapolation. Its value is unlikely to be zero, even if it can become negligible for large clusters.

VIII.2. Partition Functions of an Infinite Crystal and of a Small Phase

According to the harmonic approximation and for temperatures above the Debye temperature, the partition function per molecule of an infinite crystal of a monatomic substance can be written in the form :

$$g_o^c = \exp(-u_o/k_BT) \cdot \left(\frac{k_BT}{h\nu_o}\right)^3, \tag{39}$$

where u_o is the potential energy of an atom at rest in a lattice mode. The chemical potential of the infinite crystal is equal to the free energy of an atom in a kink position :

$$\mu_o^c = f^c = -k_BT \ln g_o^c = u_o - k_BT \ln \left(\frac{k_BT}{k\nu_o}\right)^3. \tag{40}$$

Comparison of (40) and (16) shows that u_o is equal to (minus) the differential separation work of a molecule in a kink position ϕ_o, just as is the differential entropy term

$$s_o = k_B \ln (k_BT/h\nu_o)^3.$$

This relation defines the "thermodynamic frequency" ν_o of the infinite crystal, which may not be identified with any realistic frequency of the lattice. If one pulls out of the brackets of (40) the de Broglie wave length,

$$\Lambda = \frac{h}{(2\pi mk_BT)^{1/2}}$$

(m is the mass of the molecule), one has :

$$\mu_o^c = u_o - k_BT \ln \left(\frac{k_BT}{2\pi m}\right)^{3/2} \frac{1}{\nu_o{}^3} + k_BT \ln \Lambda^3. \tag{41}$$

The term

$$\tilde{v} = \left(\frac{k_BT}{2\pi m}\right)^{3/2} \frac{1}{\nu_o{}^3} \tag{42a}$$

has dimensions of volume and is often called "mean vibrational volume". It is defined as the volume in which the probability of finding the center of the atom, vibrating harmonically with a frequency ν_o around a lattice node, is very nearly unity. In the same way, one can define the mean vibrational area $\tilde{a}$ and the mean vibrational length $\tilde{l}$:

$$\tilde{a} = \frac{k_B T}{2\pi m} \frac{1}{\nu_o^2}, \tag{42b}$$

$$\tilde{l} = \left(\frac{k_B T}{2\pi m}\right)^{1/2} \frac{1}{\nu_o}. \tag{42c}$$

Note that $\tilde{v}$, $\tilde{a}$ and $\tilde{l}$ are statistical parameters of the crystal and should not be confused with real molecular volume area or length.

The partition function of the cluster, defined in the preceding section, can be written as :

$$Q_N = \exp(-U_N/k_B T) Q_v Q_r Q_t, \tag{43}$$

where Q_v Q_r and Q_t are, respectively, the partition function of the 3N-6 modes of vibration of the cluster, the partition function of rotation of the cluster around its three principal axes, and its partition function of translation in the three-dimensional space of volume V. The separate values of the three partition functions depend on the chosen reference system (fixed center of mass, fixed atoms, etc.) but their product is reference independent. Unlike common calculations on large molecules, where fixed center of mass is prefered, we shall present our results in the reference system of fixed atoms. In order to immobilize a cluster of N atoms, having 3N degrees of freedom, we shall first fix the three positional coordinates of one arbitrary atom (which results in blocking translation), then two coordinates of a second atom (this blocks two of the rotational movements) and finally, one coordinate of a third atom (to block the third rotational movement). The determinant $|det|$ of the dynamic matrix of the cluster then has a non-zero value, related to the vibrational partition function by the expression :

$$Q_v = \prod_{i=1}^{3N-6} \left(\frac{k_B T}{h\nu_i}\right) = \Lambda^{-(3N-6)} \prod_{i=1}^{3N-6} \left(\frac{k_B T}{2\pi m}\right)^{1/2} \frac{1}{\nu_i} =$$

$$= \Lambda^{-(3N-6)} (\pi k_B T)^{(\frac{3}{2}N-3)} |det|^{-1/2}. \tag{44}$$

The rotational partition function in the chosen reference

system is that of the three fixed atoms :

$$Q_r = \Lambda^{-3}\ (4\pi r^2)\ (2\pi d)/\sigma,$$

where r is the distance between atom 1 and atom 2, while d is the distance between the line joining atoms 1 and 2, and atom 3. We can choose the three atoms to be first nearest neighbors, thus making the rotational partition function size independent for clusters larger than three atoms. σ is the so-called symmetry number, showing how many undistinguishable configurations of the cluster can be obtained by rotation. The term $4\pi r^2$ represents the smallest surface described by an atom which rotates around a nearest neighbor; similarly $2\pi d$ is the smallest rotation length of an atom around an axis joining two of its nearest neighbors. Accordingly, one can write :

$$Q_r = \Lambda^{-3}\ a_r l_r, \tag{45}$$

where a_r and l_r are weighted (through the symmetry number) elementary rotational surface and elementary rotational length respectively. The translational partition function is equal simply to

$$Q_t = \Lambda^{-3} V \tag{46}$$

Substitution of (44), (45) and (46) into (43) yields :

$$Q_N = \exp(-U_N/k_BT)\Lambda^{-3N} \cdot \prod_{i=1}^{3N-6} \left(\frac{k_BT}{2\pi m}\right)^{1/2} \frac{1}{\nu_i} \cdot a_r l_r\ V. \tag{47}$$

The excess free energy of the cluster is obtained from (47) and (41) taking into account (42a), (42b) and (42c) :

$$\Phi_N = F_N - N\mu_o^c = -\ k_BT \ln Q_N + Nk_BT \ln q_o^c$$

$$= (U_N - Nu_o) - k_BT \sum_{i=1}^{3N-6} \ln\frac{\nu_o}{\nu_i} - k_BT \ln \frac{V}{\tilde{v}} \frac{a_r}{\tilde{a}} \frac{l_r}{\tilde{l}}. \tag{48}$$

The physical meaning of (48) is that the excess energy of a small particle which can translate and rotate in a volume V, with respect to the free energy of the same number of atoms (molecules), in an immobile infinite crystal, is composed of three terms. The first is the excess potential energy, the second, the excess vibrational entropy. The third term can be called "replacement free energy" F_{rep}.[54] It expresses the variation of the free energy when the vibration of the atom 1 in the mean volume $\tilde{v}$ is replaced by a translation in the volume V, when the vibration of the atom 2 along two directions, whose amplitudes are within the mean vibrational area $\tilde{a}$ is replaced by a rotation along the surface a_r, and when the vibration of the atom 3 along the mean vibrational length $\tilde{l}$ is replaced by a rotation along the length l_r.

The first two terms of equation (48) are size dependent and might be assimilated to the surface free energy (the first term in (38)). The third term is constant and different from zero.

The latter result, fitting better than previous treatments[53] with the thermodynamic significance of the replacement free energy is due to the choice of the reference system of fixed atoms.

VIII.3. Some Model Calculations

If one can calculate all terms of eq. (48) for a particular model and construct a Φ_N vs $N^{2/3}$ plot, the graph obtained should asymptotically tend to the relation

$$\Phi_N = \gamma_o \, \Omega \, N^{2/3} + B \tag{49}$$

(Ω is a shape factor), from which one can determine γ_o and B.

Regarding the first term of (48), a semi-continuum calculation, including the real binding energy of an atom with its first coordination shell and a continuum contribution of more distant atoms, gives for the differential binding energy of an n-atomic spherical cluster with close packed structure, in the case of Lennard-Jones (6/12) pair potential, the relation

$$\phi_N - \phi_o = - \frac{\pi\psi_o}{2}\sqrt{2} \left(\frac{\kappa}{N^{1/3}} - \frac{2}{3}\frac{\kappa^3}{N} + \sqrt{\frac{3}{2}}\frac{\kappa^4}{N^{4/3}} \right),$$

where ψ_o is the energy of an atomic pair, and κ is a constant of the order of unity. The integral of the difference $\phi_o - \phi_N$ gives the excess potential energy $U_N - Nu_o$ of the cluster. Its surface being $A = (\pi r_o^2/\kappa^2)N^{2/3}$ (r_o is the equilibrium pair distance), one obtains an expression of the form :

$$U_n - Nu_o = (\varepsilon_o^\sigma - \varepsilon_1^\sigma \frac{\ln N}{N^{2/3}} - \varepsilon_2^\sigma N^{-1})\ A, \qquad (50)$$

where ε_o^σ is the potential part of the surface free energy of the infinite crystal. The values of the other potential energies, ε_1^σ and ε_2^σ, are not very different from ε_o^σ; the importance of the second and the third terms in the brackets of (50) can then be estimated. The logarithmic term decreases relatively slowly; for example when $n = 100$, $\ln N/N^{2/3}$ is still equal to 0.21,while $N^{-1} = 0.01$.

Studies of the vibrational entropy variation of crystalline clusters by numerical calculations on harmonic models have been performed for 2D and 3D lattices.[56] Relatively simple empirical formulas were ontained. For example, parallelogram shaped 2D crystals with hexagonal lattice containing N atoms have an excess vibrational entropy (entering the second term of (48); the summation in the two-dimensional space is over 2N-3 movements only) equal to*:

$$\frac{S^e}{k_B} = \sum_1^{2N-3} \ln \frac{\gamma_o}{\gamma_i} = 2.65\ N^{1/2} + \ln N - 4.159, \qquad (51a)$$

which can be written also as (50), using the total length L of the parallelogram perimeter :

$$\frac{S^e}{k_B} = (s_o^e + s_1^e \frac{\ln N}{N^{1/2}})\ L - \text{const.} \qquad (51b)$$

Here s_o^e is the vibrational part of the specific ledge entropy of a 2D-cluster of infinite size. One can notice that besides the differences due to different dimensionalities, the equations (50) and (51) are very similar, especially regarding the logarithmic terms. The constant in (51a) comes from the size independent contribution to the vibrational entropy from two of the parallelogram corners. Similar contributions are obtained for the potential part of the ledge free energy. It is interesting to compare this constant with the replacement free energy F_{rep} for the same model.

---

* In the original work of HOOVER, HINDMARSH and HOLIAN[56] the results for the vibrational entropy are presented in the fixed center of mass reference system. These results are transposed in eq. (51a) to the fixed atoms reference system.

In the two-dimensional space, the third term of (48) should be written as :

$$F_{rep} = - k_B T \ln \frac{A_r}{\tilde{a}} \frac{\ell_r}{\tilde{\ell}},$$

where A_v is the "volume" of the two-dimensional vessel. A typical order of magnitude of $F_{rep}/k_B T$ per unit "volume", all other parameters being those of a "Lennard-Jones Argon",[57] is about 40. The constant B in eq. (49) which is the sum of all constant terms, is thus neither zero, nor an unsignificant corner energy contribution. It is primarily due to the replacement free energy.

The physical reality of the above models, using continuum calculations and harmonic approximation, can be contested. For this reason, it is interesting to look for more "natural" models. Monte-Carlo simulation studies of small liquid clusters, of "Lennard-Jones Argon" at the melting point,[58] show that a fairly good linear relationship of the excess free energy with $N^{2/3}$, as expected by eq. (49), seems to be achieved for clusters of only some 20 to 30 atoms. In view of the high cost of this kind of calculations, results are scarce and it is not yet possible to assess that the fit is perfect.[59] However, the very good agreement of the experimental slope of (49) with the value of the calculated macroscopic surface tension[60] for this system shows that the capillarity approximation is, by a strange train of circumstances, not as bad as one could expect, even for cluster sizes of only few molecules.

The results of the confrontation between microscopic and macroscopic parameters can thus be summarized as follows.

(i) The excess free energy of a small phase is proportional to the area of its surface in a large size range. Important corrections to this proportionality become necessary only when the phase approaches molecular dimensions.

(ii) The dependence of the excess free energy on cluster dimensions *contains a constant term* which is due, primarily, to the replacement of some vibrational movements by translational and rotational movements of the molecules building the cluster.

Acknowledgements

The author is indebted to John Abernathey for numerous discussions and for critical revision of the manuscirpt.

References

1 . MOELWYN-HUGHES E.A., *Physical Chemistry*, Pergamon Press, Cambridge, 1961, pp. 921-981
2 . GUGGENHEIM E.A., *Thermodynamics, An Advanced Treatment for Chemists and Physicists*, North Holland, Amsterdam, 1949.
3 . FOWLER R. and GUGGENHEIM E.A., *Statistical Thermodynamics*, University Press, Cambridge, 1965; a) pp. 421-451; b) pp. 55-65
4 . AVEYARD R. and HAYDON D.A., *An Introduction to the Principles of Surface Chemistry*, University Press, Cambridge, 1973, pp. 2-29.
5 . GIBBS J.W., *The Scientific Papers*, Vol. 1, Dover Publ. N.Y., 1961, pp. 85-96
6 . McQUARRIE D.A., *Statistical Mechanics*, Harper and Row, N.Y., 1976, pp. 13-19.
7 . TOLMAN R.C., J. Chem. Phys. 16(1948)758.
8 . HONDROS E.D., Proc. Roy. Soc. (London)A286(1965)479
9 . BORN M. and STERN E., Sitz. Ber. Preuss. Akad. Wiss. 48(1919) 901.
10. HILL T.L., *Thermodynamics of Small Systems*, W.A. Benjamin Inc. N.Y., 1963, pp. 27-46.
11. MUTAFTSCHIEV B., Surf. Sci. 61(1976)93
12. STRANSKI I.N., Z. Physik. Chem. (B)38(1938)451.
13. STEFAN J., Ann. Physik 29(1886)655
14. BONISSENT A., FINNEY J.L. and MUTAFTSCHIEV B., in *Proc. Intern. Vac. Congr. and 3d Conf. Solid Surf.*, Dobrozemski R. et al. editors, Vienna, 1977, Vol. 1, p 441.
15. BONISSENT A., Ph. D. Thesis, Marseille, 1978.
16. SKAPSKI A.S., Acta Met. 4(1956)571
17. JACKSON K.A., in *Liquid Metals and Solidification*, Amer. Soc. Metals, Cleveland, 1958, p. 174.
18. CURIE P., Bull. Soc. Mineral. de France 8(1885)145
19. WULFF G., Z. Kristallogr. 34(1901)449
20. HERRING C., Phys. Rev. 82(1951)87
21. KAISCHEW R., Bull. Acad. Bulg. Sci.(Phys.) 1(1950)100
22. DASH J.G., Phys. Rev. (B) 15(1977)3136
23. SHELUDKO A., TOSCHEV B.V. and PLATIKANOV D., in *The Modern Theory of Capillarity*, Rousanov. A.I. and Goodrich F.C. editors, Khimia Publ. Leningrad, 1981, p 275.
24. KAISCHEW R., Bull. Acad. Bulg. Sci. (Phys.)2(1951)191
25. VOLMER M., *Kinetik der Phasenbildung*, Steinkopf, Dresden u. Leipzig, 1939, pp. 97-100
26. COTTERIL P. and MOULD P.R., *Recrystallization and Grain Growth in Metals*, Surrey Univ. Press, London, 1976, pp. 266-325.
27. HERRING C., in *Structure and Properties of Solid Surfaces*, Gomer R. and Smith C.S. editors, Chicago Press, 1953, p.1.
28. BURTON W.K. and CABRERA N., Disc. Farad. Soc. 5(1949)33

29. LACMANN R., in *Adsorption et Croissance Cristalline*, CNRS, Paris 1965, p 195
30. FRANK F.C., Disc. Farad. Soc. 5(1949)48
31. MULLIN W.W., Acta Met. 7(1959)747
32. MUTAFTSCHIEV B., in *Adsorption et Croissance Cristalline*, CNRS, Paris, 1965, p 231.
33. TEMKIN D.E., in *Crystallization Processes*, Sirota N.N. and Gorskii F.K. editors, Consultants Bureau, N.Y., A966, p 15
34. WEEKS J.D., GILMER G.H. and JACKSON K.A., J. Chem. Phys. 65 (1976)712.
35. BURTON W.K., CABRERA N. and FRANK F.C., Phil. Trans. Roy. Soc. (London) A243(1951)299
36. FRUMKIN A.N., Z. Physik. Chem. 116(1925)466
37. FOWLER R., Proc. Camb. Phil. Soc. 32(1936)144
38. GILMER G.H. and JACKSON K.A., in *Crystal Growth and Materials*, Kaldis E. and Scheel H. editors, North Holland, Amsterdam, 1977, p 80.
39. HEYER H., KARGE H. and POUND G.M., in *Adsorption et Croissance Cristalline*, CNRS, Paris, 1965, p 255
40. STRANSKI I.N., GANS W. and RAU H., Ber. Bunsenges. 67(1963)965
41. TOLMAN R.C., J. Chem. Phys. 17(1949)118;333
42. TAMMAN G., Physik. Z. 11(1910)609
43. THOMY A. and DUVAL X., J. de Chim. Phys. 67(1970)1101
44. BIENFAIT M., in *Phase Transitions in Surface Films*, Dash J.G. and Ruvalds J. editors, Plenum Press, N.Y., 1980, p 29
45. NELSON D.R., Phys. Rev. 18(1978)2318
46. van SWOL F., WOODCOCK L.V. and CAPE J.N., J. Chem. Phys. 73, 190 (1980)913.
47. BONISSENT A., FINNEY J.L. and MUTAFTSCHIEV B., Phil. Mag. B42 (1980)233.
48. BONISSENT A. and MUTAFTSCHIEV B., CRC Critical Reviews in Solid State and Mat. Sciences 10(1981)297
49. LOTHE J. and POUND G.M., J. Chem. Phys. 36(1926)2080
50. ABRAHAM F.F. and POUND G.M., J. Chem. Phys. 48(1968)732
51. REISS H., KATZ J.L. and COHEN E.R., J. Chem. Phys. 48(1968)5553.
52. REISS H., J. Stat. Phys. 2(1970)83.
53. NISHIOKA K., SHAWYER R., BIENENSTOCK A. and POUND G.M., J. Chem. Phys. 55(1971)5082
54. LOTHE J. and POUND G.M., J. Chem. Phys. 45(1966)630
55. MUTAFTSCHIEV B., unpublished results
56. HOOVER W.C., HINDMARSH A.C. and HOLIAN B.L., J. Chem. Phys. 57 (1972)1980
57. ABRAHAM F.F., *Homogeneous Nucleation Theory*, Academic Press, N.Y., 1974, p 217
58. LEE J.K., BARKER J.A. and ABRAHAM F.F., J. Chem. Phys. 58(1973) 3166
59. NISHIOKA K., Phys. Rev. (A) 16(1977)2143
60. MIYAZAKI J., BARKER J. and POUND G.M., J. Chem. Phys. 64(1976) 3364.

PHYSISORPTION AND CHEMISORPTION

J.P. GASPARD

Institut de Physique
Université de Liège
B 4000 SART-TILMAN (Belgium)

ABSTRACT

The nature of the adatom-substrate bond is schematically discussed in the framework of simple theories. Physisorption of inert atoms is first considered : adsorption energy of a single adatom and lateral interaction energy are calculated.

Chemisorption bond of atoms on a covalent or a metallic substrate is discussed qualitatively. A theoretical model is presented and some typical cases are considered. The role of lateral interactions is also briefly discussed.

I. PHYSISORPTION

We consider physical adsorption (physisorption) separately from chemisorption (where a true chemical bond is formed). Physisorption is defined by the van der Waals type interaction between closed shell atoms (rare gases) or molecules (e.g. CH_4) and the substrate. The energies of physisorption are of the order of 10^{-2} eV, comparable to $k_B T$ at room temperature.

Physisorption bonds are of great practical importance because they are dominant not only at submacroscopic level (interaction between colloids, powders) but also in the interaction between macroscopic entities (adhesion of solids).

There is an increasing interest during the last twenty years, in physisorbed layers, such as rare gases on graphite, because they allow studies on two dimensional (2D) condensation, and melting,

B. Mutaftschiev (ed.), Interfacial Aspects of Phase Transformations, 103–118.

polymorphism, the effect of the substrate structure on the properties of[1] the layer, as well as precise measurement of critical exponents.

From the theoretical point of view, physisorption is an interesting situation as most of the quantities can be calculated[2].

In the following two sections, a surface plasmon model is used to calculate the atom-metallic surface interaction energy and the lateral interaction energy in phisorption.

I.1. Atom - Surface Interaction

By analogy to the description of the van der Waals interaction between two closed shell atoms, (see chapter "Electronic Structure and Cohesion") we consider an adatom characterized by its polarizability α and an oscillator frequency ω_o such that $\hbar\omega_o$ corresponds to the lowest energy excitations of the atom (usually taken as the ionization energy).

Similarly the metallic surface is schematized by a surface plasmon frequency ω_{sp}. The electromagnetic interaction of the adatom with the surface gives rise to van der Waals attractive forces and a corresponding energy V_a :

$$V_a = \frac{\hbar \omega_{sp} \alpha}{2d^3} \frac{\omega_{sp}}{\omega_o + \omega_{sp}} \qquad (1)$$

where d is the distance between the atom and the planar surface (Fig. 1)

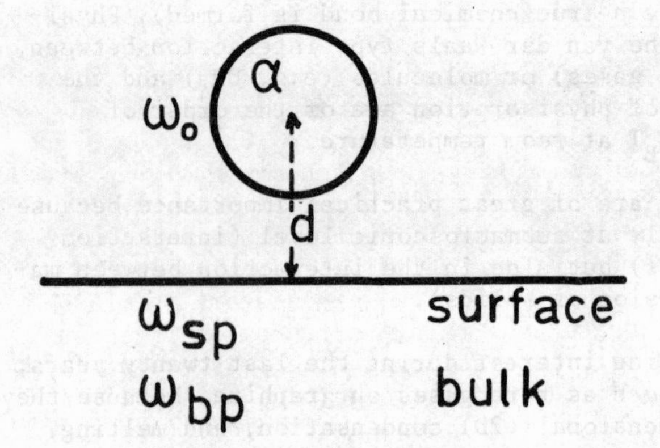

Fig. 1 : Adsorption on a flat surface.

If the surface is curved (cylinder or sphere), the van der Waals interaction is reduced for geometrical reasons; conversely adsorption on a cylindrical or a spherical hole is increased[2] (Fig. 2).

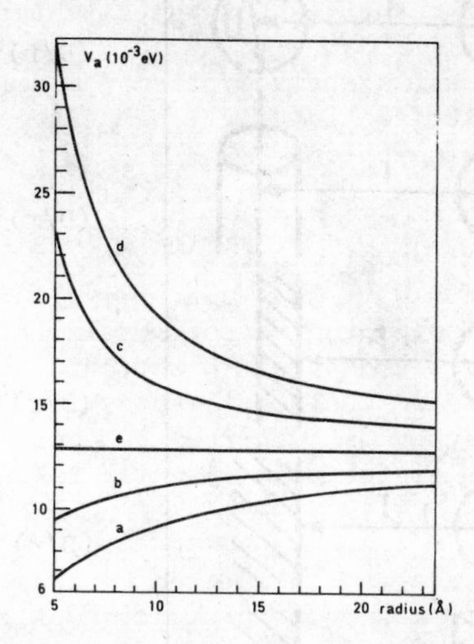

Fig. 2 : Adsorption energy of Ne on Al. Effect of curvature a) sphere; b) cylinder; c) cylindrical hole; d) spherical hole; e) plane.

As a consequence, inert atoms have a stronger tendency to attach on the surface of large grains than on smaller ones or they prefer to stay on substrate indentations. Table 1 shows the variation of the interaction energy as a function of distance r to first order in D/r (D is the typical dimension of the object)[3]

I.2. Lateral Interaction between Physisorbed atoms

In the case of van der Waals forces, it is possible to calculate the variation of the interaction between two atoms when they are brought near a metallic substrate. This variation is due to the third (and higher) order terms, as shown schematically in Fig.3.

Generally, the interaction energy between two atoms is reduced by the presence of the substrate. The relative reduction depends on the parameters and the geometry. For two argon atoms in close contact on a metallic substrate, the relative reduction is about 5 %. The reduction of the pair interaction in the presence of a substrate has consequences on the completion of the adsorbed monolayer, especially in the case of relatively small atoms (Ar).

Geometry	Power Law
Sphere + Sphere	$(D/r)^6$
Sphere + Cylinder	$(D/r)^5$
Sphere + Slab	$(D/r)^4$
Sphere + Half-Space	$(D/r)^3$
Half-Space + Half-Space	$(D/r)^2$

Table 1 : Variation of the interaction energy with distance r to first order in D/r.

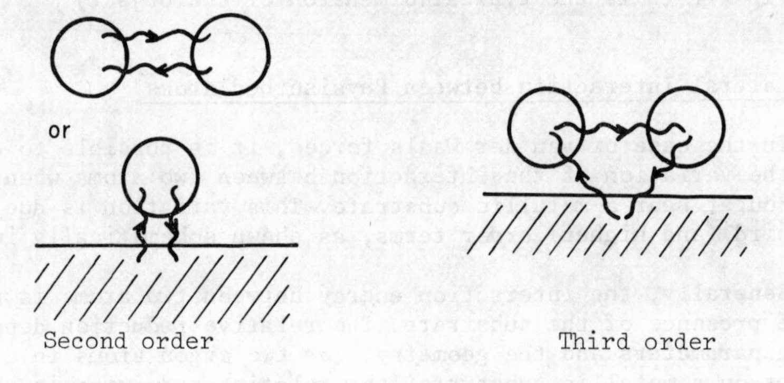

Fig. 3 : Second order (direct) and third order (indirect) interactions. The wavy arrow is the electromagnetic interaction.

Apparently due to the weakening of the lateral interaction in the monolayer of Ar, the monolayer cannot be completely close packed. This has perturbing consequences when using the BET method to determine the specific surface of catalyst by adsorption of Ar. In the literature one finds an effective surface area of the Ar atom ranging from 13.7 $Å^2$ to 18.2 $Å^2$, presumably depending on the porosity of the substrate.[2]

In conclusion, van der Waals interactions are interesting to consider, not only because they correspond to real physical situations (rare gases, inert molecules, adhesion) but because their inherent simplicity (relative to chemisorption) allows investigation of a series of situations of increasing complexity (effect of curvature, adatom interaction, sub-monolayer and multi-layer adsorption, etc.).

II. CHEMISORPTION

Chemisorption involves the formation of a true chemical bond (sharing of electrons) between an atom or molecule and the substrate. In contrast to physical adsorption (physisorption) discussed in the preceeding section, the chemisorption bond energy is of the order or greater than 1 eV; energies of 3-4 eV are commonly encountered.

It has long been believed (since Langmuir, 1918) that adsorption on solids and bonding in molecules or in solids are essentially of the same nature. Chemisorption plays a central role in the understanding of more complex processes such as epitaxial growth, corrosion, electrolysis, catalysis, etc. One must first understand the chemisorption of a single atom or a simple molecule on a clean defined surface before proceeding to more realistic and more complex situations. However most progress in the field of chemisorption is quite recent (in the last decade). More than in many other fields of surface science, experiments are ahead of theory.

The decisive progress in experiments is due primarily to the development of ultrahigh vacuum techniques (residual pressure ≃ 10^{-10} torr), that permit a reactive surface (metal, semiconductor) to remain clean for a period of an hour, and to the advent of surface sensitive techniques. In addition to the techniques of determination of the surface structure (cf. chapter "The Structure of Clean Surfaces"), angularly resolved photoemission excited by synchrotron sources[4] and the high resolution electron energy loss spectroscopy[5] have both brought an impressive amount of experimental results to occur to our understanding of the chemisorption bond.

In the following sections, we first present a relatively simple model describing the physics of the formation of the chemisorption bond. Examples are then given in the field of semiconductors and metals. Finally, a qualitative analysis of the lateral interactions between chemisorbed atoms is presented.

II.1. A Model of Chemisorption

In complicated systems, e.g. an adatom or a molecule chemisorbed on a substrate of different nature, a theory derived from simple models allows the physics to be kept transparent.

We summarize here the Friedel-Anderson theory of a magnetic impurity in a metal transposed by EDWARDS and NEWNS[6] to the chemisorption case. The hamiltonian is

$$H = H_{substr.} + H_{at} + Un_{a+}n_{a-} + Vh_{s-a} \qquad (3)$$

The first term on the right hand side describes the properties of the substrate in the absence of adatom; the second and third terms describe the atom, with energy level ε_a and correlation energy U (n_{a+} is the number of electrons of the adatom with spin up). The last term describes the interaction of the substrate and the adatom.

The energy U is the effective correlation energy originating in the Coulomb repulsion of two electrons (with opposite spins) located on the same orbital of the adatom. Fig. 4 shows a picture of the states. In this description the parameters (specially U and V) are fitted to the experimental results and/or determined by comparison with reference cases like molecules.

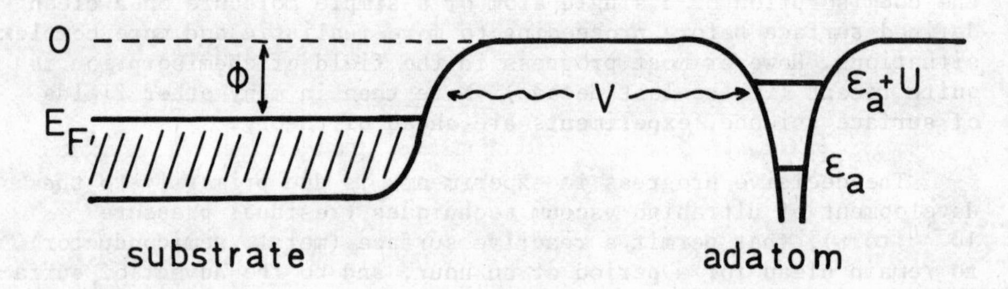

Fig. 4 : Schematic picture of the energies of the adatom and the substrate. ϕ is the work function.

If one electron occupies the adatom level, it sits at the energy ε_a, if two electrons are present they would sit at $\varepsilon_a + U$. In real cases with a substrate, the Fermi level can be between these two energies if there is a charge transfer from the substrate to the adatom . This charge tranfer is determined by the relative position of the substrate Fermi level E_F and the adatom energy ε_a, and the value of the effective U. The larger the U value, the smaller the charge transfer. This is a crucial point in chemisorption. In the hydrogen atom U is 13 eV; the presence of neighboring atoms and/or delocalized electrons cause a reduction of U (up to an order of magnitude). The U value shows how much the atom retains its atomic nature (U relatively unchanged) or how much it participates in the bulk behavior (U strongly reduced).

Let us consider first the simplest model, the limit U=0 (Hückel or tight binding model). If ε_a is within the bulk band, when V is gradually increased, one observed an evolution from a weak behavior , $V/W \ll 1$, to a strong coupling behavior, $V/W \gg 1$, (cf. Fig. 5 ; W is the substrate bandwidth).

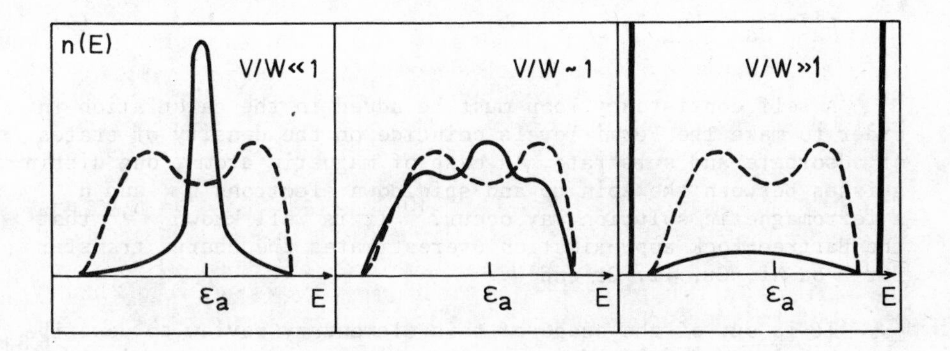

Fig. 5 : Schematic variations of the adatom density of states(—) when V/W is increased. Density of states of the substrate (---)

For small values of the adatom substrate interaction V the atomic level ε_a is weakly broadened (proportionally to V) compared to the band-width W : the adatom is weakly perturbed by the substrate. In this case, the Coulomb correlations play a very important role. When V is comparable to the bandwidth, the adatom density of states has an effective width comparable to W; if V > W, two peaks usually appear : bonding and antibonding. Due to the larger interaction with the substrate, one expects a reduction of U due to the screening by the electrons of the substrate. When V/W is large compared to unity, the bonding and antibonding states split off the substrate band. One can describe the system as a surface molecule (the adatom plus the substrate atom(s) covalently bound to it) weakly coupled to the rest of the substrate.

Most of the perturbation of the substrate due to the adatom is limited to its closest neighborhood. This description leads to simple analytic developments in the case U=0 (see e.g. ref. 6, 7). Quantitative differences occur depending on the position of the adatom (top, bridge, centered) on the substrate, on its energy ε_a relative to the substrate band, and on the details of the substrate density of states.

The introduction of U (the electron-electron effective Coulomb repulsion on the adatom orbital) considerably complicates the solution. Starting from a one-electron problem (U=0), the additional U term in the hamiltonian is a typical many-body term[8] and must be treated with the techniques of the many-body problem. One of the simplest treatment is the Hartree-Fock decoupling scheme. The one electron level is affected by its partial occupation n_a with electrons[9]:

$$\varepsilon'_a = \varepsilon_a + Un_a \tag{4}$$

A self consistency loop must be added in the calculation in order to make the Fermi levels coincide on the density of states of adsorbate and substrate. In case of magnetic atoms, one distinguishes between the spin up and spin down electrons n_{a+} and n_{a-}. A ferromagnetic solution may occur.[10] It is well known[6,9] that the Hartree-Fock approximation overestimates the charge transfer for H on Ni, Cu, Ti, Cr and W.

It is out of the scope of this elementary review to describe the techniques developed in chemisorption theory, in particular[6,8,11] the local density functional formalism. Excellent review books are now available.

II.2. Chemisorption on a Semiconductor Surface

An impressive number of both experimental and theoretical studies of chemisorption on semiconductor surfaces has recently appeared.[11-15] These studies concentrated on simple semiconductor surfaces of elemental (Si, Ge) or compound (GaAs, InSb) semiconductors and simple adatoms (H,Cl,F). A one electron model with a rough treatment of correlations was found sufficient to interpret the experimental results.

(i) H *on* Si *surfaces*. Let us consider the (111) surface of Si(Fig. 6). A dangling bond emerges from any surface atom, perpendicular to the surface when unreconstructed. A hydrogen atom chemisorbed on the surface forms a strong covalent bond by interaction of the H1s orbital and the sp^3 dangling bond. Due to their small size, the hydrogen atoms may saturate all the dangling bonds of the (111) unreconstructed surface (called the monohydride phase). The effects of H on the electron spectrum is clearly visible in photoemission (Fig. 7).

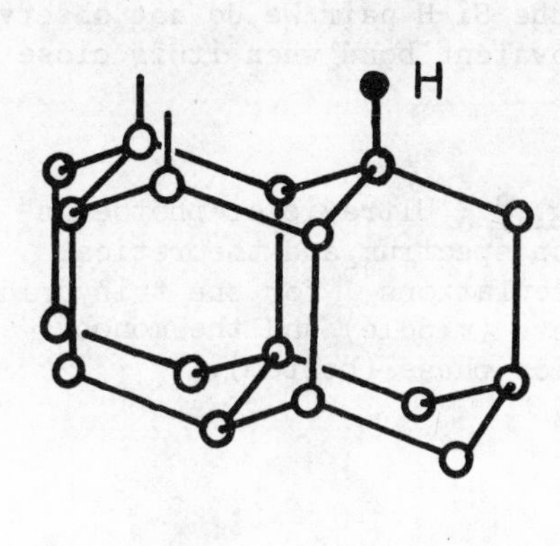

Fig. 6 : Adsorption of H on Si(111).

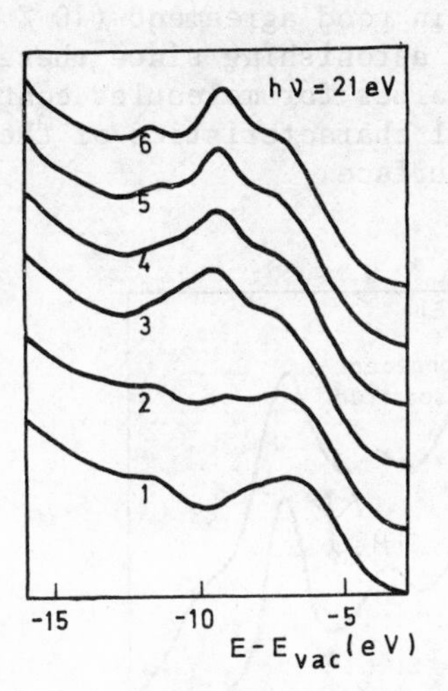

Fig. 7 : Photoemission spectra of H chemisorbed on Si.[15] The coverage by H increases from 1 to 6.

Hydrogen removes the dangling bond states from the gap and decreases the density of p states at the top of the valence bond.(*) Two hydrogen peaks appear in the density of states at -4.5 eV and -7.0 eV as shown on Fig. 7. The states at -4.5 eV are highly localized in the Si-H band, similar to the covalent bond in silane SiH_4. The state at -7.0 eV is more delocalized; it may be due to a van Hove singularity of the two dimensional chemisorbed layer. As shown in Fig. 8, there is an excellent agreement between experiment and the theory. Essentially three types of approaches were used : the self-consistent pseudopotential approach (SCPP) by APPELBAUM and HAMANN[11] the empirical tight-binding (ETB) approach of PANDEY[12] and calculations on small hydrogenated aggregates of Si, using methods of quantum chemistry developed for the study of molecules.

(*) The "cleaning" of the gap by saturation of the dangling bond orbitals with monovalent atoms is largely used now in the field of amorphous semiconductors. It gave a decisive new impetus in the field.[14] The material is produced either by glow discharge of silane or by posthydrogenation of the sample.

There is in general an excellent agreement between theory and experiment proving the validity of the one-electron approximation corrected with an averaged charge tranfer. The equilibrium distance is found to be 1.43 Å and the Si-H force constant is 0.17 a.u. again in good agreement (10 % error) with experimental data. This is not astonishing since these values are close to the corresponding values for molecules containing the Si-H pair. We do not observe special characteristics of the Si-H covalent bond when it is close to a surface.

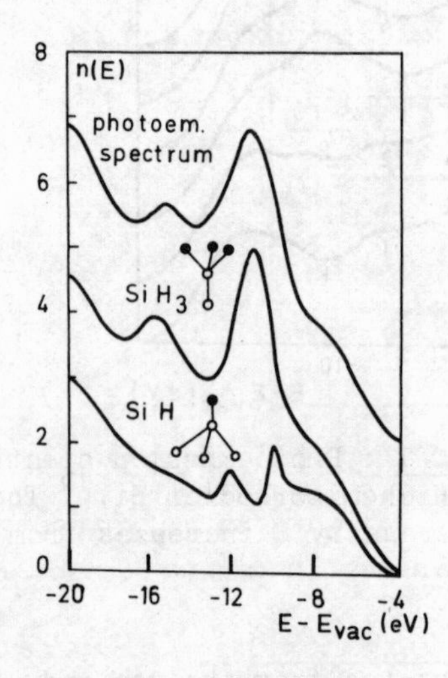

Fig. 8 : Ultraviolet photoemission spectrum and theoretical calculations[15] for the trihydride phase (middle) and the monohydride phase (bottom).

Similar conclusions can be drawn for the trihydride phase of Si(111). Additional spectral features are observed due to neighboring hydrogen atoms bound to the same Si atom.

The case of the (100) surface of Si, presenting two dangling bonds per atom in the unreconstructed geometry, is more controversial. There is no net agreement between experiment and a theory based on a simple bonding geometry. It is conjectured that a more complex H chemisorptive mode should be envisaged.[11]

(ii) *Chlorine Chemisorption* is somewhat different from hydrogen since it has seven electrons ($3s^2 3p^5$) and is more electronegative than hydrogen; consequently the bond will be partially ionic. The energy level scheme is depicted in Fig. 9.

Assuming that Cl adsorbs on top like hydrogen, one observes that one covalent bond is formed between the s and p_x chlorine orbital and the sp^3 dangling bond, while the two p_y and p_z orbitals do not interact with the dangling bond for symmetry reasons. Therefore, after the interaction, we are left with a level scheme showing four different states. The p_π doubly degenerate level gives rise to a high and narrow peak in more elaborate calculations (Fig.10) still clearly showing the three satellite peaks.

Fig. 9 : Schematic level scheme of Cl interacting with a dangling bond (sp^3) of Si.[16]

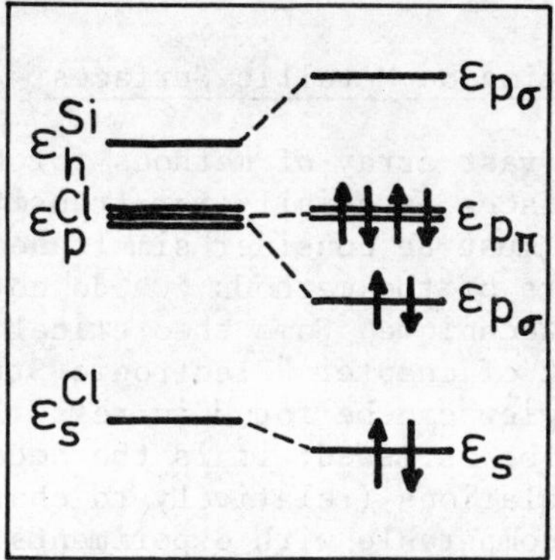

If, instead of a on site adsorption, we consider the three-fold site adsorption (chlorine interacting with three dangling bonds), we still see the p_π narrow peak but no longer the lowest energy satellites.

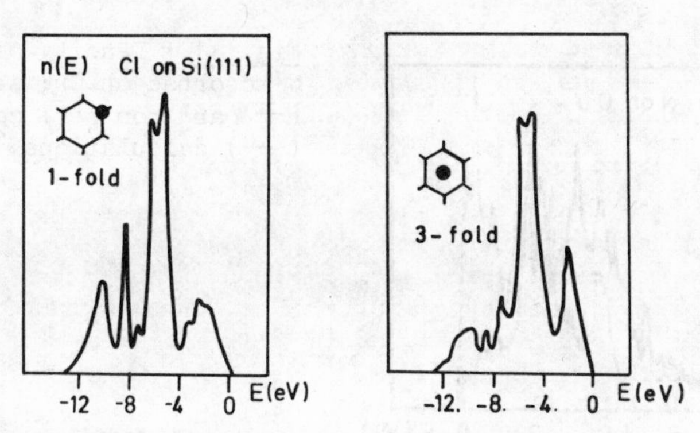

Fig. 10 : Calculated local densities of states[17] of Cl adsorbed on Si(111). Left : on top; right : three-fold adsorption.

Experimentally Si and Ge behave differently. From the photoemission spectra and the analysis of their angular variation, it is deduced that Cl on Si(111) is a on site adsorption while Cl on Ge(111) is a three-fold site adsorption. This is a clear demonstration of the intimate relations between geometry and electronic properties.

(iii) *Chemisorption of metals on semiconductors* is investigated primarily because it brings information about the formation of the metal-semiconductor interface (c.f. chapter "Chemisorption of Metals on Metals and on Semiconductors"). The calculation of the height of the rectifying Schottky barrier is still to be done from first principle. If some succes is met in some cases, a universal mechanism still to be discovered.[18,19]

II.3. Chemisorption on Metallic Surfaces

There is a vast array of methods for the study of chemisorption on metallic surfaces especially for transition metals where the s and d electrons must be consider simultaneously. Because of the technical aspects of the methods, we do not go into the details of any particular technique. Some theoretical approaches are developed in section III.2 of chapter "Electronic Structure and Cohesion". An extensive review can be found in refs. 20 and 21. One point, however, should be stressed; it is the necessity of making self-consistent calculations (relatively to charge and energy) in order to get results comparable with experiments. In order to illustrate this, Fig. 11 shows the difference in density of states between a non self-consistent and a fully self-consistent calculation for N adsorbed on Cu. For example, in the non self-consistent calculation the d band is so high in energy that it contains holes (Cu would behave like a transition metal near the surface !).

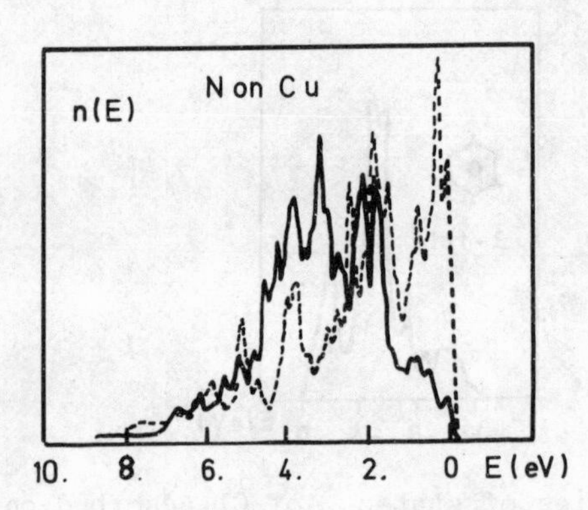

Fig. 11 : Density of states of N adsorbed on Cu; self consistent (—) and non self consistent (---) calculations.[21]

Fig. 12 shows a typical theoretical calculation compared with photoemission spectra of various adsorbates. They show that the high density of surface states is removed by the chemisorption. The state-of-the-art is at a point where most of the simple cases of adsorption on metals can be calculated using rather sophisticated (and expensive) computer programs. There is now a need to synthesize all the particular results and to develop simpler models that could be useful in the understanding of more complicated situations: e.g., larger adsorbed molecules, surface steps, amorphous surfaces.

Recall that one aim of chemisorption studies is the understanding of real industrial processes, i.e. catalysis, corrosion, electrolysis, etc., particularly in the field of metals.

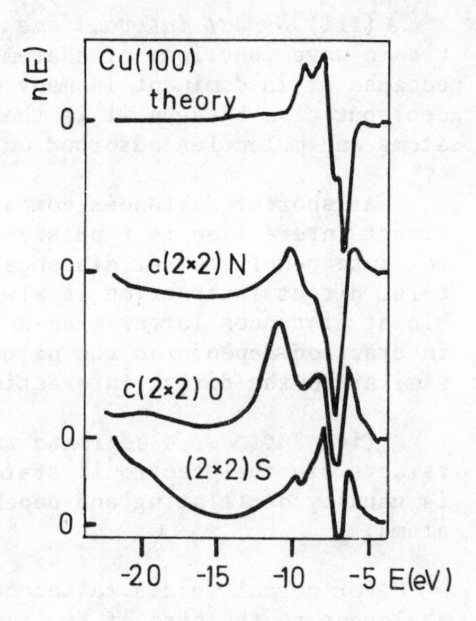

Fig. 12 : Theoretical difference density of states (bulk-surface) (top) and experimental difference spectra.[22]

II.4. Lateral Interactions in Chemisorption

By lateral interaction we mean the interaction of two adsorbed entities (atoms or molecules) distant from R in the presence of the substrate.

There are several mechanisms giving rise to lateral interactions : (i) dipole-dipole repulsion; (ii) van der Waals attraction; (iii) direct interaction (orbital overlap); (iv) indirect coupling via the substrate electronic states; (v) indirect coupling via deformation of the substrate (mattress effect) or via the phonon field mechanism. The latter may be either attractive or repulsive.

(i) *Dipole-dipole interaction*. When the adatoms and the substrate atoms differ, they often produce a permanent dipole moment $\vec{p}$ due to the charge transfer. This affects the values of the work functions. The interaction of two permanent dipoles ($U = p^2/r^3$) is usually small compared to other effect except when the ionicity of the chemisorption bond is large. In most cases dipolar interaction is negligible, e.g. when CO is adsorbed on Pd(111), the adsorption energy is $E \simeq 0.14$ kcal/mole = 6 meV, which is a negligible value. Alkali metals give a larger value, of the order of $k_B T$ at room temperature, e.g. for Na/Ni (100), $E = 0.7$ kcal/mole. The dipole-dipole interaction can then influence the geometric structure of the adlayer.

(ii) *The van der Waals lateral interaction* was treated in section I.1.

(iii) *Direct interactions*. The direct interaction of the electronic wave functions of adatoms is of primary interest, not only because it is dominant in many cases (particularly at short distances) but also because it is the key to the chemical reactivity of atoms and molecules adsorbed on a surface (heterogeneous catalysis).

At shorter distances compared to the diameter of the adatoms, direct interaction is repulsive because the electronic clouds cannot interpenetrate. At distances of the order of the atomic diameters, direct interaction is always attractive; it becomes negligible at distances larger than a few Angströms. The direct lateral interaction depends on the nature of the atoms : it is presumably similar to the direct interaction of the free atoms.

(iv) *Indirect electronic interactions*. Two adsorbed atoms interacts via the electronic states of the substrate. The interaction is usually oscillating and depends on the nature of the adsorbed atoms.

For normal metals, a second order perturbation expansion-analogous to the case of two impurities in the bulk- gives rise to an oscillating effective pair interaction (Fig. 13)

$$V_{eff}(r) \simeq \frac{\cos(k_F r)}{r^5}$$

(where k_F is the Ferm wave vector), valid for $r \simeq k_F^{-1} \simeq 1$ Å^{-1}.

If there is a partially filled band of surface states, the interaction is less damped; r^{-2} instead of r^{-5}. For transition metals, EINSTEIN and SCHRIEFFER[23] have calculated the interaction energy on various adsorption sites. $V_{eff}(r)$ is again an oscillating functions depending on the band filling and shows a strong anisotropy.

Whether direct or indirect interaction is dominant depends on the system. In the case of a bcc substrate, the direct interaction is dominant (and so the dimer is stable) on the (110) plane for nearest neighbors but for second nearest neighbors the interaction is direct on the (110) plane and indirect on the (100) plane.

Fig. 13 : Direct (——) and indirect (---) interaction between adsorbed atoms.

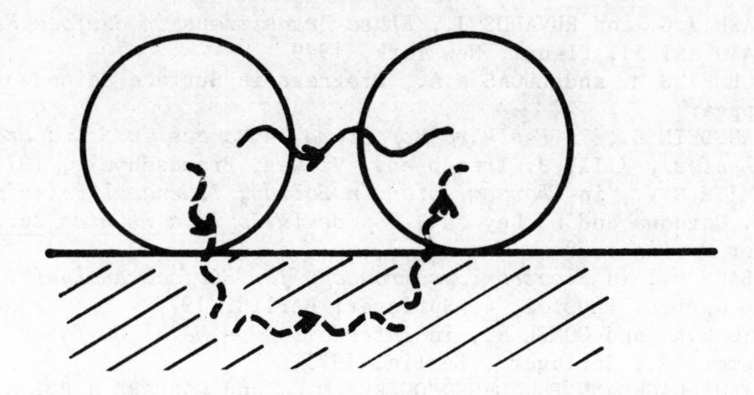

(v) Let us finally briefly mention the *indirect interaction* "mechanically" mediated by the substrate either statically (mattress effect) or via the phonon field. LAU and KOHN[24] have studied the lateral interactions due to the distortion of the substrate when adsorbing particles, in the continuum elasticity limit. They found that identical adsorbates have a tendancy to repel whilst different adsorbates may attract. The phonon field produces a weak ($\simeq 10^{-5}$ kcal/mole) attractive force varying like r^{-7}; it will be always negligible.

In summary, we are *now* able to describe in a semiquantitative way the interaction of two adatoms on a metallic substrate, at least in the asymptotic limit ($r \to \infty$). A better understanding of the interaction at short distances is of crucial importance for surface chemical reactions (dissociation, catalytic reactions, etc.). If there are more than two adatoms on the surface, the question of additivity of the pair interaction arises. Three and four (and possibly higher) body interactions could be important in the formation of islands or super-structures, as shown in preliminary estimations in the case of physical adsorption and in tight binding model calculations.

At finite coverage, the structure of the chemisorbed phase at a given pressure and temperature is determined by the two-dimensional periodic potential of chemisorption of a single atom, and the interaction potential between adatoms. One is still far from predicting the structure of the chemisorbed phase as a function of the coverage and the temperature. However our understanding of two-dimensional adsorbed phases made decisive progress in the last decade specially in the field of physisorbed layers,[1] as will be seen in the chapter "Critical phenomena in two dimensions".

REFERENCES

1 . DASH J.G. and RUVALDS J., *Phase Transitions in Surface Films*, NATO ASI 51, Plenum, New York, 1979

2 . SCHMEITS M. and LUCAS A.A., Progress in Surface Science (to appear)

3 . LANGBEIN D., in *Festkörpenprobleme (Advances in Solid State Physics)*, XIII, J. Trench ed., Vieweg, Braunschweig, 1973, p.85.

4 . SMITH N.V., in *Photoemission in Solids; I, General Principles*, M. Cardona and L. Ley ed., *Topics in Applied Physics* 26, Springer Berlin 1978

5 . IBACH H., in *Electron Spectroscopy for Surface Analysis, Topics in Current Physics*, 4, Springer, Berlin, 1977

6 . LYO S.K. and GOMER R., in *Interaction on Metal Surfaces*, R. Gomer ed., Springer , Berlin, 1975.

7 . CYROT-LACKMANN F., DESJONQUERES M.C. and GASPARD J.P., J. Phys. C., 6(1973)3077

8 . EINSTEIN T.L., HERTZ J.A. and SCHRIEFFER J.R., in *Theory of Chemisorption*, J.R. Smith ed., Springer, Berlin, 1980

9 . NEWNS D.M., Phys. Rev. 178(1969)1123

10. ANDERSON P.W., Phys. Rev. 124(1961)41 and EDWARDS D.M. and NEWNS D.M., Phys. Lett. 24A(1967)236

11. APPELLBAUM J.A. and HAMANN D.R., in *Theory of Chemisorption*, J.R. Smith ed., Springer, Berlin, 1980

12. PANDEY K.C., Phys. Rev. B14(1976)1557

13. HO K.M., SCHLUTER M.and COHEN M.L., Phys. Rev. B15(1977)3888

14. PANDEY K.C., SAKURAI T. and HAGSTRUM M.D., Phys. Rev. Lett.35 (1975)1728.

15. APPELBAUM J.A., HAGSTRUM H.D., HAMANN D.R. and SAKURAI T., Surf. Sci. 58(1976)479

16. LANNOO M., in *Electronic Structure of Crystal Defects and of Disordered Systems*, Les Editions de Physique, Orsay, 1981

17. SCHLUTER M., ROWE J.E., MARGARITONDO G., HO K.M. and COHEN M.L., Phys. Rev. Lett. 37(1976)1632

18. MELE E.J. and JOANNOPOULOS J.D., Phys. Rev. B17(1978)1528

19. SPICER W.E., LINDAU I., SKEATH P., SU C.Y. and CHYE P., Phys. Rev. Lett. 44(1980)420

20. ERTL G., in *The Nature of the Surface Chemical Bond*, Rhodin and Ertl ed., North Holland, Amsterdam, 1980

21. ARLINGHAUS F.J., GAY J.G. and SMITH J.R., in *Theory of Chemisorption*, J.R. Smith ed., Springer, Berlin, 1980

22. BURKSTRAND J.M., KLEINMAN G.G., TIBBETTS G.G. and TRACY J.C., J. Vac. Sci. Technol. 13(1976)291

23. EINSTEIN T.L. and SCHRIEFFER J.R., Phys. Rev. B7(1973)3629

24. LAU K.H and KOHN W., Surf. Sci. 75(1978)69.

CRITICAL PHENOMENA IN TWO DIMENSIONS : THEORETICAL MODELS AND PHYSICAL REALIZATIONS

Eytan DOMANY

Department of Electronics
Weizmann Institute of Science
Rehovot, Israel

I. INTRODUCTION, OUTLINE AND APOLOGY

During the past four to five years, there has been a most remarkable surge of interest in two dimensional physics.[1-7] Large numbers of workers were attracted to the field and significant progress was made in both our theoretical understanding and experimental observation of a vast range of phenomena.

A theorist working in critical phenomena in most cases tries to start out by studying what may be called a mathematical model. The model must be simple enough to entertain the hope of being able to calculate some of its properties. Thus, one encounters in theoretical papers a zoo of seemingly unphysical models. My main fascination with the field stems from the observation that no matter how imaginative we are and what models we come up with, nearly always Nature has the upper hand and provides some physical realization of even our strangest creations. This is the main point we emphasize and try to convey in this Chapter. In order to do so, we present in Section II a brief general introduction to critical phenomena,[8] where the measured (or calculated) quantities of interest are described, critical exponents are defined, and the concept of universality is introduced. Also, the special role of two dimensions (d = 2) is discussed, and some of the more exotic features of two dimensional phases and behavior are reviewed.

In Section III, various theoretical models[9-22] will be mentioned and their known properties summarized. In order to make contact with physical reality without introducing the necessary group theoretical formalism, a rather "handwaving" argument is used in Section IV, to introduce realizations of the q = 2,3,4 state Potts models[11]

B. Mutaftschiev (ed.), Interfacial Aspects of Phase Transformations, 119–141.

in d = 2 dimensions.

In Section V, the symmetry arguments[23,24] that connect microscopic models to physical systems will be put on firmer ground, and a brief summary of a classification scheme[25,26] of transitions between the liquid and commensurate solid phases of adsorbed submonolayers will be given. We also present a brief discussion of the Kosterlitz-Thouless transition and a possible physical realization by transition to an incommensurate structure. Section VI will discuss some methods of calculation of phase diagrams for adsorbed systems.

This brings us to the Apology promised in the title of this brief section. Some of the most challenging, important and interesting aspects of phase transitions in two dimensions will not receive due treatment in this Chapter, for which we apologize to the readers and the researchers working in the field. These topics include the details of the Kosterlitz-Thouless theory[21] for the XY model and melting, as well as the more recent contributions to our understanding of these phenomena.[27] Also, the fascinating problem of the commensurate-incommensurate transition[28a] were left untreated. Regarding the list of references, we decided to mainly reference review articles and/or recent work, and did not attempt to compile a comprehensive list that gives credit whenever it is due.

II. BRIEF REVIEW OF CRITICAL PHENOMENA

II.1. General Introduction and Definitions.[8]

The most basic standard manifestation of a phase transition is the observation of an "order parameter" which vanishes in the disordered phase, and has a non-zero value in the ordered phase. In many instances, the appearance of such an order parameter corresponds to spontaneous breaking of some symmetry of the system.

In order to provide a specific example, consider submonolayer of a noble gas, such as He or Kr, physisorbed on a uniform graphite substrate. The substrate provides the adsorbed atom with a triangular array of "adsorption sites". Thus the Hamiltonian of the adsorbed system has the symmetry G = P6mm of the substrate. In the high temperature (disordered) phase the adsorbate forms a gas (or liquid) phase with the same symmetry. Denoting by $n(\vec{r})$ the occupation of site $\vec{r}$,

$$n(\vec{r}) = \begin{cases} 1 & \vec{r} \text{ occupied;} \\ 0 & \vec{r} \text{ empty;} \end{cases}$$

the thermal average $\langle n(\underline{r})\rangle = n_o$, independent or $\vec{r}$. Thus a diffraction experiment will reveal Bragg peaks at the same positions as those of the substrate. However, as temperature is lowered below a critical value T_c, the adsorbate forms a superlattice of periodicity $\sqrt{3}$ times that of the substrate. That is, 1/3 of the equivalent adsorption sites is occupied with higher probability than the remaining 2/3; the symmetry of the adsorbed system is lowered, i.e. spontaneously broken. In a diffraction experiment new lines will appear,[28b] at the reciprocal lattice vector $\vec{K}$ of the superlattice. Denoting by $\vec{q}$ the momentum transfer to the scattered beam (x-rays, neutrons or electrons), the intensity $I(\vec{q})$ is given by

$$I(\vec{q}) = \frac{1}{v} \sum_{\vec{r}\vec{r}'} \exp\left[i\vec{q}\cdot(\vec{r} - \vec{r}')\right] < n(\vec{r})n(\vec{r}')>$$

which can be expressed as

$$I(\vec{q}) = \hat{G}(\vec{q}) + I^{\mathrm{Bragg}}(\vec{q}),$$

where the intensity of the Bragg peaks is

$$I^{\mathrm{Bragg}}(\vec{q}) = A(T) \sum_{\vec{K}} \delta(\vec{q} - \vec{K}).$$

The order parameter associated with this transition is

$$\psi = \sum_{\vec{r}} \exp\left[i\vec{K}\cdot\vec{r}\right] < n(r) >.$$

For $T < T_c$, we have

$$I(\vec{K}) = |\psi|^2 \sim (T_c - T)^{2\beta}.$$

The diffuse scattering part is given by

$$\hat{G}(\vec{q}) = \frac{1}{v} \sum_{\vec{r}\vec{r}'} \exp\left[i\vec{q}\cdot(\vec{r} - \vec{r}')\right] < \delta n(\vec{r})\delta n(\vec{r}') >,$$

where

$$\delta n(\vec{r}) = n(\vec{r}) - <n(\vec{r})> .$$

If we observe $\hat{G}(\vec{q},T)$ (the temperature dependence enters through $<\delta n(\vec{r})\delta n(\vec{r}')>$; see below), at $\vec{q} = \vec{K}$, we find (for $|(T-T_c)/T_c| << 1$) $G(\vec{K},T) \sim |T-T_c|^{-\gamma}$. On the other hand, keeping the temperature fixed at T_c, we have $G(\vec{q},T_c) \sim |\vec{q}-\vec{K}|^{-(2-\eta)}$.

Note that $\hat{G}(\vec{q})$ is the Fourier transform of the correlation function $<\delta n(\vec{r})\delta n(\vec{r}')>$, which measures the extent to which density fluctuations (away from the equilibrium value) at $\vec{r}$ and $\vec{r}'$ are correlated.

In ordinary phases of matter the correlation between fluctuations decays exponentially on a scale determined by the *correlation length* ξ ;

$$G(R) \sim \exp[-R/\xi] \text{ for } R \to \infty \qquad R = |\vec{r} - \vec{r}'|.$$

However as $T \to T_c$, ξ diverges;

$$\xi \sim |T - T_c|^{-\nu}$$

so that at the cirtical point itself the decay of correlations is slower than exponential. In fact, at $T = T_c$, $G(R)$ decays *algebraically*

$$G(R,T_c) \sim 1/R^{d-2+\eta}$$

In ordinary (three dimensional) systems, this type of slow delay occurs only at isolated critical points.

Finally, there is a singularity associated with the specific heat as well, which behaves near the critical temperature as

$$C \sim |T - T_c|^{-\alpha}.$$

The critical behavior of a system is characterized by the numerical values of the critical exponents $\alpha,\beta,\gamma,\nu,\eta$. Much effort has been and is currently being invested in their calculation and measurement. There are two reasons for this. One is that various scaling laws impose relations between the critical exponents, and accurate determination of their values throws light on the validity of these laws. The second reason is the so-called "universality" of critical exponents. Critical phenomena are observed in a wide variety of systems, including order-disorder, structural ferro and antiferromagnetic transitions, the liquid-gas transition, phase separation in binary liquid mixtures, the transition to the superfluid state of He, and more. Since T_c ranges from $\simeq 1°K$ to $\simeq 10^3 K$, and since the systems are so different, one expects the physical mechanisms that govern these transitions to differ widely. Nevertheless, the critical exponents fall into a relatively small number of "universality classes". According to our present day understanding of these phenomena, the universality class to which a physical system (or a theoretical model) belongs is determined by its spatial dimensionality d, the symmetry (and symmetry breaking) associated with the transition, and the range of the underlying forces. As we will deal only with models and systems with short range forces, no further mention of the effect of long range (such as dipolar) forces will be made. The effects of dimensionality and symmetry on critical behavior are quite spectacular. For dimensions $d > 4$, the exponents are classical, i.e. those predicted by mean field theory. Mean field theory neglects the effects of fluctuations, which become more important as d decreases below $d = 4$. Thus, differences between the values of exponents for different universality classes (determined by differing symmetries - see below) are rather small in $d = 3$; however, in two dimensions a *much wider variation of exponents is predicted*. This renders two dimensional critical systems an extremely sensitive tool in our investigations of universality. For example, while in $d = 3$ for a wide range of symmetries, the specific heat exponent α varies, between -0.1 and $+0.1$, in $d = 2$ exponent variation between $\alpha < 0$ and $2/3$ (or even up to $\alpha = 1$) are expected. Furthermore, in $d = 2$ there exist theoretical models (and physical systems) for which a *continuous variation of exponents* is expected. I believe that experimental observation of such *non-universal* behavior will be limited to $d = 2$. A perhaps even more striking manifestation of the fluctuation dominated aspect of the two dimensional world is the existence of *new, "critical phases" of matter*. In these phases there is *no long range order*, and the entire low temperature phase behaves like more standard systems at their critical point ! Having, I hope, conveyed the special role played by $d = 2$ in studies of critical phenomena, we can turn to more concrete statements concerning specific models and physical systems.

III. THEORETICAL MODELS

III.1. The Ising Model[9]

Denote by r, r', the sites of a lattice; with each site associate a variable $S_r = \pm 1$. The Hamiltonian of the system is given by

$$H = -\sum_{r,r'} J_{rr'} S_r S_{r'} - H \sum_r S_{r'},$$

where $J_{rr'}$ is the coupling constant of interaction between sites r and r' and H is the strenght of the applied (e.g. magnetic) field. The free energy F is a function of the temperature T and of the field strenght H :

$$F = -k_B T \ln Z,$$

where

$$Z = \sum_{\{S\}} \exp\left[-H(S)/k_B T\right]$$

is the partition sum over all S states of the system. The model (for nearest neighbor interactions only) was solved by ONSAGER ; the specific heat diverges logarithmically ($\alpha = 0$); other exponents are $\beta = 1/8$, $\gamma = 7/4$, $\nu = 1$, $\eta = 1/4$. These exponents characterize the critical behavior of a ferromagnetic Ising model on any two dimensional lattice.

III.2. Lattice-Gas Models

Consider the example of He physisorbed on graphite. The adsorption sites form a triangular lattice. When two sites, r and r', are occupied, (i.e. $n_r = n_{r'} = 1$) the two atoms that occupy them interact via a Lennard-Jones type pair potential $v(|r - r'|)$. Neglecting three-body interactions, the Hamiltonian is given by

$$H = \sum_{rr'} v(|r - r'|) n_r n_{r'} - \mu \sum_r n_{r'}.$$

In the second term μ represents the chemical potential that controls the density of adsorbed atoms. Using the transformation

$$n_r = \frac{1}{2}(1 + S_r).$$

The Hamiltonian takes the Ising form

$$H = \sum_{rr'} \frac{1}{4} v(r - r') S_r S_{r'} - \bar{\mu} \sum_r S_r .$$

Since the distance between two nearest neighbor adsorption sites is less than the "hard core diameter" of a He atom, the nearest neighbor interaction is strongly repulsive; the next nearest neighbor interaction is weakly attractive (further neighbor interactions are neglected in most calculations). This model is expressed in terms of Ising variables, and one may be tempted to conclude that the transition is in the same universality class as the ferromagnetic Ising model. This, however, is not the case. Taking the nearest neighbor repulsion to be infinite, and all other interactions zero, the model becomes the hard hexagon model, solved by BAXTER.[10] It has a phase transition as a function of μ; the critical exponents are $\beta = 1/9$, $\alpha = 1/3$ (other exponents can be obtained from scaling relations).

III.3. Potts Models[11]

Assign to each site of a lattice an integer valued variable $S_r = 1,2,\ldots q$. The Hamiltonian of the q-state Potts model is given by

$$H = - K \sum_{<rr'>} \delta(S_r, S_{r'}),$$

where $<rr'>$ denotes a nearest neighbor pair of sites, and $\delta(S,S')$ is the Kroneker delta function. In order to gain an intuitive geometrical interpretation of the model; consider the cases $q = 2,3,4$; at each site we have a vector $\vec{S}_r$ that can point into one of q directions, and nearest neighbors interact via the "standard" $- J \vec{S}_r \cdot \vec{S}_{r'}$ interaction. The assignments for $q = 2,3,4$ are given in Fig. 1. Note that for $q = 2$ the Potts model is nothing but the Ising model. BAXTER[12] has shown that for $q \leqslant 4$ these models have a continuous transition (in $d = 2$), while for $q > 4$ the transition is first order. The exponents of the model are believed to be known[11,13] and are summarized in Table 1. Note the promised wide variations of exponents. It is interesting to note that the exponents of the 3-state Potts model are identical to the exponents of the hard hexagon model; this fact is one of the strongest proofs of the statements concerning universality on which this Chapter is based.

III.4. N-State Clock Models

Assign to each site a vector $\vec{S}_r$ that lies in a plane, and can point in one of N directions, at angle

$$\Theta_r = 2\pi n_r/N \quad , \quad n_r = 1,2,\ldots N$$

with respect to some reference axis. Two neighboring vectors

Table 1 : Potts exponents; note the wide variation of α with q. The "unphysical" value q = 1 is related to the percolation problem.

q	1	2	3	4
α	-2/3	0	1/3	2/3
β	5/36	1/8	1/9	1/12

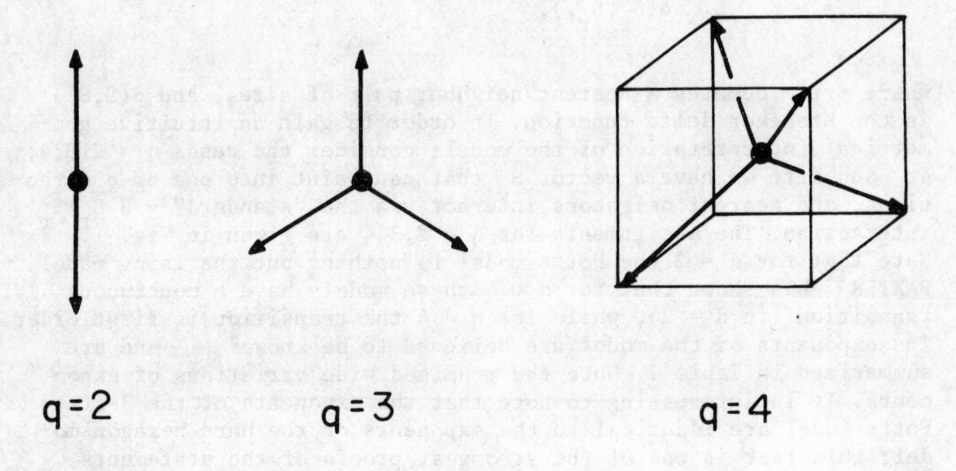

Fig. 1 : Vector representation of q-state Potts models; to each site i of a lattice a vector $\vec{S}_i$ is assigned, which can point in one of the q indicated direction. The nearest neighbor interaction is $-J\,\vec{S}_i \cdot \vec{S}_j$.

interact via a potential $V(\Theta_r - \Theta_{r'})$; one of the most commonly used forms is

$$V(\Theta_r - \Theta_{r'}) = -\cos(\Theta_r - \Theta_{r'}) = -\vec{S}_r \cdot \vec{S}_{r'} .$$

For N = 2 this is again the Ising model; for N = 3, the 3-state Potts model. For large enough values of N (i.e. $N > N_c$), and a specific form of the interaction $V(\Theta)$, ELITZUR et. al.[14] have shown that the model has a disordered high temperature phase, a low temperature phase with conventional long range order, and an *intermediate phase*, in which correlations decay algebracially ($G(r) \sim r^{-\eta}$). There is very strong evidence[15] that $N_c = 4$. At this special value, the Z(4) model has quite unique properties; it has a single transition from a disordered to an ordered phase; however, the exponents that characterize this transition are non-universal and vary continuously with a parameter.[16] This parameter is the relative "energetic cost" of opening an angle of $\pi/2$ or π between two nearest neighbors.

III.5. Ashkin-Teller and 8-Vertex Models

The Z(4) model can be represented in terms of *two* Ising spins associated with each site; s_r, τ_r. In terms of these variables, the Hamiltonian of the Z(4) model takes the form

$$H = -J \sum_{\langle rr' \rangle} (s_r s_{r'} + \tau_r \tau_{r'}) - \Lambda \sum_{\langle rr' \rangle} s_r \tau_r s_{r'} \tau_{r'} ,$$

i.e. two Ising models coupled by four-spin interactions. This is the Ashkin-Teller model. This model, however, is closely related to a different model, where two Ising models are again coupled via a four spin term; but now the two Ising models are on two different, interpenetrating square lattices, and the four-spin interactions couple the spins at the corners of basic squares of the combined lattice (see Fig. 2). This is the Ising-spin representation of the 8-vertex model, for which BAXTER has calculated[17] the variation of the exponent α with Λ(for fixed J). It is through various exact and conjectured relations between this exactly solved model and the q-state Potts model that our knowledge of the Potts exponents came about.[13,16]

III.6. The XY and Heisenberg Models

All the models mentioned so far have a common underlying feature; their Hamiltonians are invariant under some *discrete* symmetry operations (such as Z(N) for the clock models, S(q) for the Potts models, etc). Also, all the models did have a low temperature phase with conventional long range order.

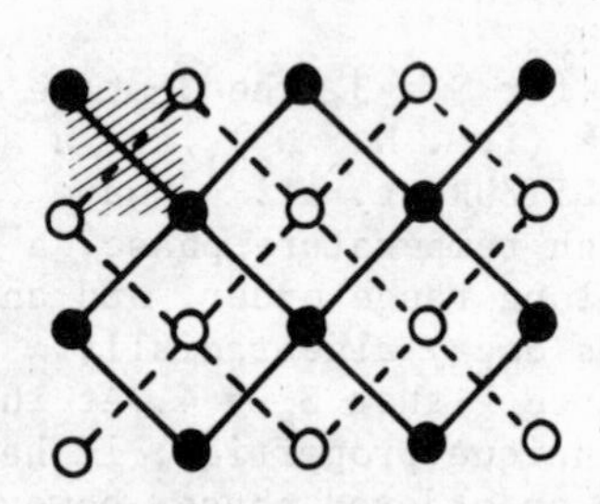

Fig. 2 : Ising-spin representation of the 8-vertex model : dots represent s_i, open circles τ_i variables. The s_i variables interact via couplings represented by solid lines; the $\tau_i\tau_j$ interactions are represented by broken lines. Four spin terms couple the variables $s_i s_j \tau_l \tau_e$ at the corners of all squares such as the shaded one.

This is not the case when the underlying symmetry is continuous; in that case, in d = 2 long range order is completely destroyed by fluctuations at any finite temperature.[18] Two such models with continuous symmetry are the Heisenberg and XY models. Both are defined in terms of a unit vector variable $\vec{S}_r$ associated with each site, and $-\vec{S}_r \cdot \vec{S}_{r'}$ nearest neighbor interactions. However, in the Heisenberg model the spin $\vec{S}$ can point in any direction in three (spin) dimensions, while it is confined to a plane in the XY model. The Heisenberg model has no transition;[19] at all finite T correlations decay exponentially. The XY model, however does have a transition; below a certain temperature correlations decay as $r^{-\eta(T)}$ with a temperature dependent exponent ! This new phase of matter is sometimes called "massless". Although the existence of a transition was known from earlier series work,[20] understanding of its nature emerged only from the pioneering work of KOSTERLITZ and THOULESS.[21] We will come back and sketch some of the essential points of their treatment at the end of Section V.

At this point we only mention that as the Kosterlitz-Thouless temperature T_{KT} is approached from above, the correlation length diverges as

$$\xi \sim \exp\left[a/(T - T_{KT})^{1/2}\right]$$

and the specific heat has only an essential singularity at T_{KT}, i.e. it is completely smooth as the temperature goes through the transition. The exponent $\eta(T)$ takes the *universal* value 1/4 at the transition.

III.7. Solid-On-Solid (SOS) and Gaussian Models

There exists an exact ("duality") transformation[15] that maps the XY model onto an SOS model. This model is defined in terms of an integer valued variable $h_r = 0, \pm 1, \pm 2, \ldots$ associated with each site; neighboring sites interact via a "potential" $V(h_r - h_{r'})$. The form

$$V(h_r - h_{r'}) = K|h_r - h_{r'}|$$

is the *linear SOS model* and describes the interface energy between coexisting (three dimensional) solid and liquid phases.[22] The phase transition of the XY model translates in this context to the "roughening transition" of such an interface. The interface, which is smooth at low temperatures, becomes rough above T_{KT}; the rough phase of the SOS model corresponds to the "massless" phase of the XY model.

When the interaction between nearest neighbor columns is quadratic,

$$V(h_r - h_{r'}) = K(h_r - h_{r'})^2,$$

one obtains the "discrete Gaussian" model. If we now take this model and relax the restriction that forces the h_r variables to take integer values only, and instead allow any continuous value $-\infty < h_r < \infty$, one obtains the trivially soluble Gaussian model, whose partition sum is given by Gaussian integrals, and for which the correlation function

$$\mathrm{Re}\,\langle e^{i(h_r - h_{r+R})} \rangle \sim R^{-\eta(T)}$$

can be calculated.[15] Note that for the Gaussian model the correlation function decays algebraically at all temperatures.

IV. PHYSICAL REALIZATIONS; INTUITIVE ARGUMENTS

To demonstrate physical realizations of some models, consider the order-disorder transition of He (or Kr) on graphite at approximately 1/3 coverage. At low T, the adsorbate occupies preferentially one out of three equivalent sublattices (see Fig. 3a). Thus the symmetry breaking associated with this transition singles out one of three completely equivalent ordered states. Now this is precisely the symmetry associated with the 3-state Potts model; therefore these order-disorder transitions to the $\sqrt{3} \times \sqrt{3}$ structure are expected to be in the 3-state Potts universality class.[29] Since the specific heat exponent shows the largest variation going from one universality class to another, thermal measurements are a natural candidate to check this prediction. Indeed, the measured values of α, ranging from[30] 0.28 to[31] 0.36, are quite consistent with the theoretical value[13,10] of 1/3.

Next, we turn to a realization of the Ising model. Consider again the graphite substrate, but this time preplated with Kr, which forms an ordered $\sqrt{3} \times \sqrt{3}$ structure. For temperatures in the 1°K range, the Kr monolayer, which orders at ≃ 100°K, can be considered as inert. Now the same amount of He is introduced into the system, and an order-disorder transition takes place as the He orders on top of the Kr layer. The triangular lattice of Kr atoms provides a honeycomb array of adsorption sites for the He atoms. He can order now into the same lattice as it did before on bare graphite; however, this ordering will cause preferential occupation of one out of two equivalent sets of (honeycomb) adsorption sites (see Fig. 3b).

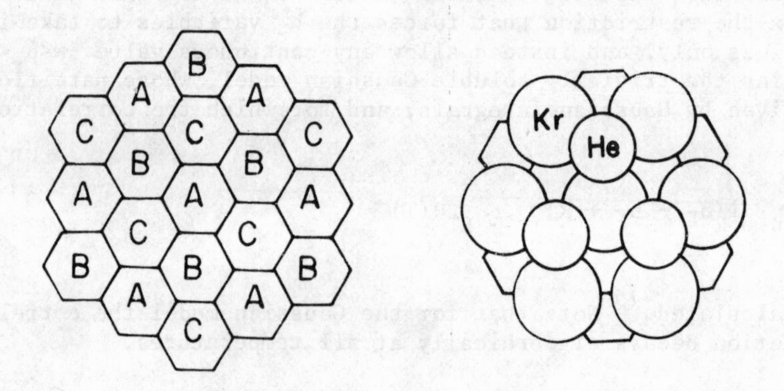

Fig. 3 : (a) Adsorption sites on graphite, He (or Kr) will order in one of three equivalent structures, with either the A,B or C sublattices occupied preferentially. (b) The $\sqrt{3} \times \sqrt{3}$ structure of Kr on graphite provides a honeycomb array of adsorption sites for He, which orders as indicated.

Thus, the symmetry associated with this transition indicates that it should be in the Ising universality class.[26] Indeed, in a beautiful experiment,[30] VILCHES and his group have found a logarithmically divergent specific heat peak for this transition, in full support of the picture of universality as determined on the basis of symmetry. Finally, turning to the 4-state Potts model; consider a submonolayer that orders into 2 × 2 structure indicated in Fig. 4, on a triangular array of adsorption sites. Evidently, this superlattice structure involves preferential occupation of one of *four* equivalent sublattices, and the transition turns out to have the symmetry of the 4-state Potts model. This is so since the symmetry operations that take sublattice A, say, into either one of B,C,or D are equivalent. A quite extensively studied physical system that has this transition is O on the (111) face of Ni. For this system exponents were measured,[32] and their values are not in agreement with those of the 4-state Potts model. This discrepancy is still not fully resolved, although a plausible explanation based on slow crossover from Ising-like behavior was recently given.[33]

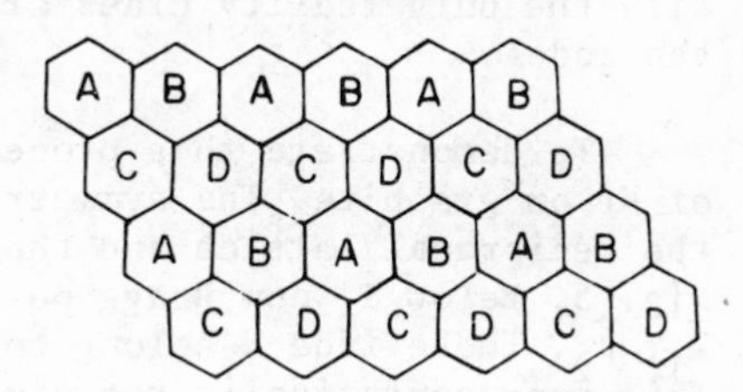

Fig. 4 : The 2 × 2 structure, which is reached by a 4-state Potts transition,is obtained by preferential occupation of one of four equivalent sublattices A,B,C or D of adsorption sites.

V. ORDER-DISORDER TRANSITIONS AND THEIR CLASSIFICATION[25,26]

We have seen that certain physical systems provide realizations of the q = 2,3,4 state Potts models; however, the arguments supporting this statement were intuitive. In this section we will first present a systematic method of establishing the correspondence between a transition observed in some physical system and the behavior of some theoretical model. We will concentrate on transitions from the disordered to commensurate phases. In such a transition, a diffraction experiment will reveal Bragg peaks associated with a superlattice,whose intensity increases from zero as the temperature decreases below some T_c; however, the position of the peaks (in $\vec{k}$-space) remains fixed (at least for a finite range of temperatures). In order to determine the universality class of the transition, one must know the symmetry group G_o of the disordered phase, i.e. that of the substrate, *and* the (lower) symmetry G of the ordered phase, its structure of diffraction pattern. In a continuous transition, sufficiently close to T_c, the dominant diffraction peaks occur at a set of wavevectors $\vec{k}_j$ that belong to one irreducible representation of G_o. That is, peaks associated with other representations may also appear but their intensity goes to zero, as T_c is approached, with a higher power than that of the dominant peaks. The occurence of these Bragg peaks means that in the ordered phase the adsorbate density, $\langle n(\vec{r})\rangle$, contains non-vanishing Fourier components $e^{i\vec{k}_j\cdot\vec{r}}$. The corresponding order parameter is $\psi_{\vec{k}_j} = \sum_r e^{-i\vec{k}_j\cdot\vec{r}} \langle n(\vec{r})\rangle \neq 0$.

Next, we identify the number of components of the order parameter as the number of independent functions in our irreducible representation.

Note that two $\vec{k}$ vectors, that differ by a reciprocal lattice vector of G_o, are not independent. Finally, we construct all possible second, third, fourth, etc. order invariants of G_o from the functions ψ_{kj}. These invariants constitute the so-called Landau-Ginzburg-Wilson (LGW) Hamiltonian of our system.[23,24] If this has the same form as the LGW Hamiltonian of some known model, we identify the universality class of our physical system with that of the model.

To demonstrate this procedure, consider the $\sqrt{3}\times\sqrt{3}$ structure of Kr on graphite. The symmetry group of the substrate is P6mm : the reciprocal lattice and the first Brillouin zone are shown on Fig. 5. Below T_c new Bragg peaks appear at the indicated positions $\vec{k}_1$, $\vec{k}_2$. These indeed belong to an irreducible representation of G_o. This representation is two dimensional; the six indicated $\vec{k}$-vectors split into two groups of three; any two vectors within each group differ by some original reciprocal lattice vector. Thus we have a two component order parameter : ψ_{k_1} and ψ_{k_2}.

There is one second order invariant, $\psi_{k_1}\psi_{k_2}$; one third order invariant, $\psi_{k_1}^3 + \psi_{k_2}^3$, and one fourth order invariant $(\psi_{k_1}\psi_{k_2})^2$. It is convenient to work with real functions, so denote $\psi_{k_1} = x + iy$; since $k_2 = - k_1$, we have

$$\psi_{k_2} = \psi_{k_1}^* = x - iy.$$

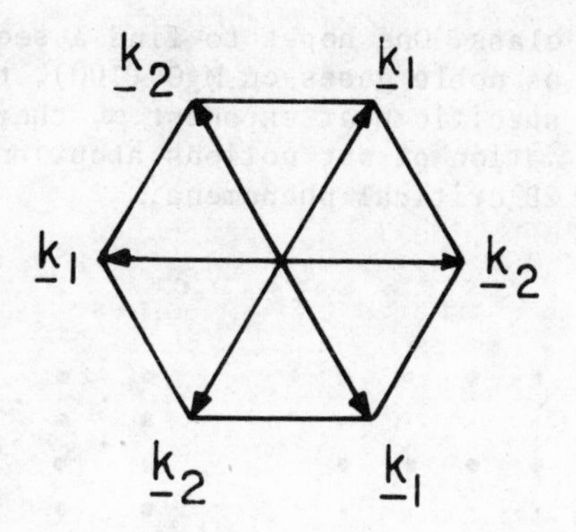

Fig. 5 : The first Brillouen zone of the graphite substrate. The $\sqrt{3} \times \sqrt{3}$ density modulation is a linear combination of $\exp[i\vec{k}_1 \cdot \vec{r}]$ and $\exp[i\vec{k}_2 \cdot \vec{r}]$. Of the six indicated $\vec{k}$ vectors, only two are independent; any two of the three indicated $\vec{k}_1$ vectors differ by a graphite reciprocal lattice vector.

In terms of x and y the invariants take the form $x^2 + y^2$, $x^3 - 3xy^2$, $(x^2 + y^2)^2$. Note that in the (x-y) plane, the third order invariant has the form $\rho^3 \cos 3\Theta$, this three-fold breaking of rotational symmetry is precisely what is obtained when the LGW Hamiltonian for the 3-state Potts model is constructed.[29]

It is of interest to determine which other universality classes may be realized in order-disorder transitions. This task is facilitated by restricting our attention to transitions to commensurate phases, with fixed periodicity. The Landau-Ginzburg theory severely restricts the $\vec{k}$ vectors that do not vary with temperature; only those representations, for which the "Lifshitz condition" is satisfied can play a role. In essence, this condition requires the $\vec{k}$ vector to be at a point of high symmetry of the Brillouin zone. Thus one can go through the two dimensional space groups, identify the $\vec{k}$ vectors for which the condition is satisfied, determine the representation, and construct the LGW Hamiltonian characteristic of the transition. Rather than going through this classification scheme, we present various superlattice structures that provide physical realizations of theoretical models that were discussed detailed classification schemes can be found in the literature. (ref. 25,26,31,35)

Ising Model : The three ordered states indicated (Fig. 6a) on a rectangular array (P2mm) (as provided by a (110) face of a simple cubic crystal) are the only commensurate structures for which the Lifshitz condition is satisfied. Transitions to all three are Ising-like. Transition to the indicated (Fig. 6b) structure on a centered rectangular array ((110) face of bcc) will also be Ising-like, as well as the structure indicated (Fig. 6c) on the (P4mm) square array, and the one shown for the honeycomb array (Fig. 6d).

The Z(4) Model : The structure shown (Fig. 7a) on the centered rectangular array belongs to the Z(4) class. This structure was observed for O on W(110). Also, the structure shown (Fig. 7b) for the square lattice belongs to this "non universal" universality class. One hopes to find a sequence of physisorbed systems, such as noble gases on MgO (100), that would reveal variation of the specific heat exponent α, thereby providing a most striking confirmation of our notions about universality and our understanding of 2D critical phenomena.

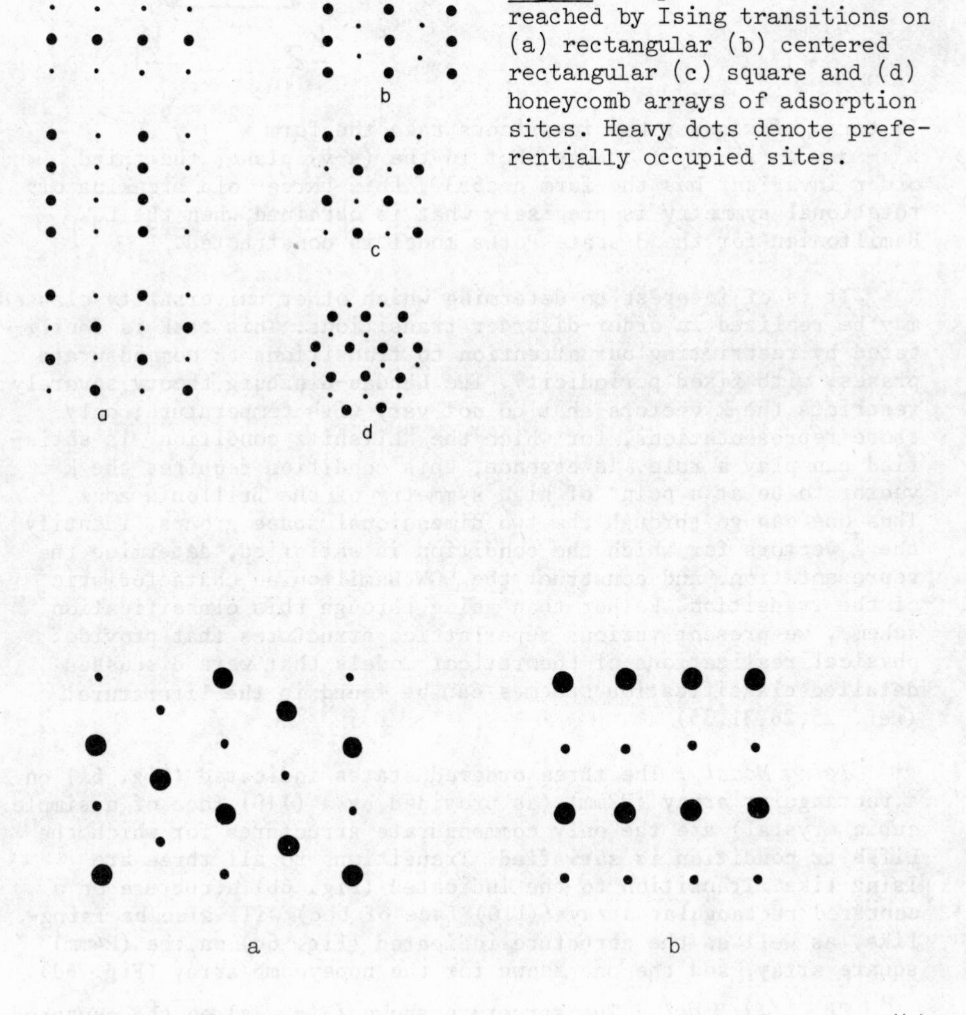

Fig. 6 : Superlattice structures reached by Ising transitions on (a) rectangular (b) centered rectangular (c) square and (d) honeycomb arrays of adsorption sites. Heavy dots denote preferentially occupied sites.

Fig. 7 : Superlattice structures reached by transitions in the Z(4) or Ashkin-Teller universality class, on (a) centered rectangular and (b) square array of adsorption sites.

The 3,4 State Potts Models : The transition to the $\sqrt{3}\times\sqrt{3}$ structure of Fig. 3a provides a realization of the 3-state Potts model. The 2×2 structure of Fig. 4 is in the 4-state Potts universality class, as well as the (2×2) structure on a honeycomb array of adsorption sites, shown on Fig. 8.

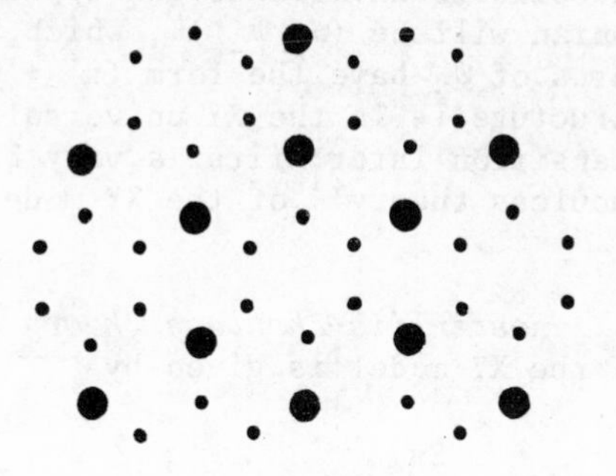

Fig. 8 : Structure reached by a transition in the universality class of the $q = 4$ state Potts model.

The XY Model : The most extensively studied realization of the XY model is the transition to a superfluid state of He^4 films. For a different realization of the XY model, presented by some theoretical models and, possibly, experimental systems, consider an order-disorder type transition on a rectangular lattice; characterized by the $\vec{k}$-vectors shown in Fig. 9a.

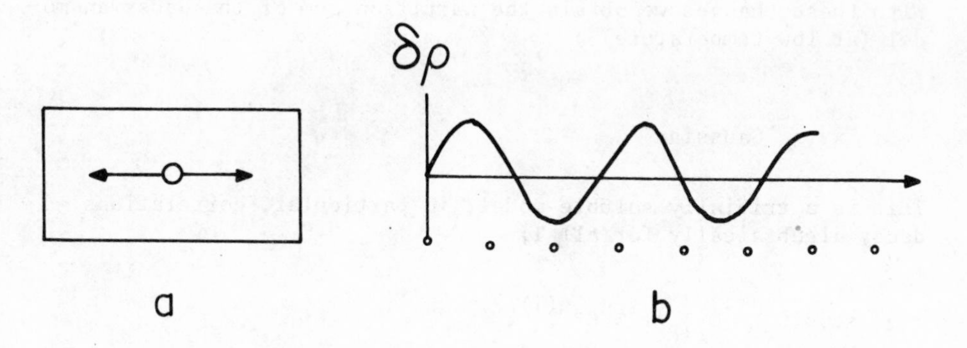

Fig. 9 : (a) An incommensurate density modulation, characterized by the $\vec{k}$ vectors indicated. (b) The density modulation is const. in the x direction and varies sinusoidaly in the y direction. Dots indicate positions of the adsorption sites.

These k-vectors correspond to an incommensurate structure (density wave), i.e., a sinusoidal density variation in one direction (see Fig. 9b). Such an ordered phase is predicted for some lattice gas models with competing nearest and next nearest neighbor interactions in one direction[36] and may occur as well in some chemisorbed system or surface reconstruction transition. The order parameter is a two component object, ψ_k, ψ_{-k}, and since k is incommensurate, no anisotropic invariants can appear. The only terms in the LGW Hamiltonian will be $(\psi_k \psi_{-k})^n$, which, in terms of the real and imaginary parts of ψ_k have the form $(x^2 + y^2)^n$. Thus the transition to this structure is in the XY universality class. We will return to this transition later; first a very brief review of the Kosterlitz-Thouless theory[21] of the XY model is given.

Kosterlits-Thouless Theory of the XY Model : The partition sum of the XY model is given by

$$Z_{xy} = \int_o^{2\pi} \prod_r d\Theta_r \exp\left[\frac{J}{kT} \sum_{<rr'>} \cos(\Theta_r - \Theta_{r'})\right].$$

Now consider a low-temperature approximation for Z. At low enough temperatures, neighboring angles will not vary much, $\Theta_r \simeq \Theta_{r'}$. We may expect only those configuration for which this approximation holds to have a significant contribution to Z. Thus we approximate

$$\cos(\Theta_r - \Theta_{r'}) \simeq 1 - \frac{1}{2}(\Theta_r - \Theta_{r'})^2$$

and also change the integration limits from $[0,2\pi]$ to $[-\infty, +\infty]$. With these changes we obtain the partition sum of the Gaussian model (at low temperature) :

$$Z_{xy} \simeq Z_{\text{Gaussian}}.$$

This is a trivially soluble model; in particular, correlations decay algebraically for all T;

$$<\cos(\Theta_r - \Theta_{r+R})> \sim R^{-\eta(T)}.$$

However, at sufficiently high temperatures, the correlation function of the XY model decays exponentially; therefore above some temperature the XY model is no longer approximated in a satisfactory manner by the Gaussian model. To estimate the temperature above which the

Gaussian approximation "breaks down", one looks for configurations which may be of importance in evaluating Z_{xy}, but which are not given the proper weight when the Gaussian approximation is used. KOSTERLITZ and THOULESS identified such configurations; they are vortices (see Fig. 10).

Fig. 10 : (a) vortex, (b) antivortex excitations of the XY model.

The main point is that by going to the Gaussian approximation, the periodic nature of the interaction $\cos(\Theta_r - \Theta_{r'})$ was lost; a vortex of size L has an energy $E_v \propto L$ when the Gaussian approximation is used, while with the proper XY interaction

$$E_v \simeq \pi J \ln (L/a_o),$$

where a_o is a characteristic core radius.

The entropy associated with such a vortex is

$$S_v = k_B \ln(L/a_o)^2,$$

and the contribution of such a vortex to the free energy is

$$F = E_v - TS_v = (\pi J - 2k_B T) \ln(L/a_o).$$

Obviously, at $T > T_{KT} = \pi J/2k_B$, vortices of unlimited size should contribute to Z_{XY}. Since in the Gaussian approximation $E_V \propto L$, such vortices are suppressed at all temperatures ! Indeed KOSTERLITZ and THOULESS have shown that for $T < T_{KT}$, the XY model is properly described by a Gaussian approximation; vortices appear only in bound vortex-antivortex pairs and correlations decay algebraically. For $T > T_{KT}$ the pairs start to dissociate, which results in an exponential decay of correlations.

Returning now to our one dimensional incommensurate density modulation (Fig. 9) we recall that the transition to this structure was predicted, on the basis of symmetry, to be in the XY universality class. If so, a natural question is the following : since vortices play a central role in the transition of the XY model, what are the vortices of our incommensurate structure ?

To answer this question, note that the one dimensional density modulation characteristic of the ordered phase associated with $\pm\vec{k}$ of Fig. 9a, can be written as

$$\langle n(x,y)\rangle = \cos\left[ky + \Theta(x,y)\right].$$

For $\Theta(x,y)$ = const. one obtains a simple translation of the density modulation along the y axis. Obviously, for $\Theta = 2$ the structure goes back to its form at $\Theta = 0$. If we represent a structure with modulated density by drawing the regions of high density as shaded thick lines, the configuration of Fig. 11 is easily interpreted as a vortex in terms of the (space dependent) phase angle Θ,

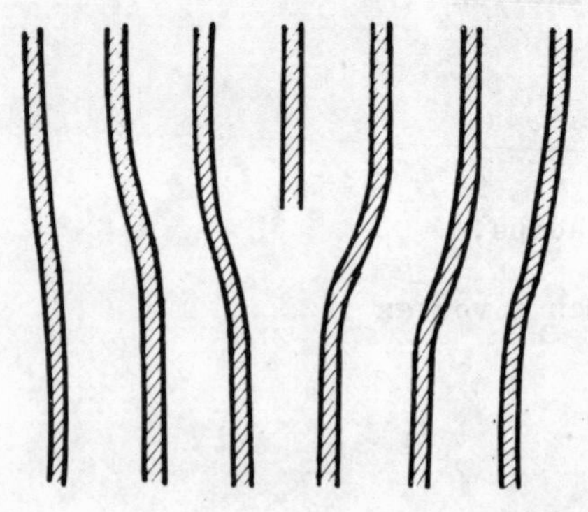

Fig. 11 : A vortex of the phase with sinusoidal density modulation.

which varies through the sequence of values $(0, \frac{\pi}{2}, \pi, \frac{3\pi}{2}, 2\pi)$ as we go clockwise aroung point 0 (the vortex "core"). Thus we see that identification of a transition on the basis of symmetry is supported even to the extent that the relevant topological excitations of the appropriate model can be easily identified.

VI. SUMMARY AND DISCUSSION

Classification of continuous transitions on the basis of symmetry arguments has proved a most useful tool to establish relations between theoretical models, as well as between these models and experimental systems. We believe that the relevance of symmetry and the concept of LGW Hamiltonians to transitions in two dimensions has been firmly established. It is important to note that arguments presented in this Chapter bear only on the universal aspects of the transitions that were discussed. Other features, such as detailed quantitative phase diagrams and thermodynamic functions cannot be derived by simple symmetry considerations. In order to obtain such (most important) information, calculations on specific (and possibly realistic) models are essential.

Two main calculational techniques have been tested and found to be quite successful when compared with experiments. These are Monte Carlo calculations[37] and position-space Renormalization Group methods.[38] A third technique, which seems to be most promising, is the transfer matrix or "phenomenological Renormalization Group" method.[39] The interested reader is referred to the literature for details.

Acknowledgements

My involvement in studies of order-disorder transitions in two dimensions was motivated mainly by the outstanding experimental effort at the University of Washington. Greg Dash, Sam Fain and Oscar Vilches were the kindest and most patient teachers of experimental systems and techniques that any theorist can hope to have. As to my theorist friends, I thank Mike Schick for sharing with me his irrepressible enthusiasm which turned work into joy and fun. I have greatly benefited from enjoyable collaborations with Eberhard Riedel, Bob Griffiths and Jim Walker, and enjoyed many helpful discussions and arguments with P. Bak, A.N. Berker, M.E. Fisher, W. Kinzel, D. Mukamel, D.R. Nelson and B. Nienhuis.

REFERENCES

1 . KOSTERLITS J.M. and THOULESS D.J., Prog. Low. Temp. Phys. 7 (1978)371
2 . RISTE T. ed., *Ordering in Strongly Fluctuating Condensed Matter Systems*, Plenum, NY, 1980.
3 . DASH J.G. and RUVALDS J. ed., *Phase Transitions in Surface Films*, Plenum, NY, 1980.
4 . BARBER M.N., Phys. Rep. 59(1980)375
5 . Kyoto Summer Inst. on "The Physics of Low Dimensional Systems", Kyoto, 1979.
6 . SINHA K., ed., *Ordering in Two Dimensions*, North Holland, 1980.
7 . COHEN E.G.D. ed., *Fundamental Problems in Statistical Mechanics V, Proc. of the 1980 Enschede Summer School*, North Holland, 1980.
8 . FISHER M.E., Rep. Prog. Phys. 30 (1967)615; STANLEY H.E., *Introduction to Phase Transitions and Critical Phenomena*, Oxford University Press, NY and Oxford, 1973.
9 . McCOY B.M. and WU T.T., *The Two-Dimensional Ising Model*, Harvard University Press, Cambridge, MA, 1973.
10. BAXTER R.J., J. Phys. A13(1980)L133.
11. WU F.X., Rev. Mod. Phys. (1982).
12. BAXTER R.J., J. Phys. C6 (1973)L445
13. DEN NIJS M., J. Phys. A12(1979)1857; NIENHUIS B., J. Phys. A (in press).
14. ELITZUR S., PEARSON R. and SHIGEMITSU J., Phys. Rev. D19(1979) 3698.
15. JOSE J., KADANOFF L.P., KIRKPATRICK S. and NELSON D.R., Phys. Rev. B16(1977)1217.
16. DOMANY E. and RIEDEL E.K., J. Appl. Phys. 49(1978)1315
KNOPS H.J.F., Ann. Phys. 128(1980)448
KADANOFF L.P. and BROWN A.C., Ann. Phys. 121(1979)318
KADANOFF L.P., Ann. Phys. 121(1979)38
117. BAXTER R.J., Ann. Phys. (NY) 70(1972)193
18. MERMIN N.D. and WAGNER H., Phys. Rev. Lett. 17(1966)1133.
19. TOBOCHNIK J. and SHENKER S.H., Phys. Rev. B22(1980)4462.
20. STANLEY H.A. and KAPLAN T.A., Phys. Rev. Lett. 17(1966)913.
21. KOSTERLITZ J.M. and THOULESS D.J., J. Phys. C6(1973)1181
KOSTERLITZ J.M., J. Phys. C7(1974)1046.
22. WEEKS J.D., ref. 2, p. 293.
23. LANDAU L.D. and LIFSHITZ E.M., *Statistical Physics*, Addison-Wesley, Reading, MA, 1969, Ch. XIII.
24. MUKAMEL D. and KRINSKY S., Phys. Rev. B13(1976)5065.
25. DOMANY E., SCHICH M., WALKER J.S. and GRIFFITHS R.B., Phys. Rev. B18(1978)2209.
26. DOMANY E. and SCHICK M., Phys. Rev. B20(1979)3828.
27. HALPERIN B.I., in ref. 5; NELSON D.R., in ref. 7.
28. (a) VILLAIN J., in refs. 2 and 6; (b) BIRGENAU R.J., HAMMONDS E.M., HEINEY P., STEPHENS P.W. and HORN P.M., ref. 6, p. 29; (c) LAGALLY M.G., LU T.M. and WANG G.C., loc. cit., p.113.

29. ALEXANDER S., Phys. Lett. A54(1975)353.
30. TEJWANI M.J., FERREIRA O. and VILCHES O.E., Phys. Rev. Lett. 44(1980)159.
31. BRETZ M., Phys. Rev. Lett. 38(1974)501.
32. ROELOFS L.D., KORTAN A.R., EINSTEIN T.L. and PARK R.L., Phys. Rev. Lett. 46(1981)1465.
33. SCHICK M., Phys. Rev. Lett. 47(1981)1347.
34. SCHICK M., in ref. 3, p. 65.
35. ROTTMAN C., University of Illinois, preprint, 1980.
36. HORNREICH R.M., LIEBMANN R., SCHUSTER H.G. and SELKE W., Z. Phys. B35(1979)91; SELKE W. and FISHER M.E., Z. Phys. B40(1980) 71. VILLAIN J. and BAK P., J. de Phys. 42(1981)657.
37. BINDER K. and LANDAU D.P., Phys. Rev. B21(1980)1941.
38. BERKER A.N., in ref. 6, p. 9.
39. NIGHTINGALE M.P., Physica A83(1976)561; KINZEL W. and SCHICK M., Phys. Rev. B23(1981)3435.

CRYSTAL-MELT, CRYSTAL-AMORPHOUS and LIQUID-VAPOR INTERFACES

A. BONISSENT

Centre de Recherche sur les Mécanismes de la Croissance
Cristalline; Campus Luminy,
13288 Marseille Cedex 9, France.

ABSTRACT

So called simple substances are those for which the molecules can be considered as having spherical symmetry. The internal energy is made up of the sum of the interaction energies of all the pairs formed by the molecules. The interaction potential follows some simple analytic law.

The structure and thermodynamic properties of the simple liquids are now relatively well known, as a number of techniques have been developed to study these problems.

The difficulties in the case of the crystal-melt interface are related to the presence of the liquid phase. The application of the methods developped for the case of the bulk liquids ,

i) dense randon packing of hard spheres(Bernal model)

ii) Monte-Carlo or Molecular Dynamics simulations,

iii) Perturbation theories,

give information concerning the structure and thermodynamics of the crystal-melt interface such as :

i) density profile

ii) interfacial specific free energy

B. Mutaftschiev (ed.), Interfacial Aspects of Phase Transformations, 143–182.

iii) structure of the liquid in the neighborhood of the crystal face.

The case of the crystal-glass and liquid-vapour interfaces is considered in the same way.

I. INTRODUCTION

I.1. The Problem

Growth from the melt is the most common method for preparation of crystals. Since the properties of the final product depend on mechanisms which occur at the interface, it is not surprising that a great deal of attention has been given to the thermodynamic properties of the interface between the crystal and the liquid.

The thermodynamic quantity which is characteristic of this two phase system is the specific interfacial free energy, i.e. the work necessary to create an interface with area unity, or to extend an existing interface by the same amount.

The most well-known studies aiming to determine the interfacial free energy between a crystal and its melt were nucleation experiments in a dispersed liquid by TURNBULL[1]. The result was that for most metals, the interfacial free energy is proportional to the heat of fusion. This is in good agreement with a calculation proposed by SKAPSKI[2], based on considerations about the environment of the molecules near the interface. However, the treatment of SKAPSKI includes the hypothesis that the structure of the liquid is not modified by the presence of the crystal surface. This assumption is hardly justified. Moreover, SKAPSKI implicitly assumes perfect wetting of the crystal by the melt. However, a few cases are known of poor wetting of the crystals by their melt; for metals like platinum[3], cadmium[4], bismuth[5] or gallium[6].

From these experimental facts, we can conclude that the crystal-melt interface cannot be treated by such simple assumptions. The structure of the liquid is perturbed at the interface by the presence of the crystal, and this affects the thermodynamic properties of the system. The first problem to be resolved is determination of the structural characteristics of the interface. These can then be used for the calculation of thermodynamic quantities.

The problems arise mainly from the presence of the liquid, for which a satisfactory theory does not yet exist. The liquid represents a state intermediate between the gas and the crystal, so that advantage cannot be taken from three dimensional periodicity, as can for the crystals, and, due to the relatively high density, the interactions between molecules cannot be neglected even in first approximation, as can be done for gases.

This chapter is devoted to a presentation of the theoretical works dealing with the structure and thermodynamic properties of the crystal-melt interface. To a large extent, they follow the basic approaches which have been successfully applied to studies of bulk liquids.

As a simplification, we shall limit our investigations to simple substances, which are composed of spherical monoatomic molecules. The most common example is that of the rare gases. For these substances, the internal energy is made up of the contributions of all the pairs formed by the molecules taken two by two (triplet and higher contributions are neglected as well as the effects of the electrons). The pair interaction is assumed to follow some simple law, which gives strong short range repulsion and attraction at intermediate range between the molecules.

The simplest potential which exhibits that kind of behavior is the hard sphere potential, although an attractive part is missing and should be taken into account in another manner. The most used "realistic" intermolecular potential is the Lennard-Jones(6-12) potential :

$$u(r) = 4\ \varepsilon \left[\left(\frac{\sigma}{r}\right)^{12} - \left(\frac{\sigma}{r}\right)^{6} \right] \qquad (1)$$

where r is the distance between two molecules, ε the energy well depth and σ the separation at which $u(r) = 0$.

The Lennard-Jones potential is a particularly good representation of the interactions between pairs of rare gas atoms. Its simple analytic form makes it ideal for computer simulations.

The principle of corresponding states tells us that two substances with the same potential law (for instance the Lennard-Jones potential), but with different constants (ε, σ) will have the same physical properties if they are expressed in reduced units, calculated as follows :

reduced temperature : $T^* = k_B T/\varepsilon$

distance : $d^* = d/\sigma$

pressure : $p^* = p/(\varepsilon/\sigma^3)$

time : $t^* = t/(\sigma\ m^{1/2} \varepsilon^{-1/2})$

etc.

Here, k_B is the Boltzmann constant; m is the mass of a molecule;

T, d, p, and t are the real absolute temperature, distance, pressure and time respectively.

I.2. Calculation of the Specific Free Energy of the Crystal-Melt Interface

This section describes how the specific surface free energy between the crystal and the melt can be calculated knowing the positions of the molecules in the crystal and in the liquid near the interface, as well as the interaction potential between the molecules. The next sections will describe how information on the positions of the molecules can be obtained. For the present time we shall assume that they are known.

The interfacial free energy can be calculated from the integrated equation of the Gibbs free energy. For a two phase solid-liquid system composed of n_s "solid" molecules with chemical potential μ_s, n_L "liquid" molecules with chemical potential μ_L and an interface with area A and specific interfacial free energy γ_{SL}, we can write :

$$G = n_s \mu_s + n_L \mu_L + \gamma_{SL} A \qquad (2)$$

Since we are dealing with two condensed phases with a small density difference, the Gibbs potential or Helmholtz potential can be used interchangeably. The Helmholtz free energy F is directly related to the canonical partition function so we shall use it preferentially, and write :

$$A \gamma_{SL} \simeq F - n_S \mu_S - n_L \mu_L \qquad (3)$$

At equilibrium (i.e. at the melting point) :

$$\mu_S = \mu_L = \mu \qquad (4)$$

and

$$\gamma_{SL} = (F - N\mu)/A \qquad (5)$$

with

$$N = n_S + n_L \qquad (6)$$

This means that the position of the interface is unimportant for the determination of the specific surface free energy. However, since the liquid is an entropy rich phase while the solid is an energy rich phase, the position of the interface determines the respective amplitude of the potential and entropic parts of the interfacial free energy.

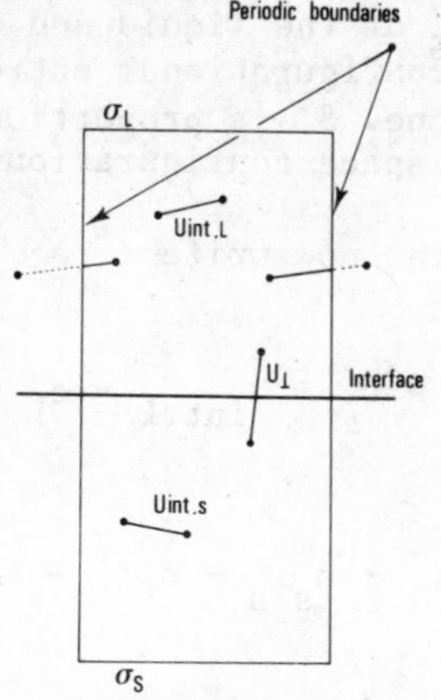

Fig. 1 : A schematic representation of a model of the crystal-melt interface, and definition of the potential energy terms which enter eq. (7)

Fig. 1 is a schematic representation of a portion of the interfacial zone, bounded by peroidic limits in the directions parallel to the interface, thus representing a quasi infinite interface. In the direction perpendicular to the interface, the model is limited by planes cutting the crystal and the liquid into zones where they are perfectly homogeneous. The so-defined interfacial zone is composed of n_L "liquid" molecules and n_S "solid" molecules. The free energy of the interfacial zone is composed of :

i) the potential energy U_L of the bonds between liquid and crystalline molecules, across the interface ;

ii) the potential energy U_{int} of the solid or liquid molecules respectively, between themselves ;

iii) the binding energy of the molecules in the interfacial zone with their respective solid or liquid adjacent bulk phases. This is approximately minus the potential part σ_S or σ_L of the surface free energy of the bulk phases (assuming that the density and structure of the liquid at a liquid-vapor interface are those of the bulk liquid);

iv) an entropy term $-TS$, in which S can be expressed as the logarithm of the free volume accessible to the molecules. Calculation of the free volume is a complex task, because of the collective character of the movements of liquid molecules. As a simplification, it can be assumed that each molecule performs essentially oscillatory movements, separated by occasional jumps from one position to another. Therefore, the corresponding configurational integral can be wirtten as a product of the local free volumes (calculated by the same methods as the mean vibrational volume in a crystal) and the number of possible space configurations of the liquid molecules. In this approximation, three terms enter the expression for the free energy of the interfacial zone : the vibrational free energies F_L^v and F_S^v of the liquid and crystalline molecules respectively, and the configurational entropy S^c of the molecules of the interfacial zone. S^c is proportional to the logarithm of the number of possible space configurations of the molecules in the system.

We can now write :

$$\gamma_{SL} = U_{\perp} + U_{int,L} - \sigma_L + U_{int,s} - \sigma_s - TS^c + F_L^v + F_s^v \quad (7)$$

$$-n_L u_L - n_s u_s - n_L f_L^v - n_s f_s^v + n_L T s_L^c$$

where u_s and u_L, f_S^v and f_L^v are the potential energy and the vibrational free energies per molecule in the liquid and solid bulk phases respectively and s_L^c is the configurational entropy per molecule in the bulk liquid. The configurational entropy is zero in a perfect crystal. The calculation of the interfacial free energy involves evaluation of all the terms of eq. (7).

II. THE MODEL APPROACH

II.1. Introduction

The model approach is mainly derived from the works of BERNAL[7] on the structure of bulk liquids. According to BERNAL, the liquid is a homogeneous, coherent and essentially irregular assembly of molecules containing neither crystalline regions nor holes large enough to admit another molecule.[8]

A basic hypothesis, later justified by the success of perturbation theory, is that the structure is determined largely by the form of the repulsive potential between the molecules. Since all real potentials are strongly repulsive at short separations, the molecules can be considered in first approximation as perfectly hard spheres.

II.2. The Bernal Model

The Bernal model of the liquids is essentially a dense random packing of hard spheres that can be considered as representing a snapshot of the liquid structure at a given time. The statistical characteristics of the geometry of the Bernal model are close to those of the liquid structure, as shown by various experiments[9]. Physical properties such as density, viscosity, heat of fusion or vaporization can be correctly predicted.

Other structural information, which cannot be obtained from experiments on real liquids, is provided from observation of the packing in the Bernal model. The space can be divided into Voronoï polyhedra, constructed around each sphere by the planes which perpendicularly bissect the segments joining the neighboring molecules. Each polyhedron contains all the points of space closer to the corresponding sphere center than to any other. Characterization of the Voronoï polyhedra shows the importance of pentagonal symmetry, which allows a high local order. The incompatibility of pentagonal symmetry with long range order is of little importance for the liquid state. It is also shown that octahedral packing, which is typical for the crystalline close packed structures is nonexistent in the Bernal model.

The shortcomings of the model are related to the fact that the geometry does not account for the real intermolecular potential, or for the thermal motion of the molecules in the liquid. In this respect, it is a representation of a glass rather than a liquid. Since it is based essentially on geometrical considerations the Bernal model is not concerned with statistical mechanics, and cannot answer questions like : why the liquid has its structure, or what is the entropy per molecule ?

The realization of the models by an experimental process as done by Bernal and coworkers[7] had some drawbacks too. One drawback is of course the need for tedious manipulations, and human intervention is the source of experimental errors. At the same time, the speed and capacity of the computers improved, so that people started to think of building virtual random packings in a computer. The first successful algorithms were those of BERNAL-MASON-FINNEY,[10] BENNETT[11] and MATHESON.[12]

The Bernal-Mason-Finney model consists of a compression of hard spheres. The starting configuration is a random assembly of hard spheres with a very low density. The volume of the system is then decreased by a homothetic reduction of all the coordinates, while preventing overlaps between neighboring spheres. This process aims to simulate what happens when a real assembly of steel spheres is compressed. The algorithm leads to models with a density equal to that of the laboratory built model, and with a very similar structure. The size of the models, is, however, limited to a few hundreds of molecules, if one wants to keep the computational time within reasonable limits.

The Bennett model is a polytetrahedral packing of hard spheres, based on the idea that the presence of three first neighbors for each sphere is a necessary condition for mechanical stability. The building process starts with a tetrahedral seed of four molecules. The new spheres are then arranged one by one, in the tetrahedral sites (pockets) formed by the previously placed molecules, taken in sets of three. The site closest to the center of the model is always chosen to be occupied first, ensuring the highest possible density. As soon as the model has a moderate size, sites appear which do not belong to the crystalline lattice. Their occupation leads to a disordered structure with a high local density.

The structure of the Bennett model is a satisfying approximation of the Bernal model. However, the spherical symmetry causes oscillations in density as a function of size.

The Matheson algorithm is similar to that of Bennett in that it places the spheres one by one in the pockets determined by previously deposited spheres. The main difference is that the spherical symmetry is replaced by a cylindrical one, in view of suppressing the density fluctuations. Comparison with the laboratory built Bernal model in terms of radial distribution function shows reasonable agreement. (fig. 2).

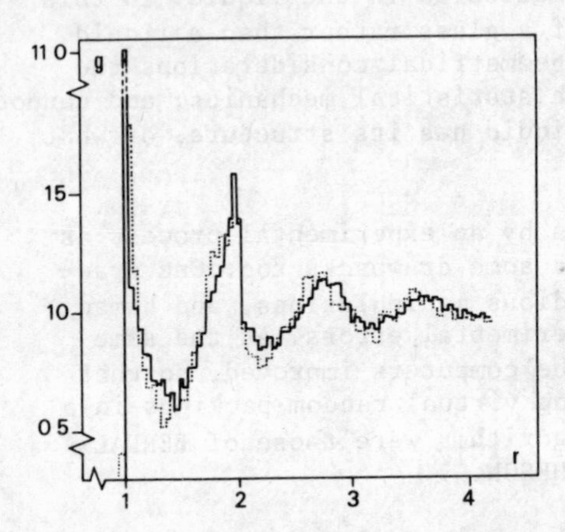

Fig. 2 : Radial distribution function for the Matheson random close packing of hard spheres, and comparison with the Bernal loose random packing, as obtained by SCOTT. (Reprinted from ref.12) (With permission from the Institute of Physics, London)

The density of the Matheson model is 5 per cent lower than that of the laboratory built model. For a hard sphere system, the density can be expressed in terms of packing fraction, which is the ratio between the volume of the spheres and the total volume of the system. The following packing fractions have been obtained :

Laboratory built model : 0.637

Matheson model : 0.606

hcp or fcc crystal : 0.7405

II.3. The Models of the Crystal-Melt Interface

The structural approach of Bernal can be applied to the crystal-melt interface. The liquid is represented by the hard sphere model, in contact with a crystal face. Because poor wetting of the crystal by the liquid usually occurs in the case of metals on faces with a hexagonal structure (111)fcc or (0001)hcp , primarily these faces have been studied by the various authors. The hexagonal close-packed two-dimensional lattice has one important peculiarity. It presents to the disposition of atoms with the same size two interpenetrating triangular sublattices of tetrahedral pockets. The occupation of a pocket of any sublattice prohibits occupation of neighboring pockets of the other sublattice. This peculiarity is responsible for the presence of stacking faults in hcp or fcc crystals.

ZELL and MUTAFTSCHIEV[13] built a model in the laboratory by physically packing some 2000 ping pong balls in a container. The model was then put in contact with a plane with hexagonal structure composed of spheres of the same size, representing the crystal face. The spheres were then removed one by one, while measuring their coordinates for further analysis. Assuming a realistic Lennard-Jones (6-12)potential between the molecules, the potential part of the adhesion energy $U_{\perp}$, eq. (7) has been calculated. The results demonstrate the possibility of a poor wetting of the crystal by the melt. The use of this model was limited however, because it did not allow the realization of very large packings.

Another interesting approach was proposed by SPAEPEN.[14] For Spaepen, the principal characteristic of the liquid structure is the absence of octahedral holes, which are non-existent in the Bernal model, while abundant in the close packed crystalline structures. Spaepen assumed that the first liquid layer in contact with the crystal consists of molecules disposed in the tetrahedral sites determined by the crystal molecules in such a way that the two systems of sites are occupied equally, that the density is maximal, and that no octahedral holes are formed. Geometrical considerations lead to estimation of the density of the first layer as some 75 % that of a crystal lattice plane. Fig. 3 gives an example of the structure of the first layer. From the same type of geometrical considerations, Spaepen concludes that there is no density deficit at the interface, and thus that the potential part of the interfacial specific free energy is zero. Statistical geometry on the two-dimensional packing at the interface also allows an estimate of the entropy of the first liquid layer at the interface, giving access to the interfacial specific free energy. The conclusion was that there must be a perfect wetting of the crystal by the melt.

The shortcomings of Spaepen's approach are that the model is limited to one liquid layer and that the basic assumption of the model, namely the non existence of octahedral packings, is not justified by thermodynamic considerations.

The last model[15] to be presented here is a computer built version of the model proposed by ZELL and MUTAFTSCHIEV.[13] It makes use of the MATHESON[12] algorithm to build a model starting from a crystal face with a hexagonal structure. Like in the Matheson algorithm, the spheres are deposited one by one in the site which is the closest to the original crystal face. If several sites have the same lowest "altitude" Z, the site to be occupied is chosen randomly among them. Each new sphere deposited prohibits occupation of sites situated at a distance smaller than the sphere diameter. It also creates new sites, in combination with its closest neighbors. The list of possible sites is updated after each sphere deposition. Periodic boundary conditions in the directions parallel to the interface allow simulation of a model with infinite size by elimination of the inhomogeneities near the boundaries. Fig. 4 shows a cross sectional view of one of these hard sphere models.

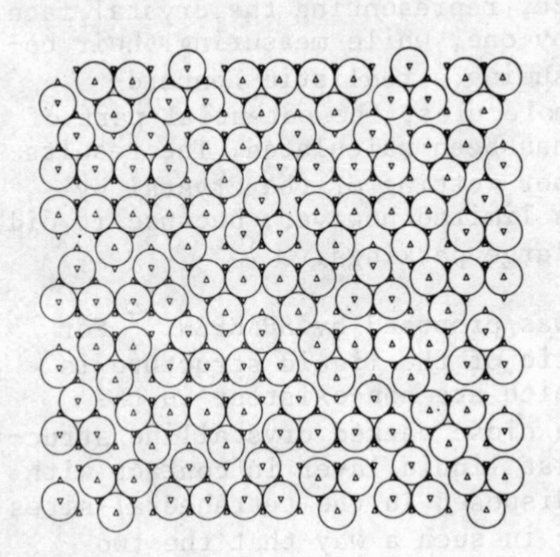

Fig. 3 : Structure of the first liquid layer in contact with the crystal, as obtained by Spaepen. The two types of triangles Δ and ∇ indicate the two interpenetrating sublattice of sites (Reprinted from ref. 15) (With permission from Taylor and Francis, London)

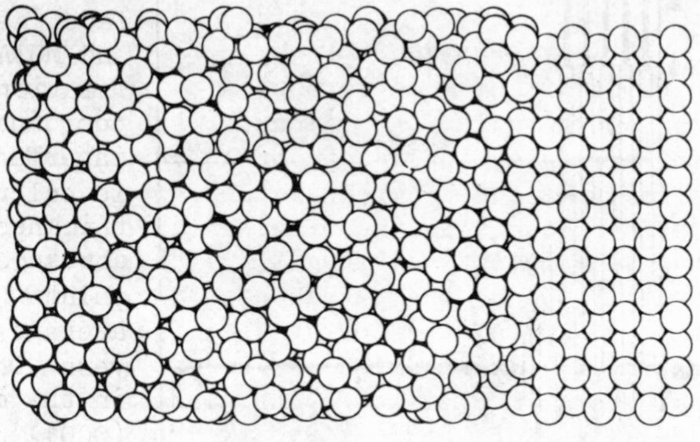

Fig. 4 : Cross-sectional view of a model of the crystal-melt interface. The crystal has an h.c.p. structure, and the interface is parallel to the (0001) direction. The plane of the figure is parallel to the (11$\bar{2}$0) direction. (reprinted from ref. 56)

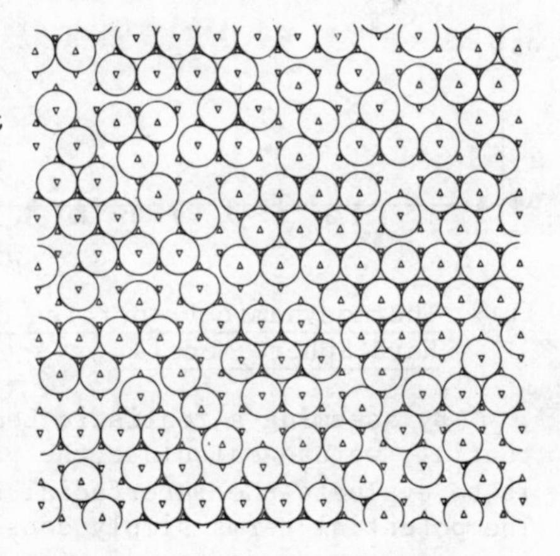

Fig. 5 : Structure of the first liquid layer in contact with the crystal, in the model of Bonissent and Mutaftschiev. The two types of triangles Δ and ∇ indicate the two interpenetrating sublattices of sites. (reprinted from ref. 15).
(With permission from Taylor and Francis, London)

Far from the interface, the computer built models are similar to those built by the algorithm of Matheson. In the first layer in contact with the interface, the equal occupation of the two sublattices of sites leads to the peculiar structure of fig. 5, formed of rafts of spheres in coherent positions, separated by channels in which it is not possible to place a new sphere in contact with the crystal face. This is the beginning of disorder, which will become complete in the bulk liquid part of the model.

Figure 6 gives the density profile of the models, in the Z

direction perpendicular to the interface. We observe a slight minimum of density near the interface. This has consequences on the thermodynamic properties, and is the origin of the poor wetting obtained from the laboratory built model of Zell and Mutaftschiev, and confirmed by the computer built model.

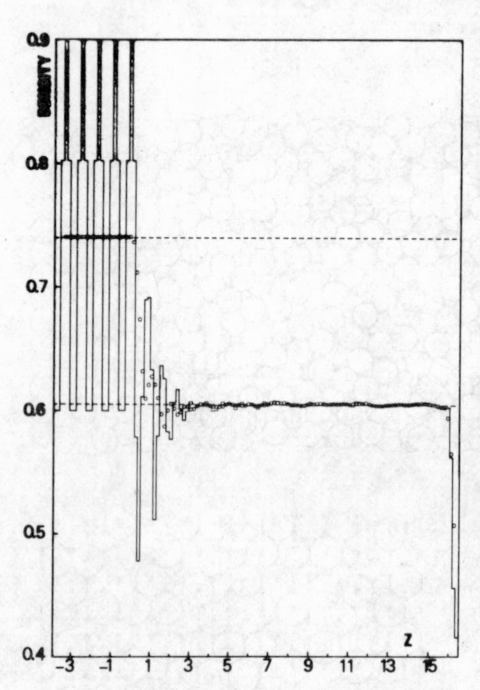

Fig. 6 : Density profile of a model of the crystal-melt interface, as obtained by BONISSENT and MUTAFTSCHIEV. The packing fraction is plotted versus z coordinate perpendicular to the interface. The packing fraction is calculated in slices with thickness 1/5 of the interlayer distance in the crystal. The points 0 represent the local average on five neighboring layers. The interface is oriented in the direction (111) f.c.c. or (0001) h.c.p. (from ref. 56).

II.4. Thermodynamic Properties of the Interface Predicted by the Hard Sphere Models.

By assuming a realistic Lennard-Jones(6-12)intermolecular potential, we can calculate the parameters entering eq. (7), and use it to evaluate the interfacial specific free energy.
The potential terms simply consists of a summation over the necessary pairs, as defined in fig. 1.

The vibration frequencies are calculated for each molecule, in the Einstein approximation, i.e. assuming that the neighboring molecules remain fixed in their equilibrium position.
The errors introduced by this approximation should be similar in the bulk or interfacial phases and are thus minimized. Figure 7 gives the values of the vibrational frequency as a function of the z coordinate.

Finally the configurational entropy must be estimated.[16] This is possible, due to the algorithm used to build the models. At any time during the building process, the computer program knows all accessible sites. The spheres placed in these sites will form the next (m^{th}) "monomolecular" layer. Since the system is disordered, the number of sites is larger than the number of spheres which will form the layer. The number of complexions of this layer is equal to the number of possibilities for occupation of the sites with a maximum number of molecules (the maximum number of molecules means the highest possible density in the final model). This number is obtained by the randon sequential filling method, as proposed originally by BAKER[17] for the two dimensional honeycomb lattice.

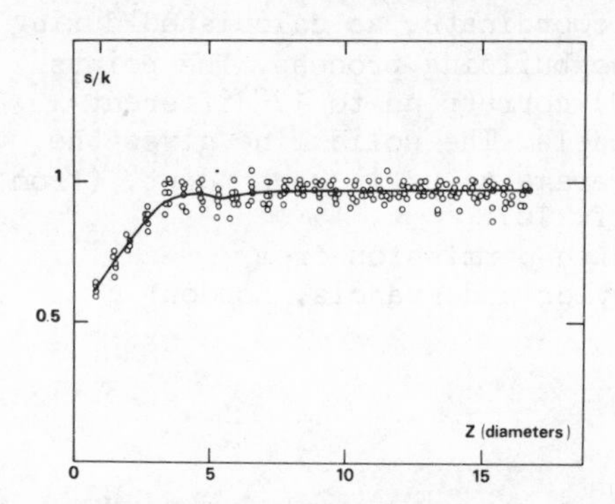

Fig. 7 : Vibration frequencies, in the Bonissent-Mutaftschiev model of the crystal-melt interface for a molecule versus its z coordinate, and for argon at the triple point. The points (0) correspond to three different models, and the curve gives the average values for the three models. (from ref. 56)

At a given moment of the building process, a random sequential filling is performed. All the sites are formed by allowing them to be occupied randomly and removing, after deposition of a new sphere, all sites whose occupation becomes impossible. The sites created thereby are not taken into consideration. After deposition of some N_m spheres, a "jamming" occurs. It becomes impossible to place a new sphere since there is no site available. The number of possibilities for placing each sphere (labelled i) is identical to the number $N_{s,m}^{(i)}$ of disposable sites. The number W'_m of complexions of the N_m spheres forming the jammed layer can be approximated by

$$W'_m = \prod_{i=1}^{N_m} N_{s,m}^{(i)} \tag{8}$$

If the spheres (molecules) are undistinguishable, the number of complexions is :

$$W_m = W'_m / N_m ! \tag{9}$$

and the entropy per molecule is :

$$s = \frac{1}{N_m} k_B \ln W_m \tag{10}$$

Figure 8 shows the configurational entropy per molecule at various distances from the interface.

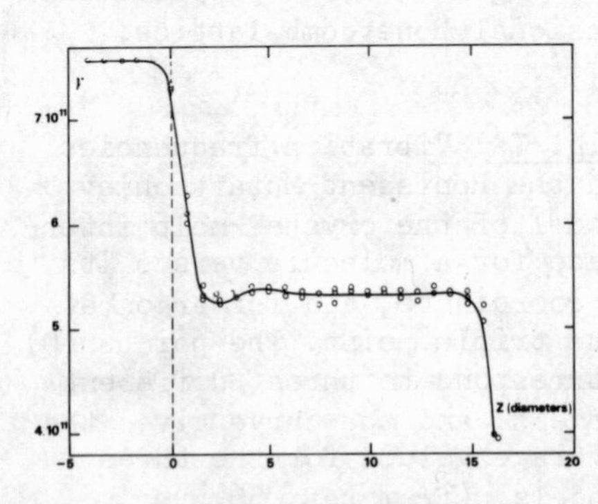

Fig. 8 : Configurational entropy per molecule as a function of the z coordinate, as calculated during the building process. The points (0) correspond to 12 different models. The solid line gives the average value in each point. (from ref. 16).
(With permission from Taylor and Francis, London)

An estimate of the specific crystal-melt interfacial free energy for Argon at the triple point by the methods outlined above is :

$$\gamma_{SL} = 11.2 \text{ mJ/m}^2 \tag{11}$$

Applying the same methods, one can estimate the crystal-vapor and liquid-vapor interfacial free energies for the same substance respectively as :

$$\gamma_S = 32.6 \text{ mJ/m}^2 \tag{12}$$

$$\gamma_L = 22.1 \text{ mJ/m}^2 \tag{13}$$

These values are approximate since they do not take into account that the structure of the solid or liquid phases near the solid-vapor interfaces are not identical to those of the bulk phases. This is consistent with other approximations used in the models, namely assumption of a constant lattice parameter in the crystal up to the solid-liquid interface, or of using a density independent diameter for the hard spheres.

The contact angle α for a liquid drop on a $[111]$ face of solid Argon is determined by :

$$\cos \alpha = \frac{\gamma_S - \gamma_{SL}}{\gamma_L} = 0.97 \qquad (14)$$

$$\alpha = 14° \qquad (15)$$

This is in qualitative agreement with the experimental finding of poor wetting of some dense faces of crystals by their own melt.[3-6] In conclusion, we see that the model approach gives a good description of the structure of the interface and permits estimation of the interfacial free energies.

The limitations of the model approach are the same as those of the Bernal model, i.e., the building procedure does not take into account the real intermolecular potential, or thermal motion. Moreover, the thermal motion of the molecules in the crystal can cause deformation of the interface or some kind of surface roughness, the character of which remains to be determined. Accordingly, it is of great interest to check (by other methods) the validity of the structure of the liquid near the crystal face as determined by the models. This will be the object of the following section.

III. COMPUTER SIMULATIONS

Computer experiments have become a very useful tool for studying bulk liquids. Their advantage is that they apply on well defined (although not necessarily real) systems. In computer experiments it is possible to vary microscopic parameters such as molecular mass and intermolecular potential, which cannot be modified in a real experiment. This is why computer experiments provide a convenient basis for verification of the theories of the liquid state. The techniques for computer simulation can be classified in two main groups; the molecular dynamics method and the Monte-Carlo method.

III.1. The Molecular Dynamics Method

The molecular dynamics technique was first applied by ALDER and WAINWRIGHT[18] to a system of hard spheres. This technique consists of integrating the coupled differential equations which describe the behavior of the system. These equations derive from the application to each molecule of the principle of Newtonian mechanics :

$$\vec{f}_i = m_i \vec{g}_i \tag{16}$$

where $\vec{g}_i$ is the acceleration of the molecule labeled i, m_i is its mass, and $\vec{f}_i$ is the force applied to it by the intermolecular potential of the neighboring molecules. As for any differential equation, initial conditions must be specified which are in this case, the positions and velocities of each molecule at the origin of time. This specification defines the temperature, but one must wait until the system is equilibrated to know its exact value since exchanges can occur between the potential energy (related to the positions of the molecules) and the kinetic energy (related to their velocities). When the system is equilibrated both of these quantities remain constant except for small fluctuations.

The systems simulated by the molecular dynamics technique have constant energy, volume, and number of molecules. A technique has been proposed[19] recently to perform molecular dynamics simulations with constant pressure, or constant temperature. This will certainly be of great interest for the future.

III.2. The *Monte-Carlo* Method

Apart form the choice of initial conditions, a molecular dynamics simulation is a completely deterministic process. This is not the case for the Monte-Carlo method,[20] which involves a probabilistic element. Successive configurations are generated by a random displacement of the molecules, usually one at a time. The configurations are accepted or rejected, depending on the difference in potential energy ΔU_{ij} between the two successive configurations labeled i and j. The new configuration is accepted with the probability exp $(-\Delta U_{ij}/k_B T)$. It can be shown[21] that this algorithm generates a Markov chain, i.e. an ensemble of states of the system, such that the occurence of a given state i is proportional to its thermodynamic probability exp $(-U_i/k_B T)$. The thermodynamic quantities can then be obtained by averaging over all states of the system in the chain.

In practice, the successive configurations are obtained by selecting one molecule at a time. A movement is then applied to this molecule, according to the procedure described above. The molecule, the amplitude, and the direction of its movement are chosen by a random process.

The Monte-Carlo simulations are normally performed at constant temperature, volume, and number of molecules. Algorithms have been developed to perform simulations at constant temperature, volume and chemical potential (the grand canonical ensemble),[22] or at constant temperature, pressure, and number of molecules (the isothermal-isobaric ensemble).[23]

No time scale is involved in the Monte-Carlo simulation, and the order in which the configurations occur has no special significance. This differs from the molecular dynamics method, which appears to be more adapted to studying systems out of equilibrium or kinetic parameters such as diffusion coefficients. On the other hand, the Monte-Carlo method has the advantage of a formalism directly applicable to the constant temperature system.

The two methods suffer from limitations due essentially to the small number of molecules which can be taken in consideration (usually less than a thousand). This is extremely small compared to the number of molecules which are involved in any real laboratory experiment. Some procedures have been proposed to extrapolate the computer simulated properties to those of an infinite system. The phenomena which do not occur because of the small size of the system, however, cannot be extrapolated. This is why it is extremely difficult to simulate liquids near the critical point. The shape of the sample, even though periodic boundaries are used, can also be a serious source of trouble, especially if a phase transition is being studied.

Another difficulty is the one of the equilibration, or in other words, when the system can be considered as being in its equilibrium state, remembering that the starting configuration is usually not one of equilibrium. This is a crucial problem because of the amount of computer time necassary for these simulations, so that one wants to know the minimum equilibration time. This can be judged only in an empirical way, by monitoring the properties of the system and the way they fluctuate in the course of the simulation. It is always possible, however, that the system remains in a metastable equilibrium. This is often a source of controversy.

The computer simulations give access to the state parameters which are not fixed in the ensemble in which the simulation is performed : pressure, internal energy, kinetic energy or temperature, and to the structure, since the position of all molecules are known at any time. However, it is not possible to directly obtain the free energy F, or entropy S. On the other hand, the derivatives of these quantities with respect to thermodynamic variables such as V and T are known. Hence, by performing Monte-Carlo calculations for a number of values of V (or T) and integrating numerically the pressure (or heat capacity), free energy differences can be evaluated. If the free energy of the system is known in a reference state, in principle it can then be obtained in any other state. The reference state can be a perfect gas (high temperature, low pressure), or a perfect crystal (low temperature). For these two systems, the free energy can be calculated by standard methods of statistical mechanics. The only inconvenience of this procedure is the amount of computer time required, as the precision of the result depends on the grid size used for the integration.

III.3. Application to the Solid-Liquid System

Computer simulations of a system composed of a liquid and a solid phase, and an interface, present some specific difficulties. One is the fact that the two phases can coexist only at the melting point. This means that the thermodynmmic conditions under which the experiment is performed must be fixed with a high precision. Even in this case, stability is not assured since, in such small systems, one can expect the fluctuation to be rather important and one of the phases could easily disappear. Nevertheless, the problems of thermodynamic equilibrium do not seem to be too crucial, probably because of the very short time over which the simulation extends. It is sufficient that the system remains in a metastable state for the length of the computer experiment. This condition can be obtained by carefully selecting the starting configuration and/or applying special techniques during the equilibration.

The first simulation of a system containing a crystal-melt interface was in fact a three phase system, as described by LADD and WOODCOCK.[24] This was a molecular dynamics simulation of a Lennard-Jones substance. The system, composed of 1500 molecules, was limited by periodic boundary conditions in two (x - y) directions. In the z perpendicular direction it was composed of a static crystal, with thickness larger than the effective range of the interatomic potential, a mobile crystal slab, a liquid slab and finally a vapor with low density. The crystal phase has a f.c.c. structure, and the interface is oriented in the [100] direction. The temperature is $T^* = 0.72$ close to the triple point temperature.[25]

The initial configuration is made up of a crystal, expanded in the z direction, so that its density is approximately that of the liquid at the triple point. Because of the presence of the static crystal, the liquid and solid phases separate and the system exhibits a cristal-liquid and a liquid-vapor interfaces.

The main conclusion of this work is that the interface is rather diffused since the density profile perpendicular to the interface exhibits a layered structure in the liquid, over some 5 atomic diameters, as shown in figure 9. No analysis was made, however, to determine more precisely the structure of the interfacial region. No density deficit was observed at the interface. The density profile of the liquid-vapor interface is given as well, and appears to be consistent with other simulations.

Another simulation of a system involving a crystal-melt

interface was done in the same period by TOXVAERD and PRAESTGAARD[26]. It is a molecular dynamics simulation on a set of some 1700 molecules interacting through a Lennard-Jones pair additive potential.

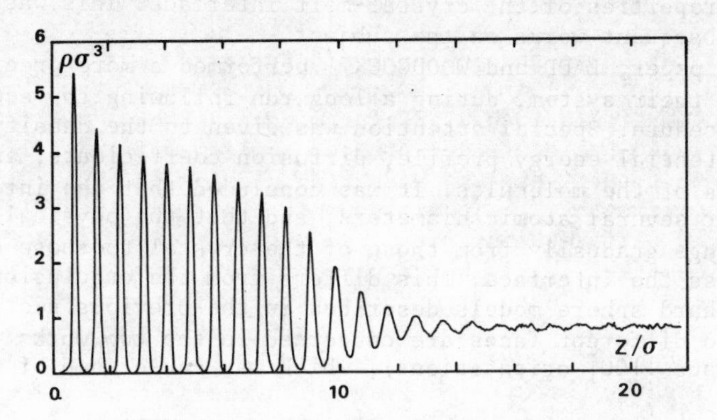

Fig. 9 :One particle density profile obtained by LADD and WOODCOCK[27] for the (1 00)crystal-melt interface of a Lennard-Jones substance. (Reprinted from ref. 27, with permission from the Institute of Physics, London)

The system consists of a "sandwich" composed of a liquid slab enclosed between two solid slabs, thus defining two solid-liquid interfaces. The system is surrounded by periodic boundary conditions in all three directions. During equilibration the system is divided into two subsystems, separated by planes which no molecule is allowed to cross. There is however no limitation to the interaction potentials, in the sense that molecules of one subsystem "feel" the effect of the molecules in the other subsystem. The velocities in each subsystem are periodically rescaled during the equilibration in order to fix the temperature at the desired value of $T^* = 1.15$. Simultaneously, the density of the liquid subsystem is adjusted by rescaling the positions in the z direction (perpendicular to the interfaces), so that the pressure is finally equal to the equilibrium pressure between crystal and liquid at the considered temperature. The separation planes between the two subsystems are then removed, and a supplementary equilibration period is run, without any temperature or pressure calibration. The state of the system at this stage is considered as the initial configuration for the simulation. Only the subsequent configuration obtained are taken into consideration to study the thermodynamic properties.

The interface is oriented in the [100] direction of the (f.c.c.) crystalline part of the system.

The conclusions are very similar to those of Ladd and Woodcock namely that the density profile perpendicular to the interface

presents oscillations over some 5 atomic diameters. The main interest of these two works was to show that it is possible to keep a two phase system in equilibrium during a time long enough to observe the properties of the crystal-melt interface. This was developed in subsequent works on the subject.
In a second paper, LADD and WOODCOCK[27] performed a more precise analysis on their system, during a long run following the equilibration procedure. Special attention was given to the density profile, potential energy profile, diffusion coefficients, and trajectories of the molecules. It was concluded that the interface extends over several atomic diameters, and that the physical properties change gradually from those of the crystal to those of the liquid across the interface. This differs from the conclusions drawn from hard sphere models described in the previous section. However, two different faces are concerned in the two works ([111] versus [100] orientations), which may cause some of the difference.

The purpose of the simulation presented by BONISSENT, GAUTHIER and FINNEY[28] was to check the validity of the assumptions on which the hard sphere Bernal model presented in the previous section is based, i.e. not taking into account the realistic interatomic potential or the thermal motion of the molecules during the building process. Comparison of the properties of the hard sphere model with those of a computer simulated realistic system should be illuminating in that respect.

The computer experiment[28] consists of a Monte-Carlo simulation, on a system composed of some 860 Lennard-Jones molecules. Periodic boundary conditions are again imposed in the x-y direction, and in the z direction there is successively a static crystal, a mobile crystal limited to two reticular plane, a liquid slab, and finally a vapor phase. The interface has the [111] orientation, and the starting configuration is realized by the hard sphere sequential deposition technique described in section I. The molecules are then softened, i.e. the hard sphere potential is replaced by a Lennard-Jones potential. The transitional period of thermal equilibration over which the potential energy exhibits significant changes appears to be very short, confirming that the structure of the realistic model is close to that of the hard sphere model. The potential energy change during the equilibration process is evaluated to 5 % of the total potential energy.

The density deficit observed with the hard sphere models remains, although of smaller amplitude, and the extent of the interfacial zone is larger (some 5 atomic diameters).

Observation of the trajectories of the molecules shows that the first liquid layer, on which the island-like structure remains has essentially the behavior of a crystal plane with a high concentration of defects. The next (liquid) layer has the behavior of a liquid, as can be deduced from the observation of the mean square displacement as a function of computer time, for molecules situated at different positions in the model. It is not clear, however, to what extent these properties are effected by the small extension of the mobile crystal in the z direction (two planes only).

The last contribution to be presented here is a comparative molecular dynamics simulation of two Lennard-Jones systems containing crystal melt interfaces, with [100] and [111] orientations respectively.[30] In both cases, the system is limited by periodic boundaries in the three directions. The interfaces are perpendicular to the z direction. The simulation starts with two crystal slabs separated by a z expanded crystal slab (this will become liquid during the equilibration period). The original positions are fixed so that the crystal and the "expanded crystal" parts of the system have the respective densities of the triple point crystal and liquid, as determined by LADD and WOODCOCK.[24] Each system is equilibrated in two stages. The atoms in the solid parts are initially held fixed at their initial lattice positions, and the atoms in the liquid are allowed to move, while rescaling the velocities to a temperature $T^* = 0.75$, well above the melting point. After a period of time sufficient to insure that the "liquid part" of the model becomes liquid, the crystalline atoms are released. The temperature is then rescaled every time step to the triple point temperature $T^* = 0.67$.

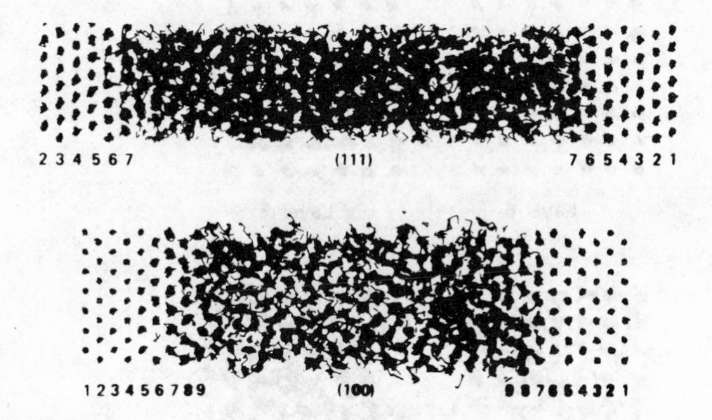

Fig. 10 : Trajectories of the molecules during the simulation in a slice perpendicular to the interface (x - z plane) for the (a) (111) and (b) (100) systems. Any atom entering the slice, at any time during the simulation, is represented so long as it remains in the slice (from ref. 30)(With permission from the American Institute of Physics)

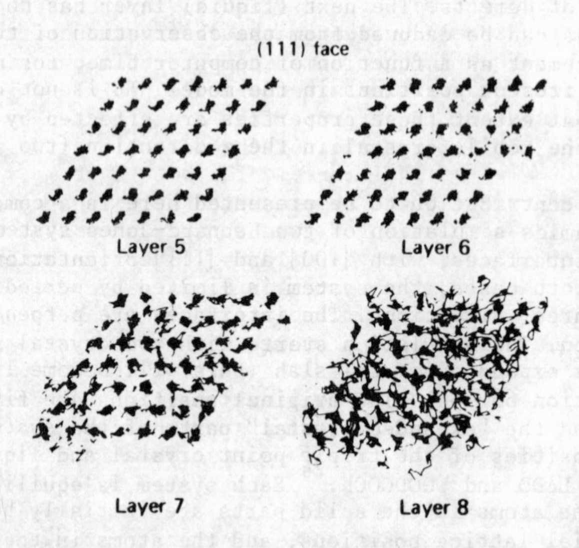

Fig. 11 : Trajectories of the molecules in layers parallel to the interface (x - y plane) using the same procedure as for Fig. 10, (111) interface. (from ref. 30) (With permission from the American Institute of Physics)

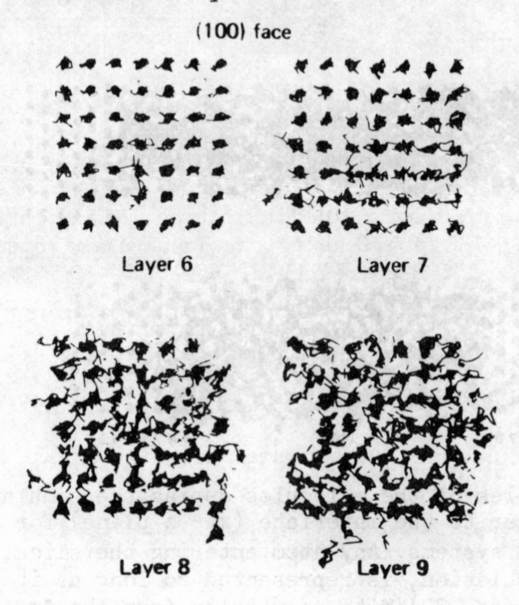

Fig. 12 : Trajectories of the molecules in layers parallel to the interface (x - y phase), (100) interface. (from ref. 30) (With permission from the American Institute of Physics)

After some 5000 time steps, the system is equilibrated and can be allowed to evolve freely, without any significant drift of the temperature over 5000 further time steps. Such a procedure creates stable interfaces, as well as a temperature profile which is flat over the whole system.
Figure 10 shows the trajectories of the molecules over the 5000 time steps performed at equilibrium, for both systems.
Figure 11 and 12 present the same movements in cross sectional views perpendicular to the plane of figure 9. They show that the two systems appear to behave very similarly. Layer 6 of the (111) system and layer 7 of the (100) system can be considered as equivalent. They represent a quasi crystalline layer, in which the amount of diffusion is not negligible, although the molecules spend most of their time in the lattice sites. The next layer up, in each case, definitely represents a two-dimensional liquid, but careful examination of the trajectories indicates a higher residence time in the positions of the ideal lattice sites. This is due to the potential minima created by the field of the layer underneath, which is virtually crystalline. The trajectory plots, therefore, indicate that the transition from crystal to liquid is rather sharp. The two-dimensional radial distribution functions and the diffusion coefficients, calculated for the molecules in the same slices, confirm that observation. We can then conclude that as far as the structural characteristics are concerned, the interface between the crystal and the melt extends essentially over two atomic diameters. The subsequent oscillations in density or related quantities, observed in this simulation as well as in others, are minor compared with the abrupt change of symmetry which occurs at the interface itself. This is in good agreement with the conclusions obtained from the hard sphere models of the (111) interface. The potential part of the interfacial free energy can be calculated by the method exposed in section I, as the potential energy of all molecular bonds is known at each moment during the simulation. The molecular dynamics simulation also gives access to the local free volume, or vibrational entropy, for each molecule. However, an important contribution to the interfacial free energy is missing, namely the configurational entropy (this is defined in section I.2). Therefore, the interfacial free energy cannot be calculated by this method.

III.4. Conclusion.

The computer simulation techniques give a great deal of information on the structure and thermodynamic properties of the crystal-melt interface. They are in some respect more powerful than the hard sphere model building procedure described in the previous section since realistic systems can be simulated. They allow simulation of interfaces with any orientation, which is up to now impossible with the model building procedure, limited to the (111) face.

It is unfortunate that these methods do not give access to the interfacial free energy. The hard sphere model is presently the only way to obtain an estimate for this important quantity. It is thus comforting that there is a reasonable agreement between the structural characteristics of the hard sphere model and those of the computer simulated models.

The main difference between the hard sphere model and the molecular dynamics simulations, concerns the density deficit which is observed in the hard sphere model at the interface, but which is absent in most of the computer simulated models. This is due partly to the fact that in the hard sphere model, the crystal is by definition perfect and static, while in the computer simulations, the molecules in the crystal, and especially in the last layer, undergo thermal vibrations. This results, as will be seen in the next section, in a spreading of the density peaks, which become wider and of smaller amplitude. The same effect applied to an eventual minimum can be sufficient to make it disappear. In this respect, it is indicative that in the Monte-Carlo simulation of BONISSENT et al.[28] the minimum is present, although of smaller amplitude than in the hard sphere model. Remember that in this simulation, only two crystalline planes at the interface are free to move. The surface molecules in this case perform movements of smaller amplitude than in the more realistic simulations. The smoothness of the Lennard-Jones potential, as compared to the hard spheres potential has a similar effect.

IV. A DIRECT APPROACH : THE ABRAHAM-SINGH PERTURBATION THEORY FOR NON UNIFORM LIQUIDS AND ITS APPLICATION TO THE CRYSTAL-MELT INTERFACE.

IV. 1. Introduction

A theoretical approach to the thermodynamic properties of the bulk liquids which has shown a considerable development during the last few years uses the principles of the statistical mechanics, without any model considerations, or computer simulation. This approach resulted in perturbation theories which take advantage of the relative simplicity of the equations for the hard sphere fluids and of the fact that the realistic intermolecular potentials are sufficiently close to the hard sphere potential. Equations of state for the Lennard-Jones substance, in good agreement with the "experimental" computer simulated results, have been derived by these methods.

Recently, perturbation techniques have been applied to the problem of a liquid near an interface.

In this section, we shall describe the Abraham-Singh

perturbation theory for non uniform liquids, and its application to the crystal-melt interface. In order to make it more clear, we shall first make an overview of bulk liquids theories.

IV. 2. The Theories of Bulk Liquids

The grand canonical partition function is defined as :

$$\Xi = \sum_{N=0}^{\infty} \frac{\exp(\beta\mu N)}{N!\ h^{3N}} \int \ldots \int \exp\left[-\ \beta\ H_N\ (r^N, p^N)\right]\ dr^N\ dp^N \qquad (17)$$

where $\beta = 1/k_B T$, N is the number of molecules, μ the chemical potential, h the Planck constant, and H_N the Hamiltonian (total energy), which is a function of the position and momentum coordinates of the N molecules, denoted r^N and $\underline{r}^N$ respectively. The integral is 6N - fold. The limits are $-\infty$ to ∞ for the momentum coordinates and the volume of the container for the position coordinates.

The link with thermodynamics is made through the relation

$$\Xi = \exp(\beta PV) \qquad (18)$$

while the probability for finding the system with precisely N particles having coordinates r^N and momenta p^N is :

$$\frac{\exp(\beta N\mu)\ \exp\left[-\beta\ H(r^N, p^N)\right]}{N\ !\ h^{3N}\ \Xi} \qquad (19)$$

Due to the quadratic character of the kinetic part of the Hamiltonian, which contains only square terms of the momentum coordinates, the integral over momenta in eq. 17 factors into 3N identical contributions $(2\pi mkT)^{1/2}/h = \Lambda^{-1}$

The activity, or fugacity, is defined as :

$$z = \exp(\beta\mu)/\Lambda^3 \qquad (20)$$

and the grand canonical partition function can be written as :

$$\Xi = \sum_{N=0}^{\infty} \frac{z^N}{N!} \int \exp\left[-\beta U_N(r^N)\right] \; dr^N \tag{21}$$

or

$$\Xi = \sum_{N=0}^{\infty} \frac{z^N}{N!} Z_N \tag{22}$$

with

$$Z_N = \int_V \exp\left[-\beta U_N(r^N)\right] \; dr^N. \tag{23}$$

Z_N is the potential part of the canonical partition function for a system of N particles in the volume V. Z_N is often called the configurational integral. In the case of the liquids, even for the simplest intermolecular potentials, the configurational integral Z cannot be separated into simple integrals. Thus, no analytic solution can be obtained. The alternative of numerical calculation is not any more realistic, because of the high order of the integral. Theories of the liquid state propose approximations which make this integral tractable, or find other methods to calculate the thermodynamic quantities.

An other important notion is the so-called density functions : $\rho^n(r^n)$, which is the probability of finding any $\underline{n}$ particles at the positions r^n in the system. The expression for ρ^n is obtained by integrating the thermodynamic probability, (Eq. (19)) on the coordinates of the (N - n) molecules whose position is not specified. In the grand canonical ensemble, the result is :

$$\rho^n(r^n) = \frac{1}{\Xi} \sum_{N>n} \frac{z^N}{(N-n)!} \int \exp\left[-\beta U_N(r^N)\right] dr_{n+1} \ldots dr_N \tag{24}$$

These functions allow a compact and complete description of the microscopic structure of liquids. Knowledge of the lowest order ones is generally sufficient to calculate most equilibrium properties of the system. The most important functions are the one and two particle density distributions. It can be shown easily [3] that for a homogenous fluid :

$$\rho^{(1)}(r) = \frac{\langle N \rangle}{V} = \rho \tag{25}$$

The distribution functions are defined as

$$g^{(n)}(r) = \frac{\rho^{(n)}(r^n)}{\prod_{i=1}^{n} \rho^{(1)}(r_i)} \qquad (26)$$

which is for a uniform system :

$$g^{(n)}(r^n) = \rho^{(n)}(r^n)/\rho^n \qquad (27)$$

The distribution functions are normalized expressions for the density functions, in the sense that they tend to unity for a large separation of the molecules. The pair distribution function depends, for a homogeneous system, only on the separation r_{ij} of the two molecules considered :

$$g^{(2)}(r_1,r_2) = g(|r_1 - r_2|) = g(r) \qquad (28)$$

The quantity g(r) is called the radial distribution function. Its importance is evident in the case of a pairwise additive potential, as the total internal energy per particle in an assembly of N molecules is :

$$\frac{U_N}{N} = 2\pi\rho \int_0^\infty g(r)\ u(r)\ r^2\ dr, \qquad (29)$$

where u(r) is the intermolecular potential.

The equation of state can be expressed by a similar expression. An alternative for the calculation of the thermodynamic properties of liquids is then to use the radial distribution function. In this respect, the first step is to calculate g(r) for a given intermolecular potential u(r).

It is beyond the scope of this course to give a complete review of the possible techniques available for evaluation of g(r). The reader will find all the desired information on this subject elsewhere.[31,32] What should be mentioned here is that an approximation has been proposed by PERCUS and YEVICK,[33] which is justified for low densities, but the validity of which extends to high enough densities that it is applicable to the liquid state. Owing to the special character of the hard sphere potential, the equations obtained in the Percus-Yevick approximation can be resolved leading to analytical expressions for the radial distribution function and the equation of state of the hard sphere fluid.

The perturbation method is a well known technique in several fields of physics. It applies to systems which are so complicated that they cannot be treated in a rigorous way. If a simplified system can be found, which is in some sense similar to the system under consideration and which is tractable, then the differences between the two systems can be treated as a small perturbation. The perturbation is applied to the properties of the simplified reference system to obtain the properties of the real system. In the case of the liquid state, the reference system is usually the hard sphere fluid, the properties of which can be obtained analytically in the Percus-Yevick approximation. The perturbation theories consist of evaluating the effect of a modification of the intermolecular potential (from the hard sphere to e.g. the Lennard-Jones potential) on the thermodynamic properties. They make use of functional derivative techniques to obtain a series expansion in functions of a parameter which represents a measure of the difference between the two potentials concerned. The result can be obtained in principle with any degree of accuracy by incorporating as many terms as necessary in the series expansion.

The BARKER and HENDERSON[34] perturbation theory consists of a division of the intermolecular potential into two parts :

$$U_o(r) = \begin{cases} U(r) & r < \sigma \\ 0 & r > \sigma \end{cases} \qquad (30)$$

which will be considered as the reference potential, and :

$$U_1(r) = \begin{cases} 0 & r < \sigma \\ U(r) & r > \sigma \end{cases} \qquad (31)$$

which represents the perturbation. Barker and Henderson show that the behavior of a fluid with a purely repulsive potential is the same as that of a hard sphere fluid, if the hard sphere diameter is correctly adjusted. The hard sphere diameter depends on the temperature but not on the density. The remaining part U_1 of the potential is then incorporated by a perturbation procedure. Development to second order is necessary to obtain sufficient accuracy. The results are then in good agrement with those obtained from computer simulation.

In the WEEKS, CHANDLER and ANDERSEN theory,[35] the intermolecular potential is devided into :

$$U_\circ(r) = U(r) + \varepsilon \qquad r < r_m$$
$$= 0 \qquad r > r_m \tag{32}$$

and

$$U(r) = -\varepsilon \qquad r < r_m$$
$$= U(r) \qquad r > r_m \tag{33}$$

where r_m is the position of the minimum in the U(r) function.

It is shown that the free energy and radial distribution function of the reference system defined by $U_\circ$ are :

$$F_\circ = F_{HS}$$
$$g_\circ(r) = \exp\{-\beta U_\circ(r)\} \; y_{HS}(r) \tag{34}$$

The subscript HS denotes the hard sphere system; y_{HS} is defined by :

$$g_{HS}(r) = y_{HS}(r) \exp\{-\beta U_{HS}(r)\}, \tag{35}$$

where $U_{HS}(r)$ is the hard sphere potential. It is interesting to note that although $g_{HS}(r) = 0$ and $U_{HS}(r) = \infty$ for $r < \sigma_{HS}$, $y_{HS}(r)$ is defined for any value of r.
The hard sphere diameter is defined by :

$$\int_\circ^{r_m} r^2 y_{HS}(r)\, dr = \int_\circ^{r_m} r^2 \exp\{-\beta U_\circ(r)\} \; y_{HS}(r)\, dr, \tag{36}$$

which expresses that the first term in the expansion of the grand canonical partition function in a series of $\exp\{-\beta U_\circ(r)\}$ is zero, i.e. that the hard sphere reference system is as close as possible to the soft repulsive system. The next step is the incorporation of the perturbation potential $U_1(r)$. To first order, this is simply the attractive potential energy of the neighbors of each molecule :

$$\Delta F = \frac{1}{2} N\rho \int U_1(r)\, g_o(r)\, dr \qquad (37)$$

The hard sphere diameter depends both on density and temperature. It must be found by iteration (eq. 36), and is therefore more difficult to compute than in the Barker-Henderson theory. However, this division of the potential has the effect of reducing the second order terms, which can than be neglected.

The ABRAHAM-SINGH[36] theory for nonuniform liquids is based on the same approximations as the Weeks-Chandler-Andersen theory of bulk liquids.

IV.3. <u>A Perturbation Theory for the Crystal-Melt Interface</u>

Perturbation theory of the crystal-melt interface has not yet been applied to the prediction of interfacial free energy or to the detailed structure of the liquid in the neighborhood of the crystal face. However, a treatment for the density profile perpendicular to the interface is known. This is the Abraham-Singh perturbation theory for nonuniform fluids[36] applied to the case of the crystal-melt interface.

In this theory, it is assumed that the interaction potential between the solid and any atom in the fluid depends only on the normal distance z between the solid face and the fluid atom. Furthermore, it is assumed that it is the repulsive nature of the potential that dictates the fluid structure neighboring the solid face. Hence, the repulsive potential of the solid face is considered to be

$$\begin{aligned} \Phi^r_w(z) &= \Phi_w(z) - \Phi_w(z_m), & z < z_m \\ \Phi^r_w(z) &= 0 & z > z_m \end{aligned} \qquad (38)$$

The grand partition function Ξ of the fluid-wall system is treated as a function of the Boltzmann factor $b^r_w(z) = \exp[-\beta\Phi^r_w(z)]$ and a Taylor expansion of $ln\ \Xi\{b^r_w\}$ is performed in powers of the ANDERSEN-WEEKS-CHANDLER[33] "blip" function $\Delta b_w(z) = b^r_w(z) - b^h_w(z)$, where $b^h_w(z) = \exp[-\beta\Phi^h_w(z)]$ is a Heaviside step function for a hard wall potential at position d_w, i.e ,

$$ln\ \Xi\ \{b_w^r\} = ln\ \Xi\ \{b_w^h\} + \int y_w^h(z)\ \Delta b_w(z)\ dz \tag{39}$$

where

$$y_w^h(z) = \frac{\delta\ ln\ \Xi\ \{b_w^h\}}{\delta\ b_w^h(z)} = \rho^h(z)\ \exp\left[\beta\Phi_w^h(z)\right] \tag{40}$$

The $\rho^h(z)$ is the single-particle density distribution of the fluid in contact with a hard wall positioned at d_w, i.e., the center of a fluid particle is excluded from the region $z < d_w$. The value of d_w is chosen such that the second term in eq.(19) vanishes, i.e.,

$$\int y_w^h(z)\ \Delta b_w(z)\ dz = 0, \tag{41}$$

and therefore

$$ln\ \Xi\ \{b_w^r\} \simeq ln\ \Xi\ \{b_w^h\} \tag{42}$$

Equations (41) and (42) may be used to determine the density distribution $\rho^r(z)$ of the fluid in contact with a soft repulsive wall, where

$$\rho^r(z) = \exp\left[-\beta\Phi_w^r(z)\right] \frac{\delta\ ln\ \Xi\ \{b_w^r\}}{\delta b_w^r(z)} \tag{43}$$

Following the arguments of ANDERSON-WEEKS-CHANDLER, ABRAHAM and SINGH[36] find that

$$\rho^r(z) \simeq \exp\left[-\beta\Phi_w^r(z)\right]\ y_w^h(z) \tag{44}$$

Equations (41) and (44) are the basic results of the Abraham-Singh theory. The values of $y_w^h(z)$ for the hard-sphere fluid in contact with a hard wall have been obtained by HENDERSON, ABRAHAM and BARKER by applying the Percus-Yevick approximation to a mixture spheres with two diameters, in which one of the components becomes infinitely dilute and infinitely large in size.

The validity of the Abraham-Singh theory has been established by comparing it with Monte-Carlo simulations of a hard sphere fluid in contact with a repulsive Lennard-Jones 9/3 wall potential.[36]

IV.4. The Crystal-Melt Interface

When applying the Abraham-Singh theory to the crystal-melt interface, two questions arise which are not relevant when the idealized soft wall system is considered. These two points are the effect of the two dimensional structure of the potential exerted by the solid on the liquid molecules and the thermal motion of the crystal molecules. Due to the periodicity of the crystal lattice, it is useful to expand the potential $\Phi_w(x,y,z)$ in a two dimensional Fourier series in the x - y plane. The leading term of the series is a function only of the z coordinate. In the present state of the Abraham-Singh theory this is the only term which can be taken into account. The higher order terms have to be neglected. ABRAHAM[39] has shown that if the expansion is performed not on the potential, but on $\exp\left[-\Phi_w(x,y,z)/k_BT\right]$, the leading term is a more occurate approximation of the exact potential. This is due to the fact that the Boltzmann factor of the wall potential plays a dominant role in the statistical mechanics of this system, through the expression for the partition function. Another reason is that the exp function is bounded by 0 and 1, while the Φ function can have sharp peaks for a molecule situated in the neighbourhood of the crystal face. Thus, the exp function is more regular and will be better approximated by its average value, which is the leading term in the Fourier series expansion.

Following Abraham, the effective potential $\Phi_w^e(z)$ is defined by

$$\exp\left[-\Phi_w^e(z)/k_BT\right] = \frac{1}{A}\int_A \exp\left[-\Phi_w(x,y,z)/k_BT\right]dxdy \qquad (45)$$

In a crystal face, it can be assumed that neighboring atoms will perform "in phase" oscillations, i.e., it is unlikely that the positions of any two, three or four neighboring surface atoms will be highly uncorrelated. Furthermore, the liquid structure near the crystal face will be dictated by the configurational distribution of surface sites on the crystal face. We can also consider that, to a good approximation, the density profile in the normal direction to a crystal plane is dominated by the vibrational modes with a wavelength significantly larger than the crystal lattice parameter (i.e., those for which an adsorption site at the interface is distorted very little). With this viewpoint, the interfacial density profile $\rho^T(z)$ is altered from the zero-temperature crystal fluid profile using the following crude approximation :

$$\rho^T(z) = \int\rho^o(z+\delta)\,P(\delta)\,d\delta \Big/ \int P(\delta)\,d\delta \qquad (46)$$

where P(δ) is the density profile in the normal direction of the crystal face, obtained from computer simulation.

Finally, the hard-sphere diameter for the fluid must be adjusted according to the interatomic potential, temperature, and density for the fluid under consideration[35]. This is done using the WEEKS, CHANDLER, and ANDERSEN theory. The results are given in figures 13 and 14 for the (111) and (100) faces respectively, and compared with the molecular dynamics results.

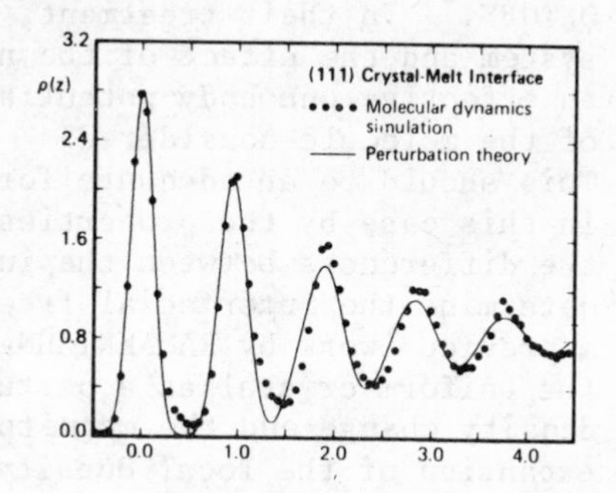

Fig. 13 : Density profile of the melt neighboring the (111) face predicted by the perturbation theory and compared with the "experimental" profile obtained by molecular dynamics simulation. The crystalline peak, around z = 0, has been fitted with a Gaussian function, as described in the text. (from ref. 30).
(With permission from the American Institute of Physics)

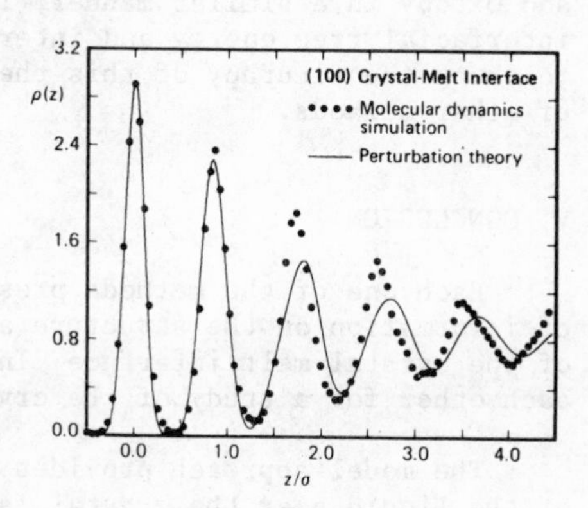

Fig. 14 : Same as Fig. 13, for the (100) face.

We see that the agreement is rather good especially if one recalls the crudeness of the approximation used. This may be due to the fact that the function considered (ρ(z)) is a two-dimensional average, thus having a weak dependence on the two-dimensional structure of the crystal face.

The agreement is, however, much better on the (111) face than on the (100) face. The reason is that the reference system is the same (hard spheres against a hard wall) in the two cases.

It has been shown by BERNAL[40] that the structure of a hard sphere system near a flat wall exhibits hexagonal symmetry. This is obviously more like the (111) interface than the (100). Since the perturbation theory modifies only the first peak of the density profile (the one which is in the range of the wall potential), it is not suprising to observe a good fit only for this peak in the (100) interface. The subsequent peaks remain those of a hard sphere fluid against a hard wall, and are more representative of the structure of the (111) interface than of the (100) interface.

Another perturbation theory has been proposed by HAYMET and OXTOBY.[41] In their treatment, the bulk liquid is the reference system and the effect of the non uniformity is treated by using an effective one body potential which depends on the environment of the molecule considered.
This should be an adequate formalism, since one is not concerned in this case by the properties of the interface itself, but by the differences between the interface and the bulk material, which determine the interfacial free energy. This treatment expands upon a previous work by RAMAKRISHNAN and YOUSSOUFF[42] which represents the uniform crystal as a perturbation of the uniform liquid. The density change and the magnitude of the Fourier components of the expansion of the local density under freezing are obtained.

The case of the crystal-melt interface is treated by Haymet and Oxtoby in a similar manner, leading to expressions for the interfacial free energy and interfacial density profile. It remains to check the accuracy of this theory by comparison with the results of other methods.

V. CONCLUSION

Each one of the methods presented here gives a limited amount of information on the structure and related thermodynamic properties of the cyrstal-melt interface. In this respect, they complement each other for a study of the crystal-melt system.

The model approach provides a useful structural description of the liquid near the crystal face and permits the evaluation of the specific interfacial free energy. The main results of the model approach are the fact that the interface is limited in extention perpendicular to the crystal face, and that the liquid poorly wets the crystal. The use of this method is, however, presently limited to the (111) direction, and its validity is the one of the approximations on which the hard sphere model is based.

The computer simulation techniques can be applied to two phase systems. It appears possible to maintain coexistence between solid and liquid during a time long enough to obtain quasi equilibrium properties of the interface. The small extension of the interfacial region in the z direction is confirmed for the (111) face, and demonstrated for the (100) interface. A characterization of the interface is obtained, in terms of the density, potential energy and configurational entropy profiles. It has not yet been possible to calculate the interfacial free energy. A limitation for this kind of study is the cost of the computer time necessary to perform the simulations.

The direct approach with the perturbation theory does not bring any new results, since the density profiles are already known from computer simulations. It must be considered instead as a self teaching process in the sense that it provides information on the validity of the approximations and thus on the physics of the system.

V.1. Further Developments

A number of questions remain to be answered, related to the nucleation and growth of crystals from the melt. One of them is : what is the activation energy in the nucleation process, or the excess free energy of formation of a crystal nucleus in the melt ? For sufficiently large nuclei, this excess free energy can be calculated if one knows the interfacial free energies of all the faces present. This requires knowledge of the interfacial free energy in many directions.

It would be interesting also to study how this free energy of formation depends on the size of the nucleus.

We have already seen that the model approach is presently the only method leading to the interfacial free energy. The way in which it can be applied to faces other than (111)f.c.c. remains to be found. Further developments in perturbation theory or in computer simulations might change that situation.

Another point is the effect of a step on the crystal surface. This is related to the existence of a lateral growth mechanism for the crystal growth versus the continous mechanism frequently assumed.[43] The value of the ledge free energy is an important parameter for this problem.[57] On the (111) face, the model approach can give an estimate of this quantity, by the difference between the free energy of a system containing a ledge, and that of a system with a flat interface. The other techniques might be applied to this system as well but it is unlikely that growth could be observed with the computer experiments because a typical simulation (expending large amounts of computer time) corresponds to a very short interval of time in the real system (about 10^{-10} sec).

VI. THE CRYSTAL-GLASS INTERFACE

Little is known about crystal-glass interfaces, and it is only during the last few years, with the technological interest in the recrystallization (or crystallization from the solid phase) that attention has been paid to the interface between the crystal and the amorphous phase. From an experimental point of view, the problem is slightly easier than that of the crystal-melt interface, in the sense that the system does not undergo any important modifications over a large temperature range. Observations at room temperature are thus possible. However, it is presently impossible to prepare a glass from a pure simple substance with central interatomic forces. Experiments are performed on two component alloys which makes the comparison with theory more difficult. Structural characterization is not microscopic, but rather macroscopic, concerned with the observation of the defects in the interfacial region.[44]

As far as computer simulation is concerned, we can mention the works of RAHMAN[45] which consist of rapidly quenching a liquid during a molecular dynamics simulation. Whether or not the system is in a glassy state after quenching is controversial[46] as the lifetime of this low temperature disordered state appears to be very short. Nucleation then occurs, followed by the growth of the crystal. The nature of the interface between the ordered and disordered phases has not yet been investigated.

Since the Bernal model is well known to represent the glass phase rather than the liquid,[47] the model approach presented in section I also represents the crystal-amorphous interface. In fact, the main criticisms against this model are irrelevant in the case of the crystal-amorphous interface, which exists at low temperature. In this case, the thermal motion of the crystal molecules can be neglected, and the glass molecules can be approximated as hard spheres.

VII. THE LIQUID-VAPOR INTERFACE

This section will deal with simple liquids only. We will again pay attention to the structure of the interface, and then to its main thermodynamic property, the surface tension or surface free energy.

VII.1. The Structure of the Liquid-Gas Interface

The liquid-vapor interface should be simpler than the crystal-liquid interface, since it involves only one condensed phase. For this reason, the question of how the structure of the liquid is perturbed by the presence of the molecules of the other phase does not arise. The "surface layer structure" observed in some computer simulations[48] has shown[49] to be an artificat of the extremely slow convergence of the density profiles, as compared to the number of configurations generated. It is now generally admitted that the density varies smoothly from that of the liquid to that of the vapor. An example of this density profile is given in fig. 15 for different temperatures.

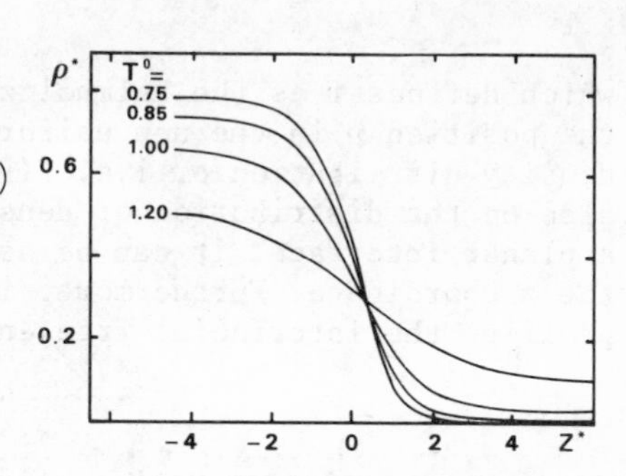

Fig. 15 : Density distribution in the liquid-gas interface zone for a Lennard-Jones substance at different temperatures (from ref. 53) (With permission from the American Institute of Physics)

The interface width extends when increasing the temperature, and becomes theoretically infinite at the critical point temperature. The effect of the system size and temperature on the density profile and surface tension have been investigated by WEEKS.[50] He concludes that the surface tension measured on a system with a macroscopie size should be lower than in a small system (like those studied in computer simulation), and that very long range correlations parallel to the interface (capillary waves) exist. In principle there is also the possibility of long range correlations perpendicular to the nominal interface and a divergent total width, but this would occur only for systems much larger than usual macroscopic distances.

VII.2. The Surface Free Energy

Computer simulation. A precise determination of the liquid vapour surface free energy has been performed by MIYAZAKI, BARKER and POUND.[51] It consists of a Monte-Carlo simulation in which the free energy required to reversibly create a surface in a bulk liquid is calculated directly. The liquid is separeted by steps into two slabs, limited by hard walls. When the distance between the two slabs is large enough, the walls are removed and the system is allowed to relax. The free energy change is integrated during the process by appropriate methods. A value of 18.3 ± 0.3 mJ/m^2

is obtained for the surface energy of the Lennard-Jones liquid at the triple point.

Perturbation Theory. The basic idea of the perturbation theory for the thermodynamic properties of nonuniform fluids is that the Helmholtz free energy F of the system can be written as :

$$F = \int_V f(r)\, dr \tag{47}$$

which defines f as the Helmholtz free energy per unit volume at the position r in the non uniform fluid. $f(r)$ is a function of the density distribution ρ, i.e. $f(r)$ depends not only on $\rho(r)$, but also on the distribution of density in the neighborhood of r. For a planar interface, it can be assumed that $f(r)$ depends only on the z coordinate. Furthermore, if f(z) is known for a given density profile, the interfacial free energy can be wirtten as[52] :

$$\gamma = \lim_{h\to\infty} \{\int_{-h}^{h} f(z)\, dz - h[f(-h) + f(h)]\} \tag{48}$$

Due to its smooth shape, the density profile $\rho(z)$ can be approximated by relatively simple analytic expressions of the tanh form, with one or two parameters. These are adjusted by minimizing the interfacial free energy.

The problem is thus to determine the local free energy function f(z) for a given density profile $\rho(z)$.

The first treatment of this problem, was proposed by TOXVAERD.[53] It is an extension of the BARKER-HENDERSON[34] perturbation theory for bulk liquids. ABRAHAM[54] has reformulated this theory within the framework of the WEEKS, CHANDLER ANDERSEN theory.[33] He concludes that the treatment of TOXWAERT is based on "local" assumptions like:

$$g(r_1, r_2, \{\rho\}) \simeq g[|r_1 - r_2|, \rho(r_1)] \tag{49}$$

where the brackets {} denote a functional dependence. Recently, SINGH and ABRAHAM[55] go beyond adopting the "local" approximations, and develop a perturbation theory which accounts for the non local contributions neglected by assuming eq. (48). Its application to the Lennard-Jones fluid leads to the value :

$$\gamma = 18.35 \text{ mJ/m}^2 \tag{50}$$

in agreement with the "exact" computer simulation results. In conclusion, it can be said that the structure and thermodynamic properties of the liquid-vapor interface are now well understood, at least in the case of a simple substance. This is significantly different from the case of the crystal-liquid interface.

REFERENCES

1 . TURNBULL D., J. Chem. Phys 18 (1950) 768
2 . SKAPSKI A., Acta Met. 4 (1956) 576
3 . KAISCHEW R. and MUTAFTSCHIEV B., Z. Phys. Chemie 204 (1955)334
4 . STRANSKI I.N. and PAPED E.K., Z. Phys. Chemie B 38 (1937) 451
5 . GLICKSMAN M.E. and VOLD C.L., Acta Met. 17 (1969) 1
6 . VOLLMER M. and SCHMIDT O., Z. Phys. Chem. Leipzig (1937) 467
7 . BERNAL J.D., Proc. Roy. Instn. 37 (1959) 355
8 . BERNAL J.D., Proc. Roy. Soc. A 280 (1964) 299
9 . FINNEY J.L., Proc. Roy. Soc. A 319 (1970) 495
10. FINNEY J.L., Journal de Physique, Coll. C2, Suppl. N°4, 36 (1975) C 2-1
11. BENNETT C.H., J. Appl. Phys. 43 §1972) 2 727
12. MATHESON A.J., J. Phys. : Solid State Physics 7 (1974) 2 569
13. ZELL J. and MUTAFTSCHIEV B., J. Cryst. Growth 13/14 (1972)231
14. SPAEPEN F., Acta Met. 23 (1975) 731
15. BONISSENT A. and MUTAFTSCHIEV B., Phil. Mag. 35 (1977) 65
16. BONISSENT A., FINNEY J.L. and MUTAFTSCHIEV B., Phil. Mag. B 42 (1980) 233
17. BAKER B.G., J. Chem. Phys. 45 (1966) 2 694
18. ALDER B.J. and WAINWRIGHT T.E., J. Chem. Phys. 27 (1957) 1 208
19. ANDERSEN H.C., J. Chem. Phys. 72 (1980) 2 384
20. METROPOLIS N., ROSENBLUTH A.W., ROSENBLUTH M.N. and TELLER A.H. J. Chem. Phys. 21 (1953) 1 087
21. WOOD W.W. and JACOBSON J.D., J. Chem. Phys. 27 (1957) 1 207
22. NORMAN G.E. and FILINOV V.S., High Temp. USSR 7 (1969) 216
ADAMS D.J., Mol. Phys. 29 (1975) 307
ROWLEY L.A., NICHOLSON D. and PARSONAGE N.G., J. Comput. Phys. 17 (1975) 401
23. WOOD W.W., in *Physics of simple Liquids*, ed. by Temperley, H.N.V., Rowlinson, J.J. and Rushbrook, G.S., North Holland, Amsterdam, 1966, Chap. 5.
WOOD W.W., J. Chem. Phys. 52 (1970) 729
ABRAHAM F.F., Phys. Rev. Letters 44 (1980) 463
24. LADD A.J.C. and WOODCOCK L.V., Chem. Phys. Letters 51 (1977) 155
25. HANSEN J.P. and VERLET L., Phys. Rev. 184 (1969) 151
26. TOXVAERD S. and PRAESTGAARD E., J. Chem. Phys. 11 (1977) 5 291
27. LADD A.J.C. and WOODCOCK L.V., J. Phys. C. Solid State Physics 11 (1978) 3 565

28. BONISSENT A., GAUTHIER E. and FINNEY J.L., Phil. Mag. B 39 (1979) 49
29. The Lennard-Jones potential does not represent exactly the behaviour of the argon molecules.
30. BROUGHTON J.W., BONISSENT A. and ABRAHAM F.F., J. Chem. Phys. 74 (1981) 4 029
31. Mc. DONALD I.R. and HANSEN J.P., in *Theory of Simple Liquids*, Acad. Press London, New York, San Francisco, 1976
32. BARKER J.A. and HENDERSON D., Rev. Mod. Phys. 48 (1976) 587
33. PERCUS J.L. and YEVICK G.J., Phys. Rev. 110 (1958) 1
34. BARKER J.A. and HENDERSON D., J. Chem. Phys. 47 (1967) 4 714
35. WEEKS J.D. and CHANDLER D., Phys. Rev. Letters 25 (1970) 149
WEEKS J.D. and CHANDLER D., J. Chem. Phys. 54 (1971) 5 237
WEEKS J.D., CHANDLER D. and ANDERSEN H.D., J. Chem. Phys. 55 (1971) 5 422
ANDERSEN H.C., WEEKS J.D. and CHANDLER D., Phys. Rev. A 4(1971) 1 597
36. ABRAHAM F.F. and SINGH Y., J. Chem. Phys. 67 (1977) 2 384
37. BONISSENT A. and ABRAHAM F.F., J. Chem. Phys. 74 (1981) 1 306
38. HENDERSEN D., ABRAHAM F.F. and BARKER J.A., Mol. Phys. 31(1976) 1 291
39. ABRAHAM F.F., J. Chem. Phys. 68 (1978) 3 713
40. BERNAL J.D., Proc. Roy. Soc. A 280 (1964) 299
41. HAYMET A.D.J.and OXTOBY D.W., J. Chem. Phys.to be published.
42. RAMAKRISHNAN T.V. and YOUSSOUF M., Phys. Rev. B 19 (1979)2 775
43. JACKSON K.A., UHLMANN D.R. and HUNT J.D., J. Cryst. Growth 1 (1976) 1
44. DROSD R. and WASHBURN J., J. Appl. Phys. 51 (1980) 4 106
45. HSU C.S. and RAHMAN A., J. Chem. Phys. 71 (1979) 4 974
46. WENDT H.R. and ABRAHAM F.F., Phys. Rev. Letters 41 (1978) 1 244
47. CARGILL G.S., J. Appl. Phys. 41 (1970) 12
48. LEE J.K., BARKER J.A. and POUND G.H., J. Chem. Phys. 60 (1974) 1976
49. ABRAHAM F.F., SCHREIBER D.E. and BARKER J.A., J. Chem. Phys. 62 (1975) 1958
50. WEEKS J.D., J. Chem. Phys. 67 (1977) 3 106
51. MIYAZAKI J., BARKER J.A. and POUND G.M., J. Chem. Phys. 64(1976) 3 364
52. ABRAHAM F.F., Physics. Reports 53 (1979) 95
53. TOXVAERD S., J. Chem. Phys. 55 (1971) 3 116
54. ABRAHAM F.F., J. Chem. Phys. 63 (1975) 157
55. SINGH Y. and ABRAHAM F.F., J. Chem. Phys. 67 (1977) 537
56. BONISSENT A., Thèse d'Etat, Marseille, France. 1978.
57. MUTAFTSCHIEV B., Materials Chemistry 4 (1979) 263.

THE INTERMOLECULAR NATURE OF LIQUID WATER - THE HYDRATION OF MOLECULES AND IONS.

Felix FRANKS

Department of Botany
University of Cambridge
Downing Street, Cambridge
United Kingdom

ABSTRACT

Most physical properties of water are anomalous and such anomalies must be accounted for in terms of the intermolecular potential and the molecular distributions. Aqueous solutions reflect some of these anomalies, usually referred to as "water structure". Several fundamentally different types of hydration interactions can be distinguished. Their sensitive balance is the main contributing factor to the maintenance of so-called native structures which are essential for the functioning of biologically active molecules.

I. INTRODUCTION

There are several inherent problems in the formulation of a molecular description of a liquid. The two most popular approaches start respectively with a dense gas or a perturbed solid. The development of X-ray and neutron scattering techniques has made the former approach more useful. The classification of liquids is based on the nature of the intermolecular forces involved, shown in Table 1. These interactions determine the PVT phase diagram and the physical properties. Table 2 is a summary of the properties of some typical members of the groups of substances shown in Table 1. Water occupies an anomalous position, characterized by the following properties;

1) the liquid is denser than the solid,
2) the liquid has a negative thermal expansion coefficient up to 4°C (11°C for D_2O),

B. Mutaftschiev (ed.), Interfacial Aspects of Phase Transformations, 183–198.

3) the latent heat of fusion is only 15 % of the latent heat of evaporation,
4) the liquid range is very large,
5) heat capacity of the liquid is very large.

Table 1. Nature of intermolecular forces in various types of liquids.

Nature of molecule	Predominating intermolecular force
Monatomic (spherical)	Van der Waals (dispersion), short range
Homonuclear diatomic	do. + quadrupole interactions
Metals, fused salts	Electrostatic, long range
Polar	Electric dipole moments
Associated	Hydrogen bonds, orientation dependent
Macromolecules	Some of the above, depending on chemical nature, also intramolecular forces
Helium	Quantum effects.

By comparison with the equilibrium properties, the transport behavior of water shows few of these anomalies but is, on the whole, typical of a liquid composed of small molecules.

Any credible structural model for liquid water must be able to account for the physical properties of all isotopic water species from the undercooled state to above the critical point.

II. THE STRUCTURE OF LIQUID WATER

Any structural calculations must take as their starting point the structure of the H_2O molecule. The calculated electron density distribution lends support to the Bjerrum four point charge model, shown in Fig. 1, in which the oxygen atom is placed at the centre of a regular tetrahedron and the fractions of charge $\pm \eta e$ are placed at the vertices at distances of 0.1 nm from the centre. The van der Waals diameter of 0.282 nm is identical to that of neon (water and neon are isoelectronic). The vertices carrying positive

	Argon Solid/Liquid	Benzene Solid/Liquid	Water Solid/Liquid	Sodium Solid/Liquid
Density ($kg\ m^{-3}$)	1636/1407	1000/899	920/997	951/927
Latent heat of fusion ($kJ\ mol^{-1}$)	7.86	34.7	5.98	109.5
Latent heat of evaporation ($kJ\ mol^{-1}$)	6.69	2.51	40.5	107.0
Heat capacity ($J\ mol^{-1}K^{-1}$)	25.9/22.6	11.3/13.0	37.6/75.2	28.4/32.3
Melting point (K)	84.1	278.8	273.2	371.1
Liquid range (K)	3.5	75	100	794
Isothermal compressibility (m^2N^{-1})	1/20	8.1/8.7	2/4.9	1.7/1.9
Surface tension ($mJ\ m^{-2}$)	13	28.9	72	190
Viscosity (poise)	0.003	0.009	0.01	0.007
Self diffusion coefficient ($m^2\ s^{-1}$)	$10^{-13}/1.6\times10^{-9}$	$10^{-13}/1.7\times10^{-9}$	$10^{-14}/2.2\times10^{-9}$	$2\times10^{-11}/4.3\times10^{-9}$
Thermal conductivity ($J\ s^{-1}m^{-1}K^{-1}$)	0.3/0.12	0.27/0.15	2.1/0.58	134/84

Table 2. Physical properties of representative types of molecular substances in condensed phases.

charge are the positions of the two hydrogen atoms, with the lone electron pair orbitals being directed towards the other two vertices. The interaction of two such molecules would have the features of what is called the hydrogen bond. With minor modifications, the Bjerrum model has been successful in accounting for the physical properties of ice and water.

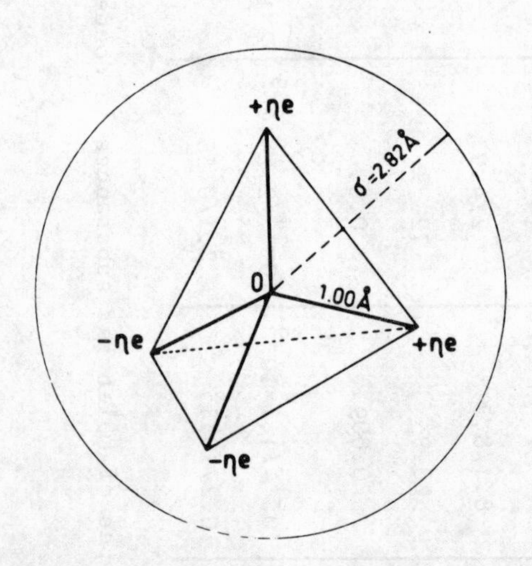

Fig. 1 : Bjerrum four-point-charge model for the water molecule, showing the van der Waals diameter.

It is difficult to calculate by *ab initio* methods the interaction details of two water molecules. Semi-empirical methods of treating the hydrogen bond lead to the conclusion that the following contributions must be considered : electrostatic effects, delocalization energy, dispersion interactions and core repulsion. All calculations predict the linear hydrogen bond to be the most stable configuration, with a dissociation energy lying between 20 and 35 KJ, and an equilibrium O-O distance between 0.26 and 0.30 nm. The covalent contribution (charge displacement) makes larger water clusters more stable than dimers. This type of interaction which depends on previous processes is described as cooperative, and hydrogen bonding in water is believed to be highly cooperative.

The importance of hydrogen bonding becomes apparent in the infrared and Raman spectra of water vapour, liquid water and ice. Thus, the characteristic O-H vibrations are severly perturbed by the proximity of other OH groups. The bending and final rupture of hydrogen bonds under the influence of temperature and pressure are also reflected in the spectra. The sensitivity of hydrogen bonded structures is best demonstrated by the P-T phase diagram of ice, shown in Fig. 2. All the polymorphs are fully hydrogen bonded, i.e. each oxygen atom participates in four bonds, but there are considerable deviations from the regular tetrahedral geometry of ice-I, as the pressure is increased.

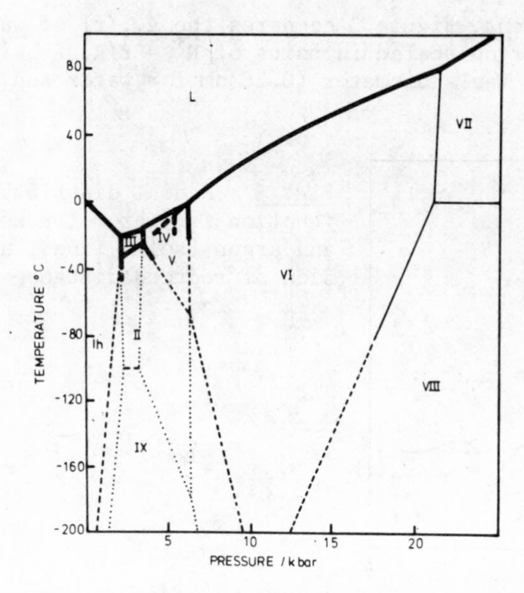

Fig. 2 : Solid-liquid phase diagram of ice. Solid lines : stable regions, broken lines : metastable regions, dotted lines : estimated or extrapolated phase boundaries.

In the liquid, the regular geometry is further perturbed by thermal motion. The application of scattering techniques permits the calculation of $g_{ij}(r)$, the radial distribution function which measures the relative probability of finding a molecule (or atom) i at a distance r from a given molecule (or atom) j. Thus, large values of $g_{ij}(r)$ imply that there is a preferred molecular spacing, rather than a random arrangement of molecules, in which case $g_{ij}(r) = 1$. To describe the molecular distributions in water completely, three radial distribution functions are required, expressing O-O, O-H and H-H distributions respectively. Only the first of these can be obtained from x-ray and the second form neutron scattering experiments. It requires yet another, independent scattering experiment (electrons ?) to determine the third distribution function.

The distribution of molecules in a fluid is governed by the interactions between them. This, in turn, depends on the molecular geometry and charge distribution. In the case of water, tetrahadral hydrogen bonding is expected to be the dominant interaction. A major shortcoming in all calculations is the necessity to express the total energy as the sum of pairwise interaction energies, thus denying the element of cooperativity. There are several formal relationships between the pair potential $u_{ij}(r)$ and $g_{ij}(r)$. The former is the more basic quantity, but the latter is more easily accessible to experiment. Figure 3 compares the $g_{ij}(r)$ of water with that of liquid argon scaled in units of $R^* = r/R$, R being the molecular van der Waals diameter (0.28 nm for water and 0.34 nm for argon).

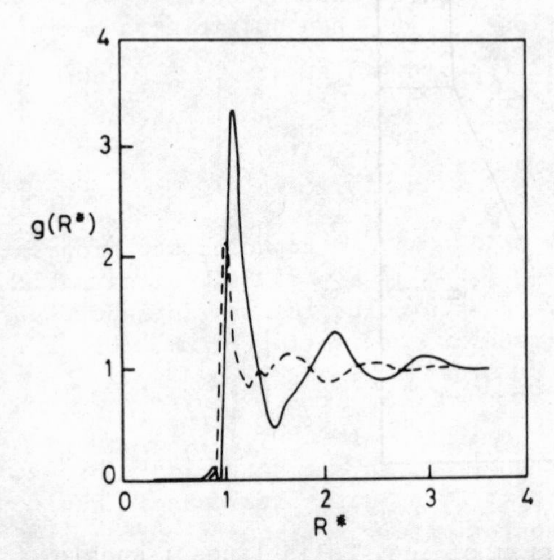

Fig. 3 : Radial distribution function for water (broken line) and argon (solid line), as function of reduced distance R^*.

The coordination number is obtained from the area of the first peak (4.4. for water, 10 for argon). For water, this number increases with rising temperature, reaching 5 at the boiling point. The second peak is at $1.63R^*$ ($2R^*$ for argon), in agreement with tetrahedral geometry. The widths of the peaks are the real indicators of structure in a liquid. Thus, a narrow peak indicates little perturbation about the mean value of R^*. By this criterion, water is more structured than argon, but such structure persists only over three molecular layers, beyond which $g_{ij}(r)$ approaches unity.

The advent of fast computers has led to the development of simulation methods for the study of liquids. With an assumed pair potential $u_{ij}(r)$, the energy of a system of n molecules is minimized and the equilibrium distribution is calculated by one of two available techniques. There are as yet severe problems, not least of which is the formulation of a credible $u_{ij}(r)$, but simulation has become a popular technique for the study of water and aqueous solutions. Both $u_{ij}(r)$ and $g_{ij}(r)$ are of intrinsic interest, because they form the bridge between structural measurements and thermodynamics, that is, the time-averaged behaviour of the liquid.

For a complete characterization of the liquid we also require a knowledge of its dynamic behaviour, as reflected in spectra and transport processes, such as self-diffusion and viscosity. Details of molecular motions are described in terms of time correlation fucntions, C(t), defined by

$$C(t) = \langle A(0) \cdot A(t) \rangle$$

where C(t) measures how long some property A persists before it is averaged out by random motions or collisions. A(0) refers to t = 0 and A(t) to the same property after time t, the average being taken over all molecules. Often C(t) can be expressed in terms of a single exponential decay function ; it is then possible to define a correlation time τ_c, where

$$\tau_c = \int_o^\infty C(t)\,dt.$$

This is then the time after which the magnitude of the property A has decayed to 1/e of its magnitude at t = 0. Nuclear magnetic relaxation and dielectric measurements provide information about τ_c, taken to be the period required for the isotropic tumbling of a molecule. For water τ_c is of the order of 1 ps.

Water has frequently been described as being ice-like. While certain of its properties lend support to this description- for instance, the X-ray and neutron scattering data- its dynamic behaviour argues against this. Any approximately tetrahedral order is of a short range nature and molecular exchange takes place on a picosecond time scale. The computer simulation results suggest that water is a random, almost completely hydrogen bonded network, but that there exists a large proportion of bent hydrogen bonds and distortation from the regular tetrahedron is common. The simulations do not reveal marked heterogeneities in the density, a conclusion which is at odds with the comparatively high compressibility, an indicator of marked density fluctuations.

III. HYDRATION AND ITS CONSEQUENCES

The properties of solutions must be considered in terms of three contributions :

Solvent - solvent interactions
Solvent - solute interactions
Solute - solute interactions.

In a dilute solution, the third of these effects can be neglected. Solute - solute interactions are measured by the second virial coefficient B in the osmotic pressure equation

$$\pi = RT\ (c/M + Bc^2 + \ldots.)$$

where c is the concentration and M the molecular weight of the solute. In other words, B measures the deviation of the solution from ideality. B is also related to $W_{ss}(r)$ and $g_{ss}(r)$, and hence to structural properties : $W_{ss}(r)$ is the potential of mean force between two solute particles, and $g_{ss}(r)$ is the solute radial distribution function. Here again, if $g_{ss}(r) = 1$, then the solute is distributed randomly in the solvent.

Nonideality is well represented by the thermodynamic excess functions. Thus, the excess Gibbs free energy change resulting from the mixing of two components is defined by

$$\Delta G^E = \Delta G_{exptl} - \Delta G_{id}$$

where ΔG_{exptl} is the experimental quantity and ΔG_{id} is the free energy calculated on the basis of an ideal mixture, viz.

$$\Delta G_{id} = RT\ (x_1\ ln\ x_1 + x_2\ ln\ x_2)$$

where x_i is the mole fraction of species i which, through Raoult's law is proportional to its partial pressure. Thus, positive deviations from Raoult's law are equivalent to $\Delta G^E > 0$. Similar excess functions can be written for the enthalpy, entropy, heat capacity, volume, tec. Mixtures can be conviently classified according to the disposition of their excess functions, as represented in Fig.4. Several significant features emerge :

(i) ΔG^E is usually symmetrical about $x_i = 0.5$,
(ii) both ΔH^E and $T\Delta S^E$ can exhibit complex dependence on x,
(iii) the sign of ΔG^E can be determined by that of ΔH^E (a, b, d) or by that of $T\Delta S^E$ (c and e).

The "normal" behaviour is that depicted in Fig. 4a, b and d, but there are classes of solute which in aqueous solution exhibit large negative ΔS^E values. They include nonpolar compounds and alkyl and aryl derivatives with only functional polar group : alcohols, ethers, amines, ketones, etc. The negative mixing entropy is explained in terms of structural change induced in water by nonpolar residues; the effect is normally referred to as hydrophobic hydration.

Structural analogues are believed to be the clathrate hydrates, in which solute (guest) molecules are contained in polyhedral cages composed of water molecules (host lattice). The water lattice resembles that of ice; each oxygen is tetrahedrally surrounded by four others. The next nearest neighbour geometry deviates slightly from that of ice, and this slight distortion can give rise to a multitude of similar cage structures. The phenomenon of hydrophobic hydration therefore arises because the introduction of an apolar molecule or residue leads to a reduction in the number of orientational degrees of freedom of the water molecules; the OH vectors can now only point along the sides of, or away from the cavity, but not towards the guest molecule. Hydrophobic hydration thus leads to a reduction in thermodynamic stability, and any process which can reverse this loss of orientational freedom will be spontaneous.

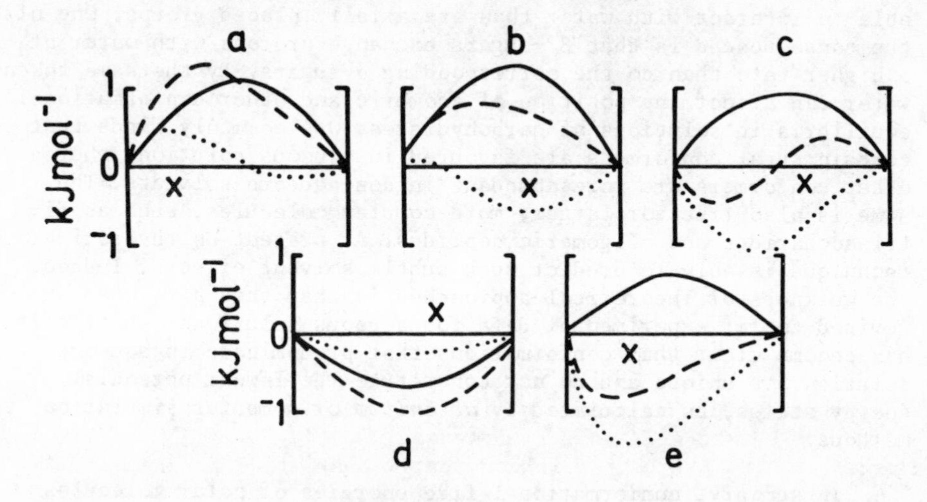

Fig. 4 : Thermodynamic excess functions for liquid mixtures: a) ethanol-benzene, b) methanol benzene, c) water-dioxan, d) water-hydrogen peroxide, e) water-ethanol. Solid lines : ΔG^E, broken lines : ΔH^E, dotted lines : $T\Delta S^E$.

Such a reversal can be achieved by the interaction of two such hydrated particles since some of the perturbed water will be able to relax back to its normal state. The net result is an apparent attraction between apolar residues in water, termed hydrophobic interaction, but this is unlike other interactions, in that it is of an entropic origin and results from the unfavourable structural arrangement which water molecules are forced to adopt in the proximity of an apolar residue. Recent statistical mechanical and simulation studies of the hydrophobic interaction suggest that $W_{ss}(r)$ is of a complex and long range nature, possibly with several minima and maxima. The biological consequences of hydrophobic interactions are the many supramolecular structures involving the large scale aggregation of apolar residues, as in micelles, viruses and biological membranes, and the complex folding processes which characterize globular proteins.

The interaction of water with polar molecules, especially those which themselves possess exchangeable protons is of an altogether different nature. The carbohydrates serve as good model compounds for such investigations. Hydration is dominated by solute- water hydrogen bonding, but there are subtle differences in the hydration behaviour of physically and chemically very similar molecules, such as a series of hexose sugars. The details of the hydration behaviour are determined by the spacings and orientations of hydroxyl groups and the compatibility of such spacings and orientations with those in liquid water. In general, equatorial hydroxy groups are better able to interact with water than are axially placed groups. One of the consequences is that β -sugars exchange protons with water at a higher rate than do the corresponding α-sugars. By the same token water can affect the position of anomeric and other conformational equilibria in solutions of carbohydrates. One commonly finds that certain sugar conformers are favoured in aqueous solution, whereas other conformers are more abundant in non-aqueous solvents. The same is also true for larger, more complex molecules, such as di-trisaccharides and oligomeric peptides. At present no theoretical technique is able to predict such subtle solvent effects. Indeed, the weakness of theoretical approaches is that they have been devised to fit experimental data for aqueous solutions, whereas it has become clear that conformations that predominate in aqueous solution are unique and do not constitute the lowest potential energy states, as calculated by *ab initio* or computer simulation methods.

In summary, conformational free energies of polar molecules in aqueous solution are markedly affected by hydration effects. These effects can lead to quite long-lived structures ($\simeq$1 ns), the detailed properties of which are intimately related to the spatial and orientational correlations between hydrogen bonding sites on the solute molecule and in water. Such subtle effects are also responsible for the tendency of carbohydrates to form gels which thermodynamically resemble supersaturated solutions. They owe their stability to complex interrelations between short range hydration, long range order of polymer domains, temperature and the method of preparation.

In principle, ions in solution can be described in molecular terms by the same techniques discussed above for non-electrolytes. The situation is complicated, because an electrolyte consists of at least two solute species : cation and anion, and this introduces even more terms into the total $g_{ij}(r)$. On the other hand, the existence of a dilute solution law in terms based on electrostatics (the Debye-Hückel theory) simplifies the extrapolation to infinite dilution. In all classical theories the solvent is treated as a continuum whose only significant property is a dielectric permittivity. Such treatments become inadequate where short range forces become important, and the past eight years have seen the development of more refined theories which allow for non-electrostatic factors, such as hydration and the overlap of ionic hydration shells which can lead to a net attraction or a net repulsion.

An important development has been the application of neutron diffraction methods to study the detailed geometry of ionic shells. The total radial distribution function of even the simplest ionic solution $M^+X^- \cdot H_2O$ contains ten $g_{ij}(r)$ terms which must be separately determined from a single experiment. Ion-water scattering contributes only up ot 10 % of the total scattering intensity, and ion-ion scattering even less. To overcome such practical problems, isotope difference scattering is now used, in which identical experiments are performed with two different isotopes of either M^+ or X^-. This reduces the number of radial distribution functions from ten to four. Figure 5 shows the radial distribution function for Ni^{++} in 1.5 molal $NiCl_2$ in D_2O.

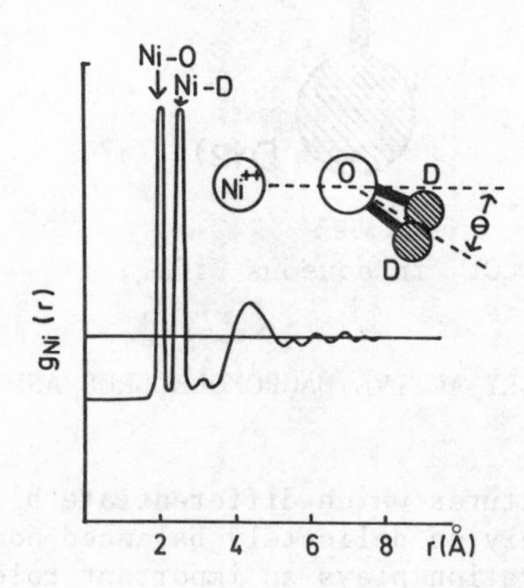

Fig. 5 : Radial distribution function for Ni in 1.5 molal $NiCl_2$ in D_2O, obtained by isotopic difference neutron scattering. Insert: hydration geometry.

The first peak corresponds to the Ni-O distance and the second to the Ni-D distance. The ratio of the areas under the two peaks is of course 1:2, and integration of the peaks yields a hydration number of 5.8 ± 0.2. From the position of the peaks and the geometry of the D_2O molecule it becomes clear that the D_2O molecules are tilted at an angle to the Ni-O axis; in very dilute solutions this angle tends to zero.

By performing a similar experiment, using two isotopes of chlorine it is possible to chart the hydration shell of Cl^-. Here again the hydration number is found to be approx. 6 and the water molecules are tilted, as shown in Fig. 6. In neither case can any structure be detected beyond 0.5 nm. Although many problems remain in the detailed interpretation of scattering data, the technique has already contributed significantly to the development of a new generation of electrolyte solution theories.

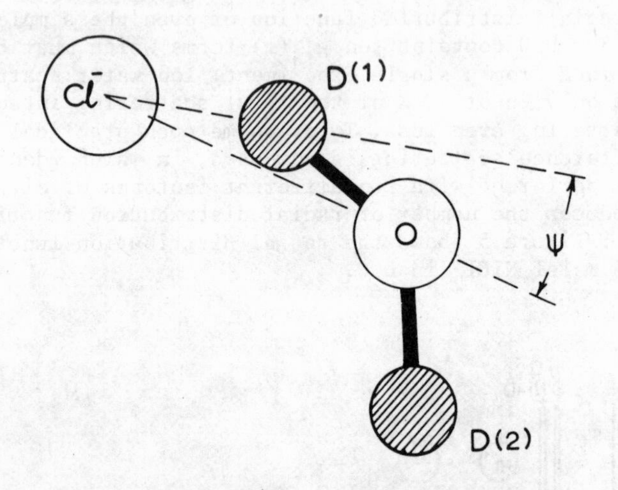

Fig. 6 : Hydration geometry of Cl^- in aqueous $NiCl_2$.

IV. THE STABILITY OF BIOLOGICALLY ACTIVE MACROMOLECULES AND SUPRAMOLECULAR STRUCTURES

The so-called native structures which differentiate biopolymers from synthetic polymers rely on delicately balanced noncovalent interactions in which hydration plays an important role. These effects are not limited to the innermost hydration shell but extend to water-water interactions at some distance from a polymer surface, so that protein crystals contain at least 40-50 % of water by weight. Even in highly idealized systems, such as an aqueous medium between two mica plates, hydration forces have been predicted and measured over separations approaching 5 nm. Beyond this distance the forces between the plates are well accounted for by classical theory (electrical double layer repulsion and van der Waals attraction). The newly discovered hydration forces introduce an additional repulsion term which decays in an exponential fashion with distance of separation, with typical correlation lengths of 2-3 molecular diameters; it is therefore a long range interaction. The role of hydration forces in maintaining native biological structures is not yet clear, but the dimensions of aqueous domains in biological systems are of the same order as the range of such forces.

Macromolecular hydration at short range has been studied by scattering and spectroscopy, and even at the macroscopic level it is gradually becoming clearer how the solvent medium affects the stability of macromolecules. This stability of a native state, which is highly marginal, is the resultant of a number of contributing factors, such as electrostatic repulsion or charged groups, conformational entropy, hydrogen bonding within the macromolecule and with the solvent, hydrophobic effects and dispersion interactions. Some of these effects are very large and they are difficult to calculate with any degree of confidence. In some cases, the interchange of just one amino acid residue in a protein can drastically alter its native state stability and solution behaviour (e.g. normal and sickle cell haemoglobin).

Just as the integrity of individual macromolecules is of crucial importance for the correct functioning of a living cell, so is the process of self-assembly of such molecules into supramolecular structures. Indeed, most enzymes consist of several -sometimes very many- subunits, each of which is a correctly folded polypeptide chain. The interactions which lead to the assembly of such large structures from individual globular polymers also involve delicately balanced hydration contributions. It is believed that the hydrophobic element plays a dominant role in self-assembly processes.

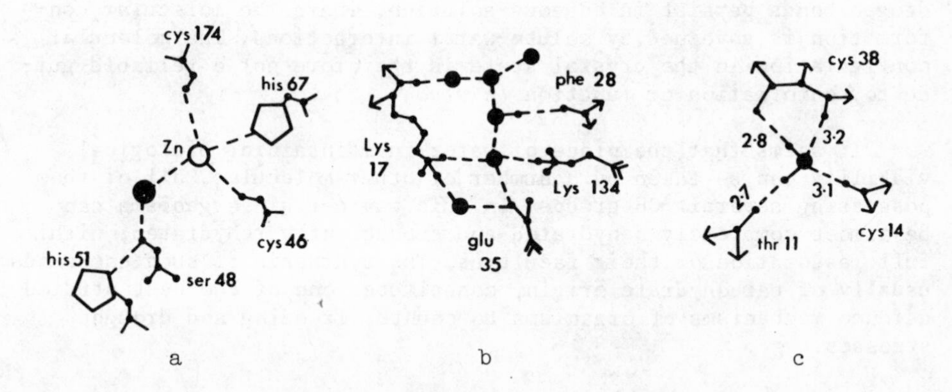

Fig. 7 : Internal water molecules in proteins : a) horse liver alcohol dehydrogenase, b) water mediated salt bridge in papain, c) four-coordinated water molecule in pancreatic trypsin inhibitor.

Water molecules in crystalline proteins are found in a number of distinct locations : internally they occur coordinated to metal ions, close to internal charged groups, such as COO^- and NH_3^+, or linking peptide bonds located at different places in the peptide chain. Figure 7 shows examples of such internal water molecules in proteins. These molecules possess a low diffusional mobility and

and can be quite well defined by diffraction methods. The opposite is the case for water molecules at the periphery of the macromolecule, but even here the existence of water bridges between protein molecules in crystals has been established; some of these bridges contain up to 12 water molecules. Other techniques which are currently contributing to an insight into protein hydration include H/D exchange kinetics, NMR relaxation and calorimetry. Attempts are also being made to treat the related problems of hydration, protein folding and native state stability by computer simulation methods, but the current state of the art is so rudimentary that such procedures cannot as yet be taken very seriously.

Most theoretical procedures as well as some experimental approaches, take as the starting point the known crystal structure of a macromolecule. The validity of extrapolating solid state conformations to a dilute aqueous solution, or even to the *in vivo* state, needs careful examination. In the case of globular proteins such extrapolations are perhaps permissible, because the structural integrity of such proteins arises mainly from intramolecular factors. However, even here, it must be remembered that interactions between molecules in the crystal are governed by packing considerations. This becomes very evident in the case of carbohydrates which in the solid state exhibit considerable hydrogen bonding, both intra and intermolecularly. It is most unlikely that such hydrogen bonds persist in aqueous solution, where the molecular conformation is governed by solute-water interactions. The molecular configuration in the crystal state is therefore not a reliable guide to conformation or function *in vivo*.

It seems that the place of water in maintaining biological viability can be taken by a number of other molecules, all of them possessing several -OH groups. In this way certain organisms can be almost completely dehydrated and subsequently rehydrated, with full restoration or their faculties. The synthesis of such compounds, usually of carbohydrate origin, constitutes one of the best studied defence mechanisms of organisms to counter freezing and drought stresses.

V. BOUND WATER

In systems possessing a large surface area/volume ratio, an appreciable proportion of the water present appears to have properties very different from those characteristic of bulk liquid water. This is of particular importance in biological and other colloidal systems, where it is well known that a proportion of the water does not freeze. Such unfreezable water can markedly affect the physical properties of the solid substrate and any chemical processes that take place in partly dehydrated systems. Bound water has been extensively studied by a variety of techniques. NMR relaxation seems to offer most promise, despite the fact that the interpretation of such relaxation data is still subject to some ambiguity. From a dynamic standpoint, three distinct populations of water molecules can generally be distinguished : the behaviour of a small fraction of the molecules is completely governed by the motions of the substrate, whether a bulk solid or a macromolecule in solution. The diffusive motions of a further, significant, proportion of water molecules are apparently perturbed, both as regards the rate and isotropy of diffusion, and finally, the motions of unperturbed water can be observed.

It must be clearly understood that unfreezable, or bound water is not primarily a manifestation of equilibrium phenomena; it originates from kinetic barriers to diffusion, rendering the system thermodynamically metastable. Nevertheless, the observed phenomena are of considerable physiological and technological importance and deserve detailed study.

All figures are reproduced, with permission of the authors, from *Water - A Comprehensive Treatise*, F. Franks ed., Plenum Press, New York, 1972-1979, Vols. 1-6.

BIBLIOGRAPHY

The properties of aqueous systems described in this article have been reviewed in depth in various volumes of *Water - a Comprehensive Treatise*, F. Franks editor, Plenum Press, New York, Vols. 1-6, 1972-1979; Vol. 7 in press. Hereafter, the relevant chapters.

KERN C.W. and KARPLUS M., The Water Molecule, Vol. I, p. 21.
FRANKS F., The Properties of Ice, Vol. I, p. 115.
BEN-NAIM A., Application of Statistical Mechanics in the Study of Liquid Water, Vol. I, p. 413.
FRANK H.S., Structural Models, Vol. I, p. 515.
FRANKS F. and REID D.S., Thermodynamics of Aqueous Solutions, Vol. II, p. 323.
ZEIDLER M.D., N.m.r. Spectroscopic Studies of Aqueous Solutions of Non-electrolytes., Vol. II, p. 529.
FRIEDMAN H.L. and KRISHNAN C.V., Thermodynamics of Ion Hydration, Vol. III, p. 1.
FRANKS F., The Hydrophobic Interaction. Vol. IV, p. 1.
SUGGETT A., Polysaccharides. Vol. IV, p. 519.
EAGLAND D., Nucleic Acids, Peptides and Proteins. Vol. IV, p. 305.

ENDERBY J.E. and NEILSON G.W., X-ray and Neutron Scattering by Aqueous Solutions of Electrolytes, Vol. VI, p. 1.
FINNEY J.L., The Organization and Function of Water in Protein Crystals, Vol. VI, p. 47.
CHAN D.Y.C. et al., Solvent Structure and Hydrophobic Solutions, Vol. VI, p. 239.
WOOD D.W., Computer Simulation of Water and Aqueous Solutions,Vol. VI, p. 279.

ON THE STRUCTURE OF GRAIN BOUNDARIES IN METALS

H. GLEITER

University of Saarbrücken,
D-6600 Saarbrücken
West Germany.

ABSTRACT

The first part of this paper reviews the atomistic model proposed for the equilibrium structure of grain boundaries in metals. These models are based on dislocation, coincidence, plane matching and polyhedral unit concepts. The basic ideas and shortcomings of the various concepts are discussed. The present understanding of the structure of dislocation and vacancies (non-equilibrium defects) in grain boundaries is summarized in the second part of the chapter. The existing evidence suggests that these defects have a similar structure in low energy grain boundaries as the analogous defects in the perfect lattice. However, in high energy boundaries, the defects may be smeared out (delocalized) in the plane of the boundary.

I. INTRODUCTION

The purpose of this chapter is to summarize the present state of understanding of the atomic structure of grain boundaries by highlighting results that provide most insight. For earlier work, the reader is referred to existing review articles and conference reports.[1-8]

II. EQUILIBRIUM GRAIN BOUNDARY STRUCTURE

II.1. Dislocation Models

The first dislocation model of a high angle grain boundary

B. Mutaftschiev (ed.), Interfacial Aspects of Phase Transformations, 199–222.

was proposed by READ and SHOCKLEY[9] for symmetrical high angle tilt boundaries. It was assumed that, if dislocations are uniformly spaced in the plane of the boundary, a low energy interface results, by analogy to the strain-field energy of a small angle tilt boundary. For all other tilt angles the boundary may be described as a boundary with a uniform dislocation spacing and a small angle tilt boundary superimposed that accounts for the deviation from the tilt angle required for uniformly spaced dislocations. This idea has also been used in subsequent years for boundary models that are not based on the dislocation concept (cf. section II.2). Two inherent limitations exist in the dislocation model developed by READ and SHOCKLEY. First, the singular behavior of elastic strain fields near dislocation centers was removed by an inner "cut-off" radius. The second deficiency is that linear superposition results in a complete neglect of the interactions among the dislocations in the array.

Li[10] first attempted to account for such effects. He proposed a hollow core dislocation model[11] of grain boundaries (Fig.1) and recognized that as the tilt misorientation increased, with its attendant decrease in dislocation spacing, the core shape changed.

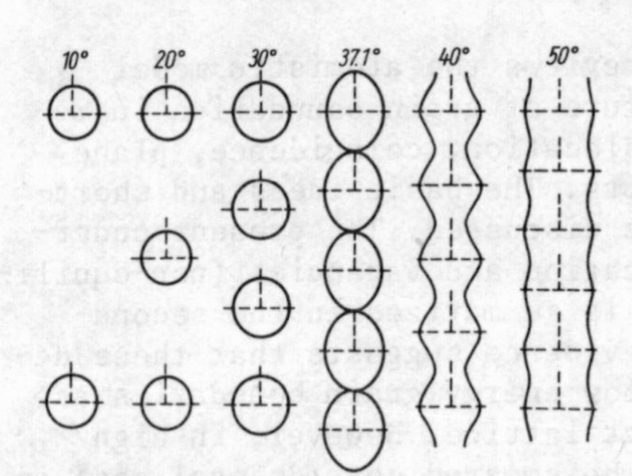

Fig. 1 : Calculated shape of the dislocation cores in tilt boundaries corresponding to different tilt angles. The tilt angles are indicated by the numbers above the figs.

Thus the traction-free core condition assumed in the model was violated. Li, therefore, extended the dislocation model by postulating that shape and radius of the actual core depended on the boundary structure. The result was an increase of the core radius and a change of the core shape with increasing misorientation (see Fig.1). To ameliorate some of the inherent deficiencies of the continuum dislocation array model, GLICKSMAN et al.[12,13] added a phenomenological core-energy term of thermodynamic origin. This led to the concept of the equilibrated "heterophase" dislocation model with a second phase in the core. A finite array of edge dislocations of this type was superimposed to obtain the elastic energy per dislocation. Included in a self-consistent manner was a chemical energy term for a core of circular cross-section. The total energy was minimized with respect to the core radius.

The linear superposition techniques used in the above analyses by definition neglect the elastic interactions occuring among the grain boundary dislocations. An attempt to include these interactions was made by MASAMURA and GLICKSMAN[13,14] using the heterophase dislocation model and solving the two-dimensional elastostatic-boundary value problem for an arbitrary periodic spacing between the dislocations. According to this model, liquid-core dislocations should be prevalent in boundaries at temperatures not too far below the equilibrium melting point. Measurements of the energy of high angle grain boundaries in bismuth at temperatures close to the melting point[12,15] seem to support this result (Fig. 2).

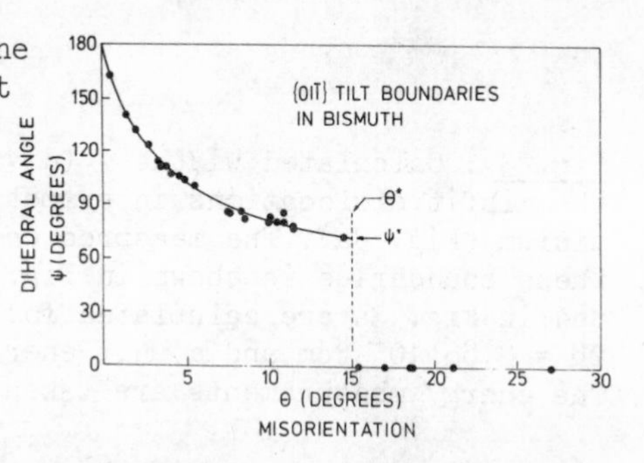

Fig. 2 : Dihedral angle (ψ) at the solid/melt interface at the point where a grain boundary of misorientation Θ terminates at that interface. The conspicuous discontinuity occuring at the cordinate (Θ^*, ψ^*) may be interpreted as a first-order phase transition in the boundary.[12]

Some of the inherent difficulties of the method used by GLICKSMAN et al have been discussed critically by JONES.[16] Recent observations[17] of interphase boundaries and the solid/liquid interphase in carbontetrabromide-hexachloroethane alloys gave little support for any discontinuity of the type shown in Fig. 2. The same conclusion was drawn from measurements of grain boundary energies in copper as a function of temperature.[18] If a phase transition due to the collapse of heterophase dislocations occurs, the boundary energy should become independent of the misorientation angle (Fig. 2) since a slab of liquid is formed in the boundary core. This was not observed.

In addition to the above models based on linear elasticity theory,non-linear dislocation models (PEIERLS model, FRENKEL-KONTOROWA model) have been proposed.[19-21] The results suggest[21] that the width of the core of a dislocation in a boundary depends on the boundary structure (Fig. 3). In boundaries of low energy, the cores of the dislocations are comparable in size to a lattice dislocation. With increasing deviation from the low energy structure the core diameter grows and, finally, spreads to infinity if the boundary energy is independent of the misorientation between the two crystals forming the interface. The physical reason for this effect will be discussed in section III.2. A number of experimental observations[12,23-33] support these views. Nevertheless, the problem is still a matter of controversy.[34-38]

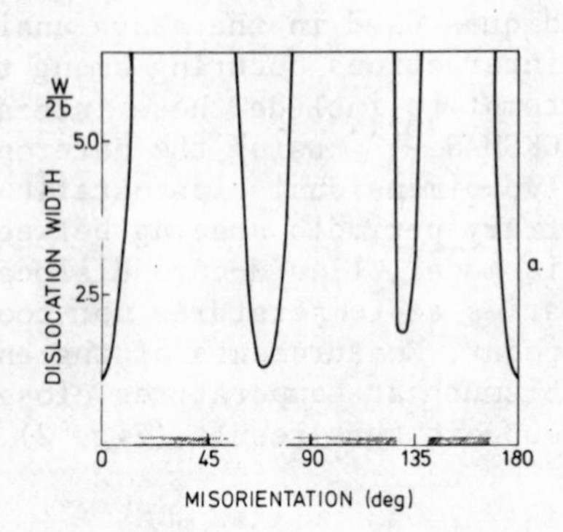

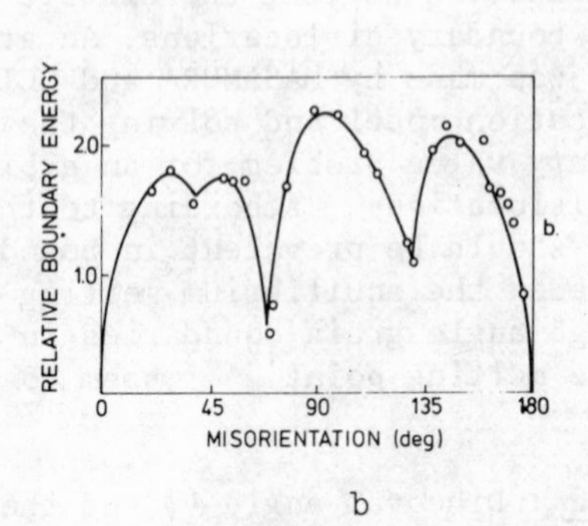

Fig. 3 : Calculated widths w (divided by the Burgers vector 2b) of the misfit dislocations in symmetric <110> tilt boundaries in aluminium (Fig. 3a). The measured energy misorientation curve of these boundaries is shown in Fig. 3b. The dislocation widths plotted in Fig. 3a are calculated for $G = 2.5\ 10^{11}$ dyn/cm^2, $2b = 2.86\ 10^{-8}$cm and a unit energy (1.0 in Fig. 3b) of 400 erg/cm^2. The energy measurements are taken from ref.22.

In addition to non-linear dislocation models, partial dislocation models[20] and disclination descriptions[39,40] have been considered to describe the structure and energy of grain boundaries (Fig. 4). Similar to all dislocation models proposed so far, the existing disclination models do not account for rigid body relaxations between the crystals (sectionII.2) forming the boundary.

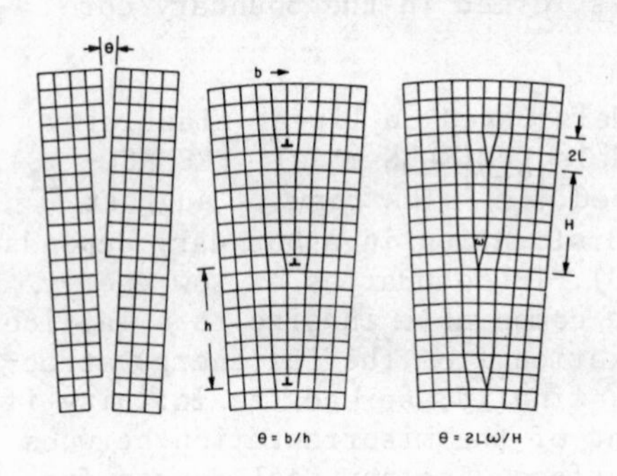

Fig. 4 : A simple tilt boundary made of dislocations or disclinations. The tilt angle is Θ and the Burgers vector of the dislocations[39] is b.

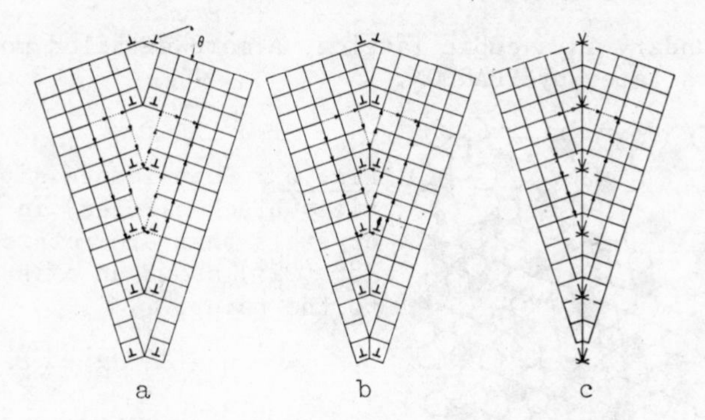

Fig. 5: a) Completely unrelaxed tilt type symmetric grain boundary of misorientation angle Θ = 36.9°;
b) partially coalesced modification of the grain boundary shown in (a)
c) total coalescence of the grain boundary[51] shown in a).

This concept was extended further by using a formalism[47-50] that describes the distorsion field, the tensor of Burgers vector density, the metric strain, and the torsion of a grain boundary by means of differential geometry. In a non-rigid lattice model[51] of a symmetric tilt boundary of arbitrary misorientation, a grain boundary was portrayed in terms of an array of voids and/or asymmetric cracks upon whose surfaces are distributed surface dislocations. The boundary shown in Fig. 5a contains no elastic energy and consists of large voids which, in turn, correspond to a high surface energy. The total amount of free surface energy can be reduced by a partial (Fig. 5b) or total (Fig. 5c) coalescence. In the case of total coalescence (Fig. 5c), the elastic distorsion is a maximum and the surface energy is a minimum. Real crystals exhibit a behaviour somewhere in between the two extremes as represented in Fig. 5b. The calculations indicate that a minimum energy configuration of the boundary is obtained if the boundary dislocations are spread out in the plane of the boundary. As pointed out above, the delocalization of boundary dislocations is also suggested by other arguments.

II.2. Coincidence Models

The pioneering work of KRONBERG and WILSON[52] led to the concept of coincidence sites in grain boundaries. This concept has since been adopted by several other workers[53-56] and developed in more detail. For example, Fig. 6 shows the coincidence site lattice formed by two hexagonal arrays of atoms. ARKHAROV[57] developed the first model of a grain boundary in which an attempt was made to interpret the structure and the properties of grain boundaries on the basis of the coincidence site concept. He considered a

[100] tilt boundary in a cubic lattice. A more detailed model was given some years later by BRANDON.

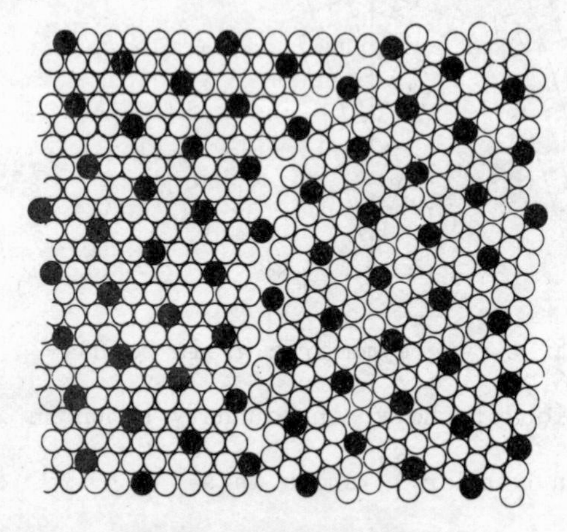

Fig. 6 : Coincidence site lattice (black circles) in two crystals that are rotated by 38° (22°) about an axis normal to the paper.[52]

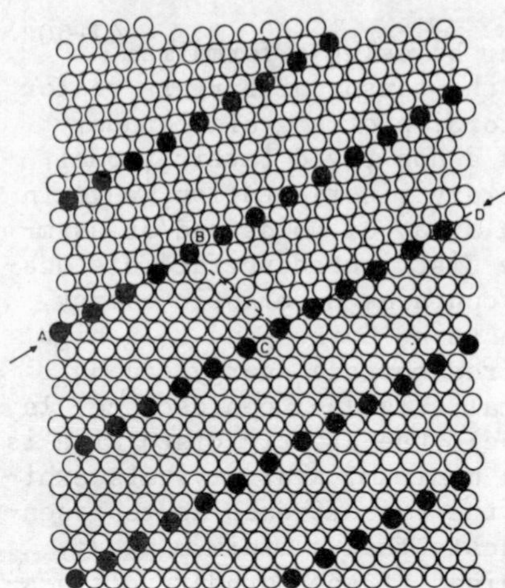

Fig. 7 : Structure of a grain boundary according to the model of BRANDON et al.[54,58] The boundary is a 50.5° <110> tilt boundary between two crystals that have b.c.c. structure. The sections AB and CD of the boundary lie in a plane of high coincidence density. (The atoms represented by black circles occupy coincidence sites.)

BRANDON's model[54,58] represents an extension and combination of the coincidence lattice concept[52] and the dislocation model[9,11] (cf. sectionII.1). The coincidence theory originally proposed (Fig. 6) brings out the dependence of the boundary structure on the orientation relationship between both crystals, but ignores the effect of boundary inclination. The effect of boundary inclination was incorporated by suggesting that boundaries constrained to lie at an angle to the densely-packed coincidence plane take up a step structure (Fig. 7). In the subsequent development of the atomistic models of grain boundaries, attention was focused essentially on the following two problems : (i)the structural characterization of low energy boundaries and (ii) the development of a general crystallographic theory of grain boundaries. These two problems led to two separate lines of development in the understanding of the

structure of grain boundaries. Let us first consider the relevant ideas to improve understanding of the characteristic features of low energy boundary structures.
The concept of atoms occupying coincidence sites (in terms of boundary or lattice coincidences) had to be abondoned after it was recognized from computer simulations of the atomic structure of grain boundaries[59-61] and from bubble raft experiments[62] that two crystals forming a boundary relax by a shear type displacement (rigid-body shear) from the position required for the existence of coincidence site atoms at the boundary (Fig. 8). This result was confirmed in subsequent years by other more sophisticated computer work[63-65] as well as by experimental observations.[66,68]

Fig. 8 : 18° symmetric tilt boundary displaying six-fold corrodinated units. Shaded atoms represent structural units.[60]

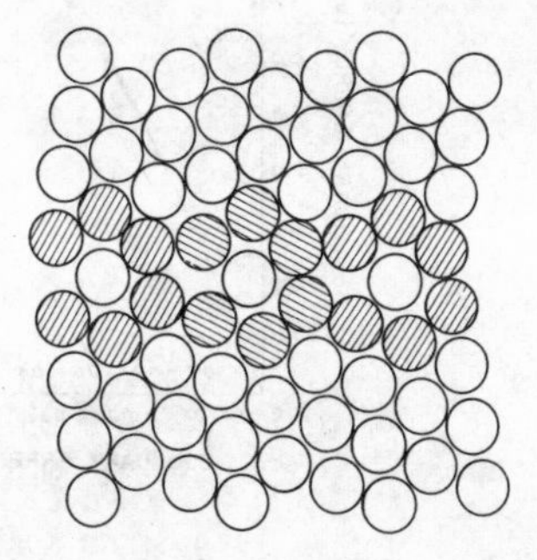

The existence of rigid body relaxations led to the conclusion[69] that boundary periodicity rather than the existence of boundary coincidence *per se* is the physical meaningful parameter. However, recent measurements[70,71] of the orientation relationships corresponding to low energy boundaries in noble metals and noble metal alloys, indicate[71,72] that the boundary periodicity may not be the only parameter that controls the energy of a grain boundary. The low energy boundaries observed experimentally suggest a division of low energy boundaries into two groups[70]: "electron sensitive" and "electron insensitive" boundaries. The "electron insensitive" boundaries are those observed in all metals and alloys of the same lattice structure irrespective of alloy composition. The "electron sensitive" boundaries are observed only in some pure metals or in alloys of certain compositions. On this basis, it was proposed that the energy (and structure) of the "electron insensitive" boundaries is controlled primarily by the geometry of the atomic arrangement in the boundary (atomic packing density similar to the lattice), whereas the energy of the second group depends primarily on the electron band structure of the material. Examples of "electron insensitive" boundaries in Ag, Au, Cu are the 70.5° [110] or the 50.5° [110] boundaries. The 59° [110] or the 81° [110] boundaries belong to the second category. The significance of the electronic contribution to the energy of grain boundaries was pointed out by several authors[73–74] on the basis of free electron calculations. In fact, SEEGER and SCHOTTKY showed (Fig. 9) that the screening of the positive charge deficit at a grain boundary by the conduction electrons may be the dominant part of the boundary energy.

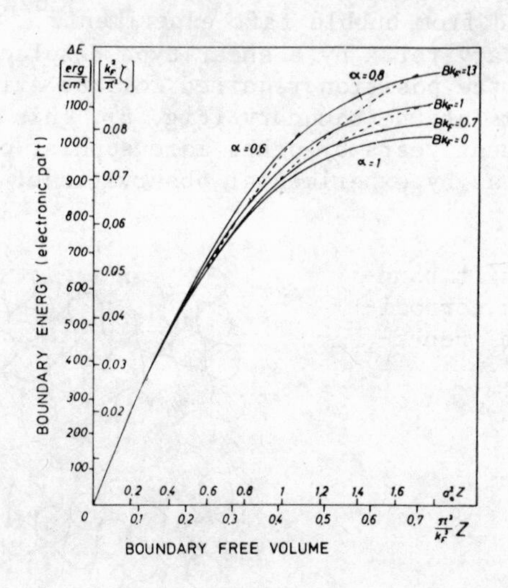

Fig. 9 : Electronic contribution (ΔE) to the boundary energy due to electron screening effects (of the positive charge deficit in the boundary) as a function of the boundary free volume (positive charge deficit) for a free electron gas model. Z is the charge per atom, a is the lattice constant, k_F is the wave vector at the Fermi energy ζ, a^2Z is the positive charge deficit per unit area of the boundary. The numbers given on the energy axis are calculated for silver ($a_o = 4.078\ 10^{-8}$ cm, $k_F = 1.204\ 10^{-8}\text{cm}^{-1}$, $m^* = m$). The boundary is represented as a square well potential of width 2B and hight U_o. The broken curves are computed[73] for various heights of the potential well ($\alpha = U_o/\zeta$).

The considerations discussed so far have neglected the effect of temperature, pressure etc. on the boundary structure. There is substantial evidence in the literature[18,75–82] suggesting *structural phase transformations* in grain boundaries similar to the well known phase transformations of free surfaces.[83] Figs. 10a and 10b show (hypothetically) how such a transformation may be visualized. Due to the smaller free volume of the boundary structure shown in Fig. 10a, the Gibbs free energy of this structure is lower at high pressures than the Gibbs free energy of a more open structure (Fig. 10b). Therefore, at a certain pressure, the boundary structure may transform from the more open structure into the densely packed structure as the pressure is increased.[18,81]

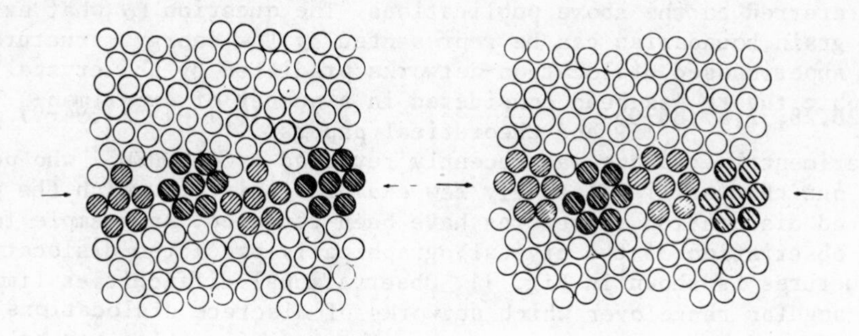

Fig. 10 : Schematic model of the atomic arrangement in a grain boundary due to a phase transformation for a boundary between two hexagonal arrays of atoms.

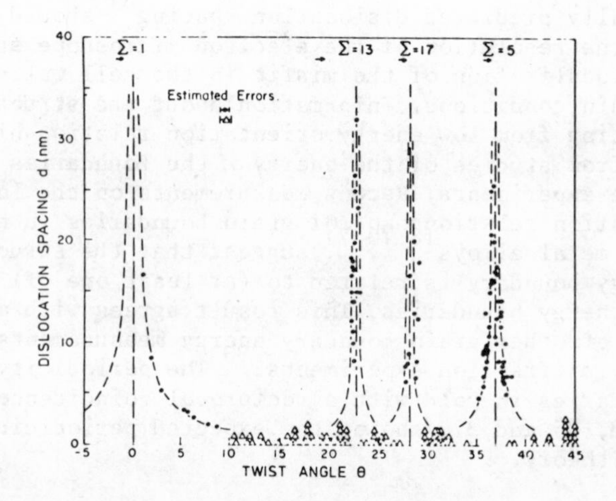

Fig. 11 : Spacing d of the secondary grain boundary dislocations found in <001> twist boundaries in gold. The dashed curves were calculated from the crystallographic theory. The filled circles represent the measured spacings. The open triangles represent cases where dislocation networks were not detected.[88]

The *crystallographic theory* of grain boundaries was pioneered by the work of BOLLMANN[84-86] as the O-lattice theory. This theory is a mathematical method for calculating the crystallographic structure of interfaces and the permissible Burgers vectors of the perfect interfacial dislocations between two arbitrary crystals in arbitrary relative orientation for any position of the boundary between them. For the derivation of relevant equations the reader is referred to the above publications. The question to what extent all grain boundaries can be represented by low energy structures and superimposed dislocation networks predicted by the crystallographic theory has been considered in a number of experimental[26,28,32,87,88-93] and theoretical papers.[10,12-15,21,94-97] The experimental evidence was recently reviewed by GOODHEW[81] who pointed out that still relatively few examples exist in which the predicted dislocation structures have been revealed. An example for the observation of the crystallographically predicted dislocation structures is shown in Fig. 11. Observational difficulties limit the angular range over which networks of discrete dislocations can be detected by electron microscopy. These difficulties may be due to the small Burgers vectors of grain boundary dislocations or to wide cores (and, thus, a weak strain field) as shown by the analysis of the contrast width.[24,28-38] The importance of the latter effect was emphasized by observations on misfit dislocations in the interfaces between α/β brass.[26] The dislocation network in such interfaces became invisible upon small rotations although the crystallographically predicted dislocation spacing[84] should have been well above the resolution of the electron microscope suggesting that a non-localization of the misfit in the cell walls may occur under certain conditions. Information about the structure of boundaries deviating from low energy orientation relationships may also be derived from studies of the energy of the boundaries and X-ray diffraction experiments. Recent measurements on the low energy misorientation relationships of grain boundaries in noble metals and noble metal alloys[18,70,71] suggest that the structure of any high energy boundary is related to (at least one of) the neighboring low energy boundaries. This result agrees with a considerable number of other grain boundary energy measurements and with recent X-ray diffraction experiments.[98] The periodicity of |001| twist boundaries in gold with a reciprocal coincidence site density of $\Sigma = 13$, 85 and 377 showed the expected periodicity from crystallographic theory.

Whether or not a given grain boundary will tend to preserve a structure characteristic of a given coincidence orientation relationship depends on the energies of alternative possible boundary structures. Two examples have been documented for $\Sigma = 5$ and $\Sigma = 9$ boundaries, where the manner in which a coincidence structure is maintained, depended on the boundary orientation.[99,100] In particular, for the $\Sigma = 5$ boundary, a change of boundary plane favored a change from minimum length Burgers vectors to crystal lattice Burgers vectors. Therefore, the simple b^2-rule (Frank-rule) is not generally applicable for dislocations in grain boundaries, although this rule has been used in the literature.

II.3. Plane Matching Models

The electron microscopy of high angle grain boundaries has shown that the images of grain boundaries frequently exhibit periodic linear features, which have been demonstrated not to be Moiré fringes. These observations prompted PUMPHREY[101] to explain the periodic line structures on the basis of a plane matching approach. The situation is shown schematically in Fig. 12. The overall array of traces shown in Fig. 12 produces a Moiré pattern consisting of parallel bands of high opacity. These bands are interpreted as poor atomic fit along the dashed lines in Fig. 12. Between these bands are corresponding bands of high transparency (associated with good atomic matching).

Fig. 12 : Slightly mismatched traces of two sets of lattice planes (dark lines) in the boundary. The two sets of planes are assumed to impinge from the two adjoining crystals.[101]

PUMPHREY suggested that atomic relaxation occurs in the vicinity of these Moiré bands. This results in strain fields that are imaged by the electron microscope as a single set of parallel lines with the same geometry as the bands. The defects that may result in grain boundaries if planar matching is dominant have been worked out by RALPH et al.[102] BALLUFFI and SCHOBER[103] have discussed the relationship between the plane matching approach and the description of a grain boundary by means of crystallographic theory (cf. section II.2). In principle, a plane matching structure may always be accounted for (at least geometrically) in terms of the boundary dislocation network given by the crystallographic theory. Geometrically, the plane matching representation may therefore not represent an independent model of the boundary structure. The difference between the two approaches lies in the physical significance associated with the boundary dislocation structure.

As an example, Figure 13[104] shows a pattern of parallel lines in a grain boundary with the misorientation 10.4°[8.5, 9.3, 1.0]. It is difficult to interpret these lines in terms of grain boundary dislocations since the boundary shown deviates more than 10° from the nearest high density coincidence orientation ($\Sigma = 19$). If the limit $\Sigma = 100$ is imposed, the closest CSL is $\Sigma = 51$ resulting in an array of $\frac{a}{102}$ [1,1,10] edge dislocations with a spacing of about one lattice constant. They are therefore probably not physically significant. It was concluded that boundaries may exist which show dislocation type defects but do not maintain a CSL interface. The observed spacing and orientation of the lines shown in Fig. 13 agrees, however, with those predicted by the plane matching model from the mismatch of the (110) planes. Recently, independent evidence for the possible existence of plane matching structures in grain boundaries has been reported for several other systems.[105,106]

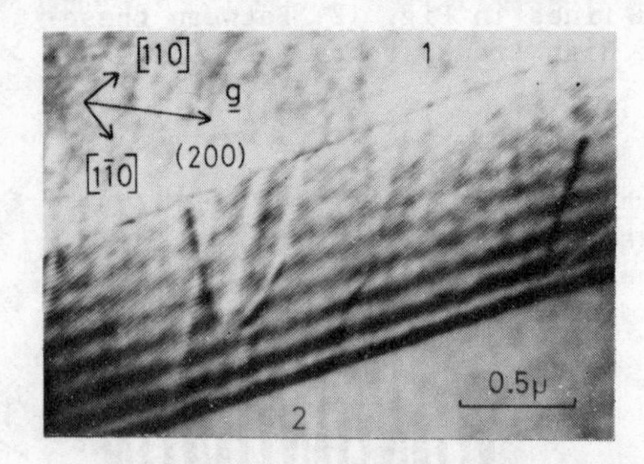

Fig. 13 : Grain boundary in recrystallized Al 0.9 wt/o Mg (cold rolled 92 %, annealed 30 min. at 350°C). The spacing of the periodic lines[104] is 185 Å.

II.4. Polyhedral Unit Models

The basic idea of the polyhedral unit models is that the structure and properties of a grain boundary may be described in terms of a two-dimensional array of one or several types of atomic configurations (also termed as atomic clusters or polyhedral units).

The idea of describing a boundary this way was first suggested on the basis of observations by field ion microscopy.[54] The low energy of such a configuration was rationalized in terms of a two-dimensional array of an atomic configuration containing a coincidence atom plus the surrounding atoms.

With the advent of computer simulations of the atomic structures of grain boundaries it was recognized[59-61] that atoms occupying coincidence sites may not exist in coincidence grain boundaries (cf. section III.2). This result prompted WEINS et al.[60] to suggest that the structure of a grain boundary may be described by a two-dimensional periodic pattern of characteristic atomic groups consisting of a central atom surrounded by five, six or seven atoms (Fig. 14). In terms of this model, the structure of a general grain boundary was interpreted as a combination of several types of these units. In recent computer calculations of the structure of [100] symmetrical tilt boundaries in aluminium, close packed triangles of atoms (which stack to give trigonal prisms) were found in low energy relaxed structures.[107] It was therefore proposed that a low energy structure of a boundary may be associated with the existence of planar, close packed groups of atoms (Fig. 15).

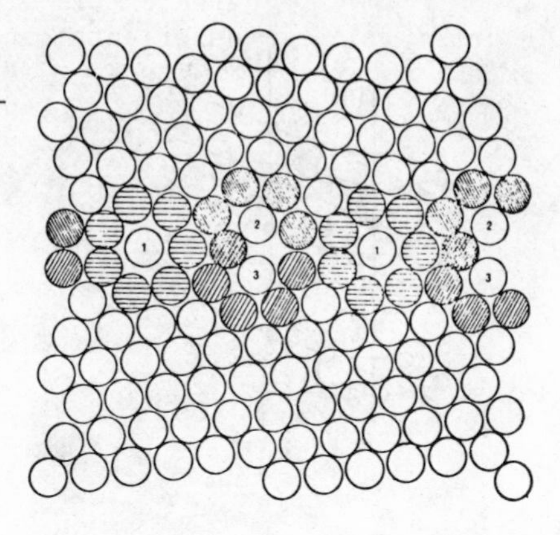

Fig. 14 : 37.8° symmetric tilt boundary displacing the coordinated (polyhedral) units (1 = sevenfold unit, 2 = five-fold unit and 3 = five-fold unit). Shaded atoms represent the structural units.[60]

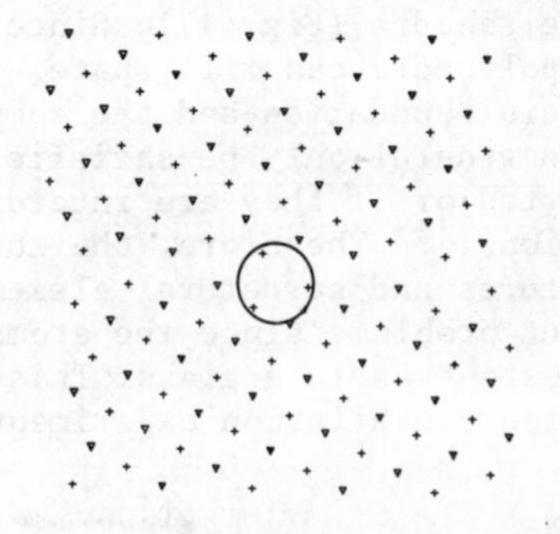

Fig. 15 : Triangular group of atoms (circle) in a (210) symmetrical tilt boundary.[107]

Similar behavior apparently occurs in a covalently bound material such as germanium. It was found, by high resolution electron microscopy,[106] that tetragonal configuration is preserved as far across the boundary as possible. Comparison of the atomic arrangements in a grain boundary with the atomic arrangements in amorphous structures was apparently, first proposed by POTAPOV et al.[108] and AARON and BOLLING.[109,110] A 38° [110] boundary was found[107] to consist of periodically arranged rings formed by five atoms with a central atom between them (Fig. 16). On the basis of these observations it was concluded that a grain boundary may be represented in terms of atomic configurations existing in amorphous metals. In the work of AARON and BOLLING,[109,110] the free (excess) volume of a grain boundary was explicitly calculated for dislocation boundaries and boundary structures formed by random close packed units. Description of the structure of a grain boundary in terms of the packing of certain of the 8 basic deltahedra that result when equal spheres are packed to form a shell, such that all spheres touch their neighbors was developed in detail by ASHBY et al.[112,113]

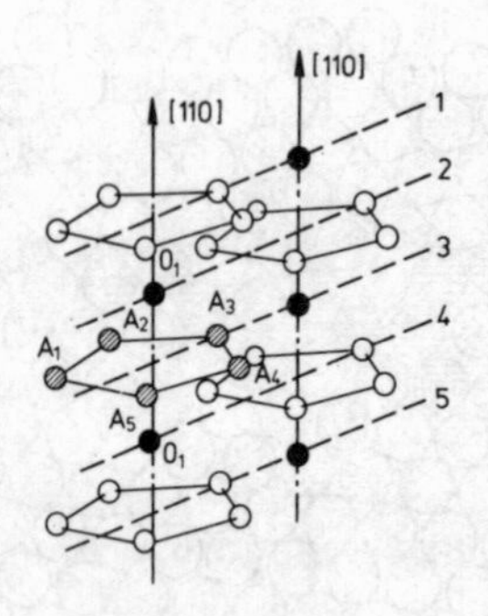

Fig. 16 : Schematic diagram of the arrangement of the atoms in a 40° <110> tilt boundary in W derived from a sequence of field ion microscopy images. The position of the <110> common tilt axis in the two grains is indicated. 1,2,3,4 and 5 indicate subsequent layers of the boundary. The polyhedral rings proposed by POTAPOV et al.[108] are labelled by letters.

It was found in all of these cases that the structure may be described completely and uniquely as nesting stacks of certain of the deltahedra (Fig. 17). Since only very special arrangements of such polyhedra can fill space, the requirement of space filling at grain boundaries and the compatibility with the adjoining grains can in general only be satisfied if the polyhedra are elastically distorted or if they are interdispersed with other atomic configurations.[114] Therefore, the comparison between grain boundary structures and structural elements of amorphous materials is not without problems since the atoms in a boundary cannot relax to the same extent as in a glass. This difference is also born out by positron annihilation experiments[115] that suggest atomic packing

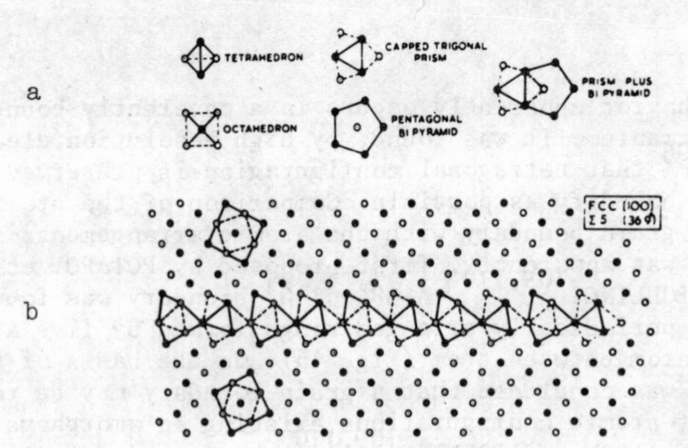

Fig. 17 : Demonstration of graphically constructed boundaries in terms of polyhedral units.[111,112] (a) Undistorted projections of the units appearing in the following diagram. Octahedra and tetrahedra are not shown in subsequent diagram. (b) A $\Sigma = 5[100]$ tilt boundary between f.c.c. crystals.

in a grain boundary is more "open" than in glassy structures. This conclusion is supported by Mössbauer measurements on grain boundaries in Zn, Fe and Al.[116-118]

III. STRUCTURAL (NON-EQUILIBRIUM) DEFECTS IN GRAIN BOUNDARIES

This section is concerned with the atomic structure of point defects and dislocations that are not part of the equilibrium structure of the boundary.

III.1. Point Defects

The attempts that have been made so far to study the atomic structure of vacancies in grain boundaries are based on computer simulation techniques[7,119-125] and dynamic hard sphere models.[7,122] Depending on the atomic structure of the "vacancy-free" boundary, two kinds of vacancy structures were obtained from computations.[123-125]

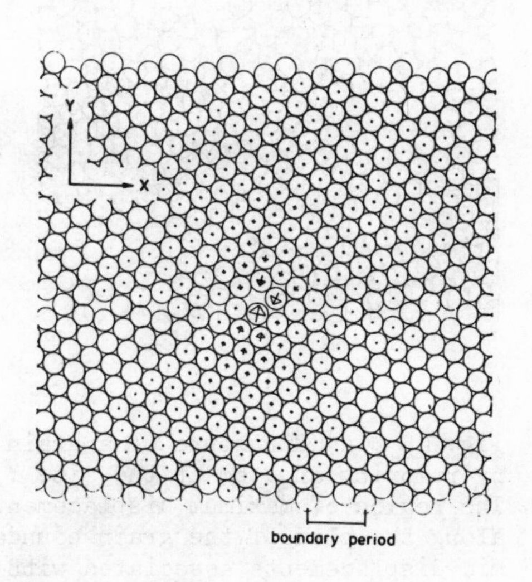

Fig. 18 : Atomistic structure of a vacancy in a symmetrical $\Sigma = 7$ tilt boundary between two hexagonal arrays of atoms. The arrows indicate the displacement of the atoms relative to the positions occupied before the vacancy was introduced. The largest and smallest arrows correspond to displacements of 25.5 % and 0.643 % of the interatomic spacing respectively. The arrows scale linearly. Third neighbours were included in the calculation.[124]

1) Vacancies in short periodic boundaries (Fig. 18) of good fit may be described as an empty site in the boundary (localized vacancy) surrounded by a displacement field consisting of cone-shaped sectors, the tips lying approximately at the site of the vacancy (Fig. 18). The relaxation decays in each cone with increasing distance from the vacancy. The structure of the boundary some lattice constants away from the vacancy remains essentially unaltered due to this decay.

2) The generation of a vacancy in such a random boundary induces

a displacement field composed of the following three components :

(a) A displacement field similar to the cone-shaped sectors observed in short periodic boundaries of good atomic fit (Fig. 18).

(b) A group comprising several boundary atoms in the "vacancy-core" rearranges by about one interatomic spacing, e.g. the group shown in Fig. 19a (marked by heavy arrows) comprised of four atoms.

(c) Displacement fields in boundary regions several interatomic spacings away from the site at which the vacancy was introduced e.g. the displacement fields on the right and left section of Figs; 19a and 19b.

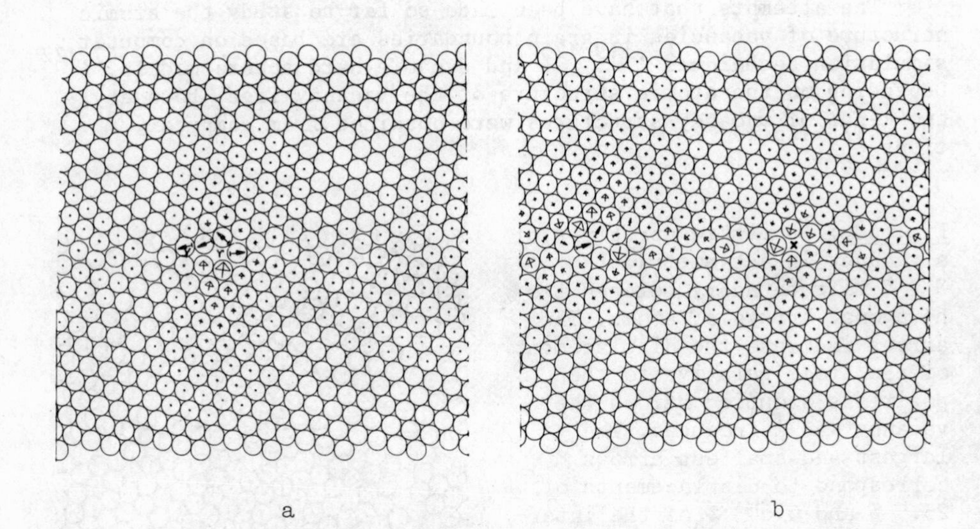

a b

Fig. 19 : a) Structure of a grain boundary vacancy that resulted by removing an atom at the site Y from the vacancy-free boundary. The region of maximum displacements ("vacancy core") is extended along the plan of the grain boundary. The arrows indicate the atomic displacements associated with the generation of the vacancy. The heavy arrows correlate directly with the displacements, the light arrows scale linearly with the displacements. The largest and smallest one corresponds to displacements of 0.164a and 0.082a (a = interatomic distance). The atomic relaxation in the vicinity of the vacancy displaces the hole generated by removing an atom from Y to Y'.[124] b) Grain boundary vacancy structure generated by removing an atom at the site *X* of the vacancy-free boundary. The regions of maximum displacements are extended over a wide boundary region far away from *X*, e.g. one region is in the vicinity of *X*, one region is in the right part of the figure. The arrows indicating the displacements scale in the same way as in (a).[124]

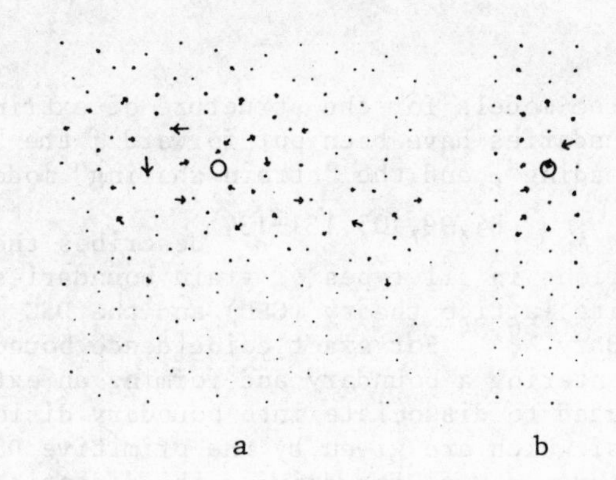

Fig. 20 : Displacement field obtained by computer simulation model around a vacancy in a [001] twist boundary ($\Sigma = 5$, $\theta = 36.87°$) in b.c.c. tungsten. (a) and (b), plan and edge-on views of atomic relaxations around the vacancy. Each atomic relaxation represented by a vector displacement projected on the plane of the paper.[122]

BRISTOW et al.[122] applied a computer simulation method to study vacancy structures in three-dimensional crystals. Fig. 20 shows the relaxation around a vacancy in a twist boundary in tungsten. In this case relatively large displacements occured around the vacancy and it became essentially delocalized. A large variety of other relaxed vacancy structures were found including cases where the vacancy was "semi-localized" or split depending upon the boundary structure,the site of the vacancy in the boundary, and the interatomic potential. These results agree with observations by DAHL et al.,[121] NABEREZHNYKH et al.,[126] and with the experimentally observed high efficiency of high energy grain boundaries as vacancy sources whereas in boundaries of good atomic fit an energy barrier was found to exist for emission of vacancies.[125,127-129]

Investigations by bubble raft and hard sphere dynamic models[122] as well as boundary modelling of colloidal systems[130] apparently support the above results.

The physical reason for the different types of vacancy structures observed in boundaries of good and poor atomic fit may be understood as follows. The free volume of a vacancy in a random grain boundary is reduced by a translational motion of the two adjacent crystals towards the boundary. This motion is made possible by atomic displacements described in the core of the vacancy and in boundary areas far away from this defect. In the case of a short-periodic boundary of good atomic fit, such displacements would destroy the good fit (low energy) structure of the boundary.

III.2 Dislocations

Essentially, three models for the structure of extrinsic dislocation in grain boundaries have been put forward : the "dissociation", the "core spreading", and the "strain sharing" model.

The *Dissociation Model*[84,99,105,131-134] describes the extrinsic boundary dislocations in all types of grain boundaries in terms of the coincidence site lattice theory (CSL) and the DSC dislocation model of a boundary.[56,84] For exact coincidence boundaries, a lattice dislocation entering a boundary and forming an extrinsic dislocation is envisaged to dissociate into boundary dislocations, the Burgers vectors of which are given by the primitive DSC lattice vectors. The major energy source for driving the dissociation reaction is believed to be the reduction of the elastic energy associated with the strain field of the dislocations.

Fig. 21 : Splitting of a lattice dislocation into five boundary dislocations in a $\Sigma = 29$ boundary in stainless steel according to the equation $1/2\,[1,1,0] = 3(1/58)\,[3,7,0] + 2(1/58)\,[10,4,0]$.

If a boundary deviates from the exact coincidence lattice orientation, the dissociation model assumes the boundary contains an (equilibrated) network of (secondary) boundary dislocations. The incorporation of an extrinsic dislocation would occur again by dissociation into boundary dislocations, the Burgers vectors of which are given by the DSC lattice. An example is shown in Fig. 21. Despite the attractive features of the dislocation model, it is not obvious that it is the most physically realistic model for all types of boundaries due to the narrow dislocation spacing and the small Burgers vectors.

The *Dislocation Core Delocalization Model* suggests that the structure of an extrinsic dislocation depends on the boundary structure. In boundaries with well defined boundary dislocations, the incorporation of an extrinsic dislocation is assumed to occur by a dislocation reaction of the type described above. However, for high energy boundaries, the reaction is proposed to occur by the widening (delocalization) of the cores of the dislocations [21,26,33,135,136] in the plane of the boundary (Fig. 22).

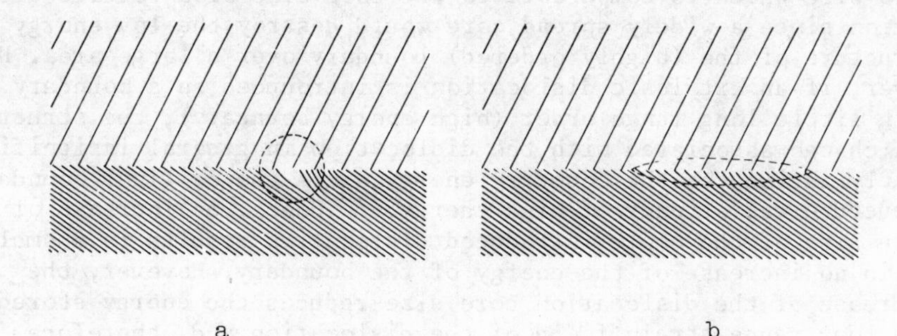

Fig. 22 : Schematic diagram showing the structure of a localized (a) and a delocalized (b) boundary dislocation. For simplicity, only one set of lattice planes of the two adjoining crystals forming the boundary is shown. The approximate size of the dislocation core is indicated by the broken lines.[135]

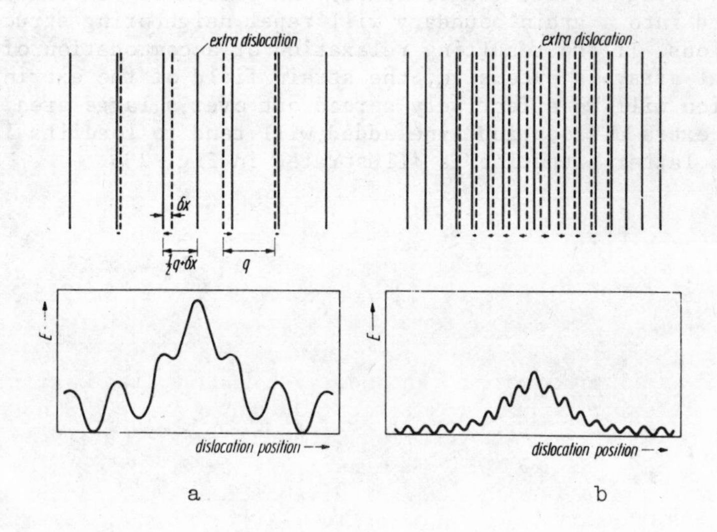

Fig. 23 : Schematic illustration[87] of the spreading of the elastic strain field (E) of an extra dislocation (marked) when added to a network of parallel boundary dislocations with the same Burgers vector. The added dislocation/final dislocation positions are shown by solid vertical lines, the original positions by broken lines. In (b) the density of structural dislocations is greater than in (a).

The physical reason for spreading of the core of extrinsic boundary dislocations seems to be the reduction of the energy of the extrinsic boundary dislocation. Hence, an extrinsic boundary dislocation in a highly ordered, low energy boundary is expected to have a core size which is comparable to the core size of a lattice dislocation since a widely spread core would destroy the low energy structure of the (highly ordered) boundary over a large area. However, if an extrinsic dislocation is introduced in a boundary with little long range order (high energy boundary), the structural change associated with the dislocation in general insignificantly influences the boundary energy since the original boundary structure was already of high energy.[140] Hence, an increase of the core size of the newly introduced dislocation results in a small or in no increase of the energy of the boundary. Hovever, the increase of the dislocation core size reduces the energy stored in the long range strain field of the dislocation and, therefore, the total energy of the system is reduced if the core of the dislocation is widely spread in the plane of the boundary (Figs. 22a,b).

The *Strain Sharing Model*[87] is based on the visiblility of extrinsic dislocations in symmetrical [110] tilt boundaries in aluminium. The observed behavior was interpreted in terms of a "strain sharing" effect (Figs. 23a,b). An extrinsic dislocation introduced into a grain boundary will repel neighboring structural dislocations. If the resulting relaxation or accommodation of the structural array is extensive, the strain field of the extrinsic dislocation will be effectively spread out over a large area. The original extra dislocation line added will tend to lose its identity. The latter situation is illustrated in Fig. 23b.

REFERENCES

1 . GLEITER H. and CHALMERS B., Prog. Mat. Sci. 16 (1971)
2 . GLEITER H., Phys. Stat. Sol. (b) 45(1971)9
3 . *The Nature and Behavior of Grain Boundaries*, H. Hu ed., Plenum Press, New York (1952)
4 . CHAUDHARI P. and MATTHEWS J.W. eds, Surf. Sci. 31(1972)
5 . *Grain Boundary Structure and Properties*, G.A. Chadwich and D.A. Smith eds., Academic Press, New York (1976)
6 . *Interfacial Segregation*, Johnson W.C. and Blakely J.M. eds., ASM, Metals Park, Ohio (1979)
7 . BALLUFFI R.W. ed., ASM Materials Science Seminar : *Grain Boundary Structure and Kinetics*, 15-16 Sept. 1979, Milwaukee, WI, ASM, Metals Park, Ohio (1980)
8 . GLEITER H., Materials Sci. and Eng., in press.
9 . READ W.I. and SHOCKLEY W., Phys. Rev. 78(1950)275
10. LI J.C.M.,J. Appl. Phys. 32(1961)525
11. FRANK F.C., Acta Cryst. 4(1951)497
12. GLICKSMAN M.E. and VOLD C.L., Surf. Sci. 31(1972)50
13. MASAMURA R.A. and GLICKSMAN M.E., Can. Met. Quart. 13(1974)43
14. MASAMURA R.A. and GLICKSMAN M.E., NRL Report 7851, Naval Research Laboratory, Washington, D.C. (1975)
15. VOLD C.L. and GLICKSMAN M.E., in ref. 4 p. 171
16. JONES D.R.H., J. Mat. Sci. 4(1974)1
17. KAUKLER W., Ph. D. Thesis, University of Toronto (1981)
18. ERB U. and GLEITER H., Scripta Met. 13(1979)61
19. SEEGER A. and HORNING R., quoted in A. Seeger, *Handbuch der Physik*, Vol. VII/1, Springer Verlag, Berlin, 1955, p. 654
20. van der MERWE J.H., Proc. Phys. Soc. A63(1950)616
21. GLEITER H., Scripta Met. 11(1977)305
22. HASSON C.G. and GOUX C., Scripta Met. 5(1971)889
23. KHALFALLA O. and PRIESTER L., Scripta Met. 14(1980)839
24. MORI T. and TANGRI K., Met. Trans. A10(1979)733
25. BALLUFFI R.W. and SCHINDLER R., Proc. V. Polish Conf. on Electron Microscopy, J.A. Kozubowski ed., Warsaw-Jadwisin(1978)51
26. KLUGE-WEISS P. and GLEITER H., Acta Met. 26(1979)117
27. FOLL H. and AST D., Phil. Mag. 40(1979)589
28. PUMPHREY P.H., GLEITER H. and GOODHEW P., Phil. Mag. 36(1977) 1099
29. VARIN R.A., Phys. Stat. Sol. (a) 51 (1979) K 189
30. VARIN R.A., Phys. Stat. Sol. (a) 52 (1979) 347
31. VARIN R.A., WYRZYKOWSKI J., LOJKOWSKI W. and GRABSKI M.W., Phys. Stat. Sol. (a) 45(1978)565
32. KUHN H., BARO G. and GLEITER H., Acta Met. 27(1979)959
33. ISHIDA Y., HASEGAWA T. and NAGATA F., Trans. Jap. Inst. Metals 9(1968)504.
34. GLEITER H., Scripta Met. 14(1980)569
35. VITEK V., SUTTON A.P., SMITH D.A. and POND R.C., Phil. Mag. A39(1979)213.

36. SUTTON A.P. and VITEK V., Scripta Met., in press.
37. WARRINGTON D.H. and POND R.C., Phil. Mag. 39(1979)821
38. PUMPHREY P.H. and GOODHEW P.J., Phil. Mag. 39(1979)825
39. LI J.C.M., Surf. Sci. 31(1972)12
40. SHIH K.K. and J.C.M. LI, Surf. Sci. 50(1979)109
41. MARCINKOWSKI M., SADANANDA K. and TSENG W.F., Phys. Stat. Sol. (a) 17(1973)423
42. MARCINKOWSKI M. and SADANANDA K., Phys. Stat. Sol. (a) 18 (1973) 361
43. MARCINKOWSKI M. and DWARAKADASA E.S., Phys. Stat. Sol. (a) 19 (1973) 597
44. MARCINKOWSKI M. and SADANANDA K., J. Appl. Phys. 45(1974)1521
45. ibid. 45(1974)1533
46. MARCINKOWSKI M.J., Archiwum Mechaniki Stosowang 31(1979)763
47. MARCINKOWSKI M.J., in *Unified Theory of Mechanical Behavior*, J. Wiley and Sons, 1979
48. KRONER E., in *Kontinuums Theorie der Versetzungen*, Springer Verlag 1958
49. MARCINKOWSKI M.J., Phys. Stat. Sol. (a) 49(1978)725
50. MARCINKOWSKI M.J., J. Mat. Sci. 14(1979)205
51. MARCINKOWSKI M.J. and JAGANNADHAM K., Phys. Stat. Sol. 50 (1978)601.
52. KRONBERG M.L. and WILSON F.H., Trans AIME 185(1949)501
53. FRANK F.C., Conf. Plastic Def. of Cryst. Solids, Mellon Inst., Pittsburgh (1950)p 150
54. BRANDON D.G., RALPH B., RANGANATHAN S. and WALD M.S., Acta Met. 12(1964)813
55. RANGANATHAN S., Acta Cryst. 21(1966)197
56. BOLLMANN W.,Phil. Mag. 16(1967)363
57. ARKHAROV V.I., Fiz Met. i Metall. 12(1961)223
58. BRANDON D.G., Acta Met. 14(1966)1479
59. WEINS M., GLEITER H. and CHALMERS B., Scripta Met. 4(1970)235
60. WEINS M., CHALMERS B., GLEITER H. and ASHBY M.F., Scripta Met. 3(1969)601
61. WEINS M., GLEITER H. and CHALMERS B., J. Appl. Phys.42(1971) 2639
62. YAMRAGUCHI Y. and VITEK V., Phil. Mag. 34(1976)1
63. BRISTOWE P.D. and CROCKER A.B., Acta Met. 25(1977)1363
64. POND R.C., SMITH D.A. and VITEK V., Acta Met. 27(1979)235
65. BRISTOWE P.D. and CROCKER A.B., Phil. Mag. A38(1978)487
66. POND R.C. and SMITH D.A., Can. Metall. Quart. 13(1974)39
67. PUMPHREY P.H., MALIS T.F. and GLEITER H., Phil. Mag. 34(1976) 227
68. MARUKAWA K., Phil. Mag. 36(1977)1375
69. CHALMERS B. and GLEITER H., Phil. Mag. 23(1971)1541
70. HERRMANN G., GLEITER H. and BARO G., Acta Met. 24(1976)353
71. SAUTTER H., GLEITER H. and BARO G., Acta Met. 25(1977)467
72. HERRMANN G., SAUTTER H., BARO G. and GLEITER H., Scripta Met. 9 (1975)357
73. SEEGER A. and SCHOTTKY G., Acta Met. 7(1959)495.

74. KUBALSKI M. and GRABSKI M.W., subm. to Phys. Stat. Sol.
75. AUST K.T., Can. M t. 9(1969)173
76. SIMPSON C.J., AUST K.T. and WINEGARD W.C., Met. Trans. 2(1971) 987
77. DEMIANCZUK D.W. and AUST K.T., Acta Met. 23(1975)1140
78. LAGARDE P. and BISCONDI M., Can. Met. Quart. 13(1974)245
79. GLEITER H., Zeitschr. f. Metallk. 61(1970)282
80. HART E.W., in ref. 3 p. 155
81. MEISSER H., GLEITER H. and MIRWALD E., Scripta Met. 14(1980)95
82. GLEITER H., Radex-Rundschau 1(1980)51
83. DASH J.G. and RUWALDS J. eds, in *Phase Transitions in Surface Films*, Plenum Press, N. Y., 1980, NATO Advanced Study Institutes Series, Senes B. Physics Vol. 51.
84. BOOLMANN W., in *Crystal Defects and Cryst. Interfaces*, Springer Verlag, Berlin, 1970
85. BOLLMANN W. and PERRY A.J., Phil. Mag. 20(1969)33
86. WARRINGTON D.H. and BOLLMANN W., Phil. Mag. 25(1972)1195
87. HORTON C.A.P., SILCOCK J.M. and KEGG G.R., Phys. Stat. Sol. (a) 26(1974)215
88. BALLUFFI R.W., KOMEM Y. and SCHOBER T., Surf. Sci. 31(1972)68
89. GOODHEW P., in ref. 7.
90. HORTON C.A.P. and SILCOCK J.M., J. Micr. 102(1974)334
91. CLARK W.A.T. and SMITH D.A., Phil. Mag. 38(1978)367
92. MORI M. and ISHIDA Y., Scripta Met. 12(1978)11
93. VARIN R.A., Phys. Stat. Sol. (a) 51(1979)K189
94. LOJKOWSKI W., KIRCHNER H.O.K. and GRABSKI M., Scripta Met. 11 (1977)1127
95. PUMPHREY P.H., Scripta Met. 9(1975)151
96. GLEITER H., Phil. Mat. 36(1977)1109
97. BOLLMANN W., Acta Cryst. A33(1977)730
98. CAUDIG W. and SASS S.L., Phil. Mag. A39(1979)725
99. BOLLMANN W., MICHAUT B. and SANIFORT G., Phys. Stat. Sol. (a) 13 (1972)637
100. VAUGHAN D., Phil. Mag. 22(1970)1003
101. PUMPHREY P., Scripta Met. 6(1972)107
102. RALPH B., HOWELL P.R. and PAGE T.F., Phys. Stat. Sol. (b) 55 (1977)641
103. BALLUFFI R.W. and SCHOBER T., Scripta Met. 6(1972)697
104. PUMPHREY P., Scripta Met. 7(1973)895
105. KEGG G.R., HORTON C.A.P. and SILCOCK J., Phil. Mag. 27(1973) 1041
106. GRONSKI W. ang THOMAS G., Scripta Met. 11(1977)791
107. SMITH D.A., VITEK V. and POND R.C., Acta Met. 25(1977)475
108. POTAPOV L.P., GOLOVIN B.F. and SHIRYAEV P.H., Fiz Met. Metall. 32(5) (1971)227
109. AARON H.B. and BOLLING G.F., Surface Sci. 31(1972)27
110. AARON H.B. and BOLLING G.F., in *Grain Boundary Structure and Properties*, G.A. Chadwich and D.A. Smith eds., Academic Press, New York, 1976, p. 107.
111. ASHBY M.F., SPAEPEN F. and WILLIAMS S., Acta Met. 26(1978)1647.

112.ASHBY M.F. and SPAEPEN F., Scripta Met. 12(1978)193
113.FROST H.J., ASHBY M.F. and SPAEPEN F., Scripta Met. 14(1980) 1051
114.POND R.C., SMITH D.A. and VITEK V., Scripta Met. 12(1978)669
115.CHEN H.S. and CHANG S.Y., Phys. Stat. Sol. (a) 25 (1974)581
116.OZAWA T. and ISHIDA Y., Scripta Met. 11(1977)835
117.ISHIDA Y. and OZAWA T., Scripta Met. 9(1975)1103
118.NASU S. and GLEITER H., to be published.
119.INGLE K.W., BRISTOWE P.D. and CROCKER A.G., Phil. Mag. 33(1976) 83
120.FARIDI B.A. and CROCKER A.G.,Phil. Mag. A41(1980)137
121.DAHL R.E., BEELER J.R., BOURQUIN R.D., in *Interatomic Potential and Simulation of Lattice Defects*, Plenum Press, 1972,p 673
122.BRIWTOWE P.D.,BROKMAN A., SPAEPEN F. and BALLUFFI R.W., Scripta Met. 14(1980)943
123.HAHN H., Diploma Thesis, Univ. Of Saarbrücken (1980)
124.HAHN H. and GLEITER H., Acta Met. 29(1981)601
125.GLEITER H., Progr. Mat. Sci., 25(1981)125
126.NABEREZHNYKY V.P., FELDMAN E.P. and IURCHENKO V.M., Metalofizika 2(1980)11
127.SIEGEL R.W., CHANG S.M. and BALLUFFI R.W., Acta Met. 28(1980) 249
128.SEGALL R.L., Acta Met. 12(1964)117
129.JAEGER W. and GLEITER H., Scripta Met. 12(1978)675
130.ISHIDA Y.,OKAMATO K. and HACHISU S., Acta Met. 26(1978)651
131.POND R.C. and SMITH D.A., Phil. Mag. 36(1977)353
132.DINGLEY D.J. and POND R.C., Acta Met. 27(1979)667
133.DARBY T.L. and BALLUFFI R.W., Phil. Mag. A37(1978)245
134.CLARK W.A.T. and SMITH D.A., Journal of Mat. Sci. 14(1979)776
135.PUMPHREY P.H. and GLEITER H., Phil. Mag. 30(1974)593
136.SMIDODA K. and GLEITER H., Zeitschr. f. Metallk. 89(1978)81
137.JOHANNESSON T. and THOLEN A., Metal Sci. J. 6(1972)189
138.ASHBY M.F., Surf. Sci. 31(1972)498
139.AARON H.B. and BOLLING G.F., Surf. Sci. 31(1972)27
140.The energy density in the core of a high energy boundary is of the order of the latent heat of fusion. This energy density presents an upper limit, and, hence, the introduction of an additional dislocation in such a high energy structure can not increase the energy significantly.[137,138] It has also be argued that the relatively low density[81,139] of the material in the boundary should result in lower restoring forces there, and thus allow the long range strain field to reduce its energy by widening the boundary dislocation core.

KINETICS OF ELEMENTARY PROCESSES AT SURFACES

Milton W. COLE
Physics Department
The Pennsylvania State University.
University Park, Pennsylvania 16802

Flavio TOIGO
Istituto di Fisica Galileo Galilei and
Unita GNSM-CNR. Università di Padova
35100 Padova, Italy.

ABSTRACT

This chapter presents fundamental aspects of kinetic processes involving atoms or molecules near surfaces. The topics include scattering, sticking, diffusion, and desorption. Emphasis is placed on results obtained with well characterized surfaces. Selected for particular attention are examples that illustrate particularly well the basic phenomena or recently developed techniques. Both chemisorption and physisorption are treated. The picture that emerges is one of a rapidly evolving field. Experimental data are becoming sufficiently comprehensive that calculations with realistic potential energy functions and phonon spectra are now needed. This is particularly true of diffusion studies and scattering measurements, which can provide detailed information about energy losses and residence times. The theoretical community has responded with some elegant calculations, but many questions and problems remain.

I. INTRODUCTION

The domain of phenomena for which surface kinetics is relevant is extraordinarily diverse, including crystal growth,[1] monolayer adsorption[2] and phase transitions,[3] and chemical reactions.[4] These lectures describe some of the simpler concepts associated with the kinetics of adsorption. This discussion treats scattering, sticking, diffusion and desorption. The breadth of subject matter requires

B. Mutaftschiev (ed.), Interfacial Aspects of Phase Transformations, 223–260.

focus on some fairly specific aspects of each topic and the choices are biased toward our view of particularly interesting developments of a fundamental nature. Our hope is that these tutorial lectures will provide stimulation and a modest degree of conceptual background.

We first address the simplest possible problem--the states of a particle adsorbed on a rigid and perfect surface. This idealization provides a basis for treating the various processes of interest. For example, a particle incident on a surface may undergo any of the following processes:scattering (elastic or inelastic, specular or diffractive) or sticking in a bound state. In some cases, we are interested in the detailed behavior (e.g., the angular dependence of the energy distribution of scattered particles), although such complete information is not usually available. There exist various possibilities (e.g., molecular dissociation) which we must omit because of lack of space.

A complete understanding of such phenomena requires knowledge of the static and dynamic properties of the surface, in addition to its interaction with the adsorbed particles. Even if these ingredients were known, substantial computational problems would remain. Nevertheless, impressive progress has been made in recent years, stimulated by a wealth of experimental data. This is particularly true of the scattering problem. The scattering technique is very powerful because it leads to information[5-8] about specific states of adsorbed or reflected particles; the more traditional tools for studying kinetics have yielded, in contrast, averages over an ensemble of states.[8,9] It will be evident, however, that the latter still have a very useful role to play in the future.

II. PARTICLES NEAR A RIGID, PERFECT SURFACE

II.1. Properties

Understanding this idealized problem helps clarify some fundamental aspects of the processes of adsorption and desorption. We distinguish between quantum and classical adsorption on the basis of the particle de Broglie wavelength, $\lambda = h/p$. λ vanishes in the classical limit, in which we may take the particle to have both a well defined position and momentum p. Its migration across the surface is inhibited by periodic barriers due to the potential energy V(r) and interaction with phonons (as discussed in Sec. VI) or surface imperfections.

In the quantum regime, the state of an adsorbed particle extends across the surface. Because $V(\vec{r})$ is periodic in the two dimensions (2D) parallel to the surface, an energy eigenfunction must satisfy a 2D Bloch theorem :

$$\psi_{\vec{K},n}(\vec{r} + \vec{l}) = e^{i\vec{K}\cdot\vec{l}}\ \psi_{\vec{K},n}(\vec{r}), \tag{1}$$

where $\vec{l}$ is any of the 2D lattice vectors associated with the surface periodicity, K is the 2D wave vector, and n is a band index. Thus the function can be written

$$\psi_{\vec{K},n}(\vec{r}) = e^{i\vec{K}\cdot\vec{r}}\ u_{\vec{K},n}(\vec{r}) \tag{2}$$

where u is 2D-periodic. The simplest possible case arises when the lateral variation of $V(\vec{r})$ is negligible, so that $V(\vec{r}) \simeq V(z)$. In this "smooth surface" limit, Eq. (2) becomes

$$\psi_{\vec{K},n}(\vec{r}) = e^{i\vec{K}\cdot\vec{r}}\ \phi_n(z), \tag{3}$$

$$-\frac{\hbar^2}{2m}\frac{d^2\phi n}{dz^2} + V(z)\ \phi_n = \varepsilon_n\phi_n \tag{4}$$

$$E_{\vec{K},n} = \hbar^2K^2/2m + \varepsilon_n, \tag{5}$$

where, in general, there is a finite number (N) of bound states ($0 \leqslant n \leqslant N-1$). Equations (3) and (4) separate the solutions into distinct motions parallel and perpendicular to the surface. The former is that of a free particle; the latter takes on a series of eigenvalues ε_n associated with particle vibration against the surface.

Equations (3) through (5) are only approximations, in general, but they are often assumed in 2D analyses of adsorbed phases. Physically adsorbed rare gases on basal plane graphite and close-packed metal facets come reasonably close to satisfying the assumptions. The reason is that the weak van der Waals forces do not attract the adatom very close to the surface; since the periodic component of $V(\vec{r})$ is associated primarily with atomicity of the top layer, this component is small in the relevant spatial region.[2]

Figure 1 illustrates this with the potential energy[10] of a single He atom above symmetry positions of a graphite surface. The coordinate z is the distance to the top C layer. Both this function and the allowed energy bands[11] have been determined from atomic scattering data[12,13] as described below. Note that the ground state of the atom (at -12.2 meV for ^{4}He) lies substantially above the minimum of the potential energy (-19 meV). The difference is the zero point energy of which about 5 meV and 1.8 meV come from perpendicular and parallel motion, respectively.[14]

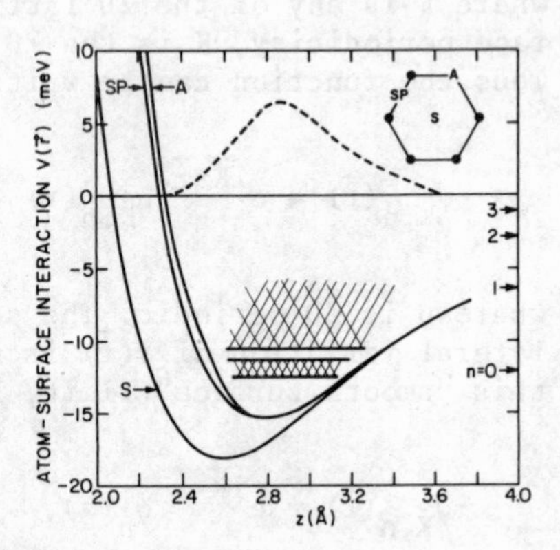

Fig. 1 : The potential energy of a He atom above various points (see upper right; the circles represent C atoms) on a graphite surface (from Ref. 10). Also shown for ^{4}He are the ground state probability density (dashed curve, from Ref. 14), the lowest allowed energy bands (from Ref. 11), and the eigenvalues ε_n (arrows at right, from Ref. 12) for vibrational motion in the laterally averaged potential.

The former value, for example, can be roughly estimated from the force constant for surface normal motion, $\kappa_z = d^2V/dz^2 \simeq 90$ meV/A^2. In a symplifying[15] harmonic oscillator model, this gives a zero point energy $\Delta E_z \simeq \hbar\omega_z/2 \simeq 4.6$ meV, where

$$\omega_z = (\kappa_z/m)^{1/2} \quad (6)$$

is about 1.4×10^{13} rad/sec for ^{4}He. Note that the ground state probability density extends beyond the potential minimum, so there is relatively little effect of the lateral force on particle translation. Perhaps the most dramatic quantum effect is that the ground state energy exceeds even the minimum potential energy above the C atom (curve A in Fig. 1); thus there is no classically forbidden region for particle motion along the surface. There are, however, residual effects of the lateral variation of the potential : (a) a small (0.05) effective mass correction, and (b) the states fall into energy bands;[11,16] the lowest are separated by a gap (of width 0.5 meV), which has been detected in specific heat data.[11,16-18]

One can estimate a characteristic site localization time[19]

$$\tau_s \simeq \hbar/\Delta \quad (7)$$

for a localized (Wannier) state constructed from the lowest band, of width Δ. For the case of Fig. 1, $\Delta \simeq 1$ meV, about 70 percent of the smooth surface value. This gives $\tau_s \simeq 10^{-12}$ sec, about twice the period of perpendicular motion corresponding to Eq. (6).

We may contrast this quite delocalized case with examples such as He in a more corrugated potential[20] or a classical gas on graphite.[21] In either case, as exemplified in Fig. 2, the substantial lateral potential barrier produces localization. This is manifested in band narrowing and a corresponding increase in τ_s in each band. Thermally excited particles will dominate in the diffusion process, except at low temperature.

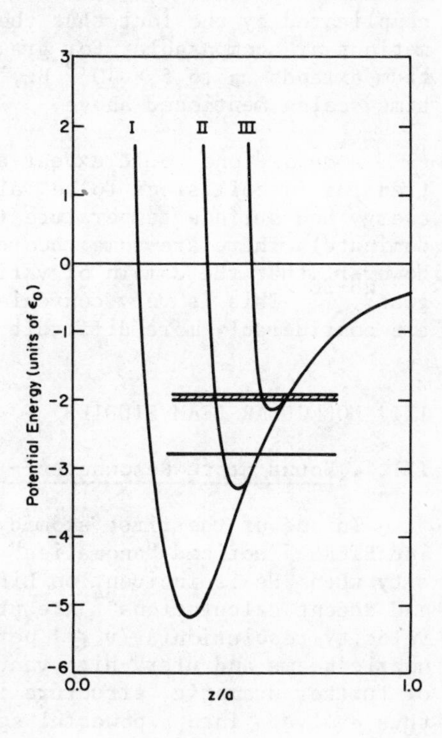

Fig. 2 : Potential energy (in units of $\varepsilon_0 = 4.16$ meV) for a ^{4}He atom as a function of z (in units of a = 3.84 Å) above an Ar crystal. The curves label adsorption site (I), saddle point (II) and on-top (III) sites. The horizontal lines represent the lowest energy bands (of width 0.01 and 0.3 meV). Results from Ref. 20.

II.2. Relevance of the Static Model

The preceding discussion refers to elastic potentials, obtained by assuming that the substrate atoms are localized at their equilibrium positions.[22] We may now ask about the validity of this assumption. The root mean square vibration amplitude of a substrate atom perpendicular to the surface, measured by inelastic neutron scattering,[23] yields for graphite[24] values varying from 0.07 Å at temperature T = 0 to 0.12 Å at T = 300 K. A Debye-Waller determination from ^{4}He specular scattering intensities agrees well with these numbers, indicating that the surface dynamical properties are not substantially different from the bulk (except for the presence of surface phonons[25]). For graphite, this conclusion is a consequence of the very weak interlayer forces. Atoms at most other surfaces are found by LEED to exhibit larger vibrations because of weaker binding at the surface than in the bulk.[26]

Even though the vibrational amplitudes at low T are a relatively small fraction of the adatom-surface separation, this motion is very important because of the steepness of the repulsive potential (Fig. 1). Thus the atom-phonon interaction represents a major mechanism for energy transfer.[27] The theory of such processes is complicated by the fact that the time scales for adatom and lattice motions are comparable. For graphite, for example, the phonon spectrum extends up to 5×10^{13} Hz,[23] which is of order the inverse of time scales mentioned above.

A priori one would expect a classical calculation of energy transfer in collisions to be valid only at high incident particle energy and surface temperature (since multiphonon processes then dominate). There are some theoretical and experimental indications, however, that the domain of validity is broader than this suggests.[28-30] This is very convenient because the quantum calculations are considerably more difficult and less certain.[31]

III. MOLECULAR BEAM STUDIES

III.1. Bound State Resonances-- Elastic and Inelastic

In one of the first atomic scattering experiments, FRISCH and STERN[32] noticed "anomalies" in the specularly reflected intensity when ^{4}He is incident on LiF. A modern version[33] of their data and recent calculations[34] are presented in Fig. 3. The very high velocity resolution($\Delta v/v \lesssim 1$ percent) available with supersonic nozzle beams and ultra-high vacuum has resulted in the appearance of further dramatic structure in the data; this technique has thus evolved into a powerful spectroscopy. The original interpretation of the anomalies by LENNARD-JONES and DEVONSHIRE[35] remains conceptually useful, although it has been refined in the last five years.[5,7,36,37] The original name for the phenomenon was "selective adsorption";more common today is "bound state resonance". To understand it, let us focus on the pronounced minima labelled 0 - 0,1 or 1 - 0,1 in Fig. 3. The idea is that at the corresponding angle (e.g., $\phi = 25°$) the incident particle can make a transition to a bound state propagating along the surface. It then has a high probability of being scattered by a phonon or imperfection, becoming lost from the specular beam. Supposing the initial transition is elastic, the incident energy E_i must satisfy the relation

$$E_i = E_{\vec{K}_b, n} \simeq \hbar^2 K_b^2/2m + \varepsilon_n, \qquad (8)$$

where the equality holds if Eq. (5) is satisfied.

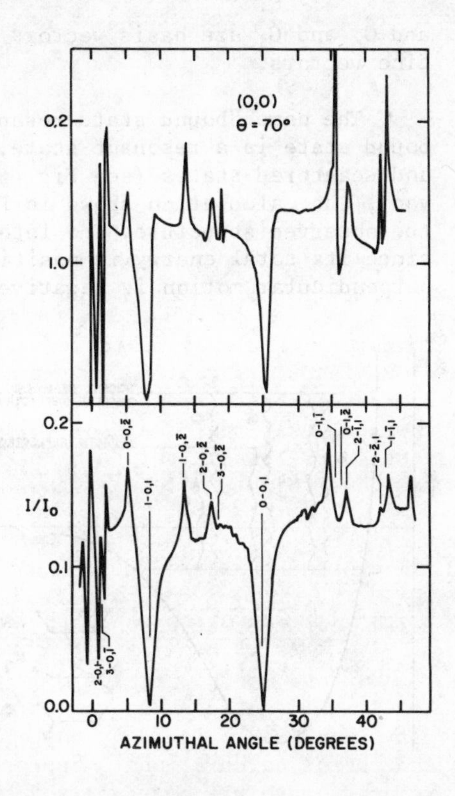

Fig. 3 : Specular intensity for ^{4}He incident with $E_i \simeq 17$ meV on LiF at $\theta = 70^o$ as a function of azimuthal angle. The labels of the features are discussed in the text. Upper half, experiment of DERRY et al.[33]; lower half, calculation of GARCIA, CELLI, and GOODMAN.[34]

The final state is bound with respect to motion perpendicular to the surface if $\varepsilon_n < 0$; this implies $\hbar^2 K_b^2/2m > E_i$. Such a transfer of incident energy to lateral translation in the bound state is allowed by the possibility of particle diffraction :

$$\vec{K}_b = \vec{K}_i + \vec{G}, \qquad (9)$$

where $\vec{G}$ is a reciprocal lattice vector of the surface and $\vec{K}_i$ is the projection on the surface of the incident wave vector. This is inherently a quantum process; it has been seen to date with H, He, H_2 and some of their isotopic variants. Equations (8) and (9) are satisfied only at specific angles of incidence because the values of $\vec{G}$ and ε_n are discrete. Various features in Fig. 3 are thus assigned values n - kl, where

$$\vec{G}_k = k\vec{G}_0 + l\vec{G} \qquad (10)$$

and $\vec{G}_0$ and $\vec{G}_1$ are basis vectors for the set of 2D reciprocal lattice vectors.

The name "bound state resonance" reflects the fact that the bound state is a resonant state, intermediate between the incident and scattered states (see Fig. 4). Inelasticity need not be relevant; the calculation shown in Fig. 3 is elastic and reproduces the observed structure. The intermediate state is not truly bound since its total energy is positive (Eq. (8)); only its energy of perpendicular motion is negative.

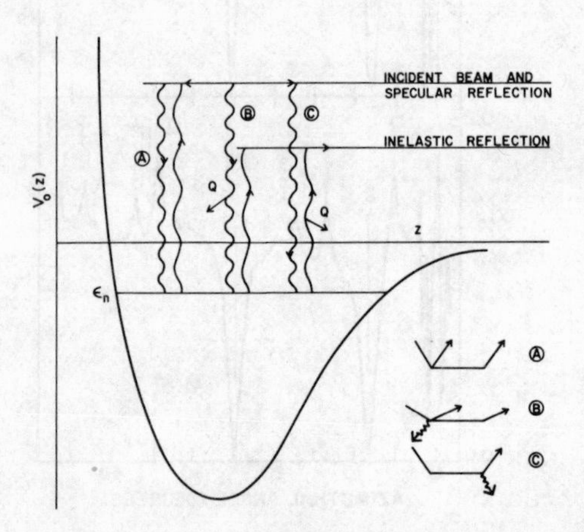

Fig. 4 : Schematic views of (elastic) bound state resonance or selective adsorption (A), phonon-assisted selective adsorption (B), and phonon-assisted selective evaporation (C). Only the energy of motion perpendicular to the surface is shown. The level ε_n is an eigenvalue for the one dimensional potential $V_0(z)$. In schematic at lower right, a horizontal line denotes a bound state and a wiggly line represents a phonon.

It is thus a quasi-stationary state with energy overlapping the positive continuum. Analogous resonances appear in quite diverse scattering experiments, e.g., the compound nucleus[38] phenomenon and surface electron resonances.[39] Their common feature is that the width of the "anomaly" in the scattering data (Fig. 3) can be related to the lifetime for decay of the quasi-stationary state.[37,40,41] In general this decay can lead to either "scattering states" (positive total energy) or truly bound states. The latter possibility corresponds to sticking; the particle will remain localized near the surface until it desorbs in a subsequent event, e.g., absorption of a phonon.

One principal application of bound state resonance studies has been to deduce the elastic part of the atom-surface interaction[6] and the energy spectrum of deeply bound states.[11] For example,

Fig. 1 was obtained from Eqs. (5) and (8), generalized to incorporate the lateral variation of the potential energy). Of possibly greater relevance is its recent use[42-48] as a probe of the atom-phonon interaction, which is intimately related to sitcking. First, we write the kinematic condition for non-diffractive scattering to a polar angle θ_f in an event involving one phonon of energy $\hbar\omega$ and wave vector projection $\vec{Q}$ along the surface :

$$\pm \hbar\omega = E_i - \frac{\hbar^2}{2m} \left[\frac{\vec{K}_i \pm \vec{Q}}{\sin \theta_f}\right]^2 \tag{11}$$

where $\vec{K}_i$ is the surface projection of the incident wave vector, of magnitude $(2mE_i)^{1/2}/\hbar$. We consider a competing two-step process of the following form (see Fig. 4b) :

1. Incident particle diffracts into a bound state, assisted by a phonon :

$$\vec{K}_b = \vec{K}_i + \vec{G} \pm \vec{Q}; \tag{12}$$

2. The bound state diffracts into the final scattered state $\vec{K}_f = \vec{K}_b - \vec{G}$ via the reverse of $\vec{G}$.

The first step is thus a modification of the previously described elastic resonance. The revised version of Eq. (8) is

$$\pm \hbar\omega = E_i - (\hbar^2 K_b^2/2m + \varepsilon_n) = E_i - \varepsilon_n - \hbar^2(\vec{K}_i + \vec{G} \pm \vec{Q})^2/2m \tag{13}$$

The simple inelastic process described by Eq. (11) is nonresonant, occurring for any phonon satisfying that relation. The second process, in contrast, involves the resonant bound state. The two possibilities can interfere in the same way that elastic, direct, specular scattering interferes with resonant specular scattering (with interference lineshape shown in Fig. 3). This inelastic interference requires simultaneous solution of Eqs. (11) and (13). For fixed E_i, ε_n, and $\vec{G}$, this corresponds to a constant value of θ_f. We expect, furthermore, from the limit where ω and Q vanish simultaneously that the angle θ_f (corresponding to $\omega = 0 = Q$) is simply the incident angle obtained from Eq. (8) for an elastic resonance. Very dramatic evidence of this phonon-assisted resonance concept appears in the tails of the specular peak, shown in Fig. 5. The data exhibit a lineshape expected for interference between the two processes.[45] As predicted, the position stays localized near $\theta_f = 57°$ for variable θ_i; this angle is indeed the incident angle for that particular resonance ($n = 0$, $G = G_{10}$ for graphite) in the elastic, specular scattering (Fig. 1 of Ref. 45). More extensive data over the interval $40° < \theta_i < 54°$ show the same effect. Also observed in the data is evidence of an inelastic double-resonance which involves two bound states. Such data give clues about the cascade process associated with sticking.

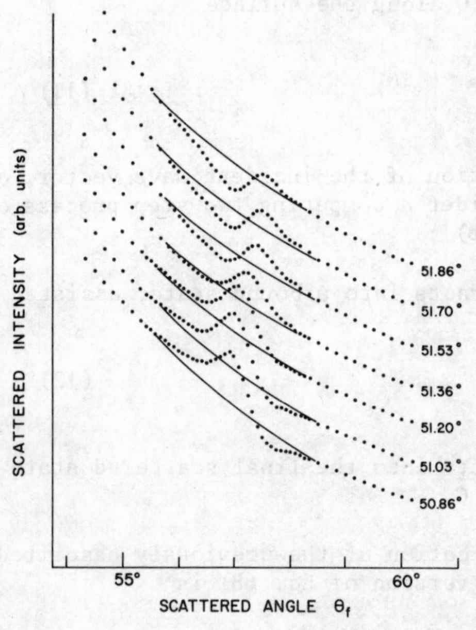

Fig. 5 : Scattered intensity showing interference pattern associated with phonon-assisted selective adsorption (see Fig. 4, case (B)), involving the n=1 bound state. The labels refer to the incident polar angles. From CANTINI and TATAREK (ref.45) for He/graphite.

These studies are obviously very important for analyzing the details of adsorption. More valuable information can be gleaned from measurements of the energy distribution of the scattered particles.[46-48] The techniques used to date are time-of-flight and diffraction by a secondary analyzer crystal. One obvious application is to deduce the dispersion relation of phonons interacting with the particle, using Eq. (11).[46-48] Of further interest for understanding adsorption kinetics is a determination of absolute intensities, which reflect the probabilities of various processes. While implementation of this program is only beginning, some remarkable results have already emerged.

Figure 6 shows the total scattering intensity measured at $\theta_f = 90° - \theta_i$ and $\phi_f = \phi_i$ for He incident on LiF.[47] Between the diffraction peaks appear several sharp features. BRUSDEYLINS, DOAK and TOENNIES[47] have interpreted these in terms of the mechanism labelled C. in Fig. 4. Called "selective evaporation", the phenomenon was anticipated long ago.[49-50] It consists of an elastic selective adsorption transition (to a state labelled by the bars in Fig. 6), followed by a phonon-stimulated desorption. The signature is a peak for the transitions near $\theta_i = 35°$ labelled by G = (11)

since the second step (forward scattering by a phonon) is expected to be a likely process.

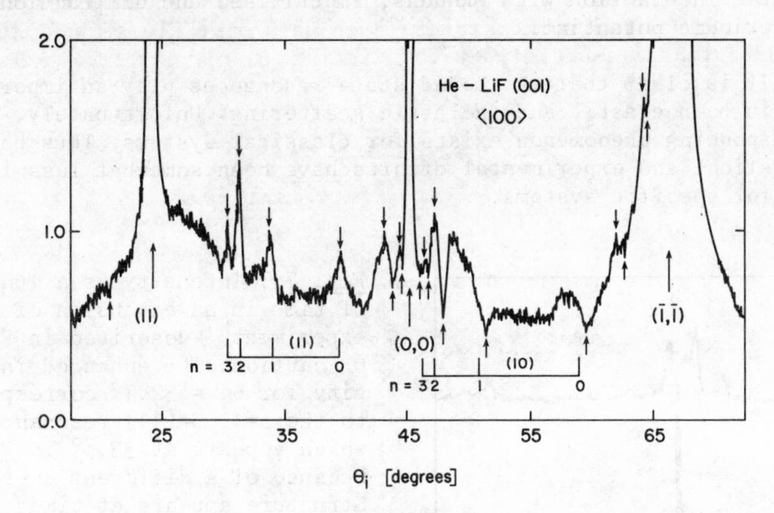

Fig. 6 : Scattered intensity (relative to specular) for He/LiF measured by BRUSDEYLINS et al.[47] at detector position $\theta_f = 90 - \theta_i$, for $k_i \simeq 6.15$ A^{-1} and variable θ_i. The very strong features are diffraction peaks. The connected horizontal and vertical bars provide assignments of selective evaporation features corresponding to a particular bound state n and G vector.

This interpretation is supported by the time-of-flight data in Fig. 7. Specifically, the selective adsorption process is kinematically allowed[51] for the n = 1 bound state, at $\varepsilon_1 = -2.46$ meV.[33] At the corresponding angle $\theta_i = 32.7°$, a substantial enhancement of the detected signal is seen. Furthermore, the arrival time of the particles at the detector in Fig. 7 can be related to the energy loss. The peak at t = 2.3 msec is associated with phonon creation; the fastest particles (t ≃ 1.55 and 1.7 msec) have gained energy by phonon annihilation.[52] BRUSDEYLINS et al. have used these data to deduce the dispersion relation of phonons created or destroyed in these events. The results agree well with predicted[53] Rayleigh wave dispersion curves, except near the Brillouin zone boundary, suggesting the need for a revised calculation.[54]

These data have been exploited[46] further to yield the lifetime τ of the bound state. Specifically, the angular width of the features in Fig. 6 corresponds, according to Eq. (8), to a width of ε_n value. For example, after correction for the small divergence in the incident beam, the n = 3 state yields $\Delta\varepsilon \simeq 0.08$ meV. By the uncertainty relation $\tau \simeq h/\Delta\varepsilon$, BRUSDEYLINS et al.[46] deduced a value $\tau \simeq 50 \times 10^{-12}$ sec for this state (with a 30 percent uncertainty).

This corresponds to a mean free path of nearly 500 Å. States of smaller n have shorter τ; their proximity to the surface enhances particle interaction with phonons, impurities, and diffraction by the periodic potential.

It is clear that the bound state resonances play an important role in both elastic and inelastic scattering. Unfortunately, no corresponding phenomenon exists for classical systems. Thus both theoretical and experimental efforts have been somewhat less intensive for specific systems.

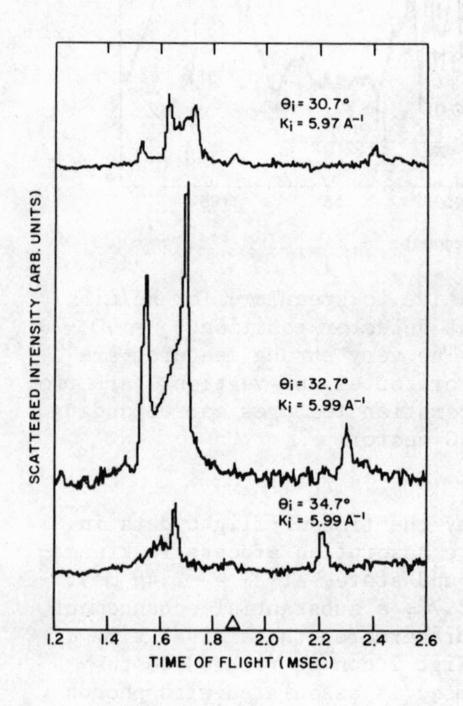

Fig. 7 : Intensity as a function of time in an extension of the experiment[47] described in Fig. 6 caption. The enhanced intensity for θ_i = 32.7° corresponds to the n=1, G=(11) resonance, which appears at 33.8° in Fig.6 because of a different energy. Structure appears at times shorter or longer than that corresponding to the elastic scattering (triangle on abscissa) because of phonon creation or annihilation.

III.2. A Chemisorption study : CO/Pd(111)

In order to illustrate some concepts and techniques relevant to the rather different kinetics of chemisorption, we consider an elegant experiment of ENGEL.[55] An initially clean Pd(111) surface was exposed over a 5 mm region to a CO beam from a supersonic nozzle source; the background pressure (10^{-10} torr) was at least 3 orders of magnitude smaller in its effective flux. The data, some of which appear in Fig. 8, correspond to a cosine distribution[56] for all temperatures studied (300 < T < 1020 K). This means that we need discuss only the detected intensity I_d at a single position. This is given by

$$I_d(\theta,T) = f\{I_\circ[1 - S(\theta,T)] + D(\theta,T)\} , \qquad (14)$$

where f is a geometrical factor, θ is the coverage, S is the sticking coefficient, and D is the desorption rate per unit area. The incident intensity $I_\circ = 1.5 \times 10^{15}$ molecules/cm^{2}sec was calibrated by monitoring the coverage with a series of flash desorption measurements.[9,57] Equation (14) assumes that the desorbing molecules emerge with the same (cosine) distribution and are detected with the same efficiency[58] as the directly scattered particles; this interpretation is partly motivated by the fact that the cosine distribution is that of particles incident from an equilibrium vapor.

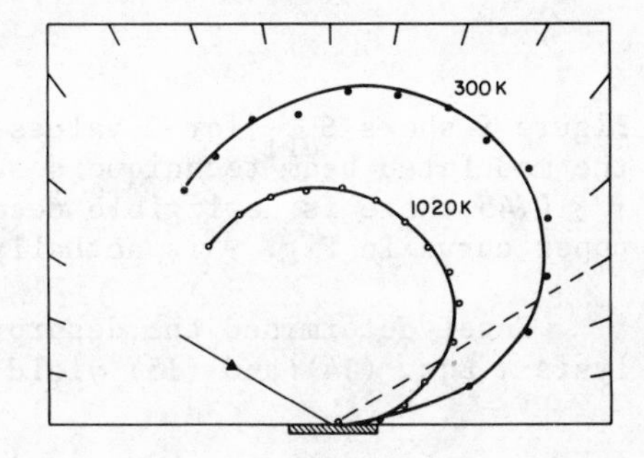

Fig. 8 : Nearly cosine angular distribution of CO scattered from Pd(111) at 2 different surface temperatures, for $\theta_i = 60^\circ$. From ENGEL, Ref. 55.

Following ENGEL[55] we define an "effective" sticking coefficient by a relation similar to (14) :

$$I_d(\theta,T) \equiv f\, I_\circ[1 - S_{eff}(\theta,T)] \qquad (15)$$

S_{eff} is obviously equal to S if there is negligible desorption; this is the case at low θ and T. After the equilibrium coverage θ_e is attained, the sticking rate equals the desorption rate :

$$D(\theta_e,T) = I_\circ S(\theta_e,T) \qquad (16)$$

so that $I_d(\theta_e,T) = f\, I_\circ$, from (14). Thus from (15)

$$S_{eff}(\theta,T) = 1 - \frac{I_d(\theta,T)}{I_d(\theta_e,T)} \qquad (17)$$

At $\theta = 0$, the left side can be replaced by S_o (the true initial sticking rate) according to the remark following Eq. (15). This allows an absolute determination of S_o.

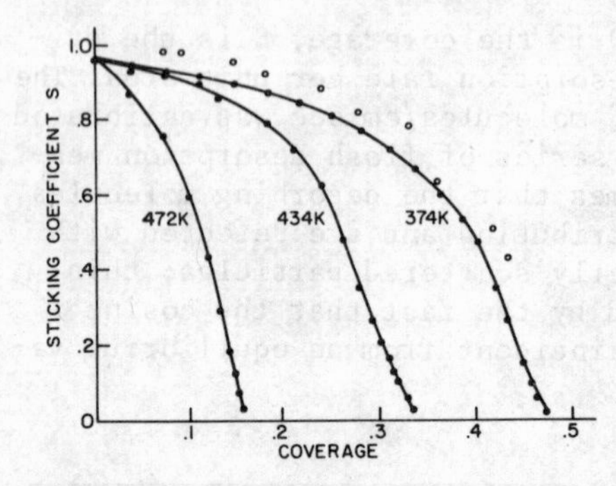

Fig. 9 : Effective sticking coefficient S_{eff} for CO incident on Pd at 3 temperatures. At $T_s = 374$ K the results are obtained from modulated beam (open circles) as well as dc (full circle) techniques. Only at this temperature does $S_{eff} = S$. From Ref. 55.

Figure 9 shows S_{eff} for 3 values of T. Numerical estimates based on the modulated beam technique described below indicate that for $\theta \leq 0.45$ there is negligible desorption for $T < 374$ K. Thus the upper curve in Fig. 9 is actually the true sticking coefficient S.

Engel determined the desorption rate from the following analysis : Eqs. (14) and (15) yield

$$D(\theta,T) = I_o \; S(\theta,T) - S_{eff}(\theta,T) \tag{18}$$

Figure 9 indicates that $S_o = 0.95$, independent of T. The modulated beam data yield no dependence on T at finite θ. Thus Eq. (18) becomes

$$D(\theta,T) = I_o\left[S(\theta,374) - S_{eff}(\theta,T)\right] \tag{19}$$

which allows a direct determination of D at finite coverage (from Eq. 17)). As is discussed in Section V, it is common to assume an activated process:[59,60]

$$D(\theta,T) = \theta\nu \exp\left[-E_d(\theta)/k_B T\right] \tag{20}$$

Modulated beam measurements give $\log \nu\,(\mathrm{Hz})^{-1} = 14.4 \pm 0.8$. The resulting heats of desorption E_d (32 kcal/mole at $\theta = 0$ and 26 kcal/mole at $\theta = 0.42$) are consistent with direct isosteric heat measurements[61] (which give $E_d \simeq 34$ and $\simeq 30$ kcal/mole at these coverages).

Furthermore, the coverages deduced by integrating the net condensation rate,

$$\theta(t) \simeq I_o \int_o^t S_{eff}(t')dt'$$

are consistent with LEED data.[61]

We describe the modulated beam measurements in terms of an assumed sinusoidal signal; the actual data were taken with a square wave (chopped beam) source using lock-in detection.[62] Assuming first order kinetics with a single characteristic time τ, the surface density changes with time according to

$$\frac{dn_s}{dt} = S\, I_o e^{i\omega t} - \frac{n_s}{\tau}$$

which has a solution[63]

$$n_s = \frac{I_o S \tau}{(\omega^2\tau^2 + 1)^{1/2}} e^{i(\omega t - \phi)}$$

$$\phi = \tan^{-1}(\omega\tau).$$

The total rate I_d scattered into the detector is the sum of the desorption rate (n_s/τ) and the rate of direct scattering. If we reference the latter with zero phase, we obtain

$$\frac{I_d}{I_o} = \frac{S}{\sqrt{\omega^2\tau^2 + 1}} e^{-i\phi} + (1 - S). \qquad (21)$$

For T > 500 K, the experiment conditions correspond to $\theta < 0.01$, so that (from Fig. 9) the second term is small. In this regime, therefore, the T dependence of I_d is dominated by that of τ. At very high T(>> 600 K); the data indicate that $\omega\tau \ll 1$ so that $\phi \simeq 0$ and one measures the dc scattering rate (from Eq. (21); see Fig. 10). For T ≈ 600 K, $\tau \simeq \omega^{-1}$, so that the first term in Eq. (21) falls to a lower value as T decreases. As T falls below 500 K, however, θ increases so that S decreases; thus the detected signal starts to rise as T falls below 500 K. These data can then be used independently of the dc measurements to deduce the θ and T dependence of S. As seen in Fig. 9, the results agree. Furthermore, the activation form (20) is confirmed.

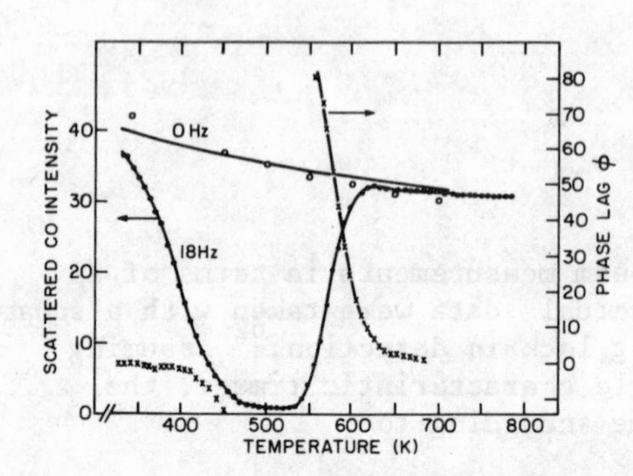

Fig. 10 : Intensity and phase lag for CO incident on Pd using dc and ac (18 Hz) methods. The dc curve takes into account the velocity dependence of the detector; the other curves are drawn to aid the eye. From Ref. 55.

The cosine dependence observed in Fig. 8 was attributed by ENGEL[56] to the absence of a direct scattering component. The incident particles were all trapped initially in a physisorbed precursor state,[9,64-66] in which they equilibrated. They subsequently either desorbed (after time τ) or fell into a chemisorbed state. The absence of a direct scattering component (which generally gives a lobular reflection pattern near the specular direction) is by no means general, as discussed below.

III.2. Rotational Energy Transfer

The exchange of internal energy is obviously an important aspect of molecular sticking, association, and dissociation at surfaces. In some cases it has been possible to study this problem without directly determining the states of individual molecules.[67-71] In fact only recently have direct measurements been made.[72-76] One detects the fluorescent light emitted by molecules which have been excited by either an electron beam or a laser. We discuss here the latter method, which offers somewhat wider applicability. Either the incident or reflected particles are illuminated by a tunable laser. When the frequency is appropriate to a transition between rotational states J and J', the total fluorescent intensity I_S is proportional (within a known factor that depends on J') to the initial state population N_J divided by (2J + 1). If the rotational states have an equilibrium distribution characterized by a temperature T_r, the intensity satisfies

$$I_S \propto \frac{N_J}{(2J+1)} \propto \exp\left[-\frac{E_J}{k_B T_r}\right] . \qquad (22)$$

Thus if $ln\ I_s$ is observed to be a linear function of E_J, the initial states had an equilibrium population distribution.

FRENKEL et al.[74] found such behavior for NO scattered from an NO covered Pt(111) surface. The value of T_r was equal to that of the surface (290 K) within experimental error. The presumption of surface equilibration is supported by the observation of a cosine angular distribution. In contrast, when the Pt surface was covered with graphite, the scattered molecules were incompletely accommodated;[73] a deficiency of low J states was observed.[77] POLANYI and coworkers have also found incomplete accommodation for CO and HF incident on LiF(100). The rotational temperature of the scattered beam is 50-100 K below the surface temperature if $T_s \lesssim 600$ K. At higher T_s values, T_r seems to saturate.

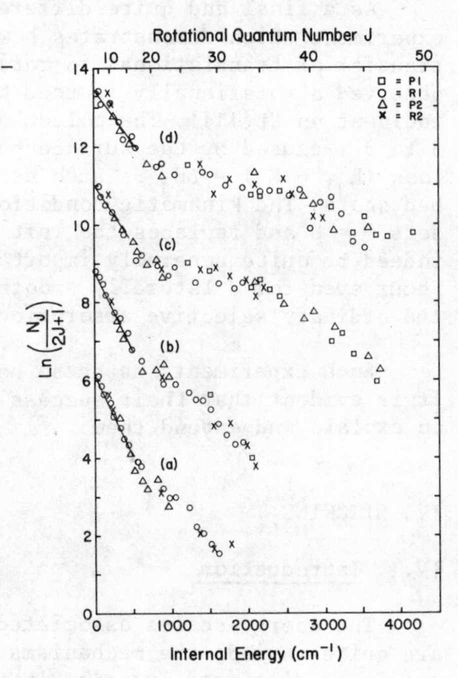

Fig. 11 : Logarithm of occupation number of rotational state J divided by 2J + 1 as a function of the energy E_J; the different symbols correspond to distinct rotational and spin-orbit transitions (see Ref. 76). The data correspond to incident energies and angles (a) 0.32 eV, 40°, (b) 0.32 eV, 15°, (c) 0.75 eV, 15°, and (d) 1 eV, 15°.

Figure 11 shows results obtained by KLEYN, LUNTZ and AUERBACH[76] for NO incident on Ag(111). The linear variation at small J is consistent with Eq. (22). The resulting temperatures, however, are not those of the surface, but depend on both incident energy and angle. It is found that T_r is proportional to the normal energy $E_n = E_i \cos^2 \theta_i$ plus a shift 0.75 ± 0.1 eV. KLEYN et al. interpreted these data as indicating a major role of the attractive potential in shifting the normal energy. The value cited above is, in fact, reasonably consistent with the known[78] 1 eV heat of adsorption.

The most remarkable feature in Fig. 11 is the deviation from Eq. (22) at high E_J. The explanation presented by KLEYN et al. is quite intriguing : a surface rotational rainbow. This is analogous to both a similar phenomenon seen in the gas phase[79] and to a surface translational rainbow effect.[80] In these cases, the cross-section for a particular event depends inversely on a derivative with respect to impact parameter; the rainbow arises from a vanishing of this derivative. In the present case, the relevant derivative is that of energy transfer with respect to orientational angle of the incident molecule. This must vanish at some particular angle, giving a large probability of scattering into the corresponding final state J. While this interpretation of the large J enhancement in Fig. 11 is plausible, calculations are necessary to validate it.

As a final and quite different example, we mention a recent experiment which demonstrates how sticking may arise from the transfer of translational to rotational energy. COWIN et al.[81] observed a rotationally induced bound state resonance for HD incident on Pt(111). The molecule makes a transition from state J to J', caused by the surface potential. The translational energy loss $(E_{J'} - E_J) = \Delta E_J$ is such as to leave it in a selectively adsorbed state. The kinematic condition coincides with Eq. (13), if one sets Q = 0 and replaces the left side with ΔE_J. Such a process may indeed be quite generally important in the sticking process. It can occur even for a laterally smooth repulsive wall, in which case the ordinary selective adsorption mechanism does not occur.

Such experiments as these have only recently been performed. It is evident that their success will stimulate considerable effort to explain and extend them.

IV. STICKING

IV.1. Introduction

The energy scales associated with the examples cited above are quite varied; the mechanisms for kinetic processes are correspondingly distinct. Trapping in the case of physisorption (binding energy $\leq$ 0.1 eV/particle $\simeq$ 2 kcal/mole) is associated with a small number of phonons and relatively little electronic excitation compared with chemisorption. To provide some focus for a discussion of sticking, we shall concentrate on the physisorption case. This is somewhat better understood conceptually and quantitatively because of both the extensive data available about the elastic potential and the weakness of the relevant interactions. It will emerge from the discussion that even this case leaves important questions unanswered. Since chemisorptive sticking sometimes commences with trapping in a physisorbed precursor state, our description will be of relevance to that problem also.

If a particle is incident with energy E_i from a solid angle Ω_i, we may define an accommodation coefficient

$$\alpha(E_i, \Omega_i, T_s) = \frac{\langle E_f \rangle - E_i}{E_s - E_i} \quad (23)$$

Here $\langle E_f \rangle$ is the mean scattered energy and E_s is the mean energy of a hypothetical particle which has been completely equilibrated with the surface. Note that the case $\langle E_f \rangle = E_s$ corresponds to $\alpha = 1$. For molecules, the internal state energy may be included in these energies, or one may define partial accommodation coefficients referring to the electronic, vibrational, rotational, and translational degrees of freedom. In the following we focus on atoms, in which case $E_s = 2k_BT_s$, the mean energy of an atom incident from a vapor in equilibrium with the surface.

One might define an averaged accommodation coefficient

$$\bar{\alpha}(T_g,T_s) = \frac{(k_BT_g)^2}{2\pi} \int dE_i \exp(-E_i/k_BT_g) \int d\Omega_i \cos\theta_i \alpha(\Omega_i,E_i,T_s), (24)$$

where the weighting factors correspond to those of a gas at temperature T_g near a surface. Instead, the conventional defintion is

$$\alpha(T_g, T_s) = \frac{T_f - T_g}{T_s - T_g}, \quad (25)$$

where $T_f = \langle E_f \rangle/2k_B$, the average being over the particles which strike the surface from a gas at T_g. The equilibrium accommodation coefficient is defined by

$$\alpha(T_s) = \lim_{T_g \to T_s} \alpha(T_g,T_s) \quad (26)$$

The value of α is determined by the probabilities and characteristics of the various processes which may occur during a collision with the surface. The most complete experiment possible (pulsed beam scattering with final state analysis) can reveal details about each individual process. As illustrated in Section III, however, considerable information can often be deduced without such a detailed study. In addition, averaged quantities such as α may suffice for characterizing the overall energy transfer properties.

The case of noble gases near W has been studied rather

extensively and exemplifies some general aspects. It is a convenient system because of ease of outgassing and the low equilibrium film coverage at room T. Experimental methods and some conclusions are reviewed by GOODMAN and WACHMAN[8,82] and KING.[9]

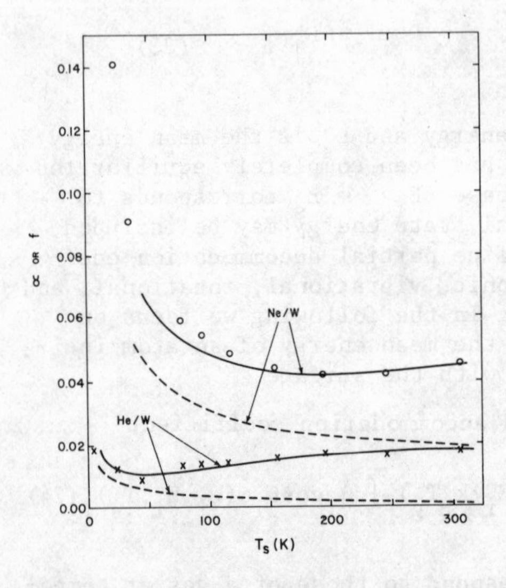

Fig. 12 : Theoretical trapping fractions (dashed curves) and equilibrium accomodation coefficients α (full curves) compared with experimental values of α for He (crosses) and Ne (circles) incident on W. Assumed well depths D/k_B were 50 K and 350 K, respectively. From Ref. 83.

We may summarize the data as follows (see Fig. 12) :

(i) for Xe and Kr, α(T) decreases monotonically with T from the value $\alpha(0) \simeq 1$; (ii) for He, Ne, and Ar, α decreases at first, has a minimum at T_m, and then increases. The values of T_m are about 40, 225, and 600 K, respectively; (iii) for He, $\alpha(T_g, T_s)$ is essentially independent of T_s; for the other gases, it decreases with T_s at fixed T_g.
Interpretation of (iii) is complicated by the finite coverage (except at high T_s). Taking this into account, the data are consistent with a conclusion $(\partial\alpha/\partial T_s) = 0$ for fixed T_g and θ. We may extrapolate this conclusion, tentatively, and treat the case $T_s = 0$. Then $\alpha = -\langle\Delta E_g\rangle/E_g$, the average fractional energy loss in a collision with a cold surface.

IV. 2. Classical Calculations

Perhaps surprisingly, classical theories have been relatively successful in treating this case.[8,9,28,80,82-84] For example, Fig. 12 shows the semi-quantitative agreement obtained with the very simple "soft cube" model. In the latter, the surface atoms are cubes, connected to the solid by a spring with force constant $\kappa = m_s\omega^2$; no momentum parallel to the surface is exchanged. The attraction is simulated by a flat well of depth D, the repulsion by an exponential. Because of its simplicity, the model yields many analytic results.[83,84] Note the correlation between α and the trapping fraction f. This is expected because a trapped atom will tend to equilibrate with the surface.

For several reasons, such agreement as is found above does not validate the model. One is that the experiment[85] was performed with a polycrystalline material, while α is surely facet-dependent.[9,86] Another is the fact that the parallel momentum transfer is found experimentally to be finite.[87,88] This points out the need for more detailed study of the scattering pattern than is provided by α. We illustrate this in Fig. 13 with data[87] for Xe scattered from Pt(111).

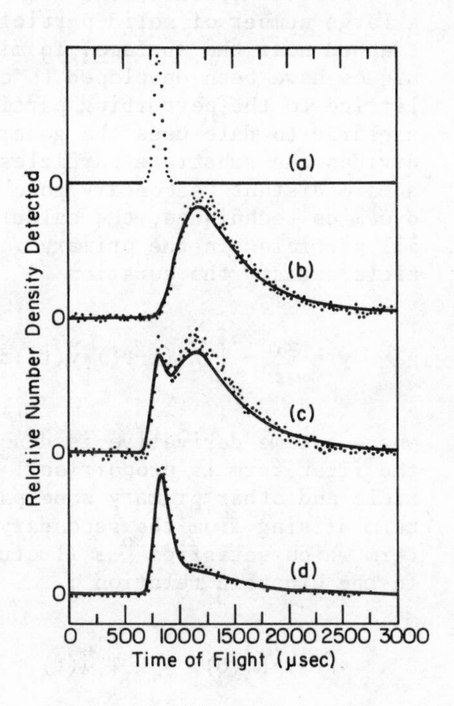

Fig. 13 : Time of flight spectra for Xe at E_i/k_B = 1615 K incident (curve a) upon Pt(111). Scattered intensities shown are for scattered angles θ_f = 0° (curve b), 45° (curve c), and 75° (specular curve d). From HURST et al. (Ref. 87). Curves are fit to 2-component model discussed in the text.

The particles may be divided into 2 components, with relative weights depending on θ_f. The slow (long time) component varies as $\cos\theta_f$ and is well described by a Boltzmann distribution employing the surface temperature (independent of beam energy, which was varied by a factor of 2); this is plausibly interpreted as a trapping-desorption component.[89] The fast component is interpreted as arising from direct inelastic scattering. Its mean energy is found to vary with both[88] E_i and T_s. A detailed study[88] for Ar on polycrystalline W finds

$$\langle E_F \rangle = B_2 \langle E_i \rangle + B_3 \langle 2k_B T_s \rangle \tag{27}$$

with constant coefficients B_2 and B_3; this corresponds to partial accommodation. These results have been interpreted with a soft-sphere model by BARKER and AUERBACH[90] and with a hard cube calculation by GRIMMELMANN, TULLY and CARDILLO.[91]

While such simple models often provide qualitative explanations of the data, one would like also to utilize and test realistic potentials. This necessitates a determination of the trajectories of particles incident on many points within a unit cell, i.e. an expensive calculation.[80] One must solve Newton's equations for a large number of solid particles, especially if the particle is trapped near the surface. To mitigate this problem, various techniques have been developed to characterize the response of the lattice to the perturbing particle.[93-101] The most extensively explored to date uses the generalized Langevin equation.[96,97] One divides the substrate particles into a nearby "primary zone" and a more distant "secondary zone". Using conventional molecular dynamics techniques, the calculation follows the trajectories of all particles in the primary zone. The coordinates of such a particle satisfy the equation[93,98]

$$\ddot{\vec{y}} = \vec{F}' - \int_0^t \overleftrightarrow{\lambda}(t-t') \cdot \dot{\vec{y}}(t')\,dt' + \vec{R}(t), \tag{28}$$

where a time derivative is denoted by a dot over a quantity. Here the first term is proportional to the force from the incident particle and other primary zone particles. The second is a friction term arising from the secondary zone. The last is a fluctuating term which satisfies[98] a fluctuation-dissipation relation analogous to the Einstein relation :

$$\langle \vec{R}(t) \cdot \overleftrightarrow{R}^{\dagger}(0) \rangle = k_B T \, \overleftrightarrow{\lambda}(t) \tag{29}$$

Equation (28) assumes only that the solid is harmonic and that the force on the incident particle from the atoms of the secondary zone can be evaluated from the equilibrium positions of the latter. The forms of λ and R are chosen to yield the principal dynamical properties of the solid.[102]

Figure 14 shows the scattering pattern predicted[92,93] on this basis using alternative models of the interaction potential. The dashed curve corresponds to a rather corrugated potential, obtained by summing Lennard-Jones interactions between Ar and Pt atoms. The solid curve corresponds to a substantially smoother potential. In the latter case, the collisions tend to impart relatively little parallel momentum, so there is a larger reflection in the specular direction and smaller sticking coefficient than in the former case. The better agreement with experiment[103] seen in Fig. 14 substantiates the belief[104] that the conduction electrons tend to smooth out potentials obtained by pairwise summation.

Note that this basis for trapping is the classical analogue of the bound state resonances entering the quantum problem. Normal momentum is converted into parallel momentum during the initial stage of the collision (this need not be phonon-assisted). Subsequent energy transfer is likely as the particle translates along the surface.

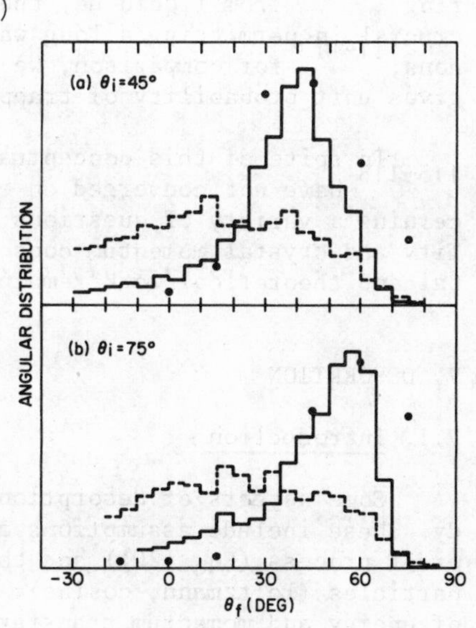

Fig. 14 : Scattered particles distribution for Ar incident on Pt(111) with E_i 60 meV. Calculations (from Ref. 92) assume alternatively a sum of Lennard-Jones potentials (dashed curve) and a smoother potential (solid curve). Data from Ref. 103, Figure from Ref. 93.

IV.3. Quantum Calculations

The basic ingredients of the quantum theory of sticking were presented by DEVONSHIRE and CABRERA et al.[105] The states $\psi_{K,n}$ described in Section II are perturbed when the solid deforms. The matrix element describing the interaction is proportional to

$$M = \langle \vec{K}, n | \nabla v | \vec{K}', n' \rangle \tag{30}$$

multiplied by coupling factors associated with the lattice motion.

These depend on the phonon spectrum; surface phonons play a particularly important role because of their localization near the boundary.[97,106] Approximations often used in the analysis include Eq. (3) (smooth unperturbed surface), a Debye spectrum for bulk phonons, and the neglect of parallel components of $\vec{M}$ (since the unperturbed potential has its largest gradient in the normal direction). Even a 1D calculation is instructive,[107] but suffers from its failure to incorporate the important inelastic bound state resonances discussed in Sec. III.1.

Early calculations[108] of the accommodation coefficient for He/W approximated the unperturbed interaction with a Morse potential. Because it is short-ranged, however, an incident low energy particle has its wavefunction almost totally reflected before it arrives within the phonon force field. This leads to a vanishing probability of trapping as T_s approaches zero, in disagreement with experiment. As in the qualitatively similar problem of He scattering[109-110] from liquid He, the long range of the potential is crucial in permitting a long wavelength particle to excite phonons;[111-113] for comparison, we note that the classical calculation[114] gives unit probability of trapping at low energy for $T_s = 0$.

In spite of this conceptual advance, recent calculations[113, 115-118] have not converged on either experiment or each other concerning a variety of questions (e.g. the roles of three-dimensionality and crystal momentum conservation). Evidently both experimental and theoretical work remain to resolve these issues.

V. DESORPTION

V.1. Introduction

Some aspects of desorption kinetics have been discussed already. These include assumptions about the activated form of a first order process (Eq. (20)) and the distribution of the desorbing particles (Boltzmann, cosine). Moreover, the underlying mechanisms of energy and momentum transfer are also those of sticking. Time reversal and detailed balance arguments can be used advantageously to relate the two phenomena. For example, Eq. (16) is valid for film-vapor equilibrium if I_0 is replaced by the intensity incident from the vapor.

Desorption may be studied by scattering (Section IIIB), by flash desorption (surface heating), or by isothermal methods (vapor evacuation). The papers in refs. (9) and (57) review these methods and interpret results for many systems. At low coverage, a first order kinetic description is appropriate (except for reactive species); only this case will be discussed here.

Denoting by R_{fi} the transition rate from state i to state f, the net rate of change of the occupation of bound state b is given by

$$\dot{n}_b = -(R_{cb} + \sum_{b'\neq b} R_{b'b}) n_b + \sum_{b'\neq b} R_{bb'} n_{b'} + \sum_c R_{bc} n_c \quad (31)$$

where c refers to a continuum state. In isothermal desorption, the last term is negligible and the rates may be calculated from an equilibrium distribution at temperatures T_s of phonons and, initially, bound particles. For flash desorption, in contrast, the last term may be important and the adsorbate and phonon temperatures differ initially.

V.2. Quantum Calculations.

GORTEL et al.[118] have solved eq. (31) to obtain

$$\frac{\theta(t)}{\theta(0)} = s_0 e^{-\lambda_0 t} \{1 + \sum_{b=0}^{N-1} \frac{s_b}{s_0} \exp[(\lambda_0 - \lambda_b)t]\} \quad (32)$$

where λ_i is the i-th eigenvalue of the transition matrix $\overleftrightarrow{R}$, λ_0 being the smallest, and

$$s_b = \sum_{i=0}^{N-1} \frac{f_i}{\theta(0)} \tilde{e}_i^{(b)} \sum_{k=0}^{N-1} e_k^{(b)} , \quad (33)$$

$$f_i = \exp[(E_i - \mu)/k_B T] , \quad (34)$$

with μ as the chemical potential. Here e_i and $\tilde{e}_i$ are right and left eigenvectors of $\overleftrightarrow{R}$. At long times, $t \gg (\lambda_1 - \lambda_0)^{-1}$ so that the sum in Eq. (32) vanishes, leaving only a single exponential with characteristic lifetime $\tau = \lambda_0^{-1}$.

Gortel et al. have used the one phonon approximation incorporating the matrix elements of Eq. (30) :

$$R_{fi} \propto \sum_p \omega_p^{-1} [M_{fi}]^2 (\nu_p + m) \delta(E_i - E_f \pm \hbar\omega_p), \quad (35)$$

$$\nu_p^{-1} = \exp(\hbar\omega_p / k_B T) - 1 , \quad (36)$$

where the sum is over phonon states and the delta function ensures energy conservation. For phonon emission/absorption, m = 1/0 and a minus/plus sign is used. Since the subsequent calculation omits lateral variation of both the unperturbed wave functions and the perturbing force, it is essentially one dimensional. The results are qualitative at best. In the case of weakly bound states and T less than the Debye temperature, the method yields activated desorption in the form of Eq. (20). The activation energy E_d agrees with that of the deepest bound state, within a few percent. The pre-exponential factor ν^{-1} ranges between 2×10^{-9} sec for He/LiF to 2×10^{-12} for He/Ar; it has no simple interpretation as an inverse attempt frequency for the desorption process. In the opposite case of deeply bound levels, a cascade of phonon processes must occur for desorption; ν^{-1} is reduced to 10^{-13} to 10^{-15} sec. This has a qualitative interpretation as arising from the larger number of desorption channels present in this case.

The neglect of parallel momentum transfer described above is surely a severe defect if we may generalize from the sticking problem.[119] GOODMAN and ROMERO[120] improve on this slightly by incorporating Umklapp processes, i.e., exchange of a 2D reciprocal lattice vector. While this allows selective evaporation to occur (Sect. III. 1), it still restricts markedly the contribution of finite wave vector phonons.

V.3. Classical Calculations

A major obstacle to computing desorption lifetimes (or rate constants $k = \tau^{-1}$) is that τ is so much longer than the lattice vibrational period (10^{-13} sec). Using molecular dynamics one needs time steps at least an order of magnitude shorter than the latter, so a conventional calculation is prohibitively expensive. Recently, however, TULLY and coworkers[121,122] have developed an efficient scheme, based on work of KECK,[123] which can surmount this difficulty. Rather remarkable results have emerged already. The rate constant may be factored into two terms,

$$k = k_{TST} F, \tag{37}$$

where k_{TST} is given by transition state theory (TST).[123] The factor F corrects for the fact that when the system crosses the activation barrier, it does not always lead, directly or otherwise, to the activated process. In the present case, F is less than one because a particle crossing a plane $z = z_0$ may make more than a single pass through this surface. This possibility can be related to the sticking probability P_s for a time-reversed trajectory; if $V(z_0) << k_B T$, $F = P_s$. Since sticking occurs more quickly than desorption, P_s can be evaluated much more readily than k. Note that F incorporates the past and future of a system's migration in phase space; such dynamical behavior is absent from the equilibrium TST.

An efficient procedure called the "compensating potential" method has been developed to expedite the evaluation of k_{TST};[121] an alternative approach is the Monte Carlo technique of ADAMS and DOLL.[101,124] Results obtained with the first method for k and k_{TST} are shown in Fig. 15 for Ar and Xe initially adsorbed on Pt(111).

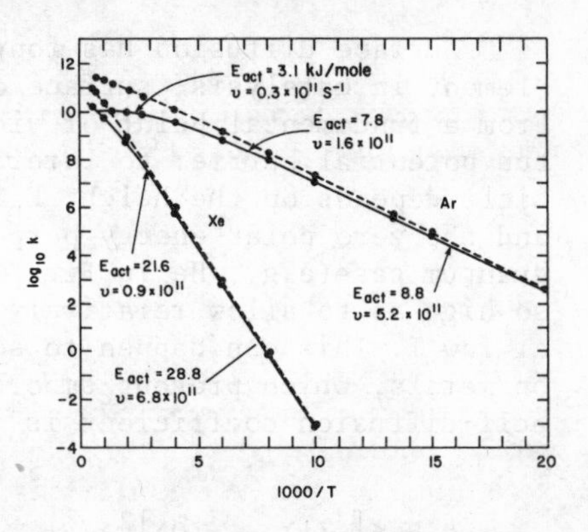

Fig. 15 : Desorption rate constants (solid curves) as a function of T for Ar and Xe on Pt(111) computed by TULLY (Ref. 122). The dashed curves denote the predictions of transition state theory. The prefactors and activation energies are shown for various regions of T.

As T approaches zero, F goes to one, so that k_{TST} and k coincide. The activation energies deduced from the slope in this regime correspond closely to the binding energies U_0 of the potential (9.2 and 29.3 kcal/mole for Ar and Xe), which was based in part on experimental values of U_0.[125,126] At finite T the variation of F is the principal source of the observed deviation of k from an Arrhenius form, Eq. (20). The much smaller deviation of k_{TST} is due[122] to the contribution of an entropy term to the activation free energy.

Note that the pre-exponential factors in Fig. 15 are comparable to vibrational frequencies of these adatoms (10^{11}Hz). Their decrease with increasing T has been interpreted by TULLY[122] as arising from two factors absent from the TST : (a) a reduction in the effective attempt frequency due to a smaller fractional energy change per vibration at high T, and (b) a reduced (non-equilibrium) occupation of weakly bound states at high T because of depletion by desorption.[127]

Among the most interesting results of this study are the distributions of angle and energy. These are obtained by time-reversing a Boltzmann and cos θ distribution of particles crossing a plane above the surface. Only those particles which stick and equilibrate[128] are counted as having desorbed; thus these contribute to the desorption distribution. Since low energy particles have a relatively high sticking probability,[129] the mean energy of the desorbing particles is less than $2k_BT$. The specific results for Ar and Xe on Pt(111) are consistent with data of the Chicago group.[87,103] A calculated deviation from cos θ dependence was too small to have been confirmed unequivocally.

VI. DIFFUSION

VI.1. Introduction

Surface diffusion has long been recognized as an essential element in catalysis, surface ordering and film growth.[3,9,130,132] From a fundamental point of view, it is of interest as a probe of the potential barrier to lateral translation. The motion of a particle depends on the height E_D of this barrier, the temperature T, and the zero point energy perpendicular to the surface.[133] In the quantum case(e.g., He in Fig. 1) the ground state energy may be so high as to allow relatively uninhibited lateral motion, even at low T. This can happen to some extent even for heavier gases on metals, which present smooth equipotential surfaces.[134] The self-diffusion coefficient is defined in two dimensions (2D) by

$$D = \langle [\vec{r}(t) - \vec{r}(0)]^2 \rangle / 4t \tag{38}$$

In this smooth surface case D should vary relatively slowly with T (e.g. a power law).[135] Metal atoms, in contrast, are usually highly localized on metal surfaces. Migration occurs by thermally activated hopping. This can be characterized as a random walk with step length l and time interval τ_D:

$$D = l^2/4\tau_D \tag{39}$$

By analogy with Eq. (20) the T dependence for this case is usually exponential :

$$D = D_o \exp(-E_D/k_BT) \tag{40}$$

E_D is of order one-fifth the binding energy, depending substantially on the particular facet. In fact the surface geometry is often so anisotropic[9,131,136] that the motion occurs along channels. A crude estimate of D_o could be obtained from eqs. (39) and (40) by multiplying the square of a lattice constant (jump distance) by the Debye frequency, i.e., $D_o \simeq 10^{-2}$ cm^2/sec. Experimental results reported for D_o may differ by several orders of magnitude.[9,131,137,138]

One of the most interesting possibilities is a transition between the two regimes described above. DI FOGGIO and GOMER observed this for the case of hydrogen on W(110).[139] At high T, Eq. (40) is satisfied, while at lower T, D varies slowly with T. Thir represents a transition from thermally activated diffusion to quantum migration.

The coefficient D defined above coincides with the Fick's law coefficient D_F only if interactions can be neglected.[140] In that case, D satisfies the Einstein relation to the mobility. This means that one can determine D by evaluating the adsorbed particle's motion in response to a lateral driving force.[141]

Diffusion coefficients have been determined by a variety of methods including field ion and electron microscopy (FIM and FEM),[9,131,136-139,141,142] quasi-elastic neutron scattering (QENS)[143] nuclear magnetic resonance (NMR),[144] the Mössbauer effect,[145] and observation of macroscopic migration.[9,130,146] Each of these has obvious advantages and limitations. For example, QENS, NMR, and Mössbauer are restricted in choice of adsorbate and the necessity to use high specific area substrates (e.g. powders). FIM is conventionally limited to refractory substrates and probes facets of quite small dimensions (< 100 Å).

In spite of these difficulties, intensive effort has yielded impressive results concerning surface diffusion. Information has been gleaned about both the relevant interactions (atom-surface and atom-atom) and the transitions between phases of adsorbed particles. The examples discussed below are chosen to exemplify such progress for cases of chemisorption and physisorption.

VI.2. An FIM Study : Si/W(110)

Diffusion is explored in the FIM study by direct observation of an adatom before and after a heating period of well-defined duration and temperature.[131,141] This must be repeated many times in order to obtain statistical validity. Si on W(110) represents a particularly interesting system because of possible covalent bond formation in the adlayer. In addition, few extensive FIM studies with non-metallic adsorbates exist;[147,148] the results described below should stimulate further efforts.

In the present case, Si is adsorbed above an isosceles triangle of W atoms. Nearest neighbor sites are separated by 2.74 Å, which is $\sqrt{3}/2$ times a W lattice constant. TSONG and CASANOVA[147] found that the Si diffusion could be fitted by Eq. (40), with log D_o = -3.5 ± 1.3 and E_D = 0.70 ± 0.07 eV. The latter is about 15 percent of the binding energy estimated by field desorption.[148] Both of these are comparable to values found for metal adatoms.

The most intriguing results of their study concern the case when several atoms coexist on the plane. Then the FIM can be used to deduce the pair correlation function and the adatoms' mutual interaction.[149,150] Specifically, if V_{ij} is the interaction between Si atoms at sites i and j, the number of pairs on these, or equivalent, sites satisfies

$$n_{ij} = c\, g_{ij} \exp(-V_{ij}/k_B T) \qquad (41)$$

where c is a normalization coefficient and g_{ij} is a weighting factor which depends on the number of possible pairs on the surface plane. Eq. (41) assumes that the coverage is sufficiently low that an adatom interacts with at most one other adatom at any time.[149]

The interaction thus obtained for Si/W(110) is an oscillatory function of separation, as predicted[150] and observed previously for metal adatoms.[149] What is new here is that the oscillations are so dramatic that they lead to a prediction of superlattice formation; the Si atoms lie in parallel chains along the $[1\bar{1}0]$ direction. This has been observed by TSONG and CASANOVA.[147,148] Work is in progress to determine the phase diagram.

The most exciting prospect is to observe the dynamics of the phase transition. We are commencing calculations[152] to describe this phenomenon. It should be noted, however, that the limited facet size will smear any singularities which would be present for an infinite system.

VI.3. CH_4 on Graphite

The motion of CH_4 on graphite is an extremely interesting problem for a variety of reasons. One is the existence of internal degrees of freedom, which can be perturbed and excited by the surface potential energy field.[153] Another is the strong incoherent scattering by neutrons (cross-section 320 barns[143]), which permits an experimental analysis of internal and translational degrees of fredom.[143,154,155] Finally, there is the fact that the motion along the surface is relatively free, so that the two-dimensional (2D) limit is nearly realized. This last is of fundamental interest because of the peculiar behavior of diffusion in 2D. As first found for a hard disk system by ALDER and WAINWRIGHT[156] the velocity autocorrelation function

$$A(t) = \frac{\langle \vec{v}_i(\tau) \cdot \vec{v}_i(\tau + t)\rangle}{\langle v_i^2(\tau)\rangle} \tag{42}$$

has a slow decay (proportional to t^{-1}) at large t. Here the average is over time τ for an arbitrary particle. This means that in 2D

$$D = \frac{1}{2}\int_0^\infty dt\, A(t) \tag{43}$$

diverges. TOXVAERD investigated this with a 2D molecular dynamics simulation for CH_4 on graphite.[157] Numerical considerations confined the upper limit of time to 5×10^{-12}sec. He confirmed, however, this t^{-1} dependence at intermediate and low coverage. There was no apparent difference in this behavior between the realistic model and an artificial 2D model without a substrate potential.

COULOMB, BIENFAIT and THOREL[143] have investigated this system with quasi-elastic incoherent neutron scattering.[158] The energy loss spectrum for 2 meV neutrons was fitted to the Lorentzian form predicted by Fick's law :

$$S(Q,E) = G(Q)\,\frac{\hbar DQ^2}{(\hbar DQ^2)^2 + \Delta E^2} \tag{44}$$

where Q and ΔE are the 2D wavevector and energy transfers, respectively, and G(Q) is a factor which is not crucial to the present discussion. The fit yields values of D which vary markedly with coverage and T. A fully 3D molecular dynamics calculation by SEVERIN and TILDESLEY[159] obtains good agreement with the magnitude and coverage dependence at high T (90 K), but not at lower T (61 K). It is conceivable that the discrepancy originates from the finite crystallite size (≈200 Å) and corresponding inhomogeneous potential fields. In fact, TABONY and COSGROVE[144] found comparable values of D, but a smaller θ dependence, in an NMR pulsed field gradient experiment which used a more heterogeneous substrate. On the theoretical side, the calculations suffer from the limited dimension (40 Å) of the periodically reproduced cell and the omission of substrate dynamics.

As in the example described in Section VI.2, some of the most interesting results of the diffusion study pertain to phase transitions. These include a sharp reduction of D when the CH_4 solidifies[143] and evidence[154] for barriers to rotation when T < 20 K. Rotational ordering is of particular interest because of the role of the substrate potential.

VII. OVERVIEW

We hope that these chapter transmits the feeling of excitement which has arisen in this field. Thanks to ingenious or detailed experiments, the essential ingredients of many phenomena have been revealed. Sophisticated theoretical techniques are being developed to quantify these and others phenomena. We are confident that similar progress will ensue in a variety of problem areas. One of these is the role of heterogeneity, to which little theoretical attention has been paid thus far. Imperfection is indeed important in all kinetic processes at surfaces. We hope, therefore, that systematic experimental investigations will stimulate a corresponding theoretical effort, or vice versa.

ACKNOWLEDGEMENTS

This paper has benefited significantly from preprints from and discussions with many individuals. We are particularty indebted to D.R. Frankl, W.A. Steele, T.T. Tsong, M. Giri, J.P. Toennies, J.C. Tully, T. Engel, D.J. Auerbach, and D. Tildesley.
One of the authors (M.W.C) acknowledges partial support by Department of Energy Contract DE-AC02-79ER10454, and by a travel grant from GNSM - Unitâ Basse Temperature, Padova.

REFERENCES

1 . See articles by D. W. SHAW and E. KALDIS, in *Crystal Growth, Theory and Techniques*, Vol. 1, C.H.L. Goodman ed., Plenum, New York, 1974, pp. 1-48 and 49-192
2 . STEELE W.A., in *The Interaction of Gases with Solid Surfaces*, Pergamon, Owford, 1974
3 . *Ordering in Two Dimensions*, 1980, Sinha S.K. ed., North Holland, New York.
4 . WATSON W.D. and SALPETER E.E., Astrophysical J. 174 (1972) 321-340
5 . COLE M.W. and FRANKL D.R., Surf. Sci. 70 (1978) 585-616
6 . HOINKES H., Rev. Mod. Phys. 52 (1980) 933-970
7 . GOODMAN F.O., CRC Critical Rev. Sol. St. Mat. Sci. 7 (1977) 33-80; CARDILLO M.J., Ann. Rev. Phys. Chem. (1981) to be published.
8 . GOODMAN F.O. and WACHMAN , *Dynamics of Gas-Surface Scattering*, Academic, New York, 1976
9 . KING D.A., CRC Crit. Rev. Sol. St. Mat. Sci. 8 (1978) 167-208
MENZEL D., in *Interactions on Metal Surfaces*, R. Gomer ed., Springer, New York, 1975, pp. 101-142.
10 . CARLOS W.E. and COLE M.W., Surf. Sci. 91 (1980) 339-357
11 . CARLOS W.E. and COLE M.W., Phys. Rev. B 21 (1980) 3713-3720
12 . DERRY G.D., WESNER D., CARLOS W.E., and FRANKL D.R., Surf. Sci. 87 (1979) 629-642
13 . BOATO G.P., CANTINI P., GUIDI C., TATAREK R. and FELCHER G.P., Phys. Rev. B20 (1979) 3959-3969.
14 . COLE M.W. and TOIGO F., Phys. Rev. B23 (1981) 3914-3919.
15 . The approximation of a parabolic potential in not quantitatively accurate; see Ref. 14 discussion of the shifted Morse potential.
16 . COLE M.W., FRANKL D.R. and GOODSTEIN D.L., Rev. Mod. Phys. 53 (1981) 199-210
17 . SILVA-MOREIRA A.F., CODONA J. and GOODSTEIN D.L., Phys. Lett. A76 (1980) 324-326
18 . BRUCH L.W., Phys. Rev. B23 (1981) 6801-6804
19 . DASH J.G., J. Chem. Phys. 48 (1979) 2820-2821
20 . NOVACO A.D. and MILFORD F.J., J. Low Temp. Phys. 3 (1970) 307-329
21 . HAGEN D., J. Chem. Phys. 56 (1972) 5413-5416; BONINO G., PISANI C., RICCA F., and ROETTI C., Surf. Sci 50 (1975) 379-387
22 . An alternative interpretation is that the potential (derived from elastic scattering) corresponds to that produced by averaging over the substrate vibrations; this is the case if the latter is much faster than the adatom motion.
23 . NICKLOW R., WAKABAYASHI N. and SMITH H.G., Phys. Rev. B5 (1972) 4951-4960
24 . CANTINI P., BOATO G., SALVO C., TATAREK R. and TERRENI S., prepint. (1981)

25 . De ROUFFIGNAC E., ALLDREDGE G.P. and de WETTE F.W.,Phys. Rev. B23 (1981)4208-4219
26 . SOMORJAI G.A. and FARRELL H.H., Adv. Chem. Phys. 20 (1972) 215-339
27 . Electronic excitation may also play a role. See BRAKO R. and NEWNS D.M., Sol. St. Comm. 33 (1980)713-715 and references therein.
28 . TULLY J.C., Ann. Rev. Phys. Chem. 31 (1980)319-343
29 . SHUGARD M., TULLY J.C. and NITZAN A., J. Chem. Phys. 66 (1977) 2534-2544
30 . See Sect. IV.
31 . See Refs. 102-114 of Tully's review (Ref. 28 here).
32 . FRISCH R. and STERN O., Z. Phys. 84 (1933) 430-442
33 . DERRY G., WESNER D., KRISHNASWAMY S.U. and FRANKL D.R., Surf. Sci. 74 (1978) 245-258
34 . GARCIA N., CELLI V. and GOODMAN F.O., Phys. Rev. B19 (1979) 634-641
35 . LENNARD-JONES J.E. and DEVONSHIRE A.F.,Nature 137(1936)1069
36 . CHOW H. and THOMPSON E.D., Surf. Sci. 54 (1976) 269-292
37 . CELLI V., in Proc. 12th Rarefied Gas Dynamics Conf., to be published.
38 . BJORNHOLM S. and LYNN J.E., Rev. Mod. Phys. 52(1980)725-932
39 . McRAE E.G., Rev. Mod. Phys. 51 (1979) 541-568
40 . HUTCHISON J.S., Phys. Rev. B22 (1980)5671-5678
41 . HAMAUZU Y., J. Phys. Soc. Japan 42 (1977) 961-970
42 . WILLIAMS B.R., J. Chem. Phys. 55 (1971) 1315-1322
43 . CANTINI P., FELCHER G.P. and TATAREK R., Surf. Sci. 63 (1977) 104-112
44 . FRANK H., HOINKES H. and WILSCH H., Surf. Sci. 63 (1977) 121-142
45 . CANTINI P. and TATAREK R., Phys. Rev. B23 (1981) 3030-3040
46 . BRUSDEYLINS G., DOAK R.B. and TOENNIES J.P., Phys. Rev. Lett. 46 (1981)437-439; FEUERBACHER B., ADRIAENS M.A. and THUIS H., Surf. Sci. 94 (1980) L171-177; YERKES S.C. and MILLER D.R., Vac. Sci. Tech. 17 (1980) 126-129
47 . BRUSDEYLINS G., DOAK R.B.and TOENNIES J.P., J. Chem. Phys. 75 (1981) 1784-1793
48 . MASON B.R. and WILLIAMS B.R., Phsy. Rev. Lett. 46 (1981)1138-1142
49 . DEVONSHIRE A.F., Proc. Roy. Soc. A156 (1936)37-44
50 . STRACHAN C., Proc. Roy. Soc. A158 (1937)591-605
51 . The corresponding peak in Fig. 6 lies at different angle(33.8°) because of a slightly different incident wave vector.
52 . The two peaks correspond to distinct values of Q and (Q)
53 . CHEN T.S., de WETTE F.W. and ALLDREDGE G.P., Phys. Rev. B15 (1977) 1167-1186
54 . A model incorporating surface relaxation has recently been proposed to explain this by BENEDEK G. and GARCIA N., Surf. Sci. 103 (1981) L143-148

55 . ENGEL T., J. Chem. Phys. 69 (1978) 373-385
56 . This behaviro is not universal, see below, Ref. 55, and CAMPBELL C.T., ERTL G., KUIPERS H. and SEGNER J., Surf. Sci. 107 (1981) 207-219 for CO/Pt(111).
57 . MADIX R.J., CRC Crit. Rev. Sol. St. Mat. Sci. 8 (1978) 143-166
58 . As described in Ref. 55, there is a velocity dependence of the detector efficiency.
59 . BRENIG W. and SCHONHAMMER K, Zeits. Phys. B24 (1976) 91-97
60 . D'AGLIANO E.G.,KUMAR P., SCHAICH W.L. and SUHL H., Phys. Rev. B11 (1975)2122-2143
61 . CONRAD H., ERTL G., KOCH J. and LATTA E., Surf. Sci. 43 (1974) 462-480
62 . The analysis in the actual case is discussed in Refs. 5, 31, and 33 of the Engel paper (Ref. 55).
63 . This neglects the coverage dependence of S; in the subsequent discussion n_s is very small.
64 . EHRLICH G., J. Phys. Chem. 59 (1955) 473-477; KISLIUK P.J., J. L. Chem. Sol. 3 (1957) 95-105
65 . KING D.A. and WELLS M.G., Proc. Roy. Soc. A339 (1974)245-269
66 . SCHONHAMMER K., Surf. Sci. 83 (1979) 633-636
67 . DRAPER C.W. and ROSENBLATT G.M., J. Chem. Phys. 69 (1978) 1465 1472 and references therein.
68 . ROWE R.G. and EHRLICH G., J. Chem. Phys. 63 (1975)4648-4665
69 . BOATO G., CANTINI P. and MATTERA L., J. Chem. Phys. 65 (1976) 544-549
70 . WOLKEN G., Chem. Phys. Lett. 23 (1973) 373-379
71 . GELB A. and CARDILLO M., Surf. Sci. 75 (1978) 119-214
72 . RAMESH V. and MARSDEN D.J., Vacuum 24 (1974) 291)294; THOMSON R.P., ANDERSON D. and BERNASEK S.L., Phys. Rev. Lett. 44 (1980) 743-746
73 . HEPBURN J.W., NORTHRUP F.J., OGRAM G.L., POLANYI F.C. and WILLIAMSON J.M., submitted to Phys. Rev. Lett.; ETTINGER D., HONMA K., KEIL M. and POLANYI J.C., submitted to Phys. Rev. Lett.
74 . FRENKEL F., HAGER J., KRIEGER W., WALTHER H., CAMPBELL C.T., ERTL G., KUIPERS H.and SEGNER J., Phys. Rev. Lett. 46 (1981) 152)155
75 . McLELLAND G.M., KUBIAK G.D., RENNAGEL H.G. and ZARE R.N., Phys. Rev. Lett. 46 (1981) 831-834
76 . KLEYN A.W., LUNTZ A.C. and AUERBACH D.J., IBM Research Report RJ-3188, (1981)
77 . This agrees with a prediction of NICHOLS W.L. and WEARE J.H., J. Chem. Phys. 66 (1977) 1075-1078
78 . GODDARD P.J., WEST J., and LAMBERT R.M., Surf. Sci. 71 (1978) 447-461
79 . SCHEPPER W., ROSS U. and BECK D., Zeits. Phys. A290 (1979) 131-141
80 . McCLURE J.D., J. Chem. Phys. 57 (1972) 2810-1822; KLEIN J.R. and COLE M.W., Surf. Sci. 79 (1979) 269-288

81 . COWIN J.P.,YU C.F., SIBENER S.J. and HURST J.E., J. Chem. Phys. 75 (1981) 1033-1034
82 . GOODMAN F.O., J. Phys. Chem. 84 (1980) 1431-1445
83 . LOGAN R.M., in *Solid State Surf. Sci.* 3, M. Green ed., Marcel Dekker, New York, 1973, pp. 1-103
84 . LOGAN R.M., Surf. Sci. 15 (1969) 387-402
85 . THOMAS L.B., in *Proc. 12th Intl. Rarefied Gas Dynamics Symp.*, Vol. 1, C.L. Brundin ed., Academic Press, 1967, p. 155
86 . WANG C. and GOMER R., Surf. Sci. 84 (1979) 329-354
87 . HURST J.E., BECKER C.A.,COWIN J.P., JANDA K.C., WHARTON L. and AUERBACH D.J., Phys. Rev. Lett. 43 (1979) 1175-1177
88 . JANDA K.C., HURST J.E., BECKER C.A.,COWIN J.P., WHARTON L. and AUERBACH D.J., J. Chem. Phys. 72 (1980) 2403-2410
89 . In fact, below T_s = 170 K, finite residence times were observed.
90 . BARKER J.A. and AUERBACH D.J., Chem. Phys. Lett. 67 (1979) 393-396
91 . GRIMMELMANN E.K., TULLY J.C. and CARDILLO M.J., J. Chem. Phys. 72 (1980) 1039-1042. This calculation uses an enhanced effective mass for the surface atom to simulate collective effects.
92 . TULLY J.C. and GRIMMELMANN E.K. : unpublished.
93 . TULLY J.C., Accounts of Chemical Research, 14 (1981) 188-194.
94 . GOODMANN F. O., Surf. Sci. 3 (1965) 386-414
95 . BARKER J.A. and STEELE W.A., Surf. Sci. 74 (1978) 596-611
96 . ADELMAN S.A. and DOLL J.C., J. Chem. Phys. 64 (1976) 2375-2388 ADELMAN S.A. and GARRISON B.J., J. Chem. Phys. 65 (1976) 3751-3761
97 . DIEBOLD A.C., ADELMAN S.A., and MOU C.Y., J. Chem. Phys. 71 (1979) 3236-3251
98 . TULLY J.C., J. Chem. Phys. 73 (1980) 1975-1985
99 . GARRISON B.J. and ADELMAN S.A., J. Chem. Phys. 67 (1977) 2379-2380
100. GARRISON B.J., DIESTLER D.J., and ADELMAN S.A., J. Chem. Phys. 67 (1977) 4317-4320
101. ADAMS J.E. and DOLL J.D., J. Chem. Phys. 74 (1981) 1467-1471
102. Relatively simple models suffice because the scattering properties are not very sensitive to fine details of the lattice dynamics;see Ref. 98.
103. HURST J.E., BECKER C.A., COWIN J.P., WHARTON L., AUERBACH K.J. and JANDA K, to be published.
104. LIFSHITZ E.M., Sov. Phys. JETP 2 (1956) 73-83
105. DEVONSHIRE A.F., Proc. Roy. Soc. A158 (1937) 269-279; CABRERA N., CELLI V., GOODMAN F.O. and MANSON J.R., Surf. Sci. 19 (1970) 67-92
106. KELLY M.J., Surf. Sci. 108 (1981) L407-411
107. GOODMAN F.O. and GILLERAIN J.D., J. Chem. Phys. 54 (1971) 3077-3083
108. GOODMAN F.O., J. Chem. Phys. 56 (1972) 6082-6088
109. EDWARDS D.V. and FATOUROS P.P., Phys. Rev. B17 (1978) 2147-2159
110. ECHENIQUE P.M. and PENDRY J.B., J. Phys. C9 (1976) 3183-3191
111. GOODMAN F.O., J. Chem. Phys. 55 (1971) 5742-5753

112. GARCIA N. and IBANEZ J., J. Chem. Phys. 64 (1976) 4803-4804
113. GARCIA N., CELLI V. and MANSON J.R., J. Chem. Phys. 72 (1980) 3436-3437
114. ZWANZIG R.W., J. Chem. Phys. 32 (1960) 1173-1177
115. KNOWLES T.R. and SUHL H., Phys. Rev. Lett. 39 (1977) 1417-1420
116. SEDLMEIER R. and BRENIG W., Zeits. Phys. 536 (1980) 245-250
BRENIG W., Zeits. Phys. B36 (1980) 227-233
117. DOYEN G., Surf. Sci. 89 (1979) 238-250
118. KREUZER H.J., Surf. Sci. 100 (1980) 178-198, and references therein; GORTEL Z.W., KREUZER H.J. and TESHIMA R., Phys. Rev. B22 (1980) 5655-5670
119. See also DE S.G., LANDMAN U. and RASOLT M., Phys. Rev. B21 (1980) 3256-3268 and JEDRZEJK C., FREED K.F., EFRIMA S. and METIU H., Chem. Phys. Lett. 79 (1981) 227-232 and Surf. Sci. 109, in press.
120. GOODMAN F.O. and ROMERO I., J. Chem. Phys. 69 (1978) 1086-1091
121. GRIMMELMANN E.K., TULLY J.C. and HELFAND E., J. Chem. Phys. 74 (1981) 5300-5310
122. TULLY J.C, Surf. Sci. (1981) in press
123. KECK J.C., Disc. Far. Soc. 33 (1962) 173-182; and Adv. Chem. Phys. 13 (1967) 85-121
124. ADAMS J.E. and DOLL J.D., Surface Science 103 (1981) 472-481
J. Chem. Phys. 74, 5332-5333.
125. For Ar, an estimate was made from Ar/W data of ENGEL T. and GOMER R., J. Chem. Phys. 52 (1970) 5572-5580
126. NIEUWENHUYS B.E., MEIJER D T. and SACHTLER W.M.H., Phys. Stat Sol. A24 (1974) 115-122
127. PAGNI P.J. and KECK J.C., J. Chem. Phys. 58 (1973) 1162-1177
128. "Sticking" is defined somewhat arbitrarily as becoming bound with less than $-3k_BT$ total energy; the results are not sensitive to this choice.
129. The reverse of this argument is appropriate to the case of activated adsorption (i.e. a potential energy maximum outside the region of attraction). See CARDILLO M.J., BALOOCH M. and STICKNEY R.E, Surf. Sci. 50 (1977) 263-278
130. LANGMUIR I. and TAYLOR J.B., Phys. Rev. 44 (1973) 423-458
131. EHRLICH G. and STOLT K., Ann. Rev. Phys. Chem. 31 (1980) 603-637
132. REED D.A. and EHRLICH G., Surf. Sci. 102 (1981) 588-609
133. HOLLENBACH D. and SALPETER E.E., J. Chem. Phys. 53 (1970) 79-86
134. For example, Ar and Xe on Pt(111) have zero point energy about 0.025 kJ/mole, or one-fourth of E_D. See TULLY, Refs. 93 and 122.
135. For non-interacting particles undergoing translation with a 2D momentum transfer lifetime τ_m, $D = kT\,\tau_m/m$. τ_m arises from phonon and impurity scattering. The associated weak T dependence of D appears in the results of Tully, Ref. 122.
136. AYRAULT G. and EHRLICH G., J. Chem. Phys. 60 (1974) 281-294
BASSET D.W. and WEBBER P.R., Surf. Sci. 70 (1978) 520-531

137. KELLOGG G., TSONG T.T. and COWAN P., Surf. Sci. 70 (1978) 485 519
138. An exceptional case ($D_0 \simeq 10^{-8}$ cm^2 for Xe/W(110)) war reported by CHEN R. and GOMER R., Surf. Sci. 94 (1980) 456-468; see BANAVAR J.R., COHEN M.H. and GOMER R.,Surf. Sci. 107(198) 113-126 for a discussion.
139. DI FOGGIO R. and GOMER R., Phys. Rev. Lett. 44 (1980)1258-1260
140. MAZENKO G., BANAVAR J.R. and GOMER R., Surf. Sci. 107(1981) 459-468
141. TSONG T.T. and KELLOGG G., Phys. Rev. B12 (1975) 1343-1353
142. GOMER R., Surf. Sci. 38 (1973) 373-393; TSONG T.T. and CASANOVA R., Phys. Rev. B22 (1980) 4632-4649
143. COULOMB J.P., BIENFAIT M. and THOREL P., J. de Physique 42 293-306
144. RICHARDS M.G., in *Phase Transitions in Surface Films*, ed. by Dash, J.G. and Ruvalds J., Plenum, N. Y., 1980, pp. 165-192; TABONY J. and COSGROVE T., Chem. Phys. Lett. 67(1979) 103-106
145. SCHECHTER H., SUZANNE J. and DASH J.G., Phys. Rev. Lett. 37 706-710
146. BONZEL H.P., CRC Crit. Rev. Sol. St. Mat. Sci. 6 (1976) 171-194; BUTS R. and WAGNER H., Surf. Sci. 63 (1977) 448-459
147. TSONG T.T. and CASANOVA R., Phys. Rev. Lett. 47 (1981)113-116
148. CASANOVA R. and TSONG T.T., 1981, preprint
149. CASANOVA R. and TSONG T.T., Phys. Rev. B22(1980)5590-5598
150. FINK H.W., FAULIAN K and BAUER E., Phys. Rev. Lett. 44(1980) 1008-1011
151. EINSTEIN T.L., Crit. Rev. Sol. St. MAT. Sci. 7(1978)261-288
152. The calculations are by BOYER J.R., NICHOLSON D. and COLE M.W.
153. SEVERIN E.S. and TILDESLEY D.J., Mol. Phys. 41 (1980)1401-1418
154. NEWBERRY M.W., RAYMENT R., SMALLEY N., THOMAS R.K., and WHITE J.W., CPL 59(1978) 461-466
155. BOMCHIL G., HULLER A., RAYMENT R., ROSEN S.K., SMALLEY M.N. THOMAS R.K., and WHITE J.W., Phil. Trans. Roy. Soc. B, to be published.
156. ALDER B.J. and WAINWRIGHT T.E., Phys. Rev. A1(1970) 18-21
157. TOXVAERD S., Phys. Rev. Lett. 43 (1979) 529-531
158. The CH_4 was adsorbed on papyex, a form of exfoliated graphite similar to grafoil. The distribution of basal plane orientations was included in the data analysis.
159. SEVERIN E.S., D. Phil. Thesis, Oxford University (unpublished) 1981, and SEVERIN E.S. and TILDESLEY D.J., to be published.

THREE DIMENSIONAL NUCLEATION

Joseph L. KATZ,
Department of Chemical Engineering
The Johns Hopkins University
Baltimore, Maryland 21218
U.S.A.

I. INTRODUCTION

The realization that phase changes, i.e. condensation of a vapor to its liquid or solid phase, boiling of a liquid, melting of a solid does not necessarily occur at equilibrium was published as early as 1724 by G. FAHRENHEIT.[1] J. T. LOWITZ in 1785 showed that acetic acid crystals are selective nuclei for the crystalization of acetic acid.[1] But it was not until 1926, when a paper by VOLMER and WEBER[2] was published, that a predictive equation for the rate of nucleation became available. Advances quickly followed. FARKAS,[3] based on a suggestion by SZILLARD, provided a kinetic scheme for calculating the unknown constant. BECKER and DÖRING[4] used this scheme but obtained the backward rate (this terminology will be explained later) from the Kelvin equation. ZELDOVICH suggested a continuum version. Many others have also made significant contributions. A very good review of much of the early history can be found in Volmer's book.[1] While this book is in some places difficult to read it is nevertheless highly recommended. Other sources are the three volume set edited by ZETTLEMOYER[6,7,8] and the book by ABRAHAM.[9]

In this chapter we shall begin with a brief review of the basic concepts. Then we shall present nucleation theory in its standard form, pointing out some weaknesses. It will then be extended to associated vapors;[10,11] this extension will simultaneously provide the opportunity to describe the continuum version. A kinetic approach to nucleation which has the great advantage that it clearly separates kinetic and thermodynamic quantities will then be presented.[12,13,14] This approach is especially valuable since it makes more simple and much more self evident the procedure for

B. Mutaftschiev (ed.), Interfacial Aspects of Phase Transformations, 261–286.

generalizing nucleation theory for more complicated situations. Two examples of such generalizations will be presented. They are : the nucleation of voids from vacancies in the presence of interstitial atoms,[15] and nucleation with simultaneous chemical reaction.[16] It will then be used to describe nucleation in dense systems, e.g. crystallization from the melt.

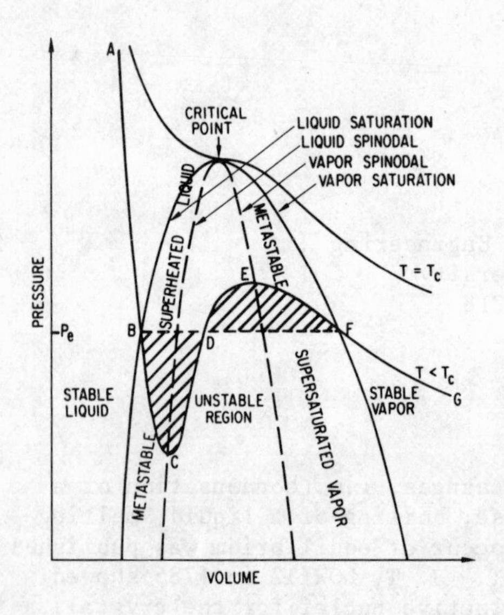

Fig. 1 : Schematic of pressure-volume isotherms for a fluid at and below the critical temperature.

II. THE SPINODAL LIMIT

Figure 1 depicts the pressure-volume diagram of a pure substance. Two isotherms are shown, one for the critical temperature and one for a subcritical temperature. The dashed line connecting B to F denotes the coexistance of the liquid and gas phases at the pressure p_e. However if no nuclei are present it is experimentally possible to temporarily continue along the line F to E and the line B to C. Such states are not stable states. It is usual to call them metastable and to call the locus of all points on the isotherm where $(\partial p/\partial V)_T = 0$ by the name, the spinodal. Certainly nucleation should occur before (or at) the spinodal since at the spinodal the system becomes unstable to even the smallest density fluctuations. The liquid spinodal turns out to be a good predictor of nucleation of bubbles in superheated liquids. Figure 2 is a plot of the temperature at which bubble nucleation occurs as a function of pressure (both quantities are in reduced units). While the van der Waals equation of state is poor, the Bertholet equation,

$$(p + a/TV^2)\ (V - b) = RT, \qquad (1)$$

does a good job. Nonetheless, the kinetic version of nucleation theory which we will present later is much better (on the scale of this figure, it would go through the points perfectly). Vapor nucleation is poorly predicted by the spinodal; it always occurs at much smaller supersaturations. We shall see that the kinetic theory is (on this scale) also an excellent predictor of vapor condensation nucleation.

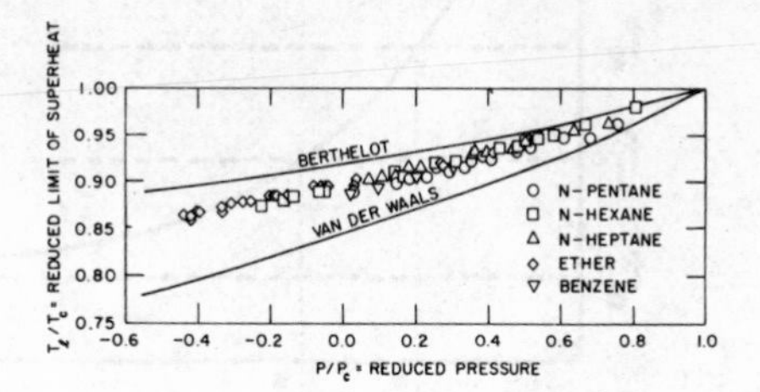

Fig. 2 : Reduced limits of superheat as a function of reduced pressure. The thermodynamic limits of superheat calculated from the Berthelot and van der Waals equations are plotted.

III. CONCEPTS

We shall now briefly discuss two concepts which are the source of much confusion in nucleation theory, the concepts of "critical size" and "the equilibrium distribution of clusters".

III.1. Critical Size

Consider a cluster of molecules. The probability per unit area, $\gamma(i)$, that a molecule will leave the cluster depends on the surrounding conditions, the temperature, on properties of the molecule, etc. *and* on the size of the cluster. This size dependence arises because a small cluster has fewer atoms which exert an attractive effect on any given atom than is the case for a larger cluster or for a flat surface. This effect increases the vapor pressure of a substance over curved surfaces. At large sizes the Kelvin equation,

$$\ln(p/p_\infty) = 2\sigma/r, \tag{2}$$

accurately describes this effect. Thus the larger the cluster the smaller is $\gamma(i)$. Figure 3 is a plot of the "backward" rate, $b(i)$, which is defined as the rate at which a cluster consisting of i molecules loses one molecule. Obviously $b(i) = \gamma(i)\ a(i)$ where $a(i)$ is the surface area of the cluster. Note that the $b(i)$ curve does not extend to very small sizes, where it is possible that certain sizes of unusually large stability occur. But beyond this initial region, the curve is smoothly decreasing. At equilibrium (e.g. in a saturated vapor) the "forward" rate (i.e. the rate at which clusters add molecules) is equal to the backward rate for a flat surface. Also shown in Fig. 3 is the equilibrium forward rate, f_e.

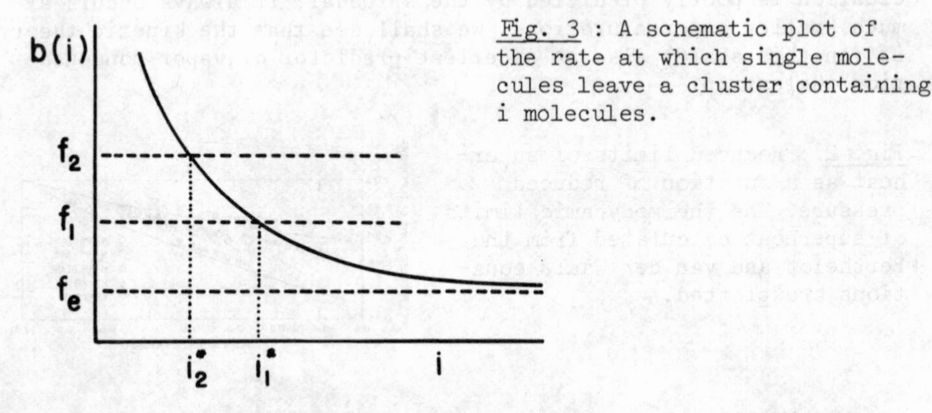

Fig. 3 : A schematic plot of the rate at which single molecules leave a cluster containing i molecules.

It is, except for a negligible geometric factor, independent of cluster size. If the mother phase is "supersaturated", then the forward rate has to be higher than the equilibrium forward rate since supersaturation means that the concentration of condensing species is higher than the equilibrium concentration. Also shown in Fig. 3 are forward rates, f_1 and f_2, for a supersaturated system and for an even more supersaturated system respectively. One can see that f_e asymptotically touches the b(i) curve (at infinity), while f_1 and f_2 cross it. The size cluster for which this crossing occurs is called "the critical size". For super-critical sized clusters, the probability for a cluster to grow is greater than is the probability that it will shrink. For sub-critical sized clusters the reverse is true. "Critical size" is a kinetic concept and it changes as any of the parameters of the system change.

III.2. The Equilibrium Distribution of Clusters

We shall use without further justification the equation

$$n(i) = N \exp\{-W(i)/kT\}, \tag{3}$$

where n(i) is the number of clusters of size i, N is a normalization factor, k is Boltzmann's constant, T is the absolute temperature and W(i) is the reversible work it would take to create a cluster of size i from the mother phase. The validity of this equation has been the subject of great controversy and continues to be so. Nonetheless, it is conventional in nucleation theory to use it. What can be rigorously derived is the equation that the reversible work required to create a critical sized cluster containing i molecules in (unstable) equilibrium with a surrounding phase (called the mother phase) is given by

$$W(i) = \sigma a(i) - [p_i - p_a]V_i, \quad (4)$$

where $\sigma a(i)$ is the surface free energy of the cluster, p_i and V_i are its pressure and volume and p_a is the ambient pressure. If the cluster size i is not the critical size, then one has to add to the right hand side of eq. (4) an extra term, with the result

$$W(i) = \sigma a(i) - [p_i - p_a]V_i + i[\mu_i(p_i) - \mu_a(p_a)], \quad (4')$$

where $\mu_i(p_i)$ is the chemical potential of the cluster at its internal pressure p_i and $\mu_a(p_a)$ is the chemical potential of the mother phase at the ambient pressure p_a. Thus for a liquid with a constant compressibility, $\kappa \equiv -(\partial V/\partial P)_T/V)$, one obtains

$$W(i) = \sigma\, a(i) - i\, k\, T \ln [p_a/p_\infty] + V_i [p_a - p_\infty] + iv\kappa [p_i - p_\infty]^2/2, \quad (5)$$

where p_∞ is the equilibrium vapor pressure and v is the molecular volume at this pressure. The last two terms in eq. (5) are both usually negligible. One therefore usually finds the statement that the concentration of clusters of size i as a function of size, for a supersaturated gas, is given by the equation

$$n(i) = N \exp[-\sigma a(i)/kT + i \ln p_a/p_\infty]. \quad (6)$$

It follows that for a saturated gas, i.e. when $p_a = p_\infty$, the equilibrium distribution of clusters is

$$n(i) = N \exp[-\sigma a(i)/kT]. \quad (7)$$

IV. CONVENTIONAL VERSION OF NUCLEATION THEORY.

Because of the rapid decrease in the number of clusters as a function of size, it is usually an excellent approximation to suppose that clusters increase or decrease in size by acquiring or losing single molecules. Only for substances whose vapor phase is significantly associated, e.g. acetic acid vapor, does this approximation fail. For the case, an appropriate generalization will be presented in the next sections.

The net rate J at which clusters of size less than i+1/2 become larger than size i+1/2 is

$$J(i) = f(i)\, n(i) - b(i+1)\, n(i+1), \tag{8}$$

where f(i), the forward rate, is the rate at which single molecules impinge on a cluster of size i, n(i) is the concentration of such clusters, and b(i+1) the backward rate, is the rate at which single molecules will leave clusters containing i+1 molecules. Figure 4 symbolically shows this process on a line in size space.

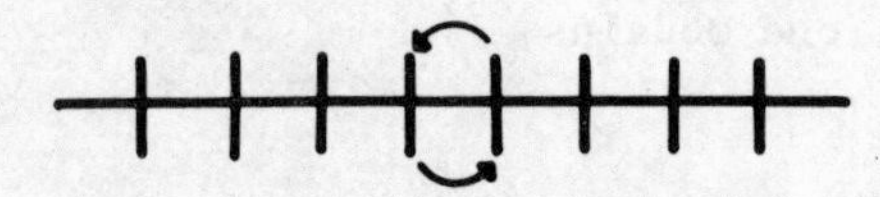

Fig. 4 : The processes by which clusters grow larger or smaller if only single molecules can be added or lost at each step.

In many nucleation processe, the forward rate is fairly well known. For example, for an ideal gas containing spherical clusters

$$f(i) = \beta a(i), \text{where } \beta = \alpha p/(2\pi m k T)^{1/2} \text{ and } a(i) = 4\pi r^2, \tag{9}$$

where α is the condensation coefficient, i.e. the fraction of arriving molecules which actually condense, p is the pressure, m is the molecular mass and r is the radius of the cluster. If one also assumes that clusters have the same density as the infinite liquid phase (would have at the same temperature T and at the equilibrium pressure p_∞), one can express the surface area as

$$a(i) = 4\pi\left(\frac{3v}{4\pi}\right)^{2/3} i^{2/3}. \tag{9'}$$

The volume per molecule in the liquid can be calculated from $v = MN_o/d$, where M is the molecular weight, d is the density of the liquid and N_o is Avogadro's number.

However, little is known about the backward rate. One can always write that $b(i) = \gamma(i)\, a(i)$ where $\gamma(i)$ is the probability per unit area that a molecule will leave a cluster of size i and a(i) is the cluster surface area, but $\gamma(i)$ still depends on i.

The conventional source of information about b(i) is the "constrained equilibrium distribution of clusters". If one could somehow constrain the mother phase to be in equilibrium while not changing any of the thermodynamic variables, e.g. pressure or temperature, then one could use eq. (8) to obtain b(i) since at equilibrium J(i) = 0 for all i. Thus, one finds that

$$b(i + 1) = f(i)n^c(i)/n^c(i + 1), \tag{10}$$

where $n^c(i)$ is the concentration of i sized clusters that would exist if such a constraint were possible. It is further assumed that $n^c(i)$ is given by eqs. (3) and (4) or equivalently, eq. (5). Substituting (10) in (8) and rearranging terms one obtains

$$\frac{J(i)}{f(i)n^c(i)} = \frac{n(i)}{n^c(i)} - \frac{n(i + 1)}{n^c(i + 1)}. \tag{11}$$

Note that the terms on the right hand side differ only by the value of their index. Thus on adding such equations for successive values of i, successive terms cancel with the result that

$$\sum_{i=1}^{\bar{i}} \frac{J(i)}{f(i)n^c(i)} = \frac{n(1)}{n^c(1)} - \frac{n(\bar{i} + 1)}{n^c(\bar{i} + 1)} . \tag{12}$$

Thermodynamic models for $n^c(i)$, i.e. eqs. (3) to (5), predict that $n^c(i)$ increases without bounds as i increases. Because n(i), the actual concentration of cluster is necessarily finite, the last term in eq. (12) goes to zero for large i. Furthermore, it is conventional to choose the monomer density in the constrained state to be the same as in the actual supersaturated vapor, i.e. $n^c(1) = n(1)$.

For processes where the mother phase is a gas, a steady state is very rapidly established. Since the time dependence of n(i) is given by the population balance equation,

$$\frac{dn(i)}{dt} = J(i - 1) - J(i), \tag{13}$$

steady state means that dn(i)/dt = 0 and therefore J(i = 1) = J(i) = constant $\equiv$ J for all i. One thus obtains for a sufficiently large value of $\bar{i}$,

$$J = 1/\sum_{i=1}^{\overline{i}} \frac{1}{f(i)n^c(i)} . \quad (14)$$

Substituting (9), (9'), (3) and (5) into (14) one obtains

$$J = \frac{2N\alpha p v}{\sqrt{2\pi m k T}} \left(\frac{\sigma}{kT}\right)^{1/2} \exp\left\{\frac{-16\pi}{3}\left(\frac{\sigma}{kT}\right)^3 \left[\frac{v}{\ln(p/p_\infty)}\right]^2\right\}^* . \quad (15)$$

In typical units this becomes

$$J = 5.5(10^{31})\left(\frac{p_\infty}{p}\right)\left(\frac{\sigma M}{d}\right)^{1/2}\left(\frac{p}{p_\infty}\right) \exp\left\{-17.56\left(\frac{\sigma}{T}\right)^3\left[\frac{M}{d \ln (p/p_\infty)}\right]^2\right\} \quad (16)$$

where σ is in erg/cm^2, M is in g/mol, d is in g/cm^3, p is in atmospheres and T is in Kelvins. p_∞ is the equilibrium vapor pressure at the temperature T.

V. ASSOCIATED VAPORS

In the last section we allowed clusters to increase or decrease in size by gaining or losing only one molecule at a time. Substances having a hydrogen bond or can form other unusually stable complexes often have vapor phases which are significantly associated.[10,11] Good examples are formic, acetic and propionic acids. The equation describing the nucleation process is just like the one in Section IV except that we now have to allow for clusters growing and shrinking several molecules at a time. On the line in size space this is simbolically shown in figure 5. The corresponding equation for the net rate J at which clusters of a size less than i + 1/2 become larger than i + 1/2 is

$$J(i) = \sum_{\alpha=1}^{m} \sum_{\beta=1}^{\alpha} \{f_\alpha(i-\beta)n(i-\beta) - b_\alpha(i-\beta+\alpha)n(i-\beta+\alpha)\} \quad (17)$$

where the index m is the size of the largest species which is present in appreciable concentration, f_α and b_α are the rates at which clusters containing α molecules arrive at or leave from clusters

---

* To actually obtain eq. (5) one also converts the summation to an integral and approximates the integral to a Gaussian. Many authors have commented on the accuracy of these steps. The definitive word on this subject was published by E. R. Cohen.[17] He showed that under almost all conditions of interest in nucleation processes, they are extremely accurate.

containing the number of molecules indicated in parentheses . For an ideal gas mixture, $f_\alpha(i) = a(i)p_\alpha(2\pi\alpha mkT)^{1/2}$ where p_α is the partial pressure of α sized clusters. Again, assuming a constrained equilibrium distribution then each term in braces in eq. (17) must equal zero, i.e.,

$$b_\alpha(i-\beta+\alpha) = f_\alpha(i-\beta)n^c(i-\beta)/n^c(i-\beta+\alpha)$$

$$\text{for } 1 < \beta < \alpha < m \quad \text{and all } i > \alpha. \tag{18}$$

Fig. 5 : The processes by which clusters grow larger or smaller if several molecules can be added or lost at each step.

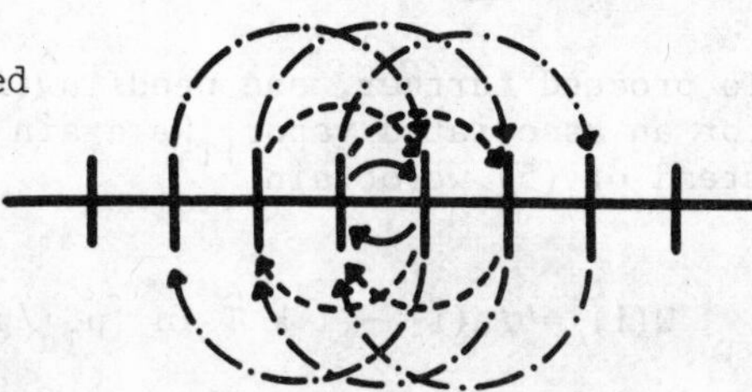

Substituting (18) in (17) one obtains

$$J(i) = \sum_{\alpha=1}^{m}\sum_{\alpha=1}^{\alpha} f_\alpha(i-\beta)n^c(i-\beta)\left[\frac{n(i-\beta)}{n^c(i-\beta)} - \frac{n(i-\beta+\alpha)}{n^c(i-\beta+\alpha)}\right]. \tag{19}$$

Expanding both the product $f_\alpha n^c$ and the term in brackets in Taylor series about i and keeping only the lowest non-zero term one obtains

$$J(i) = -n^c(i)\frac{\partial(n/n^c)}{\partial i}\sum_{\alpha=1}^{m}\alpha^2 f_\alpha(i), \tag{20}$$

which, using the ideal gas equation for $f_\alpha(i)$ given above, results in

$$J(i) = -\frac{a(i)n^c(i)}{(2\pi mkT)^{1/2}}\frac{\partial(n/n^c)}{\partial i}\sum_{\alpha=1}^{m}\alpha^{3/2}p_\alpha. \tag{21}$$

Rearranging and integrating eq. (21) one obtains

$$\frac{(2\pi mkT)^{1/2}}{\sum_{\alpha=1}^{m} \alpha^{3/2} p_\alpha} \int_1^\infty \frac{J(i)}{a(i)n^c(i)} di = -\int_1^\infty \frac{\partial(n/n^c)}{\partial i} di = -n/n^c \Big|_1^\infty = 1, \tag{22}$$

because we choose the conditions of the constrained state such that $n(1) = n^c(1)$ and because the constrained distribution increases without bounds as i increases. Again, at steady state J becomes size independent. From eq. (22), using the fact that the partial pressures $p_\alpha = X_\alpha p$ where X_α are the mole fractions and p is the total pressure, one obtains

$$J = \frac{p}{(2\pi mkT)^{1/2}} \Sigma\alpha^{3/2} X_\alpha \int_1^\infty [a(i)n^c(i)]^{-1} \, di. \tag{23}$$

To proceed further, one needs $n^c(i)$, the constrained distribution for an associated vapor.[11] We again use eqs. (3) and (4'), but instead of (5) we obtain

$$W(i) = \sigma a(i) - i\,k\,T \ln [p_{1a}/p_{1\infty}] + V_i [p_a - p_\infty] + iv\kappa[p_i - p_\infty]^2/2 \tag{24}$$

The only difference between eqs. (24) and (5) is that the argument of the logarithm is the supersaturation of the monomer species, $p_{1a}/p_{1\infty}$ and not the actual supersaturation p/p_∞. (Note that the monomer supersaturation is identical to the ratio of fugacities of the actual vapor and a saturated vapor, i.e., $p_{1a}/p_{1\infty} \equiv f_a/f_\infty$.) Using eqs. (3) and (24) and evaluating the integral in (23) one obtains (in the same units as in eq. (16))

$$J = 5.5 \times 10^{31} \left(\frac{p_\infty}{T}\right)^2 \frac{\sqrt{\sigma M}}{d} \left(\frac{p}{p_\infty}\right) \sum_{i=1}^{m} i^{3/2} X_i \times \exp\left\{-17.56 \left(\frac{\sigma}{T}\right)^3 \left[\frac{M}{d \ln (p_{1a}/p_{1\infty})}\right]^2\right\}. \tag{25}$$

Note that the initial number is 5.5×10^{31} instead of the more common 9.5×10^{25} only because the units of p_∞ were chosen as atmospheres and not mm of Hg.

VI. KINETIC VERSION OF NUCLEATION THEORY

If nucleation can occur only along a single path (binary or

multicomponent nucleation is thus excluded), as in Section IV, one finds that the net rate J at which clusters of size less that i + 1/2 become larger than i + 1/2 is

$$J(i) = f(i)\, n(i) - b(i+1)\, n(i+1). \qquad (8)$$

At this point in most derivations of nucleation theory (and in Section IV), the backward rate was eliminated from eq. (8) by relating it to the forward rate and the hypothetical "constrained" equilibrium distribution of clusters. However, the "constrained" distribution is not necessary and the backward rate need not be eliminated until much later.

A convenient equation which contains all the nucleation information can be readily derived by defining a recursive relation

$$Z(i+1) \equiv Z(i)\, b(i+1)/f(i+1) \quad \text{for } i > 1 \qquad (26)$$

and

$$Z(1) = 1. \qquad (27)$$

This recursive relation is easily solved yielding

$$Z(i) = \prod_{j=2}^{i} b(j)/f(j). \qquad (28)$$

Multiplying eq. (8) by Z(i) and summing, one obtains

$$\sum_{i=1}^{\bar{i}} J(i)Z(i) = \sum_{i=1}^{\bar{i}} \left[f(i)n(i)Z(i) - f(i+1)n(i+1)Z(i+1)\right]. \qquad (29)$$

Note that the two terms on the right-hand side are identical except for their indices. Thus, on summing, successive terms cancel and eq. (29) simplifies to

$$\sum_{i=1}^{\bar{i}} J(i)Z(i) = f(1)n(1) - f(\bar{i}+1)n(\bar{i}+1)Z(\bar{i}+1). \qquad (30)$$

If one carries out the summation to a sufficiently large value of $\bar{i}$, the last term on the right-hand side of this equation is negligible compared to the first because Z(i + 1) goes to zero as i becomes large; this follows from (28) for the following reasons. The quotient $b(i)/f(i)$ is the ratio of the number of clusters of size i that shrink to those that grow. This ratio depends on both the cluster size and supersaturation. At any supersaturation there is a critical size i^* defined as that size for which this ratio is unity. For all sizes smaller than the critical size the ratio is greater than unity, while for all sizes larger than the critical size this ratio is less than unity. Since there are only a *finite* number of sizes smaller than the critical size (and $b(i)/f(i)$ is finite for all i) the product of $b(i)/f(i)$ from $i = 1$ to $\bar{i}$ can be made arbitrarily small by making $\bar{i}$ arbitrarily large. Note that the reason that this term is negligible is a kinetic result based on the concept, the critical size, and not on the idea (expressed after eq. (12)) that the constrained distribution of clusters increases without bounds.

Thus, eq. (30) becomes

$$\sum_{i=1}^{i} J(i)Z(i) = f(1)n(1). \tag{31}$$

In most cases of interest, a steady state is very rapidly established. Since the rate of change of concentration of clusters containing i molecules is given by

$$\frac{dn(i)}{dt} = J(i - 1) - J(i), \tag{13}$$

at steady state, $dn(i)/dt = 0$ and J becomes a constant for all sizes. It can then be factored out of the summation to obtain

$$J = f(1)n(1)/\{1 + \sum_{i=2}^{\bar{i}} \prod_{j=2}^{i} b(j)/f(j)\}. \tag{32}$$

To proceed further, one needs explicit expressions for the backward rate $b(i)$ and the forward rate $f(i)$.

The same arguments and line of reasoning which were used in Section IV (see eq. (9)) apply again. But we are no longer forced to invent a constrained equilibrium distribution, e.g. eq. (10), to obtain an expression for the backward rate, $b(i)$.

The backward rate is the rate at which molecules leave a cluster. This is not known in general, but can be determined in dilute solution. For dilute systems (i.e., systems in which the number density of solute molecules is much smaller than the number density of solvent molecules) the interactions between a molecule leaving a cluster and other solute molecules are negligible compared to its interactions with solvent molecules. Therefore, in dilute systems, the backward rate $b(i+1)$, although a complicated function of temperature and cluster size, is independent of the concentration of the nucleating species, i.e., the solute. Consequently, if $b(i+1)$ can be determined at any concentration or supersaturation (since it is independent of concentration) it is known at all other concentrations. At equilibrium, that is, at the true equilibrium state, and not some hypothetical "constrained" equilibrium - the nucleation rate is zero, and eq. (8) becomes

$$0 = f^e(i)n^e(i) - b(i+1)n^e(i+1), \quad (33)$$

where the superscript e is used to denote the value of that function at the equilibrium state. This equation can be solved for $b(i+1)$ and the result when substituted into (32) gives :

$$J = f(1)n(1)/\{1 + \sum_{i=2}^{\bar{i}} \frac{f(1)n^e(1)}{f(i)n^e(i)} \prod_{j=2}^{i} \frac{f^e(j-1)}{f(j-1)}\}. \quad (34)$$

A further simplification occurs when f/f^e is independent of the size of the cluster. This is usually the case and occurs when the forward rate β, given by eq. (9), is a product of terms which are either size independent, or whose size dependence is independent of supersaturation. Equation (34) can then be written

$$J = \frac{n(1)\beta^e}{n^e(1)\beta} / \sum_{i=1}^{\bar{i}} [a(i)\beta n^e(i)(\beta/\beta^e)^i]^{-1}. \quad (35)$$

The numerator of this equation is equal to unity whenever the arrival rate β is proportional to the number density $n(1)$. This is the case for ideal gases and is also believed to be true for somewhat non-ideal gases.

Equation (35) can formally be made identical to eq. (14) if one chooses to identify $n^c(i)$ with $n^e(i)(\beta/\beta^e)^i$. This choice is suggested by the following idealization and approximations : (i) for ideal gases $\beta/\beta^e = p/p^e$, (ii) equations (6) and (7) have the same relationship to each other. However this identification is not always necessarily true. Equation (35) (actually its more general form (32)) thus provides one with the means of imporving on the classical theory of nucleation as one obtains better expressions for the equilibrium distribution of clusters than eq. (7) and as one obtains more sophisticated expressions for the concentration dependence of β/β^e. It also provides us with a means for creating a theory of nucleation for systems where the process is sufficiently complicated that it would be impossible to make guesses of what the correct form for $n^c(i)$ would be since in these systems the (β/β^e) term becomes so complicated that no thermodynamic model could possibly produce it. It is precisely because of this confusion of thermodynamic and kinetic concepts that there has been much controversy in nucleation theory in the past.

VII. NUCLEATION OF VACANCIES AND INTERSTITIALS

One very important example of such a nucleation process occurs in the claddings of fuel rods in nuclear reactors. Energetic neutrons knock atoms from their lattice sites, creating high concentrations of vacancies and of interstitial atoms. The vacancies can cluster and grow to become voids, typically 50 to 100 Å in diameter. The interstitial atoms also nucleate and form interstitial loops, that is, an extra plane of atoms growing radially. These structural defects lead to swelling and warping of the fuel rods.

The description of this nucleation phenomenon is complicated because the vacancies and the interstitials are matter and antimatter to each other. Not only can they combine directly with each other to produce a nothing (except for the release of energy), but each can also combine with clusters of the opposite species to make the cluster smaller by one unit. Furthermore, the clusters are capable of evaporating not only the self specie, but also the opposite specie.

In the limit that vacancies and interstitials arrive at a void at the same rate, no growth occurs. Attempts to model this phenomenon using previous versions of nucleation theory led to a great deal of confusion. In fact, it was their efforts[18-21] to solve this complex nucleation problem that led WIEDERSICH and KATZ[15] to develop the general formalism presented here.

Consider the nucleation of voids in a metal containing excess concentrations of both vacancies and interstitials. The forward and backward rates are now each the sum of two terms (we have changed the symbol for the cluster size from i to x to avoid confusion with the subscript i which denotes interstitial) :

$$f(x) = \left[\beta_v + \gamma_i(x)\right] a(x), \tag{36}$$

$$b(x) = \left[\beta_i + \gamma_v(x)\right] a(x). \tag{37}$$

where $a(x)$ is the surface area of vacancy cluster, β_v is the condensation rate (per unit area) of vancanies, $\gamma_i(x)$ is the rate (per unit area) at which a cluster composed of x vacancies emits an interstitial atom, β_i is the condensation rate (per unit area), and $\gamma_v(x)$ is the rate of emission of vacancies from a cluster containing x vacancies.

For the determination of the emission terms, detailed balance can be used again in an even more restricted sense. For the metal at the same temperature but with equilibrium concentrations of both vacancies and interstitials, not only is it true that $J(x) = 0$ and therefore $f^e(x)n^e(x) = b^e(x + 1)n^e(x + 1)$, but also that the fluxes in cluster space caused by vacancies and interstitials must separately be equal to zero (i.e., at equilibrium the vacancy condensation is balanced by the vacancy emission). Thus

$$\beta_v^e a(x)n^e(x) = \gamma_v^e(x + 1)a(x + 1)n^e(x + 1) \qquad (38)$$

and

$$\beta_i^e a(x + 1)n^e(x + 1) = \gamma_i^e(x)a(x)n^e(x). \qquad (39)$$

Solving eqs. (38) and (39) for $\gamma_v(x)$ and $\gamma_i(x)$, substituting into (36) and (37), and using eq. (28) for Z(x), we obtain

$$\frac{b(j)}{f(j)} = \frac{\beta_i a(j) + \beta_v^e a(j - 1)n^e(j - 1)/n^e(j)}{\beta_v a(j) + \beta_i^e a(j + 1)n^e(j + 1)/n^e(j)} \qquad (40)$$

In this case there is no cancellation of terms as in eq. (35). However the rate of nucleation can still be obtained by using (40) with (32), since it is a straightforward matter to evaluate the summation numerically.[15]

VII.1. Sample Calculations

The singificance of the effects discussed, i.e. the antinucleating effect of the interstitials, can perhaps best be appreciated by sample calculations. Since much of the interest in this subject, aside from its intrinsic scientific value, is due to its applicability to void formation in the structural materials (usually fcc stainless steels) used in fast-breeder reactors, calculations for such materials would be of greatest interest. However, the appropriate quantities such as jump frequencies and defect-formation energies are not well known for these complicated alloys. For this reason, our sample calculations have been made using material parameters appropriate to pure nickel (a major component of stainless steels that has the same fcc structure), except that a surface free energy of 1000 ergs/cm^2 was used, which we think is more appropriate for these alloys.

The interstitial flux during irradiation must be nearly equal to the vacancy flux, since interstitials and vacancies are produced simultaneously by the irradiation and must annihilate with equal rates to maintain steady state. The interstitial flux to voids must also be less than the vacancy flux, or voids would never form. Since no adequate model for the relationship of the interstitial and vacancy flux exists, it is customary to use the ratio of the rates of condensation of interstitials β_i to the rate of condensation of vacancies β_v as a parameter; the ratio β_i/β_v, called the arrival-rate ratio, has been variously estimated to be between 0.8 and 1.0. Rough order of magnitude estimates of minimum nucleation rates that are required for the observed number densities of 10^{13} to 10^{16} voids/cm^3 can be obtained from the typical time scales of exposures; i.e., months to years in reactors and tens of minutes to hours in accelerators. Thus, nucleation rates of 10^5 to 10^9 voids/cm^3/s are typical of reactor conditions, and rates of 10^6 to 10^{13} voids/cm^3 s are encountered in accelerators.

As Fig. 6 shows, the strong dependence of the nucleation rate on the supersaturation, which is a characteristic of ordinary nucleation, i.e., an arrival-rate ratio $\beta_i/\beta_v = 0$, is essentially unchanged even by large values of the arrival-rate ratio. A factor of ten increase in supersaturation causes the nucleation rate to increase from 1 to 10^{15} nuclei/cm^3 s at 600°C or to 10^{10} at 400°C, independent of the arrival-rate ratios. Higher values of β_i/β_v require a higher supersaturation, but the factor by which the supersaturation must be increased is only weakly dependent on the desired nucleation rate. At high supersaturations, the slope of the curves in Fig. 6 goes to two since the critical nucleus size is then a single vacancy. Thus, the rate of nucleation becomes the rate of formation of di-vacancies.

The effect of the arrival-rate ratio on the nucleation rate at fixed vacancy supersaturation is shown in Fig. 7. The nucleation rate decreases with an increase in β_i/β_v, at first moderately but becoming increasingly sharp as β_i/β_v approaches unity. Inspection of Fig. 7 shows that increasingly larger vacancy supersaturations are required to obtain a given nucleation rate as the arrival-rate ratio increases.

As the production rate of defects during irradiation becomes balanced by the loss rate that results from recombination and sink annihilation, a quasi-steady state is approached. The steady-state concentrations and supersaturations are determined by the temperature, displacement rate, and sink-annihilation probability. The curves labeled y = 0 in Figs. 8 and 9 show the effect of temperature on the rate of homogeneous nucleation for various values of the relevant parameters. The two curves in Fig. 8 are for a displacement rate (i.e., the rate of production of vacancies and interstitials) of 10^{-6}/s at two sink-annihilation probabilities, $p = 10^{-6}$ and $p = 10^{-3}$ ($1/p$ is the average number of jumps that a defect, a vacancy or an interstitial, makes before it annihilates at sinks such as dislocation loops, void surfaces, and grain boundaries).

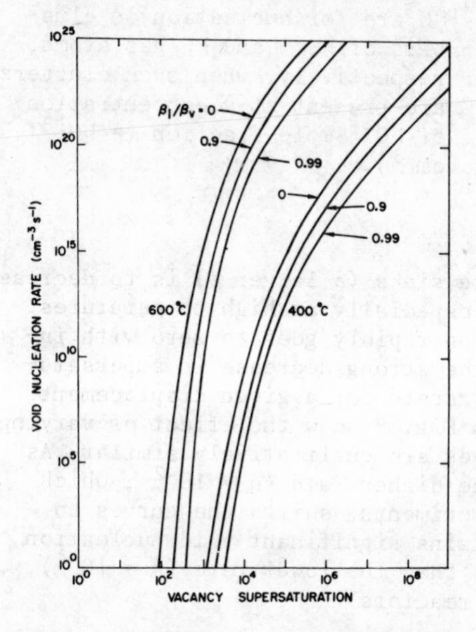

Fig. 6 : Nucleation rate as a function of the vacancy supersaturation without helium. The parameters are the arrival-rate ratio β_i/β_v and temperature.

Fig. 7 : Nucleation rates at 625° as a function of arrival-rate ratio for several vacancy supersaturations S shown as a parameter. No helium.

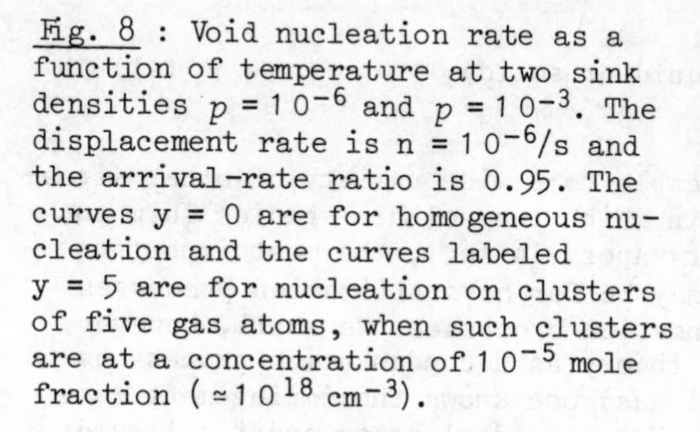

Fig. 8 : Void nucleation rate as a function of temperature at two sink densities $p = 10^{-6}$ and $p = 10^{-3}$. The displacement rate is $n = 10^{-6}$/s and the arrival-rate ratio is 0.95. The curves y = 0 are for homogeneous nucleation and the curves labeled y = 5 are for nucleation on clusters of five gas atoms, when such clusters are at a concentration of 10^{-5} mole fraction ($\approx 10^{18}$ cm^{-3}).

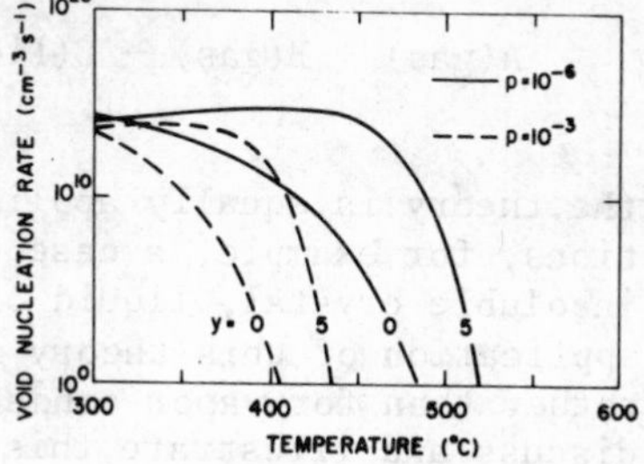

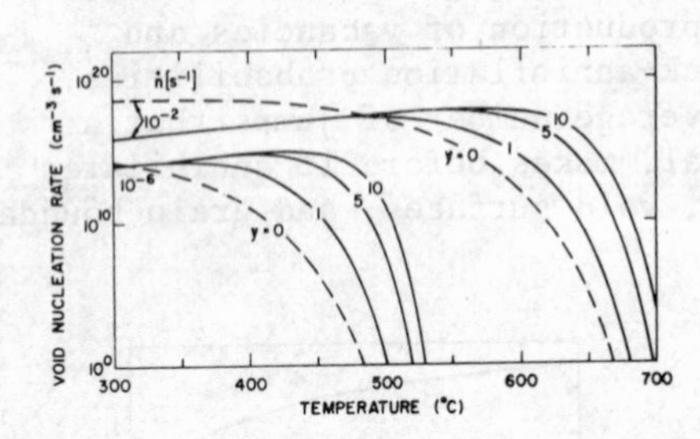

Fig. 9 : Void nucleation rate as a function of temperature. The sink density is $p = 10^{-6}$ and the arrival-rate ratio is 0.95. The curves labeled y = 0 are for homogeneous nucleation and those labeled y = 1, 5, and 10 are for nucleation on clusters of 1, 5 and 10 gas atoms, respectively, when such clusters are present at a concentration of 10^{-5} mole fraction ($\simeq 10^{18}$ cm^{-3}).

The effect of the presence of more sinks (a larger p) is to decrease the nucleation rate drastically, especially at high temperatures. Furthermore, the rate of nucleation rapidly goes to zero with increasing temperature because of the strong decrease in supersaturation with temperature at steady state for a given displacement rate. The curves labeled y = 0 in Fig. 9 show the effect of varying the displacement rate $\dot{n}$. The curves are qualitatively similar. As expected, producing defects at the higher rate ($\dot{n} = 10^{-2}$), which is typical of ion-bombardment experiments, shifts the curves to higher nucleation rates and maintains significant void nucleation rates at much higher temperatures than the lower rate ($\dot{n} = 10^{-6}$), which is typical of fast-breeder reactors.

VIII. NUCLEATION WITH SIMULTANEOUS CHEMICAL REACTION

Although we shall present and illustrate this generalization of the theory for the case of a non-volatile product being formed from gas phase reactants, i.e.

$$A(gas) + B(gas) \rightarrow C(liquid\ or\ solid), \quad (41)$$

the theory is equally applicable to a wide variety of other situations, for example, a case in which two soluble species form an insoluble crystal, liquid or vapor. In fact, the most important application of this theory may be for crystallization processes rather than for vapor condensation processes. We shall, however, discuss and illustrate this theory as a condensation process because (at least for an ideal gas) one knows the impingement rate exactly from kinetic theory. For non-ideal gases and for liquids, the equations presented for the nucleation rate are still exact, but some assumptions must be made about the impingement rate, β, before actual calculations can be made.

Since we are considering the case in which the reactants are much more volatile than the product, it is reasonable to assume that the average number of reactant molecules which are adsorbed or dissolved in the nuclei will be small, often less than one. Under these conditions, one can characterize the clusters, or nuclei, by the number of molecules of species C they contain, and treat the nucleation process as that of a single component. The net rate of growth of a cluster of size i, then, is given by the equation

$$J_c(i) = f_c(i)n_c(i) - b_c(i + 1)n_c(i + 1), \tag{42}$$

where the subscript c denotes that all the rates and concentrations are for species C.

We shall consider a nucleation process in which reaction can occur both homogeneously in the gas phase and heterogeneously on the surface of the nuclei. In other words, this is a self catalytic process since the presence of the reaction product as a condensed phase will promote further reaction.

The forward rate is the rate at which a cluster containing i molecules of species C grows to i + 1 molecules. This growth process can result from the condensation of a C molecule that was formed by the gas phase reaction, or it can result from the heterogeneous reaction of adsorbed (or dissolved) A and B molecules to form new C molecules. The forward rate is then

$$f_c(i) = \beta_c \theta_c \alpha a(i) + k[A]\ [B]a(i), \tag{43}$$

where β_c is the rate of impingement of C molecules from the vapor phase, θ_c is the fraction of the cluster surface that is covered by C molecules, [A] and [B] represent the surface concentrations of species A and B, respectively, and k is the surface reaction rate constant.

A nucleus can also lose species C by two processes : by direct evaporation and by the reverse reaction to form species A and B. The backward rate is thus given by

$$b_c(i) = \gamma_c(i)\theta_c a(i) + d'[C]a(i), \tag{44}$$

where γ_c is the rate of evaporation of C molecules per unit area, k' is the rate constant for the reverse reaction, and [C] is the surface concentration of species C. The surface concentrations (molecules per unit area) and surface fraction are related by

$$\theta_c = s_c[C], \tag{45}$$

where s_c is the area on the surface occupied by one molecule of species C and

$$s_a[A] + s_b[B] + s_c[C] = 1. \tag{46}$$

The backward rate b(i) can then be written as

$$b(i) = (\gamma_c s_c + k')[C]a(i). \tag{47}$$

Although the evaporation rate per unit area, γ_c, can still be assumed independent of supersaturation, eq. (47) shows that the backward rate is supersaturation dependent whenever [C] depends on the supersaturation. Thus one can no longer substitute b(i + 1) as was done before. Nonetheless, it is still true that at equilibrium the forward and backward processes proceed at exactly equal rates; therefore

$$0 = f^e(i)n^e(i) - b^e(i + 1)n^e(i + 1). \tag{48}$$

However, from eq. (47), the backward rate can be related to the equilibrium backward rate, i.e.

$$b(i + 1) = b^e(i + 1)\ [C]/[C]^e. \tag{49}$$

Substituting (48) into (49), one obtains

$$b(i + 1) = \frac{f^e(i)n^e(i)[C]}{n^e(i + 1)[C]^e} \tag{50}$$

which is equivalent to

$$\frac{b(i)}{f(i)} = \frac{f^e(i - 1)n^e(i - 1)[C]}{f(i)n^e(i)[C]^e}. \tag{51}$$

Substituting from eq. (43) for the forward rates gives

$$\frac{b(i)}{f(i)} = \frac{(\beta_c^e \theta_c^e \alpha + k[A]^e [B]^e) a(i-1) n^e(i-1)[C]}{(\beta_c \theta_c \alpha + k[A][B]) a(i) n^e(i)[C]^e} . \quad (52)$$

It is convient at this point to define an effective impingement rate β^*

$$\beta^* \equiv (\beta_c \theta_c \alpha + k[A][B])/[C]s_c . \quad (53)$$

Substituting (53) into (52) and then this equation, together with equation (43), into (32) gives

$$J = \frac{n(1)}{n^e(1)} (\alpha\beta_c \theta_c + k[A][B]) / \sum_{i=1}^{i} [a(i) n^e(i) (\frac{\beta^*}{\beta^{*e}})^{i-1}]^{-1} \quad (54)$$

As before, the summation can be converted to an integral, expanded around its maximum value and integrated.

To proceed further $n^e(i)$ has to be specified. Equation (7) when used with a size independent surface tension is equivalent to the standard assumption used in classical nucleation theory. Since the purpose of this paper is to illustrate the generalization of nucleation theory to allow for surface chemical reaction, (7) is used in the calculations below. Substituting eq. (7) into eq. (54), the result after integration is

$$J = N_c a_o [\frac{a_o \sigma}{kT}]^{1/2} (\beta_c \theta_c \alpha + k[A][B]) \exp[\frac{-4a_o^3 \sigma^3}{27(kT)^3 (\ln \beta^*/\beta^{*e})}] \quad (55)$$

Example calculations[16] show that there can indeed be a significant decrease in the barrier to nucleation because of the self catalytic nature of this process.

IX. NUCLEATION IN DENSE SYSTEMS

In previous sections, we used the approximation that the backward or "evaporation" rate is independent of solute supersaturation. This is an excellent assumption at very low concentrations, because the interactions between solute molecules leaving a cluster and other solute molecules are negligible. At high concentrations this assumption can no longer be valid, since molecules leaving a cluster do interact with other nearby solute molecules. For a crystal nucleating from its melt, these interactions totally dominate (there are no solvent molecules).

KATZ and SPAEPEN[13] have shown that by making what appears to be a most reasonable assumption, this problem can be overcome. This assumption concerns the variation of the evaporation rate with concentration C at constant temperature. They assumed that the effect of solute density on the probability of a molecule leaving a cluster is the same for clusters of all sizes, that is,

$$\frac{b(i) \text{ at } T,C}{b(i) \text{ at } T,C'} = \frac{b(j) \text{ at } T,C}{b(j) \text{ at } T,C'} \text{ for all } i \text{ and } j. \quad (56)$$

To use this result, it is necessary to choose the concentration C' and the size j so that b(i) at the system temperature T and composition C can be related to known quantities. The most readily obtainable values are for the equilibrium concentration, $C' = C_{sat}$, at system temperature T and for a macroscopic bulk phase ($j = \infty$). Then eq. (56) can be written

$$b(i) = b^e(i)b(\infty)/b^e(\infty) \equiv b^e(i)m(T,C), \quad (57)$$

where m is in general a function of temperature and concentration (or pressure), but is independent of i. The quantities $b^e(i)$ and $m(T,C)$ can both, in principle, be determined experimentally. Using the equilibrium condition (33), $b^e(i)$ can be determined from measurements of the concentrations $n^e(i)$ together with an appropriate expression for $f^e(i)$ from kinetic theory. The quantity $m(T,C)$ can be determined from the rate of growth of one macroscopic phase at the expense of another at supersaturated conditions.

For the dilute solution case discussed previously, m is equal to unity. In concentrated solution, m is not unity, but it is a straightforward matter to repeat the derivation and, instead of eq. (35), obtain

$$J = \frac{n(1)}{n^e(1)} \Big/ \sum_{i=1}^{\bar{i}-1} \left[f(i)n^e(i)\left(\frac{\beta}{m\beta^e}\right)^{i-1}\right]^{-1}. \quad (58)$$

This result is identical to the dilute solution result except that we have $\beta/m\beta^e$ instead of the ratio of arrival rates β/β^e.

A complication arises in condensed systems. This concerns the minimum size of a nucleus. The large density difference between a gas phase and a nucleating condensed phase makes the identification of the condensed-phase nuclei straightforward; any cluster containing two or more solute molecules is considered to be a potential nucleus. However, for nucleation in condensed systems (e.g. melt→ crystal, very concentrated solution →crystal, crystal→crystal) where the density difference between phases is small, this criterion is no longer useful. Therefore other structural characteristics which make these two condensed phases distinct from each other must be used to identify the nucleus.

For a crystal-crystal transformation, the difference in crystallography between the two phases is the obvious choice for identification of a phase transition. Specifically, an assembly of neighboring molecules can be identified as the nucleus of a new phase if the lines connecting the centers of the molecules form polyhedra that are characteristic of the new phase. For example, a fcc nucleus that forms in a bcc matrix can be identified by outlining those parts of the system which contain a specific array of regular octahedra and tetrahedra typical of the fcc structure and quite distinct from the array of distorted octahedra that make up the bcc structure.

The identification of crystalline nuclei in a melt is similar; the only difference is that the melt has even less characteristic structure than a bcc crystal. It has been shown that an excellent model for melts of noble and late transition metals is the dense random packing of hard spheres.[22-24] The polyhedra that make up the structure of such a fluid are a random mixture of tetrahedra, with a few octahedra and occasional deltahedra. The preponderance of tetrahedra, however, makes this structure distinctly different from the more ordered crystal states. This is illustrated by considering the ratio of tetrahedra to octahedra : for random packing this ratio is 15 to 1;for a fcc structure it is only 2 to 1. Therefore, since the octahedra are a necessary element of a fcc structure and only incidental to the dense random packing, this suggests that a nucleus of a fcc crystal can be identified as any part of the system that is made up of properly aligned octahedra.

The identification of crystalline nuclei in a concentrated solution is quite similar. Even though little is known about the structure of these solutions, the identification of regular polyhedra is, in principle, always possible.

This procedure for identification of nuclei in dense systems makes it clear that it is impossible to identify nuclei containing fewer than a certain number of molecules; there is no way to decide whether a pair or triplet of molecules in the dense system is actually a crystal nucleus. The minimum number of molecules is determined by the types of polyhedra that characterize the nucleating phase. For the example of fcc crystals, the minimum identifiable nucleus size is one octahedron and therefore requires at least six atoms to identify the nucleus. The few octahedra that are always present in the dense random packing of hard spheres can perhaps be considered as incipient nuclei for crystallization from the melt.

Accounting for the minimum size of an identifiable nucleus in the calculation of the nucleation rate is straightforward. Since we consider only clusters larger than some minimal size, i_o, we need only sum over the size range i_o to $\bar{i}$ in eq. (58). The result is that

$$J = \frac{n(i_o)}{n^e(i_o)} \Big/ \sum_{i=i_o}^{\bar{i}-1} \left[f(i)\,n^e(i)\,(\beta/m\beta^e)^{i-i_o}\right]^{-1}, \tag{59}$$

where $n^e(i_o)$ is the concentration of clusters of size i_o in the true equilibrium state. Note that $n^e(i)$ is the size distribution of crystal nuclei in a melt at true equilibrium : it can, in principle, be measured. However, even if no such measurements are available, it is still safe to say that $n^e(i)$ is a strongly decreasing function of i. The factor $(\beta/m\beta^e)^{i-i_o}$ is a strongly increasing function of i. This quantity can, in principle, be obtained from the growth velocity u of a planar crystal face growing into the melt, i.e.

$$u = (\beta - \gamma)v_o, \tag{60}$$

where v_o is the atomic volume and $\gamma = b(i)/a(i)$ is the rate at which atoms leave a crystal per unit area. Therefore

$$m = \frac{b(\infty)}{b^e(\infty)} = \frac{\gamma}{\gamma^e} = \frac{\beta - u/v_o}{\beta^e} \tag{61}$$

and

$$\frac{\beta}{m\beta^e} = \left[1 - \frac{u}{\beta v_o}\right]^{-1}, \tag{62}$$

since the crystal growth velocity, u, is equal to zero at equilibrium. Measurement of u, in conjunction with the knowledge of β from kinetic theory of liquids, makes this factor experimentally available. Note that $(\beta/m\beta^e)$ is always larger than 1, which makes $(\beta/m\beta^e)^{i-i_o}$ strongly increasing with i. In order for the sum in (59) to converge, $n^e(i)$ must decrease less strongly with i than $(\beta/m\beta^e)^{i-i_o}$ increases with i. If this is the case, eq. (59) can be evaluated directly by performing the summation numerically. If one whishes to obtain an analytical expression, this can be done as described in the previous section, i.e. convert the sum to an integral, expand in a Taylor series and keep only the leading terms. Apart from the factor $n(i_o)/n^e(i_o)$, which is not likely to be very different from unity in a melt for topological reasons, and a knowledge of from kinetic theory, we have thus determined the nucleation rate completely in terms of experimentally obtainable quantities; one thermodynamic : the distribution $n^e(i)$ of nucleus sizes in the equilibrium melt, and one kinetic : the growth velocity of a planar crystal at the nonequilibrium temperature and pressure.

V. CONCLUSIONS

We have presented nucleation theory as it exists today. There are many difficulties with it, only some of which we have covered. We certainly have not considered the question of how to correctly describe the equilibrium distribution of clusters (see eq. (3)). Even more fundamental is the question of what is a crystalline cluster. Can one even use those concepts which arise from gas condensation ideas for crystallization from dense systems ? When is a cluster a small crystal ? How does one distinguish it from a cluster which is incapable of growth into the macroscopic crystal ? There are no answers as yet to these questions. Perhaps you will provide them someday.

Aknowledgements

The author is grateful to the Division of Materials Sciences, Office of Basic Energy Sciences, U.S. Department of Energy, for financial support during the writing of this chapter.

REFERENCES

1 . Cited by VOLMER M., in *Kinetic der Phasenbildung*, T. Steinkopff, Dresden and Leipzig, 1939. An English translation is available from the National Technical and Information Service, Cite : ATI No 81935 (F-TS-7068-RE).
2 . VOLMER M. and WEBER A., Z. Phys. Chem. (Leipzig) 119(1926)277.
3 . FARKAS L., Z. Phys. Chem. A125 (1927)236.
4 . BECKER R. and DÖRING W., Ann. Phys. 24(1935)719.
5 . ZELDOVICH J., Acta. Physicochim. U.S.S.R., 18(1943)1.
6 . *Nucleation*, A.C. Zettlemozer, ed. Dekker, New York, 1969.
7 . *Advances in Colloid and Interface Science*, Vol. 7, 1977.
8 . *Advances in Colloid and Interface Science*, Vol. 10, 1979.
9 . ABRAHAM F.F., *Homogeneous Nucleation Theory*, Academic Press, New York, 1974.

10. KATZ J.L., SALTSBURG H. and REISS H., J. Coll. Inter. Sci. 21 (1966)560.
11. KATZ J.L. and BLANDER M., J. Coll. Inter. Sci. 42(1973)496.
12. KATZ J.L. and WIEDERSICH H., J. Coll. Inter. Sci. 61(1977)351.
13. KATZ J.L. and SPALPEN F.,Phil. Mag. B37(1978)137.
14. KATZ J.L. and DONOHUE M.D., Adv. Chem. Phys. 40(1979)137.
15. WIEDERSICH H. and KATZ J.L., in ref. 8 p. 33.
16. DONOHUE M.D. and KATZ J.L., J. Coll. Inter. Sci. (to appear).
17. COHEN E.R., J. Stat. Phys. 2(1970)147.
18. KATZ J.L. and WIEDERSICH H., J. Chem Phys. 55(1971)1414.
19. KATZ J.L. and WIEDERSICH H.,J. Nucl. Materials 46(1973)41.
20. WIEDERSICH H. and KATZ J.L., *Proceedings of the Conference on Defects and Defect Clusters in bcc Metals and their Alloys.* AIME Nuclear Metallurgy Series 18(1973)530.
21. WIEDERSICH H., BURTON J.J. and KATZ J.L., J. Nucl. Materials 51(1974)287.
22. CARGILL G.S. III, in *Solid State Physics*, H. Ehrenreich, F. Seitz, and D. Turnbull, ed., Academic Press, New York, 1975, Vol. 30, p. 227.
23. FINNEY J.L., Proc. R. Soc. 319A(1970)479.
24. BERNAL J.D., Proc. R. Soc. 280A(1964)299.

THREE AND TWO DIMENSIONAL NUCLEATION ON SUBSTRATES

R. KERN

CRMC2, CNRS
Marseille - Luminy
France

I. INTRODUCTION

In this chapter the growth of a crystal of species A is considered to take place on a substrate crystal of species B. This old problem has received new attention in the last 15 years primarily because :

- new experimental methods, coming from surface science, are available to study the mechanisms on an atomic scale

- the problem is related to the important technological aspects of the growth of epitaxial layers[1] (oriented overgrowth of a crystal A on a substrate B).

We will consider both the problem of thermodynamical stability of two and three dimensional phases, and the kinetical barriers that must be overcome when such phases start to grow. The approach uses very simple models from the point of view of thermodynamics, structure and bonding, but the essential features known from experiment are described well.

We start with the simplest model. Crystal A and B are strictly isomorphous (simple cubic lattices with the same periodicity). They are KOSSEL crystals : this means that only first neighbor additive (and positive) interaction energies (ϕ_{AA}, ϕ_{BB}, ϕ_{AB}) are considered. They also behave as Einstein solids. This model makes possible the distinction between two dimensional (2D) and three dimensional (3D) stability, nucleation and growth. Later, when these restrictions are relaxed new aspects appear such as multilayer growth, interface coherency or incoherency. The existence of three epitaxial growth

B. Mutaftschiev (ed.), Interfacial Aspects of Phase Transformations, 287–302.

modes can be deduced in accordance with known experimental facts. In all these analyses we intentionally consider the substrate B as an inert solid, in that sense that A atoms cannot enter the solid B. Alloying is not considered in spite of important ϕ_{AB} interactions, since few facts about "surface alloying" are known.

II. PHASE STABILITY ON A SUBSTRATE

Consider two KOSSEL crystals, one of species A the other of B but strictly isomorphous (their lattice parameters are $a_B = a_A$). The crystal to be deposited has an equilibrium vapor pressure :

$$P_{3\infty} = f_v/f_c \exp(-3\phi_{AA}/k_B T) \quad (1)$$

where $-3\phi_{AA}$ is the potential energy of an atom in a repeatable step site, f_v and f_c are respectively the partition functions of the monoatomic perfect gas of A and the vibrational partition function of the crystal A such that :

$$f_v = (2\pi m)^{3/2} k_B T^{5/2}/h^3 \quad ; \quad f_c = (k_B T/h\nu)^3 \quad (2)$$

where m is the atomic mass of A, ν the Einstein frequency, h and k_B the Planck and Boltzman constants.

As mentioned before, we assume that the vapor pressure of the substrate crystal B is negligible with respect to that of crystal A (inert substrate), accordingly $\phi_{BB} >> \phi_{AA}$. A (100) face of substrate B is exposed to the vapor of A at pressure P. The A particles adsorb on (100) B with a superficial concentration $n = n_S\theta$, $n_S = a^{-2}$ being the maximum concentration of adsorption sites. $\theta = 1$ means that full isomorphous monolayer of A sits on (100) B. θ is the mean degree of coverage per surface site that also represents the occupancy probability of such a site, $(1 - \theta)$ being the probability to be empty.

An isolated A particle on (100) B having an interaction energy ϕ_{AB} with the substrate has a relative partition function with respect to vapor of

$$\varepsilon(T) = (f_{ad}/f_v)\, P \exp(\phi_{AB}/k_B T) \quad (3)$$

If the particle is part of an adsorption layer, the evaluation of its partition function is a difficult task. In the Bragg-Williams mean field approximation it is simply assumed that all sites surrounding the one under consideration have a mean occupancy θ. If a given site is occupied, this atom A is in an average environment and its potential energy is $-4\phi_{AA}\theta$, since ϕ_{AA} is the lateral interaction energy of two effectively present first neighbors of A particles (four are the maximum possible). The relative partition function of this non isolated particle in now : $\varepsilon(T)\exp(4\phi_{AA}\theta/k_BT)$. If the given surface site is empty its relative partition function is unity by definition. The ratio of the two partition functions is equal to the probability ratio of this site being occupied or of being empty :

$$\frac{\theta}{1-\theta} = \frac{f_{ad}}{f_v} P \exp\left[(\phi_{AB} + 4\phi_{AA}\theta)/k_BT\right] \tag{4}$$

This equation is called the Frumkin-Fowler [2,3] adsorption isotherm. It shows some interesting features (Fig. 1).

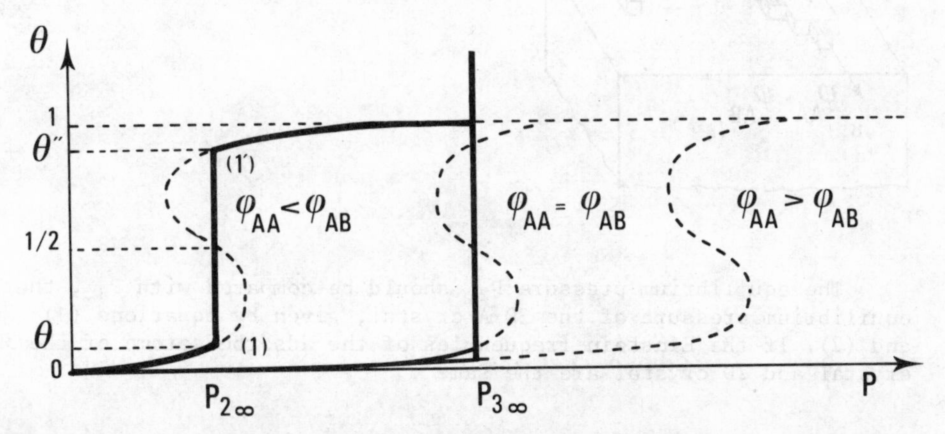

Fig. 1 : Adsorption isotherms at the same temperature T (θ degree of surface coverage, P vapor pressure of deposit) at different interaction energies ϕ_{AA}, ϕ_{AB}. Full lines are real equilibrium states.

The main result of this isotherm is a loop characteristic of an instability occuring when $\phi_{AB}>0$ and $\phi_{AA}>k_BT$. The dashed curve (on the left) represents non physical states in contrast to the full line representing the equilibrium state. The vertical segment (1) - (1') cutting the loop at $\theta = 1/2$ is the coexistence line of two phases in equilibrium, a very dilute one at concentration θ(2D gas) and a dense one at θ'' (full layer with some holes in it). Along (1) - (1') the relative proportion of the two phases changes from 0 to 100%. This phase change is called 2D condensation. It is of first order since at T, P = cte. There is a discontinuity of some properties when θ changes from θ' to θ''. This phase change takes place at temperature T and pressure $P_{2\infty}$ obtained by setting $\theta = 1/2$ in (4) :

$$k_B T \ln P_{2\infty} = -(2\phi_{AA} + \phi_{AB}) + k_B T \ln \frac{f_v}{f_{ad}} \qquad (5)$$

$(2\phi_{AA} + \phi_{AB})$ is exactly the potential energy of a particle in the kink position of the 2D-A crystal sitting on the (001) B face (Fig.2).

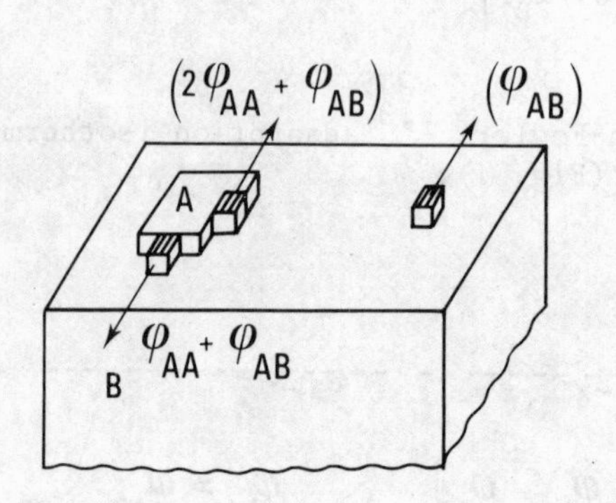

Fig. 2 : Two dimensional phase condensation on a substrate indicating the interaction energy of A particles on different sites.

The equilibrium pressure $P_{2\infty}$ should be compared with $P_{3\infty}$, the equilibrium pressure of the 3D-A crystal, given by equations (1) and (2). If the Einstein frequencies of the adsorbed atome of the 3D crystal and 2D crystal are the same :

$$P_{2\infty}/P_{3\infty} \simeq \exp(\phi_{AA} - \phi_{AB})/k_B T \qquad (6)$$

The 2D-A crystal is stable with respect to the 3D-A crystal if $P_{2\infty} < P_{3\infty}$ implying $\phi_{AA} < \phi_{AB}$. It is worth mentioning that a 3D-A crystal in the vicinity of a (001) B substrate (at the same temperatures) loses its particles and a 2D-A crystal forms on the (001)B substrate. Growth takes place at undersaturation with respect to the vapor pressure of the 3D-A crystal.

In the special case $\phi_{AA} = \phi_{AB}$ and $f_c = f_{ad}$, equation (5) gives $P_{2\infty} = P_{3\infty}$ (see figure 1 where the phase transition now takes place at $P_{3\infty}$). This is the case of the growth of a crystal on its own substrate (sometimes imporperly called autoepitaxy).

This special case answers whether, if in the case $\phi_{AA} < \phi_{AB}$ (Fig. 1), a second layer can be formed after the first one. The answer is yes, and a third .... a xth layer can grow, since for an A particle adsorbed on the very nearly full first layer, $\phi_{AB} \equiv \phi_{AA}$ and the isotherm (4), where θ now means the coverage in the second layer, gives the phase transition at $P_{3\infty}$. But clearly this property of the formation of only one layer at undersaturation is due to assumptions in the KOSSEL crystal model where only first neighbor interactions are assumed (see later a generalization).

In summary the stability conditions for 2D and 3D phases on a substrate are given in two first columns of the table

Stable Phase	Binding Conditions	Adhesion Conditions	Interfacial Conditions	Experimental Conditions
2D	$\phi_{AA} < \phi_{AB}$	$\beta > 2\sigma_A$	$\sigma_A < \sigma_B - \sigma^*$	Undersaturation
3D	$\phi_{AA} > \phi_{AB}$	$\beta < 2\sigma_A$	$\sigma_A > \sigma_B - \sigma^*$	Saturation

The third column gives another expression of the same stability conditions. For a KOSSEL crystal the surface specific energy of the (001) face is by definition $\sigma_A = \phi_{AA}/2a^2$ for the A crystal and for B isomorphous, $\sigma_B = \phi_{BB}/2a^2$.
The adhesion energy for (100) A and (100) B is by definition $\beta = \phi_{AB}/a^2$. The specific interfacial energy σ^* defined by DUPRE'S relation $\sigma^* = \sigma_A + \sigma_B - \beta$ gives the condition in the fourth column.

III. KINETIC BARRIERS TO NUCLEATION

Equilibrium conditions don't tell us about the *formation* of a phase. Since GIBBS[4] and VOLMER[5] it is known that a new phase cannot be formed at equilibrium. For producing a new phase we have to start from the smallest geometrical extension and since a small phase has a greater vapor pressure than an infinite one (GIBBS-THOMSON equation), supersaturation is a necessary condition for its formation. This is due to the fact that a small phase has an excess energy (surface free energy) referred to the same number of atoms contained in an infinite crystal. This is the origin of an energy barrier.

Let us see this closer with the same model as in section II. We build a 3D - A crystal of $m = n^2x$ atoms on the substrate (001)B. Around this crystal a vapor pressure P exists. Taking m atoms from a saturated vapor phase at pressure $P_{3\infty}$, the reversible work to be done is $mk_BT\ln(P/P_{3\infty}) = m\Delta\mu$. $\Delta\mu$ is the thermodynamic driving force (per atom) according to VOLMER. The A crystal having n^2 atoms in the base and x layers in thickness, has an excess energy E(m) with respect to a infinite crystal of vapor pressure $P_{3\infty}$. The free energy change is :

$$\Delta G(m) = -m\Delta\mu + E(m) \tag{7}$$

To evaluate E(m), take m particles from the infinite crystal A and disperse them, expending energy $m3\phi_{AA}$. Build with them an A crystal, containing $m = n^2x$ molecules, on the substrate. The energy gain is

$$- |n^2 \phi_{AB}+2(n-1)\phi_{AA}+2(n-1)^2\phi_{AA}| - (x-1)|n^2\phi_{AA}+2(n-1)\phi_{AA}+2(n-1)^2\phi_{AA}|,$$

where the first bracket corresponds to the first layer on the substrate, the other to the second layer on the first (there are (x-1) of that type). Then (7) simply gives :

$$\Delta G(m) = - m(\Delta\mu - \Delta\phi/x) + 2\sqrt{mx}\,\phi_{AA} \tag{8}$$

where $\Delta\phi = \phi_{AA} - \phi_{AB}$.

If $\underline{\Delta\phi < 0}$ we know that a 2D layer is stable. For x = 1, from (8), ΔG (m) is, at $\Delta\mu$ = constant, an increasing function going through a maximum when $\Delta\mu > \Delta\phi$, located at $m_1^* = \{\phi_{AA}/(\Delta\mu - \Delta\phi)\}^2$ (see figure 3).
This initial nucleus grows spontaneously when $m > m_1^*$ and forms a full monolayer. In order to do so, an activation barrier, $\Delta G_1^* = \phi_{AA}^2/(\Delta\mu - \Delta\phi)$ has to be overcome, whose height can be decreased if the supersaturation $\Delta\mu$ is increased. In fact $\Delta\mu$ may be negative. The only condition is that $\Delta\mu > \Delta\phi$ or the real driving force $\Delta\mu^* = \Delta\mu - \Delta\phi$ must be positive for spontaneous growth to take place.

In figure 3 the activation barriers for double (x = 2), triple (x = 3) layer crystals.

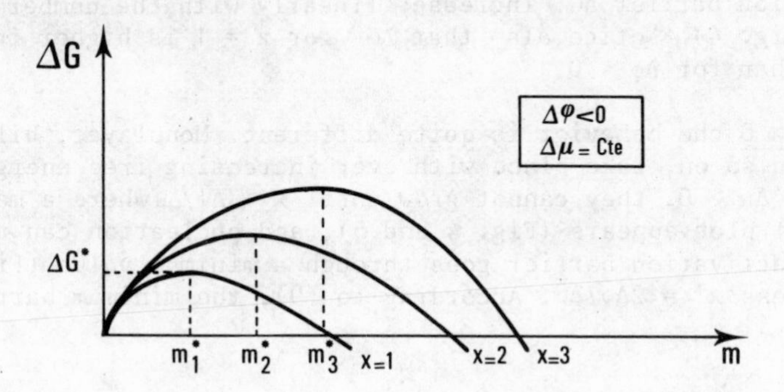

Fig. 3 : Free energy variation as a function of the number of particles m, when 2D islands of thickness x = 1, x = 2 .... are formed on the substrate. Case where $\Delta\phi < 0$. The lowest activation barrier is that of x = 1.

All these barriers are higher than for the 2D-layer. For an x-layer crystal, the energetic barrier is :

$$\Delta G^*_x = x^2\phi^2_{AA}/(\Delta\mu x - \Delta\phi) \qquad (9)$$

In fig. 4 the barrier heights as a function of x are shown. The result is that the growth of a 2D-A crystal is the most favorable among all other multilayers. The critical nucleus contains $m^* = [\phi_{AA}/(\Delta\mu - \Delta\phi)]^2$ atoms in the base.

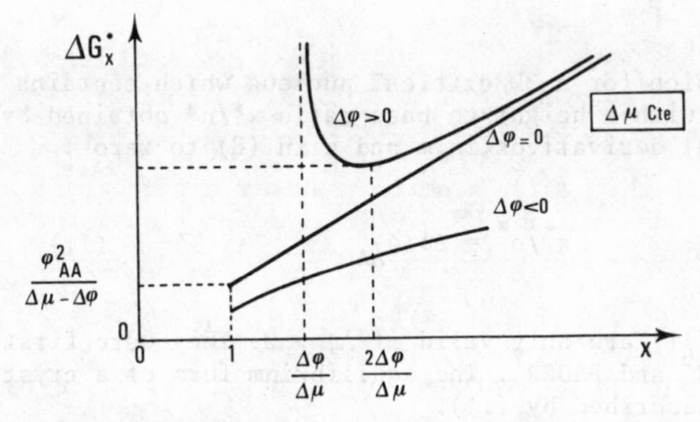

Fig. 4 : Heights of activation barriers at constant driving force $\Delta\mu$ as a function of number of layers x. Three cases : $\Delta\phi = 0$ where there is a linear increase, $\Delta\phi < 0$ with a smaller increase and $\Delta\phi > 0$ with an optimal barrier with $x > 1$.

The special case $\Delta\phi = 0$ has similar properties as when $\Delta\phi < 0$ but the activation barrier ΔG^* increases linearly with the number of layers x (fig. 4). Notice also that ΔG^* for $x = 1$ is higher in this case than for $\Delta\phi < 0$.

If $\Delta\phi > 0$ the behavior is quite different. Monolayer, bilayer crystals and so on, take place with ever increasing free energy. For a given $\Delta\mu > 0$, they cannot grow until $x > \Delta\phi/\Delta\mu$ where a maximum in the $\Delta G(m)$ plot appears (Fig. 4 and 5), and nucleation can then occur. The activation barrier goes through a minimum value (Fig. 5) at a thickness $x^* = 2\Delta\phi/\Delta\mu$. According to (9), the minimum barrier is then :

$$\Delta G^*_{min} = 4\phi^2_{AA}\Delta\phi/\Delta\mu \tag{10}$$

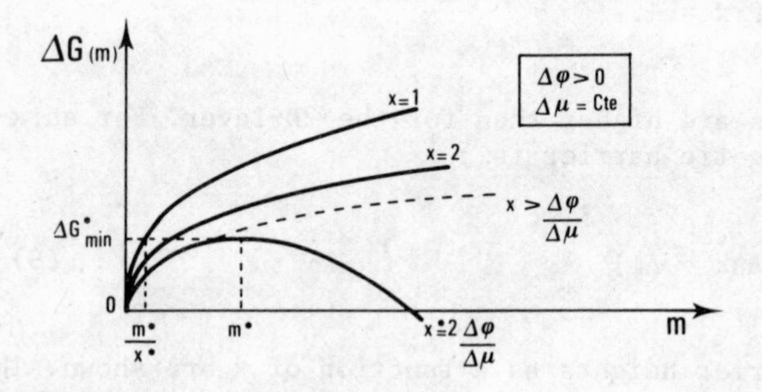

Fig. 5 : The same as figure 3 but for the case $\Delta\phi > 0$. A finite activation barrier appears only for islands whose thickness is $x \geq x^*$

This is the expression for a 3D critical nucleus which contains in its basis n^{*2} atoms with a height to base ratio x^*/n^* obtained by equating the partial derivatives in m and x in (8) to zero :

$$n^* = 2\phi_{AA}/\Delta\mu ; \qquad x^*/n^* = \Delta\phi/\phi_{AA} \tag{11}$$

Formulas (10) and (11) are only valid if $\Delta\phi > 0$. They were first obtained by KAISHEW[6] and BAUER[7]. The equilibrium form of a crystal on a substrate is described by (11).
In the absence of a substrate, $\phi_{AB} = 0$ implies $\Delta\phi = \phi_{AA}$. Then $x^*/n^* = 1$ and the equilibrium form is a cube. The activation energy is then $\Delta G^*_{hom} = 4\phi^3_{AA}/\Delta\mu^2$.
This homogeneous nucleation has its activation barrier diminished

by heterogeneous nucleation on a substrate by ratio $\Delta G^*_{het}/\Delta G^*_{hom} = \Delta\phi/\phi_{AA}$.

The question may arise how such a critical nucleus builds up at constant $\Delta\mu$. Does it increase its number of atoms m according to the curve x = constant (Fig. 5) or does it follow trajectories as x = 1, x = 2, ...? A clear answer can be given to the more precise question : What is the most probable kinetic trajectory ? When it follows x = 1, growing as a 2D building, it reaches the same energy level ΔG^*_{min} (fig. 5) as does the 3D critical nucleus of m^* atoms and thickness x^* but having now in its plane m^*/n^* atoms, exactly the same number as has the 3D critical nucleus in its basis plane. The same property is true for a double layer (x = 2) building containing in its base m^*/x^* atoms and $2\,m^*/x^*$ in its bulk.
All buildings with $x < x^*$ disappear as they form in contrast to the critical nucleus x^* which has also the possibility of growing spontaneously. Growth can take place only along x^* = constant. But clearly when a 2D building of m^*/x^* atoms is obtained, a second layer may be formed on it, then a third layer, etc... attaining progressively the critical nucleus. In figure 6 the free energy path of this process is given. When a second layer is formed on the 2D building $n^{*2} = m^*/x^*$, an activation barrier must be overcome. Its maximum corresponds when the second layer has exactly half the number of atoms $(m^*/2x^*)$ as a full second layer (use eqn. (8) with x = 1 and $\Delta\phi = 0$). The height of this barrier is $\phi^2_{AA}/\Delta\mu$. This fact was known since the early days of crystal growth theory by STRANSKI [8] and KAISCHEW [9]. When a two layer building is formed and a third layer starts, a new bump in the ΔG diagram appears (Fig. 6) etc... Evidently the trajectory x^*= constant is a better one than x = 1 hopping over all the bumps in order to reach a thickness of x^* layers. But x^* = constant is not the best trajectory. There is a minimal free energy path according to formula (8) if the constant ratio $x/n = \Delta\phi/\phi_{AA}$ is preserved not only for the critical nucleus according to (11) but also along the entire trajectory, m = 0 to $m = m^*$. Introducing this "equilibrium ratio" in formula (8) the minimum path $\Delta G_{m.p.}$ is obtained as shown in Fig. 7 as a function of n.
In the same figure it is compared to the path at x^* = constant. This path is the most probable one for the critical nucleus.

The calculations of the kinetic rate equations for nucleation on substrate follows very closely that considered in the chapter "Homogeneous Nucleation".

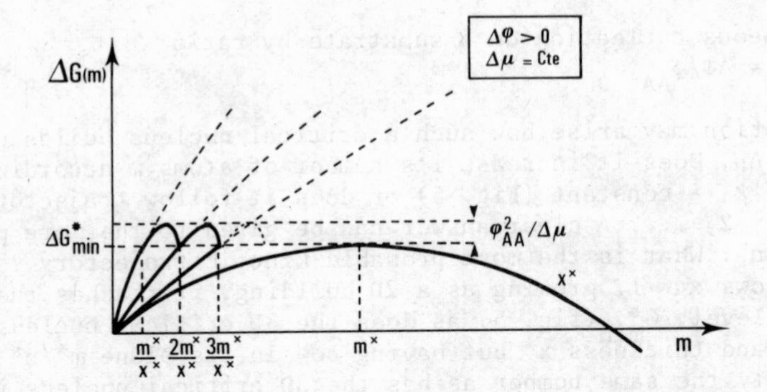

Fig. 6 : Activation energy bumps with a single, double... etc. layer building forming the critical nucleus m^*.

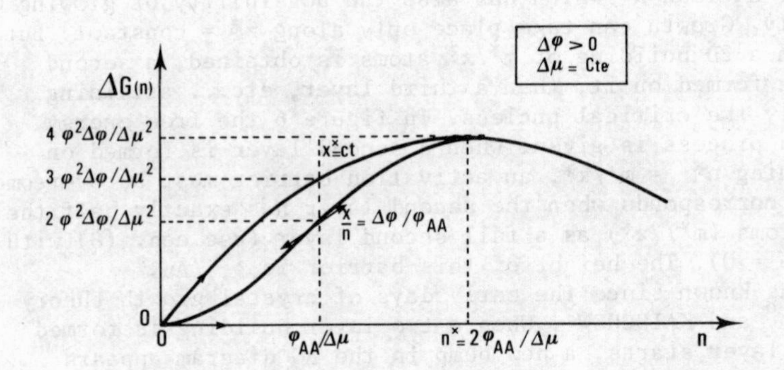

Fig. 7 : Minimum activation path compared to the x^* = constant path.

IV. SUBSTRATE INTERACTION AT LONGER DISTANCES

The simple Kossel model may be refined when including interactions between A and B beyond first neighbors. We don't apply this refinement to interactions between A atoms because nothing new is obtained other than more complicated calculations.

An isolated atom A adsorbed in a x-th layer is bound to the (x-1) underlying layers and to the substrate. The interaction energy of this atom, $\phi_{AB}(x)$, tends towards ϕ_{AA} with increasing distance x (Fig. 8).
If, as an example, the bonding is of Van der Waals type, SINGLETON and HALSEY[10] showed that for the case $\phi_{AA}(1) < \phi_{AA}$:

$$\phi_{AB}(x) = \phi_{AA} + \left[\phi_{AB}(1) - \phi_{AA}\right] x^{-3} \qquad (12)$$

where the distance x is scaled in number of monolayers.

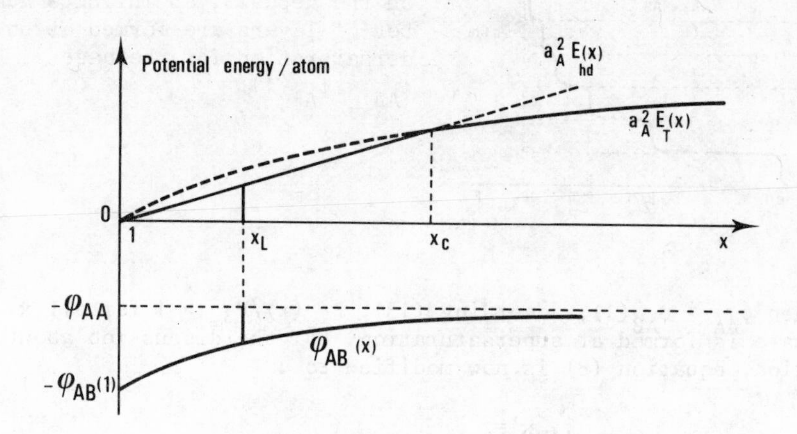

Fig. 8 : Potential energy diagram per atom adsorbed as a function of the layer numbered x above the substrate (below). Upper part : strain energy per atom for layer of increasing thickness x. Straight line homogeneous strain ; curved line : compound strain due to homogeneous strain + dislocation strain.

The contribution of the substrate in the second layer is $1/8^{th}$ that of the first layer, $1/27^{th}$ in the third layer, etc. All these contributions were zero for the Kossel crystal.

The previous conclusions are now somewhat changed. Relation (6) becomes :

$$P_{2\infty}(x)/P_{3\infty} = \exp\left[(\phi_{AA} - \phi_{AB}(1)\right] x^{-3} \qquad (13)$$

Again if $\phi_{AA} < \phi_{AB}(1)$, then $P_{2\infty}(1)/P_{3\infty} < 1$. This means that a first layer is more stable than a 3-D crystal. The inverse is true if $\phi_{AA} > \phi_{AB}(1)$. The conclusions of column 2 in the table are still valid with $\phi_{AB}(1)$ replacing ϕ_{AB}. However some new phenomena appear.

According to (13), if $\phi_{AA} < \phi_{AB}(1)$, then $P_{2\infty}(x)/P_{3\infty} << 1$ regardless of x. This means that the isotherm now has an infinite number of steps (Fig. 9) at well defined pressures $P_{2\infty}(1)$, $P_{2\infty}(2)$, etc., representing layer by layer growth in which each layer is a distinct 2D-phase.

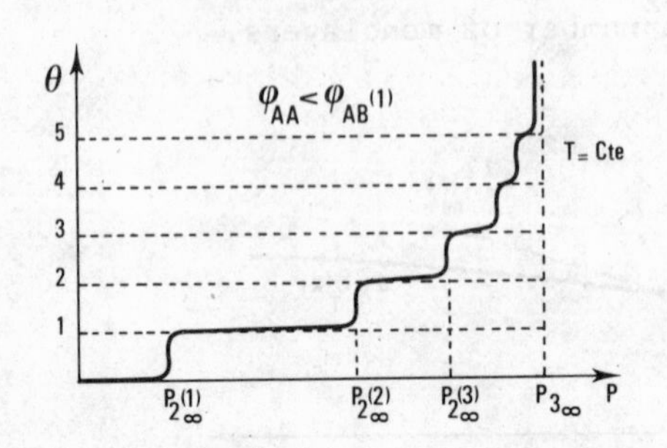

Fig. 9 : In the case of long range action of the substrate on the deposit, an infinite number of layers are formed at undersaturation for the case $\phi_{AA} < \phi_{AB}(1)$.

When $\phi_{AA} > \phi_{AB}(1)$, according (13), $P_{2_\infty}(x)/P_{3_\infty} > 1$ for all x. A 3D-phase is formed at supersaturation. For the discussion about nucleation, equation (8) is now modified to :

$$\Delta G(m) = -m\left(\Delta\mu - \frac{\Delta\phi}{x}\sum_{x'=1}^{x'=x}(x')^{-3}\right) + 2\sqrt{mx}\,\phi_{AA} \qquad (14)$$

with $\Delta\phi = \phi_{AA} - \phi_{AB}(1)$. Derivation in m or x in order to find all equivalent formulas (9) (10) (11) may be performed by assuming the variation of $\sum_{x'=1}^{x'=x}(x')^{-3}$ is small at large values of x.

Then all formulas are modified only slightly by replacing $\Delta\phi$ by $\Delta\phi \sum_{x'=1}^{x'=x}(x')^{-3}$ in (9) (10) (11). It is worth mentioning that the equilibrium form is no longer homothetic with changes of crystal size as was the case for first neighbor interactions only (Eq. 11).

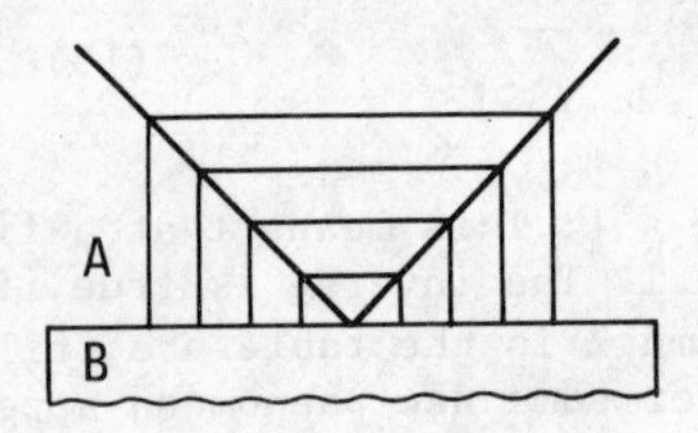

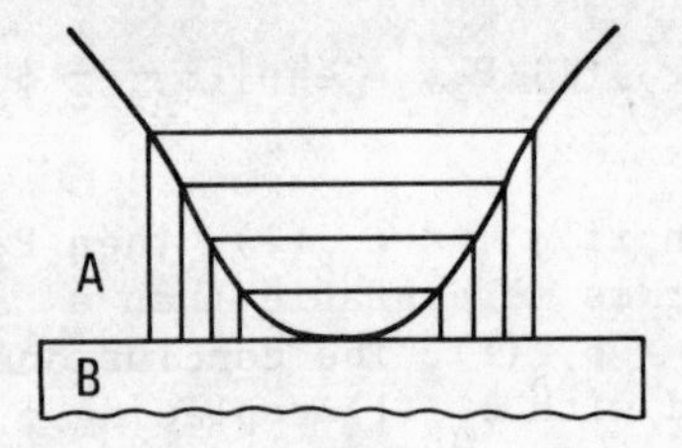

Fig. 10 : Left part : first neighbor interaction with the substrate gives homothetic equilibrium form. Right part : longer range action gives a flattening of the equilibrium form.

Equation (11) becomes $x^*/n^* \simeq \frac{\Delta\phi}{\phi_{AA}} \sum_{x'=1}^{x'=x}(x')^{-3}$ and the equilibrium form becomes flatter when the size decreases (Fig. 10). This effect, mentionned by DASH[11] could hardly have been observed because it appears only for crystals of several atomic layers thickness, the maximum effect being given when $\sum_1^\infty (x')^3 \simeq 1.2$ corresponding to $\simeq$ 20% flattening.

V. MISFIT BETWEEN SUBSTRATE AND DEPOSIT

Until now A and B were considered as strictly isomorphous ($a_A = a_B$). If $a_A \neq a_B$ there is a natural misfit f defined by

$$f = (a_A - a_B)/\bar{a} \qquad (15)$$

where $\bar{a}$ is a mean value of a_A and a_B.

On a very thick substrate the layers accomodate their misfit by changing their periodicity $a_A \rightarrow a'_A = a_B$ and increasing their energy by homogeneous deformation. For a layer of thickness xa_A, the stored energy, per unit of surface, is :

$$E_{hd}(x) = G_A \frac{1 - \nu}{1 - 2\nu} \; xa_A \; f^2 \qquad (16)$$

where G_A is the macroscopic shear modulus of the deposit and ν its Poisson's ratio. This supplementary energy per atom is represented in fig. 8 as a straight line.

Dislocation theory[12,13,14] tells that there is another way to store strain energy, especially in the A-B interface. In the case of our model two orthogonal arrays of edge dislocations could reduce the natural misfit to a smaller value f'. The number of dislocations per unit surface will be 2N, N being N = (f-f')/b, where b is their Burgers's vector. The energy stored in this way is :

$$E_{disl}(x) \simeq \frac{4\, G_A\, G_B}{G_A + G_B}\, b \left[1 + ln(xa_A/b)\right] (f-f'), \qquad (17)$$

where G_B is the shear modulus of the substrate. The remaining strain f' not accounted for by dislocations remains as a homogeneous deformation in the bulk of the deposit. It is given by (15) where f' replaces f.

The total energy of this composite state is :

$$E_T(x) = E_{hd}(x) + E_{disl}(x) \qquad (18)$$

and is drawn in figure 8. This curve intercepts the straight line of homogeneous deformation at a critical thickness x_c. This tells us the well known experimental fact that, at small thicknesses, pure homogeneous strain is prefered and the layer is coherent with the substrate. If $x > x_c$, the compound mechanism is prefered, dislocations appear in the interface, the stored deformation energy now increasing less with thickness.

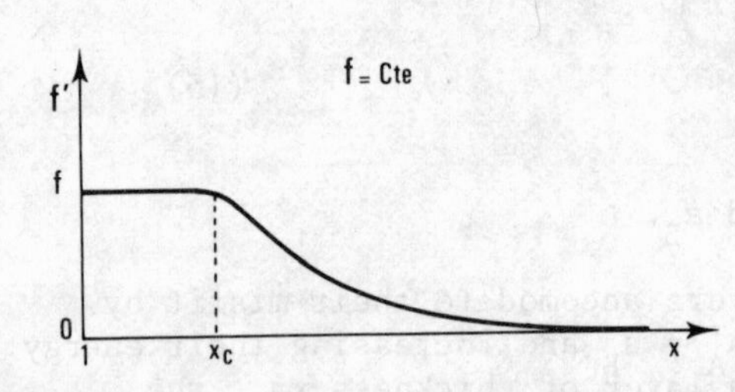

Fig. 11 : Misfit f' in layers as a function of their thickness x. f = Cte, natural misfit of the couple (A,B).

In Fig. 11 the actual misfit of such a layer for different thicknesses is given (obtained by equating to zero the derivative in f' of (18)).
The layer has the same parameter as the substrate only for $x < x_c$. This has been called a pseudomorphic layer by FINCH. For $x > x_c$, a_B asymptoticaly tends to a_B (no more strain remains in the thick layer).

The thickness x_c where this transition occurs depends on the material constants G but especially on the natural misfit. There are epitaxial couples, where pseudomorphism occurs in several layers (f.c.c. metals), others where partial incoherency appears until the second layer, and systems where dislocations appear in the first layer (rare gases on graphite).

The question arises if a presupposed layer by layer growth can occur when there is a natural misfit. Or in other words : is there a possibility of changing the growth mode (2D → 3D) ? To address this we again use formula (13) but inside the exponential add the actual deformation energy of the layer numbered x. This can be discussed with the curves of figure 8, when summing them up. Clearly there appear to be a limited number of layers x_L growing layer by layer, in contrast to the case (fig. 9) where the natural misfit was assumed zero.

The adsorption isotherm now (fig. 12) cuts the vertical line $P_{3\infty}$ and 3D-growth occurs by increasing the vapor pressure. If $x_L < x_c$, the layers are pseudomorphic, if the opposite is true there are dislocations in the interface.

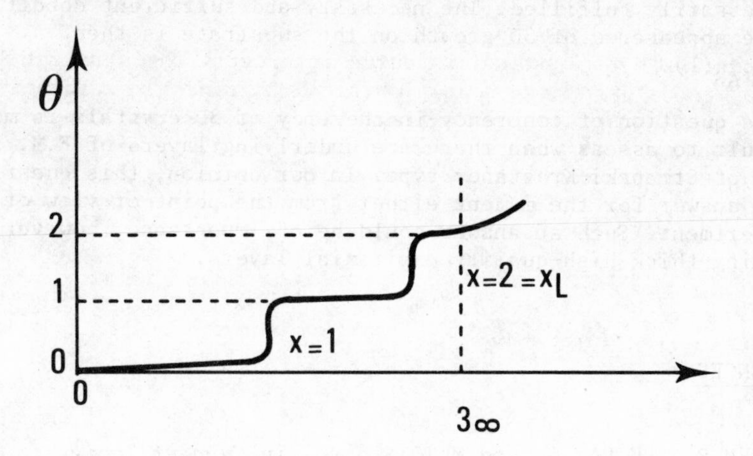

Fig. 12 : When layer growth occurs on a couple with any natural misfit $f \neq 0$, the number of layers is limited to x_L. At a pressure of the vapour phase P higher than P_{3_∞}, three dimensional growth occurs.

An important conclusion is that when A and B are non isomorphous, the layer by layer growth is necessarily relayed by three dimensional growth during thickening.

The first mode has been called Frank-van der Merwe mode, the second Vomer-Weber mode. The F.M. mode appears when $\phi_{AA} < \phi_{AB}(1) - E_{deform}(1)$, where $\phi_{AB}(1)$ is the adsorption energy of a single molecule in the first layer. It can be pseudomorphic or dislocated, depending on the ratio x_L/x_c, but it is necessarily followed by the V.W mode if x becomes greater than x_L. A third mode has been called the Stranski-Krastanov mode. It is a special case where the first layer is isomorphous and if $x > 1$, 3D-growth follows. f.c.c. metals on silicon or germanium follow this mode.

Another point to address is the coherency-incoherency of 3D-crystallites of the V.W. mode[14]. The strain energy in this case is more complicated to evaluate.
For a very thick substrates, strain also appears in the substrate in contrast to layer growth. However, qualitatively the phenomena are the same. The 3D-crystals could first form in a quasi homogeneous strained state until a critical radius r_c is reached (assuming they are hemispherical). The homogeneous strain is partly released by the introduction of dislocations in the interface. The total strain energy is again a slowly increasing function of size as it was for layer growth. The condition for a 3D-growth on the substrate is the $\phi_{AA} > \phi_{AB}(1) - E'_{deform}(r^*)$ where r^* is the size

of the critical nucleus. Clearly if $\phi_{AA} > \phi_{AB}(1)$, the above condition is necessarily fulfilled. The necessary and sufficient condition for the appearence of 3D growth on the substrate is then $\phi_{AA} > \phi_{AB}(1)$.

The question of coherency-incoherency of 3D-crystals is more difficult to assess when there are underlying layers of F.M. type or one of Stranski-Krastanov type. In our opinion, this question has no answer for the moment either from the point of view of theory or experiment. Such an answer could be of importance, however, in obtaining thick high-quality expitaxial layers.

REFERENCES

1 . KERN R., LE LAY G. and METOIS J.J., in *Current Topics in Material Sciences*, Vol. 3, ed. Kaldis, N.H. Pub. Co., 1979,p 128-419
2 . FRUMKIN A.N., Z. Physikal.Chem. 116(1925)466; 135(1926)792
3 . FOWLER R. and GUGGENHEIM E.A., in *Statistical Thermodynamics*, Cambridge Univ. Press, 1965
4 . GIBBS J.W., in *The Scientific Papers of J.W. Gibbs*, Vol. 1., Dover Pub. 1961.
5 . VOLMER M., *Kinetik der Phasenbildung*, Leipzig, 1939
6 . KAISCHEW R., Bull. Acad. Sc. Bulg. Ser. Phys. 2(1951)191
7 . BAUER E., Z. Krist, 110(1958)372
8 . STRANSKI I.N. and KAISCHEW R., Z. Phys. Chem. (B) 26(1934) 81, 114, 312
9 . KAISCHEW R., Z. Elektrochem. 61(1957)35
10. SINGLETON J.H. and HASLEY G.D., J. Chem. Phys. 52(1954)1011
11. DASH J.G., Phys. Rev. B15(1977)3136
12. FRANK F.C. and Van der MERWE J.H., Proc. Roy. Soc. A198(1949) 205; ibid. A200(1949)125
13. MATHEWS J.W., in *Epitaxial Growth*, Vol. A and B. Material Science Series, Acad. Press. N.Y., 1975
14. MATHEWS J.W., in *Dislocations in Solids*,Vol. 2, F.R.N. Nab rro Ed., N.H., Co., Amsterdam, 1979, p 461.

COALESCENCE OF NUCLEI ON SUBSTRATES

R KERN

CRMC2, CNRS
Marseille-Luminy
France

ABSTRACT

The actual formation of epitaxial layers may be far from what would be expected close to equilibrium. Considering the growth of three dimensional crystals on a substrate far from equilibrium, various kinetical processes are analyzed.

In fact, the actual crystal growth form (GF) rarely corresponds to the equilibrium form (EF). When surface self-diffusion is active on the grains, the GF tends to EF. In addition, when surface diffusion on the substrate is active, material exchange takes place between the grains (Ostwald ripening). Grains may also be mobile on the substrate. These two types of material exchange can induce either "static" or "dynamic" grain coalescence when two grains come in contact. Changes of epitaxial orientation of the grains can happen during these different processes.

I. INTRODUCTION

In the chapter "Three and Two-Dimensional Nucleation on Substrate" (further called (NS)) some fundamental points were discussed about the formation of thin layers on a substrate. In particular, the very early stages of stability and growth were discussed within the framework of thermodynamics. The kinetical point of view was not neglected but many of its aspects were intentionally avoided. Some of them will now be discussed, as material transport on the grains and between the grains by surface self diffusion or surface diffusion on the substrate (section II and III). In section IV, we discuss the more peculiar behavior of grain migration on a substrate.

B. Mutaftschiev (ed.), Interfacial Aspects of Phase Transformations, 303–314.

When grains are touching each other during growth or migration. coalescence occurs, two grains become one (section V and VI) by a sintering process on the substrate. These different phenomena may be isolated experimentally, but during the actual growth they all occur simultaneously. In addition, grains may change their epitaxial orientation during migration and during coalescence (section VII).

II. GROWTH FORMS AND EQUILIBRIUM FORM

When a flux of atoms A, impinges onto the substrate B, part of the atoms diffuse on the surface, meet each other, form embryos and critical nuclei. Critical nuclei are the cluster which by addition of one more atom are able to grow spontaneously. Surface diffusion on the substrate and grains as well as desorption back to the vapor phase, constitute the various processes occuring during the growth of the layer. Equilibrium is attained when the balance of the incoming flux is just equal to the outgoing one. The grains then show, in steady state, the behavior discussed in NS. A dynamical equilibrium is achieved and crystals show the equilibrium form predicted by thermodynamics. Their form depends only on the relative surface specific free energies and the adhesion on the substrate. In NS[1], a simple cubic model was used for simplicity but it may be generalized for every crystal structure.

In practice, special experimental conditions must be realized in order to obtain an equilibrium form. The substrate and the deposit must be enclosed in an isolated system (constant temperature and constant number of particles). Experimentally this was done by SUNDQUIST[2] for several metal-substrate systems and later done under more careful conditions by HEYRAUD et al.[3] Describing the behaviour in simple experiments should give some useful understanding.

Figure 1 represents different shapes of gold crystals with their (111) face sitting on the (0001) face of a graphite crystal. Figure 1b shows the equilibrium form obtained in an isolated system at $T \simeq 1000°C$. Fig. 1a shows what is observed when this system loses some gold atoms by evaporation (evaporation form), Fig. 1c, when more gold atoms are coming in the system than atoms going out (growth form). The EF has well defined flat {111} and {100} faces and rounded regions, the other two forms differ slightly. The evaporation form has many facets, the growth form only {111} and {100} faces but more or less straight edges and no rounded parts. The crystal shape is very sensitive to the experimental conditions. By varying these conditions, very extreme cases could be produced with the same system.

If gold atoms are evaporated on a cold (20°C) graphite substrate, very thin crystals with the (111) plane in contact with the substrate are formed showing dendritie features (see Fig. 2a). At this temperature, the gold atoms are able to supply the border of the crystals. The surface diffusion coefficient, $D_{sD}^{Au-graph}$, seems to be rate determining, but the surface self diffusion coefficient, D_{sD}^{Au-Au}, is not high enough to contribute to the thickening. Thus, the activation free energies for diffusion show the relationship $\Delta G_{sD}^{*Au-graph} < \Delta G_{sD}^{Au-Au}$.

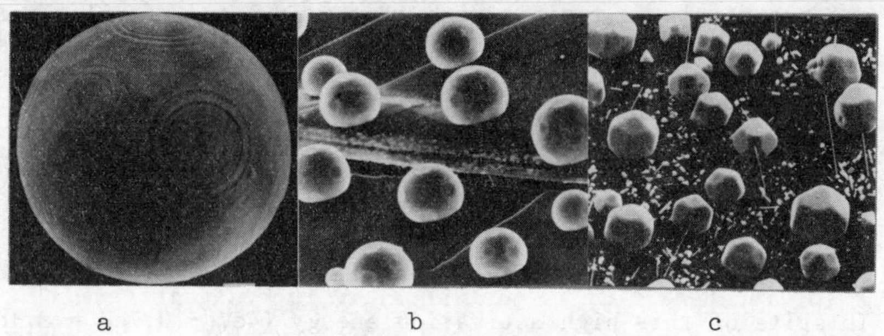

Fig. 1 : Illustration of the morphology of gold crystals on (0001) graphite. a) form after slight evaporation, b) equilibrium form, c) growth form (Courtesy J. C. HEYRAUD, J. J. METOIS).

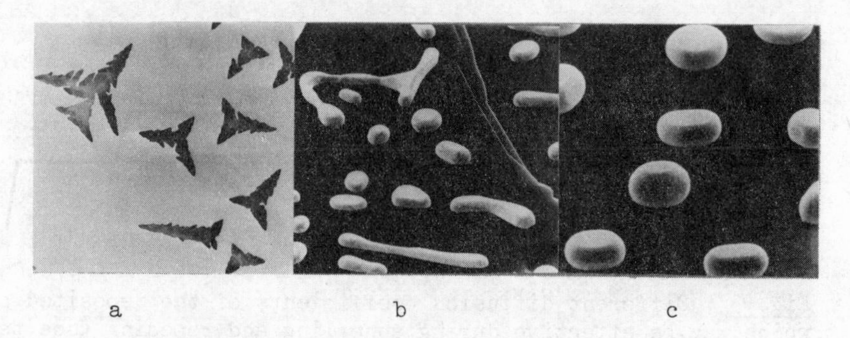

Fig. 2 : a) Gold dendrites on graphite
b) after a short annealing at 50°C
c) thickening at 450°C (Courtesy J.C. HEYRAUD, J.J.METOIS).

An annealing at 150°C for several minutes[4] rounds the dendrites without thickening (Fig. 2b). At this low temperature, isolated gold atoms are supplied singe there is a local driving force pushing them from regions of high curvature to lower according to HERRING's theorem :

$$\mu(K) = \mu(0) + a^3 \left[\sigma(\alpha) + \frac{d^2\sigma}{d^2\alpha}\right] K(\alpha), \qquad (1)$$

and the corresponding j current of atoms by Fick's law is :

$$j = - \frac{D'_{sD}}{k_B T} \, grad_s (\mu(K) - \mu(0)), \qquad (2)$$

where K is the inverse radius of curvature, $\sigma(\alpha)$ is the specific surface free energy as a function of the angular coordinate α, a^3 is the atomic volume, $\mu(K)$ and $\mu(0)$ are respectively the chemical potentials in regions with curvature K and ∞(a flat reference surface). D'_{sD} is the self diffusion coefficient of gold adatoms that must be extracted from a kink position. Its activation energy is approximately (fig. 3) :

$$\Delta G^*_{sD} \simeq \Delta G^{Au}_{evap} - \Delta G^{Au-Au}_{ads}$$

In spite of this high activation energy (4eV - 0.7eV = 3.3eV), the diffusion process is still effective at low temperature because the highest curvature change is of the order of $\Delta(K^{-1}) \simeq 10^{-6}$ cm if the dendrites have tip radii of $\simeq 10^2$Å . The surface energy gradient plays only a minor role for this system.

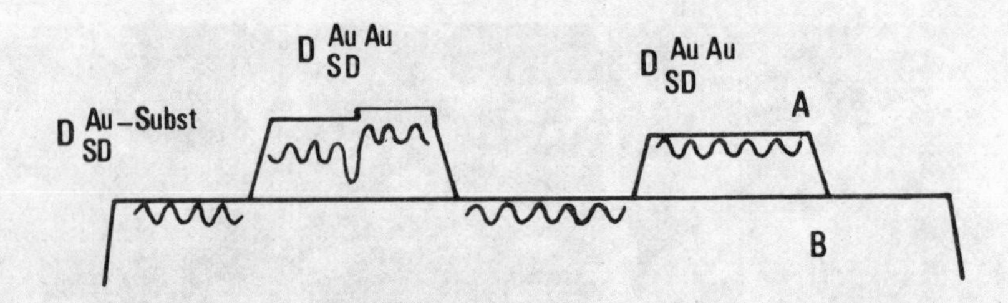

Fig. 3 : Different diffusion coefficients of the deposited atoms A which may be effective during annealing and repening (see text).

At 150°C annealing thickening is not observed, but when the dendrites are fully rounded (K $\simeq \infty$), further annealing at 450°C produces thickening. This new process has a higher activation energy than the rounding of dendrites. New layers on the flat (111) Au face have to be formed. It is likely that such small crystals (<1μ) contain no screw dislocations and the layer-by-layer growth must proceed by two dimensional nucleation with an activation barrier higher than for surface diffusion.

The thickening proceeds slowly, requiring 100 hours even at 1000°C (50°C below the melting point of gold) in order to obtain the EF. (fig. 1b).

It is clear from the discussion above that when gold atoms are deposited on a hot graphite substrate, the crystals more rapidly attain their EF. First of all, the high temperature increases the probability of passing the different activation barriers, but also the impinging gold atoms are able to diffuse more easily. The actibation energy is no longer dominated by G^{Au}_{evap} as in the annealing experiment but rather by hopping over the surfaces (which is estimated as 1/5 or 1/10 of G^{Au}_{evap}.

In summary this description shows[4] that :

a) during condensation, the 3D crystals come closer to their EF the larger the surface diffusion coefficients D_{sD} on the deposit and the substrate.

b) In order to obtain the EF, during condensation the system must be able to be heated high enough. This means the substrate must be stable enough and the deposit must not evaporate at this temperature. This is the case for gold on graphite or MgO or other metals at 600°C but not for substrates such as alkali halides. A good example of the latter type is platinum deposited on (100) NaCl at 400°C which exhibits epitaxial crystals of a {100} shape far from their EF.

III. OSTWALD RIPENING

Well separated 3D-crystallites on a substrate may achieve their EF not only by surface self-diffusion but also by exchanging atoms among themselves either by surface diffusion or through the vapor phase (if the system is closed and the temperature high enough). This phenomenon, called Ostwald ripening, is well known in the case of crystal growth from solution.

A cap-shaped crystallite on a substrate with a radius r has a vapor pressure P(r), larger than P_∞ of an infinite crystal, given by the Gibbs-Thomson equation :

$$P(r)/P(\infty) = \exp\ (2\sigma a^3/rk_BT). \qquad (3)$$

where σ is the average specific surface energy. When the deposits show a size distribution around $\bar{r}$, the mean pressure $P(\bar{r})$ produces an average concentration n of adatoms. If the concentration close to a crystallite is n(r), this crystallite grows when $n(r) < \bar{n}$, but loses atoms to the substrate if $n(r) > n$. CHAKRAVERTY[5] has formulated this problem by a Lialikov differential equation under the condition that the total mass of grains is constant (conservative system).

The analytical result is that :

i) the total number of grains varies as $t^{-3/2}$ (the result is a unique crystal at $t \to \infty$)

ii) the mean radius and the dispersion increase as $t^{1/4}$. We will not consider the details further but will consider here the real possibility of such a ripening.

A certain concentration of adatoms is necessary to assure the kinetic efficiency of Ostwald ripening. By using equ. (3) vapor tension and the degree of coverage of atoms ($\theta \ll 1$) by equations (1) and (4) of NS[1], one has :

$$ln\ \theta(r) = -(\Delta H_{evap} - \Delta H_{ad} - 2\sigma a^3/r)/k_B T \qquad (4)$$

where ΔH_{evap} is the heat of evaporation of the deposit and ΔH_{ad} the heat of adsorption on the substrate.

Consider gold crystallites as small as $r = 10$Å, $a \simeq 3$Å, $\sigma = 2400$ erg cm^{-2}, and $\Delta H_{evap} = 4$ eV. On a (100) KCl or a (0001) graphite substrate, $\Delta H_{ad} \simeq 0.7$ eV. Taking $T \simeq 1100$°K, equ. (4) yields $\theta \simeq 10^{-12}$. This degree of coverage of adatoms ($\simeq 10^3$ atoms cm^{-2}) is much smaller than the usual number of crystallites on such a substrate ($\simeq 10^{11}$ cm^{-2}). Thus Ostwald ripening is obviously excluded at this temperature (For KCl this temperature is also much too high). Taking again gold, but on a (100) MgO substrate or on (111) Si , the adsorption energy is as high as $\Delta H_{ads} \simeq 3$eV, giving a coverage of $\theta \simeq 10^{-2}$ at 1100°K. This corresponds to 10^{13} atoms cm^{-2} which is sufficient to yield appreciable ripening.

As a conclusion, Ostwald ripening must only be considered for systems where $(\Delta H_{evap} - \Delta H_{ads}) \simeq 14\ k_B T$ could be a possible experimental condition, without appreciable evaporation of the deposit.

IV. CRYSTALLITE MOTION

Ostwald ripening doesn't occur for many deposit-substrate combinations. Au/(100) KCl is among these. This is surprising since we know that by annealing at only 200°C, this system increases its grain size[5] as indicated in figure 4. The Au crystallites were condensed on a 20°C (100) KCl cleavage substrate in ultrahigh vacuum (curve 1) and then annealed for 30, 60, and 90 minutes.

New facts, necessary in order to understand this curious behavior, were given[7] by measuring the positions of a large number of gold grains on the substrate and subsequently determining the radial pair distribution function g(r). Figure 5 gives this function just after gold crystallites are deposited on the cold (100) KCL substrate. A slight subsequent annealing at only 120°C for 15 minutes shows (right part of fig. 5) that g(r) changes drastically. At this temperature the grains have not changed their mean diameter <d> or their number density ρ_o, but they have obviously moved on the substrate during annealing. Other proof of crystallite migration were shown[1], for example, by measuring migration profiles which showed that the motion is thermally activated and depends on the grain size. Direct observations of the movement with an electron microscope have been made[8].

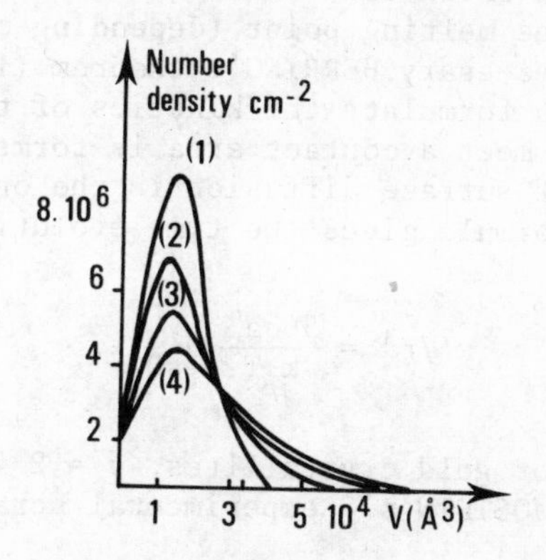

Fig. 4 : Size histogram of gold crystallites on (100) KCl for different annealing times at 200°C (1) t = 0, (2) t = 30 min, (3) t = 60 min, (4) t = 90 min.

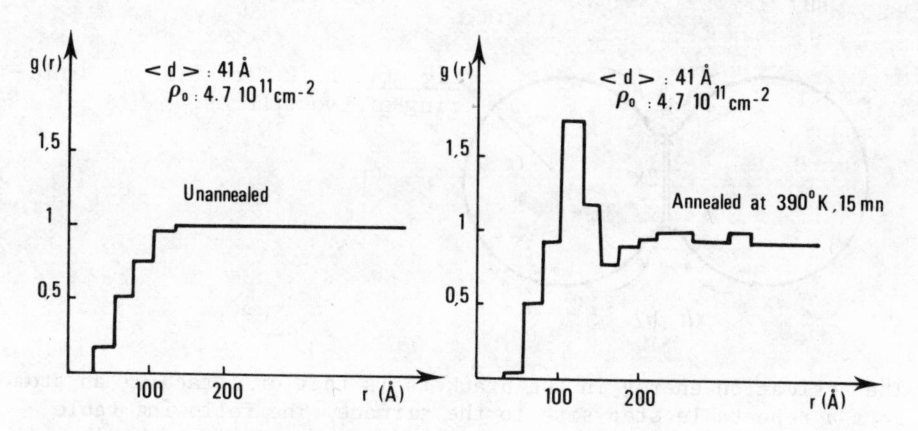

Fig. 5 : Radial pair distribution function of Au crystallites on (100) KCl before and after annealing but without change of their mean diameter and their density.

This migration behavior is quite general for couples where

adhesion energy is low and natural misfit very high. Several theories discussing the movement have been given.[9,10,11]

In the following section we will see another way to explain increasing grain size in a discontinous film during annealing if crystallites meet during their motion.

V. CONTACT COALESCENCE

Two crystallites touching each other may sinter. Sintering is an often used technique in industry, but temperatures close to the melting point (depending on the size of the solid particle) are necessary.HERRING's theorem (1) and FICK's law (2) gives the basis to formulate the kinetics of the process.When two spheres of radius r meet a contact area is formed with a neck diameter 2x (Fig. 6). If surface diffusion is the only active process, KUCZYNSKII's[12] formula gives the time evolution of the neck diameter :

$$x^7/r^3 = \frac{56\sigma a^4}{k_B T} D'_{sD}\ t. \qquad (5)$$

For gold crystallites, σ = 2 400 erg cm^{-2}. D'_{sD} can be taken from GJOSTEIN's[13] experimental scratch decay method :

$$D'_{sD} = 10^6 \exp - [54300/RT]\ \mathrm{cm^2 sec^{-1}}. \qquad (6)$$

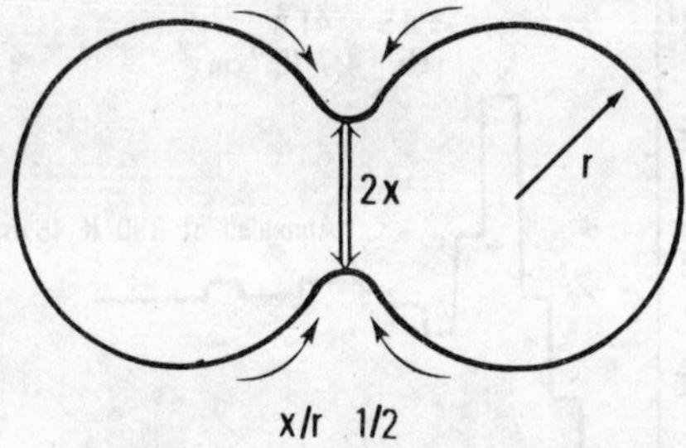

Fig. 6 : Neck formation by sintering of two spheres.

The activation energy in the brackets is that of detaching an atom from a repeatable step site to the surface. The following table gives values (in seconds) of the time $t_{1/2}$ required to obtain a relative neck radius x/r = 1/2.
Contact coalescence is effective at 200°C for crystallites with a diameter of 20 Å, and very rapid at 400°C, even for 200 Å diameter crystallites.

Table 1 : Time necessary to obtain a relative neck radius 1/2.

particle radius r Å / T °C	10^2	10
200	10^9	10^3
400	1	10^4

BASSET[14] observed such contact coalescence with an electron microscope when gold crystals are growing on MoS_2 and touch each other. At 200°C the coalescence was so quick that he called it "liquid like behavior", in analogy to coalescence of liquid droplets but is spite of the fact that an electron diffraction pattern was observed. We should mention that in this case there was an incoming flux of gold atoms so that in eq. (5) D'_{SD} must be replaced by the diffusion coefficient D_{SD} with an activation energy no longer containing the term for detachment from a repeatable step because of the presence of free atoms. If $D_{SD} \simeq 10^{-3}\exp[-15000/RT]\,cm^2sec^{-1}$; the half times $t_{1/2}$ are reduced by a factor 10^{11}! The "liquid like behavior" stops when the incoming flux of gold atoms is stopped.

HONJO[15] did not observe the same effect for the ionic crystals, NaCl, PbS, PbSe on a MgO substrate. Surface self diffusion of these species is probably much lower than for metals (the σ values are also ten times smaller).

VI. STATIC AND DYNAMIC COALESCENCE

The coalescence occuring when crystallites touch each other during growth we call "static" in contrast to "dynamic" coalescence where crystallites meet during their motion on the substrate.

VINCENT[16] developed a static coalescence theory whereby the volume of the grains on the substrate is supposed to grow with a constant rate. Grains are not supposed to influence each other when they touch. When coalescing they add their volumes instantaneously. During deposition, the density of grains varies as $t^{-2/3}$, their mean radius (and thickness) vary as $t^{1/3}$, but the size dispersion increases.

Dynamic coalescence has received several treatments similar to Smoluchowski's treatment of the coagulation of an unstable emulsion. These formulations differ somewhat in their basic hypotheses and none of them consider an incoming flux of atoms.

SKOFRONICK and PHILLIPS[17] studied the electrical conductivity of gold crystallites on an amorphous carbon film and the effect of thermal annealing between 200 and 400°C. They supposed that the grains are fixed in potential wells of depth E_o. A grain whose thermal energy is higher than E_o leaves this site with a constant velocity according its mass and can hit the other grains and quickly coalesce. The integro-differential equation is solved by calculating first and second moments of the size distribution with time. The experimental results fit well; E_o comes out to be $\simeq 1.3$ eV.

METOIS et al.[6] studied the system Au/(100)KCl of figure 4 by electron microscopy. They gave a theoretical formulation with the motion of the crystallites described by their diffusion coefficient $D_c(r,T)$. Since these coefficients have been measured independently the system is well known. The rate of binary collisions was calculated for successive collisions. In the case of 30 Å diameter grains, $D_c \simeq 7\ 10^{-14} cm^2 sec^{-1}$ at 120°C ; an initial density of $\rho_o \simeq 1.8\ 10^{11} cm^{-2}$ gives a collision rate of $D\rho_o^2 \simeq 10^9 cm^{-2} sec^{-1}$. That means that after 10^2 seconds each grain collides at least once and the density must at least drop to $\rho_o/2$. In fact the observed coalescence rate is much lower (fig. 4) than this estimation shows. The results are in quantitative agreement if an collision efficiency factor of $\delta = 10^{-6}$ is introduced. δ is independant of temperature (nonactivated) and independant of the density ρ_o. This factor probably is related to a repulsive potential between crystallites.[18] By applying an electric field parallel to the surface, the interaction between the crystallites is changed and coalescence is enhanced.[19]

A similar theory has been developed by RUCKENSTEIN et al.[20] and KASHCHIEV.[21] All crystallites are mobile with $D_c(r) = D_o r^{-n}$. Explicit solutions for the time evolution of density, mean radius, dispersion, and mean free surface area can be obtained. Qualitative agreement with SKOFRONICK's experimental results can also be obtained. HENRY et al.[22] qualitatively analyzed with this theory the coalescence behavior of very small gold crystallites (< 15 Å)

VII. ORIENTATIONAL POSTNUCLEATION PHENOMENA

Static or dynamic coalescence are postnucleation phenomena. Annealing experiments demonstrate this quite clearly. But epitaxial orientation effects also occur after nucleation takes place.

In N.S., the activation energy for 3D - nucleation on a substrate was written[10] for a Kossel crystal as :

$$\Delta G^* = 4\Phi_{AA}^2\ \Delta\Phi/\Delta\mu^2 \qquad (7)$$

The critical nuclei were assumed to be oriented parallel to the [100] direction of the substrate. For other azimuthal orientations α, $\Delta\Phi$ changes and must be replaced by $\Delta\Phi' = \Phi_{AA} - \Phi_{AB} - E(\alpha)$. The last term is a strain energy depending on the orientation. For couples with small natural misfit, $E(\alpha)$ shows a cusped minimum at $\alpha = 0$. If the misfit is high, this minimum is flat and not very deep. During nucleation there is an orientational selection of critical nuclei, the parallel orientation being favored. This phenomena is less pronounced as $E(\alpha)$ becomes smaller and its minimum flatter, especially if $\Delta\mu$ is high.

It is well known that systems of high misfit with high super saturation produce texture deposits (azimutal misorientations). Using for illustration again the Au/(100) KCl system, electron diffraction shows that Au crystals condensed at $\simeq 20°C$ on a clean (100) KCl substrate have a [111] texture. Other systems such as PbS/(100)NaCl with low natural misfit display good epitaxial orientation. When such a non epitaxial layer is slightly annealed at 150°C (where the crystallites are mobile) gradually the epitaxy (111)Au//(100)KCl, $[1\bar{1}0]$Au//$[1\bar{1}0]$KCl is established[23] and arcing of the electron diffraction rings are observed.

By annealing at higher temperatures, 250 - 300°C, this system undergoes coalescence as described in sectionVI. Au/(100) KCl,[23] among other systems, has a special behavior. During coalescence, two meeting (111) particles are replaced by a single (100) particle ((100)Au//(100) KCl ; $[1\bar{1}0]$Au//$[1\bar{1}0]$KCl). The electron diffraction pattern gradually attains fourfold symmetry. Individual coalescence events are seen by dark-field imaging. This epitaxial reconstruction is not clearly understood in terms of a three grain boundary problem.
As a matter of fact,[6] these new epitaxial crystallites are firmly anchored on the substrate. They show no motion in contrast to the (111)Au grains which are mobile at this temperature. This information has been introduced in the dynamic coalescence steps when there are no more (111) mobile crystallites.

REFERENCES

1 . KERN, LE LAY G. and METOIS J.J., in *Current Topics in Materials Science*, Vol. 3, ed. E. Kaldis, N.H. Pub. Co. 1979, p 178-419
2 . SUNDQUIST B.E., Acta Metal. 12 (1964) 67
3 . HEYRAUD, J.J. METOIS and J. of CRYST. GROWTH 50 (1980) 571; and Thin Sol. Films 75 (1981) 01
4 . METOIS J.J. and HEYRAUD J.C., Surface Science, (1981) in press.
5 . CHAKRAVERTY B.K., J. phys. Chem. Solids, 28 (1967) 2401
6 . METOIS J.J., GAUCH M., MASSON A. and KERN R., Thin Solid Films 11 (1972) 205
7 . ZANGHI J.C., METOIS J.J. and KERN R., Phil. Mag. 31 (1975) 743
8 . METOIS J.J., HEINEMANN K. and POPPA H., Phil. Mag. 35 (1977) 1
9 . REISS H., J. Appl. Phys. 39 (1968) 5045
10. KERN R.,A. MASSON and METOIS J.J., Surf. Sci. 27 (1971) 483
11. KOTZE I.A., J.C. LOMBARD and HENNING C.A.O., Thin Solid Films 23 (1974) 221
12. KUCZYNSKI G.C., Metals Trans. (1949) 169
13. GJOSTEIN N.A., in *Adsorption et Croissance Cristalline*, Ed. C.N.R.S. Paris, (1965) p 97
14. BASSET G.A., in *Condensation and Evaporation of Solids*, Goldfinger Ed., 1962, p 599
15. HONJO G., YAGI K. J. Vac. Sci. Technol. (1969) 576 and Thirty Fourth Annual EMSA Meeting (1976) p 442
16. VINCENT R., Proc. Roy. Soc.Lond. A 321 (1971) 53
17. SKOFRONICK J.C., PHILLIPS W.B., J. Appl. Phys. 38(1967)4791; 39(1968)3210
18. ZANGHI J.C., METOIS J.J. and KERN R., Surf. Sci. 52(1975)556
19. ZANGHI J.C., GAUCH M., METOIS J.J. and MASSON A., Thin Solid Films 33(1976)83
20. RUCKENSTEIN E. and PULVERMACHER B., J. Catal. 29(1973)224
21. KASHCHIEV D., Surface Sc. 55(1976)477
22. HENRY C.R., CHAPON C. and MUTAFTSCHIEV B., Thin Sol. Films 46 (1977)157
23. MASSON A., METOIS J.J. and KERN R., in *Advances in Epitaxy and Endotaxy*, Ed. Schneider, Ruth, VEB Deutscher Verlag für Grundstoffindustrie, Leipzig, 1971, p 103.

CRYSTAL GROWTH KINETICS

F. ROSENBERGER

Department of Physics
University of Utah
Salt Lake City, Utah 84112
U.S.A.

I. INTRODUCTION

In this chapter we review the various growth kinetics models ("crystal growth theories") for the steady-state formation of a solid phase from a fluid nutrient phase near equilibrium. We concentrate only on the interfacial atomic kinetics. Macroscopic growth morphologies (e.g., step bunching and macrosteps)[1,3,23,55] are not considered. Also, transport in the bulk nutrient, i.e., fluid dynamics, is only treated when it strongly modifies the interfacial transport. Similarly, the role of impurities[1,2] is indicated only from a conceptual point of view. Futhermore, it is assumed that the crystalline phase, onto which growth occurs, is large on an atomic scale.

For an ultimate comparison with experiment we are mainly interested in relations between growth rate, driving force, and temperature. In Section II-IV we discuss the various growth kinetics theories in order of increasing complexity of the underlying structural models. Emphasis is put on the physical assumptions made and the results obtained. Details of the derivations are presented only if they incorporate additional, typically restrictive, conditions. For more rigorous treatments, reference is made to the numerous original papers and earlier reviews[1-3]. Computer simulations of crystal growth[4,5] are not discussed *per se*; however, selected simulation results are used for illustrative purposes. In Section V,

B. Mutaftschiev (ed.), Interfacial Aspects of Phase Transformations, 315–364.

the applicability of the various theories and experimental evidence for their validity is discussed. It is suggested that some of the conclusions drawn from interface roughness models, with respect to the prevailing growth mechanism, are unwarranted.

The complexity of a growth process is largely determined by the structure of the crystal and the molecular configuration of the nutrient. The growth unit (i.e., the crystallographic basis that is "tacked onto" the space lattice to form the crystal structure) is rarely present in the nutrient in its final form. Only some monomeric high purity vapors offer such ideal conditions. In all other nutrients the growth units exist in some precursory form : (i) associated, as in polymeric congruent vapors, or congruent melts; or (ii) dissociated as in some compound vapors (e.g., Cd + Te_2 above CdTe) or in solutions (e.g., Na^+ + Cl^-); and/or (iii) solvated (physically bound) with a solvent (or excess component), as in solutions or incongruent melts and vapors; or (iv) chemically bound to other components, as in chemical vapor deposition or in reactive solution growth.

In addition, any growth process can consist of several stages through which growth units or precursors pass (see also Fig. 1). In general one distinguishes between :

a) Transport from or through the (bulk) nutrient to an impingement site, which is not necessarily the final ("growth") site.
b) Adsorption at the impingement site; where precursors may (partly) shed solvent molecules or react towards growth units. Hence, solvent or reaction product species (possibly going through adsorption states of their own) must be transported back into the nutrient.
c) Diffusion of growth units of (smaller) precursors as admolecules from the site of impingement to a growth site.
d) Incorporation into the lattice; for precursors after completion of desolvation or chemical reaction. Thus, the growth site may also be a source of solvent and reaction products which, again, after possibly going through adsorption state, must escape into the nutrient.

Note that all of the processes (a-d) depend on the morphology of the interfacial region. For instance, activation energies for diffusion steps involved in (a) and (c) will be determined by the local arrangement of nutrient and crystal molecules. One is tempted to assign importance of the interfacial nutrient structure only to step (a) and consider an independently derived crystal surface morphology only for steps (b-d). Such modeling efforts have yielded considerable insight into interfacial processes and are discussed throughout this chapter. We must not forget, however, that this approach is conditioned by our limited understanding of the structure of liquids, in particular of interfacial liquids. In reality, an interface morphology results from the (local) interaction between the structured, contacting nutrient and (possibly imperfect) crystal layers. Only for vapor-solid interfaces, due to the much lower mass density of the vapor, is an independent modeling approach for the crystal "surface" promising. For liquid-solid interfaces, however, we will find that the models which attempt to lump nutrient contributions to interface morphology into bulk parameters (e.g., latent heat of transition) meet only with limited success.

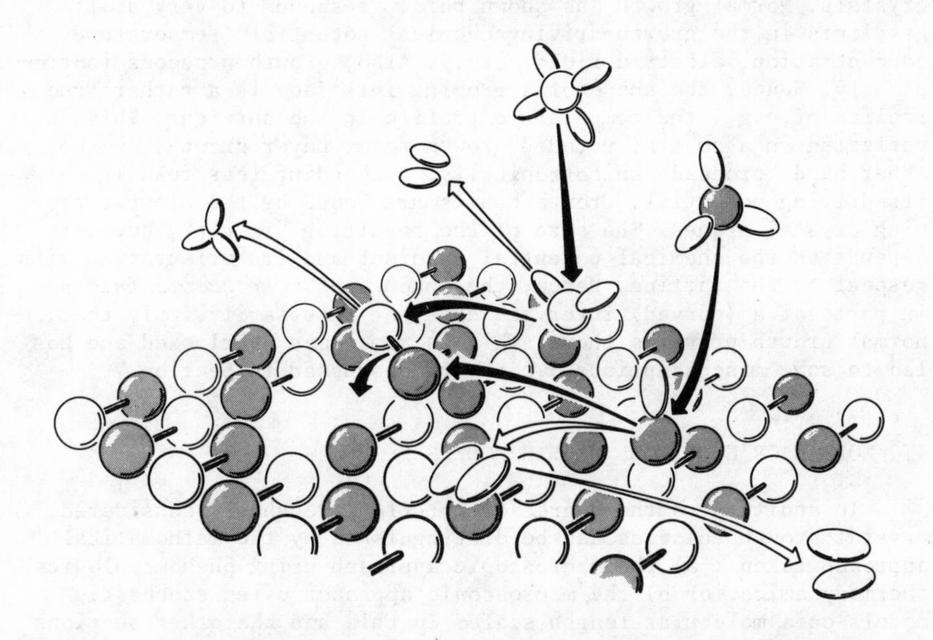

Fig. 1 : Schematic representation of molecular events in growth with reactive formation of growth unit (dark and white dumbbell) on interface.

In the following sections we circumvent this conceptual shortcoming by simply dividing interfaces into *atomically rough* and *atomically smooth* ones. The origin of a specific morphology is addressed in Section V, where we try to identify specific growth mechanisms for specific materials processes. Here we use the interface roughness criterion simply to distinguish between two groups of fundamentally different growth mechanisms. When the interface roughness is very high, i.e., when any impingement site is also a potential growth site, surface diffusion and actual surface morphology (including defects) can be ignored. The corresponding *normal* or *continuous growth* models are then solely concerned with the growth rate as the net flux resulting from impingement and rejection fluxes normal to the interface. Yet, when the interface roughness is so low that surface diffusion of adatoms must be taken into account, the surface morphology must be specified. Obviously the resulting *lateral* or *layer growth* models are more complex than the continuous growth description.

As a consequence of the different kinetics, normal and layer growth lead also to fundamental differences in the growth habit of crystals. Normal growth, as shown below, responds to very small gradients in the growth-driving chemical potential (temperature, concentration, electric field, etc.). Also, growth proceeds isotropically. Hence, the shape of a growing interface is a rather true replica of,e.g., the temperature profile in the nutrient. This is reflected in a smooth, rounded growth form. Layer growth, on the other hand, proceeds anisotropically, responding less readily to its driving potential. Growth habits are bound by the slowest growing crystal planes. The size of the resulting "facets", however, depends on the chemical potential gradient and its orientation with respect to the lattice. Hence, the absence of *macroscopic* facets in parts of a (curved) interface does not necessarily imply that normal growth prevails there. This is sometimes overlooked and has led to some misconceptions, which are discussed in Section V.

II. NORMAL OR CONTINOUS GROWTH MODELS

In addition to the degree of interface roughness considered, crystal growth theories can be distinguished by the mathematical approach taken : a) the macroscopic approach using phenomenological thermodynamics, or b) the microscopic approach using stochastic models on a molecular length scale. In this and the other sections we present first the phenomenological ("classical") models and then the stochastic descriptions.

II.1. Classical Models

Let us assume that a crystal-nutrient interface is sufficiently rough that details of its morphology can be ignored for the discussion of growth kinetics. (Thermodynamically speaking, this "ideal roughness" model implies that the surface free energy does not change with the addition or removal of a growth unit.) Also, suppose that each atom (growth unit) at the interface moves independently. Furthermore, let us view crystal growth as a reversible process in which atoms are both leaving and joining the crystal. During growth, the joining process predominates. When the leaving process predominates, the crystal etches. If the atom motion at the interface is a simple thermally activated process, by applying absolute reaction rate theory one can write for the net growth rate

$$R = R_a \exp(-Q_a/k_B T) - R_d \exp(-Q_d/k_B T), \quad (1)$$

where the first and second therm on the right hand side of (1) are the rates with which atoms arrive and depart, respectively. Thus, the arrival and departure probabilities are taken as independent of the surface configuration. Q_a is understood as some activation energy for atom movement (diffusion) in the nutrient to and from the solid. For departure, an atom also has to break the bonds with its "solid neighbors". Hence, one approximates the activation energy for departure with

$$Q_d = Q_a + L, \quad (2)$$

where L is the latent heat of the phase transition. This simple model conceals of course, some of the essential physics of the real growth process. For more realistic accounts of the morphology (e.g., density of active sites for arrival and departure) as well as energetics (loss of kinetic energy on impingement, final accomodation into the lattice, etc.) considerably more complex formulations for the pre-exponential terms and activation energies are required.

If equation (1) is applied to the growth from a pure melt at low undercooling $\Delta T = T_e - T$ (T_e is the equilibrium melting temperature) one can assume that the pre-exponential terms keep approximately their equilibrium values, i.e., $R_a = R_a^\circ$ and $R_d = R_d^\circ$. For equilibrium, where $R = 0$ and $T = T_e$, (1) and (2) yield

$$R_a^\circ/R_d^\circ = \exp(-L/k_B T_e). \quad (3)$$

From (1) and (3) one obtains

$$R = R_a^\circ \exp(-Q_a/k_B T)\left[1 - \exp(-L\Delta T/k_B T_e T)\right]. \quad (4)$$

For $L\Delta T/k_B T_e T \ll 1$, (4) reduces to

$$R = R_a^\circ \exp(-Q_a/k_B T)\,\frac{L\Delta T}{k_B T_e T}. \quad (5)$$

For a narrow temperature range (5) can be approximated with a linear growth law. Thus, for small undercoolings one can expect

$$R \propto \Delta T. \quad (6)$$

For larger variations in temperature, the variation of the "mobility term" $\exp(-Q_a/k_BT)$ must be taken into account so that

$$R \exp(Q_a/k_BT) \propto \Delta T. \quad (7)$$

Relations like (4) and (5) were derived for melt growth by WILSON[6] and FRENKEL.[7] We have followed presentations by JACKSON and CHALMERS.[8-10] WILSON set Q_a equal to the activation energy for self-diffusion in the melt, whereas FRENKEL associated Q_a with the temperature dependence of viscosity. Although both values are similar, neither of these bulk properties are necessarily characteristic of the jump probabilities at an interface governed by the atomic structure of the liquid and solid in contact. Note that with this step one implies that all atomic sites on the interface are equivalent and represent potential growth sites and/or that diffusion on the surface is infinitely rapid. In reality such ideal atomic roughness of an interface is probably never fully met. Hence, predictions of the Wilson-Frenkel growth law always represent an upper limit.

In this context it is interesting to note that the first limiting growth law (i.e., with the tacit assumption that each site is a potential growth site) was developed for growth from vapor by HERTZ[11] and KNUDSEN.[12] From kinetic gas theory, assuming no coupling between condensation and evaporation (i.e., ideal gas behavior and ideal surface roughness), they obtained the net flux of molecules per unit surface area.

$$J = (p - p_e)/(2\pi m k_BT)^{1/2}, \quad (8)$$

with p_e the equilibrium vapor pressure at surface temperature T, p the pressure of the supersaturated vapor at the same temperature, and m the molecular weight. Experimental investigations of vapor growth rates showed that in many cases (8) must be corrected with a "sticking coefficient" smaller than unity. Introducing the "relative supersaturation" $\sigma = \Delta p/p_e = (p - p_e)/p_e$ (which is the isothermal analogy of the "relative undercooling" $\Delta T/T_e$) one obtains for the growth rate a linear law similar to (6),

$$R \propto J \propto \frac{(2\pi m k_B T)^{1/2}}{p_e} \sigma \tag{9}$$

The Wilson-Frenkel law can, in general, be modified with a temperature and interface morphology dependent correction factor f to "fit" other growth mechanisms. In particular, one finds in the literature such factors for the surface nucleation and screw dislocation models (Section III). Since in both models it is assumed that growth occurs only on steps, f is taken to be proportional to the step density.

In the surface nucleation model, the number of critical nuclei is

$$N_n^* \simeq N \exp(-g_1 \gamma^2 T_e / k_B T L \Delta T), \tag{10}$$

with N the density of atomic sites, γ the step free energy and g_1 a factor dependent on the nucleus shape. Since the average distance between nuclei is approximately $1/(N_n^*)^{1/2}$, the total step density is of order $(N_n^*)^{1/2}$. Consequently one may express the growth site density factor for the surface nucleation model as

$$f_{sn} \propto N^{1/2} \exp(-g_1 \gamma^2 T_e / 2k_B T L \Delta T). \tag{11}$$

In combination, (5) and (11) predict a growth rate which is small up to $\Delta T \simeq g_1 \gamma^2 T_e / 2k_B T L$ and increases exponentially with larger supercoolings.

In the screw dislocation model, the growth sites are on the edge of a spiral. The spacing between turns of the spiral is of the same order as the radius of the critical 2D nucleus.[13] The Wilson-Frenkel law can be applied to screw dislocation growth when the spacing between turns is so large that each step grows independently of the presence of other steps, i.e., at very low supersaturation. One then obtains

$$f_d = g_2 \frac{L \Delta T}{\gamma T_e} \tag{12}$$

where g is again a geometric factor of order unity. Combining (12) with (5), one expects $R \propto \Delta T^2$ for smaller supercoolings. As discussed in Section V, both these types of functional behavior for $R(\Delta T)$ are observed. However, in view of the conceptual simplifications of these descriptions, in particular the absence of surface diffusion, it is not surprising to find differences of several orders of magnitude between measured and predicted growth rates.

II.2. Kinetic SOS Models

The main shortcomings of the Wilson-Frenkel model are (i) the constrained equilibrium assumption for the evaporation flux expressed in (3), and (ii) the lack of morphological information about the interface. These points have been addressed in more recent continuous growth treatments, which are based on the kinetic solid-on-solid (SOS) model for the crystal-vapor-interface (a restricted version of the kinetic Ising model widely used for Monte Carlo simulations of surface roughening [14,15] and growth kinetics [16]). Arrival and departure of atoms is envisioned to occur stochastically on an interface parallel to the (001) plane of a simple cubic crystal (Fig. 2), in which only nearest neighbors interact ("Kossel crystal"). The SOS restriction implies that every occupied lattice site must be directly above another occupied site. "Overhangs," and vacant sites in the "bulk" of the crystal are not allowed.

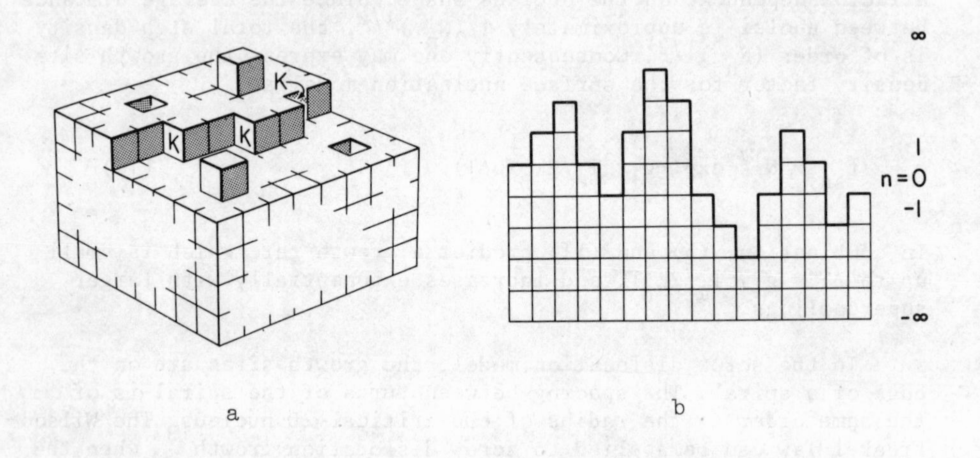

Fig. 2 : Atomic configurations on a (001) solid-vapor interface of a simple cubic crystal. (a) Perspective view : surface atoms may have up to four lateral neighbors ($0 < m < 4$), in the kink position (K) $m = 2$. (b) Section normal to interface : index n characterizes position of (incomplete) layers with respect to (arbitrarily chosen) reference plane.

This restriction permits a treatment in terms of a two-dimensional array of solid columns of varying integer heights. Growth and evaporation then occurs only through addition and annihilation of atoms on top of these columns. Atoms arrive on the surface at random with an externally imposed rate k^+. The evaporation rate of a particular atom, however, decreases with increasing number of the bonding nearest neighbors. Hence, one assumes that the evaporation rate (transition frequency) of a surface atom with m lateral neighbors (here $0 \leqslant m \leqslant 4$) and nearest-neighbor interaction energy ϕ is

$$k_m = \nu \exp(-m\phi/k_B T),$$

where ν is the evaporation rate of an isolated adatom (i.e., m = 0 and one bond to an atom below).

Based on these assumptions one obtains exact kinetic equations for the rate of change of the fraction of sites that are occupied in the n-th layer, C_n (see Fig. 2b) in the form[2,17]

$$\frac{dC_n(t)}{dt} = k^+\left[C_{n-1}(t) - C_n(t)\right] - k(n,t)\left[C_n(t) - C_{n+1}(t)\right] \quad (14)$$

where k(n,t), the effective evaporation rate of surface atoms in layer n, can be written with (13) as

$$k(n,t) = \nu \sum_{m=0}^{4} \exp(-m\phi/k_B T) f_{n,m}(t) \quad (15)$$

with $f_{n,m}(t)$ the fraction of surface atoms in the n-th layer which have m lateral neighbors.

The first and second term on the right side of (14) represent, respectively, the rates of adsorption on and evaporation from the n-th layer. The square brackets represent the fraction of sites in layer n that are available for deposition and evaporation, respectively. The SOS restriction allows for very simple formulations of these fractions. For adsorption, it is the fraction of sites occupied in the next lower layer minus the one already occupied in layer n. For evaporation, the available site fraction is the fraction of surface atoms in layer n minus the fraction occupied (i.e. covered) by atoms in the next higher layer.

Note that the essential physics of the problem lies in the functions $f_{n,m}(t)$ (at this point unspecified). Unless some specific assumptions are made concerning the clustering represented in the $f_{n,m}(t)$'s, the system of (14) and (15) cannot be solved. The following treatments of the kinetic SOS model differ, in essence, only in the approximations made to obtain a mathematically tractable form for the functions $k(n,t)$ or $f_{n,m}(t)$. Note also that one can, in principle, include surface diffusion (within the n-th layer considered) into these fonctions. At not too high surface roughness, surface diffusion results in the effective increase growth of the fraction of atoms having many neighbors and, thus, enhances growth.

The net crystal growth rate R is then the difference between the rate of adsorption in all layers, R^+, and the rate of evaporation from all layers, R^-, i.e.,

$$R = R^+ - R^- = k^+ \sum_{n=-\infty}^{\infty} P_{n-1} - \sum_{n=-\infty}^{\infty} k(n,t) P_n , \qquad (16)$$

with $P_n = C_n - C_{n+1}$, $P_{n-1} = C_{n-1} - C_n$ and the time dependence deleted in all P and C terms.
Since each column has an adsorption site on top independent of its height n, the first sum is always one. Hence, R^+ is always the imposed arrival rate k^+, irrespective of the specific model-assumption made for the effective evaporation rates $k(n,t)$. This is the main (mathematical) merit of the SOS assumption. In a real system (as in the unrestricted Ising model with overhangs) the local adsorption flux depends on the local configuration.

The simplest solution to (16) is obtained with the assumption of the Wilson-Frenkel model discussed above : complete roughness (i.e., in the SOS model all sites are kink sites with 2 lateral bonds to be broken on evaporation) and (constrained) equilibrium evaporation flux. Thus one gets

$$R_{WF} = k^+ - \nu \exp(-2\phi/k_B T_e). \qquad (17)$$

This can be rewritten as a growth law that is linear for small deviations $\Delta\mu$ of the nutrient chemical potential from equilibrium. One assumes that $k^+ = k^+_{eq} \exp \Delta\mu/k_B T$ and sets the equilibrium rate k^+_{eq} equal to the equilibrium kink evaporation rate $\nu \exp(-2\phi/k_B T_e)$. We will, however, not follow this route since it simply reduces the (then superfluous) SOS model to an approach without concern for interface morphology.

A slightly more accommodating approach was taken by TEMKIN[18] by applying *mean field theory* to (16). One assumes that all surface atoms in layer n interact laterally with the average number of nearest neighbors in that layer, $\langle m \rangle_n$. Temkin approximated $\langle m \rangle_n$ by the average number of neighbors in a random distribution of the surface atoms in layer n. If the probability of finding an atom at a given site is C_n, then $<m>_n = 4\, C_n$. Thus Temkin's approximation for k(n,t) is

$$k_T(n,t) = \nu \exp(-4\phi C_n(t)/k_B T). \tag{18}$$

Physically the mean field (Bragg-Williams) approximation implies that clustering (i.e., an increasing number of nearest neighbors) has no effect on the (local) evaporation rate of surface atoms. This is equivalent to assuming that the lateral interaction forces are infinitely long-ranged. Consequently, the same growth rates result whether one includes surface diffusion in the model or not. Such assumptions are, of course, at variance with the basic concept of a Kossel crystal. Since clustering actually reduces the evaporation rate, its effect is most pronounced in systems with not too high surface roughness. Hence, it is not surprising to find that the mean field theory always gives a range of $\Delta\mu$ (near equilibrium) for which the predicted growth rate is zero. This "metastable range" is solely due to the unrealistic interaction assumption and does not represent a nucleation barrier. The width of this range decreases with increasing surface roughness. But even above the roughening temperature of a surface, this artifact does not disappear. This is illustrated in Fig. 3. The solid curve represents a Temkin mean field solution for $\phi/k_BT = 1.5$ obtained by numerical integration of (14-16) using the mean field formulation (18) for the effective evaporation rate terms.

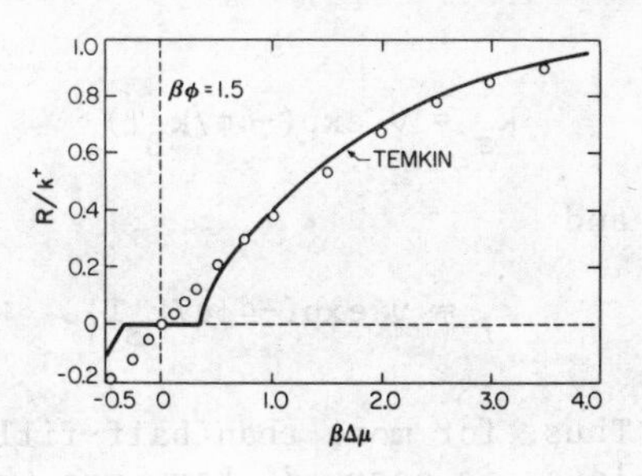

Fig. 3 : Growth rate normalized by impingement rate versus normalized growth potential. $\beta = (k_BT)^{-1}$ Comparison of Temkin mean field prediction (solid line) with Monte Carlo simulation (circles). From ref. 2 by permission of Wiley and Sons.

Note that the roughening temperature T_R of the (001) Kossel crystal is related to the interaction energy ϕ by $\phi/k_BT_R \simeq 1.6$ (see ref. 14). Hence, the system of Fig. 3 possesses a rough interface. Yet, the mean field approximation leads to a "metastable range" of considerable width. The circles in Fig. 3 represent solutions obtained by Monte Carlo simulation (MCS) which accounts for clustering. The MCS shows the expected linear growth law for small $\Delta\mu$, and no "metastable range". At higher supersaturations (i.e., at higher $\Delta\mu/k_BT$) the results of the mean field model and the MCS agree very well, and both approach the Wilson-Frenkel limiting growth law (not shown here). This can be understood in terms of kinetic roughening discussed below (Fig. 5).

The most serious shortcoming of the mean field model, i.e., neglect of clustering, is somewhat alleviated in more recent work by WEEKS, GILMER and JACKSON.[17] These authors retain some of the essential physics of the problem by setting the evaporation rate for highly filled layers (where clustering effectively reduces evaporation) equal to some "slow" constant value k_s. For less populated layers, however, the k(n,t) is set equal to a "faster" constant k_f. The dividing line between fast and slow in this *two-rate model* is drawn at half filled layers, so that

$$d(n,t) \quad \begin{cases} k_s & \text{for } 0.5 < C_n < 1 \\ k_f & \text{for } 0 < C_n < 0.5 \end{cases} \tag{19}$$

This assumption, together with other simplifications and approximations (which "tune" the model to high temperatures) allow for a quasi-continuous formulation and analytical soltuion of (13-16). Specifically, the two rate constants were choses as (ses also eqs. (13) and (18))

$$k_s = \nu \exp(-4\phi/k_BT)$$

and (20)

$$k_f = \nu \exp(-4\phi\bar{c}/k_BT).$$

Thus, for more than half-filled layers, the behavior of a completed layer is assumed. For less than half-filled layers, the parameter $\bar{c}$ represents some effective average probability per surface atom of finding a neighboring site occupied. To that extent, the approach resembles a mean field model for the less populated layers. In reality, $\bar{c}$ depends on temperature and the impingement and surface diffusion rates.

Under these provisions, the normalized growth rate can be written as

$$\frac{R}{k^+} = \frac{2\sinh(\Delta\mu/k_BT)}{\exp(\Delta\mu/k_BT) + \cosh(\alpha\phi/k_BT)} , \tag{21}$$

where $\alpha \equiv 2(1 - 2\bar{c})$. To evaluate (21), a semi-empirical choice for α (or $\bar{c}$) was arbitrarily made such that R/k_+ agrees with the MCS value at $\phi/k_BT = 4$ (i.e., below T_R) when $R/k^+ = 0.5$. Figure 4 compares the growth rates obtained from (21), after the above single point fit, with MCS results for a wide range in surface roughness (values of ϕ/k_BT) and supersaturation (values of $\Delta\mu/k_BT$). No surface diffusion was allowed in the MCS results, or indirectly through the $\bar{c}$ fit, in the two-rate model. There is excellent agreement between the two-rate model and computer experiments (MCS) for high temperatures up to very high supersaturation (deposition rate), where the original model assumptions are valid. In the region of low temperatures and supersaturation, where the continuum approximation and crude $k(n,t)$ assumptions are most in error, the model overestimates the growth rates considerably (Fig. 5).

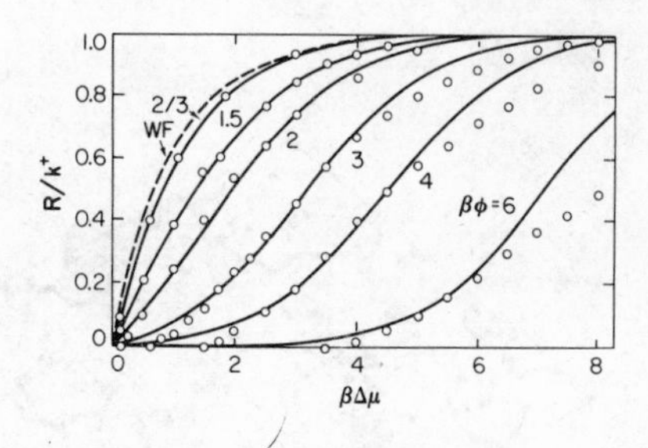

Fig. 4 : Comparison of normalized growth rate as calculated from two-rate model (solid curves) and MCS (circles) for a wide range of parameters. $\beta \equiv (k_BT)^{-1}$. Wilson-Frenkel rate, dashed curve. From ref. 2 by permission of Wiley and Sons.

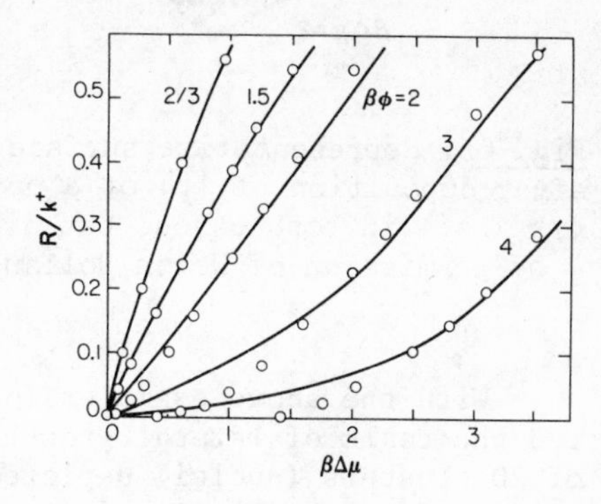

Fig. 5 : Expanded plot of the low supersaturation range of Fig. 4. From ref. 2 by permission of Wiley and Sons.

This is not surprising in view of the above fit (choice of $\bar{c}$) to conditions representing relatively high kinetic roughness (see below).

Note that the two-rate results and the MCS data for a surface with high equilibrium roughness (e.g., $\phi/k_BT = 2/3$ in Fig. 4) more closely approach the Wilson-Frenkel law the higher the supersaturation and, thus, the higher the impingement rate. This is due to the increased clustering on increase of the deposition flux. Figure 6 presents two microstates generated by MCS on an initially relatively smooth (001) face of a Kossel crystal after condensation of the same number of "atoms". At low impingement rates (Fig. 6a) an adatom has a higher chance of reevaporation before impingement on adjacent sites stabilizes it. For this case, with $\Delta\mu/k_BT = 2$, the impingement rate k^+ is 7.4 larger than the equilibrium value k_{eq}. Yet, less than 3 percent of the impinging atoms remained attached. For the case of $\Delta\mu/k_BT = 10$ (Fig. 6b) $k^+ = 2200\ k_{eq}$ and 86 percent remained attached. This kinetic roughening is obviously most important in growth onto atomically smooth interfaces ($T < T_R$), as in Fig. 6. However, the comparison in Fig. 4 of the curves with $T > T_R$ (i.e., $\beta\phi \leqslant 1.5$) shows that kinetic roughening comes to bear even on intrinsically rough surfaces.

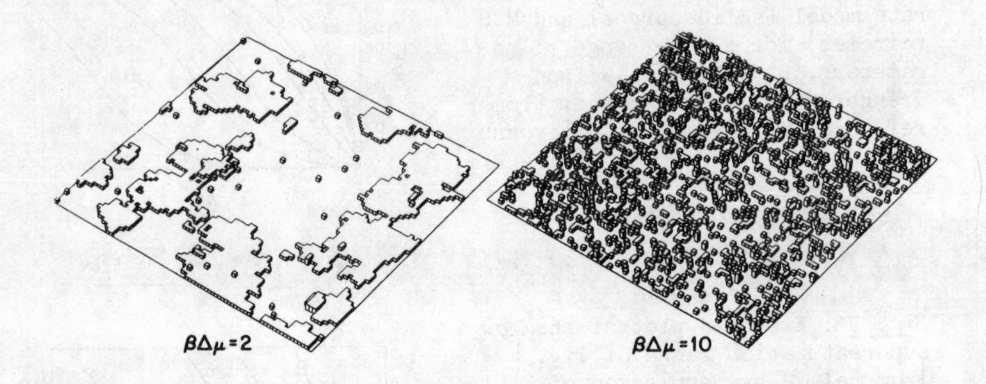

Fig. 6 : Representative surface configurations generated by MCS after deposition of 1.4 of a monolayer on a (001) face with $\phi/k_BT = 4$ in both cases, but different deposition rates. From ref. 4 by permission of North Holland Publishing Company.

With the above examples for $\phi/k_BT > 1.6$ we have clearly entered the realm of lateral growth. Note, for instance, the spreading of 2D clusters (nuclei) depicted in Fig. 6a, and the nucleation barrier (threshold value in $\beta\Delta\mu$) exhibited by the MCS data for $\phi/k_BT \geqslant 3$ in Fig. 5.

The physical consequences of the continuum approximation become particularly obvious in the MFA-CA kinetic SOS model employed by PFEIFER, HAUBENREISSER and KLUPSCH.[19] These authors solved the system of rate equations, simultaneously applying the mean field approximation (MFA) and continuum approximation (CA). Figure 7 compares their analytical results with MCS and Temkin's MFA solutions.

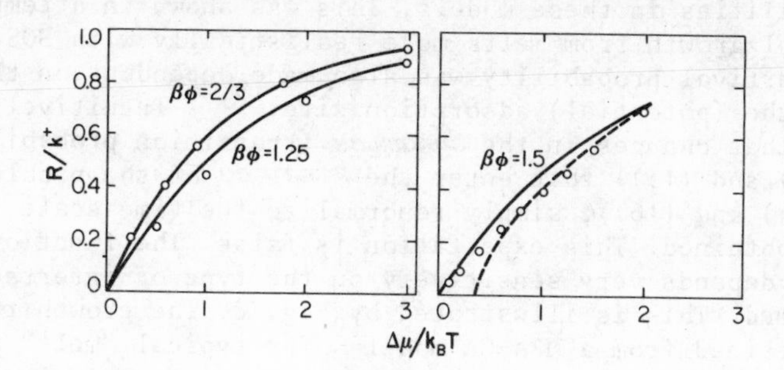

Fig. 7 : Comparison of normalized growth rate obtained from MFA-CA model (solid curves), MCS results[17,20] (circles) and Temkin's MFA (dashed curve). After ref. 19.

Most norteworthy, the combined MFA-CA approach does not result in "metastable" states. This is due to a cancellation of errors between MFA and CA. As pointed out before, the MFA ignores the consequences of clustering that are particularly important at low T and $\Delta\mu$. The CA, on the other hand, simulates an increase in the surface roughness and, thus, counteracts the "smoothing" effect of the MFA.

The (kinetic) SOS model, particularly during its early development[21,22] has frequently been applied to liquid-solid systems. This appears conceptually troublesome. In all the above vapor-solid SOS models, the arrival probability is assumed to be independent of the specific morphology of the adsorption site. Also, only solid-solid interaction $\phi \equiv \phi_{ss}$ is considered, i.e., $\phi_{vapor-s} = \phi_{vv} = 0$. Pure liquids, however, possess *about* the same density and short range order as their solids. Significant differences between the two contacting phases occur only in the long range order or symmetry. All nearest neighbor interaction terms in the system are comparable, i.e. $\phi_{liquid-s} \simeq \phi_{ll} \simeq \phi_{ss}$. Consequently, the motion of the growth units from *and to* the solid surface will depend on the strongly varying local potential at the interface; statistical knowledge of the actual morphology of the interfacial region becomes essential. Such insight can hardly come from linking of ϕ_{ls}'s and ϕ_{ll}'s, that in reality vary locally, to parameters typical for the homogeneous bulk phases, such as latent heats, via the usual liquid-solid "lattice models". It has been argued that no specific structure assumptions for the liquid are necessary in an SOS interface model[23,24]. Yet, it is difficult to see how a model liquid, particularly with nearest neighbor interaction only and essentially the same density as the solid, can "avoid" propagating the periodic structure information imposed on its boundary.

Besides the structural deficiencies in applications of the usual SOS lattice description to growth from liquids, there are serious consequences from the asymmetric formulation of the transition probabilities in these models. This was shown in attempts to model normal growth from melts more realistically with SOS models, in which the arrival probability was also made dependent on the structure of the (potential) adsorption site.[19,25] Intuitively one might expect that changes in the *dynamics* (transition probabilities, e.g., eqs (13) and (15)) that enter the *kinetics* of the problem (rate eqs. (14) and (16)), simply renormalize the time scale for the solution obtained. This expectation is false. The functional form of $R(\Delta\mu)$ depends very sensitively on the type of interface dynamics assumed. This is illustrated by Fig. 8. The growth rate functions, obtained from a MFA-CA model[19] for typical "melt" and vapor dynamics, differ drastically. For the vapor case (only the departure probability morphology dependent), $R(\Delta\mu)$ is asymmetric. For the melt case, where arrival and departure probabilities were made dependent on the local configurations, a symmetric $R(\Delta\mu)$ is obtained.

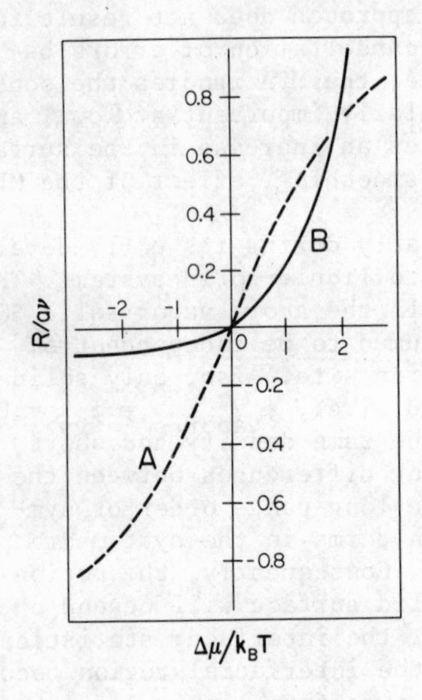

<u>Fig. 8</u> : Growth rate vs $\Delta\mu/k_BT$ for case A (melt) and B (vapor) of interface dynamics. Calculated for $2I/k_BT = 2/3$, where $I_A = 1/2(\phi_{ss} + \phi_{ll}) - \phi_{ls}$ and $I_B = \phi_{ss}/2$. After ref. 19.

Obviously these solutions cannot be mapped onto one another by simple adjustments of the time scale. A similar dependence of the interface kinetics on the form of the transition probabilities has been observed in recent "fluctuation theory" treatment (SOS with special overhangs allowed, MFA) of normal growth.[26,27]

Recent work towards a realistic description of the structure of liquid-solid interfaces without lattice models raises hope for a better understanding of melt growth kinetics in the not too distant future. These efforts range from largely heuristic, phenomenological defect models[28,29] (see also the dislocation models mentioned in Section IV) to molecular dynamics Monte Carlo simulations based on Lennard-Jones (6-12) potentials.[30,31,129]

The above stochastic models focus on the adsorption-evaporation mechanism, i.e., on the "relaxation kinetics" part of the crystal growth process. In reality, unless the interface roughness is extremely high, *surface diffusion* can considerably modify the interface morphology and expected growth rate. This has been shown by SAITO and MÜLLER-KRUMBHAAR[32] with a quasi-chemical MFA to the kinetic SOS model including exchange processes within each layer (intra-layer) and between layers (inter-layer diffusion). Such a model requires a more complex formultaion of the "Markovian" master equations than given by (14-16) which allow for intra-layer exchange only. Figure 9 shows that the incorporation of a mean diffusion length X_s of one or two lattice parameters increases the growth rate sizably even on an interface above the roughening temperature. The diffusion length is defined as $X_s = a(\nu_D/\nu)^{1/2}$; where a is the lattice parameter, ν_D the frequency with which an adatom jumps into a neighboring adsorption site (surface migration frequency) and ν is the desorption frequency. The predicted results agree very well with MCS results that include surface diffusion.[32] For "infinitely rapid" interfacial diffusion, i.e., for large values of ν_D/ν, the growth rate reaches the limiting Wilson-Frenkel rate and the interface roughness relaxes to its equilibrium value.[32,34]

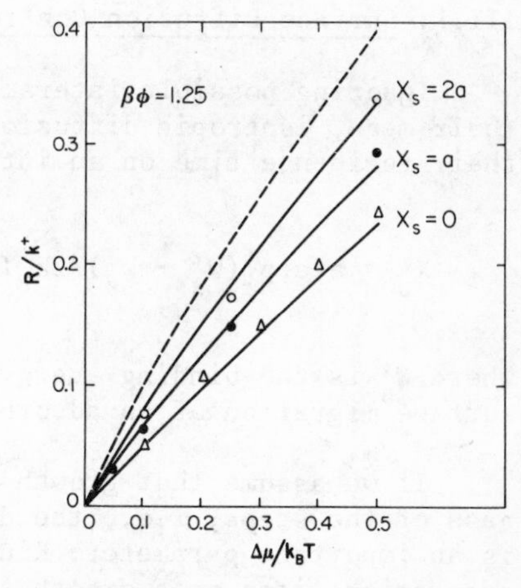

Fig. 9 : Growth rate vs $\Delta\mu/k_BT$ obtained from quasi-chemical MFA (solid curves) and MCS[33] (symbols) showing influence of surface diffusion. X_s is diffusion length in multiples of lattice parameter a. Wilson-Frenkel rate (dashed curve). After ref. 32.

Kinetic SOS-models have been extensively applied to binary and multicomponent systems[35-45] utilizing all approximations discussed above including surface diffusion and reactive interface processes.[45] The additional difficulties arising with such extensions of the monocomponent models have been well summarized in ref. 43.

III. LATERAL OR LAYER GROWTH MODELS

At lower atomic interface roughness the distance between sites that offer a higher interaction (binding) energy than the close-packed, unperturbed parts of the interface, is large with respect to molecular dimensions. Consequently, fewer molecules will be in a position to reach such energetically favorable growth sites via an elementary jump from the nutrient. Diffusion must be incorporated into the kinetics model. Diffusion to growth sites can occur either in an adsorbed state on the interface (surface diffusion) or directly through the interfacial nutrient layer ("volume diffusion"), or simultaneously through both.

In the following we first outline the surface diffusion model introduced by BURTON, CABRERA and FRANK[46] (see also 23) and extended by CHERNOV.[1] Following the classical "terrace-ledge-kink" model of KOSSEL,[47] STRANSKI,[48,49] STRANSKI and KAISCHEW,[50] and FRENKEL[51] we assume that the solid surface is a low index or singular face, without intrinsic (i.e., equilibrium) adatoms and vacancies. Growth is envisioned to occur via surface diffusion to monatomic, straight steps. Growth through volume diffusion is treated after discussion of the various sources of steps, in particular of the screw dislocation and 2D nucleation mechanisms.

III.1. Surface Diffusion Controlled Motion of Elementary Steps

Ignoring possible interactions between adsorbed molecules, their mean, isotropic diffusion path (migration distance) during their residence time on an interface is[46]

$$X_s \simeq a \exp\left[(W_s' - \varepsilon_s)/2k_B T\right] \tag{22}$$

where W_s' is the binding energy and ε_s the activation energy for surface migration of an adsorbed molecule.

If we assume that growth occurs onto steps, the atomic roughness of the steps, i.e., the density of thermally created kinks is an important parameter. Kinks offer a higher binding energy than adsorption sites on a smooth step. Note that the 2D step roughening problem (see ref. 16) is analogous to the 3D surface roughening situation. However, the bond energy of a molecule in a step is considerably lower than in a complete surface layer.

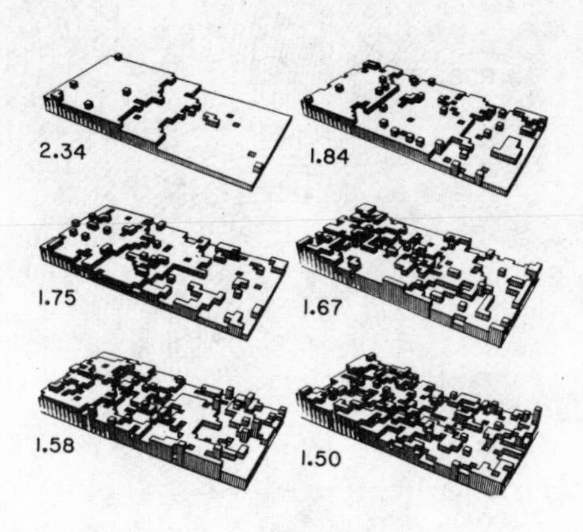

Fig. 10 : Perspective drawings of representative (20,1 ,0) surface configurations (microstates) at various values of ϕ/k_BT. (Monte Carlo simulation, Kossel crystal, SOS, no surface diffusion.) From ref. 52 by permission of North Holland Publishing Company.

Thus we can expect that (i) steps are rougher than their bounding low index faces, and (ii) that the step roughness decreases with increasing step height. The first expectation is well illustrated in Fig. 10 which presents microstates, obtained through Monte Carlo simulations, for increasing temperatures.[52,53] Long before the roughening temperature of the singular terraces is reached, the two steps show already rough profiles extending over several lattice constants. Actually, the steps lose their identity before the surface roughening temperature $T_R(\phi/k_BT_R \simeq 1.6)$[2,14,15] is attained.

Experimental evidence for the dependence of step roughness (kink density) on step height is given in Fig. 11. The steps on an evaporating (100) NaCl surface are round (i.e., isotropically vaporizing) or square (i.e., anisotropically vaporizing), respectively, for step heights of one or two atomic spacings.

The mean distance between kinks on a close-packed monatomic step (Fig. 12) ignoring possible overhangs,[1,46] is approximately

$$Y_o \simeq \frac{a}{2} \exp\left[(W/k_BT) + 2\right], \tag{23}$$

where W is the energy required to create a single kink. Since typically $W \ll W_s' - \varepsilon_s$, comparing (22) and (23), we assume here that the mean diffusion length is larger than the mean kink distance (BCF have also dealt with $X_s < Y_o$).[23,46] This allows one to consider the step as a continuous line sink. Diffusion along steps is ignored as a rate limiting process and the surface diffusion of adatoms towards the step can be viewed as a one-dimensional problem.

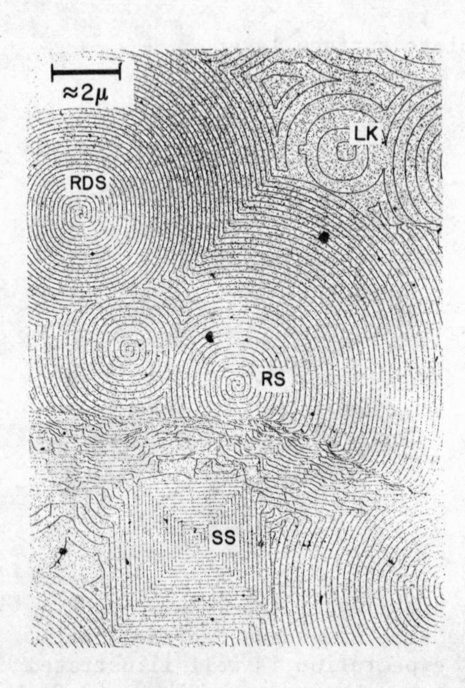

Fig. 11 : Electronmicrograph of NaCl (100) surface after high-vacuum cleaving, evaporation at 400°C for 90 min and gold decoration. Steps on round spirals (RS) and round double spirals (RDS) as well as on negative 2D nuclei ("Lochkeime", LK) are one interatomic distance (a = 2.81 Å) high. For steps with height of 2a the step roughness is sufficiently reduced that anisotropic kinetics leads to square spirals (SS). From ref. 54 by permission of McGraw-Hill.

BURTON, CABRERA and FRANK (BCF)[46] have made the following additional assumptions in their treatment of the (growth) motion of a monoatomic step : (1) Fick's law describes the adatom flux towards the step, i.e., the diffusion coefficient is independent of concentration. This restricts the treatment to low adatoms concentrations. (2) The steps are considered stationary. This is reasonable as long as the diffusion velocity towards the steps is large compared to the growth velocity of the step itself. BCF[46] have shown that this holds for low supersaturation, yet it is difficult to determine the upper limit. Only growth from adatoms onto steps is considered; contributions from moving surface vancancies or recombinations of adatoms and surface vacancies are neglected.

The transport of atoms to growth steps is considered in two consecutive stages (see Fig. 12)

a) The net flux $j_s(x)$ on the surfaces towards steps. Atoms which "land" closer to a step have a higher chance to become incorporated during their residence time. Thus a concentration gradient (depletion zone) develops around the step, which drives *net* diffusion towards the step. Consequently, the surface concentration of adatoms $n_s(x)$ increases with distance form a step until at $|x| >> X_s$ it reaches a constant value. Correspondingly, the supersaturation value $\alpha_s(x) = n_s(x)/n_{so}$ (with n_{so} the equilibrium surface concentration) reaches a constant value α far from the step.

Since it is assumed that transport in the nutrient is not rate determining, α can be set equal to the α of the bulk nutrient. Hence, the relative supersaturation

$$\sigma_s(x) = \frac{n_s(x) - n_{so}}{n_{so}} = \alpha_s(x) - 1 \qquad (24)$$

equals σ at large x.[46] The linear net flux towards a step on one bounding surface is then

$$j_s = D_s n_{so} \frac{d\psi}{dx}, \qquad (25)$$

where the potential function $\psi = \sigma - \sigma_s(x) = \alpha - \alpha_s(x)$.

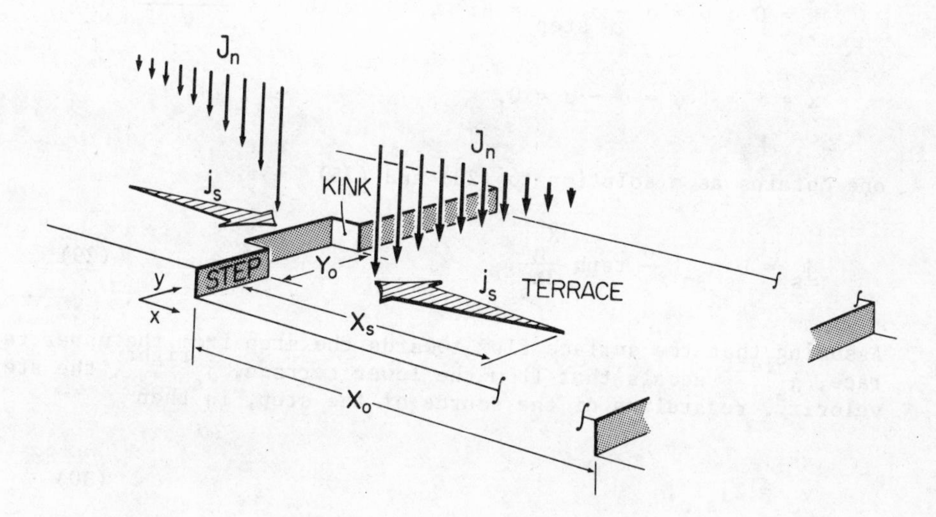

Fig. 12 : Transport of growth units from nutrient to (kinked) steps. Terrace-ledge-kink model of surface diffusion growth.

b) The net flux $J_n(x)$ from the nutrient onto the interface. Under steady state conditions j_s and J_n balance such that

$$\frac{dj_s}{dx} = J_n. \qquad (26)$$

For J_n, the difference between atoms arriving at and leaving the surface, one obtains

$$J_n = [\alpha - \alpha_s(x)] \frac{n_{so}}{\tau_s} = \psi \frac{n_{so}}{\tau_s}, \tag{27}$$

where the mean residence time τ_s is related to the mean diffusion length via Einstein's relation, $\tau_s = X_s{}^2/D_s$. Combining (25) and (27) through the conservation condition (26) yields the governing equation

$$X_s^2 \frac{d^2\psi}{dx^2} = \psi . \tag{28}$$

For the boundary conditions

$$x = 0 : \psi = \sigma - \sigma_{step} = \beta\sigma \quad , \qquad \text{with } \beta = \frac{\sigma - \sigma_{st}}{\sigma}$$

$$x = \pm \infty : \psi = \sigma - \sigma = 0,$$

one obtains as a solution to (28) and (25)

$$j_s = D_s n_{so} \frac{\beta\sigma}{X_s} \tanh \frac{X_o}{2X_s}. \tag{29}$$

Assuming that the surface flux towards the step from the upper terrace, j_s^{left} equals that from the lower terrace, j_s^{right}, the step velocity, regardless of the source of the step, is then

$$v_o = 2j_s f_o, \tag{30}$$

where f_o is the area occupied by one growth unit.

Through the introduction of the kinetic coefficient β one can accommodate special kink kinetics. For details, see ref. 3. If, for instance, the relaxation time of a growth unit at a step, τ_{st}, is short with respect to τ_s, then the concentration at the step will equal the equilibrium concentration.

Hence, the "supersaturation" $\sigma_{st} = 0$. Consequently, $\beta = 1$ and the kink or step kinetics do not modify the overall process. If, however, the exchange of adatoms between adsorbed layer and step is not rapid enough, i.e., if $\tau_{st.} \simeq \tau_s$, the concentration of growth units entering the step will remain above the step equilibrium value and $\beta < 1$. Such a retardation can result from adsorbed impurities that "poison" kinks, desolvation processes, or "steric hindrance" in the case of atomically complex growth units. Note that, possibly, the incorporation from the adjacent terraces proceeds differently, i.e., $\beta^{left} \neq \beta^{right}$, see.[55] Note that in addition to an anisotropy in β which enters (29), orientation dependence of the growth flux j_s can also originate from an anisotropic D_s and X_s.

III.2. Growth onto Steps of Specific Origin

Screw dislocations. FRANK[56] suggested in 1949 that a screw dislocation that intersects the surface of a crystal at right angles leads to a spiraling step during growth or dissolution. BCF[46] extended this concept to all dislocations with nonvanishing component of the Burgers vector $\vec{b}$ normal to the interface, ($\vec{b}\cdot\vec{n} \neq 0$). The actual shape of the spiral presents a solution to the surface diffusion field equation, possibly with $\beta < 1$. BCF assumed infinitely rapid incorporation at the steps ($\tau_{st} \ll \tau_s$, i.e., $\beta = 1$) and neglected diffusion parallel to the steps. They assumed that, under these conditions, a spiraling sequence of equidistant steps (see e.g., RS in Fig. 11 not too close to the spiral's center) can be approximated by the Archimedian spiral (polar coordinates : r, Θ)

$$r = 2r_c\Theta = 2\frac{\gamma' a}{\Delta\mu}\,\Theta \simeq 2\frac{\gamma' a}{k_B T\,\ell n(1+\sigma)}\,\Theta, \tag{31}$$

with r_c the radius of the critical 2D nucleus. The ledge free energy γ' in (31) is typically assumed to be the same as for an unperturbed lattice. In reality one may expect that γ' depends on the core strain energy of the dislocation. From (31) one obtains for the distance between steps (at low supersaturation $\ell n(1+\sigma) \simeq \sigma$)

$$X_o = 2r_c\left[(\theta + 2\pi) - \theta\right] = 4\pi r_c = \frac{4\pi\gamma' a}{k_B T\,\sigma}\,. \tag{32}$$

CABRERA and LEVIN,[57] in a more detailed discussion, found that $X_o = 19\, r_c$. (For a similar treatment of polygonized spirals see[58] and Section IV). Their paper is also noteworthy for its consideration of the core energy of the screw dislocation. They predict for dissolution (evaporation, etc.) a critical $\Delta\mu$, beyond which the spiral should develop a macroscopic hollow core (etch pit), even in pure systems. CABRERA and COLEMAN[59] and SUREK, POUND, and HIRTH[60] pointed out that the flux of adatoms to the center of the spiral is somewhat reduced due to the competing fluxed to the neighboring arms of the spiral. In their quantitative discussions of this "back-stress" effect, they approximate the diffusion field with concentric circular step sinks and show that CABRERA and LEVIN's[57] model gives too closely spaced steps, particularly at high supersaturations. The partial *ad hoc* assumptions of these theories were cicumvented in a recent analytical theory of the surface diffusion problem by Van der EERDEN.[61] His treatment yields a more complex $X_o(\sigma)$ but typically the distance between steps is considerably larger than the X_o-values predicted by the earlier back stress theories.[59,60]

The overall growth rate normal to the interface, i.e., the product of step density, step velocity v_o and step height d, is from (29-32)[46]

$$R = \frac{\beta \Omega D_s n_{so}}{X_s^2} \frac{\sigma^2}{\sigma_1} \tanh \frac{\sigma_1}{\sigma} , \qquad (33)$$

with

$$\frac{\sigma_1}{\sigma} = \frac{X_o}{2X_s} \qquad (33a)$$

($\sigma_1 = 2\pi\gamma' a/k_B T X_s$, see eq. (32)) and $\Omega = f_o a$. A schematic plot of $R(\sigma)$, according to (33) is given in Fig. 13.

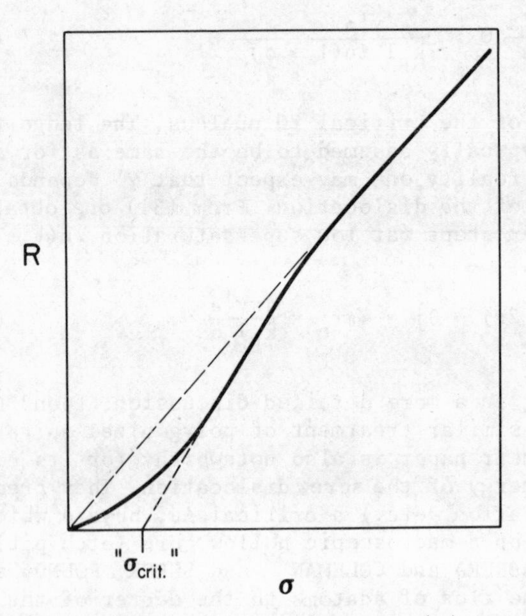

Fig. 13 : Schematic representation of $R(\sigma)$ from BCF surface diffusion theory.

For high supersaturations ($\sigma >> \sigma_1$) (33) yields a linear growth law, $R \propto \sigma$. From (33a) one sees that this occurs when the distance between steps X_o becomes shorter than the mean diffusion paths X_s. Then, the diffusion fields of adjacent steps overlap, effectively offering high surface roughness, so any adatom has a chance to find a kink. Consequently, as follows from (29-32), v_o becomes independent of σ; this makes R proportional to σ, as in normal growth.

For $\sigma << \sigma_1$, i.e., from (33a) for $X_o >> X_s$, the diffusion fields of neighboring steps do not overlap. Hence, a major fraction of adatoms has no chance of reaching a growth site during their residence time τ_s. Instead, they reevaporate from the surface. With both v_o and the step density $1/X_o$ roughly proportional to σ, one obtains R proportional σ^2. This parabolic range falls below the extrapolated linear growth rate.

If σ_1 of a system is low, due to a low γ' and/or a large X_s, the parabolic part of $R(\sigma)$ can be shifted so close to the origin that it falls below experimental resolution. On the other hand, the transition between $R \propto \sigma$ and $R \propto \sigma^2$ is very gradual (for specific examples see [23]). Hence, if values for very low σ's are lacking, one may be tempted to interpret an extrapolated transition regime curve in terms of a 2D nucleation barrier (see "σ_{crit}" in Fig. 13, and the following section).

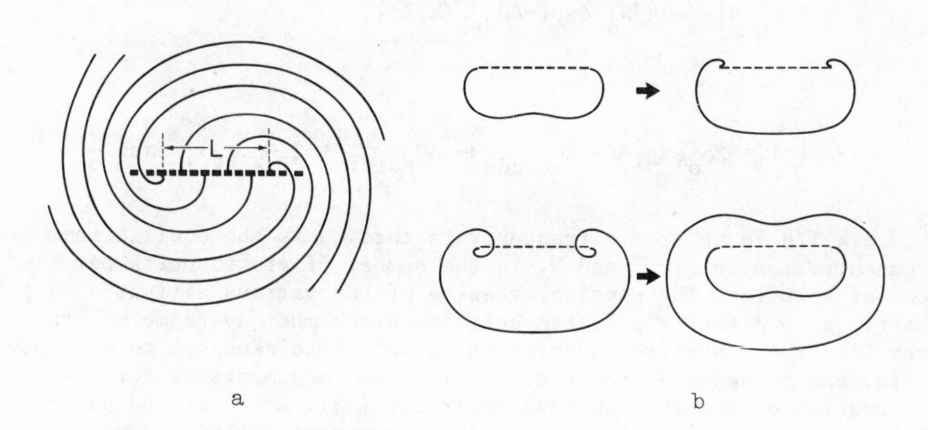

Fig. 14 : Steps from interacting growth spirals. (a) Combination of growth spirals at grain boundary, after ref. 62. (b) Closed concentric loops originating at 2 screw dislocations of opposite sign (Frank-Read source) ; after ref. 63.

It is possible that several closely-spaced screw dislocations of the same sign cooperate in the formation of growth spirals. For instance, a combination as shown in Figure 14a can result at a grain boundary. As long as the total length of n dislocations, L, is small as compared to $4\pi r_c$, the BCF law (33) holds with a new $X_o \simeq 4\pi r_c/n$, (32). Then, the combination of dislocations can be modeled with a single spiral of "n-fold strength". For combinations with $L >> 4\pi r_c$, BENNEMA[62] has shown that the surface diffusion model results in a linear growth law.

Closely spaced screw dislocations of opposite sign (Fig. 14b) can interact to form concentric closed loop steps similar to the step structure about a 2D nucleation center (LK in Fig. 11).

BCF have treated solution growth with a volume diffusion model (Section III.3). BENNEMA, however, based on thorough experiments and detailed discussions of the (desolvation) kinetics involved, concluded that solution growth will more typically proceed via a modified BCF surface diffusion mechanism. He obtained a growth law in the form[62,64]

$$R = C \frac{\sigma^2}{\sigma_1} \tanh \frac{\sigma_1}{\sigma}, \tag{34}$$

with

$$C = \beta\, c_o\, \Omega\left[\frac{k_B T}{h} \Lambda\, N_o \exp(-\Delta G_{deh}/k_B T)\right],$$

and

$$\beta = \left\{ 1 + 2c_o \exp\left[\tfrac{1}{2}(-\Delta G_{deads} - \Delta G_{sdiff} + 2\Delta G_{kink})\right] \tanh \frac{\sigma_1}{\sigma}\right\}^{-1},$$

where $k_B T/h$ is an atomic frequency factor, c_o is the equilibrium concentration in g/cm^3 and N_o is the number of growth units per cm^3 of solution. The physical meaning of the various activation barriers in β that a solvated building block must overcome on its way from the "unstirred adsorption layer" of thickness Λ to a step site, can be deduced from Fig. 15. The same arguments as for the discussion of the BCF law (33) apply for (34). A linear and parabolic law is obtained for high and low supersaturation, respectively. Due to the new, σ-dependent β-term, however, the parabolic range does not simply "retard" the linear range as in the BCF surface diffusion model. Similar to CHERNOV's volume diffusion curves (Section III.3, Fig. 19), the linear range extrapolates to a finite σ rather than through the origin.

The BCF theory for monatomic systems has also been extended to growth from binary vapors by TAKATA and OOKAWA.[65,66] In the case of incongruent vapors, the growth rate is governed by the minority component which acts as "growth unit".

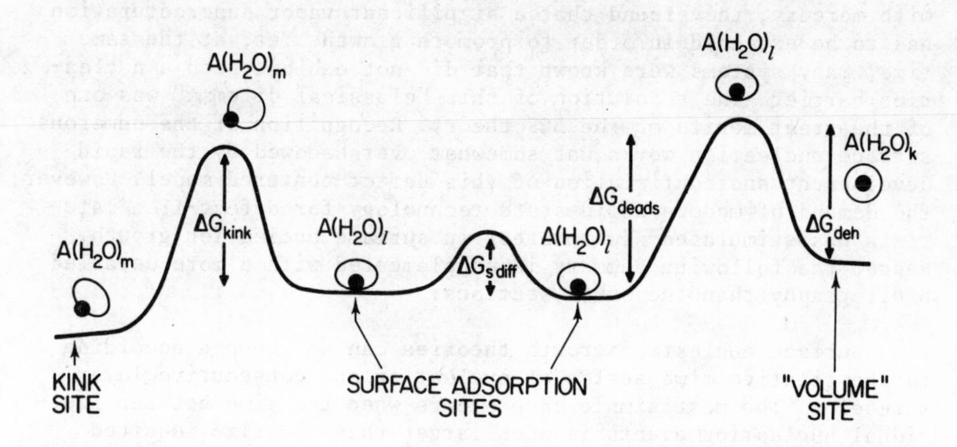

Fig. 15 : Schematized states of hydrated ions and associated potential barriers in surface diffusion model of solution growth. ($m \leqslant \ell \leqslant k$). After ref. 62.

Edge dislocations. Until very recently, pure edge dislocations ($\vec{b}\cdot\vec{n} = 0$) were assumed to influence interface kinetics only during evaporation. Negative 2D nuclei, appearing as concentric loops of steps ("Lochkeime"),[67,70] see Fig. 11, form at the emergence point of dislocation lines. Evaporation from the strained, dislocated area (possibly involving higher impurity concentrations at the dislocation) appears energetically favored over evaporation of the structurally perfect parts of the low index faces. The inverse effect, persistent growth step formation at the outcrop of pure edge dislocations, was generally assumed unlikely. However, in 1981 BAUSER and STRUNK[78] unambiguously established the growth promoting activity of edge dislocations on (001) GaAs surfaces. Growth hillocks, determined by optical interference contrast microscopy, were correlated with the location of pure edge dislocations, the character of which was identified with double crystal topography. A qualitative discussion of the phenomenon, based on the fact that surface stress energies at defects are of same order of magnitude as surface free energies, was given by FRANK.[72] He expects that preferred formation of surface nuclei at edge dislocations will only occur at higher supersaturations, approaching those values necessary for 2D nucleation at dislocation-free surface locations.

Surface nucleation. Before lattice defects were recognized as potential step sources, it was assumed that growth onto close-packed faces could only occur through surface nucleation and consecutive spreading of solid 2D clusters. This mechanism was conceived in 1921 by VOLMER and ESTERMANN.[73,74] In their pioneering experiments with mercury, they found that a significant vapor supersaturation had to be exceeded in order to promote growth. Yet, at the same time, many systems were known that did not exhibit such a nucleation barrier. The resolution of this "classical dilemma" was one of the great merits of the BCF theory. Recognition of the numerous surface nucleation works was somewhat overshadowed by the rapid development and confirmation of this defect-centered model. However, the demand of modern solid state technology for defect-free materials has stimulated new interest in surface nucleation growth. Hence, the following summary is supplemented with a more detailed bibliography than the other sections.

Surface nucleation growth theories can be grouped according to the relative time scales of nucleation and consecutive layer spreading. The most simple case occurs when the time between individual nucleation events is much larger than the time required for a step to propagate over the crystal face. Quantitatively, this requies that $1/JA \gg L/v_o$, where J is the nucleation rate per unit area, and A and L, respectively, are the area and largest diameter of the crystal face. Then, the growth rate is solely governed by the nucleation rate in the form

$$R = aJA, \tag{35}$$

i.e., the growth rate is proportional to the surface area. The kinetics of this classical *mononuclear monolayer* mechanism was developed by KOSSEL,[47] STRANSKI and KAISCHEW,[75-77] mostly based on the (001) face of a simple cubic crystal. The most general treatment, due to BECKER and DÖRING,[78,79] has been refined by KAISCHEW[80] and HIRTH,[81] utilizing the continuum approach to nucleation theory developed by FRENKEL and ZELDOVICH (see Chapter "Three Dimensional Nucleation").[80-82] The 2D nucleation rate density can be expressed as

$$J = N_s D_2^* \Gamma_2 \exp(-\Delta G_2^*/k_B T), \tag{36}$$

with N_s the concentration of adsorbed single molecules, D_2^* the rate of incorporation on single molecules into the peripheral surface of the critical 2D cluster, ΔG_2^* the Gibbs free energy of formation of the critical cluster, and Γ_2 the Zeldovich factor,

$$\Gamma_2 = \left(\frac{\Delta G_2^*}{4\pi k_B T n^{*2}}\right)^{1/2} = \left(\frac{\gamma' a^2}{2\pi k_B T}\right)^{1/2} \frac{1}{n^{*3/4}} \quad (37)$$

where n^* is the number of molecules in the critical cluster, and γ' the ledge free energy, assumed to be the same as for a macroscopic terrace. Since ΔG_2^* and n^* strongly decrease with supersaturation, (36) and (37) reflect the fact that nucleation is very unlikely below a certain ("critical") value of supersaturation, but increases exponentially when this value is exceeded. Hence, for the mononuclear monolayer mechanism to prevail with typical step propagation velocities, a very low supersaturation or, experimentally speaking, very close control of σ is required.

At practical supersaturations, multinuclear growth is more likely, unless the surface area is extremely small, since the nulceation rate (36, 37) increases with σ much more readily than the layer spreading rate (29, 30). Depending on the degree of the inequality, $1/JA < L/v_o$, a single layer may grow through concurrent nucleation, spreading and collapse of a large number of clusters (*multinulcear monolayer* growth), or, at higher supersaturation, several layers may grow simultaneously through nucleation and spreading on top of lower incomplete layers (*multinuclear multilayer* growth).

Exact analyses of simultaneous nucleation and growth in a single layer were put forth by KOLMOGOROV[83] and AVRAMI,[84] with later discussions in ref. 85. Approximate solutions for steady-state multilevel deposition were derived by numerous workers.[83-96] Transients from the initial flat interface to the steady-state multilevel "birth and spread" mode have been studied in refs. 89,90,94,97,98, 108. Monte Carlo simulations of multinuclear growth are presented in refs. 95 and 99-105.

The probability that a point on a layer becomes covered during a time interval Δt, is proportional to the density of clusters nucleated during Δt, $J\Delta t$, and to the area covered by spreading, $Gv_o^2\Delta t$ where G is a geometrical factor for the specific cluster shape assumed. Hence, the term Jv_o^2 is common to all multinuclear growth models. The differences consist in the specific idealized assumptions made for intra- and inter-layer kinetics. In general, an incomplete layer reduces the birth and spreading rate of the following layer. SOS restrictions as well as mean field approximations (MFA) have been used. The application of a MFA, i.e., the artificial introduction of "metastable states" (see Section II.2), in a model that is to reveal true metastable states (nucleation barriers) does not appear meaningful. Other shortcomings of the models are pointed out in comparison with Monte-Carlo results below.

For the steady-state growth rate most multinuclear models yield

$$R = Ca(Jv_o^2)^{1/3}, \tag{38}$$

where the model-dependent constants C fall between $1.0 \leq C \leq 1.4$. Note that the 1/3 power in (38) represents an increase in the nucleation rate because of the smaller absolute value of the negative exponent in (36). Also, R is independent of the surface area. Using the BCF notation introduced above, (38) can be rewritten[106,107] for not too high supersaturations as

$$R = A\sigma^{5/6} \exp(-B/\sigma) \tag{39}$$

with

$$A = (2\pi C_o/3)^{1/3} \; 2X_s C'/a,$$

$$B = \pi/3(\gamma'/k_B T)^2,$$

$$C' = \beta D_s n_{so} \Omega / X_s^2.$$

Here the concentration of adsorbed single atoms C_o is equal to N_s in (36), since the adsorption layer is assumed to be in equilibrium with the bulk nutrient, and γ' is the ledge free energy per growth unit on the step. A corresponding simpler formulation for growth from the melt is given by HILLIG.[88]

Comparison of predictions from 2D nucleation growth models with corresponding Monte Carlo results[24,95] reveals some interesting consequences of the simplifying model assumptions. At lower supersaturations the multinucleus theory tends to underestimate the growth rate. This results from the highly regular cluster shapes (circular, square, etc.) employed in the theory. The MCS clusters possess considerably higher step roughness and, hence, can more readily incorporate adatoms. At higher supersaturations the theory tends to overestimate the growth rate (Fig. 16). This reflects the quasi-equilibrium assumption made for the concentration of adatoms. At high supersaturations, where the cluster density is also high, the inhibition of nucleation in the depletion zones close to the clusters becomes important.

This effect is not accounted for in the nucleation theory. The importance of surface diffusion in the 2D nucleation growth mechanism is well illustrated in Fig. 16. The symbols represent MCS results for different surface migration to evaporation ratios for an atomically smooth surface ($\phi/k_BT = 4$). The solid curves were derived from a relation similar to (39)$_4$ The trend is the same as for atomically rough surfaces (see Fig. 9); surface diffusion increases the growth rate. In contrast to rough surfaces, however, diffusion leads to the largest relative increases in R at the lower supersaturation values, where the clusters are widely spaced.

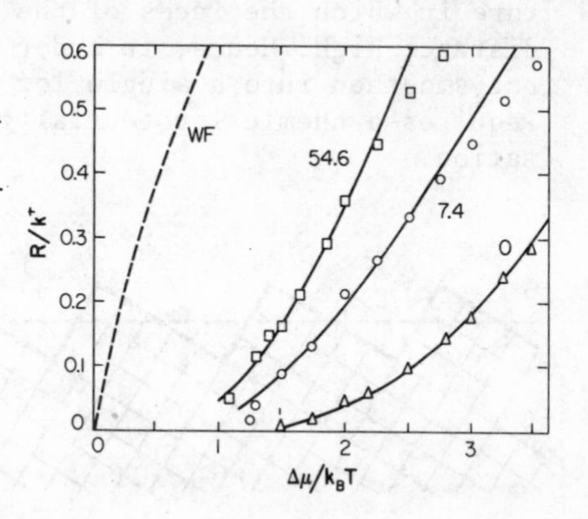

Fig. 16 : Normalized growth rate vs $\Delta\mu/k_BT$ obtained from MCS (symbols) and 2D nucleation theory (curves) on defect-free (001) surface of simple cubic crystal ($\phi/k_BT = 4$). Numbers indicate surface migration to evaporation ratio. From ref. 2 by permission of Wiley and Sons.

The models just discussed are concerned with homoepitaxial nucleation and growth, i.e., with systems in which the adsorption energy equals the nearest neighbor energy ϕ_{ss} in the lattice of same composition. Further complications arise in heteroepitaxy, i.e., for $\phi_{ads} \neq \phi_{ss}^{substrate} \equiv \phi_s$. For instance, for the often met practical case of $\phi_{ads} < \phi_s$, KASHCHIEV[94] has shown that at lower $\Delta\mu$, instead of the preferential filling of the first epitaxial layer(s), multilivel clusters ("islands") grow that eventually coalesce.

Other step sources. Besides dislocations and 2D clusters, other structural and morphological configurations have been recognized as sources for growth steps.

HAMILTON and SEIDENSTICKER[109] have proposed that twinning with two or more twin planes permeating an interface, forms an inexhaustible potential source for growth steps. These steps can nucleate in the reentrant corners which are formed by the intersection of low index faces on either side of the twin planes.

For interfaces of structurally perfect crystal, CHALMERS and MARTIUS [110] have suggested a similar mechanism as follos. Unless a crystal grows under quasi-equilibrium conditions, its macrohabit (growth form) will deviate from the equilibrium form. The crystal tends to adjust its habit to that of the growth driving temperature and/or concentration field. Consequently, the macroscopically averaged orientation of part of the interface may not coincide with those of closely packed planes. Under certain energetic conditions (Section VI) such "misoriented" interface parts are unstable and tend to break up into a "hill and valley" structure of low index faces (Fig. 17), thus lowering their surface free energy. The reentrant corners of such a structure form preferential sites for growth step "nucleation". Note that growth onto vicinal surfaces[111] is only a special case of growth onto a hill and valley structure in which the faces of the one orientation are only one atomic distance high. Hence, in order to remain stable during growth and not smoothen into a single low index plane, a vicinal surface also requires a chemical potential gradient normal to its average orientation.

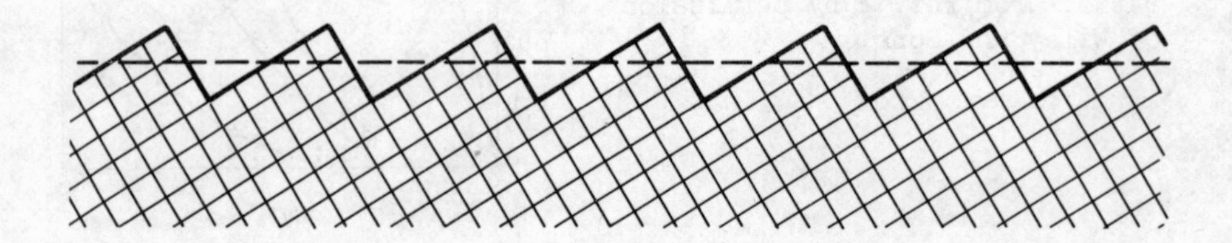

Fig. 17 : Nutrient-solid interface with finite chemical potential gradient normal to its average orientation (dashed). Hill-and-valley structure consisting of low index (low surface free energy) facets.

Recent studies of the morphology of solid surfaces have revealed another possible mechanism for the perpetual formation of growth steps : surface reconstitution (see also the Chapter "Structure of Clean Surfaces"). Driven by a possible reduction in surface free energy, the surface layer(s) may form a new crystal structure, quite different from that of the bulk. Typically these superstructures have a lower atom density than corresponding planes in the bulk. Some of the excess atoms may return to the nutrient. Energetically, however, it is more likely that surface diffusion of excess atoms sets in, leading to the formation of hillocks and ridges at which, in turn, growth steps can originate.[112] The newly spread layer can either have the ideal surface structure and then give rise to new steps via "nucleation" of the superstructure, or steps may occur during the spreading. This will depend on the relative rates of spreading and reconstruction. Though there is considerable resemblance to the 2D nucleation mechanisms, the kinetics of step creation due to reconstitution is highly speculative at this point.

III.3. Volume Diffusion Controlled Layer Growth

Growth rates in crystallization from solutions are typically rather low. To account for this, BURTON, CABRERA and FRANK (BCF)[46] assumed that surface diffusion of growth units is slow and incorporation rather proceeds via direct diffusion from the solution to kinks. BCF based their treatment on the following compound diffusion field : hemispherical diffusion about each kink was assumed and matched at the distance Y_o (for notation see Section III.1) to a semicylindrical field about the step. The individual semicylindrical diffusion fields of equidistant steps were matched at the radius X_o to a planar diffusion field which was thought to extend throughout the concentration boundary layer of width δ_c parallel to the interface. Equating the three fluxes, BCF obtained a relation between the bulk supersaturation, i.e., σ at δ_c, and the supersaturation at $r = Y_o$ from the kink in the form

$$\sigma(Y_o) = \sigma\left[1 + \frac{2\beta a\pi(\delta_c - X_o)}{X_o Y_o} + \frac{2\beta a}{Y_o} \ln \frac{X_o}{Y_o}\right]^{-1}. \tag{40}$$

Following BENNEMA,[62] we have included here the kinetic coefficient β to allow for slower kink kinetics (see the discussion after eq. (29)). Assuming steps from growth spirals with the same properties as discussed for the surface diffusion model, BCF[46] obtained for the growth rate of the volume diffusion model

$$R = \frac{D_v C_o \Omega a \sigma(Y_o)}{2Y_o r_c} = \frac{D_v C_o \Omega k_B T \sigma^2(Y_o)}{2Y_o \gamma'}, \tag{41}$$

where D_v and C_o are the coefficient for bulk diffusion and the equilibrium concentration in the bulk solution, respectively.

At low supersaturations, i.e., when the interstep distance X_o is large and, hence, comparable to the concentration boundary layer width δ_c, the logarithmic term in (40) dominates, giving rise to an approximately linear relation between $\sigma(Y_o)$ and σ. Consequently $R \propto \sigma^2$ is obtained. If, however, at high supersaturation, $X_o \ll \delta_c$, the middle term in the right hand side of (40) dominates. Then, one obtains on substitution of (40) into (41), employing (30) and (31) for X_o, $R \propto \sigma$. This linear law, when extrapolated to low σ's, passes through the origin. Thus, the BCF volume and surface diffusion models yield the same qualitative functional $R(\sigma)$ dependence. Note, however, that the transition between the linear and parabolic range should depend on the boundary layer width δ_c.

CHERNOV, in his model of solution growth,[1] also neglected surface diffusion and assumed direct diffusion from the nutrient to steps, which were considered as linear continuous sinks for building blocks (Fig. 18). This model leads also to a parabolic growth law at low σ and a linear law at high σ.

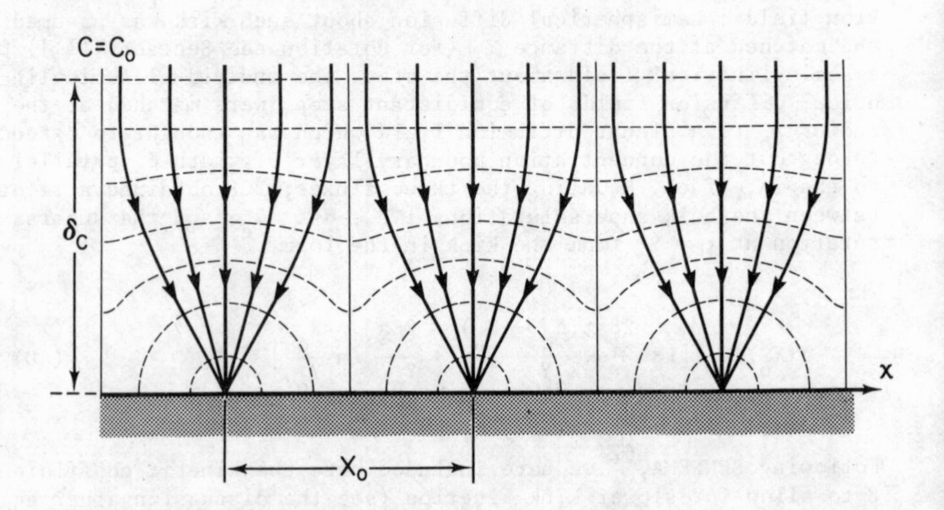

Fig. 18 : Diffusion field for solute to parallel steps in Chernov's volume diffusion model, showing isoconcentration lines and flux lines. After ref. 1 .

However, as indicated in Fig. 19., the linear regime does not extrapolate through the origin. Both this extrapolated value σ^* and the transition from $R \propto \sigma^2$ to $R \propto \sigma$ are expected to depend on δ_c.

III.4. Combined Volume and Surface Diffusion

Any interfacial growth kinetics theory requires knowledge of the concentration C_i of building blocks in the nutrient layer from which "adsorption jumps" occur. The thickness of this layer, Λ, is of the order of a mean free path. As long as the kinetics on the interface are much slower than the transport kinetics to and through this layer, C_i is approximately equal to the nutrient's bulk concentration C_b. Assuming a negligible "volume" transport impedance, BCF[46] set the supersaturation in the adsorbed layer, $\sigma(x)$ in (24), far from a step equal to the bulk supersaturation. Similarly, in the 2D nucleation model, the concentration of adsorbed molecules,

N_s in (36) and C_o in (39), was taken equal to C_b.

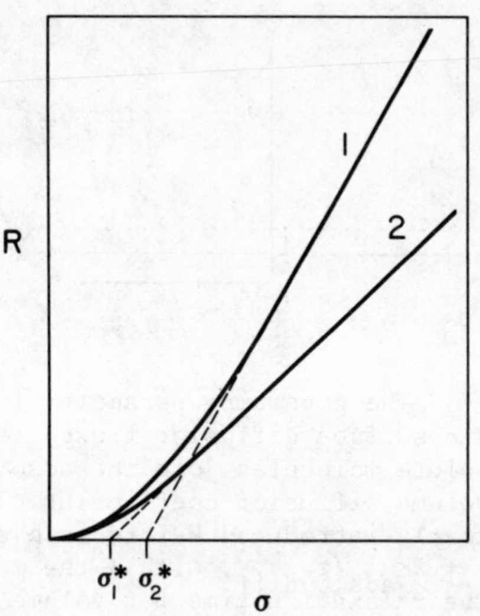

Fig.19 : Schematic plots of growth rate vs supersaturation obtained in Chernov's volume diffusion model for two different concentration boundary layer width, $\delta_1 < \delta_2$. After ref.1.

The transport in the bulk nutrient, however, may be slow. Also, the access to the adsorbed surface layer may be impeded by adsorbed impurities or solvent molecules. Consequently, the "volume" kinetics may become comparable to or slower than the surface kinetics; the assumption that $C_i \simeq C_b$ then becomes unrealistic.

For a more realistic description of the growth-driving concentration in the Λ-layer, one must simultaneously treat surface and volume transport. For instance, in a correspondingly modified step growth model, the exchange current J_n in (26) no longer originates from a fixed C_i value (equal ot C_b in the BCF treatment) but rather from C_i's which depend on the locally varying J_n (see Fig. 12). The larger J_n, the more C_i rops below C_b.

Combined volume and surface diffusion was treated by BENNEMA[64] and GILMER, GHEZ and CABRERA.[113,114] Steps were again considered to be line sinks, reducing the transport problem to two dimensions (Fig. 20). Solutions were obtained to the low concentration approximations[115] of the equation for diffusion on the surface (combined (25) and (26)) and diffusion in the nutrient, coupled by a continuity equation for J_n for the volume surface exchange process. Both groups assumed that the bulk value is attained at the outer edge of an unstirred layer of width δ_c (Fig. 20). GILMER et al.[113]

allowed for variations of C_i with the distance x from a step. BENNEMA[64] assumed an x-independent C_i, which greatly simplifies the solution of the problem. Physically, this is equivalent to putting the BCF model in series with a diffusion impedance layer of width δ_c. The uniform C_i obtained at $y = \Lambda$ corresponds to the average, macroscopic growth rate R.

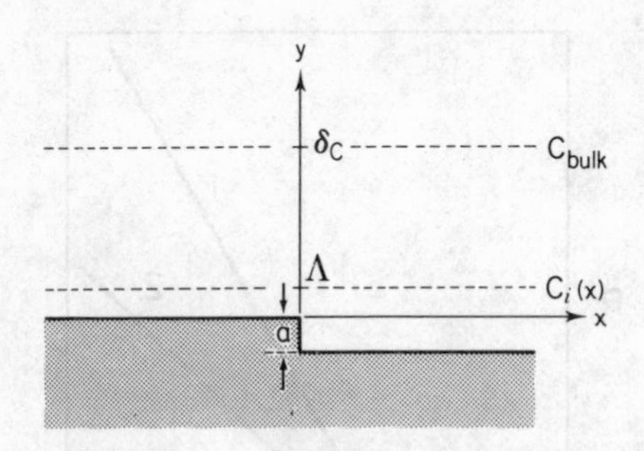

Fig. 20 : Definition sketch for parameters in combined volume and surface diffusion growth model.

The governing parameter in the coupling equation between volume and surface diffusion fluxes is the "drift velocity" D_v/Λ with which solute molecules join the adsorbed surface layer. Here D_v is the volume diffusion coefficient. The length Λ, which we have qualitatively introduced before (see also (34)), is defined as $\Lambda = \lambda\tau_{ads}/\tau_{vdiff}$, with λ the mean free path in the solute and τ_{vdiff} the relaxation time for volume diffusion. The τ_{ads} is the relaxation time (inverse jump frequency) for adsorption jumps. Thus Λ represents the distance through which solute molecules can diffuse during a τ_{ads}. Note that only when τ_{ads} equals τ_{vdiff} does Λ equal λ. More typically, for instance when a desolvation reaction occurs at the interface ($\tau_{ads} \equiv \tau_{desolv}$[65,115]), $\tau_{ads} > \tau_{vdiff}$ and, consequently, $\Lambda > \lambda$.

The general solution of GILMER et al.[113] is rather complex. However, if the mean diffusion path on the surface, X_s, is short with respect to Λ, i.e., if $b \equiv X_s/\Lambda \to 0$, the solution to the combined volume and surface diffusion growth model simplifies to

$$R = \sigma_v C_o \Omega D_v \{\Lambda + \delta_c + \Lambda\Lambda_s \frac{X_o}{X_s^2} + \Lambda[\frac{X_o}{2X_s}\coth(\frac{X_o}{2X_s}) - 1]\}^{-1}, \quad (42)$$

where σ_v is the bulk supersaturation and $\Lambda_s = a\tau_{kink}/2\tau_{sdiff}$ is the phenomenological coefficient describing the kinetics of particle exchange between the step and the adsorbed surface layer. Each term in the denominator of (42) represents an impedance to R. The first term can be regarded as the impedance of the adsorption reaction, the second term as the impedance of the unstirred layer, the third as the impedance for entering a step, and the fourth term as the surface diffusion impedance. GILMER et al.[113] have also shown that (42) is a good approximation for all systems in which $b < 0.1$, which, as discussed in 116, is typical for most solution growth situations. The functional behavior of (42) is very similar to that of the BCF relations (33) and (34). At low supersaturations $R \propto \sigma_v^2$ and at high supersaturation $R \propto \sigma_v$. As in (34) the linear range does not extrapolate through the origin. The slope of the curve is considerably reduced with increasing values of δ_c/Λ, i.e., with increasing impedance from volume diffusion. For $\delta_c/\Lambda \to 0$, (42) converges towards the BCF relations (33) or (34), depending on the specific kinetics involved. Bennema's solution[46] is equivalent to (42). This is to be expected since, with $X_s/\Lambda \to 0$, the x-dependence of C_i becomes insignificant.

The influence of volume diffusion on multinuclear multilayer growth is also outlined in ref. 113.

IV. OTHER CONCEPTS

The thermodynamics of irreversible processes[117,118] is usually only applied to bulk phases. There one finds that, not too far from equilibrium, phenomenological equations hold in which the fluxes are set equal to the gradients in the corresponding field multiplied by a transport coefficient. BCF[46] have used the same approach for their surface diffusion model. A detailed discussion of the applicability of irrevesible thermodynamics to interfaces, however, has been given only recently (see refs. 119,120 and references therein). On this basis, MÜLLER-KRUMBHAAR, BURKHARDT and KROLL[121] have applied a generalized time-dependent Ginzburg-Landau theory to the description of the motion of an interface in an anisotropic medium. They obtained a generalized kinetic equation for crystal growth, which can, in principle, accommodate any anisotropy in the transport (diffusion) coefficient and the growth-driving free energy gradient ledge free energy, chemical potential gradient including strain energy, etc.). With appropriate morphological formulations for these parameters, any growth mode can be described for which the growth rate varies continuously with $\Delta\mu$. This excludes nucleation growth modes because of the nucleation barrier. MULLER-KRUMBHAAR et al.[121] have applied their phenomenological theory in particular to anisotropic spiral growth. They found that spirals become polygonized, with the deviation from the round shape increasing with distance from the center, when the anisotropy in the surface diffusivity exceeds that of the ledge free energy. Diffusion along the steps, and increased temperatures (i.e., higher step roughness) counteract polygonization.

In all above models growth proceeds through the attachment of individual growth units. Recently, an alternative concept for growth from melts has been based on the dislocation model of the liquid state. A melt is essentially considered as a crystal saturated with dislocations. Melting occurs when the free energy of glide dislocation cores becomes negative. When this happens, dislocation cores are generated to fill the solid to capacity so that each atoms lies within a dislocation core. The total dislocation energy necessary to achieve this state has been shown[122] to agree well with observed heats of fusion. A summary of the development of the dislocation theory of melting, which has roots in works of BRAGG and SHOCKLEY, is given in ref. 123. The models to explain solidification in terms of a discontinuity in dislocation density at the interface have been reviewed by KAMADA.[124] In what is probably the physically most realistic model, dislocations are driven, in the free energy gradient that prevails under non-equilibrium conditions, across the interface, back into the melt. There, the excess conventration anihilates, giving rise to the heat of solidification. Quantitative estimates of the maximum dislocation "segregation velocities," possible in such a model, agree well with observed maximum growth rates for dislocation-free metal crystals. Solidification rates in excess of these values appear to lead to highly dislocated growth.[124]

Before closing this section, reference should be made to some general kinetics arguments put forth by COE, BRICE and TANSLEY.[125] All growth kinetics models discussed so far involve some thermal energy barriers; e.g., the activation energies for the adsorption jump to the solid, for surface diffusion, critical nucleus formation, step incorporation, etc. Traditionally, the probability of overcoming such a barrier is formulated in terms of Boltzmann statistics, i.e., the incorporation of molecules will have a reasonable chance of occurring only if their thermal energy, $k_B T^*$ is of the same order as the respective barrier height, ΔE. COE et al.[125] have pointed out that the incorporation kinetics can also be governed by a more complex statistical dependence that allows for "tunneling", i.e., for a significant transition probability even for $k_B T^* < \Delta E$. In analogy to the behavior of Schottky barriers, these authors have derived a growth law in the form

$$\ln R = A + B \ln(1 + \sigma), \tag{43}$$

where A increases with temperature and B is only very weakly T-dependent. The physical implication of (43) is that above some supersaturation, all molecules in the nutrient (or in the adsorption layer) have a finite probability of incorporation. This probability, in turn, is a function of the thermal energies of the molecules involved. COE et al.[125] found several solution growth systems that seem to follow (43) much better than other growth laws. All these systems involved complex molecules, such as ethylene diamine tartrate, which face considerable steric hindrance before incorporation.

V. APPLICATIONS AND COMPARISON WITH EXPERIMENTS

The literature on growth rate studies and direct morphological evidence for certain growth mechanisms is too extensive to be covered here. A detailed review will occur elsewhere.[126] Hence, the rest of this paper is restricted to a critical discussion of the applicability of the above growth laws and some problems met when comparing to experimental findings. First, we discuss the validity of the current interface roughness concepts for predictions of the prevailing growth mode. Then we refer to select experimental investigations of some of the growth laws.

V.1. Surface Roughness Criteria

We have seen above that the prevalence of normal or lateral growth depends on the atomic roughness of an interface. In recent years, considerable efforts have been made in the theoretical description of interface roughness and its correlation with macroscopic, thermodynamic parameters of a system (for reviews see refs. 5,14,15). To what extent can the current roughness theory be used to predict the growth mode for real systems ?

Probably the physically most realistic modeling of the roughening transition is obtained from Monte-Carlo simulation and low-temperature series expansion works. For solid-vacuum interfaces of a Kossel crystal with SOS restriction (see Section II.2) these analyses predict that the roughening temperature, T_R and the latent heat of the phase transition (here sublimation), L, not too far from equilibrium are related by

$$\alpha_{hkl} \equiv \xi_{hkl} L/k_B T_R \simeq 3.2, \qquad (44)$$

where the factor ξ_{hkl}, as introduced by JACKSON,[126,127] represents the ratio of the number of first nearest neighbors in a complete monomolecular lattice plane parallel to the interface to the bulk coordination number. The ξ factor can also be interpreted as the ratio of the energy associated with the bonds in that plane to the total crystalization energy.[23] Note that for the sublimation of a Kossel crystal, $L = 3\phi_{ss}$, and for its (001) face, $\xi = 2/3$. Thus (44) can be reduced to $\phi_{ss}/k_B T_R \simeq 1.6$, as introduced in connection with Fig. 3. This value is close to the roughening condition $\phi_{ss}/k_B T_R = 1.76$ deduced by BCF from the Onsager solution of the 2D Ising model.

For most solid-vapor systems, (44) predicts T_R's above the melting point, thus defying experimental determination of T_R. Correspondingly, most solid-vapor interfaces exhibit strongly faceted morphologies. Fortunately there are rotationally disordered molecular solids ("plastic crystals") which, due to their lower heat of sublimation and higher melting point than those of other (organic) crystals,[130-132] undergo the roughening transition distinctly below T_m. Quantitative determinations of T_R have so far been carried out for only two systems.[131] Both yielded $L/k_BT_R \simeq 16$. Even if one allows for the dynamic nature (i.e., off-equilibrium condition) of the experiments and the possible influence of impurities,[132] it appears that the predicted T_R's for solid-vapor interfaces are far too high. More experiments are needed to ascertain this point.

In melt growth one finds for numerous pure materials[21] and eutectics[133] with very few exceptions,[134] that $L/k_BT_R \simeq 2$ (in fortuitous agreement with predictions of the mean-field single-layer roughness model[127,128]). Thus, the predicted T_R's for solid-melt interfaces are considerably lower than those observed.

In solution growth there is also some morphological and kinetic evidence of rough interfaces occuring for certain solvent-solute combinations.[135-136] An actual transition from faceted to non-faceted growth with increasing temperature has been observed in the crystallization of Zn from a Zn-In solvent.[137] A quantitative interpretation of the roughening transition in solution growth is very difficult due to the complexity resulting from the solute-solvent interaction and possible adsorption of the solvent on the interface, etc. BENNEMA and VAN DER EERDEN[24] have given a very detailed, phenomenological discussion of this situation. Under highly idealized conditions their more general relations reduce to a modification of (44) for solution growth in the form

$$\alpha_{hkl} = \xi_{hkl}(L/k_BT - \ell n X_{seq}), \tag{45}$$

where L is the heat of fusion of the solute and X_{seq} is the fraction of solute in the solution at equilibrium. The trends expressed by (45) have been confirmed in experiments with biphenyl employing different solvents and, hence, solubilities.[136]

From the above one sees that the α-factor criterion for layer vs. normal (faceted vs. non-faceted) growth can supply valuable, semi-quantitative guidance within groups of materials. For quantitative predictions of interface roughness, however, the underlying models, currently exclusively based on bulk properties, must be supplemented with atomistic features of the "interface phase" that forms between the bulk phases.[30,31,129]

Before referring to some specific growth rate studies, two frequently met arguments concerning interface kinetics should be addressed :

1) In comparing the growth from various nutrients, a case is often made for a similarity between vapor and solution growth. In contrast to melt growth, it is claimed,[124] that both solutions and vapors represent a "dilute environment" for the "relevant" molecules (builiding blocks). Such analogies have led to the expectation that similar mechanisms prevail in growth from vapors and solutions. It is overlooked, however, that in practice solvents are chosen for their intimate interaction (solvation) with the solute, which results in the desired high solubility. Hence, the solute is not simply diluted, but strongly modified in its (interfacial) kinetics (see eq. (34)). With respect to its influence on the interface-structure and the resulting surface roughness, an adsorbed concentrated solvent layer may in fact much closer resemble a melt than a vapor. Of course, there can be strong coordination between components in vapors that suggest analogies to solution growth, but which differ from the "dilute environment" model.

2) The anisotropic growth of crystals is often discussed in terms of the varying interface roughness of differently oriented parts of the interface. Based on the anisotropy factor ξ in the α-criterion (Eq. 44), it is claimed that in a given system the *thermal* interface roughness increases as ξ decreases, i.e., as a part of an interface takes on higher index orientations. This claim has been supported, for instance, with the observation of higher undercoolings necessary to drive the same growth rate on off-facet parts of a germanium-melt interface.[138] Theoretical support is cited from Monte-Carlo simulation (MCS) of growth rates on (111), (001) and (011) faces of the face centered cubic lattice with nearest neighbor interaction only.[4]

As has been pointed out before,[139] such conclusions can be conceptually erroneous. Off-facet or curved parts of a macroscopically faceted interface need not be atomically rough. The anisotropy factor ξ supplies only a geometrical argument void of the *overall* energetics of the system. If the surface free energy of the facets (cusps in Wulff's plot) is low enough, then (not too far from equilibrium) even the "rounded" parts of an interface may actually consist of low-index terraces ("hill and valley" structure[140]). The size of the facets will depend, among other parameters, on the chemical potential gradient normal to the interface; for the dependence on temperature gradients see Ref. (41). From the foregoing discussion of growth kinetics, however, we know that as soon as the terrace width X_o becomes comparable to or larger than the mean surface diffusion length X_s, layer growth will prevail over normal growth. One should note that a representative range for X_s is 10-1000 Å ! Hence, facet sizes below the resolution of ordinary microscopy can result in layer growth and the lack of macroscopic facets does not necessarily imply high *thermal* roughness.

The above mentioned observation of smaller growth-driving

gradients can readily be understood in terms of reentrant corners of a hill-and-valley morphology, which facilitate the formation of growth steps (Fig. 17). But what about the computer experiments that did not show faceting, in spite of the strong anisotropy in surface free energy inherent to the underlying Kossel-type lattice ? To answer this question one must consider the stability of a given macroscopic surface with respect to faceting. This problem has been analyzed by HERRING[142-143] and MULLINS and SEKERKA;[144] for a summary see Ref. 14 . These considerations show that the Kossel surfaces employed in the MCS growth studies are intrinsically fully faceted, namely on the length scale of their interaction forces : one interatomic distance ! Hence, there is no thermodynamic driving force for "macro-faceting". In addition to these equilibrium arguments, kinetic considerations show that for most systems with some singular faces of low thermal roughness, higher index faces with potentially higher thermal roughness will disappear from the growth form. This was well illustrated in a recent review by MUTAFTSCHIEV.[145]

V.2. Experimental Identification of Growth Mechanisms

Normal growth. From the α-criterion we can expect high thermal interface roughness, even on low index planes (the prerequisite for overall normal growth) in melt growth of most metals and a few organic components. Corresponding reviews of melt growth have been given in refs. 21 and 146. Since metal melt mobilities are high (low Q_a in Eq. (4), (6), (8)), very rapid growth rates (R = 5-50 cm/sec for 1 degree undercooling) are predicted. In experiments[147] with metal dendrites, such high velocities have been observed. With macroscopic crystals, however, heat transfer limitations typically occur before kinetics limitations. Or, correspondingly, at more reasonable R's, the driving undercooling becomes so small that it is difficult to measure accurately.[146,148,150] With systems of higher melt vicosity such as glass forming materials and some organic compounds, kinetically limited normal growth is more readily accessible. The validity of Wilson-Frenkel "linear" growth laws, including the reduction of R due to an increasing viscosity at lower temperatures, appears to be well established for low-α, high purity systems.[21,146,148] The presence of impurities, however, can lead to R(Δμ)'s that resemble surface nucleation growth behavior in intrinsically low-α systems.[146]

The few known vapor-solid and solution-solid system with some evidence for rough interfaces were mentioned in Section V.1. No quantitative growth rate studies are yet available for these vapors systems. In solution growth of hexamethylenetetramine (HMT), however, a linear growth law was found for crystallization from an aquous solution, whereas a BCF like growth law governed crystallization from an alcoholic solution.[135] This behavior scales with low and high α-values, respectively, where, as mentioned above, the assignment of absolute values meets conceptual difficulties.[24]

Layer growth. Growth by lateral propagation of steps on otherwise atomically rather smooth interfaces is expected for all nutrient-solid systems with high α-factors. This is, in general, experimentally well confirmed.

For growth from the melt, faceted growth and corresponding BCF growth laws are typically observed above $\alpha = 2$ (refs. 21,146). Quantitative studies of surface nucleation controlled growth rates from melts are rather scarce, due to several experimental difficulties :

a) The limited purity of the materials with respect to particles that on incorporation can lead to spiral growth. (Note that the microfiltration of melts is much more difficult than that of low temperature solutions).

b) The low critical shear stress of materials at high temperatures allows also for ready dislocation formation during growth.

c) Heat flow can control the growth rate. CHERNOV,[1] has treated the heat tranfer from steps to the melt in a similar manner as for solute volume diffusion discussed in Section III.3. His analysis suggests that heat tranfer control of the growth process will simulate a $R \propto \Delta T$ dependence, irrespective of the actual growth mechanism, unless the material has a high thermal conductivity and a narrow thermal boundary layer is maintained at the interface.

Probably the best characterized rate studies of layer growth have been performed in crystallization from solutions. Beginning with the pioneering work of BENNEMA,[151,152] a large body of results[107,153-158] has accumulated, giving strong evidence that growth from many solutions follows the modified surface diffusion BCF relation (34). (See also the Chapters on "Aqueous and Non-aqueous Solution Growth".) Besides this functional agreement, more support for desolvation as the rate limiting step comes from kinetics considerations[64] and experimental observations.[159] The complexity of the interfacial kinetics in solution growth becomes particularly apparent in view of the strong habit dependence on the pH-value of a solution.[156] Direct volume diffusion to steps can be excluded when the growth rate shows little response to changes in the interfacial fluid dynamic conditions, i.e., in δ in (41). It appears that the combined volume and surface diffusion model (Section III.4) gives the best fit for growth rate data obtained for HMT in alcoholic solutions.[135] The deformation of BCF-like growth rate curves due to volume diffusion resistance makes a clear distinction from 2D nucleation behavior difficult. An optimum experimental approach to this problem has been discussed.[160]

The closest control of supersaturation and interface morphology in solution growth has been obtained with electrocrystallization.[161,162] Both mononuclear monolayer and multinuclear multilayer growth have been elegantly confirmed in these investigations (see also the Chapter "Electrocrystallization"). Good agreement with 2D nucleation curves has also been obtained in more conventional solution growth systems.[107,163,164]

Vapor growth systems were long considered as ideally suited for growth kinetics studies. Particularly in the early days of crystal growth modeling it was thought that vapor-solid systems most closely matched the idealized assumptions in the theoretical description. These early hopes have not materialized. Numerous revealing results have been obtained for the growth and etching kinetics on a molecular level; see the Chapter "Studies of Surface Morphology on an Atomic Scale". Well-defined macroscopic growth rate studies, however, are very scarce. There are several reasons for this : a) The volume transport phenomena that prevail in real, impure and incongruent vapor systems proved to be much more complex than anticipated based on the idealized concepts; b) Exact control of the nutrient composition is difficult. In particular, at low supersaturations, the residual gas pressure in a system can readily become comparable to the "solute" partial pressure. c) The actual surface temperature and, thus, σ, is difficult to determine. The latent heats involved in vapor-solid transitions are typically much higher than in solidification. Hence, the interface temperatures of source and crystal must be measured directly. Quantitative values of σ deduced from bulk temperatures, in particular at significant growth rates, are not too meaningful. Unfortunately, this limitation applies to all bulk vapor growth rate studies known to this author. Qualitatively, good agreement with BCF-type $R(\sigma)$ curves has been obtained by several workers.[165-168] A quantitative comparison, however, will have to wait for better defined experiments, including careful characterization of the vapor composition.

VI. CONCLUSIONS

The modeling of crystal growth kinetics has reached an impressive level of sophistication and supplies valuable, semi-quantitative insight. Quantitative modeling will require more realistic (i.e., theoretically more difficult to handle) input on the actual structure of the interacting solid and nutrient "layers".

Basically different growth mechanisms can result in the same type of growth rate curve, $R(\Delta\mu)$. Linear behavior at low supersaturations, for instance, can be expected from (a) normal growth; (b) layer growth onto steps from a group of screw dislocations of a certain spacing (Fig. 14); and (c) layer growth onto steps that originate in closely spaced reentrant angles of a hill-and-valley surface (Fig. 17). Hence, the analytical form of R(Δμ) alone can seldom give unambiguous proof of a specific growth mechanism. Though the predicted magnitudes for R differ distinctly between models, a quantitative comparison is troublesome because of the large practical uncertainties in some of the governing parameters (ledge free energy, concentration of adatoms, influence of impurities, etc.). Therefore, supplementary insight on the interfacial kinetics and morphology and, if applicable, the fluid dynamics of the process is needed for conclusive results.

Well defined growth kinetics experiments have become available for melt and solution systems. In vapor growth, however, quantitative work is badly needed.

Acknowledgments

The author is grateful for support under NSF Grant DMR 79-13183 and NASA Grant NSG-1534 during the writing of this review.

REFERENCES

1 . CHERNOV A.A., Sov. Phys. Usp. 4 (1961) 116.
2 . WEEKS J.D., GILMER G.H., in *Advances in Chemical Physics*, Prigogine I., Rice S.A., eds., Wiley, New York 1979, Vol. 40, pp. 157-228.
3 . CHERNOV A.A., in *Crystal Growth and Characterization*, Ueda R., Mullin J.B., eds., North Holland, Amsterdam 1975, pp. 33-52.
4 . GILMER G.H., JACKSON K.A., in *Current Topics in Materials Science*, Kaldis E., ed., North Holland, Amsterdam 1977, Vol. 2, Chapter 1.2, pp. 79-114.
5 . VAN DER EERDEN J.P., BENNEMA P., CHEREPANOVA T.A., Progr. Crystal Growth Characteris. 1(1978)219.
6 . WILSON H.A., Philos. Mag. 50 (1900) 609.
7 . FRENKEL J., Phys. Z. Sowjetunion 1 (1932) 498.
8 . JACKSON K.A., CHALMERS B., Canad. J. Phys, 34 (1956) 473.
9 . JACKSON K.A., in *Treatise of Solid State Chemistry*, Hannay N.B., ed., Plenum Press, New York 1975, Vol. 5, pp. 233-282.
10 . JACKSON K.A., in *Crystal Growth and Characterization*, Ueda R., Mullin J.B., eds., North Holland, Amsterdam 1975, pp. 21-31.
11 . HERTZ H., Ann. Phys. 17 (1882) 177.
12 . KNUDSEN M., Ann. Phys. 29 (1909)179.
13 . HILLIG W.B., TURNBULL D., J. Chem. Phys. 24(1956)914.
14 . LEAMY H.J., GILMER G.H. and JACKSON K.A., in *Surface Physics of Materials*, Blakely J.M., ed., Academic Press, New York 1975, Vol. 1, pp. 121-188.

15 . MÜLLER-KRUMBHAAR H., in *Current Topics in Materials Science*, Kaldis E., ed., North Holland, Amsterdam 1977, Vol. 2, Chapter 1.3, pp. 115-139.
16 . MÜLLER-KRUMBHAAR H., in *Current Topics in Materials Science*, Kaldis E., ed., North Holland, Amsterdam 1978, Vol. 1, Chapter 1, pp. 1-46; and in *Topics in Current Physics*, Vol. 7, Monte Carlo Methods, Binder K. ed., Springer Verlag, Berlin 1979, pp. 261-299.
17 . WEEKS J.D., GILMER G.H. and JACKSON K.A., J. Chem. Phys. 65 (1976) 712.
18 . TEMKIN D.E., Soviet Phys. Crystallogr. 14 (1969) 344.
19 . PFEIFFER H., HAUBENREISSER W. and KLUPSCH TH., Phys. Stat. Sol. (b) 83 (1977) 129.
20 . DEHAAN S.W.H., MEEUSSEN V.J.A., VELTMAN B.P., BENNEMA P., VAN LEUUWEN C. and GILMER G.H., J. Crystal Growth 24/25 (1974) 491.
21 . JACKSON K.A., UHLMANN D.R. and HUNT J.D., J. Crystal Growth 1 (1967) 1.
22 . JACKSON K.A., J. Crystal Growth 24/25 (1974) 130.
23 . BENNEMA P., GILMER G.H.,in *Crystal Growth*, Hartmann P., ed., North Holland, Amsterdam 1973, p. 263-327.
24 . BENNEMA P., and VAN DER EERDEN J. P., J. Crystal Growth 42 (1977) 201.
25 . MÜLLER-KRUMBHAAR H., Phys. Rev. B 10 (1974) 1308.
26 . BAIKOV YU.A., ZELENEV YU.V., HAUBENREISSER W., Phys. Stat. Sol. (a) 55 (1979) 123.
27 . CHISTYAKOV YU.D., BAIKOV YU.A., SCHNEIDER H.G., RUTH V., Acta Metall. 29 (1981) 415.
28 . FLETCHER N.H., J. Crystal Growth 35 (1976) 39.
29 . COTTERILL R.M.J., J. Crystal Growth 48 (1980) 582.
30 . BROUGHTON J.W., BONISSENT A., ABRAHAM F.F., J. Chem. Phys. 74 (1981) 4029.
31 . ABRAHAM F.F., J. Chem. Phys. 68 (1978) 3713.
32 . SAITO Y., MÜLLER-KRUMBHAAR H., J. Chem. Phys. 70 (1979) 1078.
33 . GILMER G.H., BENNEMA P., J. Appl. Phys. 43 (1972) 1347.
34 . PFEIFFER H., Phys. Stat. Sol. (b) 101 (1980) K 117.
35 . ZELENEV YU.V., BAIKOV YU.A., MOLOTKOV A.P., Kristall Technik 14 (1978) 389.
36 . PFEIFFER H., Phys. Stat. Sol. (b) 93 (1979) K 149.
37 . PFEIFFER H., HAUBENREISSER W., Phys. Stat. Sol. (b) 96(1979) 287.
38 . PFEIFFER H., HAUBENREISSER W., Phys. Stat. Sol. (b) 101(1980) 101.
39 . BAIKOV YU.A., ZELENEV YU.V., HAUBENREISSER W.and PFEIFFER H., Phys. Stat. Sol. (a) 61 (1980) 435.
40 . CHEREPANOVA T.A., Phys. Stat. Sol. (a) 58(1980) 469.
41 . CHEREPANOVA T.A., Phys. Stat. Sol. (a) 59(1980) 371.
42 . CHEREPANOVA T.A., J. Crystal Growth 52(1981) 319.
43 . BAIKOV YU.A., ZELENEV YU.V. and HAUBENREISSER W., Phys. Stat. Sol. (a) 59(1980) K 155.

44 . TEMKIN D.E., J. Crystal Growth 52 (1981) 299.
45 . PFEIFFER H., Phys. Stat. Sol. (a) 65(1981) 637.
46 . BURTON W.K., CABRERA N. and FRANK F.C., Phil. Trans. Roy. Soc. (London) A243(1951)299.
47 . KOSSEL W., Nachr. Ges. Wiss. Göttingen, Math.-Physik. Kl.IIa, 135(1927).
48 . STRANSKI I.N., Z. Physik. Chem. 136(1928)259, and B11(1931)342.
49 . STRANSKI I.N., Z. Physik. Chem. B38(1938)451.
50 . STRANSKI I.N. and KAISCHEW R., Z. Kristallogr. 78(1931)373 and 88(1934)325.
51 . FRENKEL J., J. Phys. USSR 9(1945)302.
52 . LEAMY H.J., GILMER G.H., J. Crystal Growth 24/25(1974)499.
53 . VAN LEEUWEN C., MISCHGOFSKY F.H., J. Appl. Phys. 46(1975)1056.
54 . BETHGE H., in *Molecular Processes on Solid Surfaces*, Drauglis E., Gretz R.D., Jaffee R.I. eds., McGraw-Hill, New York 1969, pp. 569-585.
55 . SCHWOEBEL R.L., J. Appl. Phys. 37(1966)3682 and 40(1969)614; VAN LEEUWEN C., et al., Surface Sci. 44(1974)213.
56 . FRANK F.C., Disc. Faraday Soc. 5(1949)48 and 67.
57 . CABRERA N. and LEVIN M.M., Phil. Mag. 1(1956)450.
58 . BUDEVSKI E., STAIKOV G., BOSTANOV V., J. Crystal Growth 29 (1975) 316.
59 . CABRERA N., COLEMAN R.V., in *The Art and Science of Growing Crystals*, Gilman J.J., ed., Wiley, New York 1963, p. 21.
60 . SUREK T., HIRTH J.P. and POUND G.M., J. Crystal Gwoth 18(1973) 20, and Surface Sci. 41(1974)77.
61 . VAN DER EERDEN J.P., J. Crystal Growth 53(1981)305 and 315.
62 . BENNEMA P., J. Crystal Growth 1(1967)278 and 287.
63 . AMELINCKX S., *The Direct Observation of Dislocations*, Academic Press, New York 1964, p. 6.
64 . BENNEMA P., J. Crystal Growth 5(1969)29.
65 . TAKATA M., OOKAWA A., J. Crystal Growth 24/25(1974)515.
66 . TAKATA M., OOKAWA A., J. Crystal Growth 42(1977)35.
67 . BETHGE H., Phys. Stat. Sol. 2(1962)3, 775.
68 . BETHGE H., Surface Sci. 3(1964)33.
69 . KELLER K.W., Phys. Stat. Sol. 36(1968)557.
70 . KROHN M., BETHGE H., in *Current Topics of Materials Science*, Kaldis E., eds., North Holland, Amsterdam, 1977, Chpt. 1.4, pp. 142-164.
71 . BAUSER E., STRUNK H., J. Crystal Growth 51(1981)362.
72 . FRANK F.C., J. Crystal Growth 51(1981)367.
73 . VOLMER M., ESTERMANN I., Z. Physik 7(1921)13.
74 . VOLMER M., Physik. Zeitschr. 22(1921)646.
75 . STRANSKI I.N., Z. Physik. Chem. 136(1928)259, (B)11(1931)342, (B)17(1932)127.
76 . STRANSKI I.N., KAISCHEW R., Z. Physik. Chem. (B)26(1934)100, 114,312 and 317.
77 . STRANSKI I.N., KAISCHEW R., Physik. Zeitschr. 36(1935)393.
78 . BECKER R., DÖRING W., Ann. Physik (Leipzig) 24(1935)719.
79 . BECKER R., Disc. Faraday Soc. 5(1949)45.

80 . KAISCHEW R., Acta Phys. Hung. 8(1957)75.
81 . HIRTH J.P., Acta Met. 7(1959)755.
82 . GUTZOW I., TOSCHEV S., J. Cryst. Growth 7(1970)218.
83 . KOLMOGOROV A.N., Izv. Akad. Nauk. Ser. Math. 3(1937)355.
84 . AVRAMI M., J. Chem. Phys. 7(1939)1103; 8(1940)212; and 9(1941) 177.
85 . BEWICK A., FLEISCHMANN M., THIRSK H., Trans. Faraday Soc. 58 (1962)2200.
86 . CHERNOV A., LJUBOV B., Growth of Crystals U.S.S.R. 5(1963)11.
87 . NIELSEN A.E., *Kinetics of Precipitation*, Pergamon, Oxford 1964, Chpt.4.
88 . HILLIG W.B., Acta Met. 14(1966)1868.
89 . VETTER K., *Electrochemical Kinetics*, Academic Press, New York 1967, p. 322.
90 . BOROVINSKII L.A., TSINDERGOZEN A.N., Sov. Phys. Cryst. 13(1969) 1191.
91 . RANGARAJAN S.K., J. Electroanalyt. Chem. 46(1973)125.
92 . HAYASHI M., J. Phys. Soc. Jap. 35(1974)614.
93 . HAYASHI M., SHICHIRI T., J. Crystal Growth 21(1974)254.
94 . KASHCHIEV D., J. Crystal Growth 40(1977)29.
95 . KASCHIEV D., VAN DER EERDEN J.P., VAN LEEUWEN C., J. Crystal Growth 40(1977)47.
96 . VAN LEEUWEN C., VAN DER EERDEN J.P., Surface Sci. 64(1977)237.
97 . ARMSTRONG R., HARRISON J., J. Electrochem. Soc. 116(1969)328.
98 . HARRISON J., THIRSK H., in *Electroanalytical Chemistry*, New York 1971, Vol. 5, p. 67.
99 . BERTUCCI U., Surf. Sci. 15(1969)286.
100. BERTUCCI U., J. Electrochem. Soc. 119(1972)822.
101. OLDFIELD J.W., Electrodep. Surf. Treatm. 2(1973/74)395.
102. CLARK M., HARRISON J., THIRSK H., Z. Physik. Chem. (NF) 98 (1975)153.
103. HARRISON J., RANGARAJAN S., Faraday Symp. 12, No. 6 (1978) 70.
104. GILMER G., J. Crystal Growth 49(1980)465.
105. MICHAELS A.I., POUND G.M., ABRAHAM F.F., J. Appl. Phys. 45 (1974)9.
106. GILMER G.H., BENNEMA P., J. Crystal Growth 13/14(1972)148.
107. BENNEMA P., BOON J., VAN LEEUWEN C. and GILMER G.H., Krystall und Technik 8(1973)659.
108. CALVERT P.D., UHLMANN D.R., J. Crystal Growth 12(1972)291.
109. HAMILTON D.R., SEIDENSTICKER R.G., J. Appl. Phys. 31(1960) 1165 and 34(1963)1450.
110. CHALMERS B., MARTIUS U.M., Phil. Mag. 43(1952)686.
111. CABRERA N., Disc. Faraday Soc. 28(1959)16.
112. VAN VECHTEN, J. Crystal Growth 38(1977)139.
113. GILMER G.H., GHEZ R., CABRERA N., J. Crystal Growth 8(1971)79.
114. GHEZ ., GILMER G.H., J. Crystal Growth 21(1974)93.
115. WESTPHAL G.H., ROSENBERGER F., J. Crystal Growth 43(1978)687.
116. JANSSEN-VAN ROSMALEN R., BENNEMA P., GARSIDE J., J. Crystal Growth 29(1975)342.

117. DEGROOT S.R., MAZUR P., *Non-equilibrium Thermodynamics*, North-Holland, Amsterdam 1962.
118. FITTS D.D., *Nonequilibrium Thermodynammics*, McGraw-Hill, New York 1962.
119. GHEZ R., Surface Sci. 20(1970)326.
120. BEDEAUX D., ALBANO A.M., VLIEGER J., Physica 99A(1979)293 and 102A(1980)105.
121. MÜLLER-KRUMBHAAR H., BURKHARDT T.W. and KROLL D.M., J. Crystal Growth 38(1977)13.
121. KUHLMANN-WILSDORF D., Phys. Rev. 140A(1965)1599.
123. COTTERILL R.M.J., J. Crystal Growth 48(1980)582.
124. KAMADA K., Progr. Crystal Growth Charact. 3(1981)309.
125. COE I.M., BRICE J.C.,TANSLEY T.L., J. Crystal Growth 30(1975) 367.
126. ROSENBERGER F., *Fundamentals of Crystal Growth II : Kinetic and Morphological Concepts*, Springer Verlag (in preparation).
127. JACKSON K.A., in *Liquid Metals and Solidification*, Am. Soc. Metals, Cleveland 1958, p. 174.
128. JACKSON K.A., in *Growth and Perfection of Crystals*, Doremus R.H., Roberts B.W., Turnbull D. eds., Wiley and Sons, New York 1958, p. 319.
129. BROUGHTON J.W., Paper at Fifth International Conference on Vapor Growth and Epitaxy, San Diego, 1981.
130. PAVLOVSKA A., NENOW D., J. Crystal Growth 12(1972) 9 and 39 (1977)346.
131. JACKSON K.A., MILLER C.E., J. Crystal Growth 40(1977)169.
132. PAVLOVSKA A., J. Crystal Growth 46(1979)551.
133. HUNT J.D., JACKSON K.A., Trans. Met. Soc. AIME 236(1966) 843.
134. AYERS J.D., SCHAEFER R.J., GLICKSMAN M.E., J. Crystal Growth 37(1977)64.
135. BOURNE J.R., DAVEY R.J., J. Crystal Growth 34(1976)230, 36 (1976)278 and 287, 39(1977)267, 43(1978)224, 44(1978)613.
136. HUMAN H.J., VAN DER EERDEN J.P., JETTEN L.A.M.J., ODERKERKEN J.G.M., J. Crystal Growth 51(1981)589.
137. PASSERONE A., EUSTATHOPOULOS N., J. Crystal Growth 49(1980)757.
138. BRICE J.C., WHIFFIN P.A.C., Solid State Electron. 7(1964)183.
139. ROSENBERGER F., DELONG M.C., GREENWELL D.W., OLSON J.M., WESTPHAL G.H., J. Crystal Growth 29(1975)49.
140. CHALMERS R., KING R., SHUTTLEWORTH R., Proc. Roy. Soc. (London) A193(1948)465.
141. WILCOX W.R., J. Crystal Growth 7(1970)203.
142. HERRING C., Phys. Rev. 82(1951)87.
143. HERRING C., in *Structure and Properties of Solid Surfaces*, Gomer R., Smith C.S. eds. University of Chicago Press, Chicago 1952.
144. MULLINS W.W., SEKERKA R.F., J. Phys. Chem. Solids 23(1962)801.
145. MUTAFTSCHIEV B., in *Dislocations in Solids*, Nabarro F.R.N. ed., North Holland, Amsterdam 1980, Vol. 5., Chpt. 19.
146. OVSIENKO D.E., ALFINTSEV G.A., in *Crystals*, Freyhardt H.C.ed., Springer Verlag, Berlin 1980, Vol. 2, pp. 119-169.

147. TEMKIN D.E., Sov. Phys. Doklady 5(1960)609.
148. VERGANO P.J., UHLMANN D.R., Phys. Chem. Glasses 11(1970)39.
149. KRAMER J.J., TILLER W.A., J. Chem Phys. 42(1965)257.
150. RIGNEY D., BLAKELY J., Acta Met. 14(1966)1375.
151. BENNEMA P., Phys. Stat. Sol. 17(1966)555 and 563.
152. BENNEMA P., in *Crystal Growth*, Peiser H.S., ed., Pergamon, Oxford 1967, p. 413.
153. REICH R., KAHLWEIT M., Berichte Bunsenges. Phys. Chem. 72 (1968)66 and 70.
154. KAHLWEIT M., J. Crystal Growth 5(1969)391.
155. GARSIDE J., MULLIN J.W., Trans. Inst. Chem. Engrs. 46(1968)T11.
156. MULLIN J.W., AMATAVIVADHANA A., CHAKRABORTY M., J. Appl. Chem. 20(1970)153.
157. GARSIDE J., in *Current Topics in Materials Science*, Kaldis E. ed., North-Holland, Amsterdam 1977, Vol. 2, pp. 483-513.
158. For additional references, in particular on high-temperature solution growth see : ELWELL D., SCHEEL H.J., *Crystal Growth from High-Temperature Solutions*, Academic Press, London 1975, pp. 171-176.
159. BLIZNAKOV G., KIRKOVA E., NIKOLAYEVA R., Kristall u. Technik 6(1971)33.
160. GARSIDE J., JANSSEN-VAN ROSMALEN R., BENNEMA P., J. Crystal Growth 29(1975)353.
161. KAISCHEW R., BUDEVSKI E., Contemp. Phys. 8(1967)489.
162. BOSTANOV V., OBRETENOV W., STAIKOV G., ROE D.K., BUDEVSKI E., J. Crystal Growth 52(1981)761.
163. GARSIDE J., MULLIN J.W., DAS S.N., Ind. Eng. Chem. Fundam. 13 (1974)299.
164. SOHNEL O., GARSIDE J., JANCIC S.J., J. Crystal Growth 39(1977) 307.
165. KITCHENER S.A., STRICKLAND-CONSTABLE R.F., Proc. Roy. Soc. A245(1958)93.
166. BRADLEY R.S., DRURY T., Trans. Faraday Soc. 55(1959)1848.
167. HEYER H., in *Crystal Growth*, Peiser H.S. ed., Pergamon, Oxford 1967, p. 265.
168. HONIGMANN B., HEYER H., Z. Electrochemie 61(1957)74.

PHYSISORPTION : STRUCTURES AND INTERACTIONS

M.B. WEBB and L.W. BRUCH

Departement of Physics,
University of Wisconsin,
Madison,
WI 53706, U.S.A.

ABSTRACT

We survey simple inert gas adsorption emphasizing surfaces where registry forces play an unimportant role. Such systems allow a detailed analysis of the structural and thermodynamic properties in terms of adatom-adatom and adatom-substrate interactions. We discuss the experimental techniques, primarily those giving structural information. Comparison of these data with results of statistical mechanical models allows quantitative assessment of the contributions to the interactions.

I. INTRODUCTION

Physisorption refers to those cases where gas atoms adhere to surfaces because of weak van der Waals or dispersion forces rather than because of the formation of primary chemical bonds. The latter case is referred to as chemisorption. There is no sharp division between the two cases, but the typical energy scales are tenths of an eV in the former and up to a few eV in the latter case. This difference means that both the experimental and theoretical approaches are quite different.

Physisorption is a very large topic, exciting from many points of view, and here we can deal with only a small part of the subject. Our emphasis will be on the simplest systems and on those techniques which give microscopic information. We will discuss in some detail the results of such experiments for inert gases adsorbed on smooth surfaces and then how these data can be analyzed in terms of thermodynamics and statistical mechanics to learn about the interactions

B. Mutaftschiev (ed.), Interfacial Aspects of Phase Transformations, 365–409.

between the adatoms within the film and between the adatoms and the substrate. We hope to show some of the separate effects of the modification of the interatomic interactions by the substrate and of the dimensionality.

The study of physisorption has a long history which we won't attempt to review. In earlier times there were challenges to theorists to understand and calculate the weak forces between atoms and between atoms and surfaces which arise from the correlated instantaneous moments ; there were questions of formulating thermodynamics for strongly nonuniform systems ; and there were problems of devising microscopic models which would account for the wide variety of adsorption isotherms. Experimentally most work was on finely divided adsorbents of high specific surface area and most measurements were of macroscopic thermodynamic properties. There was considerable technological interest in using physisorption to estimate surface area and physisorption was important in many aspects of vacuum technology.[1,2,3]

A period of renewed interest and progress began roughly a decade ago stimulated by both the demonstration that various preparations of graphite provided very uniform substrates[4,5,6,7] and by the development and application of current techniques of surface science[8] which allowed studies on well characterized single crystal surfaces. These developments reduced the extrinsic effects of heterogeneities and finite sizes. Surface thermodynamic methods such as measurements of adsorption isotherms and specific heats allow the identification of gas, liquid and both commensurate and incommensurate solid two dimensional phases and the characterization of the transitions among them. Field emission microscopy early demonstrated sensitivity of the adsorption to the crystallographic orientation of the surface.[9,10] Diffraction of low-energy electrons from single crystal surfaces[8,11,12,13,14,15] and of neutrons[16,17] and x-rays[18] from high specific area samples give detailed structural and phase transition information. Atomic scattering and diffraction give information about the atom-substrate potential and its variation along the surface plane.[19,20] Measurements of NMR relaxation times give information on atom mobilities.[21,22] The electronic structure is investigated by contact potential differences[9,12], angularly resolved photoemission[23] and surface reflectance.[24] The combination of these experimental developments now gives microscopic as well as thermodynamic information making possible a close and definite connection with quantum mechanical and statistical mechanical analyses.

In addition, theoretical questions arose concerning "2D physics" which may conveniently be approached in physisorption systems.[25,26,27,28] These include the possibility and nature of long range order in two dimensions; the related question of the nature of 2D melting; the commensurate-incommensurate transitions and the influence of registry forces generally; the critical behavior in 2D and the associated symmetry requirements. Thus physisorption presents opportunities to approach a number of interesting fundamental problems.

Further, studying these weak adsorption systems should be useful in understanding some features of the closely related chemisorption problem. Because of the weak atom-substrate interaction it is convenient to observe the system in equilibrium with 3D gas which then specifies the adsorbate chemical potential. This is generally not possible in chemisorbed systems. Also the adatom-adatom interaction energies are a larger part of the total energy and so can be studied in considerable detail. While the intralayer interactions in the chemisorption case are certainly more complicated than the dispersion-like forces, their magnitudes, inferred from phase diagrams,[29,30] are of the same order of magnitude as in physisorbed systems and so a detailed understanding must take into account those contributions which are dominant in physisorption but also present in chemisorption.

Finally, again because of the weak interactions, one is confident that the physisorbed atom perturbs the substrate much less than in the chemisorbed case. This makes the process simpler to understand, but it also provides the possibility of using physisorption as a probe of the adsorbent surface properties and there is increasing work in this direction.[31,32]

From all this it is clear that we can deal with only a small part of the topic in the time and space available. Earlier work has been reviewed in excellent books by STEELE[3] and DASH.[1] Much of the current work is covered in Conference proceedings of references 26,27,28. Prof. DOMANY will discuss phase transitions, critical phenomena and long range order and other theoretical topics in these proceedings.

The structure of physisorbed layers depends on both the adatom-adatom interactions and the periodic component of the interaction of a single atom with the substrate. Effects of the latter are manifested in the occurence of commensurate overlayer structures and in the orientational alignment of the overlayer with the substrate. These effects have been extensively studied in the adsorption of inert gases on graphite. For all the inert gases (He[33], Ne[34], Ar[15],Kr[14,35,36], and Xe[36]) one or both of these registry effects are observed. At the expense of slighting this extensive and interesting work, we will emphasize cases where the registry forces are unimportant. We further will emphasize the more microscopic information, primarily structural, that has recently become available. Such simple systems afford the clearest opportunity for comparisons of detailed theory of the interactions and the statistical mechanics of the 2D submonolayer, monolayer and multilayer films. While the simple systems do not display the richness of phenomena arising from the registry forces, the topics we do consider illustrate the potential of the techniques and serve as an introductory background for considering more complicated aspects -- both those already being investigated and those yet to be discovered.

II. EXPERIMENTS AND MEASURED QUANTITIES.

II.1. Thermodynamic Measurements

One of the most common experiments in physisorption studies is the measurement of adsorption isotherms, i.e., the quantity of gas adsorbed, θ, as a function of 3D gas pressure at constant temperature. For simple gases on uniform substrates such isotherms often display a series of sharp steps, i.e., risers, indicating nearly discontinuous increases in θ, separated by treads where θ rises only slowly with pressure. The risers indicate coexisting phases. The risers are first order transitions from a phase of smaller to one of larger density. During the transition, the chemical potential and thus the pressure are independent of total θ. The risers indicate the adsorption of successive monolayers, each of which has nearly distinct thermodynamic properties because of their different holding potential to the substrate. The spacing of pressures at which the steps occur is related to the shape of the atom-substrate potential, a point we will return to later.[37]

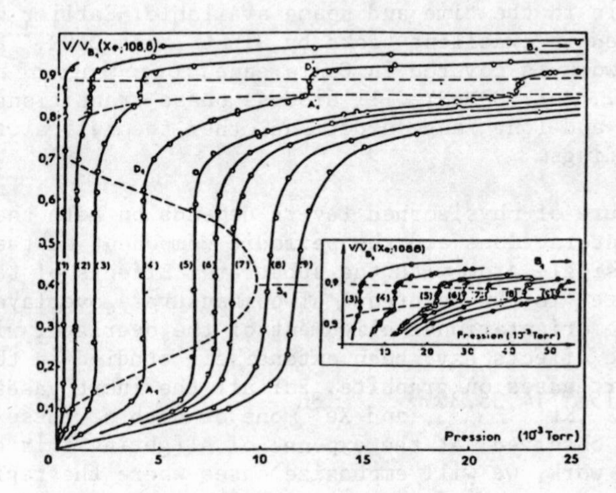

Fig. 1 : Isotherms for the adsorption of Xe on exfoliated graphite in the region of the first monolayer. (1) 101.0 K; (2) 105.6K; (3) 108.8 K; (4) 112.6 K; (5) 114.8 K; (6) 115.9 K; (7) 116.6 K; (8) 117.6 K; (9) 118.1. Reprinted from THOMY and DUVAL, ref. 6.

Fig. 1 shows in more detail a small portion of isotherms near the submonolayer region, i.e. the first layer step, for Xe on graphite from the work of THOMY and DUVAL.[6] It shows that for isotherms at temperatures below some critical temperature there are two vertical risers before completion of the monolayer. By analogy to 3D systems, this suggests the sequence of 2D gas, 2D liquid, and 2D solid phases. Boundaries of two-phase regions are indicated by dashed lines; the 2D triple point and gas-liquid critical point are evident. Heats of adsorption are obtained from pressures for corresponding points on different isotherms. Experiments such as these stimulated a great deal of subsequent work.

Heat capacity measurements, pioneered by DASH and coworkers,[1,33] are a second common thermodynamic measurement. The measurements, made as a function of temperature, show characteristic signatures for various phase transitions. Because of the trajectory in the phase diagram, essentially at constant θ, specific heat measurements are particularly convenient for locating triple lines and critical points and characterizing the nature of transitions. The temperature dependence in a single phase probes the thermal excitations of the phase. For an ideal 2D gas, for example, the specific heat C equals Nk_B independent of T; deviations from this value probe the non-ideality of the gas due to interatomic interactions and the atomic band structure of the atoms as they move in the periodic substrate potential. In the 2D incommensurate solid the low temperature specific heat varies as T^2 in the manner of a 2D Debye solid.

A particularly beautiful example of a specific heat study is illustrated in Fig. 2 which shows contours of C/Nk_B on the θ, T plane for He^4 adsorbed on grafoil.[20,7,33,38]

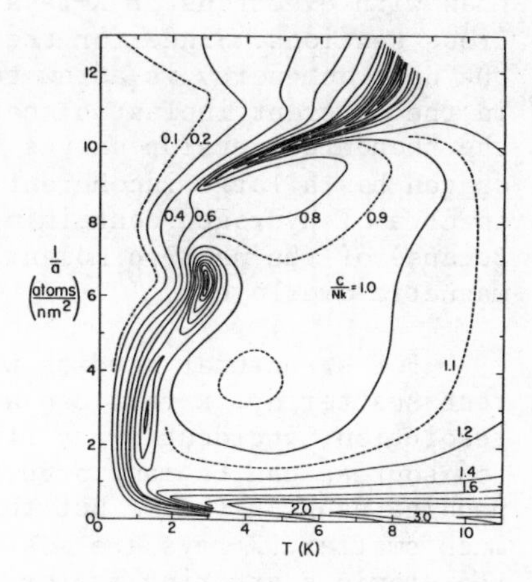

Fig. 2 : Contours of constant specific heat for ^{4}He adsorbed on Grafoil in the density-temperature plane. The upper left region corresponds to an incommensurate solid. For $a^{-1} \simeq 6.2\ nm^{-2}$ and $T \lesssim 3$ K, the atoms form a $\sqrt{3} \times \sqrt{3}$ phase in registry with the hexagonal substrate. The second layer starts forming at approximately 11 atoms nm^{-2}. Reprinted from ELGIN and GOODSTEIN, ref. 38.

From these data and from vapor pressure measurements a very complete thermodynamic characterization of this system has been constructed. This same system has been studied by He atomic beam scattering. Selective adsorption in the scattering gives the energies of the first several vibrational levels for motion normal to the surface. Dispersion in the selective adsorption as a function of the momentum parallel to the surface gives the matrix elements in the band structure of the atoms moving in the periodic potential along the surface. It is then possible to determine the gas-substrate potential and to calculate the specific heat of the 2D gas phase. The results agree well with the data of Fig. 2.

II.2. Structural measurements

From thermodynamic measurements the existence and several properties of the monolayer phases can be deduced. Verification of these identifications and much additional information is available from techniques giving micrococscopic structural information. Scattering experiments using atoms, neutrons, x-rays and electrons have recently given a wealth of such information. Low energy electron diffraction is the most common and historically the earliest of these and we will discuss it in most detail. However we begin with a brief account of neutron[17] and x-ray diffraction[18] studies which, like the thermodynamic measurements, must use samples of high specific area.

As for bulk studies, thermal neutron diffraction has many desirable features. X-ray and electron diffraction determine the pair correlation function, $G(\underline{r})$, whereas with thermal neutrons the dynamical pair correlation function $G(\underline{r},t)$ can be determined. For structural studies several light atoms may be located more readily than with electrons or X-rays because of their nuclear coherent cross sections. Since,for the appropriate de Broglie wavelengths the neutron energy is a few tens of meV, the energy lost or gained in the coherent inelastic scattering can be resolved readily and the phonon dispersion curves $\hbar\omega$ vs $\underline{q}$ can be mapped directly. Hydrogen has a large incoherent cross section so the vibrational spectra of hydrogen containing adsorbed molecules can be studied.[39] Because of the neutron moment there is diffraction from ordered magnetic overlayers.[40]

For structural studies which depend only on the elastic coherent scattering, x-rays can also be used. For similar experimental resolution, incident x-ray fluxes using rotating anode or synchroton sources can be many orders of magnitude larger than the corresponding neutron fluxes but the illuminated volume of the sample is much smaller. X-rays are most suitable for high Z elements since the atomic scattering factor is proportional to Z^2. X-ray experiments give neither inelastic nor magnetic data. Information about thermal vibrations is available from the Debye-Waller factors.

Neutron and x-ray scattering cross sections are small and their diffraction is not "surface sensitive". The experiments require substrates of high specific area which limits the applicability. The common substrate has been some form of graphite. The use of such substrates has two important consequences.

First the sample consists of basal plane flakes with a rather narrow angular distribution of their c-axes but with random orientation about this axis. The reciprocal lattice of the 2D overlayer consists of reciprocal lattice rods. (There is no periodicity normal to the plane and so there is no third Laue condition.) The diffraction geometry for two of the flakes is illustrated in Fig. 3.

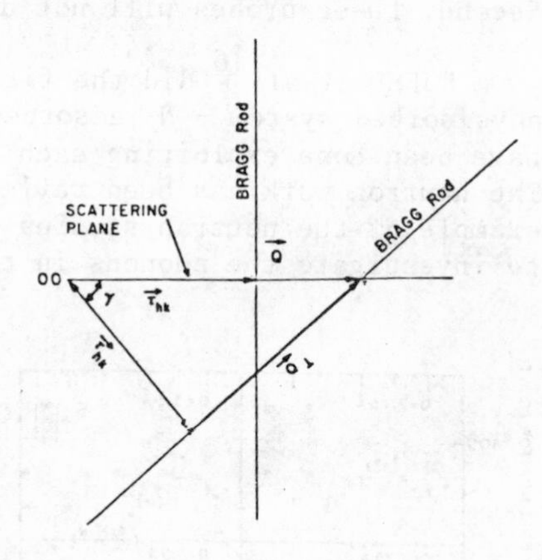

Fig. 3 : Diffraction geometry in reciprocal space for two flakes, one oriented parallel to the scattering plane and the other tilted through an angle γ.

The scattering plane is parallel with the nominal basal plane of the flakes. $\underline{Q}$ is the diffraction vector equal to $\underline{k} - k_o$ where $\underline{k}$ and k_o are wave vectors of the diffracted and incident beams respectively. For a flake whose c axis is inclined by an angle γ, the (10) reflection will occur at $\underline{Q} = \underline{G}_{h,k}/\cos\gamma$. This gives a diffraction peak which has a saw-tooth shape. The sharp rise with increasing $\underline{S}$ or scattering angle is governed by the instrumental resolution and the physical broadening of the reflection from a typical flake. The physical broadening is primarily due to the finite coherence length, L, of the overlayer. The slower fall at larger angles is governed by the distribution in tilt angles, $P(\gamma)$. The observed line shape is then fit by model parameters giving information about L, $P(\gamma)$, and the overlayer lattice parameter.

A second consequence of a high specific area sample is a large substrate background scattering. Fortunately, the basal plane lattice parameter of graphite is smaller than that of the inert gas overlayer so that the desired overlayer diffraction appears inside the nearest (hk0) substrate reflection. There remains background from the (002) reflections from those substrate flakes whose c-axis happens to lie near the scattering plane and from diffuse scattering from imperfections. For x-rays there is also a background from the thermal diffuse scattering which in a neutron experiment could be eliminated by energy analysis. The background problems can be reduced by using the most uniform substrates and best instrumental resolution so that a narrower signal competes with background in a smaller window.

The small scattering cross sections, while forcing the use of high specific area samples, provide at least two advantages compared to electron diffraction. First, the experiments can be made at high 3D gas pressures making more of the phase diagram accessible. Second, these probes will not disturb the adsorbed film.

KJEMS et al.[16] did the first neutron diffraction study of a physisorbed system - N_2 adsorbed on grafoil. Since then experiments have been done exploiting each of the advantages mentioned above. The neutron work has been reviewed by NIELSEN.[17] Here we show one example of the neutron studies illustrating their unique ability to investigate the phonons in the 2D monolayer solid.

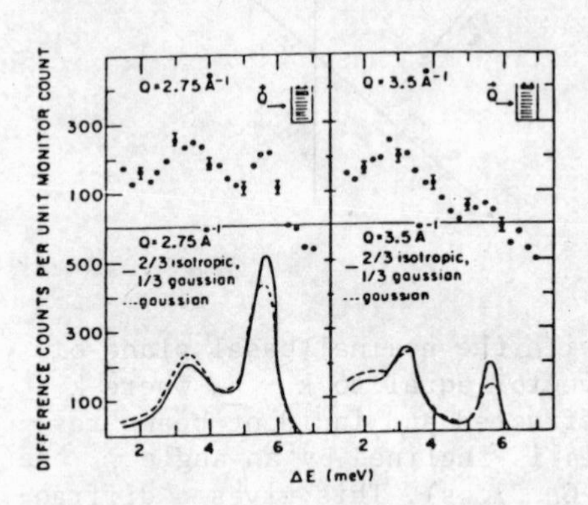

Fig. 4 : Inelastic neutron spectra from Ar monolayers on Grafoil. T = 5 K. Q is parallel to the plane of the Grafoil discs. Top panels are the data ; bottom panels are computed spectra. From ref. 39.

Fig. 4 illustrates the scattering from the incommensurate Ar solid monolayer as a function of energy transfer at two fixed momentum transfers.[39] Q lies in the nominal basal plane. For the powder average, one measures ΔE as a function of the magnitude of |q| rather than phonon wave vector q. The bottom panels are the theoretical expectations from calculations of the lattice dynamics of the monolayer using Lennard-Jones interaction parameters appropriate for Ar. The peaks at 3 meV and 5.5 meV are due to the longitudinal and transverse zone-boundary phonons. For diffraction geometry with Q normal to the c-axis one sees the Einstein-like oscillator modes with vibrations normal to the surface. The effects of coupling between the vibrations of the Ar and the graphite, not included in these calculations, have been considered in ref. 41.

As an example of structural studies we cite the x-ray diffraction results of HAMMONDS et al.[42] for Xe adsorption on ZYX exfoliated graphite and show diffraction angular profiles in Fig. 5.

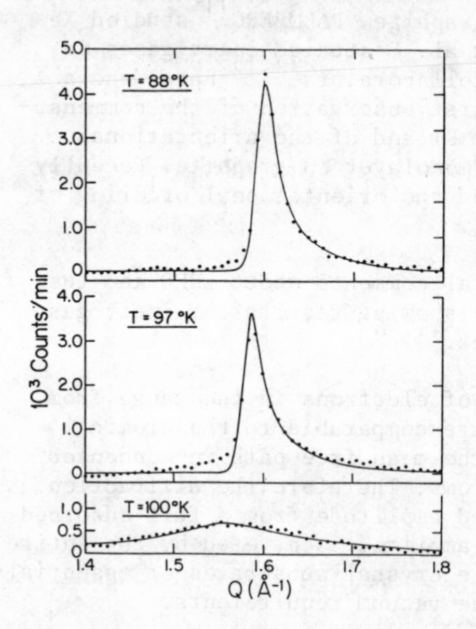

Fig. 5 : X-ray scans of the Xe(10) peak region with the graphite background subtracted. The Xe coverage is fixed at 0.72 monolayers. The solid lines are theoretical fits to the data. From ref. 42.

These panels show the Xe(10) diffraction profiles at three temperatures near the 2D triple point. The data are for a fixed total coverage of 0.72 monolayers. The lower panel shows the liquid diffraction with a Lorentzian line shape. On cooling to 97 K there is an abrupt change to a solid diffraction line shape. Xe forms a hexagonal close packed incommensurate 2D solid phase at these temperatures. Such a first order transition is seen for all smaller total coverages. For higher total coverage one sees either mixed liquid and solid or all solid diffraction confirming the existence of the triple point in the phase diagram obtained by THOMY and DUVAL.[6] The solid diffraction peaks were theoretically fit assuming an average particle size of 500 Å and a mosaic spread of 28°. The Xe interatomic separation at 80 K is 4.538 ± 002 Å in agreement with more precise than earlier measurement by VENABLES et al.[43] using high energy electron diffraction. It was shown further that the Xe second layer atoms sit in the centers of the triangles formed by first layer atoms, as would be expected.

Though beyond the limited scope of this discussion, the Kr commensurate-incommensurate transition has been studied in considerable detail using x-rays from a synchrotron source which allows considerably improved instrumental resolution.[44]

The earliest direct structural information about physisorbed films was from low-energy electron diffraction experiments. The pioneering work was by LANDER and MORRISON[8] who surveyed a variety of gases adsorbed on basal plane graphite. PALMBERG[11] studied Xe adsorption on Pd(100). CHESTERS et al.[12] studied inert gases on several metal surfaces. FAIN and collaborators[14,15] have done a number of studies including the first observation of the commensurate-incommensurate transition for Kr and of the orientational ordering of the incommensurate Ar monolayer on graphite. Recently CALISTI and SUZANNE[34] have observed the orientational ordering of Ne on graphite.

Here we first make some general comments about LEED and then later illustrate its capability by showing results for inert gas adsorption on smooth metal surfaces.[45,46]

The scattering cross section of electrons in the range from a few tens to several hundred eV are comparable to the atomic geometrical cross section and so the mean free path in condensed matter is the order of a few Angstroms. Therefore, the diffraction is surface sensitive; the scattered amplitude from a full adsorbed monolayer is roughly equal to the amplitude scattered by the entire substrate. This allows using single crystal substrates of essentially any material compatible with the vacuum requirements.

Very importantly, low energy electron diffraction is a powerful tool to characterize the substrate prior to adsorption. With present diffractometers the substrate diffraction profile allows the determination of step densities[47] and other imperfections[48] and the extent of the long range order if it is less than about 1000 Å. The same apparatus allows for Auger analysis of surface composition and relative work function measurements.

Because of the strong interactions, multiple scattering or dynamic diffraction effects are important, making a complete LEED analysis complicated in general. This, however, is generally not a serious problem in physisorption studies. First, there is very little multiple scattering within the monolayer because the atomic scattering factor is peaked in the forward direction. Second, determination of the lateral periodicity depends only on the conservation of momentum to within a surface reciprocal lattice vector, no matter how complicated the scattering may be. Dynamic scattering is a more serious problem when determining the structure normal to the surface or when dealing with structures such as adsorbed molecules with more than one atom per surface unit cell.

Presently lattice constants parallel to the surface can be determined to somewhat better than 0.01 Å in careful experiments and the overlayer substrate spacing can be determined to about 0.1 Å.[13,49]

Since generally LEED is used in nearly backscattering geometry, there is a large component of the diffraction vector normal to the surface. Thus LEED is sensitive primarily to the thermal vibrations perpendicular to the surface. The Debye-Waller factor and thermal diffuse scattering are easily observed, but dynamic diffraction and the finite penetration make the lattice dynamic information available from LEED considerably less complete and precise than that from neutron diffraction.

The same strong interactions which limit the mean free path can also cause desorption by the incident beam, particularly for the lighter adsorbates.[50] One avoids serious difficulties either by using very small incident currents and correspondingly sensitive detectors or by doing experiments at 3D gas pressures and temperatures where the beam desorption is negligible compared to the thermal desorption rate at equilibrium. A limitation of LEED is the maximum 3D pressure which can be tolerated because of electron scattering in the gas phase. This limit can be minimized by using a collimated gas flux incident on the sample.[49] Still, the practical limit to the effective pressure is less than perhaps 10^{-3} torr, and generally speaking this limits studies to the solid and dilute 2D gas phases below the triple and critical points. In our view this is the most serious limitation of the technique.

II.3. Inert Gases on Ag(111)

As an example of the use of the technique and the information which can be obtained we discuss a LEED study of the inert gases on the close-packed Ag(111) surface. Later we will use these results to discuss the intralayer interactions between the adatoms. Though the studies have been done for Xe, Kr and Ar, we illustrate them mainly with the Xe data.[13,49,51]

In these experiments the scattered electrons were detected and energy analyzed in a small aperture moveable Faraday collector. A collimated gas beam directed at the sample was used to expose the surface either to reproducible doses or to an accurately constant flux; gas atoms either not sticking or thermally desorbed were cryopumped away before they could return to the surface.

The close-packed (111) surface of the fcc Ag substrate gives a hexagonal array of diffraction beams. The substrate diffraction showed that the average separation between surface steps was greater than 1000 Å. When the surface was at temperatures between 10 and 50 K and was exposed to small Xe gas doses, a radially sharp ring of diffracted intensity developed. This indicates the growth of islands of incommensurate 2D solid with random azimuthal orientation and with long range order within the islands extending at least 500 Å. If this film is annealed at about 55 K the ring coalesces into a hexagonal array aligned with the substrate beams. The alignment following this recrystallization is thought to be due to surface steps which reflect the substrate orientation and is not an intrinsic property. This is suggested by the differences in annealing kinetics on different substrate preparations. The alignment is in fact 30° away from that which would be expected from registry forces. In all the experiments there is no evidence of registry forces, so the adsorption appears to be on a smooth uniform dielectric continuum. Fig. 6 illustrates the diffraction pattern and the angular profile of the diffracted intensity.

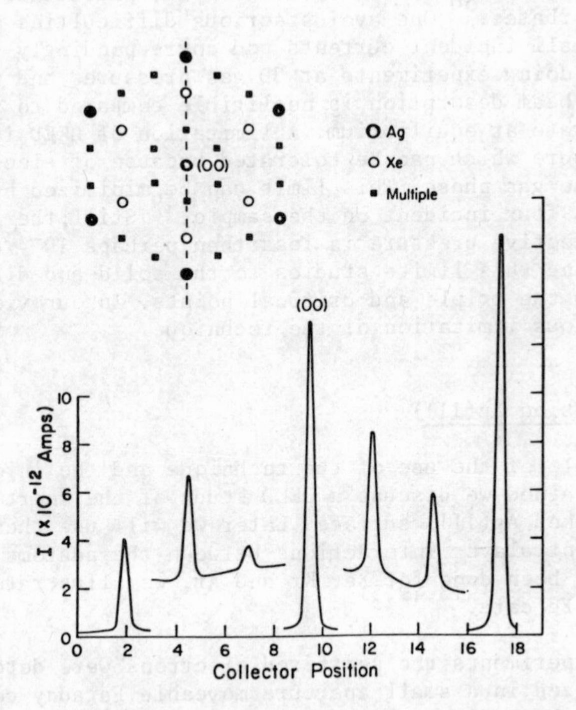

Fig. 6 : At the top is a schematic representation of a segment of the diffraction pattern from Xe covered Ag. Shown are the (00) and the Ag and Xe first order diffraction beams along with multiple scattering beams which arise from the sequential scattering from Ag and Xe. At the bottom are data showing the current measured by the Faraday collector as it moved along the dashed line shown in the diffraction pattern. The data are for a standard unconstrained monolayer at T = 25K and an electron energy of 468 e V. The Xe and multiple scattering spot intensities have been enlarged by a factor of ten.

The overlayer-substrate separation L and its mean square vibrational amplitude are determined from the interference between the overlayer and substrate scattered amplitudes. This interference is clearly seen in the ratio of the specular diffracted intensities from the clean and covered surfaces. The result is verified by dynamic LEED calculations. For Xe, $L_{Xe-Ag} = 3.45 \pm 0.1$Å (the uncertainty is primarily due to lack of knowledge of the inner potential). This can be compared to the arithmetic mean of the spacing in the two bulk strucutres of 3.61Å. If the atoms are considered hard spheres the Xe separations would be at 3.20 Å when it was in the threefold hollow, at 3.31 Å at the saddle point and at 3.61 Å on top of a substrate atom. The vibration amplitude is $\langle u_z^2 \rangle /T = 1.83 \pm 0.3 \times 10^{-4}$ Å^2/K, compared to the corresponding values of 6.6×10^{-5} Å^2/K for the clean silver surface and 9.5×10^{-4}Å^2/K for bulk Xe.

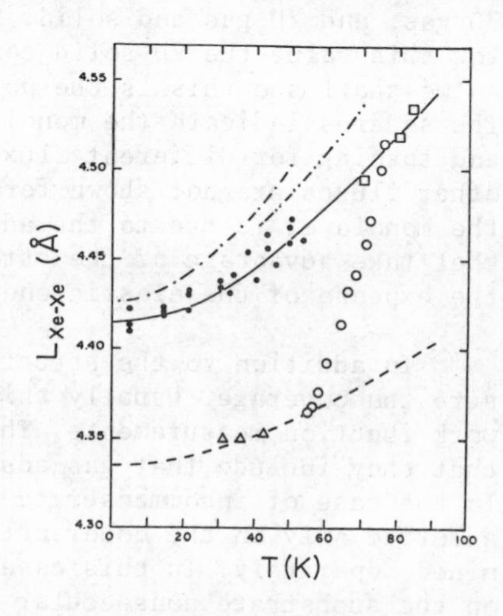

Fig. 7 : The interatomic spacing is shown as a function of temperature for : unconstrained monolayers (filled-in circles) a monolayer in equilibrium with a constant flux (open circles), the initial appearance of monolayers at various fluxes (squares), and a 20 monolayer thick film (triangles). The dashed line is L_{Xe-Xe} for bulk Xe. The (.-.-.) line is the quasi-harmonic approximation; the (..-..-..) line is the cell model approximation.

Relative interatomic lateral spacings within the overlayer solid phase can be determined to within 0.008Å by precise comparison of angular profiles of the diffraction peaks. These relative measurments are put on an absolute scale with essentially the same precision by comparing them with angular profiles from bulk inert gas crystals subsequently grown on the surface and measured under identical conditions. The results for the Xe-Xe spacing as a function of temperature are shown in Fig. 7. Two types of experiments are shown.

The first, shown by the solid circles, are data for what we refer to as an unconstrained layer. These measurements are made for coverages less than a monolayer and with no 3D gas present. The adsorbate remains on the surface at these temperatures only because of its near infinite residence time. Under these conditions the spreading pressure is negligible. For comparison the dashed line is the Xe-Xe spacing in 3D bulk crystals. Near 0 K the monolayer spacing is expanded by 1.9 % from the bulk value. (Incidentally the Ag-Ag spacing is 2.878 and the $\sqrt{3} \times \sqrt{3}$ registered phase spacing would be 4.984. The misfit is ≃ 11 %.)

In the second type of experiment, the lattice parameter is measured as a function of temperature while a constant flux of gas is directed at the surface. This is essentially an equilibrium experiment done at constant pressure.[52] The open circles are for a particular flux corresponding to an effective 3D gas pressure of 3×10^{-8} torr. Above the data point at 75K there is no 2D solid phase. The highest data point at 75K represents the coexistence of 3D gas, and 2D gas and solid. As the temperature is decreased below this value the 2D solid contracts rapidly to the bulk spacing. As we shall see this is the point where the bilayer solid forms. The squares indicate the monolayer solid condensation temperature and spacing for different fluxes. The remaining data for these other fluxes are not shown for clarity. The rapid contraction of the monolayer is due to the addition of atoms in the monolayer that take advantage of the strong substrate holding potential at the expense of the elastic energy to compress the film.

In addition to the structural information one wants to measure the coverage. Usually this is done with Auger spectroscopy or work function measurements. Those methods have the disadvantage that they include that gas adsorbed at heterogenieties and so forth. In the case of incommensurate overlayers, the coverage of that gas adsorbed only on the coherently diffracting substrate can be determined separately. In this case, the only effect of the adsorbate on the susbstrate nonspecular diffracted beams is to attenuate them. This is shown in Fig. 8 where the Ag(10) diffracted beam intensity is measured as a function of temperature while a fixed 3D gas flux is directed at the sample. The monolayer, bilayer and multilayer condensations are evident; condensation of further layers can not be resolved in these experiments. From these data we know the temperature at which this flux of gas is in equilibrium with the bulk inert gas crystal and thus we know the effective pressure corresponding to this flux from the 3D phase diagram.

The effect of the Ag Debye-Waller factor can be removed from the data using the results of a separate measurement. Incidentally it is interesting that the overlayer does not measurably change the Debye-Waller factor of the Ag surface. The attenuation of the substrate beam can be calibrated in terms of coverage from the size of the monolayer step and the lateral spacing measurement. An example of the resulting coverage measurement as a function of temperature at fixed pressure is shown in the upper panel of Fig.9.

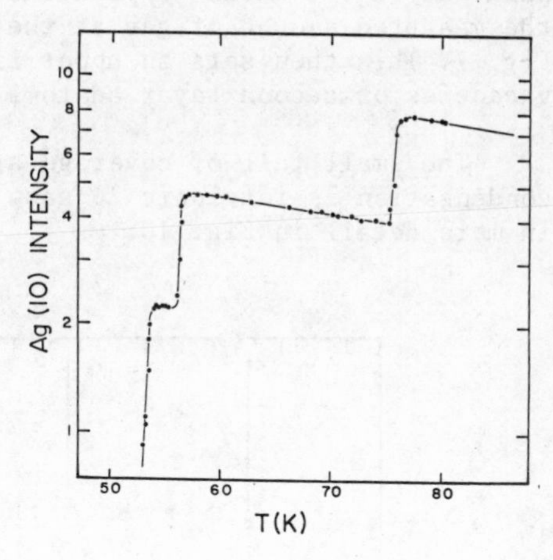

Fig. 8 : The Ag(10) intensity is shown for a fixed flux of gas ($P_{eff} = 3 \times 10^{-8}$ Torr) as the temperature is reduced.

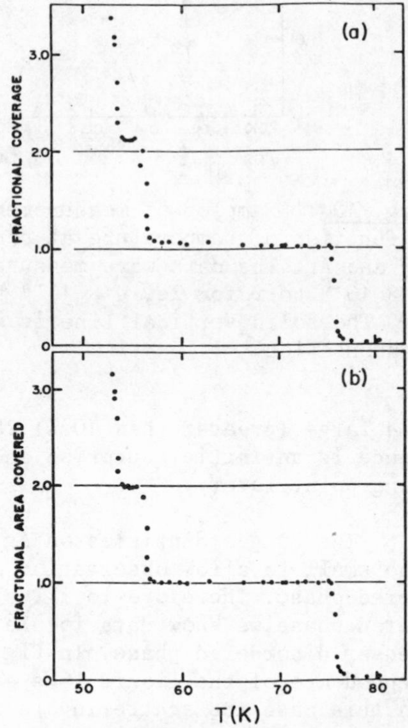

Fig. 9 : (a) Fractional Xe coverage calculated from the data in Fig. 8. (b) The fractional area covered by the Xe in (a), found by using the lattice constant from Fig. 6.

The slope of the first plateau is due to the thermal contraction of the monolayer mentioned above. This is demonstrated in the lower panel of Fig. 9 where we plot the area which would be occupied by the measured amount of gas at the interatomic separation shown in Fig. 7. This then sets an upper limit to the concentration of vacancies or second layer adatoms of less than 1%.

The small tail of coverage at temperatures above the monolayer condensation is intrinsic 2D gas. We show this part of the data in more detail in Fig. 10.

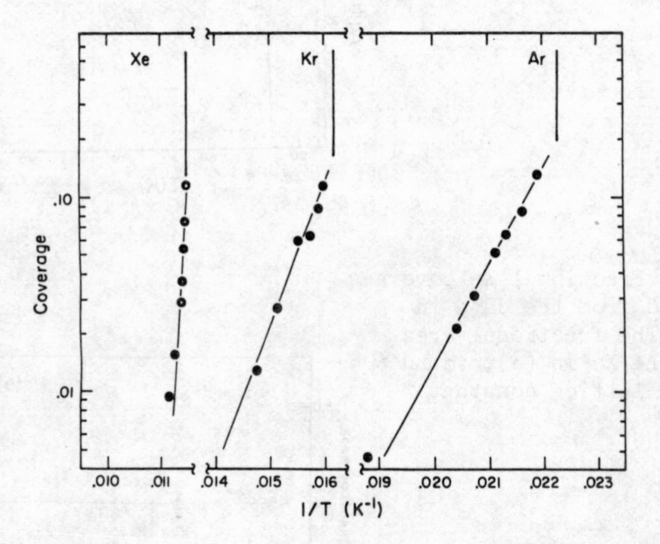

Fig. 10 : Examples of measurements of disordered phase coverage as a function of temperature at constant 3D gas flux are shown for Xe, Kr and Ar. The data were measured for effective pressures of 3×10^{-6} torr for Xe, 9×10^{-6} torr for Kr and 13×10^{-6} torr for Ar. The solid vertical line is at the temperature of monolayer condensation.

The large (greater than 10 %) 2D gas density at monolayer coexistence is an initial surprise and we will return to this interesting point later.

The 2D gas densities on Ag at the accessible temperatures are too small to allow observation of the scattering from the disordered phase. Therefore to illustrate that scattering from a disordered phase,we show data for Xe on Pd(100) where there is a much denser disordered phase. In Fig. 11 we show portions of the angular dependence of the liquid-like scattering for various Xe coverages. In this case the scattering is suggestive of a repulsive fluid.

As coverage is increased the integrated intensity rises, the ring moves outward as the average interatomic separation decreases and the peak narrows as there is less free area. Xe on Pd(100) is much more complex than the adsorpiton on Ag(111) and as yet is not understood.

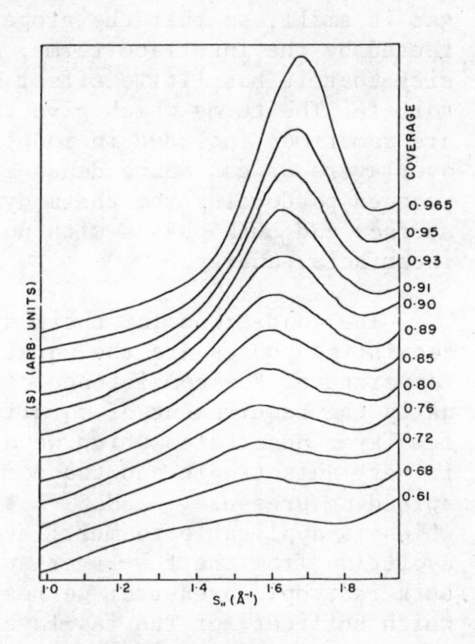

Fig. 11 : Radial intensity distribution across the liquid-like diffraction for Xe adsorbed on Pd(100) for various Xe coverages.

III. DERIVED QUANTITIES

The basic thermodynamic data for the physisorbed systems are in the form of coverage as a function of the pressure and temperature of the 3D gas. From these data we can derive heats of adsorption and other thermodynamic functions of state for the adsorbed layers. Some valuable insights into the properties of the layers are obtained simply by use of thermodynamic relations among the heats of adsorption, and we begin our discussion of the derived quantities with a thermodynamic analysis.[53,54] We follow this with an enumeration of mechanisms which contribute to the effective interaction among the adsorbed atoms and the results which are obtained from statistical mechanical calculations with models of the adatom interactions.

III.1. Thermodynamics

Viewed as a single thermodynamic system, the physisorbed system at low 3D gas pressure is an exceedingly nonuniform one, with the adsorbed layers generally having small interparticle spacings in two directions and the 3D gas having large interparticle spacings in three directions. In principle, in attempting to regard the system as composed of coexisting 3D gas and adsorbed phases there are interfacial energies, proportional to the adsorbing area. Such terms give rise to the gas-surface virial coefficients studied extensively by HALSEY and co-workers.[55,56] In our experiments, however, the ratio of the adsorbing surface to the total volume of gas is small, so that the properties of the 3D gas are little affected by the interface terms, and the 3D gas is at such low density that it has little effect on the internal energy of the adsorbate. The terms which give the gas-surface virial coefficients are sometimes included in model calculations[57] by treating dilute overlayers of gas above dense adsorbed layers. However here we proceed to develop the thermodynamic description of coexisting surface and bulk phases with no explicit reference to the (small) interfacial energy.

The conditions for thermodynamic equilibrium[1] between the coexisting phases are the equality of chemical potential and of temperature. For coexistence of the surface phases there is additionally the requirement of equality of spreading pressure. For the few-layer adsorbates which we discuss the total pressure tensor is very anisotropic and the dominant mechanical stress is the spreading pressure. STEELE[3] has presented a thermodynamic formalism which is applicable to multilayers and permits one to trace the evolution from the few-layer spreading pressure coordinate to the bulk isotropic pressure; we restrict ourselves to the formalism which suffices for the few-layer case.

With our description there are three thermodynamic coordinates, the 3D gas pressure p, the spreading pressure ϕ, and the temperature T. It is an instructive preliminary to the presentation of the phase diagrams for the adsorption systems to review the counting procedure for the Gibbs phase rule in this context. Suppose the system has C components in P (bulk and surface) phases. The conditions of chemical potential equality among the phases give $C(P-1)$ equations. To satisfy these equations one has the 3 thermodynamic coordinates and the $P(C-1)$ fractional compositions of the phases. As a result the combined system has

$$F = P(C-1) + 3 - C(P-1) = C - P + 3 \quad (1)$$

degrees of freedom. For the one-component systems which we discuss the number of degrees of freedom is

$$F = 4 - P \quad (C = 1); \quad (2)$$

this is the Gibbs phase rule for the adsorption system.

Therefore, the coexistence of 3D gas with a single surface phase has 2 degrees of freedom in the thermodynamic coordinate space; when projected on a p-T diagram the coexistence region is a plane. The coexistence of 3D gas with two surface phases, a condition which we describe as a 2D phase transition, has one degree of freedom and is represented by a line in the p-T diagram. The coexistence of 3D gas with three surface phases, which amounts to a triple point for the surface phases and a coexistence of a total of 4 phases, is represented by a point in the p-T diagram.

As a matter of convenience in presenting the data so that phase boundaries are approximately linear we use phase diagrams in the form of $\ln p$ vs $1/T$ plots. The basic adjustable parameters in our equilibrium experiments are the pressure of the 3D gas and temperature, and the area is constant. We shall describe later how the spreading pressure at a point in the diagram is evaluated.

The phase diagram constructed from our measurements[49] of the Xe/Ag(111) system is shown in Fig. 12.

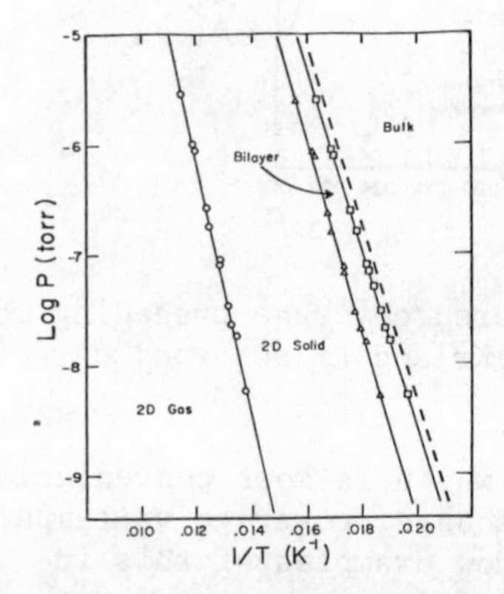

Fig. 12 : Phase diagram of Xe on Ag(111). The dashed line is the vapor pressure of bulk Xe. The squares are our data for the onset of third layer formation.

The (approximately) straight lines mark the monolayer and bilayer condensation risers observed in the experiments such as Fig. 9. The dashed curve line marks the onset of bulk 3D Xe and is constructed from published data for the 3D solid.[58] The path of a constant 3D pressure experiment, constructed from a constant flux experiment such as shown in Fig. 9, is a horizontal line in this diagram. The states of the 2D solid indicated by the open circles in Fig. 7, for decreasing temperature at constant 3D gas pressure, occur along the isobar moving from left to right beginning at the monolayer condensation line. We do not show a 2D melting line in this diagram, since the 2D liquid state has not been observed at the temperatures and pressures of the Xe/Ag(111) experiments. The structure of the phase diagram for gases adsorbed on graphite is more complex than this example, with additional phase boundaries marking commensurate phases.

In addition to the phase boundaries, it is conventional to mark isosteres on the phase diagram.[54] These are lines of constant 2D coverage and are constructed from systems of isobars such as shown in Fig. 9 or from adsorption isotherms such as shown in Fig. 1. We show examples of isosteres for the 2D gas region of the phase diagram in Fig. 13.

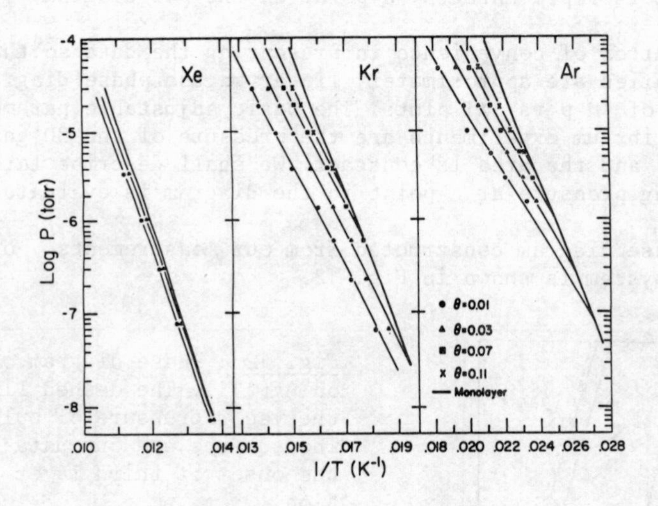

Fig. 13 : Phase diagrams of the disordered coverage preceding monolayer condensation are shown for Xe, Kr and Ar adsorbed on Ag(111).

Another presentation of the data, which is most convenient for the 2D gas state, is a family of isobars, coverage vs temperature at constant 3D gas pressure. We show examples of this in Fig. 10.

Several heats of adsorption can be defined in terms of slopes of these lines in the ln p vs 1/T diagram.[1,54] The latent heat of adsorption is the slope of the coexistence line (2D phase transition) :

$$q_i = k_B \partial ln p/\partial(1/T)\big|_{i^{th}\ \mathrm{coex.}} ; \qquad (3)$$

it is sometimes termed the "integral heat of adsorption". The isosteric heat is the slope of a line of constant coverage :

$$q_{st} = - k_B \, \partial \ln p/\partial(1/T)\big|_n ; \tag{4}$$

it is sometimes termed the "differential heat of adsorption". The isobaric heat is the slope of an isobar such as shown in Fig. 9 :

$$\tilde{q} = k_B \, \partial \ln n/\partial(1/T)\big|_p . \tag{5}$$

The heats q_i and q_{st} are approximately temperature independent since they are primarily determined by the holding potential and the internal energy (for dense 2D phases) and are large compared to the thermal energy k_BT. Values for the heats for gases adsorbed on silver, and the corresponding values for incommensurate phases adsorbed on basal plane graphite, are presented in Table 1.

Table 1 : Heats of Adsorption (meV/atom)

		Xe	Kr	Ar
Bulk	q_{bulk}	161	115	82
Ag(111)	q_1	225 ± 5	151 ± 5	99 ± 7
	q_2	173 ± 5	118 ± 4	89 ± 4
	q_{st} (gas)	215 ± 15	110 ± 10	74 ± 7
	$\tilde{q}$ (gas)	1000 ± 100	140 ± 20	100 ± 10
	$q_1 - \varepsilon_{lat}$	168	108	66
Graphite	q_1	239 ± 4	172 ± 2*	119 ± 2
	q_2	≃165	≃120	≃85
	q_{st}	163	118	93

Compiled from data in references 51 and 61

* : commensurate 2D solid.

As a point of methodology, we note that the isosteric heats for the 2D solids on silver were actually constructed from structural data such as are shown in Fig. 7. From the thermal expansion of the lattice constant along the monolayer solid phase boundary, α, and the thermal expansion along an isobar, γ, we evaluate the ratio[49]

$$q_{st}/q_1 = 1/[1 - (\alpha/\gamma)] \tag{6}$$

(where q_1 is the latent heat of adsorption of the monolayer). In this way one bypasses the problem that the temperature range for which points of constant coverage are available in the solid is frequently too small to construct precise isosteres.

There are also several heat capacities which can be defined for the adsorption system. DASH[1] has reviewed the definitions and the thermodynamic relations among them. Generally a measured specific heat includes a desorption term for the conversion of adsorbate to 3D gas.

Thermodynamic calculus. From basic data, spreading pressures for the adsorbed phases and the lateral compressibilities can be constructed, in addition to the heats of adsorption just defined. To show this and to obtain relations among the heats of adsorption, we now develop the thermodynamic calculus of the chemical potentials and their differentials.[59]

For the idealized adsorption system composed of homogeneous one-component bulk and surface phases, the chemical potential is the Gibbs energy per particle. For the 3D gas the chemical potential is then

$$\mu_3 = u - T_s + pv, \tag{7}$$

where u is the internal energy, s the entropy, and v the volume per particle. The differential of the chemical potential is

$$d\mu_3 = - sdT + vdp. \tag{8}$$

In most of the physisorption experiments on few-layer systems the 3D gas is at such low densities that it is nearly ideal, with the equation of state

$$p = (k_B T/v). \tag{9}$$

Combining Eqs (7) to (9) then gives

$$d(\mu_3/T) = h_3 d(1/T) + k_B d\ln p, \tag{10}$$

where h_3 is the enthalpy per particle,

$$h_3 = u + pv. \tag{11}$$

For each adsorbed phase the chemical potential is

$$\mu_x = u_x - Ts_x + (\phi/n_x), \tag{12}$$

where ϕ is the spreading pressure and n_x is the number density per unit area of the adsorbed phase. The differential of the chemical potential is

$$d\mu_x = -s_x dT + (1/n_x)d\phi, \tag{13}$$

and combining this with eq. (12) gives

$$d(\mu_x/T) = h_x d(1/T) + (1/Tn_x)d\phi, \tag{14}$$

where the 2D enthalpy is

$$h_x = u_x + (\phi/n_x). \tag{15}$$

A Legendre transformation of Eq. (13) to

$$d|\mu_x - (\phi/n_x)| = -s_x dT - \phi(1/n_x) \tag{16}$$

shows that there is a Maxwell relation

$$\partial\phi/\partial T|_{n_x} = \partial s_x/\partial(1/n_x)|_T. \tag{17}$$

To use these forms in the definition of the isosteric heat, Eq. (4), we note that at equilibrium between the 3D gas and a single adsorbed phase "x" of density n_x the chemical potentials are equal

$$\mu_3 = \mu_x. \tag{18}$$

The slope of the isostere at this point in the p-T plane is obtained by combining eq. (18) with the differential

$$d\mu_3 = d\mu_x|_{n_x} = \text{const.} \tag{19}$$

The isosteric heat is then expressed in terms of the thermodynamic functions of the coexisting phases as

$$q_{st} = (h_3 - h_x) + (T/n_x)\left(\frac{\partial\phi}{\partial T}\Big|_{n_x}\right) = h_3 - \frac{\partial}{\partial n_x}(n_x u_x)\Big|_T, \tag{20}$$

For the latent heat corresponding to the coexistence line between two adsorbed phases x_1 and x_2 we use the equations

$$d(\mu_B/T) = d(\mu_{x1}/T) = d(\mu_{x2}/T) \tag{21}$$

to eliminate the spreading pressure differential and have

$$q_i = h_3 - [(n_{x1}u_{x1} - n_{x2}u_{x2})/(n_{x1} - n_{x2})]. \tag{22}$$

In one application of this result the coexisting surface phases are 2D solid ($x_1 = 2s$) and 2D gas ($x_2 = 2g$). If the density of 2D gas is much less than the density of the 2D solid, a good approximation to Eq. (22) is

$$q_1 \simeq h_3 - u_{2s}. \tag{23}$$

Then the differance in slopes of the isostere and the monolayer condensation line at the point where the two lines intersect in the $ln p - 1/T$ diagram is

$$q_{st} - q_1 \simeq - n \frac{\partial u}{\partial n}\Big|_T, \tag{24}$$

which gives a measure of the lateral stresses in the solid monolayer.

The formal expression of Eq. (22) also applies to the latent heat of bilayer formation, q_2, where the coexisting surface phases are monolayer and bilayer solid. For adsorbed layers which are not in registry with the substrate, the lateral lattice constants of the coexisting monolayer and bilayer are frequently nearly equal; experimentally they have been indistinguishable.[49,51] If we make the simplifying approximation, then that the bilayer has twice the areal density of the monolayer, the expression for q_2 becomes

$$q_2 \simeq h_3 - (2u_{bi} - u_{mo}). \qquad (25)$$

The spreading pressure of a single surface phase is evaluated from the Gibbs adsorption isotherm relation.[1] This relation is obtained by equating the differentials in Eqs. (10) and (14) at constant T :

$$d\phi = k_B T\, n_x\, d\,ln p \qquad (dT = 0). \qquad (26)$$

SHAW and FAIN[15] evaluated the spreading pressure in the monolayer solid of incommensurate argon on graphite at several area densities by integrating this equation.

The isothermal lateral compressibility of a surface phase is defined by

$$K_T = [n_x \partial\phi/\partial n_x|_T]^{-1}. \qquad (27)$$

From the Gibbs adsorption isotherm relation, Eq. (26), we have

$$k_B T\, \partial\, ln p/\partial n_x|_T = (n_x^2 K_T)^{-1}, \qquad (28)$$

which was used by SHAW and FAIN[60] to evaluate the lateral compressibility for argon on graphite. Another expression is obtained from relating the partial derivatives of three variables :

$$\frac{\partial x}{\partial y}\Big|_z = -\frac{\partial x}{\partial z}\Big|_y \frac{\partial z}{\partial y}\Big|_x. \qquad (29)$$

We use

$$\partial \ln p/\partial n_x|_T = -\frac{\partial \ln p}{\partial(1/T)}\Big|_{n_x} \Big/ \frac{\partial n_x}{\partial(1/T)}\Big|_p \qquad (30)$$

to obtain

$$K_T = (T/q_{st})\, \partial(1/n_x)/\partial T|_p . \qquad (31)$$

For the low temperature monolayer solids observed in the low energy electron diffraction experiments there are very few thermally activated vacancies and only a small density of second layer gas. Therefore, the number density of the solid is given in terms of the nearest-neighbor spacing L by

$$n_x = (2/L^2\sqrt{3}). \qquad (32)$$

The lateral compressibility is then

$$K_T = (\sqrt{3})\, \frac{LT}{q_{st}}\, \frac{\partial L}{\partial T}\Big|_p . \qquad (33)$$

The last factor is the slope of the monolayer lattice constant along an isobar; thus the slope of the line of open circles in Fig. 7 is simply proportional to the lateral compressibility of the monolayer solid.[59]

The lateral compressibility also enters in the comparison of the isosteric and isobaric heats of adsorption.[61] The calculation is analogous to that in Eq. (30) :

$$\tilde{q} = k_B \frac{\partial\, \ln\, n_x}{\partial(1/T)}\Big|_p = -k_B \frac{\partial\, \ln\, p}{\partial(1/T)}\Big|_{n_x} \Big/ \frac{\partial\, \ln\, p}{\partial\, \ln\, n_x}\Big|_T \qquad (34)$$

$$= q_{st}\,(k_B T\, n_x K_T).$$

Values for the heat of adsorption difference $q_{st}-q_1$ and for the isothermal lateral compressibility K_T of several monolayer solids are shown in Figs. 14 and 15.

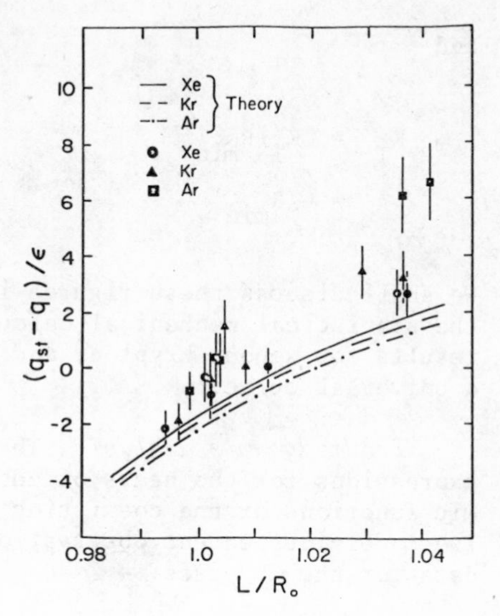

Fig. 14 : The values of the calculated and measured monolayer isosteric heats of adsorption with the measured monolayer latent heats subtracted are shown as a function of lateral spacing. The calculated isosteric heats were derived from the lateral interaction models of BRUCH and PHILLIPS, ref. 64.

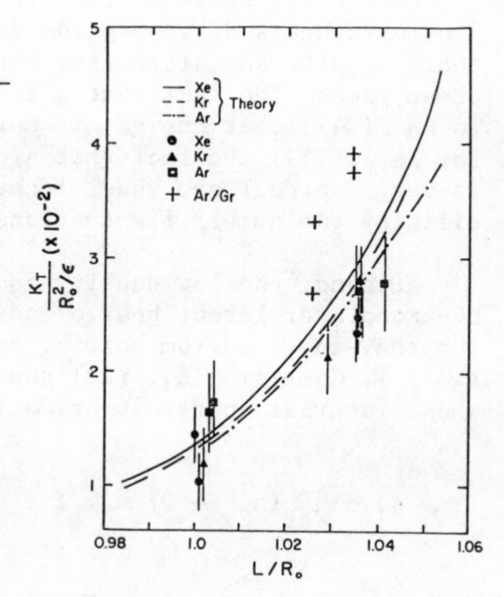

Fig. 15 : Calculated and measured isothermal lateral compressibilities for Xe, Kr and Ar monolayers are shown as a function of lateral spacing. Values are given in terms of corresponding states variables. Calculated values are from BRUCH and PHILLIPS, ref. 64. Ar on graphite data are from SHAW, ref. 60.

To emphasize the basic similarity of the 2D solids of the heavier inert gases, in the absence of significant registry effects, the data are plotted in a form suggested by the law of corresponding states.[62] The depth ε and spacing R_{min} of the 3D pair potential minimum of the inert gas[63] are used to form dimensionless ratios

$$q^* = (q_{st} - q_1)/\varepsilon$$

and

$$K_T^* = \varepsilon K_T/R_{min}^2$$
$$l = L/R_{min} \tag{35}$$

We shall discuss these figures in more detail after introducing the statistical mechanical calculations; here we note that the results for xenon, krypton, and argon on Ag(111) roughly follow a universal curve.

Thermodynamic analysis. The thermodynamic calculus gives expressions for the heats of adsorption in terms of the thermodynamic functions of the coexisting phases. We now use them to make two model-independent observations about the implications of the data for the 2D gases.

First the sense of the difference between the isobaric and isosteric heats of adsorption for the 2D gases in Table 1 shows that significant attractive lateral interactions are present in these gases. The fact that $\tilde{q}$ is greater than q_{st} shows, according to Eq. (34), that the gas is more compressible than an ideal gas. For Xe/Ag(111) the fact that $\tilde{q}/q_{st}$ is large shows that this gas is very nonideal and that, because of the relation between compressibility and number fluctuations, it has large density fluctuations.

Second, the low density isosteric heats of the 2D gases and the monolayer latent heat of adsorption q_1 give consistent values for the single adatom holding energy of Ar and Kr adsorbed on Ag(111). Combining Eq. (23) and Eq. (20) for a low density gas whose internal energy is primarily kinetic energy we have

$$q_1 - q_{st}(n_{2g} \to 0) \simeq k_B T - u_{2sl}(T)$$
$$\simeq - k_B T + \varepsilon_{1at}. \tag{36}$$

The internal energy u_{2sl} is the lateral contribution to the internal energy of the 2D solid and ε_{1at} is the lateral ground state energy of the 2D solid. The difference between these is approximately the thermal energy $2\,k_B T$ of a 2D harmonic solid. In Table 1 we subtracted values of ε_{1at} obtained in model calculations for the 2D solids from q_1.[51,64] However even a 25 % uncertainty in ε_{1at} causes less than 10 % uncertainty in $q_1 - \varepsilon_{1at}$ so that for Kr and Ar we can get consistent and essentially model independent results for the single atom holding energy obtained from q_1 and from the isosteric heat of the low density 2D gas.

III.2. Statistical Mechanics

The interactions of adsorbed atoms. The concepts in the theory of the interactions of physically adsorbed atoms are quite similar to those for the interactions of pairs of closed shell atoms in three dimensions.[65] At large separations, in the absence of permanent multipole moments, there is a van der Waals attraction which arises from the correlations of the fluctuating electric dipoles on the interacting species. The computation of the long range attraction is readily expressed, using perturbation theory, in terms of the dynamic optical response (polarizability or dielectric function) of the interacting species. Hence it can be evaluated with experimental data for the response of the separate species and is a fairly well known part of the total interaction. At short separations, the interaction is primarily an exchange repulsion arising from the overlap of the charge clouds of the species. It is termed the exchange-overlap repulsion or the Slater repulsion and can be calculated in quantum chemistry with Hartree-Fock wave functions. The minimum in the potential energy of two closed shell atoms or in the holding potential of a physisorbed atom is thought to be formed as a balance between the van der Waals attraction and the exchange-overlap repulsion. However a direct accurate computation of the potential minimum is very difficult, even for as small a system as a pair of helium atoms.[66] Consequently the interactions of pairs of inert gas atoms are mostly known from empirical determinations. Estimates of the interactions in physisorbed systems are usually obtained by adding estimates of adsorption-induced processes to the empirically determined pair potentials.

For physisorbed atoms it is convenient to divide the interactions of the adsorbed atoms into terms for the interaction of a single atom with the substrate and terms for the interactions among the adatoms. Processes arising from the interference of one adatom in the interaction of another adatom with the substrate are then included in the effective lateral interactions among adatoms.

The interaction of atoms with the substrate is probed by gas-surface virial coefficient measurements[55] and by atom-surface scattering experiments,[19,20] as well as by analysis of thermodynamic data for multilayer adsorption.[37] In spite of its importance for the understanding of gas-surface kinetics, there are few cases where the one-atom holding potential is known for a wide range of atom-surface separations. For helium adsorbed on basal plane graphite, the holding potential has been constructed from the energy levels observed by the selective adsorption in the scattering of molecular beams from surfaces.[20] For argon adsorbed on silver, the holding potential was calculated with an electron-density formalism by LANG.[67] This is the only self consistent calculation of the adatom-surface complex and of the resulting holding potential and charge densities. The silver was replaced by a jellium model but the electron densities were calculated self-consistently so that the total electron density was not simply a sum of the densities of the interacting species. The minimum potential in this calculation was in good agreement with the holding energy derived from the experiments for Ar/Ag(111); the dipole moment of the adatom-substrate complex was also calculated.

For large atom-surface distances, the interaction is a van der Waals attraction[68] which varies inversely with the cube of the distance until such large distances that relativistic retardation effects become important. For a multilayer adsorption isotherm, then, the pressures at which the higher layers condense should vary as the inverse cube of the layer index. Such considerations give the Singleton-Halsey[37] equation. For the adsorption of Xe, Kr, and Ar on graphite, step-wise condensation has been resolved up to the fifth layer.[5,69] Data follow the Singleton-Halsey equation but the coefficient of the inverse-cube dependence on the layer index is approximately twice as large as the value obtained from the theory of the van der Waals attraction.[70]

Another method of constructing the adatom-substrate holding potential is the summation of pairwise interactions between the adatom and the substrate atoms. This has been used particularly for the inert gases adsorbed on graphite. STEELE[71] showed how to separate the result conveniently into a laterally averaged holding potential and terms with the lateral periodicity of the substrate. For the case of helium on graphite, where molecular beam experiments have given the amplitude of the leading periodic term, the use of a spherically symmetric pair potential between the helium and the carbon atoms apparently underestimates the amplitude of the periodic variation.[20] There is also some uncertainty in this approach about the magnitude of nonpair-additive effects in the atom-substrate energy. A calculation by GERSBACHER and MILFORD[72] based on a model of LUCAS[73] showed substantial cancellations among appreciable higher-order terms, but a later treatment of the Lucas model by RENNE[74] showed a much better convergence of the contributions of higher-order terms.

As already mentioned in the introductory remarks, the largest part of the adatom-adatom interaction energy is the sum of the pair potentials determined for the gases in 3D.[63] The determinations, which lead to values of the pair potential minimum accurate to about 5 % in the depth and to 1 % in the spacing, are based on a combination of information from atom-atom scattering experiments, spectroscopy of inert gas dimers, thermodynamic data for the 3D gases and solids, and the perturbation theory of the van der Waals attraction. The results are expressed in the form of multiparameter potential functions; BARKER reviewed[63] calculations with these functions for the 3D inert gas solids. Later work of AZIZ[75] gave somewhat simpler functional forms. A striking feature of the potential minimum for like pairs of inert gas atoms is that there is a common shape. That is, the potential energy has the form hypothesized in the theory of corresponding states :

$$V(r) = \varepsilon f(r/R_{min}), \tag{37}$$

where members of a series differ only in the energy and length scales, ε and R_{min}.

For the adsorbed inert gases on silver and graphite, the largest correction to the isolated pair potential appears to be an energy term proposed by SINANOǦLU and PITZER[76] and by McLACHLAN.[77] It is similar to the van der Waals attraction in its origin : for the adsorbed pair, at large lateral separations, the images in the substrate of the fluctuating adatom dipoles are included in the perturbation theory of the pair. Thus, there is an idealization of the substrate as a flat dielectric continuum and ideas of electrostatic images are used at length scales approaching atomic dimensions.[78,79] To evaluate the strength coefficients of the interaction, dielectric data for the substrate are used; because the fluctuating dipoles on the inert gas atoms are at quite high frequencies an appreciable part of the substrate response is due to interband transitions. The adatom frequencies are much larger than the plasma frequency of the substrate. In the language of the perturbation theory of intermolecular forces, this substrate-mediated energy can be described as a three-body energy term, where the triple consists of the adatom pair and an extended third body (the substrate). In fact, the leading term in an expansion of the Sinanoǧlu-Pitzer- McLachlan energy in powers of the polarizability of the substrate atoms is the same as that obtained by summing the Axilrod-Teller-Muto triple-dipole energy over triples composed of the adatom pair and a substrate atom.[80]

Next there are terms reflecting the structure of the adatom-substrate complex. There are the electrostatic interactions of the adsorption-induced static dipole moments,[11,12,81] which include depolarization terms arising from the polarizability of the complex.[82] Also, there should be a short-range overlap between the complexes. The only estimate for this is a density functional approximation of FREEMAN[83] for argon on graphite which gave a term that appeared to be consistent with the McLachlan energy at separations near the pair potential minimum.

The interaction of the adatom with the substrate also deforms the substrate lattice. The indirect interaction of adatoms through the elastic distortion of the substrate was formulated by KOHN and LAU[84] and by KAPPUS.[85] The elastic anisotropy of the substrate is included in their treatments. In estimates for xenon, the contribution of this term in the 2D solid is less than the contribution of the interaction of the adsorption dipoles.

For adsorption on metals there is also an oscillatory long range interaction which is similar to the Friedel oscillation in bulk metals.[86] The wave number of the oscillation is 2 k_F where k_F is the fermi wave number of the metal. While this process can be quite important for the lateral interactions of chemisorbed atoms, in a physisorption case where it was estimated it was smaller than the McLachlan energy at intermediate separations and smaller than the interaction energy of the permanent dipoles at large separations.

Even when the 2D solid is incommensurate with the substrate, periodic components of the adatom-substrate interaction can contribute to the energy of the 2D solid. The orientational alignment of the 2D solid relative to substrate axes predicted by NOVACO and McTAGUE[87] arises from such an energy term. When included in calculations of the lateral compressibility, it gives important contributions at lattice constants which are notably different from commensurate values.[51]

As for the 3D inert gas solids, there are non-pair-additive energies among the adsorbate atoms. The only one to be included in the monolayer calculations, and even it is relatively small, is the Axilrod-Teller-Muto[88] triple dipole energy for the correlated fluctuating electric dipole moments of triples of adsorbed atoms.

When the monolayer solid is treated as a solid of atoms in mathematical two dimensions, the effective "in-plane" interaction includes an averaging over atomic vibrations perpendicular to the adsorption plane. For the small root-mean-square vibration amplitudes inferred from Debye-Waller factor data, this has been estimated[89] to be a negligible contribution to the effective interactions. In treatments of the monolayer solid which include the vibrations explicitly[57,90] this term does not occur.

Finally, the ground state energy of the monolayer solid includes the energy of the zero-point motion. For the heavy atoms such as krypton and xenon this is a small part of the total energy. It is a significant term in the ground state energy of monolayer argon and neon and, of course, becomes a very large term for the quantum solids of helium and molecular hydrogen.

Modelling. There are many quantitative similarities among the data for the 2D solids of inert gases adsorbed on silver[49,51] and between these data and the data for the corresponding incommensurate 2D solids adsorbed on graphite.[15,60,36] We have already emphasized the similarities for the adsorption on silver by the graphical presentation of the data in the form of corresponding states plots, Figs. 14 and 15. If the 2D solids of a series of adsorbates have lateral interactions which are of the form of Eq. (37) and quantum effects are small, the thermodynamic functions can be obtained from each other by scaling the thermal energy by ε and the specific area by R^2_{min}. In constructing Figs. 14 and 15 we used the energy and length scales of the BARKER[63] series of 3D inert gas pair potentials; this is appropriate when the effective interaction of an adsorbed pair is dominated by the interaction of the isolated 3D pair. For helium and neon, some departures from the corresponding states scaling are anticipated[62] on the basis of quantum (mass-dependent) effects; however for argon, krypton and xenon the mass effects are small. The data do indeed appear to follow universal curves. Model calculations show[51] that some departures from the universal behavior should be anticipated because of departures of the 3D pair potentials from the common functional form of Eq. (37). Also, the zero point energy for argon and nonscaling behavior of the substrate-dependent interactions cause visible departures from a common behavior on the scale used for Figs. 14 and 15. However the principal conclusion to be drawn from the presentation in Figs. 14 and 15 is that the monolayer solids of Ar, Kr, and Xe on Ag(111) are a family which follow corresponding states scalings probably as well as the 3D solids do.

The heats of adsorption for condensation of incommensurate monolayer and bilayer solids of Ar, Kr, and Xe on basal plane graphite are included in Table 1. They are very similar to the heats of adsorption on silver. Additionally the isosteric heats of the low density 2D gases on graphite, obtained by HALSEY and co-workers[55] are close to the values we infer for the single adatom holding energies on Ag(111). Thus the lateral internal energies of the corresponding 2D solids are quite similar.[91]

The adsorption of Ar on graphite has been modelled[57] with effective interactions similar to those used for the 2D solids on Ag(111). PRICE and VENABLES[54] had previously discussed the incommensurate solids of Kr and Xe on graphite with models which differed from these mainly in the details of evaluating the Sinanoğlu-Pitzer-McLachlan energy. Departures of the lateral compressibility of Ar on graphite from the corresponding states scalings for Ag(111) can largely be rationalized in terms of the Novaco-McTague orientational registry energy of incommensurate solids. Thus the 2D solids on Ag(111) and on graphite can be quantitatively understood with models which are quite similar. Although registry energy terms have quite visible effects in the adsorption of inert gases on graphite, the analysis of the incommensurate monolayer solids shows that for adsorption on graphite as well as for Ag(111) the total lateral energy depends little on the discrete structure of the substrate.

In these comments on the common features of the adsorption of the inert gases on Ag(111) and on basal plane graphite we have anticipated the results of the statistical mechanical evaluation of properties of the adsorbed solids. We have done this to emphasize that there are many experimentally accessible systems for which modelling the role of the substrate by a smooth dielectric continuum is adequate to understand the principal features of the monolayer and bilayer adsorption. Additionally, with the modelling we can make comparisons with the properties of the 3D solids and gain a better appreciation as to which are the surprising new effects in the observations of the adsorption system. We can compute the effects of the change in effective spatial dimension, from 3 to 2, and can estimate what new substrate-induced adatom interactions occur. We also, in the case of the dense 2D gases observed on Ag(111) get an understanding of how large departures from expectations based on experience with the 3D solids can occur.

Once a model of the adatom interactions is specified, the thermal properties of the monolayer and bilayer adsorbates are evaluated using statistical mechanics. The methods are basically the same as for the inert gas systems in 3D.[92] Low density gases are treated by virial series. Low temperature solids are treated by quasiharmonic theory, evaluating the frequencies of harmonic lattice vibrations. The solids at intermediate temperatures are treated with cell-model approximations, including quantum corrections when necessary. For extremely anharmonic solids and for dense disordered phases, Monte Carlo computer simulations are used.[93,94] The domains of application of these methods are determined by finding overlapping regions of validity. The extreme anisotropy of the few-layer adsorbates between directions in the plane and perpendicular to the plane of adsorption leads to some technical problems in the use of quasiharmonic and cell model approximations, but these have been overcome.[57,90]

The unconstrained 2D solids of Ar, Kr, and Xe on Ag(111) have nearest-neighbor distances which are expanded by 1 to 2 % from the values for the 3D solids at the sublimation curve. The expansion has been analysed most extensively for the low temperature solid of Xe/Ag(111).[89] Calculations with the Lennard-Jones (12/6) pair potential give an expansion of 2 % due to dimensionality alone. The expansion depends on the power law of the repulsion in the potential; for power laws of $1/r^9$ and $1/r^{15}$ the expansions are 3 %, and 1.5 % respectively. For Xe the Axilrod-Teller-Muto triple-dipole energy accounts for 10 % of the cohesive energy of the 3D solid. For the monolayer, the relative contribution of this term is greatly reduced, since many fewer triangles per atom can be drawn in a planar than in a 3D lattice. With the full model of the Xe interaction, about half of the expansion in the lattice constant comes from dimensionality in the two-body and three-body sums and half from the contribution of the (substrate-mediated) Sinanoğlu-Pitzer-

McLachlan interaction.

In early work[13] on Xe/Ag(111) the thermal expansion of the monolayer solid was found to be much closer to the thermal expansion of bulk Xe than to that of the Ag substrate. This was used in support of the contention that the discrete structure of the Ag has little effect on the adsorbed layer because the thermal expansion depends sensitively on anharmonic effects in the solid. Calcuations of the thermal expansion of the 2D solid at its sublimation curve have been made with quasi-harmonic and cell model approximations and with Monte Carlo simulations. As a matter of methodology, the thermal expansion for a given potential model can be calculated accurately up to the melting of the monolayer solid.[94] The calculated thermal expansion of the monolayer is larger than for the bulk. More precise measurements show this to be the case for the gases on Ag(111), although some discrepancies between the calculations and the observations remain at the highest temperatures of the data.

Comparisons of the observed and calculated lattice constants for the Xe/Ag(111) are shown in Figure 7. We emphasize that the interaction model was not adjusted to achieve this agreement. As discussed in the enumeration of components of the interactions of the adsorbed atoms, the pair potential of Xe atoms determined in 3D is the largest part of the effective interaction. The substrate-mediated dispersion energy is 15 % of the energy calculated with the 3D pair potentials. The interaction energy of the adsorption-induced dipoles is 8 %. Other thermodynamic functions calculated with this model, and the corresponding models for Kr/Ag(111) and Ar/Ag(111), are shown in Figs. 14 and 15. The energy terms in the solid ground state are shown in Table 2.

The transition from the monolayer solid to the bilayer solid occurs with very small change in the lateral lattice constant; experimentally, no change has been detected at the transition.[49,51] The transition from the nearest-neighbor spacing of the unconstrained monolayer to that of the bulk solids is almost complete by the time the bilayer condenses. The transition from the monolayer to the bilayer solid is now understood in terms of the condensation of the thermally excited second layer gas above a monolayer solid to form a registered second layer of solid. Statistical mechanical calculations[57,90] reproduce the observation that the lattice constant discontinuity at the transition must be very small; however, what would be an observable discontinuity was calculated with an interaction model which had appeared to give a fair account of the Kr/Ag(111) monolayer. The calculations give values for the latent heat differences, $q_2 - q_1$, which agree with the observed differences. The lattice constant of the bilayer is expected to be much closer to the bulk value than the unconstrained monolayer value because the atoms of the bilayer have out-of-plane near-neighbors and

Table 2 : Contributions to the Calculated Lateral Energy

Rare Gas	Xenon	Krypton	Argon
Energy Term (meV/atom)			
Pair Potential	-77.0	-55.1	-39.0
Substrate-Mediated	12.4	8.9	5.7
Permanent Dipole	6.3	2.3	0
Triple Dipole	1.5	0.9	0.5
Zero Point Energy	2.6	3.0	3.6
Total Lateral Energy	-54.2	-40.0	-29.2
0 K Lattice Constant (Å)			
Calculated	4.425	4.075	3.840
Measured	4.415±0.01	4.045±0.01	3.780±0.01 *)
Bulk	4.334	3.992	3.755

*) Derived from isosteric heat measurements

because the effect of substrate-mediated interactions is much smaller for pairs of second layer atoms. However the small lattice constant discontinuity and the fact that in observations the bilayer lattice constant has been indistinguishable (to 0.01 Å) from the bulk value are both sensitive to details of the effective interactions. This can be demonstrated[51] with a lattice gas model used by de OLIVEIRA and GRIFFITHS[95] where approximate analytical results show the dependence of the lattice constant discontinuity on the effective interactions of first and second layer atoms.

We have already used the comparison of the isobaric and isosteric heats for the 2D gas of Xe/Ag(111) to infer that the 2D gas is quite non-ideal. The coexistence of the 2D gas and 2D solid of Xe/Ag(111) has been calculated at 90 K using Monte Carlo simulations of both phases.[61] The density of gas at the coexistence is found to be 14 % of the 2D solid density; this is two orders of magnitude larger than the 0.24 % relative density of gas to solid at the triple point of 3D Xe. It is also one order of magnitude larger than the 2 % relative density calculated at the triple point of a 2D Lennard-Jones (12,6) system.[93,94] Our resolution of this puzzle is shown in Fig. 16 where we have plotted the calculated chemical potential of the 2D gas of Xe/Ag(111) as a function of density at 90 K. At low densities, it varies rapidly with density, but at intermediate densities it varies much more slowly, corresponding to a system with high compressibility and large density fluctuations. The severe clustering is shown in Fig. 17.[61] In that range, relative small changes in the chemical potential of the 2D solid give large changes in the coexistence density of the 2D gas. The difference from the results for the 2D Lennard-Jones system occurs as a consequence of the substrate-mediated repulsions between adatoms which cause a raising of the energy of the solid by 10 to 15 %. It is this increase which stabilizes the dense 2D gas against condensation and permits the occurrence of a low temperature gas with large clustering effects.

Combining an analysis of the 2D gas states for Kr/Ag(111) and Ar/Ag(111) with attempted fits to the properties of their 2D solids leads to a proposal of a soft but appreciable repulsion between adatoms which may result from the overlap of the adatom-substrate complexes.[61] Preliminary data for the adsorption of Xe on Pd(100) have also given evidence that large positive contributions to the adatom-adatom energy, which vary slowly with interparticle spacing[11,96] may occur for inert gases adsorbed on metals. The identification of such a process is still largely speculation, but there are indications that in some cases the van der Waals attraction between adsorbed atoms can be severely disrupted.

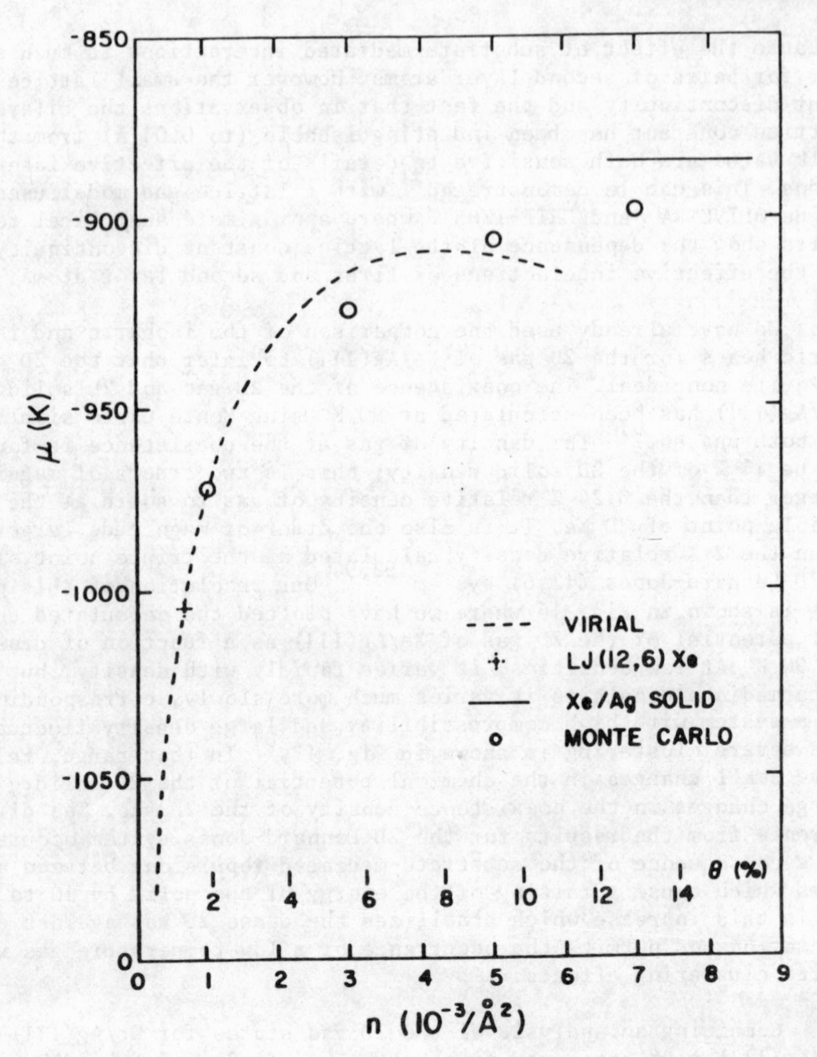

Fig. 16 : The chemical potential calculated for the 2D gas of Xe/Ag(111) is shown as a function of density (and percent of the 2D solid density) at 90 K. The values derived from the computer simulation are shown as circles; the chemical potential calculated with the truncated virial series is shown as the dashed line. The horizontal line at −885 K denotes the value of the chemical potential of the 2D solid at the approximate spreading pressure for coexistence with the 2D gas. The values of the density and chemical potential for the coexistence of the 2D Lennard-Jones system at 90 K with Xe parameters (ε = 230 K and R_{min} = 4.5 Å) are shown by dotted lines.

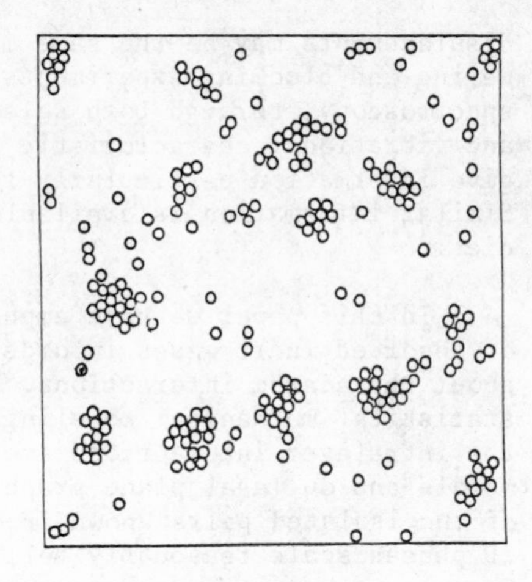

Fig. 17 : Clustering in the 2D gas of Xe/Ag(111) at 90 K. The particle positions are shown after 4000 Monte Carlo moves per particle in a gas of density 0.005/Å^2. There are 224 particles in the basic unit of the periodically extended system. On average, in this configuration each particle has nearly 3 nearest neighbors.

IV. SUMMARY AND DISCUSSION

The development of highly uniform substrates and of techniques allowing investigation of single crystal surfaces has given a large base of detailed thermodynamic and precise structural data on physisorbed systems. Three aspects of the data are noteworthy : the substrates have minimized the effects of heterogeneities and finite coherence length so that contact can be made to theories of the intrinsic properties. Second, the precision of the data has been significantly improved in recent years. This is important since many aspects of the physics manifest themselves in small ways in the structural and thermodynamic properties. Finally, there are data for several adsorbates on the same substrates so that systematic trends can be evaluated. So far the most extensive data are for the inert gases on basal plane graphite and on metals.

One expects continued improvement in both the substrate perfection and the experimental resolution of the electron and x-ray experiments.[44] Such improvements may, for example, elucidate the nature of the long-range order and melting in two dimensions. In addition one expects additional techniques to be employed in physisorption studies. We note the beautiful grazing-incidence electron microscopy experiments with dark-field illumination using overlayer diffracted beams which have been so interesting in studying reconstructed semiconductor surfaces.[97] Such experiments might resolve the extended defects in the film which presumably play an important role in melting. Atom diffraction from the overlayer with its absolute surface sensitivity and suitability to investigate lighter adsorbates is beginning to be used on physisorption systems. It also has the capability of inelastic studies. Rutherford backscattering may give detailed information particularly very near the commensurate-incommensurate transitions where displacements may be the same magnitude as the shadow cone in channeling and blocking experiments.[98] High resolution electron loss spectroscopy, through both selection rules to identify the symmetry and vibrational characteristic losses to identify the bonds should give information particularly for adsorbed molecular species.[99] Similar information is available in a few cases from neutron studies.[17]

In this paper we have emphasized the incommensurate structures of adsorbed inert gases in order to review what can be learned about the adatom interactions. From the detailed data and from statistical mechanical modeling, a rather satisfactory picture of the intralayer interactions emerges for the adsorption on simple metals and on basal plane graphite. The primary interaction is that of the isolated pairs known from 3D physics. Indeed, the inert gas 2D phases scale reasonably well with corresponding states ideas. There are, however, important modifications, the main one being the Sinanoğlu-Pitzer-McLachlan substrate mediated interaction. This repulsive interaction reduces the depth of the attractive well considerably and is much more important than the permanent dipolar repulsions and other contributions. These modifications of the adatom-adatom interactions have the consequence that the 2D gas density prior to crystallization is very large. This very interesting gas phase is exceedingly non-ideal with very large density fluctuations. Its counterpart in 3D would be a gas near its critical point.

It is perhaps surprising that, for Ag(111) and graphite substrates, the 2D phase properties can be accounted for quite satisfactorily with model potentials which take no account of the detailed electronic structure of the adatom-substrate complex or of direct overlap interactions between such complexes. Indeed there is evidence that the intralayer potentials are profoundly different for the adsorption on Pd(100) where there is a large d-band density of states at the Fermi level and where "chemistry" is obviously important. Such cases are not yet understood.

The details of the adatom-substrate interaction at distances near the equilibrium separation are more difficult to understand.

For He and H some details of this potential and its corrugation along the surface are known from selective adsorption in atom scattering. For other adsorbates one infers the depth of the potential well from thermodynamic data; there are a few measurements of the vertical position of the potential minimum. The consequences of the corrugation of the potential are evident in the commensurate-incommensurate transitions and the orientational ordering of overlayers, but no quantitative analysis of these data to give the potentials have been made.

A reasonable description of the adatom-graphite potential can be made by summing pairwise interactions with substrate atoms. This, however, does not give the magnitudes of the corrugations correctly, and anisotropic interactions have been invoked to explain the atomic scattering data. Further evidence that this and similar approaches are inadequate is the permanent dipole moment accompanying inert gas adsorption. This latter clearly indicates a distortion of charge density from the unperturbed atom plus substrate density.

The only quantum mechanical fully self-consistent calculation of the adatom substrate potential and the distortion of the charge density is that by Lang. This calculation uses a density functional formalism and a local density approximation for the exchange and correlation and it treats the adsorption on "jellium" as is appropriate for simple metals. As such the calculation cannot recover the long-range van der Waals interaction which is inherently nonlocal or the corrugation of the potential; however, it does give a good account of the well depth, spacing, permanent dipole moment, and core level binding energy shifts. Electron density-difference contours for equilibrium separations show increased density in the region between the atom and the metal," .. an electron in the valence shell shows a preference for being on the metal side the exchange-correlation hole that forms around it is more effective in lowering its energy".[67] This calculation then gives some conceptual insight and hopefully will stimulate further calculations. The results also may provide a starting point for more approximate estimates of the effect of the charge distortion on the intralayer adatom-adatom interactions.

It is evident that there has been a very constructive interplay between theory and experiment. With increasing experimental breadth and precision and with increasing theoretical sophistication it is likely to continue.

Work supported, in part, by the National Science Foundation through grants DMR-7920400 and DMR-7809429.

REFERENCES

1 . DASH J.G., in *Films on Solid Surfaces*, Academic Press, New York, 1975
2 . HOBSON J.P., CRC Critical Reviews in Solid State Sciences 4 (1974) 221-245
3 . STEELE W.A, in *The Interaction of Gases with Solid Surfaces*, Pergamon Press, New York, 1974
4 . THOMY A. and DUVAL X., J. Chim. Phys. 66 (1969) 1966-1973
5 . THOMY A. and DUVAL X., J. Chim. Phys. 66 (1970) 286-290
6 . THOMY A. and DUVAL X., J. Chim. Phys. 67 (1970) 1101-1110
7 . BRETZ M. and DASH J.G., Phys. Rev. Lett. 26 (1971) 963-965
8 . LANDER J.J. and MORRISON J., Surf. Sci. Vol. 6 (1967) 1-32
9 . EHRLICH G., HEYNE H. and KIRK C.F., in *The Structure and Chemistry of Solid Surfaces*, G.A. Somorjai ed., Wiley, New York, 1969,p 491
10. ENGEL T. and GOMER R., J. Chem. Phys. 52 (1970) 5572-5580
11. PALMBERG P.W., Surf. Sci. 25 (1971) 598-608
12. CHESTERS M.A., HUSSAIN M. and PRITCHARD J., Surface Science 35 (1973) 161-171
13. COHEN P.I., UNGURIS J. and WEBB M.B., Surf. Sci. 58 (1976)429-456
14. CHINN M.D. and FAIN Jr. S.C., J. Vac. Sci. Technol. 14 (1977) 314-317. Phys. Rev. Lett. 39 (1977) 146-149
15. SHAW C.G. and FAIN Jr. S.C., Surf. Sci. 83 (1979) 1-10
16. KJEMS J.K., PASSELL L., TAUB H., DASH J.E. and NOVACO A.D., Phys. Rev. B 13 (1976) 1446-1462
17. NIELSEN M, Mc TAGUE J.P. and PASSELL L., in *Phase Transitions in Surface Films*, Dash J.G., Ruvals J. ed., Plenum, New York, 1980, p 127
18. BIRGENEAU R.J., HAMMONDS E.N., HEINEY P., STEPHENS P.W. and HORN P.M., in *Ordering in Two Dimensions*, Sinha S.D. Ed., North Holland, New York, 1980.
19. BOATO G.P., CANTINI P., GUIDI C., TATAREK R. and FLECHER G.P., Phys. Rev. B 20 (1979) 3957-3969
20. COLE M.W., FRANKL D.R. and GOODSTEIN D.L., Rev. of Mod. Phys. 53 (1981) 199-210
21. ROLLEFSON R.J., Phys. Rev. Lett. 29 (1972) 410-412
22. RICHARDS M., in *Phase Transitions in Surface Films*, Dash J.G. and Ruvalds J. ed., Plenum, New York, 1980, p 165
23. HORN K., SCHEFFLER M. and BRADSHAW A.M., Phys. Rev. Lett. 41 (1978) 822-824
24. CUNNINGHAM J.A., GREENLAW D.K. and FLYNN C.P., Phys. Rev. B 22 (1980) 717-733
25. *Monolayer and Submonolayer Helium Films*, Daunt J.G. and Lerner E. eds., Plenum, New York, 1973
26. Colloq. Int. du CNRS., *Two Dimensional Adsorbed Phases*, Bienfait M. and Suzanne J. eds., J. Physique Paris, Vol. 38, 1977, Colloq. 4.

27. *Ordering in Two Dimensions*, Sinha S.K. ed., North Holland, New York, 1980
28. *Phase Transitions in Surface Films*, J.G. Dash and J.Ruvalds eds. Plenum, New York, 1980
29. WANG G.C., LU T.M. and LAGALLY M.G., J. Chem. Phys. 69 (1978) 478-489
30. CHING W.Y., HUBER D.L., FISHKIS M. and LAGALLY M.G., J. Vac. Sci. Technol. 15 (1978) 653-654
31. KÜPPERS J., WANDELT K. and ERTL G., Phys. Rev. Lett. 43(1979) 928-930
32. CONRAD E. and WEBB M.B. (private communication)
33. BRETZ M., DASH J.G., HICKERNELL D.C., Mc LEAN E.D. and VILCHES O.E., Phys. Rev. A 8 (1973) 1589-1615
34. CALISTI S. and SUZANNE J., Surf. Sci. 105 (1981) L255-L259
35. FAIN S.C., Jr. CHINN M.D. and DIEHL R.D., Phys. Rev. B 21 (1980) 4170-4172
36. SCHABES-RETCHKIMAN P.S. and VENABLES J.A., Surf. Sci. 105(1981) 536
37. SINGLETON J.H. and HALSEY G.D., Jr., Can . J. Chem. 33 (1955) 184-192
38. ELGIN R.L. and GOODSTEIN D.L., Phys. Rev. A 9 (1974) 2657-2675
39. TAUB H., CARNEIRO K., KJEMS J.K, PASSELL L. and McTAGUE J.P., Phys. Rev. B 16 (1977) 4551-4568.
40. NIELSEN M. and Mc TAGUE J.P., Phys. Rev. B 19(1979) 3096-3106
41. NOVACO A.D. and Mc TAGUE J.P., Phys. Rev. B 20 (1979)2469-2474
42. HAMMONDS E.M., HEINEY P., STEPHENS P.W., BIRGENEAU R.J. and HORN P., J. Phys. C 13 (1980) L301-L306
43. VENABLES J.A., KRAMER H.M. and PRICE G.L., Surf. Sci. 55 (1976) 373-379
44. MONCTON D.E., STEPHENS P.W., BIRGENEAU R.J., HORN P.M. and BROWN G.S., Phys. Rev. Lett. 46 (1981) 1533-1536
45. WEBB M.B. and LAGALLY M.G., Solid State Phys. 28 (1973) 301-405
46. PENDRY J.B., in *Low-Energy Electron Diffraction*, Academic, New York, 1974
47. HENZLER M., in *Topics in Current Physics 4*, Ibach H. ed., Springer, 1977
48. WELKIE D. 1981 Ph.D. Dissertation, University of Wisconsin-Madison (unpublished).
49. UNGURIS J., BRUCH L.W., MOOG E.R. and WEBB M.B., Surf. Sci. 87 (1979) 415-436
50. FARRELL H.H., STRONGIN M. and DICKEY J.M., Phys. Rev. B 6 (1972) 4703-4710
51. UNGURIS J., BRUCH L.W., MOOG E.R. and WEBB M.B., Surf. Sci. 109 (1981)522
52. As in most ultra-high vacuum adsorption experiments the sample is not at the same temperature as the chamber walls, but this effect should be negligible since the excess heat delivered by the gas to the sample is small compared to the incident thermal radiation and the excess kinetic energy of the atom is accommodated by the substrate within a few atomic vibrations.

53. LARHER Y.,J. Phys. Chem. 72 (1968) 1847 and J. Colloid. Interface Sci. 37 (1971) 836
54. PRICE G.L. and VENABLES J.A., Surf. Sci. 59 (1976) 509
55. SAMS J.R.Jr., CONSTABARIS G. and HALSEY G.D.Jr.,J. Phys. Chem. (1960) 1689
56. SAMS J.R.Jr. and HALSEY G.D. Jr.,J. Chem. Phys. 36 (1962) 1334-1339
57. WEI M.S., Ph.D.Thesis, Univ. Wisconsin-Madison, 1981; WEI M.S. and BRUCH L.W., J. Chem. Phys. 75(1981)4130.
58. *Rare Gas Solids*, M. L. Llein and J.A. Venables eds., Academic, New York, Vol. II., 1977
59. BRUCH L.W., UNGURIS J. and WEBB M.B., Surf. Sci. 87 (1979) 437-456
60. SHAW C.G. and FAIN S.C.Jr., Surf. Sci. 91 (1980) L1-L6
61. UNGURIS J., BRUCH L.W., WEBB M.B. and PHILLIPS J.M., Surf. Sci. (in press)
62. POLLACK G.L., Rev. Mod.Phys. 36 (1964) 748-791
63. BARKER J.A., in *Rare Gas Solids*, Vol. I, M.L. Klein and J.A. Venables eds.,Academic, New York, 1976
64. BRUCH L.W. and PHILLIPS J.M., Surface Sci. 91 (1980) 1-23
65. CERTAIN P.R. and BRUCH L.W., MTP International Review of Science Physical Chemistry Series One, Vol. 1, Theoretical Chemistry, ed. W.B. Brown, Butterworths, London, 1972
66. BERTONCINI P. and WAHL A.C.,Phys. Rev. Lett. 25 (1970) 991-994 STEVENS W.J., WAHL A.C., GARDNER M.A. and KARO A.M., J. Chem. Phys. 60 (1974) 2195-2196
67. LANG N.D., Phys. Rev. Lett. 46 (1981) 842-845
68. MAVROYANNIS C., Mol. Phys. 6 (1963) 593-600; DZYALOSHINSKII I.E. LIFSHITZ E.M. and PITAEVSKII L .P, Adv. in Phys. 10 (1961) 165-209
69. GILQUIN B., Thesis, Nancy, 1979.
70. BRUCH L.W. and WATANABE H., Surf. Sci. 65 (1977) 619-632
71. STEELE W.A., Surf. Sci. 36 (1973) 317-352
72. GERSBACHER W.M.Jr and MILFORD F.J., J. Low Temp. Phys. 9 (1972) 189-201
73. LUCAS A.A., Physica 39 (1968) 5-12
74. RENNE M.J., thesis (Utrecht, unpublished). 1971
75. AZIZ R.A., Mol. Phys. 38 (1979) 177-190
76. SINANOĞLU O. and PITZER K.S., J. Chem. Phys. 32 (1960) 1279-1288
77. McLACHLAN A.D., Mol. Phys. 7(1964) 381-388
78. LANG N.D. and KOHN W., Phys. Rev. B 7 (1973) 3541-3550
79. ZAREMBA E. and KOHN W., Phys. Rev. B 13 (1976) 2270
80. MacRURY T.B. and LINDER B., J. Chem. Phys. 54(1971) 2056-2066
81. KOHN W. and LAU K.H., Solid State Commun. 18 (1976) 553-555
82. ANTONIEWICZ P.R., Phys. Status Solidi (b) 86 (1978) 645-652
83. FREEMAN D.L., J. Chem. Phys. 62 (1975) 941-949 and 4300-4307
84. LAU K.H. and KOHN W., Surf. Sci. 65 (1977) 607-618; and LAU K.H Solid State Commun. 28 (1978) 757-762
85. KAPPUS W., Z. Phys. B 29 (1978) 239-244

86. LAU K.H. and KOHN W., Surf. Sci. 75 (1978) 69-85 and EINSTEIN T.L., Surf. Sci. 75 (1978) L161-L167
87. NOVACO A.D. and Mc TAGUE J.P., Phys. Rev. Lett. 38 (1977) 1286-1289. and Mc TAGUE J.P. and NOVACO A.D.,Phys. Rev. B 19(1979) 5299-5306
88. AXIBROD B.M. and TELLER E., J. Chem. Phys. 11 (1942)299-300 and MUTO Y., Proc. Phys. Math. Soc. Japan 17 (1943)629
89. BRUCH L.W., COHEN P.I. and WEBB M.B., Surf. Sci. 59 (1976) 1-16
90. BRUCH L.W. and WEI M.S., Surf. Sci. 100 (1980) 481-506
91. The frequencies of perpendicular vibrations, derived from Debye-Waller factor data, for Kr/Ag(111) and Xe/Ag(111) are close to the values for these gases on graphite, derived from analysis of Henry's law data. See ROSS S. and OLIVIER J.P., in *On Physical Adsorption*, Interscience, New York, 1964, sec. VIII.1.F.
92. *Rare Gas Solids*, Vol. I., M.L. Klein and J.A. Venables eds., Academic, New York, 1976
93. BARKER J.A., HENDERSON D. and ABRAHAM F., Physica 106A (1981) 226-238
94. PHILLIPS J.M., BRUCH L.W. and MURPHY R.D.,J. Chem. Phys. 75 (1981)5097
95. De OLIVEIRA M.J. and GRIFFITHS R.B., Surf. Sci. 71 (1978)687-694.
96. MOOG E.R., BRUCH L.W. and WEBB M.B. : private communication.
97. OSAKABE N., TANISHIRO Y., YAGI K and HONJO G., Surf. Sci. 102 (1981) 424)442.
98. FELDMAN L.C. and STENSGAARD I., Progress in Surf. Sci. (to be published).
99. FROITZHEIM H., in *Topics in Current Physics 4*, Ibach H. ed., Springer, Berlin, 1977.

CHEMISORPTION OF METALS ON METALS AND ON SEMICONDUCTORS

E. BAUER

Physikalisches Institut
Technische Universität Clausthal
3392 Clausthal-Zellerfeld
Federal Republic of Germany

I. METALS ON METALS

I.1. Introduction

There are only five noble gases but some sixty five metals. Noble gases have a simple electron configuration which makes their theoretical treatment simpler. They have high vapor pressures which invites their study under equilibrium conditions. Not so the metals : many have complicated electron configurations with several groups of electrons participating in bonding and many of them have low vapor pressures so that quasi-equilibrium studies can be done only with vapor beams impinging on surfaces heated to high temperatures. Condensed metal vapors have the unpleasant property of producing electrical shorts, emission centers on fluorescent screens and other disturbances, and hot surfaces frequently tend to be contaminated by carbon. Nevertheless, a large amount of information has been accumulated of which only a small amount can be presented in the form of typical examples. For additional information see recent reviews.[1-3]

This review is organized according to the atomic roughness of the substrate because experience has shown that this factor is more important for the structure of the adsorbate than the electronic structure of the substrate. Three types of surfaces will be distinguished : (i) smooth surfaces which contain 2 or 3 close-packed directions (hcp(0001), fcc(111), (100) and bcc(110) surfaces); (ii) one-dimensionally rough surfaces containing one close-packed direction, e.g. fcc(110) and bcc(112); (iii) two-dimensionally rough surfaces which contain no close-packed direction, e.g. bcc(100) and (111) . For each of these types of surface the adsorbates can be

B. Mutaftschiev (ed.), Interfacial Aspects of Phase Transformations, 411–431.

characterized by a few phenomenological parameters : (i) surface free energies of adsorbate (in its bulk state, γ_A) and substrate (γ_S); (ii) electronegativity of adsorbate (X_A) and of substrate (X_S); (iii) atomic diameter of adsorbate (d_A) and substrate (d_S). $\gamma_S - \gamma_A$ determines to a large extent whether an adsorbed layer or three-dimensional crystals are formed initially. The amount of electron transfer between adsorbate and substrate, i.e. the charge on the adsorbed atom important for lateral interaction is determined by $X_S - X_A$. The misfit between adsorbed layer and substrate, which is important for the orientation of the layer and its stability upon adsorption or condensation of additional material is determined by $d_S - d_A$. The first three sections are followed by a fourth in which alloy- and compound-forming adsorbate-substrate pairs (excluded in the preceeding sections) are discussed. The paper concludes with a brief review of metal atom adsorption on semiconductors.

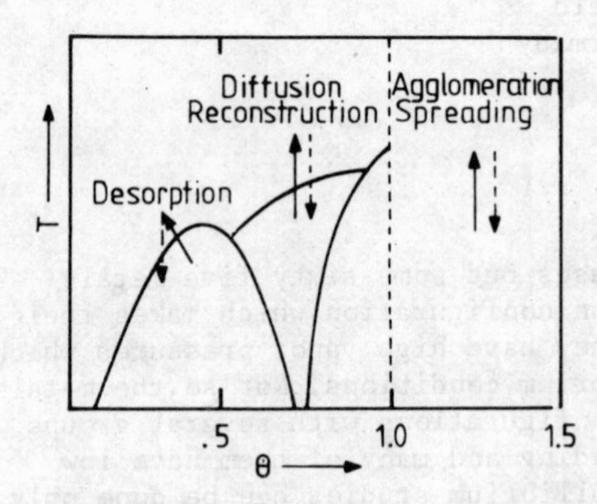

Fig. 1 : Schematic phase diagram of an adsorbate.

It is customary to present the structural information on adsorbed layers in (two-component) phase diagrams (see Fig. 1). The two components are the adsorbed atom and the vacancy. Instead of giving their concentrations it is usual to use the coverage Θ. Two definitions for Θ are used : in one, $\Theta = 1$ when the adsorbate has reached its monolayer saturation coverage (Θ_A) which may contain more or fewer atoms than the substrate surface; in the other, $\Theta = 1$ when the number of adsorbate atoms equals that of the substrate atoms (Θ_S). Because it is difficult to attain the equilibrium configuration at low temperature T, the "phase transitions" are frequently irreversible. Therefore, such "phase diagrams" should not be interpreted in the usual manner.

I.2. Smooth Surfaces

Very little work has been done on hcp(0001) surfaces; as a result, no comparisons between different A-S combinations are possible. Therefore only the adsorption of Cd on Ti(0001) will be mentioned, the only system for which a complete LEED structure analysis of a metal monolayer and of multilayers was performed. All layers formed a (1 x 1) structure ; in the first layer the Cd atoms took fcc sites, that is C sites in the ABAB... stacking sequence,[4] but the following Cd layers continued in hcp fashion (CACA...).

For fcc (111) surfaces the following examples will be considered : (i) ($X_A << X_S$) : Na on Ni and Al ; and (ii) ($X_A \simeq X_S$) : Pb on Cu[7-9] and Ag[9-11], Ag on Cu and Cu on Ag.

When $X_A << X_S$, there is considerable charge transfer to the substrate in the first group, irrespective of the electronic structure (free electron (Al) or d-electron (Ni) metal). At low coverage the dipole moment resulting from the charge transfer is large so that bonding to the substrate is strongly ionic and lateral interactions are repulsive.

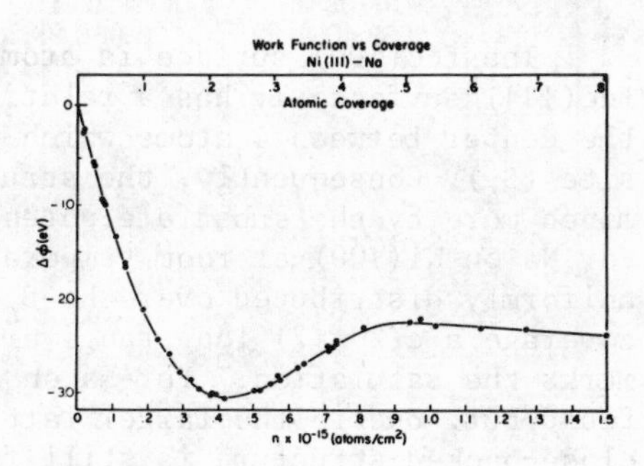

Fig. 2 : Work function change as a function of coverage for Na on Ni(111). The coverage is given in Θ_S units.[5]

With increasing coverage depolarization occurs; finally the atoms come so close that the layer becomes metallic. These changes can be seen clearly in the work function change (Fig. 2), in the desorption energy decrease and in the LEED pattern. The structure of the layer is determined by the repulsion between the atoms and by the potential modulation of the substrate. On Ni($X_A << X_S$) the strong repulsion causes a uniform distribution of the atoms without long range order at low coverage ($\Theta_S < 1/3$), at least at room temperature. At high coverage ($1/3 < \Theta_S < 1/2 = \Theta_{saturation}$) the atoms form an incommensurate hexagonal layer whose interatomic distances are continuously decreasing with coverage[5]. On Al ($X_A < X_S$) the weaker repulsion does not manifest itself so strongly so that the substrate periodicity is more apparent. The first structure seen is a ($\sqrt{3} \times \sqrt{3}$)R30° structure which saturates at $\Theta_S = 1/3$ and transforms into a (2 × 1) structure saturating at $\Theta_S = 1/2$.[6] This structure is not a densely packed layer of Na atoms as on Ni but has in one direction rather large interatomic distances, which is difficult to understand.

In case (ii) the only diffraction feature from the adsorbed Pb seen on Cu(111) is a (4 × 4) coincidence pattern that can be attributed to 2.7 % contracted Pb(111) plane,[7-9] while on Ag(111) a (√3 × √3)R30° structure is formed initially; from this densely packed Ag(111) monolayer islands develop which are rotated 4°- 5° with respect to the substrate ("rotational epitaxy").[9-11] On the other fcc(111) surfaces studied (Al and Au), these rotated islands form without preceeding superstructure. With Ag as an exception it can be said that in general two-dimensional densely packed islands form which means that the interactions are attractive. The adsorbate-substrate pair Ag-Cu has been somewhat puzzling because according to the simple $\gamma_S - \gamma_A$ criterion Ag should form an adsorption layer on Cu but not Cu on Ag while the opposite has been reported for room temperature studies.[12-14] A recent very careful re-investigation of this system with AES on very clean surfaces has shown however, that the systems behave according to the $\gamma_S - \gamma_A$ criterion[15]; Ag forms on Cu 2D densely packed islands which grow together to a monolayer while Cu on Ag grows in 3D islands from the very beginning.

The fcc(100) surface is atomically somewhat rougher than the fcc(111) surface. It has a relatively deep potential minimum in the center between 4 atoms which provides a preferred adsorption site (S_4). Consequently, the structure of the adsorbate is determined more by the substrate potential than on the (111) surface. For Na on Ni(100) at room temperature, the atoms are initially uniformly distributed over the S_4 sites and develop with increasing coverage a c(2 × 2) long range ordered structure. Its completion marks the saturation.[5] For Na on Al(100), a similar sequence is found but, due to the larger lattice constant of Al, a hexagonal close-packed structure is still formed at higher coverages.[6] Why such a densely packed monolayer is not seen on the (111) surface is not clear. The c(2 × 2) structure of Na on Al(100)[16,17] and in particular on Ni(100)[18-21] is one of the few metal adsorbate structures subjected to a complete structure analysis with LEED[16-19] APD[20] and (NPD)[21] (see Chapter "The Structure of Clean Surfaces" further called SCS) with the result that the Na atom is positioned as expected on the basis of atomic radii and potential minimum considerations : about 2.1Å and 2.2Å above to topmost Ni plane in the S_4 position.

The deeper potential wells on the (100) surface as compared to the (111) surface introduce more ordered structures only in some systems with $X_A \simeq X_S$ but not in others.An example is Pb on Cu(100); this forms only a c(4 4) structure for which several models were suggested[7,22,23] but which is probably a hexagonal packing distorted into a coincidence layer[3]. Similarly, the adsorption layer of Ag on Cu(100) is a close-packed layer distorted to achieve coincidence with the substrate.[24]

The bcc(110) surface is intermediate in roughness between the fcc(111) and (100) surfaces. Accordingly, an adsorbate behavior similar to that of both faces is expected. A comparison with the alkali adsorption data on fcc surfaces is difficult, however, because none of that work was done at sufficiently low temperatures or under sufficiently clean conditions, which are extremely important factors. An example is Na on W[25] and Mo.[26] While in room temperature adsorption studies on Mo(110) no ordered phase was found,[26] a whole series of phases was observed on W at T 100K, corresponding to optimum coverages Θ_S = 1/6, 1/4, 1/3, 3/8, 2/5 and an incommensurate hexagonal close-packed layer (Fig. 3).[25]

Fig. 3 : Ordered structures of Na on W(110) for (a), Θ_S = 1/6 (b), 1/4 (c), 1/3 (d), 3/8 (e), 2/5 (f) and the incommensurate close-packed layer (g); (h) is the multilayer structure.[25]

In the first three phases (b-c), all Na atoms are assumed to be in the potential minima of the surface, in the second two (e,f) they are on inequivalent sites and in the last one (g) their location is determined by the requirement of closest packing. A detailed analysis of the lateral interactions necessary to produce such a sequence of structures shows that the interaction cannot be simple dipolar ($\propto \frac{1}{r^3}$) but must be non-monotonic, e.g., the interaction with the 6th-nearest neighbor must be weaker than that with the 7th-nearest neighbor.[27] Therefore it is assumed that indirect interactions via the substrate electrons (which are oscillatory in nature) make a significant contribution. For indirect interactions to be strong, the electronic density of states at and near the Fermi energy must be large because these electrons mediate the interaction. The Fermi surface is anisotropic and consequently the indirect interactions are also anisotropic. Deviations from an isotropic distribution due to repulsive lateral interactions such as in Fig. 3(b) may therefore not only be caused by the anisotropy of the surface potential but also by the anisotropy of the indirect interactions. At present too little is known about indirect interactions (see, e.g., some reviews[3,28] for further discussion of this subject.

It should be mentioned that at sufficiently low temperature and with clean conditions there is no reason why on fcc(111) and in particular on fcc (100) surfaces a whole set of ordered structures should not also occur because W (and Mo) do not excel with a particularly large density of states at the Fermi energy. These experimental conditions have been fulfilled nearly exclusively by the Kiev group and in all cases which they have studied (mainly alkali and alkaline earths on W and Mo(110) and (112)) several ordered structures have been found. There is an excellent review of this work and its theoretical implications.[29] Only an example of a typical phase diagram, that of Sr on W(110),[30] will be shown here (Fig. 4).

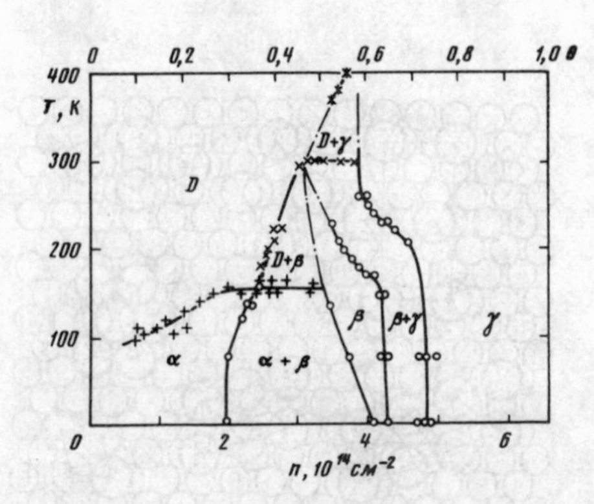

Fig. 4 : Phase diagram of Sr on W(110). The coverage is given in Θ_A units (Θ_A = 1 for 6.6×10^{14}, Sr atoms/cm^2.[30]

Here α is a hexagonal phase with coverage-dependent unit mesh dimensions, β is a commensurate phase, γ a close-packed and D is a disordered phase. The phase transitions at constant coverage caused by T changes are reversible.

Another interesting example is Be.[31] It is the smallest metal atom; although the free atom has a high ionization energy and the basal plane of crystalline Be has a high work function (5.1 eV), it behaves on W(110) like an electro-positive adsorbate with a pronounced work function minimum. For several reasons, such as small atomic size (so that the atom sits particularly "deep" in the surface potential minima), strong bonding to the substrate and probably strong indirect interactions, several ordered phases – (8 × 1), (1 × 1), (6 × 1), c(9 × 2) and c(13 × 2) – are formed with increasing coverages up to Θ_A = 1.25 (Θ_A = 1 when 22×10^{14} atoms/cm^2 (atomic density of the Be(0001) plane) are adsorbed). The desorption energy oscillates as a function of coverage (Fig.5) with maxima approximately at the optimum coverages of the (8 × 1) and (6 × 1) structures and with a minimum for the (1 × 1) structure where one would expect a maximum on the basis of misfit considerations. Apparently electronic interactions believed to be responsible for the formation of the (8 × 1) and (6 × 1) structures are, in this case, more important for binding than geometrical fit. The c(9 × 2) and c(13 × 2) structures are coincidence structures between a distorted (0001) and (10$\bar{1}$0) plane, respectively, and the substrate. It is interesting to note that unlike most other systems, a thick Be layer does not have its closest packed plane parallel to the substrate but rather a prism plane which gives a small misfit in one direction.

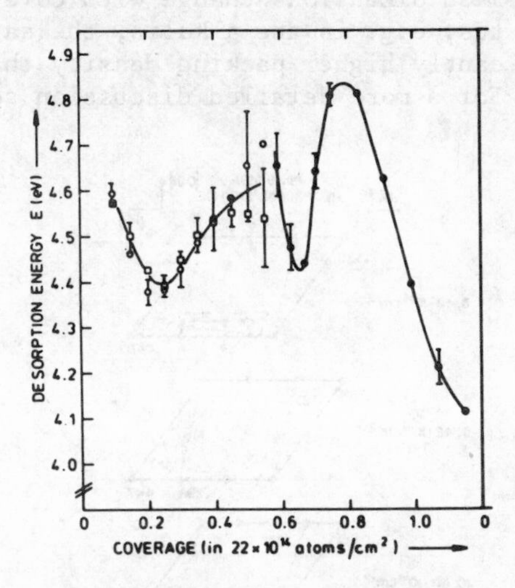

Fig. 5 : Desorption energy of Be from W(110) as a function of coverage.[31]

The metals with X_A X_S adsorb quite differently on bcc(110) surfaces, the general characteristics being the formation of densely packed islands in the early adsorption stages and the absence of coincidence structures. This is a consequence of attractive lateral interactions between the adsorbed atoms. FIM studies show that in the very early stages of adsorption the "islands" may be linear along the densely packed directions in the surface. At room temperature these islands are usually quite small so that the complete monolayer is frequently quite disordered unless the layer is annealed. Examples for such adsorbates are Pd and Au whose layers have to be annealed in order to obtain sharp diffraction spots. In the absence of dominating dipole-dipole and indirect electronic interactions the adsorbate structure is largely determined by misfit minimization. Furthermore, the first monolayer often does not screen the substrate forces enough to suppress adsorption of a second or even third layer.

Illustrations for these general statements are Cu, Pd and Pb on W. The first monolayer of Cu is pseudomorphous. A second layer is adsorbed on top of it. While this layer grows, it reorganizes the first layer into a common one-dimensionally incommensurate structure which may be considered to be a distorted fcc(111) layer.[32] Pd forms only one truly adsorbed layer. This is probably due to a specific interaction between Pd and refractory metal substrates, which causes a quite different adsorption behavior on Pd monolayers as compared to bulk Pd. For example, the Pd monolayer does not adsorb CO. The packing density and structure of the first layer depends upon temperature and upon the presence of additional Pd.[33] Pb also forms only one monolayer, but in contrast to Cu and Pd has a coverage-dependent packing density. Fig. 6 indicates how the unit mesh dimensions change with coverage.[34] Like in many other adsorbates, e.g. in the alkalis, the saturated monolayer has a significantly higher packing density than the Pb(111) plane in the bulk. For a more detailed discussion see Ref. 35.

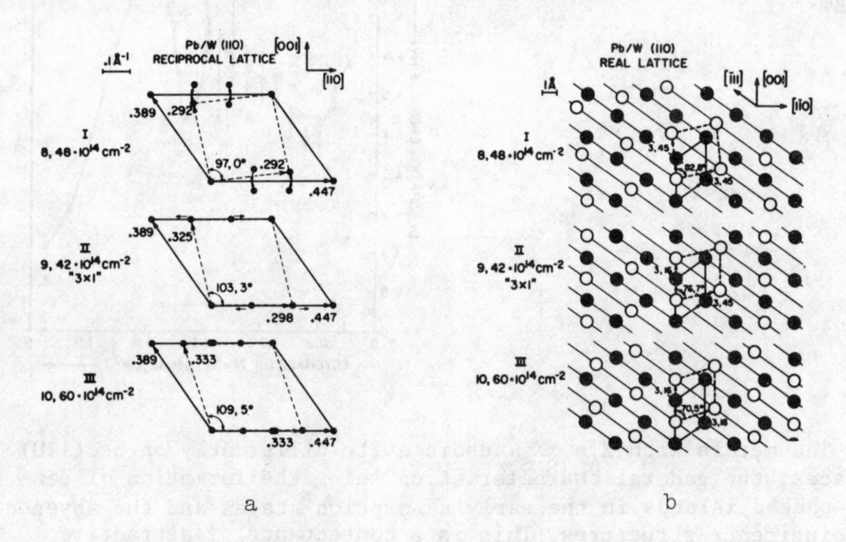

Fig. 6 : Change of the reciprocal (a) and real (b) lattice unit mesh of Pb on W(110) with coverage.[34]

I.3. One-Dimensionally Rough Surfaces

The one-dimensional roughness of those surfaces which consist of ridges and troughs parallel to the close-packed directions is considerable, even taking the contraction of the topmost layer into account (see SCS). Mobility along the troughs is large, across the ridges small. Furthermore, the atomic roughness along the troughs is small. As a consequence, a wide range of packing possibilities exists within the troughs, but in general only multiples of the trough distance occur for the packing in the perpendicular direction. The adsorbate rows in the various troughs may be shifted relative to each other as indicated in Fig. 7 for two coverages of Na on Ni(110), depending upon the relative magnitude of the lateral interactions parallel and perpendicular to the troughs. In the case of Na on Ni(110)[5] and on W(112)[36] these seem to be rather isotropic (dipole-dipole interactions). In other cases, e.g. Li on W(112)[37] and Mo(112)[38] they are very anisotropic leading initially to equidistant rows perpendicular to the troughs. This has been attributed to the dominance of indirect electronic interactions over the dipole interactions.

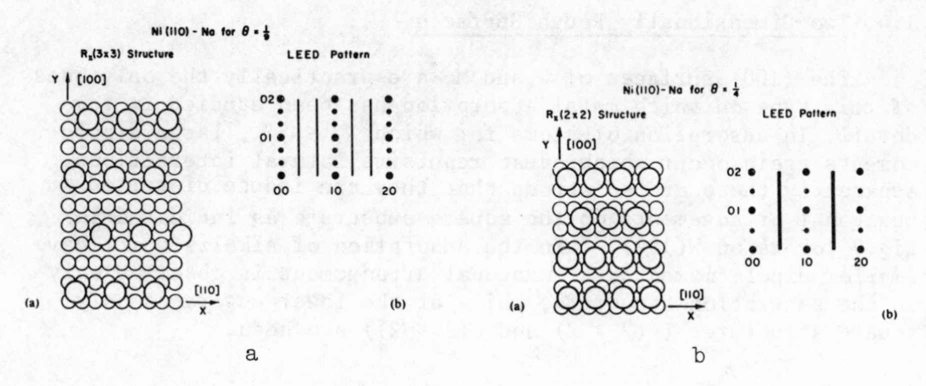

Fig. 7 : Structure models and corresponding LEED patterns for two coverages, (a) $\Theta_S = 1/9$ and (b) $\Theta_S = 1/4$ of Na on Ni(110).[5]

In adsorbate-substrate combinations in which $X_A \simeq X_S$, a much smaller variety of structures has been observed and these have been studied in much less detail. Examples are Pb on Cu(110)[22] and Ni(110)[39] for which a c(2 × 2) structure is observed at $\Theta_S = 1/2$ and several coincidence lattices ((3 × 1), (4 × 1) and (5 × 1)) corresponding to increasing packing density along the rows at higher coverages. Saturation occurs in all cases when a densely packed monolayer is reached. Usually annealing is necessary in order to obtain long range order.

Fig. 8 : Elastic interaction model for c(2 × 2) structure on a fcc (110) surface.

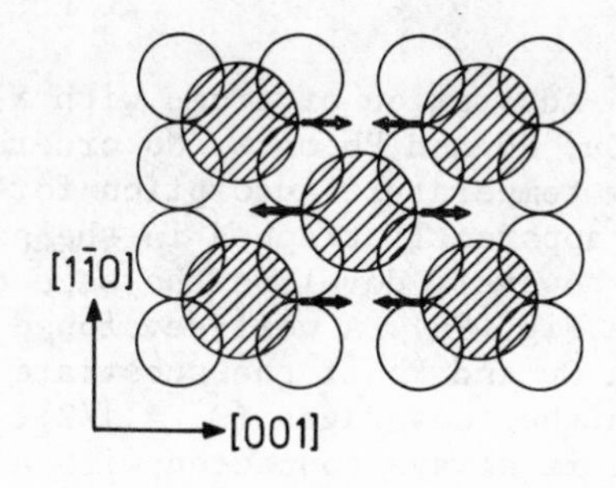

The appearance of a c(2 × 2) structure on (110) surfaces is in contrast with the adsorption behavior on smooth surfaces on which densely packed islands are formed from the beginning, Pb on Ag excepted. This suggests stronger indirect interactions on the rougher surface which appears plausible because the atoms "sink" deeper into the substrate, or elastic interactions (see Fig. 8).

I.4. Two-Dimensionally Rough Surfaces

The (100) surfaces of W and Mo are practically the only ones of this type on which metal adsorption has been studied in some detail. In adsorption of atoms for which $X_A \ll X_S$, large dipole moments again occur which cause repulsive lateral interactions. Apparently these are so strong that they can induce distorted hexagonal arrangements on the square substrate as indicated in Fig.9 for Ba on W(100).[40] In the adsorption of alkalis which have smaller dipole moments, a hexagonal arrangement is observed only in the saturation structure, while at the lower coverages only square structures (p(2 × 2) and c(2 × 2)) are seen.

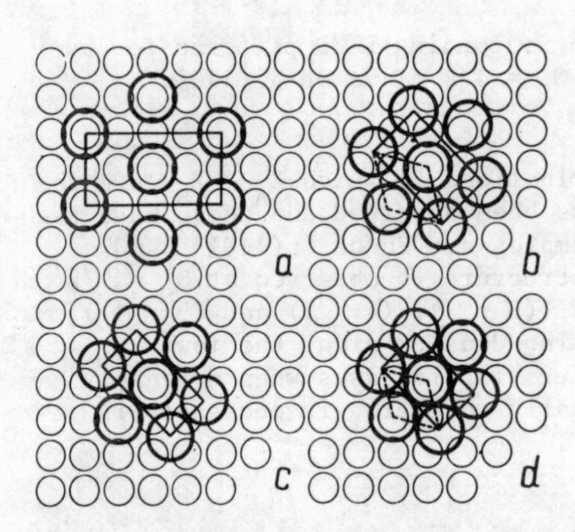

Fig. 9 : Structure models for various coverages of Ba on W(100). a) c(4 × 2), $\Theta_S = 1/4$ b) $\Theta_S = 13/30$; c) $\Theta_S = 1/2$; d) $\Theta_S = 0.58$ ($\Theta_A = 1$). Ref. 40.

Adsorption of atoms with $X_A \simeq X_S$ will again be illustrated by Cu, Pd and Pb on W. No ordered structures are obtained during room temperature adsorption for the smaller atoms (Cu, Pd) which are apparently trapped in the potential minima of the substrate. Pb, however, develops the same c(4 × 2) structure as Ba at $\Theta_S = 1/4$ (see Fig. 9a). A well-developed c(2 × 2) structure can be obtained with Cu and Pb if the substrate is heated during or after adsorption at higher coverages ($\Theta_S \simeq 1/2$). The formation of this ordered structure is always connected with a considerable work function decrease so that it is not clear at present if the surface reconstructs in this process or not.[32,34] In Pd on W(100) the work function decreases, too, but no c(2 × 2) structure is formed. Here ordered structures are observed only at higher coverages of which the first two monolayers are truly adsorbed[41] just as in the case of Cu[32] while only one Pb monolayer is adsorbed.[34]

I.5. Alloying and Compound-Forming Adsorbate-Substrate Systems

A considerable amount of work has been done with such systems but there are considerable discrepancies between the various results which have to be clarified. At the beginning of this chapter the criterion for two-dimensional phase formation was simplified by omitting the "interfacial" energy γ_{AS}. This quantity contains the strain energy and the chemical affinity of S and A aotms which can be characterized by the heat of mixing. If the heat of mixing is large, its contribution to γ_{AS} may become sufficiently large so that $\gamma_S - \gamma_A + \gamma_{SA} > 0$ although $\gamma_S - \gamma_A < 0$ so that a two-dimensional A phase may be formed. The question is : does this really occur or is a surface alloy formed from the beginning ? If no alloy is formed because the temperature is too low for the necessary place exchanges, does a two-dimensional or three-dimensional A phase then form ?

Before embarking on this discussion the simpler case $\gamma_A < \gamma_S$ in which two-dimensional A phase formation would be expected, also without γ_{SA} contributing, will be discussed. A good example, though not for alloying but for compound formation, is Pb on Au which has been studied extensively.[42-44] On all surfaces studied, that is (111), (100), (110) and the vicinals (11,1,1), (911), (711) and (511) of the (100) surface, Pb is adsorbed without alloying until a densely packed monolayer is formed. Only thereafter compound formation commences with initially two-dimensional growth. This compound has been shown on the (100)[43] and on the (111)[44] surface to be Au_2Pb and not $AuPb_2$ as suggested originally.[42] It is interesting to note that in the submonolayer range Pb on Au behaves like combinations which do not form compounds.

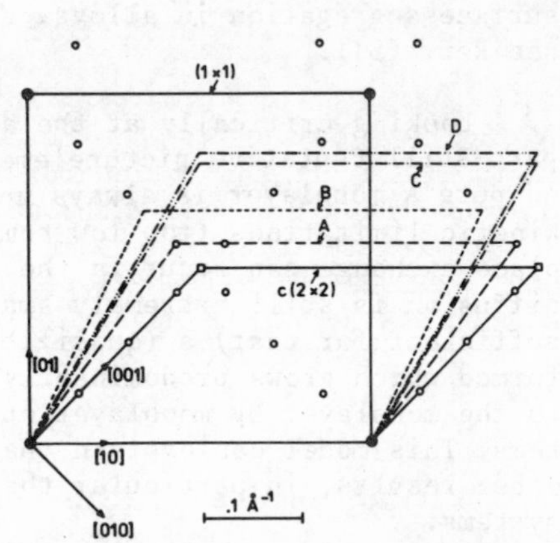

Fig. 10 : Reciprocal unit meshes of Pb on Au(100) observed in the submonolayer (Θ_A) coverage range. A) $\Theta_S = 4/7$, B) $\Theta_S = 2/3$, C) $\Theta_S = 4/5$, D) $\Theta_S = 5/6$.[43]

As an example, a whole series of coincidence lattices is formed at coverages above $\Theta_S = 1/2$ which can be understood as transitional structures between the square symmetry of the c(2 × 2) structure at $\Theta_S = 1/2$ and the hexagonal closepacking at $\Theta_A = 1$ (Fig. 10). Repulsive lateral interactions must be invoked again to understand this behavior. In contrast to Pb whose (sub) monolayer structures are obtained in well-ordered form only after heating (without alloying) Sn has been reported to form a two-dimensional surface alloy with the composition AuSn at temperatures as low as -100°C.[45] The evidence presented for this interpretation is, however, insufficient and more detailed studies are necessary. Two last examples for $\gamma_A < \gamma_S$ are the adsorption of Ag on Pd(111)[40] and of Au on Cu (100)[24]. Ag condensation takes place monolayer by monolayer without alloying up to 200°C, above 250°C a surface alloy is formed. Au on Cu(100) forms a surface alloy at room temperature, but a pure Au monolayer at -50°. Thus the situation for $\gamma_A < \gamma_S$ in alloying systems is not clear at present.

This is even more true in the case $\gamma_A > \gamma_S$. Here Pd on Ag,[47] Pd on Cu[15,48,49] and Cu on Au[15,50] are examples. Pd has been reported to be adsorbed in monolayer islands on Ag(111) at 150°C but the islands observed in transmission electron micorscopy (TEM) may as well be alloy islands. Pd condenses on Cu(111) at room temperature monolayer by monolayer according to ref. (50) but other work indicates for this surface[48] or proves[15] that alloying occurs from the very beginning of "adsorption". The change in surface lattice parameter with film thickness is then not to be interpreted in terms of elastic strains in the Pd layer but simply by the change of lattice constant with alloy composition. Also on Cu(100) and (110) there is no doubt that alloying occurs at room temperature.[48] The adsorption of one monolayer of Cu on Au(111) before three-dimensional Au crystals appear[50] probably must be interpreted in the same way and not as a Cu monolayer. Adsorption may occur not only from the outside of the crystal but also from within as in the case of surface segregation in alloys. For a recent review of this subject see Ref. (51).

Looking critically at the available evidence the following, in part still tentative picture emerges. In the case $\gamma_A > \gamma_S$, an adsorbed pure A monolayer is always unstable and can exist only due to kinetic limitations (too low temperature). At temperatures at which place exchange can occur in the surface region but in which bulk diffusion is still extremely small (room temperature is in general sufficient for that) a (quasi) two-dimensional surface alloy is formed which grows predominantly laterally but also inward similar to the monolayer by monolayer outward growth in non-alloying systems. This model can explain the TEM, AES, "misfit" strain and other results, in particular the two-dimensional growth in these systems.

II. METALS ON SEMICONDUCTORS

II.1. Elemental Semiconductors

Although metal adsorption on semiconductors produced some of the first stunning LEED patterns nearly 20 years ago[52] the large scale study of this subject began only a few years ago in connection with the attempt to understand the metal-semiconductor interface, except for the study of alkali adsorption which began in the early years of LEED, however, with initial disappointments : no new ordered structures were found during room temperature adsorption of Na, K and Cs on Si and Ge(111) and (100) surfaces.[53,54] Adsorption of these atoms weakened only the LEED pattern of the clean surface (c(8 × 2) and (7 × 7) for Ge and Si(111)) ending in a faint (1 × 1) pattern on a strong background. Na on Ge(111) was an exception by converting the c(8 × 2) structure in a strong well-developed (1 × 1)-structure, i.e. it re-reconstructed the surface at saturation into the lateral periodicity of the bulk.[53] Similarly, the intensity of the (2 × 1) pattern of the (100) surfaces is enhanced by saturation with K and Cs.[54] A detailed later study[55] of Cs adsorption on well-prepared Si(100) surfaces actually revealed two ordered phases, a (2 × 3) structure at intermediate coverage and a (2 × 1) structure at saturation (Θ_S = 1/2). The variation of the relative intensities of the fractional order LEED beams and of the work function with coverage and oxygen co-adsorption leads to the structure model shown in Fig. 11.

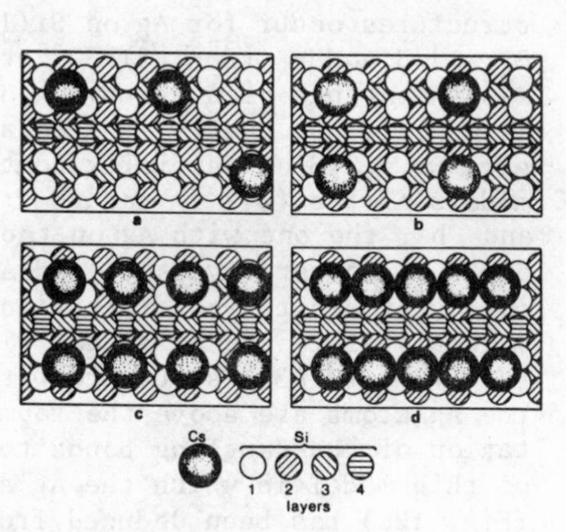

Fig. 11 : Structure models for Cs on Si(100) for various coverages.[55] For details, see text.

At low coverages Cs is adsorbed only on top of the (asymmetric) dimers of the clean Si(100) - (2 × 1) structure (see SCS) without correlation between sites on neighboring dimer rows (Fig. 11a). At Θ_S = 1/6, correlation within and between rows is enforced by the repulsive interactions (Fig. 11b). Between Θ_S = 1/6 and 1/3 the sites in between the atoms in (b) are occupied (Fig. 11c). Finally, beyond the work function minimum, condensation into a one-dimensional metal with one Cs atom per Si dimer occurs (Fig. 11d).

Long-range order can be obtained also on (111) surfaces by heating the adsorbate. Thus a (3 × 1) structure occurs in Na on Ge(111)[53] and a (6 × 1) structure in K on Ge(111) at $\Theta_S = 0.4$.[56] Although at room temperature no long-range order occurs on (111) surfaces, work function[54,57] and thermal desorption[57] measurements indicate the presence of several distinct binding sites. In the work function change ($\Delta\phi$) measurements the sites appear in linear segments as function of coverage, separated by sharp breaks at multiples of 1/8 on Ge (with c(8 × 2) structure) and of 1/7 on Si (with (7 × 7) structure). Apparently, the surface maintains its superstructure during adsorption providing different substrate sites. Similar breaks also occur on the Ge(111) - (2 × 1) structure and have been associated with specific configurations of atoms adsorbed on identical substrate sites.[58] These two interpretations are not necessarily in contradiction, because of the difference between the two surfaces and the higher stability of the c(8 × 2) structure.

With most other metals the problem to be immediately confronted is the question of pure metal adsorbate versus compound formation. Two much studied examples, Ag and Au on Si(111) illustrate this. Au easily forms metastable alloys with Si while Ag shows much less (if any) alloying tendency. Therefore there is little doubt that a Ag adsorption layer consists only of Ag while Au could form a two-dimensional surface alloy from the very beginning. Thus, in the case of Ag mainly the structure is disputed, in the case of Au also the composition of the adsorbed layer. Three ordered structures occur for Ag on Si(111) : a(6 × 1) and a (3 × 1) at $\Theta_S = 1/3$ and a $(\sqrt{3} \times \sqrt{3})R30°$ structure at $\Theta_S = 1$. This latter assignment has been a subject of discussion for a long time but it appears now that $\Theta_S = 1$ on a very clean surface while $\Theta_S < 1$, e.g. $\Theta_S = 2/3$ on partially contaminated surfaces.[59-61] Of the two models of the (3 × 1) and (6 × 1) structure shown in Fig. 12a and b , the one with Ag on top of Si atoms (b) is favored on the basis of comparative AgSi and $AgSi_3$ molecule calculations.[62] The molecular orbital considerations basic to these calculations also make plausible the hexagonal ring model (Fig. 12c), proposed for the $(\sqrt{3} \times \sqrt{3})R30°$ structure with coverage $\Theta_S = 2/3$. In this model the Ag atoms are above the topmost Si plane because of the orientation of the dangling bonds to which the Ag atoms link. A variant of this model in which the Ag atoms are immersed into the Si surface (Fig. 12d) has been deduced from LEIS experiments (see SCS)[63] and work function measurements[64] and supported by a quasi-kinematical LEED intensity analysis.[65] These deductions are, however, not unambiguous as are also the theoretical arguments used for the on-top hexagon.[62] It has been pointed out[66] that the maximum overlap principle used in Ref. (62) would favor trimer formation as is postulated in the oldest[52] $(\sqrt{3} \times \sqrt{3})R30°$ structure model for $\Theta_S = 1$ (Fig. 12 e). Although possible, it appears unlikely that there

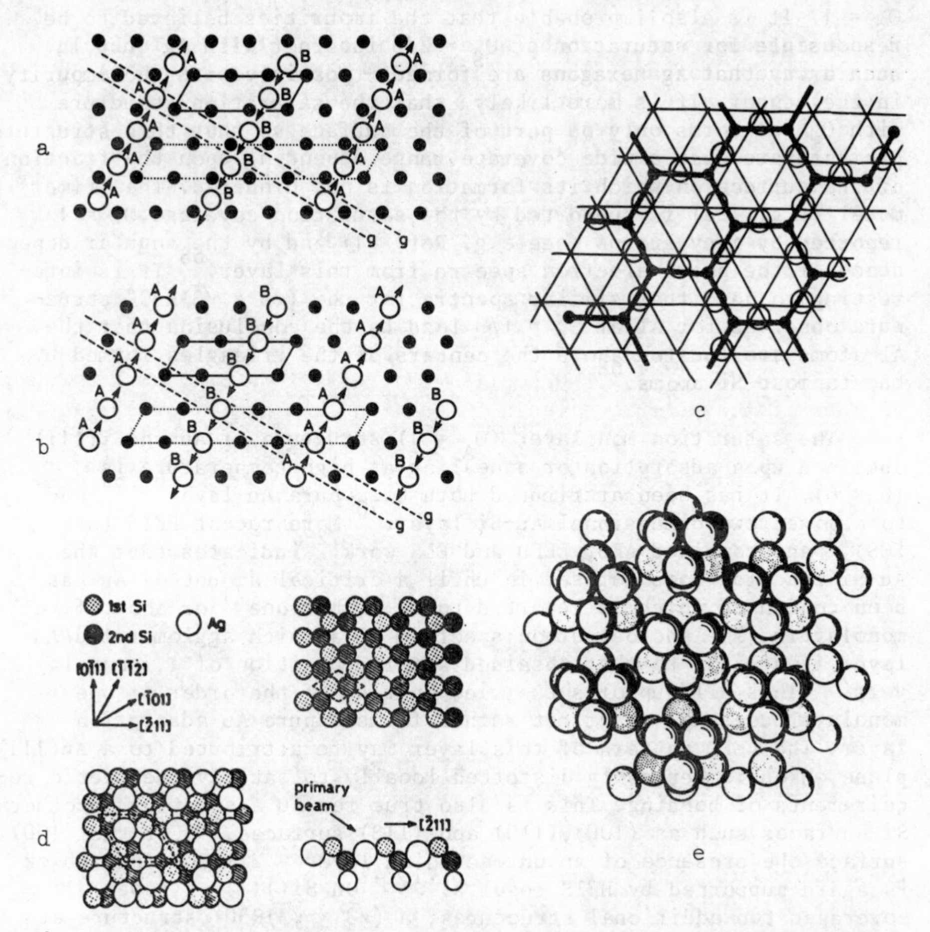

Fig. 12 : Structure models for Ag on Si(111).(a,b) : (3 × 1) and (6 × 1) structure for threefold hollow (a) and on-top adsorption (b); the arrows indicate possible displacements which transform the (3 × 1) into a (6 × 1) structure.[59] c) : ($\sqrt{3} \times \sqrt{3}$)R30° structure; the Si atoms are located at the corner of the triangles, the open circles indicate Si dangling bonds bent towards Ag atom pairs.[62] d): ($\sqrt{3} \times \sqrt{3}$)R30° structure, immersed Ag model ; top: Si(111) - (1 × 1), bottom : top and side view of Ag distribution[63] e) : ($\sqrt{3} \times \sqrt{3}$)R30° structure trimer model.[52]

are two well-defined structures corresponding to $\Theta_S = 2/3$ and $\Theta_S = 1$. It is also improbable that the impurities believed to be responsible for saturation at $\Theta_S = 2/3$ interact with Ag just in such a way that Ag hexagons are formed - possibly with the impurity in the center. It is more likely that the saturation structure with $\Theta_S = 1$ forms only on part of the surface so that this structure can saturate over a wide coverage range depending upon the fraction of the surface on which its formation is not hindered. The trimer model (Fig. 12e) is supported by the saturation coverage $\Theta_S = 1$ reported by many groups (see e.g. Ref. 61) and by the angular dependence of the photo-electron spectra from this layer.[68] It is interesting to note that similar spectra for the $(\sqrt{3} \times \sqrt{3})R30°$ structure observed for Al at $\Theta_S = 1/3$ lead to the conclusion that the Al atoms are located above the centers of the triangles formed by the topmost Si atoms.[68]

The saturation monolayer ($\Theta_A = 1$) structure of Au on Si(111) obtained upon adsorption or annealing at high temperature is (6 × 6). It has been attributed both to a pure Au layer[67,69] and to a mixed two-dimensional Au-Si layer.[70] More recent HEIS (see SCS)[71] and combined AES, LEED and ELS work[72] indicates that the Au-Si reaction does not set in until a critical amount of Au has been condensed which is reported to be either one[71] or about five[72] monolayers. The photoelectron spectrum of a thick agglomerated Au layer is similar to that obtained after deposition of 1.5 monolayers.[73] This all supports the view that Au of the order of one monolayer does not react but rather forms a pure Au adsorption layer. The LEED pattern of this layer may be attributed to a Au(111) plane which is certainly distorted locally to satisfy the steric requirements of bonding. This is also true for Au adsorption on other Si surfaces such as (100), (110) and (113) surfaces.[69] For the (100) surface the presence of an unreacted Au layer ≃ 2 monolayers thick is again supported by HEIS results.[71,74] On Si(111) at lower coverages two additional structures, a $(\sqrt{3} \times \sqrt{3})R30°$ structure at $\Theta_S = 1$ and a (5 × 1) structure at $\Theta_S = 2/5$ appear which can be interpreted in a way similar to the Ag adsorption structures. It should be noted that both for Ag and Au, elevated temperatures are necessary in order for the surface to transform from the reconstructed clean surface structure into the coverage-dependent equilibrium structure. As to the growth of the ordered structures, much should be learned by high resolution reflection electron microscopy as preliminary work[75] indicates.

In addition to the work on Ag, Au and Al there is a huge amount of work on the interaction of transition metals such as Pd, Pt, Ni, Fe but most of it was directed at the understanding of silicide formation. Very little reliable quantitative information on the adsorption process in the range $\Theta \leqslant 1$ can be extracted from the published data. At higher coverage there is always silicide formation which may occur even at liquid nitrogen temperature. There have also been a number of studies of the adsorption of less aggressive adsorbates such as Pb, Sn, Sb, Ga, in which in general leads to $(\sqrt{3} \times \sqrt{3})R30°$ structures on Si(111) as mentioned above for Al. For metal adsorption on Ge, only two examples should be mentioned briefly, Sn on Ge(111)[76] and Ag on Ge(111).[77] The first work is remarkable because in addition to a metastable $(\sqrt{3} \times \sqrt{3})R30°$ structure (with Θ_S = 1/3-1) stable (7 × 7) (Θ_S = 0.3 -0.8) and (5 × 5) (Θ_S = 0.5 - 0.9), structures were found similar to structures on Si(111). These may be important for the understanding of the Si(111) surface structures. Ag adsorption on a slightly misoriented (0.9°) (111) surface (which is stable in the clean state) produces not only several ordered structures on this plane but also causes the formation of (544) facets.

II.2. Compound Semiconductors

Cs adsorption has been studied in great detail on GaAs(110)[58,78,79] and less extensively on GaAs(100), (111), $(\bar{1}\bar{1}\bar{1})$, GaP and GaSb(100), (110), (111) and $(\bar{1}\bar{1}\bar{1})$ surfaces. For GaAs(110), multilayer formation has been reported[78] but here only the first layer will be discussed. This is completed on all (110) surfaces at about Θ_S = 1/2 and has a c(4 × 4) structure which is a compressed Cs(110) layer with $[001]_A \| [110]_S$. At room temperature no ordered structures are seen at lower coverages except on GaP where linear double spaced arrangements parallel to the GaP chains ($[1\bar{1}0]$) form. Adsorption at elevated temperature or annealing produces also on GaAs and GaSb ordered structures which can be attributed to ordered linear adatom distributions along the $[1\bar{1}0]$ directions with various coverages (1/6, 1/2). In these adsorption sites the bonding is considerably stronger than in the close-packed structure as clearly evident in ad- and desorption kinetics. $\Delta\phi$-measurements show sharp breaks at characteristic coverages similar to Cs on Ge(111) which are attributed to specific adsorbate configurations.[58]

Although much work has been done on many metal-compound semiconductor interfaces, information on the coverage range from 0 to about 1 monolayer is scarce for metals other than alkalis. Photoemission studies of Au on GaAs (110),[80] GaSb and InP (110),[81] for example, show that Au is initially present in atomic form before massive intermixing with the substrate occurs but nothing is known about the structure of the initial layer. Adsorption of column III elements Al and Ga on GaAs (110) does not change the lateral periodicity of the surface. The atoms adsorbed at room temperature might form two-dimensional, not necessarily ordered islands ("rafts"), which are sufficiently densly packed to produce a free-electron like UPS spectrum, e.g. for Ga,[82] or randomly adsorbed on top of the surface, e.g. for Al.[83,84] However, if a crystal covered with >1/2 monolayer of Al is annealed at 450°C, Al replaces the Ga atoms in the second atomic layer beneath the surface, according to a detailed LEED intensity analysis.[83,84] In this replacement reaction the structure of the topmost GaAs layer of the clean surface is retained, but with a relaxation of 0.1Å towards the second AlAs layer (see also Ref. 85). In contrast to the column III atoms, the column V atom Sb leads even at room temperature to a major change of the surface structure without changing the lateral periocicity.[82]

On II-IV semiconductor surfaces (of which the zincblende (110) and wurtzite (10$\bar{1}$0), (11$\bar{2}$0) surfaces were studied) LEED studies suffer from the dissociation of the surface by the electron beam. Photoemission, work function and electrical studies indicate however, that massive in-diffusion of Au, Ag, Cu, Al and In occurs into Cd chalcogenides at room temperature even at very low coverages. This is a field in which a considerable amount of information will certainly become available in the next few years.

III. CONCLUDING REMARKS

Although some metal adsorbates on metal and semiconductor surfaces such as the alkalis have already been studied rather thoroughly, the present understanding of the adsorption of most other metals is still quite rudimentary. Even in alkali adsorption the lateral interactions responsible for the appearance of ordered structures are known only very qualitatively. As a consequence, general statements about metal adsorption are necessarily of a rather sweeping type while specific statements based on the rather limited (and in many cases not very reliable experimental data) are apt to be overturned by future experiments. In the hope that this may not already be the case at the time of printing, the author has, therefore, tried to stay away from such statements.

REFERENCES

1 . KERN R., LELAY G. and METOIS J.J., in *Current Topics in Materials Science*, Vol. 3, 1979, pp.131-419
2 . BIBERIAN J.P. and SOMORJAI G.A., J. Vac. Sci. Technol. 16(1979) 2073-2085
3a. BAUER E., in *The Chemical Physics of Solid Surfaces and Heterogeneous Catalysis*, King. D.A. and Woodruff D.P. eds., Vol. IIIa, Elsevier, Amsterdam, to be published.
b. BAUER E., Appl. Surface Sci., to be published.
4 . SHIH H.D., JONA F., JEPSEN D.W. and MARCUS P.M., Phys. Rev. B15 (1977)5550
5 . GERLACH R.L. and RHODIN T.N., Surf. Sci. 17(1969)32-68 Surf. Sci. 19(1970)403-426
6 . PORTEUS J.O., Surf. Sci. 41(1974)515-532
7 . HENRION J. and RHEAD G.E., Surf. Sci. 29(1972) 20-36
8 . BARTHES M.G. and RHEAD G.E., Surf. Sci. 80(1979)421-429
9 . RAWLINGS K.J., GIBSON M.J. and DOBSON P.J., J. Phys. D11(1978) 2059
10. TAKAYANAGI K., KOLB D.M., KAMBE K. and LEHMPFUHL G., Surf. Sci. 100(1980)407-422
11. TAKAYANAGI K., Surf. Sci. 104(1981)527
12. VOOK R.W., HORNG C.T. and MACUR J.E., J. Cryst. Growth. 31 (1975)353-357
13. VOOK R.W. and HORNG C.T., Phil. Mag. 33(1976) 843-861
14. HORNG C.T. and VOOK R.W., J. Vac. Sci. Technol. 11(1974)140-143
15. BAUER E. and POPPA H., 1981, unpublished.
16. HUTCHINS B.A., RHODIN T.N. and DEMUTH J.E., Surf. Sci. 54(1976) 419-433
17. Van HOVE M., TONG S.Y. and STONER N., Surf. Sci. 54(1976)259-268
18. ANDERSON S. and PENDRY J.B., Solid State Commun . 16(1975)563-566
19. DEMUTH J.E., JEPSEN D.W. and MARCUS P.M., J. Phys. C8(1975) L25
20. WOODRUFF. D.P., NORMAN D., HOLLAND B.W., SMITH N.V., FARRELL H.H. and TRAUM M.M., Phys. Rev. Letters 41(1978)1130
21. WILLIAMS G.P., LAPEYRE G.J., ANDERSON J., CERRINA F., DIETZ R.E. and YAFET Y., Surf. Sci. 89(1979)606-614
22. SEPULVEDA A.and RHEAD G.E., Surf. Sci. 66(1977)436-448
23. ARGILE C. and RHEAD G.E., Surf. Sci. 78(1978) 115-124
24. PALMBERG P.W. and RHODIN T.N., J. Chem. Phys. 49(1968)134-146
25. MEDVEDEV V.K., NAUMOVETS A.G. and FEDORUS A.G., Sov. Phys. Solid State 12(1970)301
26. THOMAS S. and HAAS T.W., J. Vac. Sci. Technol. 9(1972)840-843
27. KABURAGI M. and KANAMORI J., J. Phys. Soc. Japan 43(1977)1686
28. BAUER E., in *Phase Transitions in Surface Films*, edit. by Dash J.G. and Ruvalds J., Plenum Press, N. Y. 1979, pp. 267-315

29. BOL'SHOV L.A., NAPARTOVICH A.P., NAUMOVETS A.G. and FEDORUS A.G. Sov. Phys. Usp. 20(1977)432
30. KANASH O.V., NAUMOVETS A.G. and FEDORUS A.G., Sov. Phys. JETP 40(1977)903
31. SCHLENK W. and BAUER E., Surf. Sci. 94(1980)528-546
32. BAUER E., POPPA H., TODD G. and BONCZEK F., J. Appl. Phys. 45 (1974)5164-5175
33. SCHLENK W. and BAUER E., Surf. Sci. 93(1980)9-32
34. BAUER E., POPPA H. and TODD G.,Thin Solid Films 28(1975)19-36
35. BAUER E., J. de Physique (PARIS) 38(1977)C4-146-C4-154
36. CHEN J.M. and PAPAGEORGOPOULOS C.A., Surf. Sci. 21(1970)377-389
37. MEDVEDEV V.K., NAUMOVETS A.G. and SMEREKA T.P., Surf. Sci. 34 (1973)368-384
38. GUPALO M.S., MEDVEDEV V.K., PALYUKH B.M. and SMEREKA T.P.,Sov. Phys. Solid State 21(1979)568
39. PERDEREAU J. and SZYMERSKA I., Surf. Sci. 32(1972)247-252
40. GORODETSKY D.A., MELNIK YU. P., SKLYAR V.K. and USENKO V.A., Surf. Sci. 85(1979)L503-L507
41. PRIGGE S., ROUX H. and BAUER E., Surf. Sci. 107(1981)101-112
42. BIBERIAN J.P., Surf. Sci. 74(1978)437-460 and references therein
43. GREEN A.K., PRIGGE S. and BAUER E., Thin Solid Films 52(1978) 163-179
44. TAKAYANAGI K., Proc. 39th Ann. Meet. Electr. Micr. Soc. Am., (1981)204-207
45. BARTHES M.G. and PARISET C.,Thin Solid Films 77(1981)305-312
46. GUGLIELMACCI J.M. and GILLET M.,Thin Solid Films 68(1980)407-416
47. YAGI K., TAKAYANAGI K., KOBAYASHI K., OSAKABE N., TANISHIRO Y. and HONJO G., Surf. Sci. 86(1979)174-181
48. FUJINAGA Y., Surf. Sci. 84(1979)1-16
49. CHAO S.S., VOOK R.W. and NAMBA Y., J. Vac. Sci. Technol. 18 (1981)695-699 and references therein
50. MACUR J .E and VOOK R.W., Thin Solid Films 66(1980)311-324, 371-379
51. ABRAHAMSON F.F. and BRUNDLE C.R., J. Vac. Sci. Technol. 18 (1981)506-519
52. LANDER J.J. and MORRISON J., Surf. Sci. 1(1963)125-164 Surf. Sci. 2(1964)553-565; J. Appl. Phys. 36(1965)1706
53. PALMBERG P.W. and PERIA W.T., Surf. Sci. 6(1967)57-97
54. WEBER R.E. and PERIA W.T., Surf. Sci.14(1969)13-38
55. HOLTOM R.and GUNDRY P.M.,Surf. Sci. 63(1977)263-273 and references therein
56. WEBER R.E. and JOHNSON A.L., J. Appl. Phys. 40(1969)314
57. SURNEV L. and TIKHOV M.,Surf. Sci. 85(1979)413-431 and references therein
58. CLEMENS H.J., von WIENSKOWSKI J. and MÖNCH W., Surf. Sci. 78 (1978)648-666

59. ICHIKAWA T. and INO S., Surf. Sci. 97(1980)489-502 and references therein
60. LELAY G., MANNEVILLE M. and KERN R., Surf. Sci. 72(1978)405-422 and references therein
61. KERN R. and LELAY G., J. de Physique 38(1977)C4-155
62. BARONE V.,DEL RE G., LELAY G. and KERN R., Surf. Sci. 99(1980) 223-232
63. SAITOH M., SHOJI F., OURA K. and HANAWA T.,Jap. J. Appl. Phys. 19(1980) L 421-L424
64. OURA K., TAMINAGA T. and HANAWA T., Solid. State Commun. 37 (1981)523-526
65. TERADA Y., YOSHIZUKA T., OURA K. and HANAWA T., Jap. J. Appl. Phys. 20(1981)L333-L336
66. GASPARD J.P., DERRIEN J, CROS A. and SALVAN F.,Surf. Sci. 99 (1980)183-191
67. BAUER E., J. Japan. Assoc. Cryst. Growth. 5(1978)49-71
68. HANSSON G.V., BACHRACH R.Z., BAUER R.S. and CHIARADIA P., J. Vac. Sci. Technol. 18(1981)550
69. GREEN A.K. and BAUER E., Surf. Sci. 103(1981)L127-L133 and references therein
70. LELAY G. and FAURIE J.P., Surf. Sci. 69(1977)295-300
71. NARUSAWA T., KINOSHITA K., GIBSON W.M. and HIRAKI A., J. Vac. Sci. Technol. 18(1981)872-875
72. OKUNO K., ITO T., IWAMI M.and HIRAKI A.,Solid State Commun. 34 (1980)493-497
73. ABBATI I., BRAICOVICH L.,FRANCIOSI A., LINDAU I., SKEATH PR SU C.Y. and SPICER W.E., J. Vac. Sci. Technol. 17(1980)930-935
74. NARUSAWA T., KINOSHITA K. and GIBSON W.M., Proc. 4th Intern. Conf. Solid Surf. and 3rd Europ. Conf. Surf. Sci. (1980)673-676;
75. YAGI K., OSAKABE N., TANISHIRO Y. and HONJO G., ibid. pp. 1007-1010, 1980; Surf. Sci. 97, pp. 393-408
76. ICHIKAWA T. and INO S., Surf. Sci. 105(1981)395-428
77. SULIGA E. and HENZLER M., 1981 to be published
78. DERRIEN J. and ARNAUD D'AVITAYA F., Surf. Sci. 65(1977)668-686
79. Van BOMMEL A.J. and CROMBEEN J.E., Surf. Sci. 93(1980)383-397 and references therein
80. LINDAU I., SKEATH P.R., SU C.Y. and SPICER W.E., Surf. Sci. 99(1980)192-201
81. CHYE R.W., LINDAU I., PIANETTA P., GARNER C.M., SU S.Y. and SPICER W.E., Phys. Rev. B18(1978)5545-5559
82. SKEATH P., SU C.Y., LINDAU I. and SPICER W.E., J. Vac. Sci. Technol. 17(1980)874-879
83. KAHN A., KANINI D., CARELLI J., YEH J.L., DUKE C.B., MEYER R.J. PATON A. and BRILLSON L., J. Vac. Sci. Technol. 18(1981)792-796
84. DUKE C.B.,PATON A.,MEYER R.J.,BRILLSON L.J., KAHN A., KANINI D; CARELLI J.,YEH J.L.,MARGARITONDO G. and KATNANI A.D.,Phys.Rev. Letters 46(1981)440-443
85. HUIJSER A.,van LAER J and van ROOY T.L.,Surf. Sci.102(1981)264 270 and references therein.

CHEMISORPTION ON SINGLE CRYSTAL AND HETEROGENEOUS CATALYSIS

J. OUDAR

Laboratoire de Physico-Chimie des Surfaces,
Ecole Nationale Supérieure de Chimie de Paris,
11, rue Pierre et Marie Curie,
75231 Paris Cedex 05, France.

ABSTRACT

In the first part of this paper, the equilibria of adsorption and segregation are treated in a symmetrical manner, referring to the Fowler-Guggenheim isotherms.

In the second part, examples of first order transition on well defined surfaces (single crystals) are given for

i) sulphur adsorption on silver

ii) carbon segregation on nickel.

In the last part, the interest of heterogeneous catalysis on single crystals is illustrated by two examples : CO methanation on nickel and CO oxidation on platinum.

I. ADSORPTION AND SEGREGATION EQUILIBRIA

When a metallic surface at high temperature is in contact with a reactive gas such as oxygen, sulphur or a halogen, two types of interactions must be considered.

If the pressure is higher than the dissociation pressure of the most stable bulk compound, (oxide, sulphide or halide) this compound can be formed. The reaction may go on until consumption of the metal is complete, or, it may give rise to a surface layer whose growth is limited by diffusion.

B. Mutaftschiev (ed.), Interfacial Aspects of Phase Transformations, 433–451.

When the pressure is lower than the dissociation pressure of the stable compound, interactions result in the formation of adsorption layers. Due to the high affinity between gas and metal resulting from the unsaturated metallic surface bonds, the adsorption bonds are of chemical nature and of high energy. Later we will see that these bonds can be even stronger than in the bulk compound. Adsorption stops when the whole surface is completely covered. In the process of oxidation, the adsorption stage can be considered as the precursor stage of bulk compound formation. The adsorption layer can be obtained not only from the gas phase but also by segregation of atoms dissolved in the bulk phase. However, in that case, the temperature must be high enough to allow diffusion to the surface. A unified treatment of adsorption and segregation phenomena can be made using equilibrium concepts. The only distinction between the two is in the expressions for the chemical potential of the adsorbed species. For adsorption, the chemical potential is referred to the vapor phase pressure. In case of segregation, it is referred to the bulk concentration in the condensed phase. Equilibrium segregation concepts can be extended to other types of defects such as grain boundaries, dislocations and vacant lattice sites (Fig. 1). A new distribution of active species between phases in equilibrium occurs when defects are present. It essentially results in a lowering of the total free energy of the system.

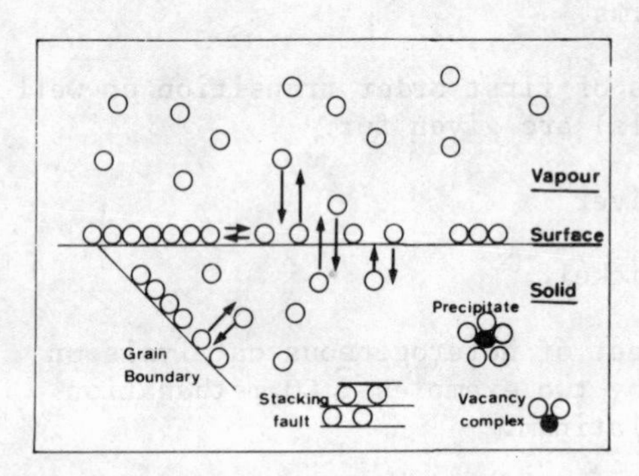

Fig. 1 : Adsorption or segregation on external and internal surfaces and on various defects.

As for the treatment of solid solution-surface equilibria, only dilute binary systems will be considered here (systems in which dissolved atoms are in low enough concentrations not to interact each other in the bulk of the crystal).

A useful way of studying such systems is to construct adsorption isotherms. The superficial concentration of adsorbed atoms is measured as a function of the chemical potential, expressed in terms of gas pressure or of bulk concentration.

We will first show how different interactions between adsorbed atoms influence the properties of the theoretical isotherms. Later, comparison of these results with real examples will be made.

II. ADSORPTION AND SEGREGATION ISOTHERMS

To obtain an analytical expression for an adsorption isotherm, the chemical potentials in both phases, vapor and superficial, for the various constituents are set identically equal. The segregation isotherm is obtained in the same way by referring not to the gas but to the condensed phase. We shall restrict ourselves here to the special cases of a vapor phase behaving as a perfect monoatomic gas and a condensed phase behaving as a dilute solid solution.

According to statistical thermodynamics[1], the chemical potential of the atoms of a monoatomic gas is given by the relation :

$$\mu_g = k_B T \, ln \, P \, (k_B T)^{5/2} \, (2\pi m)^{3/2} \qquad (1)$$

where P is the gas pressure, m is the mass of the atom, and k_B is the Boltzmann constant.

Assuming that dissolved atoms have access to only one kind of surface site, their chemical potential, in the dilute solid solution, is given by the equation :

$$\mu_b = E_b + k_B T \, ln \, X - k_B T \, ln \, f_b \, (T) \qquad (2)$$

where E_b represents the potential energy of a dissolved atom, X is the atomic fraction of dissolved atoms, and f_b (T) is the vibrational partition function of the dissolved atoms.

A simple model to describe the adsorbed phase is based on the following assumptions :

i) adsorbed atoms are localized on well defined and equivalent sites of the substrate;

ii) interactions are restricted to nearest neighbors;

iii) there is a random distribution of adsorbed atoms on substrate sites (regular solutions).

The degree of coverage θ of the surface is defined as the ratio between the number of adsorbed atoms and the number of adsorbed atoms when all adsorption sites are occupied. Let E_s be the potential energy of a surface site and let w be the energy of interaction between adsorbed nearest neighbors with

$w > 0$ in the case of a repulsive interaction,

$w < 0$ in the case of an attractive interaction, and

Z = the maximum number of nearest neighbors (two dimensional coordination number).

In the Bragg-Williams approximation, the term $Zw\theta$, which represents the energy of interaction weighted by the number of neighbors $Z\theta$, is added to the chemical potential of the adsorbed atoms. The chemical potential of the adsorbed atoms then becomes :

$$\mu_s = E_s + Zw\theta + k_B T \, ln \frac{\theta}{1-\theta} - k_B T \, ln \, f_s(T)$$

where $f_s(T)$ is the vibrations partition function of an adsorbed atom. Setting $\mu_s = \mu_g$ and $\mu_s = \mu_b$, the isotherms of adsorption and of segregation respectively take the following forms :

$$P = \frac{\theta}{1-\theta} \alpha(T) \exp \frac{E_s + Zw\theta}{k_B T}$$

$$X = \frac{\theta}{1-\theta} \alpha'(T) \exp \frac{Zw\theta - E}{k_B T}$$

with $E = E_b - E_s$.

These relationships are known as FOWLER-GUGGENHEIM isotherms.[1]

A convenient representation of the isotherms is to plot θ as a function of $ln[P(\theta)/P(1/2)]$ or $ln[X(\theta)/X(1/2)]$ for different values of $Zw/k_B T$. This shows the symmetry of the curves with respect to $\theta = 1/2$ (Fig. 2). For $Zw/k_B T < -4$ the isotherms show a shape with the two loops characteristic of a phase transition. The dotted portions of the curves represent unstable parts of the isotherms. For $Zw/k_B T = -8$, for example, a surface equilibrium exists between a dilute phase with degree of coverage θ_A and a dense phase with degree of coverage θ_B. The relative proportion of each of these phases is given by the rule of moments applied to the three points, θ, θ_A and θ_B. The critical temperature at which the phase change occurs is $T_c = - Zw/4k_B$, corresponding to an isotherm with a vertical tangent of inflection (Fig. 2). Finally, the Langmuir isotherm $Zw/k_B = 0$ separates the isotherms into two types depending on the type of interaction, attractive or repulsive. One of the chief merits of the Fowler-Guggenheim isotherm is that it makes it possible to predict phase changes in the adsorbed layer in the case of attractive interactions.

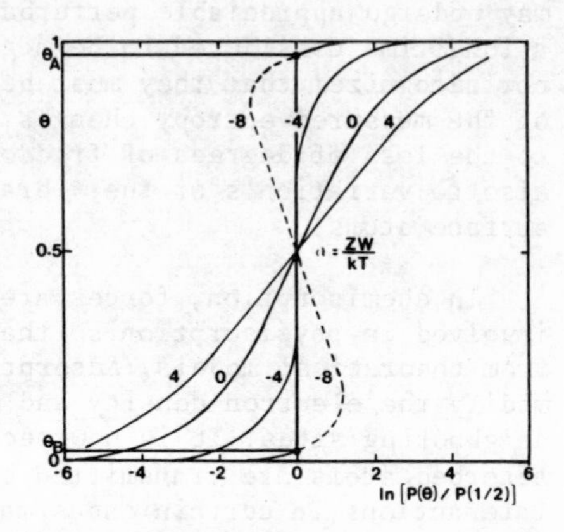

Fig. 2 : Fowler-Guggenheim isotherm of adsorption. The shape of the isotherm of segregation is the same if log $X(\theta)/x(1/2)$ is plotted instead of log $P(\theta)/P(1/2)$

III. ADSORPTION AND SEGREGATION ON WELL-DEFINED SURFACES

III.1. General Considerations

The above adsorption isotherm model is based on the assumption of a homogeneous surface (only one kind of adsorption site). In order to experimentally approach this ideal situation, it is necessary to use either bulk single crystals or lamellar compounds with all crystals having a preferential orientation. Such compounds have been widely used in studies of physisorption. With such substrates it is possible to achieve both the homogeneous character of the surface and a large specific area. Conventional techniques for determining coverage of adsorbed gases, such as volumetry or gravitmetry, can then be used.

Physisorption studies allow us to formulate very general concepts concerning phase changes in adsorbed layers. The field has been widely developed since the first significant results were obtained for rare gases adsorbed on exfoliated graphite.[2] Since then, a large number of systems with very different substrates and gases have been studied by volumetry and calorimetry. More recently, the field of low pressures on monocrystals has been explored by means of Auger spectroscopy and LEED.[3] From the set of isotherms, various phase transitions have been observed such as 2D gas $\rightleftarrows$ 2D liquid or 2D liquid $\rightleftarrows$ 2D solid.[2a] Changes in enthalpy and entropy associated with phase changes have been measured precisely from adsorption isotherms or directly by calorimetry.[4]

Attempts have been made to explain the experimental results of physisorption with simple models such as the localized model with lateral interactions, described above, or a Van der Waals type fluid. The agreement with experiment of these two approaches is rather disappointing. One of the main difficulties when treating the problem theoretically comes from the fact that the substrate may undergo appreciable perturbations. Such perturbations were for a long time considered to be negligible in physisorption. It is now recognized that they must be considered in the interpretation of the measured entropy changes. These changes are due not only to the loss of degrees of freedom of the adsorbed molecules, but also to variation's of the vibrational modes of the substrates surface atoms.

In chemisorption, forces are generally greater than those involved in physisorption so that reality is even farther away from theoretical models. Adsorption on a site may significantly modify the electron density and consequently the reactivity of neighboring sites. It is now recognized that interactions between adsorbed atoms are transmitted by conduction electrons. These interactions,in certain cases,may operate over several interatomic distances, as shown by the variations in work function caused by adsorption.

On a homogeneous surface, for example that of perfect {111} or {100} faces of an f.c.c. metal, different adsorption states characterized by different binding modes for the same adsorbate (bridge position, linear, etc.) may follow each other as a function of the degree of coverage, or they may even coexist. Additional complications may come from the dissociation of the molecule or from the restructuring of the surface by adsorption (e.g. faceting).

Contrary to physical adsorption, reversibility is attained in chemisorption mostly at high temperatures and under very low pressures of the reactive gas. In addition to the difficulties related to observation at high temperatures, other phenomena such as dissolution of the adsorbed atoms in the interior of the metal, or evaporation of the metal itself, may be superimposed on the adsorption-desorption phenomenon.

III.2. Adsorption of Sulphur on Metals (Ni, Fe, Cr, Mo, Pt)

Adsorption of sulphur on transition metals (Fe, Ni, Mo, Pt, Pd) and IB metals (Cu, Ag, Au) has been extensively studied by BENARD, OUDAR and coworkers.[5,6] Structural and kinetic data were obtained by means of Low Energy Electron Diffraction (LEED) and sometimes Reflection High Energy Electron Diffraction (RHEED). Thermodynamical data were obtained by plotting the adsorption-desorption isotherms of H_2S-H_2 mixtures.

Superficial concentrations of adsorbed atoms were measured by using radioactive sulphur (^{35}S) and Auger electron spectroscopy. The calibration of the AES was based on a radiotracer method. The relative precision of the con entration measurements was better than 5 percent.

In most investigations, a low index face is placed under a low hydrogen sulphide partial pressure. The first adsorption state observed comes from dissociation of hydrogen sulphide when it is put in contact with the clean metallic surface. At this stage of the reaction, sulphur atoms are mobile and randomly distributed on the surface. In a first approximation, interactions between adsorbed atoms can be neglected. When a critical concentration is reached, a new dense phase appears.

Growth proceeds, at medium temperatures, by nucleation of islands and can be characterized by LEED. Thin spots can be observed at a coverage much lower than that of the completed structure. Such a phase is sometimes called bidimensional (2D) sulphide.

It is possible (at least theoretically) to distinguish between the two adsorption states described above by varying the sulphur partial pressure, at a given temperature. Figure 3 represents the theoretical adsorption-desorption isotherms that shows the existence of the two adsorption states. Both states are in equilibrium with each other at pressure P_1, and their transformation into one another is first order.

Fig. 3 : Theoretical adsorption isotherm with first order transition. (Note that contrary to fig. 2, P is plotted rather than log P).

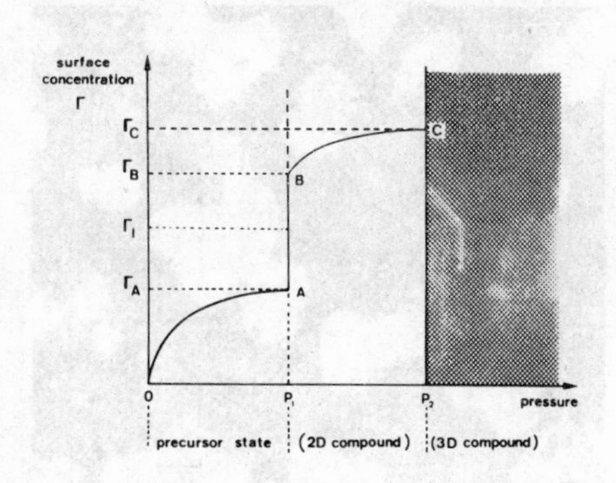

Γ_A and Γ_B represent the compositions of the equilibrium phases at a given temperature. The sulphur concentration of the initial state may vary form zero up to Γ_A, and the 2D sulphide from Γ_B to Γ_C. In the latter concentration range the 2D sulphide is non-stoichiometric. The difference between P_2 and P_1 if expressed in terms of free energies indicates the excess stability of the bidimensional sulphide compared to the tridimensional.

To illustrate the nucleation process, we will consider adsorption of sulphur on {100} Ag. The isotherm[7] obtained at 190°C is very close to the one represented in figure 3. The composition Γ_A of the dilute phase corresponds to approximatively 1 sulphur atom for every 4 superficial silver atom; and Γ_B (of the dense phase) to 1 sulphur atom for every 2 silver atoms. For concentrations between Γ_A and Γ_B, these two phases are both present and can be "frozen in" by quick quenching under vacuum (their respective proportions are given by the rule of moments). These phases can be seen by a decoration technique based on selective electrodeposition of silver from an $AgNO_3$ solution[7]. The micrograph of figure 4 corresponds to a mean coverage of 3S/8Ag, i.e. approximately to identical proportions of both phases. The regions covered by silver deposit are black. In those regions, the local sulphur concentration is that of the initial dilute adsorption state (1S/4Ag). The electrodeposition is, however, totally inhibited in the regions covered by the dense phase (1S/2Ag) which remain white.

If we could carry out systematic experiments to extablish adsorption desorption isotherms at different temperatures, we would get essential data concerning phase transitions in adsorbed layers. However, because of the strong affinity of transition metals for sulphur, the conditions in which adsorption is reversible are difficult or even impossible to obtain. It is very difficult to obtain gaseous mixtures of very low hydrogen sulphide concentration, even with a hydrogen-hydrogen sulphide mixture.

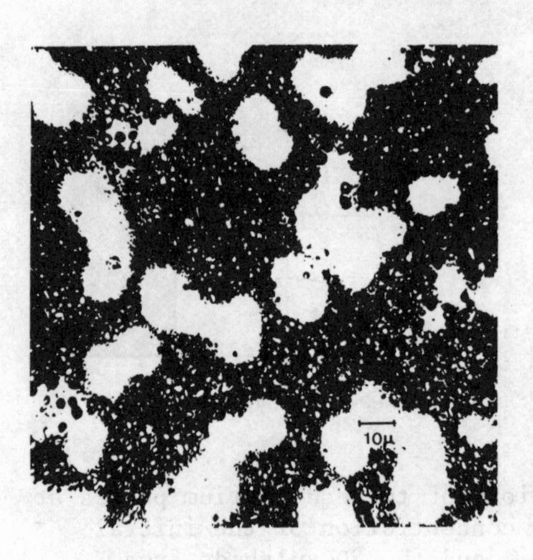

Fig. 4 : 2D sulphide nuclei (white region) shown by selective electrodeposition of Ag (after Rousseau et al.[7]).

These experimental difficulties can be overcome, on metallic monocrystals, for some specific systems such as sulphur-silver. In many cases, only the upper part of the isotherms was observed.

This permits the characterization of the saturation state when the surface is totally covered with the 2D sulphide, but does not give any information concerning its equilibrium with the precursor state.

In addition, even in the most favorable cases mentioned previously, the shape of the isotherms is very sensitive :

i) in the lower region, to the presence of crystalline defects, monoatomic steps for example[8] which are more active than the rest of the surface,

ii) in the upper region, to the existence of several 2D adsorbed sulphides.[9]

The coexistence of these different phenomena makes the interpretation of results more complex, especially considering the rather poor precision involved in measuring the isotherms.

Another way to study reversible equilibria between the precursor state and the surface compound requires conditions such that the surface can be assumed to be an isolated chemical system. To do this, exchanges between the adsorbed layer and the vapor phase, and between the adsorbed layer and the solid solution must be minimized. This can be realized by carrying out experiments under ultra high vacuum because adsorption states have a very high stability. Moderate temperatures are required in order to avoid any phenomenon of solubility. The system is then entirely defined by the temperature and the superficial concentration of adsorbed atoms. With a certain mobility of the adsorbed atoms, which can be checked by the absence of any hysteresis phenomenon, realistic diagrams of equilibrium between adsorbed phases can be obtained. The method is based on the use of Low Energy Electron Diffraction data and has been described in detail elsewhere.[10]

Studying sulphur segregation at the surface of metals requires great care because of the highly tensioactive character of this element. As an example, for a temperature of approximatively 800°C, a critical concentration of a few ppm on iron or copper, is enough to reach the upper plateau of the isotherm. On the other hand, the collection of thermodynamical data concerning both adsorption and dissolution phenomena permits, by their difference, deduction of thermodynamical data concerning the segregation phenomenon. Fig. 5 shows the different energy levels for copper. One sees here (and in several other systems such as S-Ag,[6a], S-Ni, S-Ru[11]) that sulphur is more tightly bound in the surface phase than in the bulk. In polycrystalline copper, energy of segregation of sulphur at the grain boundaries has also been measured[12] (Fig. 5)

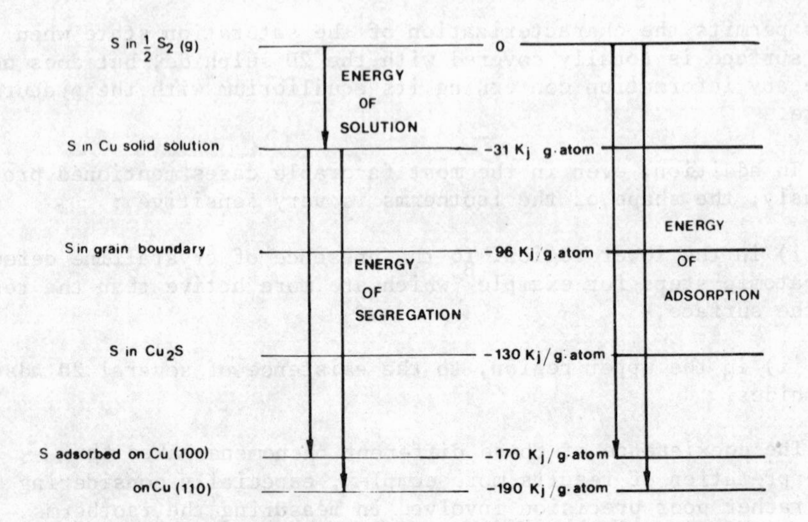

Fig. 5 : Energy-level diagram for copper-sulphur system.[52]

As far as crystal orientation is concerned, it is observed that the more rough the face is (on an atomic scale) the more tightly bound the sulphur is. This phenomenon seems to be a rather general rule.

Provided the partial pressure of the reactive gas is sufficient, a 3D compound (oxide, sulphide or halide) can appear. Numerous works were carried out between 1950 and 1960, in particular by Bénard and his coworkers, to study the process of nucleation in the growth of oxide, sulphide and halide layers.[13,14] The most favorable experimental conditions are high temperature and low reactive gas pressure. The growth of oxide or sulphide nuclei can be easily explained by surface diffusion phenomena, whereas the origin of nucleation phenomena is rather poorly understood. In some cases, metal defects such as dislocations are at the origin of the growth process of the 3D compound.[15] If it is assumed that such defects play a favorable role, especially in conditions of weak supersaturation, they still do not justify the determining role of pressure and crystal orientation on the number of nuclei.

It is well known that 3D nuclei are formed after the completion of the adsorbed layer. It may then be asked what influence the crystalline fefects of the adsorbed layer have on the nucleation of the 3D compound. It can be assumed that some of these defects are preferential paths of diffusion of the metal. In this case they would play a role comparable to that of grain boundaries or lines of dislocation in the growth of thin films of oxide.[16]

Recently, the formation of NiO on single crystals of Ni by an anodic process has been extensively studied. It appears that OH- anions play an important role as a precursor for oxidation in acid solution.[17] Sulphur adsorbed on the surface in the range of one monolayer can suppress the adsorption of OH^- anions and consequently the formation of the passive film of NiO.[18,19]

III.3. Carbon Segregation at a Nickel Surface

The surface segregation of carbon in single crystals of nickel[20,21,22] has been precisely studied using LEED and AES. On Ni(111), it was found that sharp changes in coverage versus temperature occured as shown in fig. 6

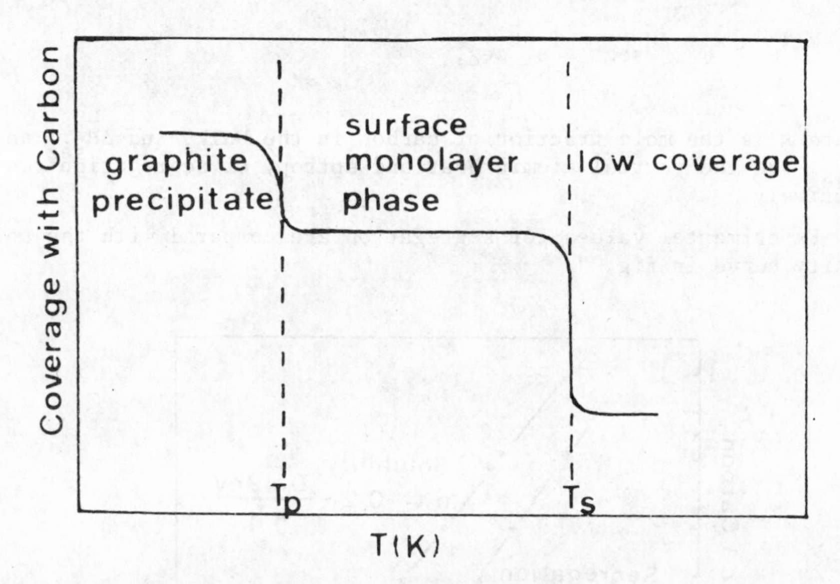

Fig. 6 : Schematic equilibrium temperature dependence of carbon coverage on the (111) surface of a carbon doped nickel single crystal. A phase transition from a low coverage to a condensed state occurs at the segregation point T_s. Graphite precipitation starts at T_p (after BLAKELY et al.[20-22]).

Three different states can be identified. In the high temperature region, the fractional coverage is very low and may be approximately equal to the bulk atomic fraction (dilute phase). In the intermediate temperature range, a surface condensed phase (interpreted as a carbon monolayer) exists and has a range of stability of 100 K. At lower temperatures, the surface is covered by a thick precipitate of graphite. The sharp change in carbon coverage at temperature T_s may be described as a first order phase transition. The origin of the transition was attributed to strong adsorbate-adsorbate interactions and to the excellent geometrical fit of the ordered phase to the metal substrate. It is possible to deposit a hexagonal carbon monoatomic layer on Ni(111) with the same arrangement and similar interatomic distances as the basal plane of graphite.

Results have been analysed by assuming that the chemical potential of carbon in the dilute bulk solution is equal to that in the two-dimensional phase at T_s and to that in bulk graphite at T_p. With such conditions, the variation of the phase transition temperature T_s with bulk composition can be represented by the following equation :

$$k_B T_s \ln x = \Delta\bar{H}_{seg} - T_s \Delta\bar{S}_{seg},$$

where x is the mole fraction of carbon in the bulk, and $\Delta\bar{H}_{seg}$ and $\Delta\bar{S}_{seg}$ are the partial atomic heat and entropy of segregation respectively.

Experimental values for segregation are compared with the solubility curve in fig. 7.

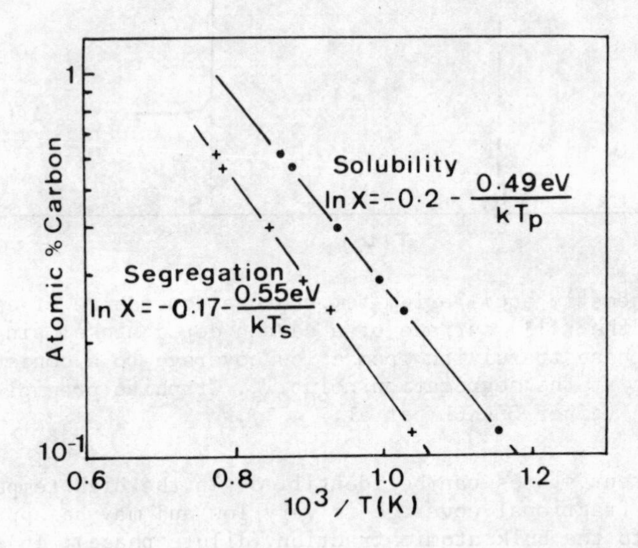

Fig. 7 : Solubility curve of C in Ni. The experimental values of $1/T_p$ that determine the doping levels of the various sample are marked on this curve. The corresponding values of $1/T_s$ determine the segregation curve. The slope of the two curves yield the heat of solution and heat of segregation respectively.(After Blakely et al.[20-22])

From the value of $\Delta\bar{H}_{seg}$ (-0.55 eV) one can conclude that the binding energy for a carbon atom in the monolayer phase is approximately 10 % greater that for bulk graphite. Other orientations of Ni (low index and vicinal planes) as well as other metals have been studied.[22] Depending on the system, the same behavior as on Ni(111), or behavior fitted to a Langmuir model was observed. In some cases, the carbon exhibits no detectable preference to accumulate on the surface.

IV. CATALYSIS ON SINGLE CRYSTAL SURFACES

IV.1. Interest of this Way of Studying Catalysis

The increased understanding of catalytic reactions on single crystal surfaces provided new interest in heterogeneous catalysis. Modern techniques such as LEED, AES, XPS, UPS, SIMS are now used for surface characterization. They permit determination of chemical composition and structure of the surface layers with a much better precision than on dispersed substances. In addition, the surface structure of a metal varies with its orientation and thus its effect on catalytic activity can be studied. In particular, experiments have been carried out on vicinal surfaces with defects such as monoatomic steps and kinks whose nature and densities can be controlled. Such an approach makes it possible to specify the nature of the sites responsible for catalytic activity. It has been shown by using such surfaces, that on platinum, monoatomic steps are mainly responsible for breaking of C-H and H-H bonds, and kinks for the breaking of C-C bonds.[23] Such mechanisms are involved in reforming and catalytic cracking, which are fundamental to the petroleum industry.

One could speculate whether catalysis experiments with single crystals are realistic or not. For industrial purposes, experiments are carried out on divided substances, possibly dispersed, on a support, having a large specific active surface. Most reactions sensitive to the surface structure are also sensitive to crystallite size. A common explanation for this is based on the fact that a change of crystallite size modifies the respective proportion of the different active sites. In particular, for crystallites of smaller size, the proportion of atoms of low coordination number in edge or vertex positions becomes much higher. These size effects are most significant for crystallite sizes between 10 and 50 Å. In this range, the proportion of atoms of high and low coordination number varies considerably. The influence of the surface structure on catalytic activity can be shown unambiguously on single crystal surfaces of various orientations presenting surface atoms whose coordination number can be varied at will. However, for highly dispersed catalysts, the influence of the support must also be taken into account.

Two types of experiments have recently appeared in the studies of catalysis on single crystals. The first type uses low experimental pressures. The structure and the composition of the surface may be determined *in situ* by previously mentioned techniques. The second type uses middle or high pressures. The initial and final states of the surface can be characterized only before and after reaction when the sample is under ultrahigh vacuum. The first type of study gives essential information about the mechanisms, but it requires experimental conditions which are far from being realistic in catalysis. Numerous examples are now well known which prove that the nature of the reaction intermediaries and, consequently, the involved mechanisms are very dependent on the reactive pressures.

In addition to the fundamental structural aspects, the chemical composition of the surface is also an essential factor. To interpret results in terms of kinetics, knowledge of the nature, the concentration of adsorbed atoms or molecules, of their arrangement, and their mobility is important and sometimes absolutely needed. For example, consider the apparently simple case of a bimolecular reaction :

$$A_{(g)} + B_{(g)} \rightarrow AB_{(g)}$$

It is well known that one of the main functions of a catalyst consists of adsorbing one or both reactive species, thus lowering the activation energy of the reaction compared to the reaction in a homogeneous gaseous phase. Let us imagine that A and B may be simultaneously adsorbed :

$$A_{(g)} + B_{(g)} \rightarrow A_{ad} + B_{ad} \rightarrow AB_{(g)}$$

Further suppose that A_{ad} is much more tightly bound than B_{ad} so that it will be in greater proportion on the surface. Two possibilities must be considered :

1) Islands of A_{ad} are formed on the surface and B_{ad} is adsorbed in the surface regions between the islands. Consequently, the reaction will take place on the edge of the islands, provided that B_{ad} is mobile enough.

2) A mixed layer of A_{ad} and B_{ad} is formed. The reaction will take place anywhere on the surface where bimolecular collisions occur, provided that A_{ad} or B_{ad} (or both) are mobile enough.

It is clear in both cases that the kinetics of the reaction, and in particular its order with respect to one or the other reactants, will be slightly different. It may be more complex if the adsorption of one reactant modifies the adsorption energy of the other one, or even totally inhibits its fixation and/or dissociation on the surface. Complex changes in the kinetics are then observed when the degree of coverage of the various adsorbed species varies. These different situations have been correlated to kinetic data in the case of the oxidation of CO on palladium.[24]

In certain reactions involving hydrocarbons, a carbonaceous deposit is formed on the surface of the catalysts. It comes from the partial or total decomposition of a reactant or of an intermediary of the reaction.[23] This deposit has been characterized for the reaction of methanation of CO on nickel. Its character depends on experimental conditions and it can be correlated to the catalytic activity of the metal (to be discussed later).

Finally, any correlation between structure and catalytic activity must necessarily take into account the possible modifications of the structure of the metal in the course of a reaction. It is recognized that the surface of a metal may be transformed by adsorption phenomenon. Some crystallographic planes may become unstable, undergo restructuring, and change into planes of different crystallographic orientation stabilized by adsorption.[25-28] The coalescence of steps, which occurs due to surface diffusion of the metal, is one of the usual processes which can account for this type of structural modification.

Monoatomic steps can also change into steps with different crystallographic orientation. It is now becoming accepted that these phenomena (known as two or one-dimensional faceting), are particularly important in heterogeneous catalysis. They can change the nature of the sites of the surface and explain the existence of some transient mechanisms observed in the beginning of reactions.

We will restrict ourselves to analysis of two examples of model catalysis on single crystals. Both examples underline the importance of good surface characterization to understand mechanisms of heterogeneous catalysis.

IV.2. <u>CO Methanation on Nickel</u>

The formation of methane by carbon monoxide hydrogenation on nickel has been the subject of a large number of studies using both divided substances and single crystals. This metal can dissociate CO according to the following dismutation reaction :

$$2\,CO \rightarrow CO_2 + \underset{\downarrow}{C}.$$

Experiments with nickel (100) have clarified the influence of the so formed carbon on the reaction kinetics. The experimental conditions were a temperature between 450 and 700 K under a total pressure of 120 Torr and a ratio H_2 : CO = 4 : 1.[29]

Two types of deposits were distinguished by their Auger[29b] spectrum (Fig. 8). The first is associated with a surface active to the methanation reaction and can be obtained by heating at 600K under CO gas. It corresponds to an amount of carbon which does not exceed one monolayer and its spectrum is analoguous to that of bulk nickel carbide (hence the name "carbodic"). It can be easily reduced at 700 K under hydrogen.

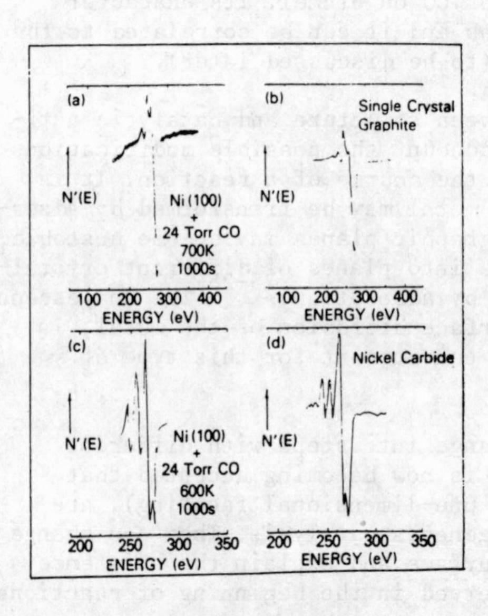

Fig. 8 : Comparison of AES Carbon signals on Nickel (100) crystal with those from single crystal Graphite and Nickel Carbide. a) Following 1000 sec heating at 700 K in 24 Torr CO; b) Single crystal Graphite; c) Following 1000 sec heating at 600 K in 24 Torr CO; d) Nickel Carbide (After Kelley et al.[29b])

The other deposit corresponds to several graphite carbon monolayers and can totally poison the surface. It is formed by heating at 700 K in a CO atmosphere. It is, however, much less easily reduced by hydrogen. Others authors have shown that the active carbon is formed and is reacting during the entire methanation reaction. This carbon can be assimiled as a true intermediary compound of the reaction.[30,31]

It is tempting to compare these carbon deposits with the ones that are formed by surface segregation when cooling a carbon-nickel alloy (see above). In both cases, a surface phase and a bulk phase can be obtained depending on the temperature.

Finally note that the catalytic activity, defined as the ratio between the number of methane molecules and the number of surface sites (Turnover Number), is very similar on both single crystals and on divided metals.[29] This is a good example of a model catalysis reaction on single crystals. Analogous studies should allow a better understanding of the formation of carbon-carbon bonds. They always come from a CO and H_2 mixture and lead to the synthesis of hydrocarbons of higher atomic weights (Fischer-Tropsch reaction). It is also interesting to find additive compounds which could prevent graphite formation without completely inhibiting the formation of active carbon.

IV. 3. S Poisoning of the CO Oxidation

It has long been recognized that sulphur is a drastic poison of numerous hydrogenation or oxidation metallic catalysts. The oxidation of CO on platinum (110) was the subject of a study by BONZEL and KU[32] which provided a better understanding of the poisoning mechanisms for that reaction. It was carried out with the aim of getting a better comprehension the sulphur desactivation of platinum catalysts, utilised in the purification process of exhaust gases. The production of CO as a function of the amount of adsorbed sulphur was continuously measured with a quadrupole mass spectrometer. At the experimental temperature (260°C), sulphur is quasi-irreversibly adsorbed and does not react with oxygen. The results, presented in figure 9, show that above a critical sulphur concentration of about 1 S per 3 superficial platinum atoms, CO_2 is no longer formed, although CO adsorption continues. The authors concluded that sulphur inhibits oxygen dissociation. On the basis of a model proposed for the structure of the sulphur adsorption layer corresponding to the critical degree of coverage, some hypotheses were made on the nature of the sites favorable to oxygen dissociation.[6b,15] They point out the existence of two effects : a blocking geometrical effect due to sulphur atoms and an electronic effect which induces the disactivation of sites neighboring the sites occupied by sulphur atoms.

Most poisoning effects result from the superimposition of these two mechanisms which cannot always be separated from each other. Work in progress in our laboratory shows that sulfur can have an opposite effect on the rate of equilibration in the H_2-D_2 exchange on Pt (111) depending on the surface concentration.[33]
At low concentration, the rate is increased. The presence of adsorbed sulphur seems to change the vibrational properties of the activated complex. (electronic effect).
At medium and high concentration, the rate decreases and the surface is entirely poisoned at a critical sulphur concentration (S/Pt = 0.66).

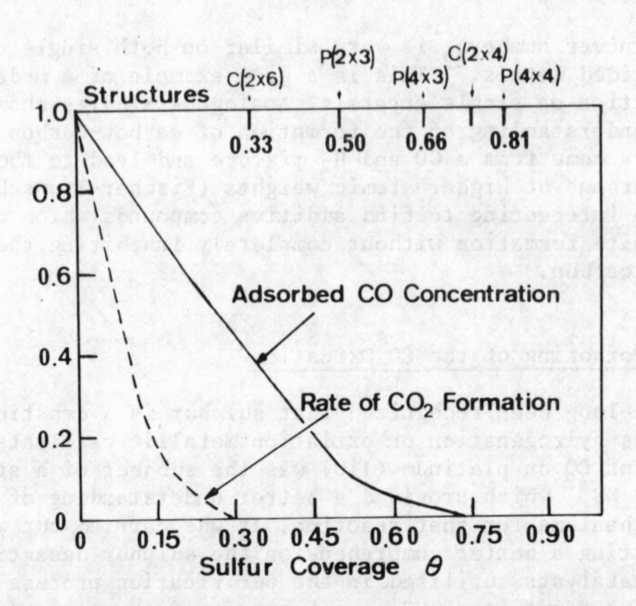

Fig. 9 : Sulphur poisoning of the CO oxidation on Pt (110). Surface normalized quantities of adsorbed CO are also shown. (Sulphur coverage θ is the ratio between the number of adsorbed sulphur atoms and the number of metallic atoms in the first layer of metal); (After Bonzel and Ku[32]).

REFERENCES

1 . FOWLER R.H. and GUGGENHEIM E.A., *Statistical Thermodynamics*, Cambridge University Press, 1939, p 429

2 . a) THOMY A. and DUVAL X., J. Chim. Phys. 66 (1969) 1966
b) THOMY A. and DUVAL X., J. Chim. Phys. 67 (1970) 286 and 1101

3 . BIENFAIT M., Surface Sci. 89 (1979) 13

4 . Colloque International CNRS, J. de Physique 10 (1977) CA

5 . BENARD J., Catal. Rev. 3 (1969) 93

6 . a) OUDAR J., Materials Science and Engineering 42 (1980) 101-109
b) OUDAR J., Catal. Rev 22 (1980) 171-195

7 . a) ROUSSEAU R., Thesis Paris, CNRS Paris n° Av. 11665
b) ROUSSEAU R., DELESCLUZE P., DELAMARE F., BARBOUTH N. and OUDAR J., Surf. Technology 7 (1978) 91

8 . BENARD J., OUDAR J. and CABANE-BROUTY F., Surface Sci. 3 (1965) 359

9 . OUDAR J., Réf. 4, p 77

10. DOMANGE J.L., J. of Vac. Sci. Tech. 9 (1971) 657

11. Mc CARTY J.G. and WISE H., H. Chem. Phys. 74 (1981) 5877
12. MOYA-GONTIER G.E. and MOYA F., SCR. Metall. 9 (1975) 307
13. BENARD J., in *L'oxydation des métaux, processus fondamentaux*, Paris, Gauthier-Villard, 1962
14. Proc. Conf. on *Processus de nucléation dans les réactions gaz-métal*, 1963, Paris, Centre Nationale de la Recherche Scientifique, Mém. Sci. Rev. Métall., 1975, p 62
15. OUDAR J., Ref. 14
16. BORIE B.S., SPARKS C.J. and CATHCART J.V., Acta Met. 10 (1962) 691-97
17. MARCUS P., OUDAR J. and OLEFJORD I., J. Microsc. Spectrosc. Electron. 4 (1979) 63-72
18. OUDAR J. and MARCUS P., Applications of Surface Sci. 3 (1979) 48-67
19. MARCUS P., OLEFJORD I. and OUDAR J., Materials Science and Engineering 42 (1980) 101-109
20. SHELTON J.C., PATIL H.R. and BLAKELY J.M., Surface Sci. 43 (1974) 493
21. EIZENBERG M. and BLAKELY J.M., Surface Sci. 82 (1979) 228
22. BLAKELY J.M., C.R.C. Critical Reviews in Solid State and Material Science 7 (1978) 333
23. SOMORJAY G.A., Advances in Catalysis 26 (1977) 1-59
24. a) ENGEL T. and ERTL G., J. Chem. Phys. 69 (1978) 1267
b) CONRAD H., ERTL G. and J. KUPPERS, Surface Sci. 76 (1978)323
25. MOREAU J. and BENARD J., J. Chim. Phys. 53 (1956) 787
26. BENARD J., Fortschr. Min. 38 (1960) 22
27. RHEAD G.E. and MYKURA H., Acta Metall. 10 (1962) 843
28. HONDROS E.D and MOORE A.J.W., Acta Metall. 10 (1962) 578
29. a) MADEY T.E., GOODMAN D.W. and KELLEY R.D., J. Vac. Sci. Technol. 16 (1979) 433-434
b) KELLEY R.D., GOODMAN D.W., MADEY T.E. and YATES Jr. J.T., *Conference on catalyst desactivation and poisoning*, Lawrence Berkeley Laboratory, Berkeley, CA, 24-26 May, 1978
30. WENTRCEK P.R., WOOD B.J. and WISE H., J. Catal. 43 (1976) 366
31. ARAKI M. and PONEC V., J. Catal. 44 (1976) 439
32. BONZEL H.P. and Ku R., J. Chem. Phys. 59 (1973) 164
33. OUDAR J., BERTHIER Y. and PRADIER C.M., C.R. Acad. Sci., Paris, (1981) 577.

GROWTH BY VACUUM EVAPORATION, SPUTTERING, MOLECULAR BEAM EPITAXY AND CHEMICAL VAPOR DEPOSITION.

R. CADORET

Laboratoire de Cristallographie et
Physique des Milieux Condensés
Université de Clermont II
63170 Aubière, France.

ABSTRACT

Thickness homogeneity and chemical purity of thin film deposition by vacuum evaporation with point and planar sources are discussed. High pressure, low pressure, and assisted sputtering methods (deposition of atoms having 10 to 10^3 times higher energy than evaporated atoms) are presented and their relative advantages discussed. The discussion of molecular beam epitaxy is focused mainly on surface kinetic studies performed to gain more insight of the GaAs deposition process. Finally, the chemical vapor deposition of {100} GaAs in a Ga - $AsCl_3$ - H_2 system and of {111}Si from silane diluted in hydrogen at reduced pressures are analysed on the basis of surface reactions.

I. GROWTH BY VACUUM EVAPORATION

I.1. Vacuum Evaporation : General Considerations

Evaporation techniques are often employed for the production of thin films when high structural quality and chemical purity is not required. The requirements of semi-conductor epitaxial films for electronic devices, however, often demand the use of chemical vapor deposition (CVD), liquid phase epitaxy (LPE), or molecular beam epitaxy (MBE). This leads us to discriminate MBE from vacuum evaporation. The molecular beam is defined as a directed ray of molecules produced by heating a source material in an effusion cell. No appreciable collisions should occur among the beam molecules and between the beam and the background vapor. Therefore ultra-high vacuum is required. In classical thin film deposition

B. Mutaftschiev (ed.), Interfacial Aspects of Phase Transformations, 453–488.

the source material is transformed into a gaseous state by heating it in a crucible. The vapor expands into the evacuated space between the evaporation source and the substrate eventually condensing on the substrate, the other cold surfaces, and the walls of the vacuum system. There is no directed beam. An appreciable increase of the background pressure occurs during the evaporation time.

I.2. Thickness Distribution of Vacuum Evaporated Layers

A homogeneous thickness distribution over large areas can be realized by designing evaporation sources with specific characteristics. A great variety of techniques[1,2] are used for the heating methods (resistive, flash evaporation, arc-evaporation, laser-evaporation, R.F., electron bombardment...). The thickness can be monitored[3] and measured by different methods[4]. The simplest evaporation sources are the point and small area type. The former is obtained by bombarding the tip of a wire of the source material with electrons. A small drop of melted material is then formed under the striking electron beam and acts as an evaporation source. The molecular evaporation flux per unit solid angle, time, and radial section is spherically symmetric. The number of molecules evaporated per unit time in a narrow beam of solid angle $d\omega$ is then

$$d\dot{n}_s = \dot{n}_s \, d\omega/4\pi \tag{1}$$

Where $\dot{n}_s$ is the total number of atoms (molecules) leaving a point source per unit time, given by the well known Knudsen formula :

$$\dot{n}_s = \frac{P}{(2\pi m \, k_B T)^{1/2}} \tag{2}$$

(P is the partial pressure of the molecules, m is the mass per molecule, k_B and T the Boltzmann constant and absolute temperature respectively). For a receiving element of area $dA_r = r^2 d\omega/\cos\theta$ of a planar substrate located at a point of spherical coordinates (r,θ,ϕ) with respect to the source (Fig. 1), the number of molecules adsorbed per unit time and area is

$$\frac{d\dot{n}_a}{dA_r} = \alpha \, \frac{\dot{n}_s \cos\theta}{4\pi r^2} \tag{3}$$

provided that there is no activation adsorption (usually the case for atomic condensation) and that condensation is complete. α (called the sticking coefficient) is the probability for an atom to be trapped when striking the surface. In the case of chemical bonds, $\alpha \simeq 1$.

For molecules (3) includes a proportionality factor $\exp -\varepsilon_a/k_B T$ where ε_a is the activation energy of adsorption. This activation energy represents the potential barrier met by molecules in the vicinity of the surface. The complete condensation hypothesis assumes first that the desorption flux can be neglected (if the substrate is cold enough). With this condition the adsorbed molecular flux is proportional to the source pressure as seen from (1) (2) and (3). Introduction of the temperature dependent desorption flux modifies this result in the sense that the adsorption flux becomes proportional to the difference between source pressure and the equilibrium pressure at the temperature of the substrate. This point is discussed in IV. 1.

A further assumption of the complete condensation hypothesis is that the substrate plane is a kinked face, or a flat face with a free path x_s of the adsorbed molecules shorter than the mean distance between steps or between clusters. The case of incomplete condensation on a surface containing steps with average equidistance $y_\circ$ introduces a corrective term proportional to $\left[(2x_s/y_\circ) \tanh (y_\circ/2x_s)\right]$ due to surface diffusion[5]. Incomplete condensation due to two-dimensional nucleation leads to a non-linear dependence of the adsorbed flux on pressure[6]. This subject will be developed in the chapter "Interface Kinetics in Nucleation and Growth".

Fig. 1 : Geometric parameters of a planar substrate with respect to a parallel planar source.

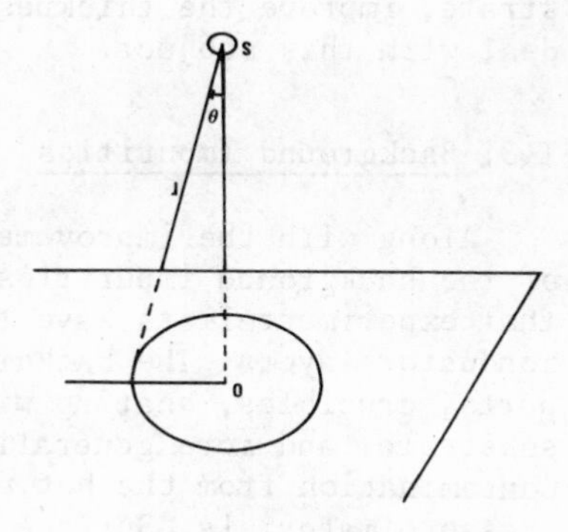

In most experiments the substrate is cold and the desorption flux may be neglected. The number of molecules condensed is then according to (3) :

$$\frac{d\dot{n}_c}{dA_r} = \frac{d\dot{n}_a}{dA_r} = \alpha \frac{\dot{n}_s d}{4\pi r^3} \quad (4)$$

where d is the source-substrate distance. Thus, thickness is proportional to the inverse third power of r.

Crucibles usually have the evaporation characteristics of a small planar source. Assuming a planar substrate parallel to the source, the molecular evaporation flux in the solid angle $d\omega$ of direction (θ,ϕ) (fig. 1) and the number of molecules adsorbed per unit time on the receiving element dA_r are respectively :

$$d\dot{n}_s = \dot{n}_s \cos\theta \frac{d\omega}{\pi} \quad (5)$$

$$\frac{d\dot{n}_c}{dA_r} = \alpha\dot{n}_s \frac{\cos^2\theta}{\pi r^2} = \alpha \frac{\dot{n}_s d^2}{\pi r^4} \quad (6)$$

The thickness decrease (with increasing r) is more rapid than for the point source. An increase of the source-substrate distance reduces the thickness inhomogeneity, but also the deposition rate. Therefore source configurations with an inherently greater uniformity of evaporation characteristics than displayed by point and small area sources have been researched. The use of several ring sources, as well as the use of rotating shutter or rotating substrate, improve the thickness homogeneity and many papers[1-4,7,8] deal with this subject.

I.3. Background Impurities

Along with the improvement of thickness homogeneity reduction of the background impurities is one of the most important problems that experimentalists have to face, particularly for thin semiconductor layers. The background impurities may come from any hot parts, crucibles, heating wire, inside the bell jar, from the substrate, and more generally from the residual gas environment. Contamination from the hot parts can be reduced by using high vapor pressure materials. Surface preparation either by in situ chemical attack, flash evaporation, or cleavage is necessary to eliminate the usual contaminated top layer of the substrate. Impurity diffusion from the substrate to the growing layer can be avoided by using the lowest substrate temperature compatible with the quality desired (transitions from monocrystalline to textured, then polycristalline to amorphous film occur by decreasing the substrate temperature). Contamination from a residual gas component, proportional to its partial pressure, can be minimized by improving vacuum. Two types of contamination processes should be considered :

(i) Adsorption of impurities. This decreases the fraction of surface sites available for autoadsorption (kinetically manifested by a decreased sticking coefficient). Impurity atoms can be trapped during the transition from a dilute to a dense monolayer particularly when lateral interactions inside the layer are not too strong.

(ii) Incorporation of impurities into kink sites of the growing layer. This produces residual doping and structural defects.

The effect of background impurities on layer properties and growth rate is not limited only to vacuum evaporation methods, but is common to all other methods of growth from the vapor phase. Surface inhibition by the gas species is one of the peculiarities of chemical vapor deposition so this question will be addressed in part IV of this chapter.

II. SPUTTERING

II.1. Sputtering Process

Sputtering is the process of ejection of atoms from the surface of a material (target) by bombardment with energetic particles. When bombarding particles are positive ions the process is referred to as "cathodic sputtering". Material erosion for SIMS and Auger analysis, thinning of material for transmission electron-microscopy, and condensation of the sputtered atoms to form a thin film are some of the technological applications. This section is devoted to the last case. Several methods of condensation have been developed during the last thirty years with low pressure deposition by ion beam methods being the most recent and powerful for research purposes. The sputtering process has been subjected to a great number of experimental and theoretical works and several excellent review papers dealing with the sputtering mechanism or applications to thin film deposition have been published.[1,9-15] For a good understanding of these papers it is first desirable to review the properties of a glow discharge by reading the book by J.D. COBINE[16] devoted to electrical conduction in gases.

The main characteristic of sputtered atoms is their high kinetic energy, which spans a range of a few to 100 eV instead of the few 10^{-1} eV of the evaporated molecules. The energy distribution is Maxwellian below the peak, with a long tail which increases in length and area with increasing energy while the peak is only slightly shifted towards higher energies.

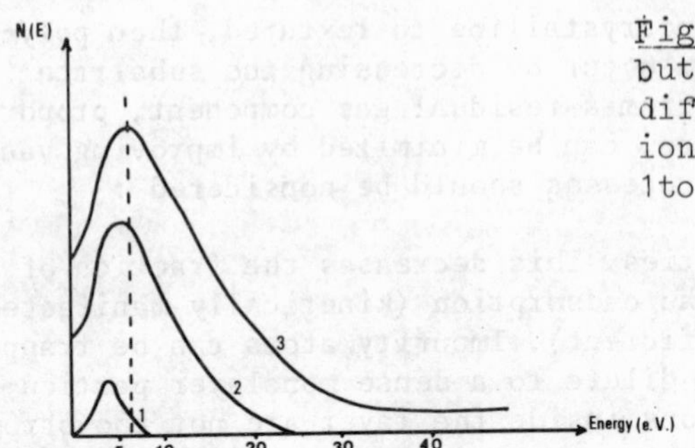

Fig. 2 : Typical energy distributions of sputtered atoms for different energies of bombarding ions. Ion energies increase from 1 to 3.

Fig. 2 represents typical energy distributions for increasing energies of bombarding ions. The peak position and tail length depend on the crystallography of the target and on the angle of ejection with respect to the normal of the target surface. Along the more closely packed directions the mean energy value may be about 3 to 4 times and the peak energy two times higher than along the other directions.[15] The peak energy[11] increases with the angle of ejection up to 60° then decreases.

The ejected particles are mainly neutral atoms or clusters of atoms.[11] With a copper target, Cu and Cu_2 molecules were detected with 2 KeV Ar ion bombardment. With an aluminium target, molecules up to Al_7 were detected with 12 KeV Ar ion bombardment and clusters of Al atoms up to Al_{18} were observed with 12 KeV Xe ion bombardment. About 1 per cent of ejected particles from metals, germanium and silicon are charged. With a III-V target material, neutral molecules are ejected for GaSb while 99% Ga and As neutral atoms and 1% neutral molecules GaAs are emitted with a GaAs target with 0-140 eV Ar ion bombardment.[11] It then appears that the nature of the ejected particles depends on the target and on the energy and nature of ions used for bombardment. The effect of the nature of ejected particles on the structural quality and the composition homogeneity of the deposited film can be studied by a mass spectrometry analysis.

The sputtering yield, defined as the average number of atoms ejected from the target per ion, increases

i) with increasing mass and energy of ions,

ii) with increasing angle between the normal to the target surface and the ion beam direction,

iii) with the closeness of atomic packing of the crystalline target along the beam direction,

iv) strongly with decreasing heat of vaporisation of the target material,

v) with the target temperature.

A sputtering threshold between 5 and 25 eV exists for most metals. A saturation of the sputtering yield occurs at high energies, the higher the heavier are the bombarding ions. Current theories consider sputtering as a momentum transfer process (see ref. 9-15) of a primary knock-on atom at the target surface which loses energy by collisions with other lattice atoms, when moving several atomic layers into the target. A fraction of these atoms diffuse to the surface and leave it as sputtered atoms.

II.2. High Pressure Sputtering Methods

The well-known phenomenon of glow discharge[16] due to an applied electric field between two electrodes in a gas at low pressures provides the simplest source of ions for sputtering.

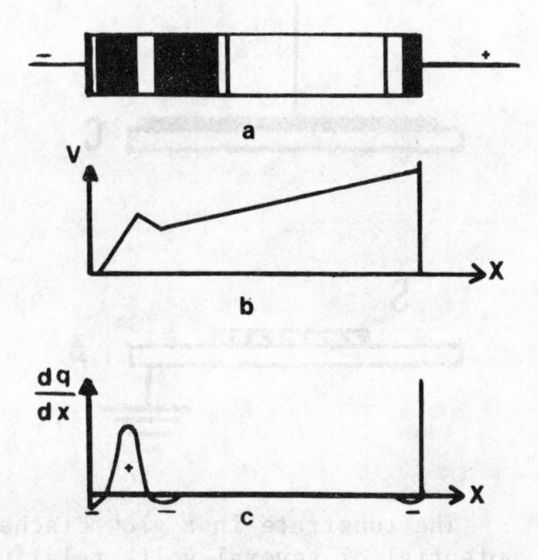

Fig. 3 : Glow discharge : a) glows, dark spaces and positive column in a tube, b) voltage distribution c) charge density distribution.

Fig. 3a represents the gas discharge in a tube operating at a few torrs. Immediately in front of the cathode is the very short "Aston dark space" followed by the "cathode glow", the "cathode dark space" called Crookes or Hittorf dark space, the "negative glow", the "Faraday dark space", the "positive column, the "anode glow",and the "anode dark space". The negative glow is the brightest of the glowing regions. By reducing the space between the electrodes or decreasing the gas pressure, the negative glow and the Faraday dark space expand at the expense of the positive column which finally disappears. When the cathode is closed to the anode there is no more discharge in the tube. This property prevents glow discharge between the back side of the cathode and the neighboring support materials.

Figs. 3b and 3c represent respectively the voltage distribution and the net charge density in the glow discharge tube. Notice that positive ions are produced in the cathode dark space, accelerated by the electric field that prevails between this region and the cathode, and strike the cathode at a high velocity. Sputtering of the cathode results from this bombardment. The simplest experimental system for deposition of films by sputtering is the diode arrangement shown shematically in Fig. 4. The substrate is placed on the anode. Sputtering of the back side of the cathode is prevented by using a grounded shield close to it. Typical conditions are 1 to 5 KV applied potential at 1 to 10 mA/cm^2 current density.

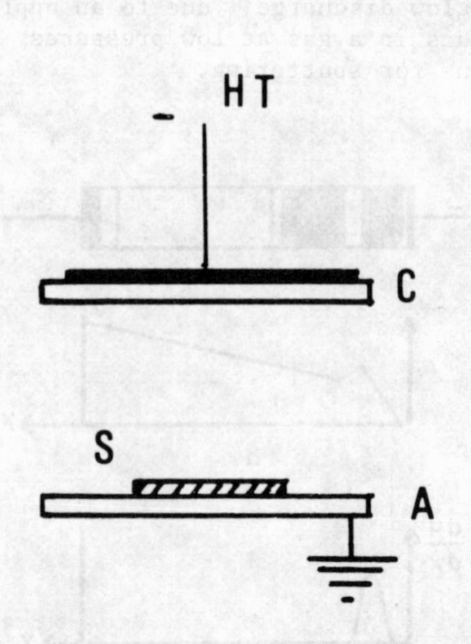

Fig. 4 : Diode system for deposition by sputtering. C : cathode; A : anode; S : substrate.

The substrate in a glow discharge system acquires a negative potential of several volts relative to the anode, attracting gas ions and impurities. An asymetric alternating applied current instead of dc allows removal of the adsorbed gas during one half cycle. Due to the asymmetry, less material is removed on one half cycle than is deposited on the other cycle. Sometimes a precoating retained the oxide layer which is thought to improve the bonding of the deposited film. This method is called "ac sputtering". The "bias" method reduces the contamination from the environment by inserting an insulator between the anode and the substrate. The substrate is then held at a sufficiently negative potential to allow steady ion bombardment throughout its growth. This removes the adsorbed gases which are trapped in the film in the conventional dc glow discharge method. The effect of sputter cleaning of a film is used in the "ion plating" method. With this method, the film is obtained by evaporation from a heated filament which forms the anode of a diode. Electron bombardment of the anode aids evaporation, while the ion bombardment of the substrate, placed on the cathode, cleans the film during its growth.

Another method of reducing impurity contamination from the environment is called "getter sputtering". Here two cathodes of the material to be sputtered are symetrically located with respect to a Ni anode on which the substrate is placed. Then, with respect to the "glow discharge" method, another cathode is used behind the anode to maximize the backside anode gettering of active gases which may enter from the pump.

II.3. Low Pressure Sputtering Methods.

All the methods described above are more or less improved "glow discharge" methods. A gas pressure of 10^{-2} to 10^{-1} Torr is necessary to produce sufficient sputtering and deposition rate. The expected advantage of low pressure systems is the reduction of film contamination by impurities. To obtain sufficient sputtering rate, one can increase the sputtering gas ionization by raising the ionizing efficiency of the available electrons, by increasing the electron densities, or by using an ion beam source.

The electron ionizing efficiency may be increased with a longitudinal magnetic field in a diode geometry. There is no effect on the electrons moving parallel to the field, but all other electrons are displaced towards the system axis and their path length is increased. A transverse magnetic field can also be used but since it concentrates the discharge on one side of a planar cathode it is normally used with cylindrical cathode. An increase in the sputtering rate by a factor of 30 at a pressure of 10^{-5} Torr can be obtained by these methods.

Ionizing electrons supplied thermoionically from a filament and accelerated by the anode potential give rise to "triode sputtering". The ionizing efficiency is further enhanced by adding a transverse magnetic field with respect to the filament-anode direction. Such a system allows fine control of the sputtering rate. (Fig. 5)

Insulator sputtering by the afore-mentioned methods encounters great difficulties because of the buildup of positive surface charges repelling the bombarding ions. This disadvantage can be overcome by bombarding the insulator with an electron beam or placing a metal grid near the surface target. The best and most commonly used method is to apply a radio frequency alternating potential between the electrodes. The insulator capacity produces a dc voltage on the front of the corresponding electrode. The electrons oscillating between the electrodes collide with atoms increasing the random component of their velocity until they build up sufficient energy to make an ionizing collision with a gas atom. Appreciable potentials develop symetrically on the front of each electrode if they are made of conducting materials, while a dc voltage appears in front of the smaller area electrode if there is a capacitor in the external circuit.[12] The ions, having a mass greater than electrons, do not oscillate but follow the mean dc gradient voltage and bombard the target. Insulating material may then be sputtered. Conducting material may also be sputtered if either a capacitor is placed in the external circuit or an rf potential superimposed on the dc voltage of a "glow discharge" system. A magnetic field parallel to the rf field can improve the sputtering yield by concentrating the electrons between the electrodes.

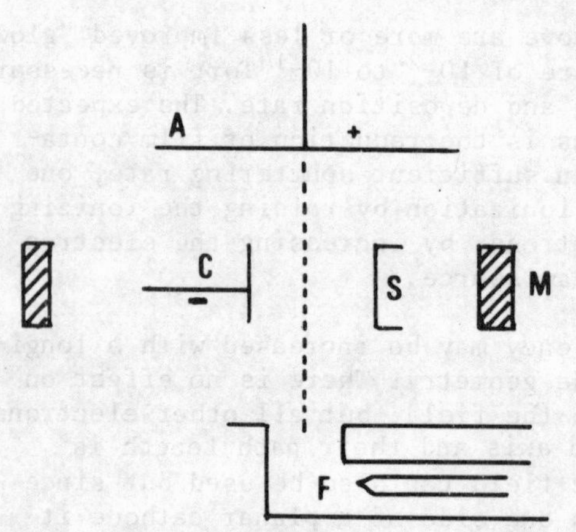

Fig. 5 : Triode sputtering. F : filament; A : anode; C : cathode; S : substrate; M : magnet.

II.4. Ion beam sputtering

The low pressure methods described in the previous section operate over 10^{-5} Torr. This vacuum is not sufficiently good to enable optimal deposition of high quality material. It prevents the utilization of *in situ* analysis methods such as Low Energy Electron Diffraction (LEED), Reflection High Energy Electron Diffraction (RHEED), Electron Spectroscopy for Chemical Analysis (ESCA), Auger Electron Spectroscopy (AES) which require at least a 10^{-8}Torr vacuum. In the last few years beam methods which can be divided into three groups[17] have been developed.

i) Ion-beam sputtering method. An ion beam produced in a high pressure chamber, then introduced into a high vacuum chamber through an aperture with the aid of suitable electron and ion optics, is used to sputter particles from the target.

ii) Low-Energy ion implantation method. The target is bombarded by an ion beam gun for sputtering, while the growing film, fed by the sputtered atoms, is subjected to a variable energy ion source used for element implantation. Silicon nitride $Si_3\ Ni_4$ has been produced by this method. Diamond-like carbon layers were also produced by directly condensing carbon ions with 40 eV energy generated in a special gun.

iii) Condensation of low energy ions method. The activation energy required for the growth of a thin film material is supplied by producing ions accelerated to an energy that spans a range of 5 to 100 eV. The ion neutralization is carried out by electrons emitted from hot filaments.

II.5. One Component Film Deposition

Metal, alloy and semi-conductor layers have all been grown by sputtering methods. One of its greatest advantages is to enable the deposition of refractory materials such as Pt,Nb, Zr, Ta, Mo, W.

The normal growth rate is generally much lower (a few hundred to thousand Å $hour^{-1}$) than in an evaporation or chemical vapor deposition process (a few to hundred microns $hour^{-1}$). The film adhesion is usually higher than in the other vapor condensation methods and is better if the material has a high-melting point temperature. Perfect cleaning of the substrate is necessary for good adhesion. In addition to the necessary chemical cleaning, a low-energy ion bombardment or a substrate heating should be performed. The formation of an intermediate oxide between a metal layer and a glass substrate, as well the use of high energies, improve the film adhesion to a great extent. Adhesion improvement with high energy sputtering is also obtained for metal on metal deposition, for example Cd on Fe.[13] Systematic studies of gold nucleation on rock salt[14] by sputtering and evaporation methods have shown that the exponential increase of the saturation cluster density with the reciprocal substrate temperature is 3 to 4 times higher for sputtered than for evaporated gold. On the other hand a decrease in the lowest substrate temperature necessary to have an epitaxial layer is observed at low rates. This comportment has been studied particularly for the case of Ge homoepitaxy.[14] Polycrystalline to monocrystalline and monocrystalline to amorphous transition temperatures are much less growth rate dependent than polycrystalline to amorphous ones. The substrate cleanliness has a strong effect on these transitions,an intermediate oxide layer increasing the epitaxial transition temperatures.

In the high pressure methods, the sputtered atoms suffer many collisions with gas atoms before reaching the substrate, leading to back-sputtering onto the target surface and a decrease of their ejection energy. The atomic flux arrives at the substrate with a wide range of incident angles giving rise to a "back coating" effect. The ion bombardment on the target, the electron bombardment, and energy dissipated by ion-electron recombination at the substrate surface cause heating, particularly for an insulating substrate. This increase of deposition temperature may change the epitaxial conditions during growth. Out-gassing of the walls occurs as a result of the plasma-discharge heating and can contaminate the deposited materials. The ion beam methods must be chosen when the film properties are very sensitive to impurity incorporation as in electronic compounds. Contamination of Si deposited layers with W, Fe and Ta originating from the cathode, the anode, and the extraction electrodes up to a concentration of 10^{19} to 10^{20} cm^{-3} has been observed even in the ion-beam sputtering method.[18] Ar implantation at the same level also occurs, probably due to the backscattering of bombarding ions on the target. This implantation cannot be prevented by electrostatic or magnetostatic lenses since the backscattered ions are neutralised. Ion source impurities are thought to be prevented by coating the electrodes with silicon. Intentional doping remains at a low level when the doping atoms are high pressure components like phosphorus, an additionnal source is then necessary to supply the atoms. With respect to MBE, ion beam sputtering has the advantage of allowing implantation of low sticking coefficient atoms. This is because of the high energy of condensing atoms. Structural defects, however, are also inherent to the high energy of sputtered atoms. The ion beam methods seem to be very attractive in the future for a better understanding of the processes of sputtered atom deposition.

III. MOLECULAR BEAM EPITAXY (MBE)

III.1. General Considerations

Originally molecular beam epitaxy was considered as an improved method of material deposition from pure vapor. The improvement originates from using Knudsen effusion cells as material sources and ultra high vacuum (UHV) in order to minimize collisions among beam molecules and between the beam and background vapor. An UHV environment makes possible the use of *in situ* analysis methods such as LEED, RHEED, AES, SIMS as in ion beam sputtering (see II.4.) The term MBE should not be considered limited to condensation of elementary molecules, but more generally as a deposition of material from an elementary or chemical species beam onto a substrate held at a suitable temperature. Nevertheless, most experimental results published so far deal with thin film condensation from the deposited compound or its constituent elements. This subject has been discussed in several review articles[19-28] in the last few years, dealing with III-V compounds[19-22], II-VI compounds[25] or II-VI, IV-VI and III-V compounds[28], experimental design[24], surface phenomena[26], and semiconductor superlattices.[27] The object of this part of the chapter is not to describe the basic technology of the MBE process, but to report on the growth mechanisms and surface kinetics. They have been particularly well established in the case of GaAs[29-34] so the discussion will be focused only on GaAs.

III.2. Experimental Processes

The characteristics of thin film MBE are

i) a growth rate of about 600 Å to 1.8 μ $hour^{-1}$ for GaAs. Such a low value gives the possibility of producing a layer thickness of only 50 Å ± 5 Å.

ii) a low growth temperature, which allows reduction of the dopant diffusion

iii) the ability of abruptly changing the dopant concentration as well as the composition of the layers by using several Knudsen effusion cells.

Thermal beams are generated in an UHV chamber from materials heated in Knudsen effusion cells with temperatures accurately controlled to ± 0.02%. Each source is provided with an externally controlled shutter. For surface kinetic studies, modulation of the incident or desorbed beam is easily obtained by using rotating shutters.[29,31-33] The experimental system is schematically illustrated in Fig. 6.

The intensity of a beam impinging on a substrate under an angle θ is

$$J_i = \frac{a P_i}{\pi d^2 (2\pi m_i k_B T_i)^{1/2}} \cos\theta \left[\frac{\text{molecules}}{\text{cm}^2 \text{ sec}}\right]. \qquad (7)$$

P_i is the equilibrium vapor pressure of the considered species i in the ith Knudsen effusion cell at the temperature T_i, m_i is the molecule mass, d is the substrate-cell aperture distance, a is the area of the cell aperture.

A requirement for the source furnaces is a very low production of impurities originated either from outgassing or from reaction of the crucible walls with the source material. Pyrolitic BN is prefered to refractory oxides with gallium or aluminum because of the reduction properties of these compounds. The group III elements' flux is always monatomic, while the group V elements can be either diatomic or tetratomic depending on the source material. For example, As_2 and As_4 molecules are produced with gallium arsenide and arsenic sources. The molecular flux of each species, arsenic, gallium, dopants, impinging on the substrate, are controlled by adjusting the temperature of the appropriate effusion cells. It is essential that the effusion cells have good thermal isolation from the rest of the vacuum chamber, thus reducing outgassing from the walls to the lowest level possible. This is achieved with a liquid nitrogen shroud which surrounds either groups of cells or individual cells. The latter arrangement requires a larger space between the individual cells resulting in a larger angle of incidence distribution of the beams and, therefore, in a loss of homogeneity. Larger distances (> 100 mm) between substrate and cell orifices can be used to compensate this effect. Higher cell temperatures for keeping the growth rate at the same level are then required.

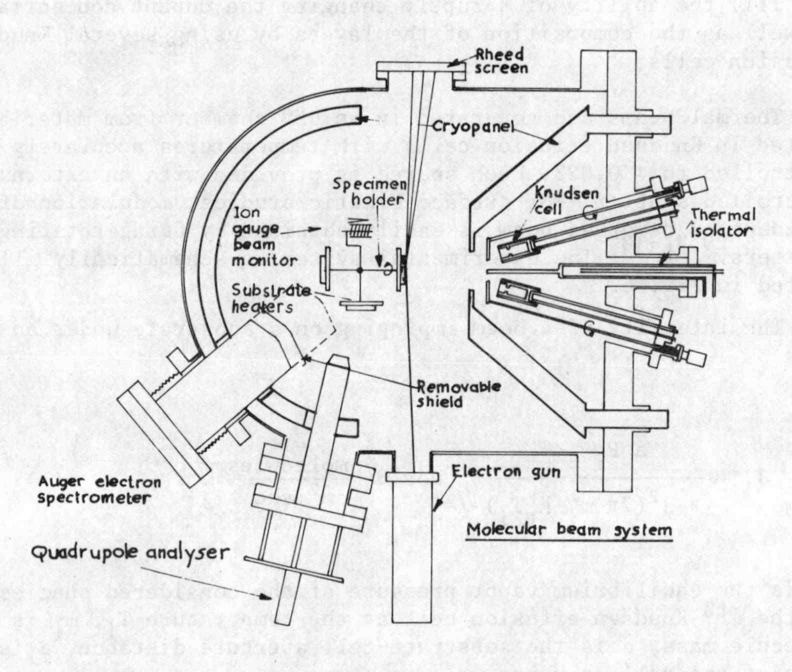

Fig. 6 : Schematic cross-section of a typical molecular beam epitaxy system for GaAs deposition after G. LAURENCE.

It is essential for the substrate to be carefully prepared before deposition by polishing and then *in situ* cleaning. After the polishing stage[8] the substrate, for example {100} GaAs, is contaminated principally by carbon and oxygen impurities (characterized by AES) after its introduction in the UHV chamber. The outermost oxygen layer is intentionally produced by a special etching to passivate the GaAs surface against further carbon contamination during loading into the bell jar. This outermost layer is easily removed by simply heating the substrate to 525-535°C. The carbon contamination is generally limited to less than 0.1 monolayer. Nevertheless it may induce faceting at a coverage as low as 6×10^{-2} monolayer[35] and affects Hall mobilities of very thin layers. Though the required purity may be achieved by a careful chemical etching, the variations observed from sample to sample have led to employment of Ar^+ ion sputter cleaning. An annealing at 350°C is then necessary to recover the crystallinity of the damaged surface. Another method to remove carbon consists of heating the substrate at 580-610°C in 10^{-6} Torr partial pressure of water vapor or oxygen. The oxide formed during this treatment can be easily removed by heating in UHV as described above.

III.3. Surface Kinetic Studies

In view of the potential, applications forGaAs MBE for electronic devices, a considerable effort has been devoted to the studies of the fundamental processes that interfere in the $(111)_A$ (gallium side), $(111)_B$ (arsenic side), and particularly in the (100) growth of this compound. The latter direction is generally preferred for thin film deposition.

Changes in the structure of the outermost layer of atoms was first observed from reflection electron diffraction patterns, RHEED and LEED. A variety of surface reconstructions have been reported for the {100} GaAs surface during simple heating and MBE growth[19,22]. The two principal stable structures for practical conditions of growth and substrate heating are the $C(2 \times 8)$ and the $C(8 \times 2)$. The first structure, called "As stabilized" occurs in the low temperature and/or high As_2/Ga flux ratio ranges (for example 740-860K with As_2/Ga = 5 to 10). The second structure called "Ga stabilized" appears in the high temperature and/or low As_2/Ga flux ratio range (for example 770 to 860K with As_2/Ga = 1 to 5). A 20% decrease in the ratio of the low energy As to Ga Auger peak heights results in transition from As stabilized to Ga stabilized structure for both {100} and $\{111\}_B$ surfaces.

Using flash desorption and mass spectrometric measurements, J.R. ARTHUR[30] has correlated this surface structure change with a loss or gain of arsenic. The time dependence of the As_2 flux leaving an initially Ga stabilized $\{111\}_B$ or {100} face exposed to an As_2 beam of 3×10^{13} molecules $cm^{-2}sec^{-1}$ was measured at 845K and 450K. On both surfaces the As stabilized structure was obtained after 20 to 30 seconds with a gain of 1.7×10^{14} As_2 molecules cm^{-2}. Once the beam was turned off, this structure remained at 450K while the surfaces reverted to the initial (845K) structure with a loss of arsenic corresponding to $1.6 \times 10^{14} As_2$ molecules cm^{-2}. Adsorption and desorption rates did not follow a first order rate law. Flash desorption from the two As stabilized and Ga stabilized surfaces heated at a constant rate gave additional information on the surface processes. The Ga desorption rate was considerably lower than the As_2 one and independent on the initial structure. On the As stabilized surface and for a {100} substrate heated at 4 deg/sec., the desorption rate is weak below 600K, increases continously in the range 650-750K, then remains constant up to 875K (which is the temperature giving complete conversion to the Ga stabilized structure), and finally increases sharply (Fig. 7). On the Ga stabilized surface, the onset of the As_2 desorption occurs in the range 800-850K and increases sharply,while the AES does not show evidence of any change in surface composition. This means that stationary bulk desorption occurs through an intermediate surface state. Ga droplets appear with further heating.

C.T. FOXON and B.A. JOYCE[33] have measured the gallium and arsenic desorption rates of {100} GaAs as a function of the intensity of an incident As_2 beam and of the substrate temperature, with the modulated molecular beam technique, using mass spectrometric detection[31]. The As_2 desorbed flux increases from 300 to 600K remaining constant with further increasing temperature, while the As_4 desorbed beam intensity increases from 300 to 450K then decreases up to 600K. Above this temperature there are no more As_4 molecules in the desorbed flux. The As_4 desorption rate was first order with respect to incident As_2 flux. The surface life time of As_4 molecules was about 1. By modulating incident and desorbed fluxes the same authors showed that below 600K the desorbed beam arises from the adsorbed molecules produced by the impinging flux, while in the range 600-900K contribution of the As_2 desorption from the GaAs bulk progressively appears. This leads to the Ga stabilized surface and to an increase of the As_2 sticking coefficient towards unity. The increase of the latter parameter with Ga surface atom population was assessed at 600K as a function of a gallium flux.

C.T. FOXON and B.A. JOYCE[20,21,31-32] using the modulated molecular beam technique, have measured the thermal accomodation, sticking coefficients, surface life time, desorption activation energy and desorption reaction order of As_4 molecules adsorbed on {100} GaAs as a function of As_4 and Ga incident beam intensities and substrate temperatures. In the range 300K-450K, the thermal accomodation coefficient was unity, the desorption activation energy was 0.38 eV and the surface lifetime varying from 5×10^{-3} to 10^{-4} sec, in the absence of Ga flux. The predeposition of Ga atoms at 10^{15} atoms cm^{-2} corresponding to about one monolayer, increases the surface lifetime of As_4 ten times without sensitive change at the desorption activation energy (0.36 eV). Similar results were obtained with simultaneous Ga and As_4 fluxes. The As_4 sticking coefficient was not measurable without Ga adatoms population, and the desorption process was first order. With a modulated As_4 beam and an unmodulated Ga beam impinging simultaneously on the substrate, the As_4 sticking coefficient α_{As_4} increases with the reciprocal temperature reaching a value close to unity at 300K. As_2 molecules were never observed as desorbed species.

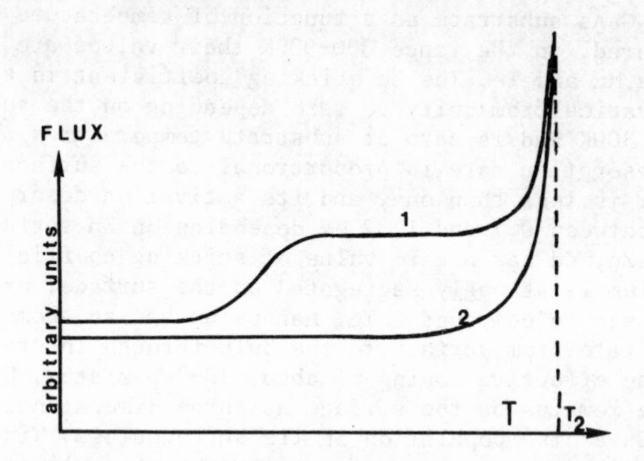

Fig. 7 : Typical behaviour of As_2 flash desorption as a function of T according to (34) : 1) As stabilized surface structure, 2) Ga stabilized surface structure. T_2 is the temperature of the structure change ≈ 875K.

In the range 450-600K, the sticking coefficient α_{As_4} was zero, increasing towards 0.5 with an increasing unmodulated Ga beam, but remaining independent of the substrate temperature. Up to a relative arrival rate J_{As}/J_{Ga} of 1, each Ga atom supplied consumes an As atom. For J_{Ga}/J_{As} values exceeding unity, the α_{As_4} value is 0.5, but stopping the Ga beam did not lead to its zero value until an amount of arsenic molecules sufficient to account for each Ga adatom had been supplied. The desorption reaction was second order with respect to the As_4 beam intensity in the J_{As}/J_{Ga} <<1 region and first order in the J_{As}/J_{Ga} >>1 region.

In the range 600-900K, the sticking coefficient α_{As_4} always has a measurable value, even if there is no Ga flux to the surface. It increases with T without exceeding 0.5 (highest value observed) The variation of the order of the desorption process with respect to the As_4 flux to the surface has the same behaviour as in the 450-600K temperature even without a supply of Ga adatoms. As_2 molecules were detected in the desorbed flux so that the essential Ga adatoms are probably supplied from bulk As_2 desorption.

At temperatures in the range 870-950K the Ga surface lifetime on {111} GaAs faces varies from 1 to 10s, corresponding to an activation desorption energy of about 2.5 eV. Below 750K its sticking coefficient on {100} and {111} is unity. Regardless of the substrate temperature the Ga desorbed flux remains at a relatively low level with respect to As_2 or As_4 fluxes.

The sticking coefficients of several dopant elements, Sn, Mg, Mn, Zn and Te on a GaAs substrate as a function of temperature have also been measured. In the range 300-900K their values are close to unity for Sn, Mg, Mn and Te. The Zn sticking coefficient is time dependent and varies from unity to zero depending on the surface composition at 300K and is zero at substrate temperatures above 373K. The Zn desorption rate is proportional to the surface coverage, its order is less than one, and its activation desorption energy varies between 0.7 and 1.37 eV depending on Zn surface coverage. Like Zn, Cd has a zero value of sticking coefficient at $T > 373K$. Tellurium is strongly segregated on the surface, presumably by forming the stable compound GaTe. Magnesium has an anomalously high diffusion rate from surface to the bulk through interstitial sites, giving an effective doping of about 10^{-5} per atom. Up to 820K, manganese remains on the surface as three dimensional islands, depleting the Ga atoms' population in its surroundings. Tin shows a tendency to segregate to the surface with a surface limited incorporation rate.

III.4. Kinetic Models

The value close to one of the Ga sticking coefficient on GaAs at all growth temperatures leads to a growth rate R in the |100| direction that depends only on the gallium flux according to the relation

$$R = \Omega \, J_{Ga} \tag{8}$$

where Ω is the volume of a GaAs molecule in the lattice and J_{Ga} is deduced from (7).

The behaviour of As_2 and As_4 molecules, thermal accomodation and sticking coefficients, surface lifetimes, desorbed fluxes and their reaction order with respect to incident fluxes have been analyzed with the aid of the precursor state model.[31-33] This model[36] assumes a two layer surface adsorption. The first layer corresponds to a shallow potential well (less than 5 Kcal per mole) where molecules are physisorbed. The second one corresponds to deep potential wells where molecules are trapped in a chemisorption state. The characteristics of such a model are :

i) a low value of the adsorption coverage and a short life time in the physisorbed layer

ii) a possible high value of the adsorption coverage and a long life time in the chemisorbed layer.

The physisorbed flux does not depend on the contamination of the chemisorbed layer. Applying this model to As_2 adsorption[31,33] gives the following reactions and relations

$$As_{2g} + V_1 \rightleftarrows As_2 \qquad (9)$$

$$As_2 + 2\,V_2 \rightleftarrows 2\,As \qquad (10)$$

$$As_b \rightarrow As \qquad (11)$$

Subscripts g and b mean respectively gaseous and bulk states. V_1 is a vacant site in the first physisorbed layer, V_2 is a vacant site in the second chemisorbed layer. Further :

$$\dot{n}_{a1} = \theta_{v1} J_{As_2} \simeq J_{As_2} \qquad (12)$$

$$\dot{n}_{e1} = \frac{n_1}{\tau_1} = n_1 \nu_1 \exp -E_1/k_B T \qquad (13)$$

$$\dot{n}_{a_2} = k_{a_2}\, \theta^2_{v2}\, n_1 \qquad (14)$$

$$\dot{n}_{e_2} = k_{e_2}\, n_2^x \qquad (15)$$

$$\dot{n}_3 = k_3 \qquad (16)$$

θ_{v1} and θ_{v2} are respectively the surface coverages with vacant sites in the first and second layers ($\theta_{v1} \simeq 1$). Subscripts a and e denote respectively adsorbed and evaporated or desorbed flux, ν_1 is an entropy term arising from the molecule vibration frequencies in the first layer, τ_1 is the surface lifetime in the first layer, n_1 and n_2 are respectively the number of As_2 molecules and As atoms in the first and second layers per unit area, $\dot{n}_{ai}$ and $\dot{n}_{ei}$

are respectively the adsorption and desorption fluxes of the ith layer per unit area.

The sticking coefficient α is defined as the ratio between the chemisorbed and the impinging fluxes[36]

$$\alpha = \frac{\dot{n}_{a1} - \dot{n}_{e1}}{J} = 1 - \frac{\dot{n}_{e1}}{J} \tag{17}$$

The order of reaction x depends on the limiting stage of the desorption reaction. When this stage is the association of two As atoms occupying random sites as considered by FOXON and JOYCE[33], x = 2. The rates of change $\dot{n}_1$ and $\dot{n}_2$ of adsorbed As_2 molecules and As atoms are

$$\dot{n}_1 = \dot{n}_{a1} + \dot{n}_{e2} - \dot{n}_{a2} - \dot{n}_{e1} \tag{18}$$

$$\dot{n}_2 = 2\,\dot{n}_{a2} + \dot{n}_3 - 2\,\dot{n}_{e2} \tag{19}$$

In the steady-state approximation $\dot{n}_1 = \dot{n}_2 = 0$.

This set of equations has been used to describe flash desorption experiments and the change of surface structures[30] by assuming a lateral interaction between As atoms. FOXON and JOYCE[33] have improved this model by considering the reactions

$$As_2 + As_2 \rightarrow As_4 \tag{20}$$

$$As_4 \rightarrow As_{4g} \tag{21}$$

below 600K, and the reaction

$$As + Ga \rightarrow AsGa \tag{22}$$

above 600K.

In the latter case, equation (18) is valid and $\dot{n}_3$ must be replaced by $-J_{Ga}$ with a sticking coefficient of unity.

$$\dot{n}_2 = 2\,\dot{n}_{a2} - J_{Ga} - 2\,\dot{n}_{e2} \quad (23)$$

In this case, with the steady-state approximation we have[33]

$$\dot{n}_{e1} = J_{As_2} - (J_{Ga}/2) \quad (24)$$

In the flash desorption case[30] $\dot{n}_{a1} = 0$, thus

$$\dot{n}_{e1} = \dot{n}_3 = k_3 \quad (25)$$

Below 600K, ARTHUR[30] found agreement with experiments by assuming $\dot{n}_3 = 0$, and $\dot{n}_{e2} = 0$, (i.e. negligible As_2 desorption flux from the bulk and rate of association of As atoms). FOXON and JOYCE[33] obtained an As_4 desorption rate equal to half the value of the impinging J_{As_4} flux by considering that below 600K the association in As_4 molecules occurs in place of the As_2 dissociation in 2 As_2 atoms, and that $\dot{n}_{e1} \ll \dot{n}_{a1}$.

Two basically different mechanisms for the interaction of As_4 and Ga on GaAs surfaces were considered by FOXON and JOYCE,[32] one below 450K and one above. Below 450K, As_4 chemisorption without dissociation was considered following a precursor physisorbed state. The rate of chemisorption was assumed to be controlled by As_4 surface diffusion, an As_4 molecule requiring an encounter with a Ga atom during its lifetime on the physisorbed layer to become chemisorbed. This model explains the increase of As_4 sticking coefficient α_{As_4} towards unity with increasing Ga beam intensity. The relation between α_{As_4} and the activation energies for As_4 desorption from and As_4 diffusion on the physisorbed layer, E_D and E_λ is that proposed by G. EHRLICH.[36]

$$\frac{\partial \ln \alpha_{As_4} (1 - \alpha_{As_4})}{\partial (1/k_B T)} = (E_D - E_\lambda)\left(1 - \frac{\alpha_{As_4}}{2} + \frac{\alpha_{As_4}}{3} \ldots\right) \quad (26)$$

The values $E_D = 0.38$ eV and $E_\lambda = 0.24$ eV were found to be in agreement with the experimental results.

Above 450K, the observed one half maximum value of the sticking coefficient α_{As_4}, and the fact that the kinetics are second order in the region of low α_{As_4} value, suggested that As_4 chemisorption was dissociative following a reaction.

$$4\ V_2 + 2\ As_4 \rightarrow 4\ As + As_{4_g} \qquad (27)$$

The dissociation was then considered a result of a pairwise interaction between Ga sites and two As_4 physisorbed molecules, with one As_4 molecule desorbed and one dissociated into two As_2 molecules. Such a model is in agreement with the absence of species other than As_4 in the desorbed flux.

When the physisorbed $\dot{n}_{e1}$ and chemisorbed $\dot{n}_{e2}$ desorption rates are respectively much lower than the physisorption $\dot{n}_{a1}$ and the chemisorption $\dot{n}_{a2}$ rates, we have

$$\dot{n}_{a1} - \dot{n}_{e1} \simeq \dot{n}_{a1} = 2\,\dot{n}_{a2} \quad \text{and} \qquad (28)$$

$$\alpha_{As_4} = \dot{n}_{a2}/J = 1/2 \qquad (29)$$

from.[17]

In addition, we have $\dot{n}_{a1} = J_{As_4}$, $\dot{n}_{e1} = k_{e1}\, n_{As_4}$ and $\dot{n}_{a2} = k_{a2}\, n^2_{As_4}$. At a zero value of the sticking coefficient α_{As_4}, $\dot{n}_{a1} = \dot{n}_{e1}$, the As_4 desorption flux is $\dot{n}_{a2} = k_{a2}\, J^2_{As_4}/k^2_e$. This is in agreement with the second order kinetics observed.

IV. GROWTH MECHANISMS IN CHEMICAL VAPOR DEPOSITION

IV.1. Experimental Analysis

Chemical vapor deposition is a common and reliable process for growing thin films of various materials. Its main characteristic is that the vapor species and the condensing species are not the same. The first problem to solve is then the determination of the vapor composition as a function of temperature and partial pressures of the introduced species. Once the deposition reaction is known, it remains to measure the effects of the gas flow dynamics and the reaction kinetics near the surface or on the surface. Industrial systems are always open systems (Fig. 8). The chemical species are diluted by a gas flowing through a tube where the substrate is placed. Most of the systems operate at atmospheric pressures but there is a trend now to perform deposition at reduced pressures. Due to the requirement of high quality layers for electronic devices, the silicon and gallium arsenide depositions especially have been studied, providing a good understanding of the surface reactions involved in the growth processes.

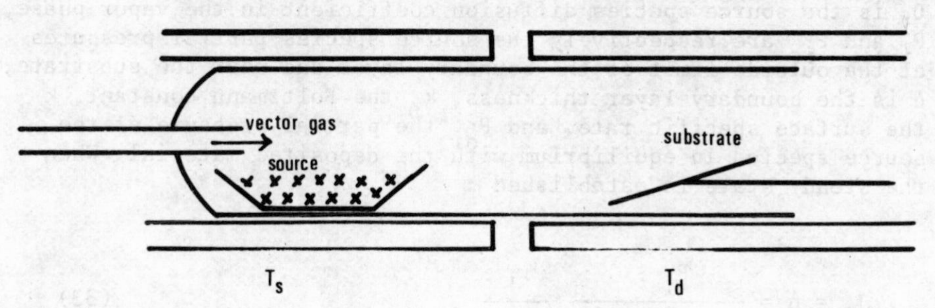

Fig. 8 : Schematic illustration of a GaAs chemical vapour deposition system.

Since the deposition reactions are well known for these systems it remains to understand the gas flow dynamics. Generally a simple boundary layer model is introduced to account for the flow and its effect on surface kinetics. There are several boundary layers to consider :

i) the viscous boundary layer arising from the viscous drag exerted by the substrate on the flow,

ii) the diffusion boundary layer caused by depletion of source species and accumulation of the vapor species from the deposition reaction near the surface,

iii) the thermal boundary layer arising in front of the heated substrate in cold wall reactors for endothermic deposition reactions. We shall consider the simplest case of the diffusion boundary layer with only one source species. This case can easily be applied to the systems of silicon deposition from silane or chlorosilanes diluted in hydrogen. Assuming a negligible effect of thermo-diffusion and a first order surface reaction we have :

$$J_v = \frac{D_i}{k_B T} \frac{P_i - P_{is}}{\delta} \tag{31}$$

for the diffusion flux in the vapor phase per unit area, and

$$\dot{n} = k\,(P_{is} - P_{oi}) \tag{32}$$

for the number of molecules deposited per unit area.

D_i is the source species diffusion coefficient in the vapor phase, P_i and P_{is} are respectively the source species partial pressures at the outside limit of the boundary layer and near the substrate, δ is the boundary layer thickness, k_B the Boltzmann constant, k the surface specific rate, and P_{oi} the partial pressure of the source species in equilibrium with the deposited material. When the steady state is established :

$$J_V = \dot{n} = \frac{P_i}{k_B T(\frac{\delta}{D_i} + \frac{1}{k})} \frac{\gamma}{\gamma + 1} \quad (33)$$

$$\text{with } \gamma = \frac{P_i}{P_{oi}} - 1 \quad (34)$$

called the relative supersaturation of the system.

At this stage we have to discriminate between low supersaturation ($\gamma \ll 1$) and high supersaturation systems. For the former, the condensation flux per unit area and the growth rate are respectively

$$\dot{n} = \frac{P_i}{k_B T(\frac{\delta}{D_i} + \frac{1}{k})} \gamma \quad (35)$$

$$R = \dot{n}\,\Omega$$

where Ω is the molecular volume in the solid state.

The dependence of the growth rate on the substrate temperature T mainly originates from the supersaturation and the specific rate. When the surface is not inhibited by impurity adsorption, the growth rate increases with 1/T for exothermic deposition reactions and this increase does not depend on the limiting step diffusion in the vapor or surface kinetics. GaAs growth from arsenic and GaCl molecules diluted in hydrogen is an example of such a system[37-40] (Fig. 9). With the same conditions, a decrease in R with increasing 1/T values is expected for endothermic reactions, however most endothermic systems are highly supersaturated.

For highly supersaturated systems :

$$\dot{n} = \frac{P_i}{k_B T(\frac{\delta}{D_i} + k)} \tag{36}$$

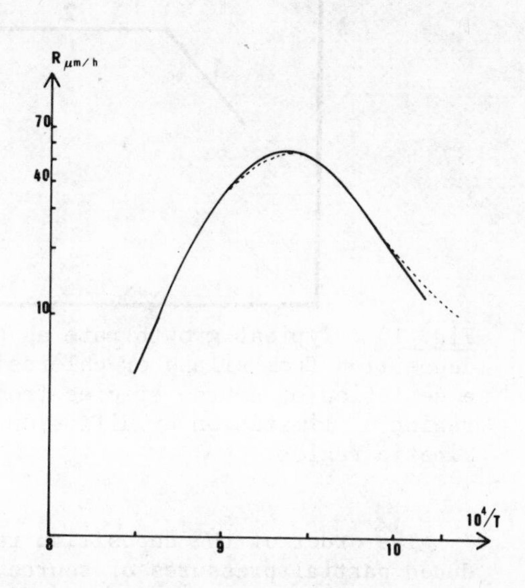

Fig. 9 : {100} GaAs growth rate R as a function of 1/T in a $Ga/AsCl_3/H_2$ system.
Full line : experimental curve after Shaw[37].
Dotted line : theoretical curve after ref. 40;
$P_{GaCl} = 7.53 \times 10^{-3}$ atm,
$P_{HCL} = 9.66 \times 10^{-4}$ atm,
$P_{As_4} = 2.14 \times 10^{-3}$ atm.

The growth rate dependence on the substrate temperature originates only in the specific rate which generally decreases with increasing values of 1/T for endothermic systems while it increases for exothermic systems if there is no surface inhibition. A non dependence of R on 1/T is generally characteristic of a deposition process limited by diffusion in the vapor, although an independence of k with 1/T may be expected in some cases.[43] Silicon deposition from silane and chlorosilanes[41] and GaAs deposition from trimethylgallium (TMG) and arsine[42] are two examples of highly supersaturated endothermic systems showing regions of independance of R on 1/T. In the case of silicon deposition, an R decrease due to k is observed at the lowest values of the substrate temperature[41,44,45] (Fig. 10 region 3)

The variation of R as a function of 1/T reported for first order deposition reactions may be considered as qualitatively unchanged for other orders.

Since D_i is proportional to 1/P and δ is proportional to $1/\sqrt{P}$, when the thickness of the boundary layer is less than the substrate-reactor wall distance, $D_i/\delta \propto 1/\sqrt{P}$. In the opposite case, δ is independent of P and $D_i/\delta \propto 1/P$. When $(D_i/\delta) \ll k$ the deposition process is limited by diffusion in the boundary layer and $R \propto D_i/\delta$.

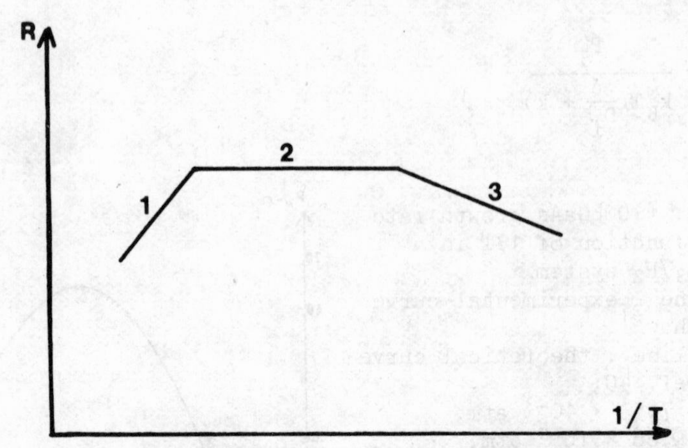

Fig. 10 : Typical growth rate as a function of 1/T in silicon deposition from silane or chlorosilanes. The increase 1 is due to a depletion of source species from homogeneous nucleation. 2 is the region of limitation by diffusion in the boundary layer. 3 is the kinetic region.

The order of the deposition reaction with respect to the introduced partial pressures of source species, is one in a vapor diffusion limited process, as observed in silicon deposition from silane and chlorosilanes[41] and in GaAs deposition from TMG and arsine.[42] The reaction order is also one in a deposition process limited by surface adsorption or diffusion towards steps of the source species[43-46] ($D/\delta >> k$). Here the γ^2 dependence arising from surface diffusion did not occur, the supersaturation having generally too high a value. The deposition reaction order is always more than one when nucleation without complete condensation occurs as observed in the growth rate of silicon from silane at reduced pressures.[45] The activation energy is increased for this process with respect to a growth controlled either by surface adsorption or diffusion.[43,45] The adsorption of an impurity on a surface leads to a negative order of deposition reaction with the corresponding impurity partial pressure. An order of reaction of -1/2 with respect to hydrogen and of one with respect to silane has been observed in the case of {111} silicon deposition from silane.[44,45] The apparent activation energy was 12 Kcal mole^{-1}. To assess the hydrogen inhibition, a careful study of monocrystalline and polycrystalline deposition at reduced pressures was performed on silicon and silicon nitride substrates[45,46] respectively. *In situ* and post analyses were carried out. The polycrystalline deposition reaction was order 1.5 with respect to silane and -1 with respect to hydrogen. The apparent activation energy was 40 Kcal mol^{-1}. The difference between monocrystalline and polycrystalline growth rates was accounted for by nucleation processes[43]. Surface inhibition from the GaCl source species was encountered for GaAs deposition from GaCl and arsenic molecules in a ($Ga/As\ Cl_3/H_2$) system.[39,40,42] As a result, the growth rate increased with P_{GaCl} at low GaCl surface coverage, then decreased when this coverage was greater than 1/2 by increasing P_{GaCl}[37-40,42] (Fig. 11).

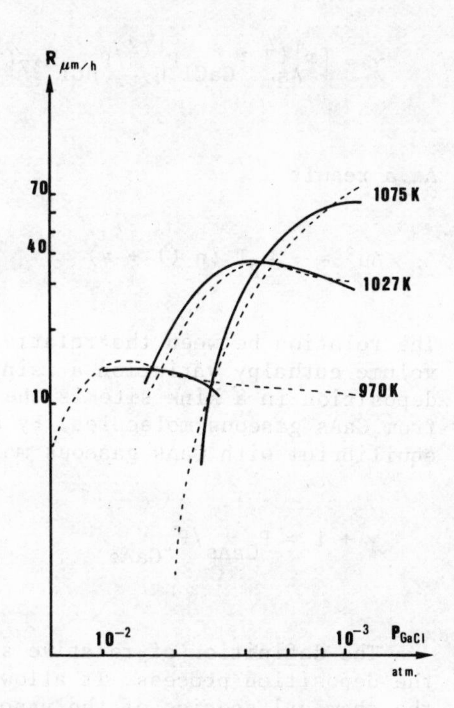

Fig. 11 : {100} GaAs growth rate R as a function of P_{GaCl}. Full line : experimental curve after Shaw.[37] Dotted line : Theoretical curve after [40]. $P_{HCl} = 9.66 \times 10^{-4}$ atm, $P_{As_4} = 2.14 \times 10^{-3}$ atm.

IV.2. Kinetic Model of {100} GaAs Growth in a $Ga/AsCl_3/H_2$ System.

The vapor species diluted in H_2 are mainly As_4, GaCl, HCl and As_2. The deposition reaction is[39,40,42] :

$$\frac{1}{4} As_{4\,g} + GaCl_{\,g} + \frac{1}{2} H_{2\,g} \rightleftharpoons GaAs + HCl \qquad (37)$$

The chemical potential variation produced by the deposition of a GaAs molecule in a kink site is

$$\Delta\mu° = - k_B T \,ln \,(P^{1/4}_{As_4}\, P_{GaCl}\, P^{1/2}_{H_2} \,/\, P_{HCl}\, K_{37}(T)) \qquad (38)$$

$K_{37}(T)$ is the equilibrium constant of the reaction (37), P_i the partial pressure of the i species.

The relative supersaturation γ is defined as :

$$\gamma = [P_{As_4}^{1/4} P_{GaCl} P_{H_2}^{1/2} / P_{HCl} K_{37}(T)] - 1 \tag{39}$$

As a result

$$\Delta\mu° = - k_B T \, ln \, (1 + \gamma) \tag{40}$$

The relation between the relative supersaturation and the free volume enthalpy variation arising from the GaAs chemical vapor deposition in a kink site is the same as in a condensation process from GaAs gaseous molecules. By applying the reaction (37) to an equilibrium with GaAs gaseous molecules we obtain :

$$\gamma + 1 = P_{GaAs} / P°_{GaAs} \tag{41}$$

The definition of relative supersaturation is independent of the deposition process. It allows for surface adsorption of all the chemical species of the vapor phase.

The decreasing growth rate with increasing values of 1/T, (Fig. 9) is characteristic of a surface inhibition. The decrease of R with increasing values of P_{GaCl} observed at the lowest substrate temperatures and highest P_{GaCl} values shows that the surface inhibition results from GaCl adsorption. Limitation of the growth rate by the chlorine desorption reaction was then assumed.[39,40,47] This assumption was assessed by a careful surface analysis.[39,48] It should be emphasized that the kinetic analysis of experimental R curves was so accurate that chlorine desorption could be ascribed to H_2 molecules, resulting in hydrogen adsorbed atoms and HCl molecules for the desorption process. The observed proportionality of growth rate with P_{H_2} was in agreement with this conclusion and led to the supposition that adsorption of hydrogen atoms did not inhibit the deposition process, taking place in an intersticial site or remaining unaffected by the type of site, Ga or As. Fig. 12 shows the {100} surface state schematically. The Ga atoms are considered to be the reference surface state. A Ga atom is bound to two arsenic atoms. The reference surface state has kink sites for As adsorption only. The As adsorbed atoms are bound to two Ga atoms. The assumption that the deposition process is limited to Cl desorption implies fast arsenic and GaCl adsorption kinetics with respect to Cl desorption. Equilibrium between adsorbed and gaseous molecules was then supposed :

$$V + \frac{1}{4} As_{4\,g} \rightleftarrows As \qquad (42)$$

$$As + GaCl_{g} \rightleftarrows AsGaCl \qquad (43)$$

$$V + \frac{1}{4} As_{4\,g} + GaCl_{g} \rightleftarrows AsGaCl \qquad (44)$$

Fig. 12 : Schematic illustration of the {100} GaAs surface state.

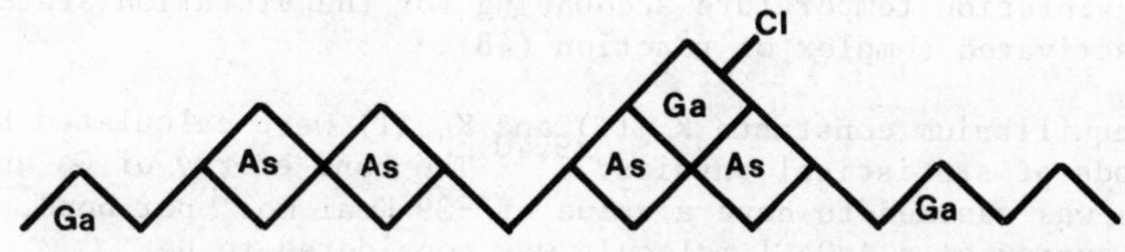

Atoms or AsGaCl molecules are considered randomly adsorbed on the reference surface state. The degree of coverage with vacant sites is then :

$$\theta_v = 1 - \theta_{As} - \theta_{AsGaCl} \qquad (45)$$

$$\theta_{As}/\theta_v = P^{1/4}_{As_4}/K_{42}(T) \qquad (46)$$

$$\theta_{AsGaCl}/\theta_v = P^{1/4}_{As_4} P_{GaCl}/K_{44}(T) \qquad (47)$$

where $K_i(T)$ is the equilibrium constant of reaction i.

At the As_4 partial pressures typically used (10^{-3} to 10^{-2} atm) computation of θ_{As} from partition functions show that regardless of the P_{GaCl} value, (10^{-3} to 10^{-2} atm), the fraction of adsorbed As atoms per Ga atom is close to one. The Ga atoms of adsorbed AsGaCl molecules were considered bound to two As atoms. Such a treatment is valid only for sufficiently high partial pressures of As_4.

The number of GaAs molecules produced per unit area and time, by Cl desorption, $\dot{n}_a$, and the number of Cl atoms adsorbed on Ga surface atoms per unit area and time are deduced from the opposed reactions $\longrightarrow$ and $\leftarrow$- by applying the theory of rate processes :

$$AsGaCl + H_{2g} \underset{\leftarrow --}{\longrightarrow} GaAs + H + HCl_g \tag{48}$$

The resulting growth rate is

$$R_{\mu m\ h^{-1}} = 6.34 \times 10^7 (T_E^*)^{-8}\ T^{15/4}\ P_{HCl} P_{H_2}^{1/2}\ \theta_v \gamma \exp(-\varepsilon/k_B T) \tag{49}$$

$$\text{with } \varepsilon = \varepsilon_e - \varepsilon_{H_{2g}}/2 + \varepsilon_H \tag{50}$$

ε_e is the activation energy of Ga chlorination from HCl (reaction 48 towards $\dashrightarrow$), ε_H the adsorption energy of H atoms. T_E^* is an apparent vibration temperature accounting for the vibration state of the activated complex of reaction (48).

The equilibrium constants $K_{42}(T)$ and $K_{44}(T)$ were calculated by methods of statistical physics.[39,40] The bond energy of Ga and As atoms was assumed to have a value of -39 Kcal mol^{-1} per bond. The bond energy of a AsGaCl molecule was considered to be

$$\varepsilon_{AsGaCl} = -4 \times 39 - 114.5 + \Delta\phi_o + a\ \theta_{AsGaCl} \tag{51}$$

in Kcal mol^{-1}. The GaCl molecule energy in the gaseous state is -114.5 Kcal mol^{-1}. $\Delta\phi_o$ is the change of a GaCl bond energy between the adsorbed and gaseous state, a is a lateral interaction term introduced to account for the decrease of $R(P_{GaCl})$ (Fig. 11). The Einstein approximation was supposed valid for all the vibration frequencies. The resulting vibration temperature (200K) was deduced by matching the experimental and theoretical expressions of $K_{37}(T)$.

The values of $\Delta\phi_o$ and a were deduced from SHAW'S experimental values[37-38] of $\partial \text{Ln } R/\partial \text{Ln } P_{GaCl}$, giving $\Delta\phi_o$ = 29.9 Kcal mol^{-1} and a = - 5.2 Kcal mol^{-1}.
By fitting the theoretical and SHAW'S experimental R(1/T) curves[37,38] the T_E^* and ε values were obtained.

$$T_E^* = 131.3\text{K} \qquad \varepsilon = -14.9 \text{ Kcal mol}^{-1}$$

The activation energy ε_e of Ga chlorination from HCl must be positive. Considering that $\varepsilon_{H_{2g}}$ = -104.2 Kcal mol^{-1} gives

$\varepsilon_H < -67$ Kcal mol-1

Measurements by GREGORY and SPICER[49] indicate that H atoms are strongly chemisorbed.

IV.3. Kinetic Model of {111} Si Growth from SiH_4 at Reduced Pressures

Silicon substrates were 3° disoriented, so that the surfaces were formed by {111} terraces, $y_\circ = a_\circ\sqrt{3}/6 \tan 3°$ wide, bordered by $a_\circ\sqrt{2}/6$ high steps ($a_\circ$ is the lattice parameter). Silane was diluted in hydrogen. Experiments were carried out between 1 to 100 Torr at temperatures in the range 1100-1400K. The substrate only was heated. First order in silane, -1/2 in hydrogen and 12 Kcal mol-1 activation energy were measured for the monocrystalline growth rate.[44-46] The deposition reaction is

$$Si\,H_{4\,g} \rightleftarrows Si + 2H_{2\,g}\,. \tag{52}$$

The relative supersaturation is

$$\gamma = \frac{P_{SiH_4}}{P_{H_2}^2\,K_{52}(T)} - 1\,. \tag{53}$$

At 1100K, $P_{SiH_4}/P_{H_2} = 6 \times 10^{-3}$, $\gamma \simeq 5 \times 10^6$ when $P_{H_2} = 10$ Torr and $\gamma = 6 \times 10^4$ when $P_{H_2} = 1$ atm. This system is highly supersaturated. Silicon nucleation in the vapor never appears at reduced pressure. although it should, according to the results at atmospheric pressures. The existence of a very thin thermal boundary layer is probably the explanation. Pyrolisis of silane in the thermal boundary layer is not to be expected in a reduced pressure system, although it probably occurs in an atmospheric pressure system. This difference could explain why a value of 40 Kcal mol-1 with a reaction order of one with respect to silane, is observed at 1 atm. This value was observed with a reaction order 1.5 with respect to silane and only in the polycrystalline deposition region at reduced pressure. From this it can be concluded that silane molecules impinge on the surface before dehydrogenation and diffuse towards the steps . Silane molecules lifetime on the surface depends on the activation energies of their desorption and dehydrogenation. By considering a unit sticking coefficient and a one layer model, the adsorbed flux is

$$\dot{n}_a = \theta_v\, P_{SiH_4}\,(2\pi m_{SiH_4}\, k_B T)^{-1/2}\, \exp(-\varepsilon_{a\ SiH_4}/k_B T) \tag{54}$$

where ε_{aSiH_4} is the activation energy of SiH_4 admolecules. Assuming that the desorption frequency is not lower than the dehydrogenation frequency[43,45], the deposited flux is

$$\dot{n} = \dot{n}_a - \dot{n}_e = \dot{n}_a \frac{\gamma}{\gamma + 1} = \dot{n}_a. \tag{55}$$

By taking into account surface diffusion towards the steps, the following relation is obtained :

$$R = \Omega \dot{n}_a \left[\frac{2x_s}{y_\circ} \tanh \frac{y_\circ}{2x_s}\right] \tag{56}$$

where x_s is the mean free path on the surface.

The -1/2 reaction order with respect to hydrogen supposes hydrogen adsorption at a surface coverage close to one, so that

$$\theta_v = 1 - \theta_H. \tag{57}$$

Computations of θ_v as a function of the bond energy of the H adsorbed atoms and of x_s as a function of the difference between the activation energy for desorption of silane ε_e and activation energy for surface diffusion U_s, permit construction of theoretical curves with the parameters ε_{aSiH_4}, $(\varepsilon_e - U_s)_{SiH_4}$, and ε_H. Agreement with experimental growth rates was obtained with $\varepsilon_{aSiH_4} = 0$, $(\varepsilon_e - U_s)_{SiH_4} \simeq 16$ Kcal mol^{-1}, $\varepsilon = -74$ Kcal mol^{-1}. The latter value is the same as that calculated by A.A. CHERNOV.[50,51] The mean free path of silane admolecules lies between $0.15y_\circ$ and $0.07y_\circ$ for temperatures ranging from 1100 to 1500K.

When admolecules coverage is not negligible, R. GHEZ[52] has established that the mean free path $x_s = (D_s\tau_s)^{1/2}$(D_s is the surface diffusion coefficient and τ_s the surface lifetime) must be replaced by a corrected mean free path

$$\lambda = x_s \left[(1 - \theta_\circ)/(1 + \gamma\theta_\circ)\right]^{1/2} \tag{58}$$

where $\theta_\circ$ is the degree of coverage with admolecules at $\gamma = 0$.

The generalization of the theoretical treatment of GHEZ to a surface mainly recovered with hydrogen gives for {111} Si :

$$\lambda_s = x_s \left[(1 - \theta_\circ - \theta_H)/(1 + \gamma\theta_\circ - \theta_H)\right]^{1/2} \simeq x_s \tag{59}$$

The GHEZ correction arises from the fact that the impinging flux depends on the vacant site surface coverage θ_v, while the desorption flux depends on admolecules coverage. When the surface is mainly recovered with foreign molecules the desorption and adsorption fluxes depend on the same way as when $\theta_v \simeq 1 - \theta_H$, and the correction term disappears. From the hydrogen reaction order, the mean free path x_s appears independent of θ_v. The diffusion coefficient expression through vacant sites is generally given as

$$D_{sj} = a_o^2 \, k_{sj} \, \theta_v \tag{60}$$

where a_o is the jump length and k_{sj} the jump frequency of the j species.

According to (60) a (-3/4) order with respect to hydrogen should be expected. The departure from this expected order must be researched in the definition of D_s[43]. For this purpose, let us consider a one-dimensional model. The number of j type molecules jumping per unit length from their position x to a neighboring vacant site $x + a_o$ is

$$\dot{n}_{j+} = k_{sj} \, a_o C_{j+}(x) \tag{61}$$

$c_{j+}(x)$ is the number of j molecules located at x per unit area having a vacant site at a jump distance a_o

$$C_{j+}(x) = N_s \, \theta_j(x) \, \theta_v(x + a_o) \tag{62}$$

N_s is the number of surface sites per unit area, $\theta_j(x)$ is the surface coverage of j molecules at x, and $\theta_v(x + a_o)$ is the surface coverage of vacant sites at $x + a_o$.

Similarly, the number of j type molecules jumping per unit length from their position $x + a_o$ to a neighboring vacant site x is

$$\dot{n}_{j-} = k_{sj} \, a_o \, C_{j-}(x + a_o) \tag{63}$$

$C_{j-}(x + a_o)$ is the number of j molecules located at $x + a_o$ having a vacant site at a jump distance $(- a_o)$

$$C_{j-}(x + a_o) = N_s \, \theta_j(x + a_o) \, \theta_v(x) \tag{64}$$

As a result the flux of j type molecules at x is

$$J_j = \dot{n}_{j+} - \dot{n}_{j-} = N_s k_{sj} a_o \left[\theta_j(x)\,\theta_v(x + a_o) - \theta_j(x + a_o)\,\theta_v(x)\right]$$

$$= N_s\, k_{sj}\, a_o^2 \left[\theta_j(x)\,\frac{d\theta_v}{dx} - \theta_v(x)\,\frac{d\theta_j}{dx}\right] \qquad (65)$$

$$\text{where } \theta_v(x) = \text{constant},\; J_j = - D_s\,\frac{dN_j}{dx} \qquad (66)$$

with $D_s = \theta_v k_{sj} a_o^2$ according to (60).

The relation (60) is valid when the equilibrium concentration of vacant sites is reached. That is not the case of a growing flat surface since a depletion of molecules occurs near the steps. An independence of D_{sj} on θ_v occurs when the molar fraction of the j species is constant over all the surface. From the definition $\theta_v(x) = 1 - \Sigma_j \theta_j(x)$ and from (65) is obtained the relations :

$$J_j = -N_s\, k_{sj}\, a_o^2 \left[\frac{d\theta_j}{dx} - \frac{d\theta_j}{dx}\,\sum_i \theta_i(x) + \theta_j(x)\,\sum_i \frac{d\theta_j}{dx}\right] \qquad (67)$$

$$\text{and } J_j = - k_{sj}\, a_o^2\,\frac{dN_j}{dx} \qquad (68)$$

$$\text{when } \frac{\theta_j(x)}{\sum_i \theta_i(x)} = \frac{\dfrac{d\theta_j}{dx}}{\sum_i \dfrac{d\theta_i}{dx}} = \frac{\theta_j(x + dx)}{\sum_i \theta_i(x + dx)}\,. \qquad (69)$$

As a result

$$D_{sj} = k_{sj}\, a_o^2 \qquad (70)$$

This probably explains the independence of D_{sj} on θ_v. A $-3/4$ rather than a $-1/2$ order of the growth rate with respect to hydrogen pressure should be observed if equation (70) was not valid.

Acknowledgement: The author would like to acknowledge G. Gautherin and A. Vapaille of I.E.N., Université Paris XI at Orsay (France) and G.Laurence of L.E.P. at Limeil-Brevannes (France), for very helpful discussions on Ion Beam Sputtering and Molecular Beam Epitaxy.

REFERENCES

1 . K.L. CHOPRA, in *Thin film phenomena*, Mc Graw Hill, New-York, 1969, p 10-82
2 . R. CLANG, in *Handbook of thin film technology*, Mc Graw Hill, New-York, 1970, chp.10
3 . G. HASS and R.E. THUN, in *Physics of thin films*, vol. 3, Academic Press, New-York, London, 1966, p 1-59
4 . W.A. PLISKIN and S.J. ZANIN, in *Handbook of thin film technology*, Mc Graw Hill, New-York, 1970, chp. 3.
5 . W.K. BURTON, N. CABRERA and F.C. FRANK, Phil. Trans. Roy. Soc. 243 (1951) 299
6 . D. KASHCHIEV J. of Crystal Growth 40 (1977) 29
7 . A. ROTH, in *Vacuum Technology*, North Holland Publishing Company Amsterdam, New-York, Oxford, 1976
8 . I.H. KHAN, in *Handbook of thin film technology*, Mc Graw Hill, New-York, 1970, chp. 10
9 . G.K WEHNER, Advan. Electron. Phys. 7 (1955) 239
10. L. HOLLAND, in *Vacuum deposition of thin films*, John Wiley and Sons Inc., New-York, 1956
11. G.K. WEHNER and G.S. ANDERSON, in *Handbook of thin film technology*, McGraw Hill, New-York, 1970, chp. 3
12. L. MAISSEL, in *Handbook of thin film technology*, Mc Graw Hill, New-York, 1970, chp. 4
13. L. MAISSEL, in *Physics of thin films*, Academic Press, New-York London, 1966, vol. 3, p 61-129
14. M.H. FRANCOMBE, in *Epitaxial Growth*, A 109-81 Ed. Matthews, John Wanchope New-York, 1975
15. M.W. THOMPSON, B.W. FARMERY and P.A. NEWSON., Phil. Mag. 18 (1968) 361-425
16. J.D. COBINE, in *Gaseous conductors*, Dover, New-York, 1958
17. G.GAUTHERIN and C.H.R. WEISSMANTEL, Thin Solid Films 50 (1978) 135-144
18. C. WEISSMANTEL, G. HECHT and H.J. HINNEBERG., J. Vac. Sci. Technol. 17 (1980) 812-816
19. A.Y. CHO and J.R. ARTHUR., Progress, Solid State Chem. 10 (1975)
20. T.FOXON, Acta Electronica 16 (1973) 323-329
21. B.A. JOYCE and C.T. FOXON, Inst. Phys. Conf. Ser. 32 (1977) 17-37
22. K. PLOOG, in *Crystals growth, properties, applications*,. Ed. H.C. Freyhard Springer Verlag, Berlin, Heidelberg, New-York 3, 1980, p 73-161
23. R.F.C.FARROW, in *Crystal growth and materials*,Eds. E Kaldis and H.J. Scheel, North Holland Publishing Company, 1977, p 237

24. P.E. LUSCHER and D.M. Collins Progr. Crystal Growth Charact. 2 (1979) 15-31
25. H. HOLLOWAY and J.N. WALPOLE, Progr. Cryst. Growth Charact. 2 (1979) 49-93
26. R.Z. BACHRACH Progr. Crystal Growth Charact. 2 (1979) 115-143
27. L.L. CHANG and L. ESACKI, Progr. Cryst.Growth Charact. 2 (1979) 3-13
28. D. SMITH, Progr. Cryst. Growth Charact. 2 (1979) 33-42
29. J.R. ARTHUR, J. Appl. Phys. 39 (1968) 4032
30. J.R. ARTHUR, Surface Sci. 43 (1974) 449-461
31. C.T. FOXON, M.R. BONDRY and B.A. JOYCE, Surf. Sci. 44 (1974) 69-94
32. C.T. FOXON and B.A. JOYCE, Surf. Sci. 50 (1975) 434-450
33. C.T. FOXON and B.A. JOYCE, Surf. Sci. 64 (1977) 293-304
34. M. PETROFF, A.C. GASSARD, W. WIEGMANN and A. SAVAGE., J. Cryst. Growth 44 (1978) 5
35. LAURENCE G., SIMONDET F., SAGET P., Appl. Phys. 19 (1979) 63
36. G. EHRLICH, J. Phys. Chem. Solids 1 (1956) 1-13
37. D.W. SHAW, J. Electrochem. Soc. 115 (1968) 405
38. D.W. SHAW, J. Electrochem. Soc. 117 (1970) 68
39. J.B. THEETEN, L. HOLLAN and R. CADORET, in *Crystal growth and materials*, ed. E. Kaldis and H.J. Scheel, North Holland Publishing Company 2, 1977, p 195-236
40. R.CADORET, in *Current Topics in Material Science*, ed. E. Kaldis, North Holland Publishing Company 2, 1980, p 219-277
41. J. BLOEM and L.J. GILING, in *Current topics in material science*, ed. E. Kaldis, North Holland Publishing Company 1, 1978, p 147-342
42. L.HOLLAN, J.P. HALLAIS and J.C. BRICE, in *Current topics in material sciences*, ed. E Kaldis North Holland Publishing Company 1, 1980, p 1-217
43. R. CADORET and F. HOTTIER, to be published
44. J.P. DUCHEMIN, M. BONNET, F. KOELSCH, J. Electrochem. Soc. 125 (1978) 637
45. F. HOTTIER and R. CADORET, J. of Cryst. Growth 52 (1981) 199
46. F. HOTTIER and R. CADORET, to be published
47. R. CADORET and M. CADORET, J. Cryst. Growth 31 (1975) 142-450
48. J.B. THEETEN and F.HOTTIER, J. Electrochem. Soc. 126 (1979)
49. P.E. GREGORY and W.E. SPICER,Surf. Sci. 54 (1976) 229
50. A.A. CHERNOV, J. Cryst. Growth 42 (1977) 55
51. A.A. CHERNOV, Soviet Phys. Crist. 22 (1977) 18
52. R. GHEZ, J. Cryst. Growth 22 (1974) 333

MELT GROWTH

R.F. SEKERKA

Department of Metallurgical Engineering
and Materials Science
Carnegie-Mellon University
Pittsburgh, Pennsylvania 15213
U.S.A.

ABSTRACT

Fundamental aspects of the growth of crystals from the melt are discussed with the aim of providing a quantitative link between the structure and dynamics of the solid-liquid interface and phenomenological solutions of the transport equations that describe solidification. The results of simple statistical models and computer simulation are used to motivate kinetic laws for the local growth rate and nonequilibrium solute segregation. Examples include the solidification of a pure crystal of spherical shape, the unidirectional solidification of a binary alloy, and a brief description of morphological stability.

I. INTRODUCTION

This chapter is concerned with fundamental aspects of the growth of crystals from the melt. The many experimental techniques and apparatus for melt growth will not be covered here but may be found in several reviews.[1-3] Instead, we will be concerned with a mathematical description of the solidification process with emphasis on those aspects that relate to the structure, properties, and dynamics of the solid-liquid interface. We shall see that a mathematical description of the solidification process may be given in terms of the solution to phenomenological partial differential equations; but the boundary conditions to be satisfied by these

B. Mutaftschiev (ed.), Interfacial Aspects of Phase Transformations, 489–508.

equations at the solid-liquid interface depend on its structure and the atomic (or molecular) dynamics in its immediate vicinity.

II. THERMODYNAMIC CONSIDERATIONS

The thermodynamics of solid-liquid equilibrium sets the stage for more complex kinetic considerations. Solidification is a first order phase transition which means that at least some of the first derivatives of G, the Gibbs free energy, are discontinuous at the melting point. Let Δ applied to any quantity indicate its value in the liquid minus that in the solid.
Then $\partial(\Delta G)/\partial T = -\Delta S$ and $\partial(\Delta G)/\partial P = \Delta V$ where T is temperature, S is entropy, P is pressure and V is volume. Derivatives with respect to composition will be considered later.

II.1. Latent Heat

The change in entropy ΔS is positive and reflects the fact that the liquid state is more disordered than that of the solid. At any fixed temperature, $\Delta G = \Delta H - T\Delta S$ where H is the enthalpy. For a pure substance at its melting point $T = T_M$, we have $\Delta G = 0$ so

$$\Delta H = T_M \Delta S \tag{1}$$

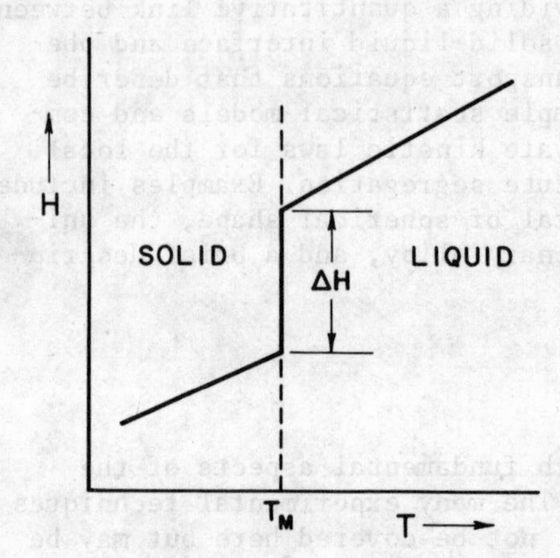

Fig. 1 : Sketch of enthalpy, H, versus temperature, T, for a pure substance near its melting point, T_M.

This change in enthalpy, ΔH, is more commonly known as the latent heat of fusion and corresponds to a jump discontinuity in a graph of enthalpy vs. temperature such as depicted in Fig. 1. The slope of the H vs. T curve above or below T_M is just the specific heat of the corresponding phase. The existence of ΔH means that there must be long range transport of heat away from a solidifying interface in order for solidification to take place. On an atomic scale, the existence of ΔH provides an energetic bias in favor of the solid which is just balanced at the melting point by a larger frequency of attempts for an atom to enter the liquid (see eq. (6)).

II.2. Volume Change

The quantity V may be either positive (usual case) or negative (water, many semiconductors) and represents a change of specific volume (or density) on solidification, as depicted in Fig. 2.

Fig. 2 : Sketch of the molar volume, V, versus temperature for a pure substance near its melting point. (a) Usual case for metals, (b) Unusual case, for water and many semiconductors.

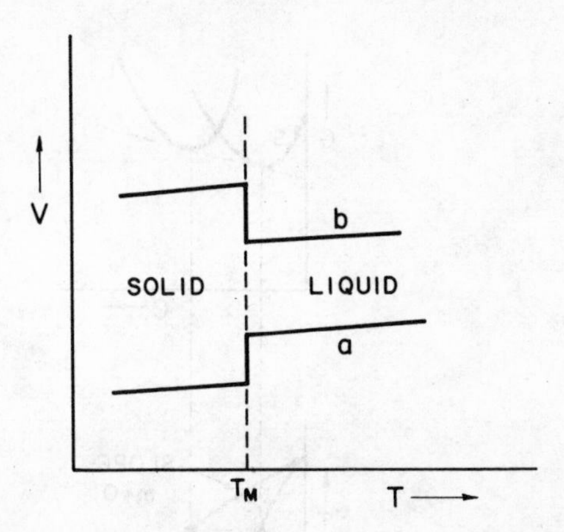

Such a volume change is usually a few percent and can have effects of great importance such as the need in many practical systems for overall fluid flow (to prevent shrinkage cavities) but probably plays a very minor role in solidification processes at the solid-liquid interface.

II.3. Composition Change

When an alloy or impure substance solidifies, there is usually a difference in the equilibirum composition of the liquid and solid. Such differences are discernable from phase diagrams, a subject too complex to be surveyed here. An important case that sheds much light on this matter is, however, a dilute binary alloy of concentration C. Figure 3 depicts typical free energy curves and the corresponding phase diagrams for the two possible cases of interest. These cases are distinguished by the value of the equilibrium distribution coefficient

$$k_E \equiv (C_S/C_L)_{\text{Equilibrium}} \tag{2}$$

in the sense that $k_E < 1$ for case (a) and $k_E > 1$ for case (b). Generally, k_E is a function of temperature unless the solidus and liquidus lines on the phase diagram are straight. Macroscopically, the need for composition change during solidification necessitates long range interdiffusion of chemical species, possibly assisted by convection. Microscopically, it represents an intrinsic bias of the system for the selective incorporation or rejection of certain chemical species at the solid-liquid interface.

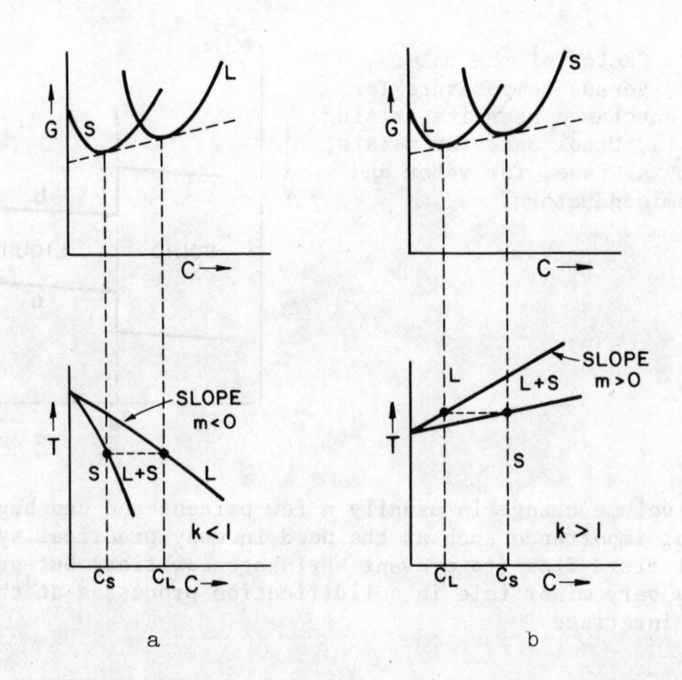

Fig. 3 : Curves of Gibbs free energy, G, versus concentration, C, and corresponding portions of phase diagrams for dilute binary alloys. (a) $k < 1$ and $m < 0$, (b) $k > 1$ and $m > 0$.

II.4. Interface Curvature

If the liquidus lines in Fig. 3 are straight, then the freezing temperature of a liquid of concentration C is just $T_M + mC$ where m is negative for $k < 1$ and positive for $k > 1$. This result is only valid, however, for the case of bulk solid and liquid phases separated by a planar interface. If the interface that separates solid and liquid is curved and if the solid-liquid surface tension γ is isotropic, then the equilibrium freezing temperature is

$$T_E = T_M + mC - T_M \Gamma K \quad (3)$$

where Γ is a capilliary constant (equal to γ/L where L is the latent heat per unit volume) and K is the mean curvature ($K = R_1^{-1} + R_2^{-1}$ where R_1 and R_2 are principal radii of curvature). The sign convention for K is such that $K = +2/R$ for a solid sphere of radius R. Typically, Γ is the order of atomic dimensions so the last term in (3) would appear to be negligible; however, T_M must be taken to be the absolute melting temperature and the term $T_M \Gamma K$ can be comparable to the term mC for dilute alloys for radii of curvature of the order of 10 μm. The curvature correction $T_M \Gamma K$ is, therefore, negligible for macroscopic considerations but essential to the understanding of nucleation and of crystal morphology on the micron scale, for example the phenomenon of morphological instability and solid-liquid interfaces that are cellular or dendritic.

In addition to the effect of interface curvature on the melting point, there can be an effect of interface curvature on the equilibium distribution coefficient which will be of the form

$$k_E(K) = k_E(0)\ (1 + \Gamma' K) \quad (4)$$

to first order in K. Here, $k_E(0)$ is the value of k_E at zero curvature (defined by eq. (2)) and Γ' is a capillarity constant, similar to Γ, the value of which depends on thermodynamic detail. A detailed derivation is given in a recent paper by MULLINS.[4] The effect of the curvature term in (4) has not been widely recognized or treated.

When the surface tension γ is anisotropic, eqs. (3) and (4) must be modified to include angular derivatives of the principal radii of curvature.[5] The corresponding modification of equation (4) for the case of anisotropic γ has not been fully explored and might be quite interesting, especially in the case of faceted interfaces for which HERRING[6] has discussed the need to interpret discontinuities in interface slope with great care.

III. KINETIC CONSIDERATIONS

Although thermodynamics sets the stage for the solidification process, kinetic considerations are also important and become increasingly important for large departures from equilibrium. If the temperature, T_I, of the solid-liquid interface is exactly equal to the equilibrium temperature, T_E, then there should be local equilibrium and neither solidification nor melting should take place.

III.1. Growth Rate

Let U be the local normal growth rate at the solid-liquid interface (positive for solidification). Then one would expect U to be some function of $\delta T \equiv T_E - T_I$ that has the same sign as δT and vanishes whenever $\delta T = 0$. This function will generally also depend on local surface orientation that we symbolize by a unit vector $\hat{n}$, referred to axes fixed with respect to the crystal. Thus, we adopt a general kinetic law of the form

$$U = f(\delta T, \hat{n}) \tag{5}$$

where $f(0, \hat{n}) = 0$, but otherwise the nature of the function f is yet to be determined. It is possible[7] to postulate functional forms more general than equation (5), but so little is known about them that we adopt the simpler equation (5) for illustrative purposes.

Much ignorance prevails regarding the functional form of f. Yet this is an important area of reserach because it is only through this type of kinetic law that the phenomenological description of solidification and the atomistics of interfacial structure and dynamics is made compatible. To understand the nature of this law from a very elementary viewpoint, we examine briefly a very simple model of solidification of a pure substance in which the solidification rate U is represented as the difference between the rate of freezing ($U_{L\to S}$) and the rate of melting ($U_{S\to L}$). In the spirit of absolute reaction rate theory, we take

$$U_{L\to S} = u_{LS} \exp(-Q/k_B T) \tag{6a}$$

and $$U_{S\to L} = u_{SL} \exp(-Q/k_B T) \exp(-h/k_B T) \tag{6b}$$

where u_{LS} and u_{SL} are rate parameters, Q is the energy of an activated complex, and h is the latent heat of fusion per atom (or molecule, for a molecular crystal) and k_B is Boltzmann's constant. Then

$$U = U_{L\to S} - U_{S\to L} = u_{LS} \exp(-Q/k_B T)\{1 - \frac{u_{SL}}{u_{LS}} \exp(-\frac{h}{k_B T})\}. \tag{7}$$

Equation (7) must be subject to the condition that U = 0 when $T = T_M$ which may be satisfied by taking

$$\frac{u_{SL}}{u_{LS}} = q(T)\ \exp\left(\frac{h}{k_B T_M}\right) \qquad (8)$$

where $q(T_M) = 1$. Then Eq (7) may be written in the form

$$U = u_{LS}\ \exp(-Q/k_B T)\left\{1 - q\ \exp\left[-\frac{h(T_M - T)}{k_B T\, T_M}\right]\right\} \qquad (9)$$

which is sketched in Fig. 4 for the case where u_{LS} is independent of T and q = 1.

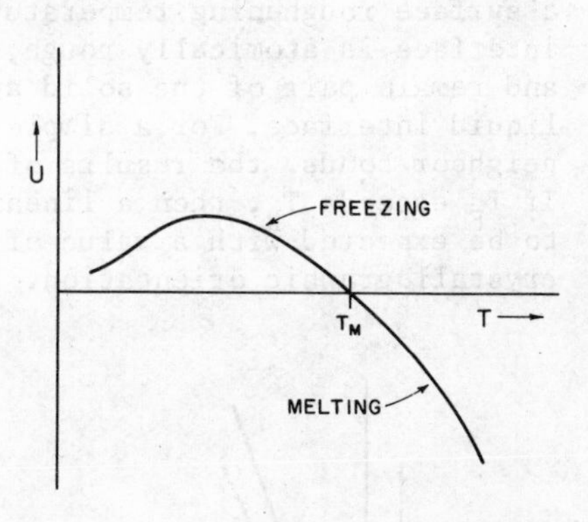

Fig. 4 : Sketch of the normal growth rate, U, versus temperature according to Eq (9). The reduction of U at low values of T stems from the factor $\exp(-Q/k_B T)$.

Note especially the asymmetry of U relative to T_M; as T departs from T_M in either direction, the driving force for phase change increases but the kinetic factor $\exp(-Q/k_B T)$ causes enhancement of rate for $T > T_M$ but reduction of rate for $T < T_M$. Although the above model is quite simple, this asymmetry phenomenon is quite general.

For T close to T_M, the exponential in (9) may be expanded (still with q = 1) to give approximately

$$U \simeq u_{LS}\ \exp(-Q/k_B T_M)\ \frac{h\Delta T}{k_B T_M^2} \qquad (10)$$

where $\Delta T = T_M - T$ is the undercooling. Then for u_{LS} independent of T, U is proportional to ΔT which is a simple linear kinetic law.

The corresponding form of equation (5) would be

$$U = \mu(\hat{n})\ \delta T \qquad (11)$$

where μ is a linear kinetic coefficient that may depend on surface orientation but depends only weakly on temperature.

Laws of the form of eqs. (9) and (10) were proposed many years ago by WILSON[8] and FRENKEL.[9] In fact, however, the factors u_{LS} and q may depend on temperature and crystallographic orientation in a complicated way. Moreover, actual crystallization may take place at preferred sites on the solid-liquid interface so it might not even be meaningful to treat the interface as being homogeneous, which is implicit in the derivation of equation (7). Much insight into this problem has been provided by JACKSON[10] by means of analytical models and especially by JACKSON[11] and GILMER[12] by means of computer simulation. Of critical importance is the concept of a surface roughening temperature T_R above which the solid-liquid interface is atomically rough, thus allowing liquid atoms to become and remain part of the solid at practically any site of the solid-liquid interface. For a simple cubic crystal with only nearest neighbor bonds, the results of computer simulation give $T_R \simeq 0.2h/k_B$. If T_E exceeds T_R, then a linear kinetic law of the form of (11) is to be expected with a value of $\mu(\hat{n})$ that depends rather weakly on crystallographic orientation.

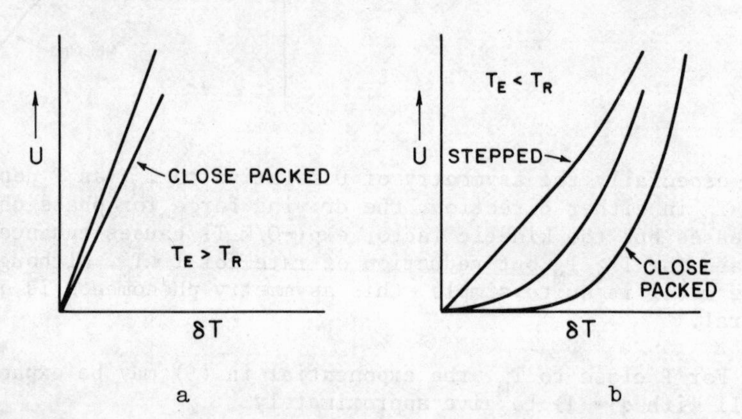

Fig. 5 : Sketches of the normal growth rate, U, versus interface undercooling, δT. (a) Equilibrium temperature, T_E, above the roughening temperature, T_R, (b) Equilibrium temperature below the roughening temperature.

This is illustrated schematically in Fig. 5a. For T_E less than T_R, the solid-liquid interface is atomically smooth, to a first approximation, but with distinct features such as raised or lowered islands of atoms, steps of atomic height, or spiraled steps associated with the presence of screw dislocations that intersect the solid-liquid interface. Thus for $T_E<T_R$, the growth rate U is greatly reduced (relative to that for a rough surface) and strongly dependent on crystallographic orientation because only atoms that join the crystal near the perimeters of islands or at steps will have a high probability of being incorporated permanently in the solid. The corresponding growth rate for $T_E< T_R$ is sketched in Fig. 5b. Note from Fig. 5b that the growth rates of close packed surfaces without screw dislocations are most strongly reduced relative to rough surfaces; at low values of δT, the dependence of the growth rate on temperature is approximately of the form

$$U \propto \exp(-\frac{A}{T}) \exp(-\frac{B}{T\delta T}) \tag{12}$$

which may be derived by calculating the probability of formation of a two-dimensional nucleus (island) on an atomically smooth surface. At high values of δT, the curves in Fig. 5b tend toward increased values of U and there is much less anisotropy, i.e., much less dependence of the growth rate on surface orientation. GILMER[12] attributes this increased rate and isotropy to a sort of dynamical roughening brought about by frequent random impingement of liquid atoms at high driving forces. As also illustrated in Fig. 5b, the growth rates of surfaces on which steps occur naturally (e.g., the (011) face of FCC) are not reduced as much as those of close packed surfaces because the steps themselves provide atomic attachment sites and there is no great need for thermally activated defects. JACKSON[12] has suggested that the factor

$$\alpha = \xi \frac{h}{k_B T}, \tag{13}$$

where ξ is the fraction of bonds parallel to the solid-liquid interface, should indicate similar modes of growth for various substances and orientations; this simple result seems to agree with the results of computer simulation for reasons not yet fully appreciated. Finally, the intermediate case in Fig. 5 can either illustrate an orientation of intermediate packing density (e.g., (001) of FCC) or a close packed surface intersected by a screw dislocation. For such cases, the growth rate may be approximated by a power law of the form

$$U = \mu' (\hat{n}) (\delta T)^m \tag{14}$$

where the exponent m has a typical value of 2.

III.2. Solute Incorporation

The incorporation of solute in a growing crystal may, for kinetic reasons, be different from that of a crystal in equilibrium with its melt. Just as in the previous section, the reason for this can be related to the atomic structure of the interface and to the processes that take place there. This problem can be addressed quantitatively by relating the actual distribution coefficient

$$k = (C_S/C_L)_{Actual} \tag{15}$$

during crystal growth to the equilibrium distribution coefficient defined by eq. (2) for the case of planar interfaces or eq. (4) for the case of curved interfaces. In the spirit of equation (5), we can, therefore, write

$$k = k_E \; g(\delta T, \hat{n}) \tag{16}$$

where the functional form of g is yet to be determined. Since $\delta T=0$ corresponds to equilibrium, we must have $g(0,\hat{n}) = 1$. Moreover, as $\delta T \to \infty$ one would expect U to become so large that no segregation could occur, i.e., $g(\delta T,\hat{n}) \to 1/k_E$ and $k \to 1$ as $\delta T \to \infty$.

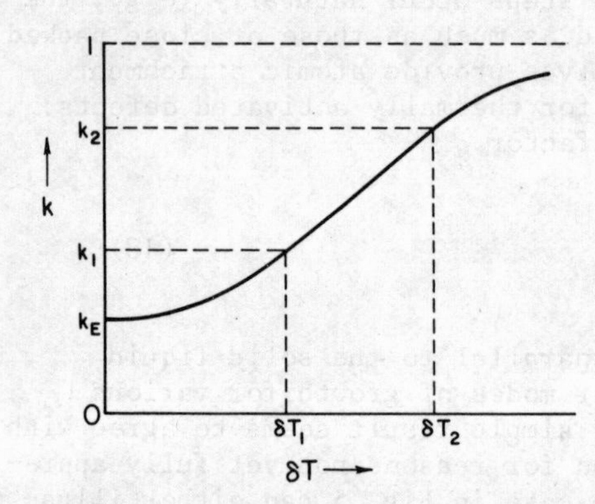

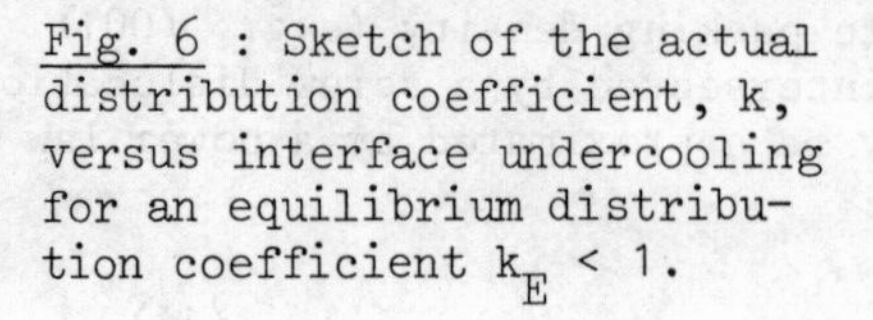
Fig. 6 : Sketch of the actual distribution coefficient, k, versus interface undercooling for an equilibrium distribution coefficient $k_E < 1$.

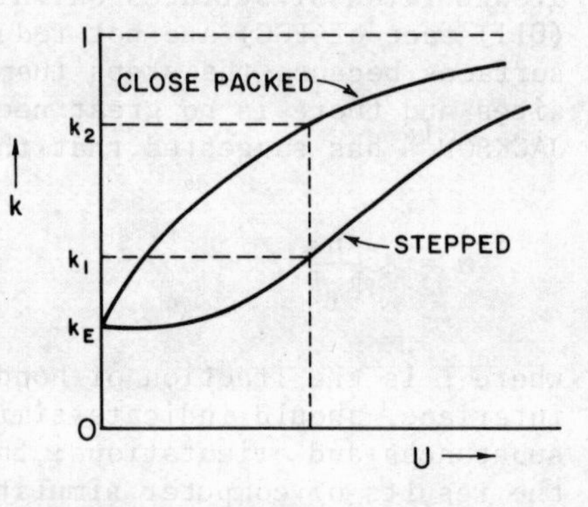

Fig. 7 :Sketch of the actual distribution coefficient versus the normal growth rate under conditions where the growth rate is anisotropic.

The variation of k with δT is sketched in Fig. 6 for the case $k_E < 1$; whether or not k gets very close to 1 at practically attainable values of δT is , however, open to question. The actual shape of the curve is not well known but simple statistical models, such as that of CHERNOV[13], and especially computer simulation by GILMER and JACKSON,[12] have shed some light on the matter. A limited number of results from computer simulations seem to indicate that the dependence of k on $\hat{n}$ is weak, i.e., that curves such as Fig. 6 do not vary strongly with crystallographic orientation. Thus, a sketch of k versus the normal growth speed U, which may depend strongly on orientation as sketched in Fig. 5b, may show a strong dependence on orientation such as sketched in Fig. 7 for a stepped and a close packed surface. Hence, a close packed surface (possibly a facet) and a stepped surface (off facet) that coexist during crystal growth at a common pulling rate may exhibit very different values of the distribution coefficient because of the much higher undercooling at the close packed surface (compare Figs. 6 and 7 with respect to k_1 and k_2).

It is noteworthy that this strong and anisotropic dependence of k on U might be quite important during the rapid solidification of hyperfine metal powders or during laser glazing where solidification rates of the order of meters per second have been calculated from heat flow considerations.

IV. TRANSPORT EQUATIONS AND BOUNDARY CONDITIONS

The temperature and solute fields that accompany crystal growth without liquid convection may be described by the partial differential equations

$$\nabla^2 T = \frac{1}{\kappa} \frac{\partial T}{\partial t} \tag{17}$$

and

$$\nabla^2 C = \frac{1}{D} \frac{\partial C}{\partial t} \tag{18}$$

where κ is the thermal diffusivity and D is the diffusivity of solute of concentration C. Eq. (17) with an appropriate value of κ must be applied in both solid and liquid for which thermal diffusivities are usually comparable. Eq. (18) is usually needed only in the liquid because diffusivities in the solid are many orders of magnitude smaller. If there is convection with velocity $\vec{v}$, then $\partial/\partial t$ in (17) and (18) must be replaced by $\partial/\partial t + \vec{v}.\nabla$ and another equation, e.g., the Navier-Stokes equation, must be solved simultaneously to determine $\vec{v}$. Such convective flow is of great practical importance for crystal growth but does not couple directly to interfacial processes because of the stagnation of fluids near solid surfaces. The interested reader is referred to standard reviews[1-3] for details.

Solutions to the above transport equations must obey boundary conditions at the solid-liquid interface. The conservation of heat may be expressed in the form

$$(k_S \nabla T_S - k_L \nabla T_L) \cdot \hat{n} = LU \tag{19}$$

where k_S and k_L are thermal conductivities of the solid and liquid, respectively, and L is the latent heat per unit volume. Conservation of solute is expressed by

$$(- D \nabla C) \cdot \hat{n} = (1 - k)\, CU \tag{20}$$

where the distribution coefficient k should be the actual distribution coefficient compatible with U and $\hat{n}$ (through (5) and (16)). The temperature, $T_I = (T_E - \delta T)$ at the solid-liquid interface must also be chosen to satisfy eq. (5).

Clearly, the above equations and boundary conditions are sufficiently complicated that one must resort to simplifications and approximations in order to obtain tractable problems.

V. EXAMPLE SOLUTIONS

In this section, we discuss two simple examples for which a mathematical description of growth is tractable. The first is the free growth of a sphere of a pure substance with emphasis on the dependence of the radius on time; the second is the unidirectional solidification at constant velocity of a binary alloy.

V.1. Sphere of Pure Substance

Even through a spherical shape is almost the opposite of what one thinks of when regarding a freely growing crystal, a great deal can be learned about the relative roles of heat flow, capillarity and interface kinetics by treating the simple case of the solidification of a sphere of a pure substance from an undercooled melt.[14] We shall, furthermore, employ the quasi-steady-state approximation according to which the growth is assumed to be so slow that the temperature field can be determined from a solution to the Laplace equation

$$\nabla^2 T = 0 \quad (21)$$

instead of eq. (17). A fully time dependent treatment for simpler boundary conditions than we consider here has been given by FRANK[15] but is much more involved and reduces to the results given here for low driving forces.

The spherically symmetric solution to (21) that reduces to the bath temperature T_∞ as $r \to \infty$ and to the interface temperature T_I at the solid-liquid interface ($r = R$) is

$$T = (T_I - T_\infty)(R/r) + T_\infty \quad (22)$$

where T_I and R depend slowly on time. Equation (22) applies in the melt ($r > R$); inside the crystal, T is uniform at temperature T_I. As boundary conditions, we take

$$T_E = T_M - 2T_M\Gamma/R \quad (23)$$

$$\frac{dR}{dt} = \mu(T_E - T_I) \quad (24)$$

and $$\frac{dR}{dt} = -\frac{k_L}{L}\left(\frac{dT}{dr}\right)_{r=R} \quad (25)$$

with μ independent of orientation. Eq. (23) is just (3) with $C = 0$ and $K = 2/R$ while eq. (24) is the linear kinetic law (11) for the case of isotropy. Eq. (25) comes from (19) for spherical symmetry.

Substitution of (22) into (25) gives

$$\frac{dR}{dt} = \frac{k_L}{L}\frac{(T_I - T_\infty)}{R} \quad (26)$$

which, where equated to the righthand site of (24) yields

$$T_I = \frac{T_E + (R_o/R)\, T_\infty}{1 + (R_o/R)} \quad (27)$$

where $R_o = k_L/(L\mu)$. According to (27), T_I will lie between T_∞ and T_E depending on crystal size R relative to R_o, a size determined by the kinetic parameter μ. Substitution of (27) into (26) with the use of (23) to obtain the dependence of T_E on R yields

$$\frac{dR}{dt} = \frac{P}{2R} \frac{\left(1 - \frac{R^*}{R}\right)}{\left(1 + \frac{R_o}{R}\right)} \tag{28}$$

where $R^* = 2T_M\Gamma/(T_M - T_\infty)$ can be identified as the nucleation radius and $P = 2k_L(T_M - T_\infty)/L$ will be called the parabolic rate constant for reasons that we shall soon see. P has the dimensions of a diffusivity and the quasi-steady-state approximation is valid if the thermal diffusivity $\kappa \gg P$.

It is useful to examine equation (28) and its integral in several limits. In the hypothetical limit of no capillarity ($\Gamma \to 0$) and extremely rapid kinetics ($\mu \to \infty$), we have $R_o = R^* = 0$ and (28) becomes

$$\frac{dR}{dt} \simeq \frac{P}{2R} \tag{29}$$

with integral

$$R^2 = Pt + \text{CONST.} \tag{30}$$

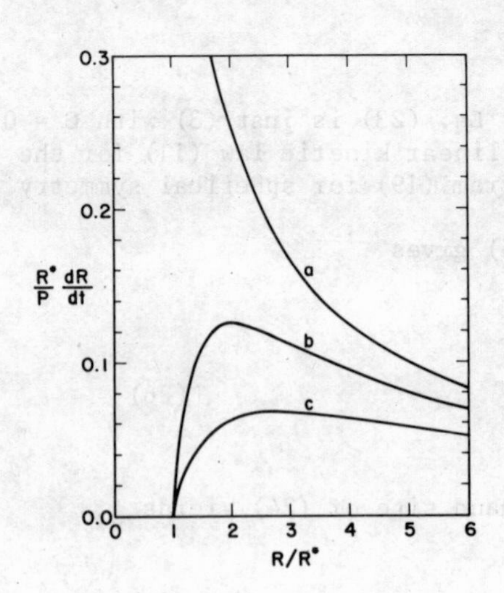

Fig. 8 : Graph of the growth rate, dR/dt, of a pure spherical crystal versus radius, R. R^* is the nucleation radius and P is the parabolic rate constant. (a) $R^*/(2R)$, neglect of kinetics and capillarity, (b) $[R^*/(2R)][1-(R^*/R)]$, neglect of kinetics only, (c) equation (28) with kinetics and capillarity both included and $R_o = 2R^*$ for illustrative purposes.

Equation (30) is a classical parabolic growth law and indicates that the growth rate is controlled by the long range transport of heat. According to (29), the growth rate becomes infinite as $R \to 0$. Of course, this singularity is fictitious and not present in equation (28). This is illustrated in Fig. 8 where we plot $(R^*/P)(dR/dt)$ according to Eq. (28) along with some of its limiting cases. We see that both capillarity and kinetics reduce the growth rate, the former to zero at the nucleation radius. If we omit the capillary correction to (28) by setting $R^* = 0$, we obtain

$$\frac{dR}{dt} \simeq \frac{P}{2} \frac{1}{(R_o + R)} \tag{31}$$

which gives constant (interface kinetic limited) growth for $R \ll R_o$ and heat flow limited growth for $R \gg R_o$. The integral of eq. (31) is just

$$R^2 + 2RR_o = Pt + \text{Const.} \tag{32}$$

which for $R \ll R_o$ gives R linear in t as expected.

The complete integral of (28) may be obtained if we are careful to begin the integration at $R = R^*(1 + \varepsilon)$ where ε is a small positive quantity. A finite value of ε is needed because the nucleation radius is a point of unstable equilibrium and a small departure is required to obtain growth in a finite time. The result is

$$(R + R^* + R_o)^2 - (2R^* + R_o)^2 + 2R^*(R^* + R_o)\, ln\left(\frac{R - R^*}{\varepsilon R^*}\right) = Pt \tag{33}$$

and eqs. (30) and (32) are seen to be special cases. The overall growth kinetics are dominated by capillarity for $R \simeq R^*$, by kinetics for $R \simeq R_o$ and by heat flow for $R \gg R_o$ or R^*. The maximum growth rate occurs at radius

$$R_{max} = R^* + \left[(R^*)^2 + R^*R_o\right]^{1/2} \tag{34}$$

where it has a value

$$(\frac{dR}{dt})_{max} = \frac{P}{2} \frac{R^*}{(R_{max})^2} . \quad (35)$$

For $R_o << R^*$, we have $R_{max} \simeq 2R^*$ and a maximum growth rate of $P/(2R^*)$ is limited by capillarity. For $R_o >> R^*$, we have $R_{max} \simeq (R^* R_o)^{1/2}$ and a maximum growth rate of $P/(2R_o)$ is limited by interface kinetics.

V.2. Unidirectional Growth of Binary Alloy

To see some of the effects of solute in a simple case, we consider the unidirectional solidification of a binary alloy at constant velocity V. This can be accomplished by means of relative motion between the sample and its thermal environment (e.g., by a traveling furnace) and approximates many situations in crystal growth. We describe the system with respect to a coordinate system that is moving uniformly with velocity V such that the plane z = 0 is always at the solid-liquid interface. If the growth is not too rapid, Eq. (21) applies and the temperature field is given approximately by

$$T = T_I + G_L z, \quad z > 0 \quad (36a)$$

$$T = T_I + G_S z, \quad z < 0 \quad (36b)$$

where G_S and G_L are temperature gradients in the solid and liquid, respectively. To find the concentration field, we transform equation (18) into our moving coordinate system and assume that C is independent of x and y; the result is

$$\frac{\partial^2 C}{\partial z^2} + \frac{V}{D}\frac{\partial C}{\partial z} = \frac{1}{D}\frac{\partial C}{\partial t} \quad (37)$$

In this reference frame, we assume that a steady state has been reached so that $\partial C/\partial t = 0$; the corresponding solution to (19) is, therefore,

$$C = C_\infty + C_\infty\left[(1-k)/k\right] \exp(-Vz/D) \quad (38)$$

where C_∞ is the concentration of bulk liquid ($z \to \infty$) and (20) has been used as a boundary condition (with $U = V$). The temperature and concentration fields that correspond to eqs. (36) and (38) are sketched in Fig. 9 for $k < 1$.

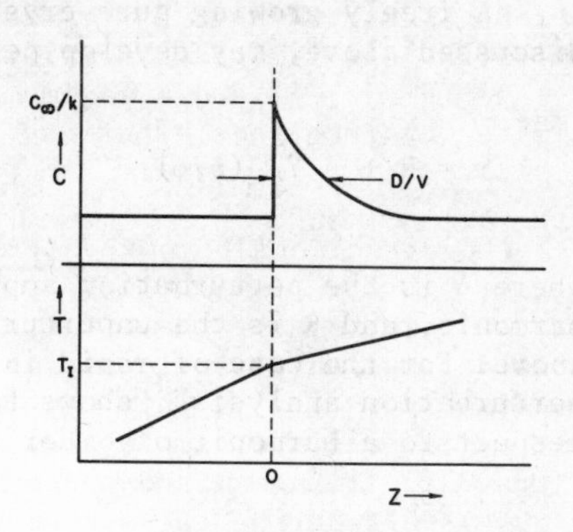

Fig. 9 : Sketches of concentration and temperature profiles during unidirectional solidification of a binary alloy. The solid-liquid interface is located in the plane $z = 0$ and is moving uniformly at velocity V to the right.

Note especially the boundary layer of height C_∞/k that decays in a distance $\simeq D/V$. From (3) we have $T_E = T_M + m\,C_\infty\,/k$ and T_I can be found from (5) if the function f is known. For example, if equation (11) applies, then

$$T_I = T_M + m(C_\infty/k) - V/\mu(\hat{n}). \qquad (39)$$

Equation (39) shows how the interface temperature is depressed below T_M by the presence of solute ($m < 0$ for $k < 1$) and by interface kinetics, which may be anisotropic. Furthermore, since k should be the actual distribution coefficient, (16) should be used to relate to equilibrium parameters. We see that even this very simple problem is beset with many complications.

VI. MORPHOLOGICAL INSTABILITY

For the sake of completeness, we consider briefly the phenomenon of morphological instability that is known to occur during melt growth. Morphological instability occurs whenever the shape of a growing crystal develops projections, depressions or undulations, usually on a length scale of several μm, because of the tendency of the transport of heat and/or solute to enhance local variations. This enhancement is related to the "point effect" of diffusion, according to which the isoconcentrates near a projection are compressed so that steeper gradients exist near a projection than at a smooth surface. These steeper gradients lead to local preferential growth via equations such as (19) and (20). Capillarity provides a stabilizing force that tends to counteract this "point effect." Details are available in several reviews,[14,16-20] so we confine ourselves here to a few simple cases.

A freely growing pure crystal in the shape of a sphere, as discussed above, may develop perturbations of a form such that

$$r = R + \delta\, Y_{lm}(\theta,\phi) \tag{40}$$

where δ is the perturbation amplitude, $Y_{lm}(\theta,\phi)$ is a spherical harmonic, and R is the unperturbed spherical radius discussed above. For the case of rapid interfacial kinetics ($R_o \to 0$), a perturbation analysis[21] shows that the sphere will be unstable with respect to a harmonic of order $l > 1$ whenever

$$\frac{R}{R^*} > \frac{1}{2}\left[(l+1)(l+2) + 2 + l(l+2)(k_S/k_L)\right] \tag{41}$$

where R^* is the nucleation radius defined above. For $l = 2$ and $k_S/k_L = 1$, this criterion gives $R > 11\,R^*$, so the spherical shape becomes unstable soon after leaving the realm where its growth rate is controlled by capillarity (see Fig. 8).

For the case of unidirectional solidification of a binary alloy, as discussed above, the solid-liquid interface may be subjected to perturbations of the form

$$z = \delta \sin(\omega x) \tag{42}$$

where δ is the amplitude of a perturbation of the initially planar interface (z = 0) and ω is a spatial frequency. Analysis[22] shows that instability obtains whenever

$$\frac{G^*}{mG_C} < S \tag{43}$$

where

$$G^* = \frac{k_S G_S + k_L G_L}{k_S + k_L}, \tag{44}$$

S is a stability function, practically equal to unity except for very large growth velocities, and G_C si the gradient of concentration at the unperturbed solid-liquid interface. Differentiation of (38) and evaluation at z = 0 gives

$$G = - C_{\infty} V(1 - k)/(kD) \tag{45}$$

so eq (43) may be written in the form

$$\frac{G^*}{V} < \frac{(1 - k)}{kD} (-mC_{\infty}) S \tag{46}$$

from which instability is seen to occur at a critical value of G^*/V that depends on the solute content, C_{∞}, of the bulk melt. Under conditions of instability, the planar interface can give way to cellular or dendritic growth.

ACKNOWLEDGEMENT

The author is grateful to the Division of Materials Research of the National Science Foundation under grant DMR78-22462 for financial support for much of the research on which this article is based. Gratitude is also expressed to S. R. Coriell and W. W. Mullins for helpful technical criticism of the manuscript, to Mrs. Carolyn Hendrix for preparing the manuscript and to Mr. Robert E. Miller for preparing the figures.

REFERENCES

1 . BRICE J.C., in *The Growth of Crystals from the Melt*, North Holland, Amsterdam, 1965
2 . PAMPLIN B.R.,ed., Cryst. Growth., Pergamon Press, 1975
3 . HURLE D.T.J., in *Crystal Growth : An Introduction*, P. Hartman ed. , North Holland, 1973, pp. 210-247
4 . MULLINS W.W., 1981, to be published in the Proceedings of an International Conference on Phase Transformations, August 1981, Pittsburgh, Pennsylvania.
5 . CAHN J.W., Cryst. Growth, H.S. Peiser ed., Pergamon, Oxford, (1967) pp. 681-690
6 . HERRING C., in *The Physics of Powder Metallurgy*, W.E. Kingston ed., Mc Graw-Hill, 1951, pp. 143-179
7 . CORIELL S.R. and SEKERKA R.F., J. Cryst. Growth. 34 (1976) 157-163

8 . WILSON H.A., Phil. Mag. 50(1900) 238
9. FRENKEL J., *Physikalische Zeitschrift der Sowjetunion* 1 (1932) 498
10. JACKSON K.A., J. Cryst. Growth 5 (1969) 13-18
11. JACKSON K.A., J. Cryst. Growth 24/25 (1974) 130-136
12. GILMER G.H. and JACKSON K.A., in *Crystal Growth and Materials*, E. Kaldis and H. Scheel eds., North Holland, 1977, pp. 80-113
13. CHERNOV A.A., Soviet Physics-Uspekhi 13 (1970) 101
14. SEKERKA R.F., in *Crystal Growth : An Introduction*, P. Hartman ed., 1973, pp. 403-443
15. FRANK F.C., Proc. Roy. Soc. A 201 (1950) 586-599
16. DELVES R.T., in *Crustal Growth Vol. 1*, B.R. Pamplin ed., Pergamon Press, Oxford, 1974, pp. 40-103
17. SEKERKA R.F., J. Cryst. Growth. 3,4 (1968) pp. 71-81
18. LANGER J.S., Reviews of Modern Physics 52 (1980) 1-28
19. CHERNOV A.A., J. Cryst. Growth 24/25 (1974) 11-31
20. WOLLKIND D.J., in *Preparation and Properties of Solid State Materials*, W.R. Wilcox ed., Marcel Dekker, 1979, pp. 111-191
21. MULLINS W.W. and SEKERKA R.F., J. Appl. Phys. 34 (1963)323-329
22. MULLINS W.W. and SEKERKA R.F., J. Appl. Phys. 35 (1964)444-451.

CRYSTAL GROWTH IN AQUEOUS SOLUTIONS

Blaise SIMON

Laboratoire de Dynamique et Thermophysique des Fluides
Université de Provence, Centre de St Jérôme
13397 Marseille Cedex 4, France.

ABSTRACT.

The interfacial aspects of crystal growth from aqueous solutions are described from different points of view.

A survey of interfacial and adhesion energies is given, with their consequences on morphology and growth mechanisms. High interaction energies between solute and solvent molecules are responsible for the apparition of faces which do not belong to the equilibrium form in the vapour phase. Methods of calculating surface roughness all suggest that crystals in aqueous solutions may be not very far from the roughening transition. This is confirmed in certain cases. The structural changes of the adsorbed layer of solution are responsible for anomalies of the growth rates in certain temperature ranges. Finally a survey is given of the various experimental methods for the determination of surface kinetics. It is sometimes even difficult to distinguish between spiral and two dimensional growth mechanisms. Distinguishing between direct integration of molecules in kinks and integration preceded by surface diffusion is shown to be impossible in many cases.

I. INTRODUCTION

Solution growth is the formation of a crystal from a liquid phase having a composition different from that of the crystal. Consider the most simple case of a pure crystal A growing from a binary mixture A_xB_{1-x}. When x=1 no diffusive mass transport in the liquid occurs (i.e. melt growth). When x=1, A grows from its melt

B. Mutaftschiev (ed.), Interfacial Aspects of Phase Transformations, 509–520.

with B as an impurity; only a few ppm of B can reduce the growth rate of A by a factor ten[1]. This is a surface "poisoning" effect of B, probably by adsorption in kinks. B can be called a solvent when $x < 0.95$, and it is still observed that an increase in the B concentration decreases the growth rate of A. In general this is a volume effect, A diffusing through B before being integrated into the crystal. Crystallization in solution then takes place by a double mechanism : melt growth and diffusion, the second being more important for crystals with low solubilities. The differences between melt and solution growth are thus not sharply defined.

Compared to vapour growth in multicomponent systems, with diffusion coefficients $\simeq 10^{-2}\ cm^2\ s^{-1}$, solution growth is characterized by low transport rates (diffusion coefficients $\simeq 10^{-5}\ cm^2 s^{-1}$) so that low growth rates may be expected. This is partially compensated by the higher concentration of solute. Another difference is the stronger solute-solvent interactions. The separation of solvent molecules from solute molecules and from kinks is difficult, thus producing slow growth rates. These interactions are stronger in aqueous solution (15 Kcal/mol for the $H_2O - Na^+$ bond) than in organic solutions where hydrogen bonds are of the order of 5 Kcal/mol.

II. ENERGETICS OF THE CRYSTAL-SOLUTION INTERFACE

The energetic and structural characters of the crystal-solution interface affect the crystal morphology (equilibrium and growth form, surface roughness) and the growth kinetics. The interfacial energy is defined by

$$\gamma_{SL} = \gamma_S + \gamma_L - \beta \quad (1)$$

where γ_S and γ_L are the surface free energies of solid and liquid respectively, β the adhesion energy. The latter is probably close to the hydration energy (about 15 Kcal/mol for cations, 5 Kcal/mol for anions).

Adhesion energies have been calculated, a priori, for water molecules on different sites of the {100} face of alkali halides.[2] The water molecule was considered a dipole with a length of 3.8 Å calculated from the molar volume. The strongest interaction energies are on cationic surface sites (10 Kcal/mol), those on other sites are about two times smaller. The chosen size of the water molecule does not allow a simultaneous adsorption of water molecules on contiguous cation and anion sites. The isosteric heat of adsorption measured from adsorption experiments at low coverage of water is in good agreement with this model regarding water molecules on cationic surface sites.

A more complete set of calculations of all quantities in (1) has been made by BIENFAIT et al.[3] The water molecule was considered a dipole with a length 2.75 Å and assumed to be adsorbed on cations and on anions, with the corresponding orientation. A correction was introduced in the γ_L term to take into account the difference between this structure and the structure of bulk water. The calculated values were $\gamma_S \simeq 150$ erg/cm^2, $\gamma_L \simeq 80$ erg/cm^2 and $\beta \simeq 220$ erg/cm^2, so that the interfacial energy is about 20 erg/cm^2, which is 7 times lower than γ_S.

Direct estimates of interfacial energies are possible from nucleation experiments. The values found are of the order of 10 erg/cm^2 (gypsum $\simeq$ 15 erg/cm^2, potassium alum $\simeq$ 2 erg/cm^2, NaCl $\simeq$ 5 erg/cm^2 sodium perborate $\simeq$ 5 erg/cm^2).

The consequences of these high interaction energies are important. Even if γ_S values of different faces of a given crystal are quite different, high β values can increase the differences in γ_{SL}, so that the equilibrium form in aqueous solution can have more faces than in vapour. This effect is well known for alkali halides.[3] In the vapour phase the only equilibrium and growth form is {100} . Smooth {111} faces limited by only one kind of ion cannot exist in this structure, because an ion of either sign close to such a surface would be at an infinite potential. In aqueous solution the growth of NaCl spheres[4] have shown {111} faces, which should be considered as equilibrium faces growing by a layer by layer mechanism. Furthermore, crystallization of alkali halides at high supersaturation always leads to habits including {111} faces. These appear as equilibrium form in aqueous solution probably because their surface, instead of being atomically smooth, is composed of microscopic {100} facets with sizes of only a few atoms. Such an ordered rough surface presents, to the water molecules, sites with very high adsorption energy. The strong water adsorption would make the presumed structure thermodynamically stable, and explain the layer by layer growth of this type of face. The same phenomenon is found in many other crystals[5].

III. SURFACE ROUGHNESS IN SOLUTION

The first computer simulation of surface roughness for a Kossel crystal in a lattice-like melt[6,7] have shown that the key parameter for the roughness of a (100) face is the factor

$$\alpha = (2\phi_{SS} + 2\phi_{LL} - 4\phi_{SL})/k_B T \qquad (2)$$

For $\alpha<3$, the surface is rough and the growth rate R is a linear function of the supersaturation σ; spirals and two dimensional nuclei do not influence the growth rate. For $\alpha>3$, the face is smooth and the growth is possible at low supersaturation only if spirals are present on the face; two dimensional nucleation is operative only at high supersaturation. A second parameter is the supersaturation; for a given α value an increase in supersaturation results in a smooth to rough transition.

In aqueous solutions with high adhesion energies, low values of α may be expected, indicating rather rough surfaces, and high growth rates. As ϕ_{SS}, ϕ_{LL}, ϕ_{SL} in (2) are not known separately, it is interesting to calculate α from macroscopic quantities. For a dilute ideal solution, α given by (2) is the enthalpy corresponding to the transition of a molecule from a kink to the surface. This quantity is proportional to the heat of dissolution, to the entropy of dissolution, and to the interfacial energy. These three macroscopic quantities can be evaluated for any crystal-solution system (non KOSSEL crystals, non ideal solutions). BOURNE and DAVEY[8] have compared the α values calculated from these three quantities for hexamethylene tetramine growing from aqueous and alcoholic solutions respectively. They found lower α values in the former case, indicating a higher roughness. This results in higher growth rates in aqueous solution, which has been observed. In fact, the calculation of α from the dissolution enthalpy in water was not possible, this being negative.

This difficulty, which seems to ruin the whole argument, was resolved by BENNEMA et al.[9]. They calculated the probability for the occurence of a given configuration in a model where solute molecules are free to move and rotate in the solution. The liquid phase is really a solution and no lattice model is involved. An equation was obtained that actually is the equality of chemical potentials at equilibrium for particles in the solid and in the liquid

$$k_B T \, ln \frac{v_L}{F_L} + k_B T + (6\phi_{SL}^{L} + 3\phi_{LL}^{L}) = k_B T \, ln \frac{v_S}{F_S} + 3\phi_{SS} \qquad (3)$$

where $1/v_L$ and $1/v_S$ are the numbers of solute particles per unit volume of liquid and solid. F_L and F_S are the free enthalpies of the solute in the liquid and in the solid. One has

$$F_S/F_L = (\pi^{0.5} e^{1.5}/\sigma_1)\,(T^3/\theta_A\,\theta_B\,\theta_C)$$

where σ_1 is a number close to 1 and θ_A, θ_B, θ_C are the characteristic temperatures for the rotations of a molecule around three directions. If $\theta_A \simeq \theta_B \simeq \theta_C \simeq T$, which holds for spherical molecules one has $ln\, F_S/F_L \simeq 1.5$. ϕ_{SL}^{L} and ϕ_{LL}^{L} are the potential energies of solute and solvent particles in the liquid, considered a saturated solution. With $x = v_L/v_S$ equation (3) becomes :

$$3\phi_{SS} + 3\phi_{LL}^{L} - 6\phi_{SL}^{L} = k_B T(1.5 - \ln x)$$

Now expressing ϕ_{LL}^{L} and ϕ_{SL}^{L} as functions of interaction between pairs and of the concentration x, one obtains :

$$\alpha = (1-x)^2\ (1.5-\ln x) \qquad (4)$$

This formula, valid only for spherical molecules, gives the solubility, expressed in mol/unit volume, as the key parameter for the surface roughness. High solubilities correspond to low α values, to high roughness and to high growth rates. The last conclusion is usually verified in practice [10,11] and can be a good guide for the prediction of growth rates.

The correspondence between high linear growth rates R and high solubilities may nevertheless be due to other reasons than only the α factor. For instance, linear growth rates may occur with high surface diffusion. This was demonstrated even in computer simulation, where the introduction of surface diffusion results in more linear R (σ) curves. It may be unwise to systematically ascribe linear R (σ) curves only to rough surfaces. It should be kept in mind that flat smooth surfaces usually obtained from solution indicate a layer by layer growth and cannot be considered rough.

There are certain cases where crystal surfaces are not very far from the roughening transition. The first indication has been given by BOISTELLE et al. [12] in the study of adsorption of cadmium ions on {100} of NaCl at equilibrium in saturated solution. It was found that crystals prepared from different conditions have a different adsorption behaviour. This implies that the surface structure depends on the mode of preparation (crystallization kinetics were very rapid and not controlled). Furthermore, the adsorption isotherms can be analyzed only with a model involving three kinds of adsorption sites, with three different adsorption energies. It is tempting to ascribe the highest energy to kink sites, the next lower energies to edge and surface sites respectively. The relative number of these sites are about in the ratios 1 : 5 :15(depending on the preparation conditions), and indicate a rather rough surface.

Another argument in favor of high surface roughness in this system is a kinetic one [3]. The morphology of alkali halides in aqueous solution at a medium supersaturation is a combination of {100} and {111} faces, all flat and smooth. For higher supersaturation the {100} faces do not stay flat and a roughening is clearly visible. To our knowledge, this is the only example of the roughening transition in aqueous solution.

IV. STRUCTURE OF THE SOLUTION NEAR THE INTERFACE

Water adsorbed on solid surfaces has properties differing from those of bulk water, with temperature anomalies corresponding to anomalies in crystal growth rates.

CHERNOV et al.[13] have investigated a great variety of substances, in conditions where the resistance of the volume diffusion was reduced to a minimum. Growth rates at constant supersaturation were measured as a function of temperature. The growth rates were monotonic functions of the temperature between about 15° and 40°C, but anomalies occured at 10° and 40°C. The interpretation was that absorbed water has certain stable structures, each corresponding to a given temperature range, and that the transition from one structure to another corresponds to a loosening of the adsorbed layer of solution. At the transition temperatures, the residence time of a growth unit traversing this layer would be smaller. This hypothesis was confirmed by the addition of ethanol to the solution, which is known to destroy the bulk structure of water. The anomalies of growth rates were actually not detected with these conditions.

There is no known property of bulk water producing anomalies at the temperatures mentioned above. CHERNOV et al. have made a careful study of the adsorbed layer of solution, by nuclear magnetic resonance which gives information on dipole orientation and mobility. Bulk water and solutions did not show anomalies in the NMR peaks. The fraction of adsorbed water in the sample was then increased by the preparation of slurries, i.e., small crystals stacked together in the saturated solution. It was found that the NMR peaks were broadened, indicating a lower mobility of protons. The plot of this broadening vs temperature shows anomalies exactly corresponding to those observed in the crystal growth rates. Insoluble salts do not show anomalies, which accordingly may be not a property of the water itself, but rather of solutions. Also there are no anomalies for solutions on amorphous substrates, such as teflon. The statement that there must be some epitaxial structure of the solution imposed by the crystal substrate is correlated to the observation that anomalies for different faces of a given crystal do not occur at the same temperature.

Such an ordering in the adsorption layer may have unexpected consequences, particularly for the diffusion coefficients : for long molecules, an ordering of the solution due to concentration gradients can enhance the transport rates by two orders of magnitude[14].

V. INTERFACIAL KINETICS DURING GROWTH FROM AQUEOUS SOLUTIONS

Crystal growth from solution is a succession of two processes. One is a surface reaction, with kinetics depending on the surface structure, the time necessary for the integration into kinks, and the concentration c_i at the interface. The other is the volume diffusion in the bulk phase, which feeds the surface reaction. Crystallization is ideally volume diffusion limited when the surface kinetics are so rapid that $c_i = c_s$ (saturation concentration), and the growth rate can be calculated *a priori* from diffusion equations. No example of such limitation is known. The opposite case is when the interfacial kinetics are very slow: This is the case for perfect crystal faces in a solution with a supersaturation insufficient for two dimensional nucleation to occur. In the intermediate cases, the interfacial concentration c_i is between c_s and c (solution concentration far from the growing crystal). It is always possible, by stronger stirring of solution, to reduce the volume diffusion resistance, and to give c_i a maximum value (depending on the interfacial kinetics) producing a higher growth rate.

V.1. Determination of the Interfacial Kinetics

The problem is to extract the interfacial kinetics from the overall growth rate. Three methods have been proposed.

In the first method[15] the growth rate is measured at a constant supersaturation as a function of the velocity u of the solution past the crystal. A theoretical model shows that the thickness of the concentration boundary layer δ (where the volume diffusion occurs) is proportional to $u^{-0.5}$. (The proportionality factor is unknown in most experimental configurations). The growth rate corresponding to volume diffusion can be written as $R = k\,u^{0.5}(c - c_i)$ where k is a constant. This growth rate is also given by the equation for the interfacial kinetics, where $R = K(c_i - c_s)^n$ (K and n are other constants). Elimination of the unknown c_i gives

$$Ru^{-0.5} = k(c - c_s) - kK^{-1/n}\,R^{-1/n}$$

A plot of the experimental values of $Ru^{-0.5}$ vs $R^{-1/n}$ should give a straight line intercepting the $R^{-1/n}$ axis at the maximum value of R, where the volume resistance has been reduced to zero. This kind of study was done by GARSIDE et al.[16] for potassium alum. It is noteworthy that the extrapolated values of R were not much affected by the choice of n($n \simeq 2$). Another important result was that these values are, at least at high supersaturation, much higher than the experimental growth rates. Thus, it is very difficult in practice to eliminate the volume diffusion resistance.

In the second method, the growth rate is measured by the rotating disc technique [17]. Here the exact value of δ is known as a function of the diffusion coefficient, the viscosity, the angular rotation rate of the disc, so that the knowledge of only one value of the growth rate enables the calculation of the interfacial concentration from the volume diffusion equation.

In the third method [18], the growth and dissolution rates of a face are measured, in the same hydrodynamic conditions, as a function of the concentration difference $(c_s - c)$. The growth rate is given by the system of equations $R = k'(c - c_i)$ and $R = K(c_i - c_s)^n$ (bulk diffusion and surface reaction respectively). If the dissolution rate R_D is found experimentally to be higher than the growth rate, and also to be a linear function of the concentration difference, it is assumed that the dissolution rate is completely limited by volume diffusion and is written as $R_D = k'(c_s - c)$. k' is calculated from the dissolution experiment and reported in the first equation for the growth rate. Then c_i can be obtained from the experimental growth rate, and a plot of R vs. $(c_i - c_s)$ is obtained. Such a calculation has been made by GARSIDE et al. [18] for potassium sulphate. The overall growth rates were parabolic functions of the concentration difference, with an activation energy of 7 Kcal/mol. The dissolution rates, about ten times more rapid than the growth rates, were linear functions of the concentration difference, with an activation energy of 4 Kcal/mol. The significant result is that the interfacial kinetics are not a parabolic function of the supersaturation, but are rather described by

$$R \propto (c_i - c_s)^{2.4}$$

Also the corresponding activation energy 12.5 Kcal/mol, is much higher than the activation energy for the overall growth rate. For these kinetics, the only possibility is to fit a two dimensional mechanism. A spiral model cannot represent the surface kinetics (but represents well the overall kinetics). This example illustrates the difficulties met in the identification of a growth mechanism. The main problem regarding this method is whether the dissolution is exactly limited by volume diffusion. It is also not certain that growth and dissolution have been measured for the same crystal morphologies.

V.2. Two Dimensional Nucleation vs Spiral Growth

Two dimensional nucleation in aqueous solutions has been directly observed only in the case of electrocrystallization of silver in capillary tubes. Other indications are indirect and based on the analysis of R (σ) curves.

In one kind of R (σ) curve, there is first a supersaturation "dead zone" where the crystal does not grow. At higher supersaturation, the curve can be fitted by an expression

$$R = A \sigma^{n} \exp(-B/\sigma)$$

with $n \simeq 1$. The edge energy can be calculated from the slope of a linear plot of $ln(R \sigma^{-n})$ vs σ^{-1}. Results regarding three different faces of palmitic acid in aqueous solution[19], and the {100} face of sodium triphosphate hexahydrate[20] have been reported. The dubious point[10] is that it is not certain whether the solution was free from impurities, the effect of which can result in an apparent two dimensional nucleation. This results from the CABRERA mechanism[21], where the growth rate is zero at low supersaturation, even for spiral growth. BELYUSTIN et al.[22] found that growth rates of faces artificially ground at an angle to singular faces also exhibit an apparent nucleation growth due to the effect of unknown impurities.

A better indication of a two dimensional nucleation is the much more rapid growth at high supersaturation than expected from an extrapolation of the growth rates at low supersaturation. The systematic deviations towards higher growth rates show that nucleation is occurring together with a spiral mechanism.[23,24]

Spiral growth has often been demonstrated, a posteriori, by optical inspection of faces. In other cases, the indication of this mechanism is indirect. In two experiments, this mechanism has received a convincing confirmation : the growth of ADP[25] and sucrose.[26] In the latter case, the crystals were first submitted to a slight dissolution causing curved surfaces. The growth rates measured on faces of very small area, followed four distinct curves. Each curve, corresponding to a given activity of spirals, was fitted to an equation

$$R = C \sigma_1^{-1} \sigma^2 \tanh(\sigma_1/\sigma)$$

where C is a constant and

$$\sigma_1 = 9.5 \gamma a/(\varepsilon k_B T x_s)$$

γ is the edge energy, a the lattice parameter x_s is the mean free path on the surface, ε is an integer. The values of C are the same for all curves. The values of σ_1(0.317,0.176,0.114,0.084) are in the ratios 1 : 1/2 : 1/3 : 1/4, which beautifully illustrates the spiral model.

It is not always possible to assign an experimental R (σ) curve to a given growth mechanism. Sometimes, due to the scatter in experimental growth rates, either fitting can be possible, as in the case for potassium alum[16], and of hexamethylene tetramine[27]. The other reason for an impossible fitting could be that the crystal actually do not grow by the simple mechanisms of the theory, because of the presence of macrosteps, or of grain boundaries.

V.3. Direct Integration vs Surface Diffusion

The most abundant of the two fluxes of molecules coming simultaneously into kinks, namely the direct flux from the bulk phase and that coming from surface diffusion, controls the overall surface reaction rate.

The concentration of molecules is so low, in vapour phase, that growth by direct impingement onto kinks is probably very slow. It has been shown that surface diffusion is effective in vapour growth (e.g. the majority of molecules arriving on the lateral faces of a H_g platelet have first diffused on the upper face of the platelet[28]) This surface diffusion mechanism was formulated by BURTON, CABRERA and FRANK.[29] They have pointed out that since the typical concentration of growth units is much higher in condensed phase than in the vapor growth by direct integration is more probable in solution, especially for crystals with a high solubility. The same authors also indicate that the mean free path in solution is probably much smaller than in the vapour because of contacts with the solvent molecules. The corresponding growth mechanism has been formulated by BURTON, CABRERA, FRANK[29] and by CHERNOV[30]. Sometimes a surface diffusion dominates growth in solution as in the case of whisker growth. The very high growth rate of the whisker tip is caused by integration of growth units that first diffused on the lateral perfect faces. Negative wiskers (long channels) are sometimes found during dissolution. It is unlikely that in such narrow channels the solution can be rapidly replaced by fresh solution, thus surface diffusion in the channels could cause the observed high dissolution rates.

But in general, no direct experimental proof of active surface diffusion is available and the identification of the integration mechanism is in general not possible. (Arguments based on morphology (rounded or linear terraces) are not conclusive). The calculated growth rates for the two mechanisms are very similar. There is a parabolic region at low supersaturation for the surface diffusion, with a transition towards a linear R (σ) curve passing through the origin. This type of curve has never been observed for aqueous solution. For the direct integration, the theory predicts also at low supersaturation a nearly parabolic curve, which shows a transition towards a linear curve not passing through the origin. Experimental curves generally are of this type.

For the direct integration mechanism BENNEMA[31] shows that the transition from parabolic to linear curves must occur at a supersaturation of about 10^{-2} or 10^{-3} (calculated from BCF and CHERNOV model). He used this argument[32] to eliminate the possibility of a direct integration for potassium alum and sodium chlorate at very low supersaturation ($\sigma < 10^{-4}$) where the experimental R (σ) curves are linear. The only remaining possibility is a surface diffusion. Fitting of a BCF law gives a mean surface free path of about 400 interatomic distances.

In other experiments the parabolic to linear transition occurs at about $\sigma \simeq 10^{-2}$ and both fittings are possible. This was shown by LEVINA et al.[33] and BELYUSTIN et al.[34] for the growth of $MgSO_4\ 7H_2O$ where interface kinetics were determined, along with activation energies. A careful discussion showed that a definite conclusion was not possible. Both fittings gave reasonable orders of magnitude for diffusion coefficients and mean free path on the surface. The authors concluded that too many terms in the equations are unknown and that knowledge only of the R (σ) curve does not sufficiently reveal the true kinetics.

A surface diffusion mechanism is more probable for R (σ) curves with a small parabolic part. When curves are parabolic in a wide range of supersaturation, a surface diffusion mechanism may be expected only if the mean free path on the surface are a few interatomic distances in length. In these conditions surface diffusion is in fact not very different from direct integration.

REFERENCES

1 . BORISOV V.T., GOLIKOV I.N. and MATVEEV Y.E., J. Cryst. Growth 6 (1969) 72
2 . BARRACLOUGH P.B. and HALL P.G., Surf. Sci. 46 (1974) 393
3 . BIENFAIT M., BOISTELLE R. and KERN R., in *Adsorption et Croissance Cristalline*, CNRS, Paris, 1965, pp 515-531
4 . NEUHAUS A., Z. Krist. 68 (1928) 15
5 . KERN R., Bull. Soc. Fr. Min. Crist., 78 (1955) 461-474
6 . GILMER G.M. and BENNEMA P., J. Cryst. Growth. 13-14 (1972) 148
7 . GILMER G.M. and JACKSON K.A., in *Current Topics in Materials Science*, vol. 2, KALDIS ed., North Holland, Amsterdam, 1977,p80
8 . BOURNE J.R. and DAVEY R.J., J. Cryst. Growth, 36 (1976) 278
9 . BENNEMA P. and van der EERDEN J.P., J. Cryst. Growth 42 (1977) 201
10. SIMON B. and BOISTELLE R., J. Cryst. Growth 52 (1981) 779

11. BOURNE J.R. and DAVEY R.J., J. Cryst. Growth 44 (1978) 613
12. BOISTELLE R., MATHIEU M. and SIMON B., Surf. Sci. 42 (1974)373
13. CHERNOV A.A. and SIPYAGIN V.V, in *Current Topics in Materials Science*, vol. 5, KALDIS ed., North Holland, Amsterdam, 1980, pp 279-333
14. PRESTON B.N., LAURENT T.C., COMPER W.D and CHECKLEY G.J., Nature 287 (1980)499-503
15. BRICE J.C., J. Cryst. Growth 1 (1967) 161
16. GARSIDE J., Janssen van ROSMALEN R. and BENNEMA P., J. Cryst. Growth 29 (1975) 353
17. BOMIO P., BOURNE J.R. and DAVEY R.J., J. Cryst. Growth 30 (1975) 77
18. GARSIDE J., MULLIN J.W. and DAS S.N., Ind. Eng. Chem. Fund 13 (1974) 299
19. MICHAELS A.S. and COLVILLE A.R., J. Phys. Chem., 64 (1960) 13
20. TROOST S., J. Cryst. Growth,13-14 (1971) 449
21. CABRERA N. and VERMILYEA D.N., in *Growth and Perfection of Crystals*, DOREMUS, ROBERTS, TURNBULL ed., Wiley, New York, 1962, pp 393-408
22. BELYUSTIN A.V. and KOLINA A.V., Sov. Phys. Cryst 23 (1978)230
23. BENNEMA P., KERN R. and SIMON B., Phys. Stat. Sol. 19 (1967)211
24. JANSSEN van ROSMALEN G. M,DAUDEY P.J and MARCHEE W.G.J., J. Cryst. Growth 52 (1981) 801
25. DAVEY R.J., RISTIC R.I. and ZIZIC G., J. Crystal Growth 47 (1979) 1
26. VALCIC A.V. and NIKOLIC S.N., ECCG 2 (Lancaster), Abstract booklet (1979) C 44
27. BOURNE J.R. and DAVEY R.J., J. Cryst. Growth 36 (1976) 287
28. VOLMER M. and ESTERMAN I., Z. Phys. 7 (1921) 13
29. BURTON W.K., CABRERA N. and FRANK F.C., Phil. Trans. Roy. Soc. 243 (1951) 299
30. CHERNOV A.A., Sov. Phys. Uspekhi 4 (1961) 116
31. BENNEMA P., J. Cryst. Growth 1 (1967) 278
32. BENNEMA P., J. Cryst. Growth 1 (1967) 287
33. LEVINA I.M., BELYUSTIN A.V., MORDERER V.I. and KHAUSTOVA Z.G., Sov. Phys. Cryst. 19 (1974) 191
34. BELYUSTIN A.V. and LEVINA I.M., Sov. Phys. Cryst. 19 (1974) 194

CRYSTALS AND CRYSTALLIZATION IN BIOLOGICAL SYSTEMS

Felix FRANKS

Department of Botany
University of Cambridge
Downing Street, Cambridge
United Kingdom.

ABSTRACT

There are few instances of real crystals in living systems, although liquid crystalline structures exist in abundance. The formation of gall and kidney stones takes place because of metabolic distrubances. Blood clotting is a defense reaction, biochemically initiated. The crystalline lens is the prime example of an *in vivo* crystal structure. Finally, the crystallization of ice in tissues and the resistance of living organisms to freezing is the most extensive occurrence of *in vivo* crystallization.

I. CRYSTALLIZATION FROM SUPERSATURATED SYSTEMS

Real crystals *in vivo* are rare; where they do occur, they are usually associated with a pathological condition. On the other hand, liquid crystals are common and will be discussed below. The best documented crystallization process *in vivo* is the growth of gall stones from supersaturated solutions of cholesterol in bile. The stones consist of cholesterol monohydrate and their precipitation and growth takes place in five stages, as shown in Fig. 1.

The first stage is of a genetic/biochemical nature and consists of the accumulation of supersaturated bile in the gall bladder. During the second stage cholesterol is accumulated; its solubility depends on the bile salt/phospholipid ratio, which in turn controls the type of micelle that is formed, and the saturation ratio. The equilibrium phase diagram is shown in Fig. 2.

B. Mutaftschiev (ed.), Interfacial Aspects of Phase Transformations, 521–529.

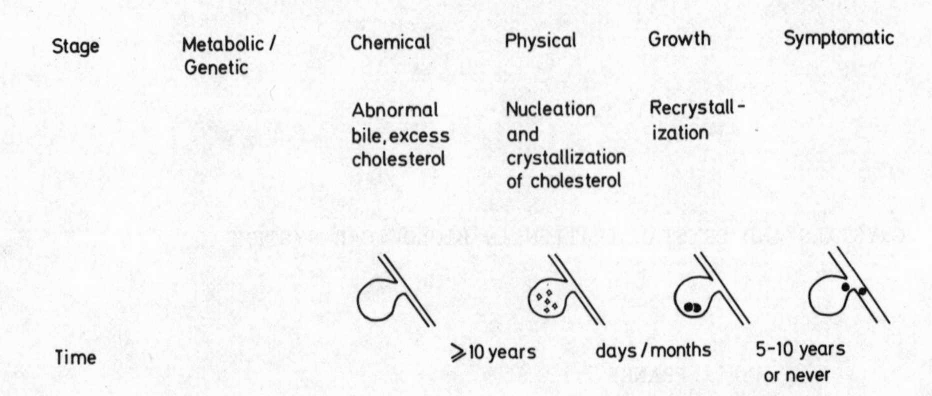

Fig. 1 : The stages of cholesterol gall stone formation, and the periods for which each of the stages persists.

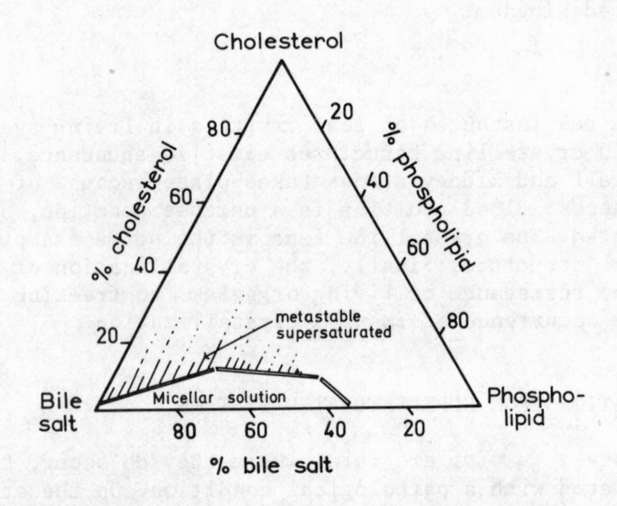

Fig. 2 : The bile salt (sodium taurocholate) - phospholipid - cholesterol phase diagram for a total solid content of 10 %.

In ethanolic solutions, supersaturation ratios of up to 13 have been measured before homogeneous nucleation of cholesterol takes place. In bile salt the ratio is approximately 3, so that cholesterol is always nucleated by a heterogeneous mechanism. The nature of the nuclei is still uncertain; calcium, bile acids, mucus and bacteria have been suggested. It is also possible that in "healthy" organisms antinucleating substances inhibit the formation of gall stones by maintaining the stability of the supersaturated cholesterol solution.

Similarly, during the recrystallization stage, growth inhibitors may retard the formation of macroscopic stones. This is also the case with kidney stones which crystallize from supersaturated urine under the action of calcium. Thus, acid urine gives rise to uric acid crystals and alkaline urine to phosphate or oxalate crystals. Nothing is known about the mechanisms which perturb the stability of the supersaturated solution and promote nucleation.

II. BLOOD CLOTTING - A BIOCHEMICAL CRYSTALLIZATION.

Haemostasis and its abnormalities have been studied extensively at a phenomenological level and a large number of essential intermediate steps have been identified. The final product is the protein fibrin which is polymerized and cross-linked to provide the protective barrier to bleeding. The precursor of fibrin, fibrinogen, is a complex protein which exists as a dimer of uncertain structure. Two suggested structures are shown in Fig. 3; both of them can be made to fit electron microscopic and hydrodynamic data, although the globular shape is now believed to be more probable.

Fig. 3 : Proposed shapes for the fibrinogen molecule. (Reproduced, with permission, from Brit. Med. Bull. 33, 245(1977).)

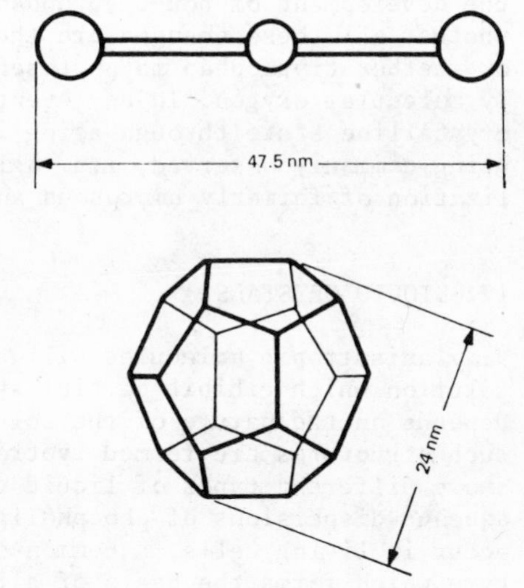

Each subunit consists of three peptide chains, Aα, Bβ and γ, with molecular weights 67000, 58000 and 47000 respectively. The Aα and γ chains of each dimer unit are linked by S-S bonds near the N-terminal, and there are also other disulphide links with greater or lesser stability; Fig. 4 shows the complex linking near the N-terminal. The production of the fibrin clot involves a reaction of thrombin with the A and B chains. Simultaneously fibrinogen is modified by yet another protein, plasmin, to produce cross-linked fibrin.

III. THE LENS : A BIOLOGICAL CRYSTAL

The lenses of all animals consist of epithelial cells which are laid down to form fibres. The lens has no vascular system, so that oxygen and nutrients are transported by diffusion. New fibres are laid down at the periphery of the lens, and no aged material is removed from the centre. The lens has the highest protein content of any tissue : 35 % of the wet weight and nearly 100 % of the dry weight are made up of three proteins, the crystallins which are crystalline in the soluble state, but which on aging become insoluble and amorphous. Little is known about the crystalline organization of the soluble proteins or the reason why the lens is transparent.

During the aging process, the crystallins become insoluble and increase in molecular weight, probably by covalent cross-linking, not all of which is due to S-S bonds. An advanced state of aging is the formation of a cataract, which is also accompanied by the racemization of aspartate, the production of acidic peptides, and the development of non-tryptophan fluorescence. It is not certain whether all these changes are the result of direct photooxidation or whether tryptophan is photosensitized and subsequently oxidized by molecular oxygen. In any event, this destruction of an *in vivo* crystalline state through aging is a unique process, the opposite being commonly observed, i.e. aging involves the gradual crystallization of formerly amorphous and hydrated tissues.

IV. LIQUID CRYSTALS

Many anisotropic molecules will adopt ordered structures in aqueous solution which exhibit partial symmetry. The particular structure depends on the nature of the solute molecule and the water content; such structures are termed lyotropic liquid crystals. Figure 5 shows different types of liquid crystal structures formed by aqueous dispersions of phospholipids. Some of these structures occur in living cells, a common one being the lipid bilayer structure which forms the basis of all biological membranes.

Liquid crystals are characterized by their distinctive optical properties : they exhibit birefringence, dichroism and optical activity. A detailed analysis of the optical properties permits the identification of many different types of liquid crystalline phases.

Such phases are also formed by the packing of biopolymers or polymers which, although of synthetic origin, possess a secondary structure (internal hydrogen bonding). These liquid crystalline phases can occur spontaneously. For instance, the homopolypeptide poly(benzyl-glutamate) exists as α-helix in solution. The helical molecules assemble into spherulitic supramolecular structures. In living organisms this type of self-assembly is common : complex structures which exhibit the properties of liquid crystals are built up from proteinaceous sub-units. They include flagella, photoreceptors (chloroplasts and visual systems), fibrous structures and effectors (muscle and nerve cells).

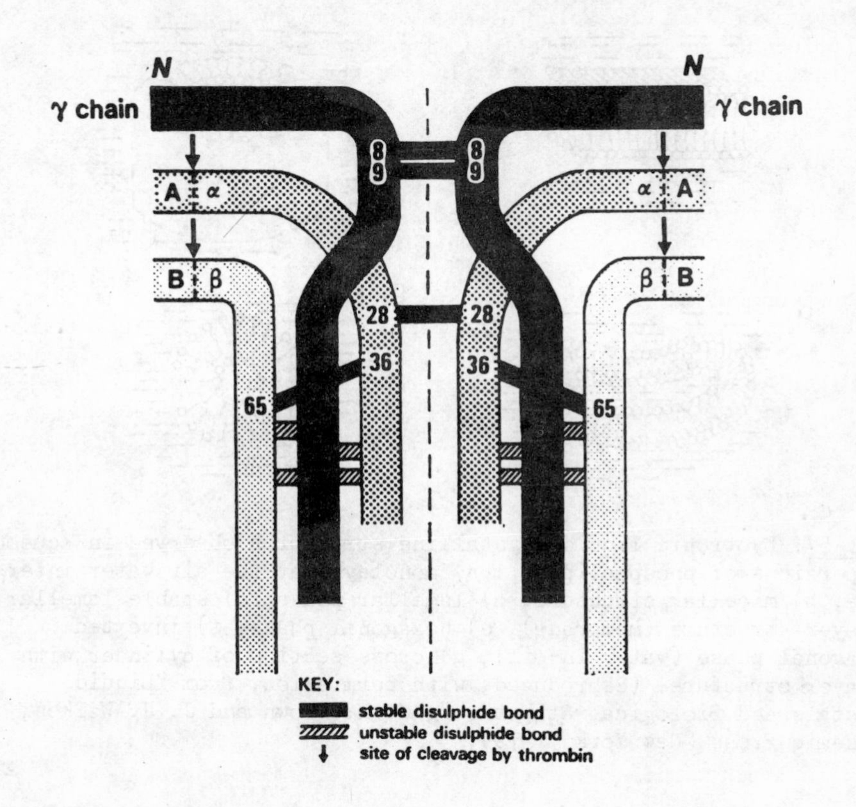

Fig. 4 : The N - terminal disulphide linked portion of the dimeric fibrinogen molecule. The bonds between the γ-chains at positions 8 and 9, and those between the Aα chains at position 28 are stable. The other disulphide links shown are metastable and can exchange. (Reproduced with permission, from Brit. Med. Bull.. 33, 245 (1977)).

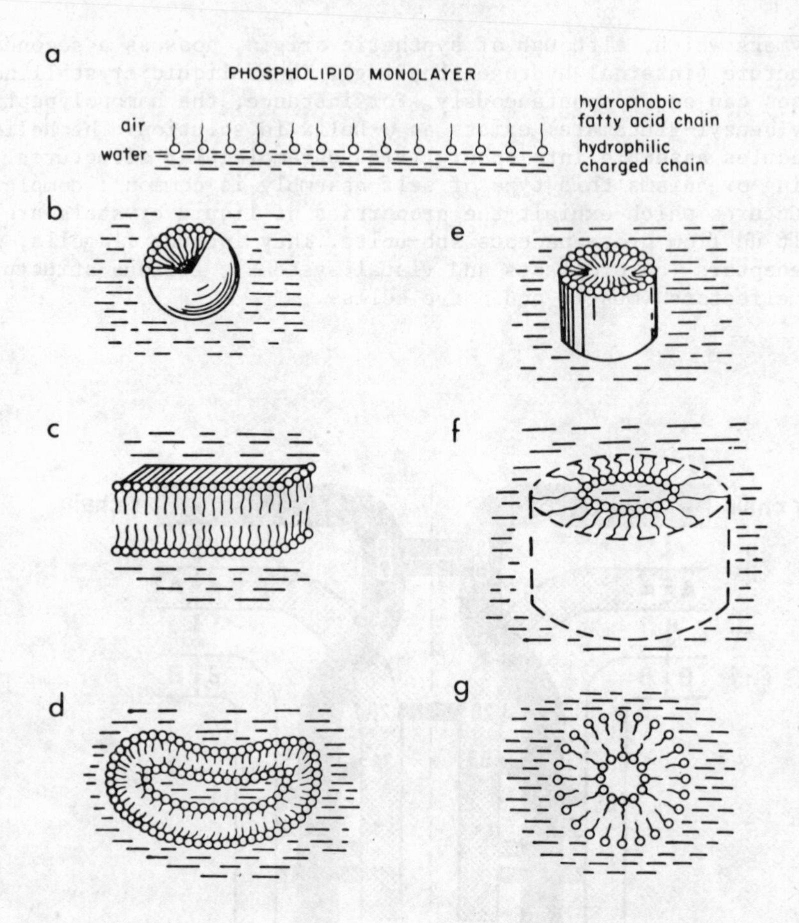

Fig. 5 : Lyotropic liquid crystalline structures observed in aqueous dispersions of phospholipids : a) monolayer at the air water interface, b) micellar structure, c) lamellar phase, d) stable lamellar bilayer structure (membrane), e) hexagonal phase, f) inverted hexagonal phase (water-in-oil), g) cross section of cylinder with bilayer structure. (Reproduced, with permission, from "Liquid Crystals and Biological Structure", G. H. Brown and J. J. Wolken, Academic Press, New York, 1979).

Fig. 6 shows the assembled structure of muscle which is built up from two structural proteins (myosin and actin) and two regulatory proteins (tropomysin and tropinin). Actin is of particular interest, because it shows how globular sub-units can give rise to aligned fibrous structures. The electron micrograph shows the elements of the two dimensional hexagonal structure. Contractile protein systems are found in a large variety of animal and plant cells.

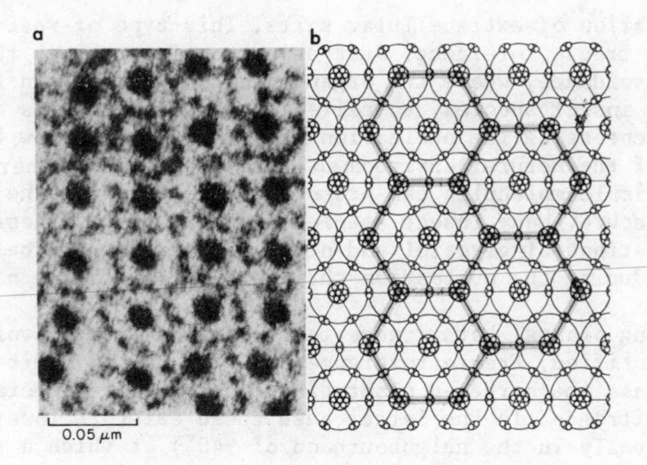

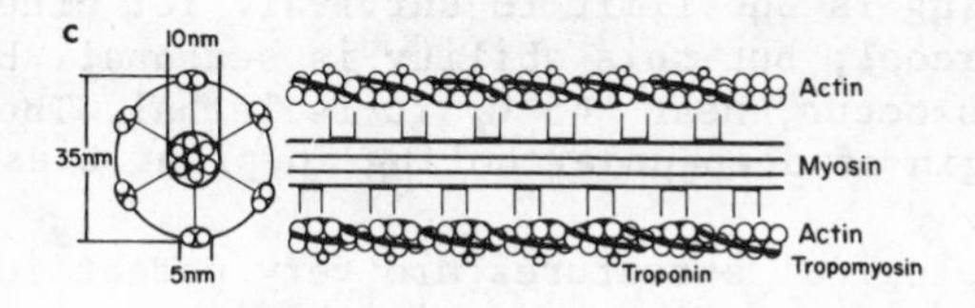

Fig. 6 : a) End-on view of the actomyosin myofibril, showing the hexagonal array. b) Schematic drawing of a), showing the positions of actin (thin filaments) and myosin (thick filaments). c) Longitudinal section of a myofibril. Reproduced, with permission from "Liquid Crystals and Biological Structures", G. H. Brown and J. J. Wolken, Academic Press, New York, 1979.

V. FREEZING STRESS AND ITS AVOIDANCE

Water is essential for life, but the freezing point of water lies at the centre of the temperature range experienced by living organisms on this planet (-40 to +40°C). Living organisms which are seasonally exposed to subzero temperatures must therefore possess survival mechanisms that enable them either to tolerate freezing or to avoid freezing. Freezing stress does not arise from the crystallization of ice directly, but from the osmotic imbalance between the cell and its environment when the extra-cellular medium freezes. In response to the osmotic stress, the cell becomes dehydrated with a comcomitant increase in the concentrations of the cell components.

Several survival mechanisms rely on the minimization of the osmotic stress produced by freezing. This is usually achieved by the synthesis of so-called compatible solutes which counter the concentration of extracellular salts. This type of resistance is based on freeze tolerance. Just as common is survival through freeze avoidance, where the organism can undercool. In Antarctic fish and insect species, so-called antifreeze proteins are implicated in depressing ice nucleation temperatures to below the temperature of the natural environment. The exact method whereby these rather simple molecules can promote undercooling of the blood is still obscure. Many insect species are known to synthesize high concentrations of glycerol and other polyols during the autumn, again reducing the nucleation threshold of ice in the haemolymph.

Among plants, both freeze tolerance and freeze avoidance have been identified. Freeze tolerance can be total or limited. In the former case the hardened plant tissue can survive immersion in liquid nitrogen. In the latter case there exists a lower temperature (usually in the neighbourhood of -40°) at which a second freezing event occurs in the tissue, possibly within the cells; this second freezing is the limit to survival. Yet other plants are able to undercool, but this ability is seasonal. Here again, when freezing does occur, near -45°C, it is lethal. The chemical and physical origin of deep undercooling in plant tissue is unknown.

Many biological structures are very potent ice nucleators, although it is not known whether this ability to nucleate ice at high subzero temperatures is implicated in the survival process. Ice crystallization *in vivo*, freezing stress and its avoidance and the biochemical and biophysical adaptation to low temperatures are of great importance from scientific and technological points of view, but the basic principles that govern survival are still quite obscure.

I wish to thank the Leverhulme Trust and the Agricultural Research Council for grants which support our work on biophysics of living organisms at subzero temperatures.

SUGGESTED READING LIST

Gall stone formation

CAREY M.C. and SMALL D.M., J. Clin.Investig. 61(1978)998.
MAZER N.A., CAREY M.C., KWASNICK R.F. and BENEDEK G.B., Biochemistry 18(1979)3064.

Blood Clotting

Haemostasis. Brit. Medical Bull. 33, No. 3, September 1977.

The Crystalline Lens, Cataract Formation

BARBER G.W., Exp. Eye Res. 16(1973)85.
BETTELHEIM F.A., ibid. 14(1972)251.
van HEYNINGEN R., ibid. 13(1972)136.
Mechanism of Cataract Formation in the Human Lens, ed. D.E. Duncan, Academic Press, London, 1981.
ZIGLER J.S. and GOOSEY J., TIBS, May 1981, p. 133.

Liquid Crystals

BROWN G.H. and WOLKEN J.J., in *Liquid Crystals and Biological Structures*, Academic Press, New York, 1979.

Physiological Freezing Stress

FRANKS F., *Scanning Electron Microscopy*, Vol. II, 1980, p. 349.

CRYSTAL GROWTH FROM NON AQUEOUS SOLUTIONS

R. BOISTELLE

Centre de Recherche sur les Mécanismes de la Croissance Cristalline - CNRS -
Campus de Luminy, case 913
13288 Marseille cedex 9 - France

ABSTRACT

Growth from non aqueous solutions often allows realization of special crystallization aims by control of the solution properties. Solvents are classified to further understand the solubility of electrolytes and non electrolytes. Several solvent and solubility effects are then pointed out such as the effect on the nucleation, which is easier when solubility is higher, on the crystallization of polymorphs, and on the surface morphology of the crystals. Solvents also change growth rates altering properties of the adsorption layers. It is shown that accurate estimates of surface roughness are difficult to achieve especially for crystals with high dissolution entropy and high structural anisotropy. Examples are also given of attempts to understand growth mechanisms by using crystallization energies.

I. INTRODUCTION

There is no essential difference between crystallizations from aqueous and from non aqueous solutions. Water is the solvent most commonly used, especially in industry, because it is cheaper than other solvents. Non aqueous solutions offer, however, a much larger choice of solvent molecular properties such as dielectric constant, dipole moment, size plus conformation and even the type of bonding with the solute or crystal molecules. Thus, large differences in nucleation and growth rates of crystals can result from the proper choice of the solvent-solute system. Experiments concerning interface effects on crystal growth in non aqueous solutions, however are quite limited. In addition, theoretical models have traditionally not included special features of non aqueous solutions, such as

B. Mutaftschiev (ed.), Interfacial Aspects of Phase Transformations, 531–557.

large molecule size. This paper deals with examples characteristic of crystal growth from non aqueous solutions. After a short review of solvent-solute systems, the relationship of solubility and solvent effects to nucleation rates and to polymorphic transformations are discussed. More closely linked to interface problems is the influence of the solvent on the surface morphology of the crystal. However the most important problems are those of the surface roughness, the adsorption layer, and the growth kinetics. These three points have been widely investigated during the last decade and the main conclusions drawn from the different studies are summarized in the sequel.

II. SOLVENTS AND SOLUBILITY

With the use of non aqueous solvents it is sometimes possible to fit the solvent-solute system to some special aims such as modification of nucleation or growth rates, modification of habit and morphology, or even simple use of high or low solute concentrations. All these phenomena result from solvent-solute or crystal-solvent interactions, the former being obviously related to solubility. Although in this paper we are only interested in solubility effects, we should still mention that solvent-solute interactions affect reaction rates, conductance, acid and base strength, etc.[1,4]

The classical principle "like dissolves like" can be explained in the following way[1,2]. There are three main classes of solvents. Hydrogen donors (methanol, formamide) are protic solvents. Solvents with dielectric constants greater than 15 but which cannot donate suitably labile hydrogen atoms to form hydrogen bonds with the solute, are dipolar aprotic solvents (acetonitrile, nitrobenzene). Finally, if the dielectric constant is weak, the solvent is non polar, and therefore also aprotic (pentane, benzene). Four strong solvent-solute interactions can also be distinguished : ion-dipole, dipole-dipole, complex forming and hydrogen bonding. Sometimes solvation also depends partially on steric effects.

II.1. Solubility of non Electrolytes.

In non electrolytic solutions the molecules of the solute are not dissociated during the dissolution process and three cases arise depending on the nature of the solvent.

In dipolar protic solvents, the solvent molecules are associated by hydrogen bonds and the solute molecules must be able to break these bonds. For dissolution to occur, the solute must therefore have a certain degree of basicity, or even it must have hydrogen bonds. In that case, the solvent-solvent and solute-solute interactions are replaced by solvent-solute interactions having nearly the same energy. If the solute is non polar(aprotic) and not basic the dissolution in a protic solvent is quite impossible.

In dipolar aprotic solvents, a material is soluble if it is also dipolar and aprotic. The dipole-dipole interactions of the solvent and of the solute are then replaced by dipole-dipole interactions between solvent and solute. If the solvent is basic, a protic solute can be very soluble because very strong hydrogen bonds between solvent and solute replace the hydrogen bonds in the solute and the dipole-dipole interactions in the solvent. If the solvent is not basic, the protic solutes have a low solubility because the strong hydrogen bonds of the solute must be broken and replaced by weaker dipole-dipole interactions between solvent and solute.

In non polar aprotic solvents all non polar materials are soluble since both solvent and solute interact by van der Waals forces. On the other hand, polar solutes (dipole-dipole interactions) and protic solutes (hydrogen bonds) are generally not very soluble, except if there is a tendency to form complexes.

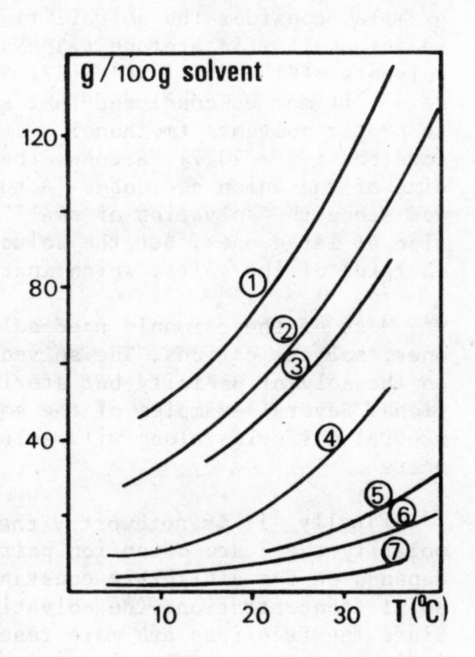

Fig. 1 : Solubility of 1,4-di-t-butylbenzene in different solvents (after ref.[5]); (1) benzene, (2) n-hexane, (3) n-heptane, (4) acetone, (5) t-butanol, (6) isopropanol and (7) methanol.

As an example of solvent effects on the dissolution of a non electrolyte, consider the solubility of DTBB (1,4-di-t-butylbenzene) in different solvents[5]. As shown in fig. 1, the solubility of DTBB ($C_{14}H_{22}$), which is a non polar aprotic solute, is high in non polar solvents (benzene) whereas it is low in polar solvents (methanol). In the series benzene, n-hexane, n-heptane, acetone, t-butanol isopropanol and methanol there is obviously a correlation between solubilities (Fig. 1) and dielectric constants ε even with the gaps which are attributed to steric effects. At 25°C[4,6] the series of ε values is : 2.3, 1.9, 1.9, 20.4, 10., 20.1 and 32.6. The solubility curves (Fig. 1) change slightly if mole fractions are used instead of concentrations, but their relative ordering does not change.

II.2. Solubility of Electrolytes

The molecules of the solute are dissociated during the dissolution process. If the solvent is dipolar protic, the anions are strongly solvated by ion-dipole interactions with superimposed hydrogen bonding which is greatest for small anions. If the solvent is dipolar aprotic, solvation also takes place by ion-dipole interactions. There is no important contribution of hydrogen bonding but an interaction is superimposed due to the mutual polarizability of the anion and the solvent molecules which is greatest for large anions. The solvation of anions by dipolar aprotic solvents decreases in the reverse order to that for protic solvents. As an example, consider the solubility series of potassium salts at 25°C in methanol or in acetone[7]. The same ordering is found in both solvents : KI > KBr > KCl > KF. From this result, and from other data[7], it may be concluded that salts are generally more soluble in protic solvents (methanol ; ε = 32.6) than in aprotic ones (acetone ; ε = 20.7). Second, the solubility decreases when the size of the anion decreases. Actually the reverse should be observed since the solvation of small anions is higher than the solvation of large ones. But the solubility also depends on the lattice energies of the salts, which increase when the anion size decreases.

Most of the commonly used solvents, especially highly polar ones, solvate cations. The solvent-solute association also depends on the solvent basicity but steric effects may hinder the interactions. Several examples of the solvation of anions, cations, and neutral molecules along with solubility values can be found elsewhere[8].

Finally, it is noteworthy that in solvents of intermediate polarity there are often ion pairs in solution. Their stability depends on the dielectric constant, the solvent, the ion size, the solute concentration, the solvation power, and the temperature. Since the free ions are more reactive than the ion pairs, some influence on the nucleation and growth kinetics is expected.

III. SOLUBILITY AND SOLVENT EFFECTS ON NUCLEATION

A correlation was made between the solubility power of different solvents and the maximum attainable undercooling of the solution[5].

The solute was DTBB ($C_{14}H_{22}$) and the solvents were those indicated in Fig. 1. The experiments were made in the following way. The solutions were saturated at $T_s = 30°C$ and cooled to the crystallization temperature T_c at a constant cooling rate. The results at a cooling rate of 0.1°C/mn are displayed in Table 1; ΔT_{max} and ΔC_{max} are the maximum attainable undercoolings expressed in terms of temperature and concentration differences respectively.

Solvent	C_s at 30°C (g/100g)	ΔT_{max} (°C)	ΔC_{max} (g/100g)	σ
Methanol	4.8	7.3	1.6	0.50
Isopropanol	11.3	9.4	3.9	0.51
t-Butanol	17.7	8.3	3.0	0.20
Acetone	38.7	6.5	11.5	0.37
n-Heptane	67.7	–	–	–
n-Hexane	82.3	4.6	13.6	0.11
Benzene	109.7	5.8	20.9	0.16

Table 1 : Solubilities C_s of DTBB in different solvents estimated from Fig. 1. Relationship of the maximum attainable undercooling of the solutions in terms of temperature and concentration differences (after [5]) and in terms of relative supersaturation.

There are significant differences in the maximum attainable undercooling C_{max} when the solvent changes from polar to non polar.[5] There is also an obvious correlation between ΔC_{max} and solubility. However, in order to avoid misleading interpretations, it is better to use the supersaturation values defined by $\sigma = (X_s - X_c)/X_c$ where X_s and X_c are the mole fractions of the solute at T_s and T_c respectively. Notwithstanding some dispersion, it is clear from table 1 that the higher the solubility, the lower the supersaturation at which nucleation occurs. This is in agreement with the well-known experimental fact that the highest critical supersaturation for nucleation are reached in very dilute solutions.

The crystallization of cyclonite (1:3:5-trinitro-1:3:5-triazacyclohexane) from acetone-water mixtures[9] borders between growth from aqueous and non aqueous solutions but it gives some interesting information about the correlation between solubility, supersaturation, surface energies, and nucleation rates. The results are summarized in table 2.

Solvent composition (Wt % of acetone)	C_s at 25°C (g/100 g)	C/C_s	γ (ergs/cm^2)
52.4	0.642	2.7	9.6
61.2	1.338	2.4	9.0
66.6	1.992	2.2	6.0
70.3	2.496	1.7	5.2

Table 2 : Solubility C_s of cyclonite in different acetone-water mixtures. Relationship of critical supersaturation ratios C/C_s necessary to get about 500 nuclei $cm^{-3}s^{-1}$, and with the crystal surface free energies γ.

Since the solute has a better affinity for acetone (ε = 20.7) than for water (ε = 80.1) the solubility of cyclonite increases with the acetone content of the solution. There is also a sharp decrease of the critical supersaturation ratios C/C_s necessary to achieve the same nucleation rate in the different solutions when the solubility increases. In other words, nucleation is easier when the solution is more concentrated. The crystal surface free energies γ change with solubility+concentration in the same way as the supersaturation ratios. Therefore, nucleation rates are affected by both C/C_s and γ due to their dependence on solubility and solvent effects. This is evident if a simple form of the nucleation rate equation [10-12] is considered :

$$J = N_o \nu \exp - \frac{\Delta G}{k_B T} = N_o \nu \exp - \left[\frac{c' \Omega^2 \gamma^3}{(k_B T)^3 (ln\, C/C_s)^2} \right] \quad (1)$$

ΔG is the activation free energy for nucleation. N_o is the number of monomers per unit volume of the solution (solubility) and ν is the frequency at which a critical nucleus turns into a supercritical one. The shape factor c' depends on the crystal, and Ω is the volume of a molecule inside the crystal. This equation illustrates why the nucleation rate increases when solubility and supersaturation increase and when the crystal surface free energy decreases.

IV. SOLVENT EFFECTS ON POLYMROPHIC TRANSFORMATIONS

Related to nucleation and growth kinetics, the solvent may also have an important effect on polymorphism which is exhibited by many substances growing from non aqueous solutions. The occurrence of polymorphs depends on kinetic factors even though the most stable phase is that which has the lowest solubility. Two cases have been widely investigated in non aqueous solutions : copperphtalocyanine (CuPc) and stearic acid (SA).

IV.1. The Polymorphism of Copperphtalocyanine

CuPc ($[C_8H_4N_2]_4Cu$) is an important material used in the pigment industry to color inks, paints, etc. The chemical and physical properties related to crystal size and structure have been reviewed recently[13,14]. Earlier papers[15,16] deal more with phase transformations. There are five polymorphic modifications named $\alpha, \beta, \gamma, \delta$ and ε. Since the crystal structures are different the solubilities are different too. The solubility of β, which is the most stable polymorph, is about an order of magnitude lower than the solubilities of α and γ.

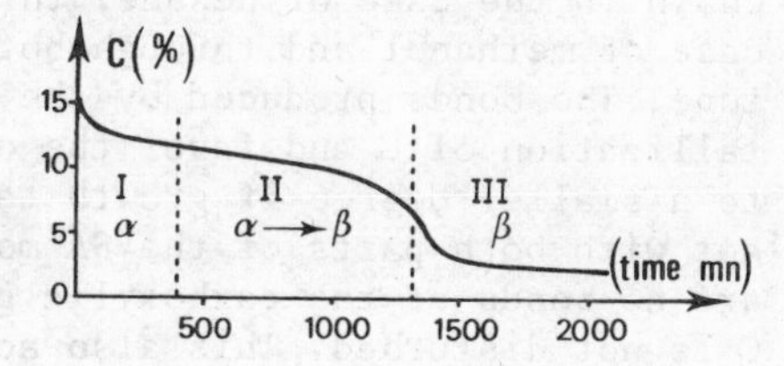

Fig. 2 : Solute concentration of copper phtalocyanine in an acetophenone solution at 40°C as a function of time showing the spontaneous α-β polymorphic transformation. (after ref. 16)

Due to differences in nucleation rates, α primarily crystallizes first but the final state is always the β form. Figure 2 shows a spontaneous transformation of α into β in an acetophenone solution at 40°C, followed by measuring the solute concentration. In region I there is a slight Ostwald ripening of the α crystals accompanied by the occurrence of β nuclei. The solute concentration decreases slowly. In region II there is dissolution of the α crystals and growth of the β ones. In region III, the crystals of the β phase continue to grow up to the moment when no residual supersaturation remains. The solute concentration reaches the saturation value with respect to the β phase. The rate of the transformation process depends on temperature, stirring, seeding, and on the nature of the solvents used[15,16].

IV.2. The Polymorphism of Stearic Acid

The polymorphism of SA ($CH_3 - (CH_2)_{16} - COOH$) has been investigated in many different non aqueous solvents. There are three polymorphs named A, B and C which can be grown from solution but only the C form grows from the melt. As only small differences are to be expected in the lattice energies, it is likely that the solubilities of the three phases are nearly identical. This is perhaps at the origin of many contradictions in the experimental results. From earlier works[17,18] no definite conclusions can be drawn. From observations made by growing SA from ethanol, benzene, cyclohexane and n-hexane[19] it turns out that the polarity of the solvent is perhaps not the most important factor determining the phase. Very recently it was confirmed that the crystallization of SA is very complex[20,21]. Nucleation and growth were performed from hexane, benzene, acetone, and ethanol, with special attention paid to supersaturation, stirring, and cooling rates[21]. The general trend is that the nature of the solvent-solute interaction directs the crystallization towards the B and C forms when the external disturbances are weak (low supersaturation and low stirring and cooling rates). When the external disturbances increase, the trend is to crystallize B and even A. The external disturbances probably also account for the discrepancies between other studies. In order to qualitatively explain the appearance of SA polymorphs, we will consider the solvent-solute interactions. The SA molecule is a non polar aliphatic chain terminated by a polar carboxylic group. The solvent-solute interactions are realized through the aliphatic chain in the case of hexane, through the carboxylic group in the case of methanol and through both in the case of benzene and acetone. The bonds produced by the carboxylic group disturb the crystallization of C and favor the occurrence of B. This is observed to a smaller degree if growth takes place from solvents which interact with both parts of the SA molecule. On the other hand, if there are no bonds at the carboxylic group as in hexane, the growth of C is not disturbed. This also accounts for the fact that only C can be grown from the melt. This interpretation seems to be confirmed by experiments where emulsifiers are introduced into the solution[22].

V. SOLVENT EFFECTS ON THE SURFACE MORPHOLOGY OF THE CRYSTAL

We have just seen that growth can be disturbed by interactions between the molecules of the solvent and of the solute. These interactions also exist at the crystal-solution interface and both can be invoked to explain the surface features of n-pentacontanol grown from different solvents[23]. Here, an attempt is made to correlate the height of the spiral growth steps and the molecular association in solution. When growth takes place from non polar solvents (xylene, petroleum ether, carbon tetrachloride) there are simultaneously single and double molecules in the solution. As a consequence the crystals exhibit growth steps of mono and bimolecular height. Actually, there are even other complexes in solution. If the growth rate is low, they do not participate to growth since they have time enough to be dissociated into mono and bimolecular complexes. If the growth rate is high, the large complexes take part in growth, the regularity of which is completely disturbed. If growth occurs from polar solvents (amyl acetate, dioxan), the n-pentacontanol molecules are not associated and the steps of the growth spirals are of monomolecular height. There is a peculiar case if growth occurs from alcohols (isopropyl, n-propyl, n-butyl, n-amyl) which are also polar solvents. Here, both mono and bimolecular steps can be found but the proportion of the latter decreases when the solvent is changed from propyl to butyl alcohol. Bimolecular steps do not exist in n-amyl alcohol. These results are explained on the basis of steric effects[23]. In solution, two pentacontanol molecules may be linked by hydrogen bonding with a solvent molecule. This results in the formation of a pseudo double molecule which accounts for the existence of bimolecular steps. When the size of the solvent molecule increases, this becomes more difficult. The larger molecule complexes also have more difficulty reaching the crystal surface. Consequently, only monomolecular steps are observed.

VI. SURFACE ROUGHNESS AND GROWTH MECHANISMS

In recent years, an attempt was made to explain or even to predict the growth mechanism of a crystal face by calculating the so-called JACKSON α factor[24]. For solutions, α gives a crude estimate of the surface roughness. Different values of surface roughness correspond to different growth mechanisms. From computer simulations[25] performed on a (001) face of a KOSSEL crystal, growth is continuous if $\alpha<3.2$ (high roughness), whereas it arises by two dimensional nucleation if $3.2<\alpha<4$ or by a spiral mechanism if $\alpha>4$. The expression for α applied to solution growth of the (001) face of a KOSSEL crystal[25,26] is :

$$\alpha = \frac{4\varepsilon'}{k_B T} = 2\,\frac{(\phi_{ss} + \phi_{ff} - 2\phi_{sf})}{k_B T} \tag{2}$$

where the ϕ terms represent the energies of the broken bonds between solid-solid (ss), fluid-fluid (ff) and solid-fluid (sf) particles. No distinction is made between the solid-fluid bonds in the bulk and at the crystal-solution interface. This lattice model yields the following equations :

$$\Delta H_d = 3\,(\phi_{ss} + \phi_{ff} - 2\phi_{sf}) \tag{3a}$$

$$\Delta H_{des} = (\phi_{ss} + \phi_{ff} - 2\phi_{sf}) \quad (3b)$$

$$\Delta H_{ks} = 2(\phi_{ss} + \phi_{ff} - 2\phi_{sf}) \quad (3c)$$

The enthalpy of dissolution ΔH_d is the sum of the enthalpy ΔH_{des} to transfer an adsorbed molecule from the surface to the solution and of the enthalpy ΔH_{ks} to remove a molecule from a kink position towards the surface.

Taking eqs. (2) and (3) into account we have :

$$\alpha = \frac{\Delta H_{ks}}{k_B T} = \frac{2}{3} \frac{\Delta H_d}{k_B T} \quad (4)$$

For solutions, $\Delta H_d = \Delta H_{fus} + \Delta H_{mix}$ which are the enthalpies of fusion and of mixing. If the solution is ideal :

$$\alpha = \frac{n_s}{n_t} \frac{\Delta H_d}{k_B T} \sim \frac{n_s}{n_t} \frac{\Delta H_{fus}}{k_B T} \quad (5)$$

where n_s and n_t are the number of bonds of a molecule within the slice (face) and the crystal respectively. In a non KOSSEL crystal, where the bonds are anisotropic, the ratio n_s/n_t must be replaced by ξ_{hkl}, the surface anisotropy factor of the hkl face under consideration. This factor is defined later in eq. (9). These ideas were applied to real systems [27,28] in order to explain the growth of glyceryl tristearate and palmitate from glyceryl trioleate and carbon tetrachloride solutions and the growth of hexamethylenetetramine form aqueous and ethanolic solutions. It was then shown that the use of eq. (5) was very limited and failed especially in the case of negative values of ΔH_d. In addition, eq. (5) does not clearly take the solubility of the crystal into account and it is very unlikely that the surface roughness is the same in very dilute as in more concentrated solutions. It was therefore proposed to replace the term ε' in eq.(2) by λ the specific edge free energy of the steps spreading over the crystal face. With some rough approximations, reasonable values of α were found. [27] In order to overcome the difficulties quoted above and to account for non ideality, the dissolution entropy ΔS_d was used in eq. (5) instead of ΔH_d :

$$\Delta S_d = \frac{\Delta H_{fus}}{T} - \frac{R}{x} \left[x \ln x + (1-x) \ln (1-x)\right] \quad (6)$$

where x is the mole fraction of the solute at the temperature T. Nevertheless, when applied to the growth of glyceryl tristearate grown from glyceryl triolate and carbon tetrachloride solutions, eq. (5) modified with eq. (6), leads to similar values of α for both systems. This implies that there is no solvent effect, contrary to the experiment where the growth rate is much higher when growth takes place from carbon tetrachloride.

During the same period, another model was proposed from computer simulations[30]:

$$\alpha = \xi_{hkl} \, (1-x_{seq})^2 \, (\Delta f_{sf} - \ln x_{seq}) \tag{7}$$

The term Δf_{sf} is the difference in internal free energies of solid and fluid particles. The term x_{seq} takes solubility into account since it is the ratio between the numbers, per unit volume, of solute molecules in the solution and of solid molecules within the crystal. The value of x_{seq} changes from 0 to 1 as a function of solubility. Nevertheless, eq. (7) is also open to question and Δf_{sf} is difficult to estimate. For this reason, another equation for α was established[31]:

$$\alpha = \xi_{hkl} \left(\frac{\Delta H_{fus}}{k_B T} - \ln x_{seq}\right) \tag{8}$$

$$\xi_{hkl} = E^{sl}_{hkl}/E_{cr} \tag{9}$$

E_{cr} and E^{sl}_{hkl} are the energies per molecule in the kink and inside the slice hkl respectively. Depending on the system investigated (vapour, melt, solution) both energies must be defined with reference to sublimation, fusion or dissolution. The different definitions and approximations necessary for the calculation of α have been discussed[32]. The most important approximations is perhaps that the energy ratio in eq. (9) is the same regardless of the system under consideration. Finally, we have also :

$$E_{cr} = E^{sl}_{hkl} + E^{att}_{hkl} \tag{10}$$

where E^{att}_{hkl} is the attachment energy per molecule on the face hkl. As an example, for the (001) face of a KOSSEL crystal, $E_{cr} = 3\phi$,

$E^{sl}_{hkl} = 2\phi$, and $E^{att}_{hkl} = 1\phi$ if ϕ is the bond energy between two first neighbors. Equation (8) was used for the case of biphenyl crystals growing from methanol, toluene and from the melt[31]. For these three systems, the α values are higher with respect to the (001) face (7.8, 6.4 and 5.2) than with respect to the (100) face (2.7, 2.2 and 1.8). It was therefore concluded that (001) faces should be smooth and properly facetted. A spiral growth mechanism is to be expected. On the other hand, (100) should be rough and no facetted growth is expected. Another example of α calculation concerns the growth of tetraoxane from acetone.[33] For (001), (100) and (111) faces respectively, the α values are 4.75, 4.30 and 2.5. The low value for (111) is in agreement with the experiment which shows that this face is rough and grows rapidly.

From all these considerations it is clear that estimation of crystal roughness by means of the α factor is not very simple and full of problems[34]. Here we can make two additional remarks. In order to calculate α, and to verify eq. (10), it is implicitly supposed that an adsorbed molecule has the same crystallographic position when it is adsorbed on (hkl) as when it already belongs to the slice (hkl). The second point is that only enthalpy terms are used for the calculation of α and both these conditions are probably not satisfied when growth concerns crystals with anisotropic structures and/or high entropies of dissolution ΔS_d. To illustrate this failure of the theory we consider the case of two monoclinic normal alkanes, hexatriacontane ($C_{36}H_{74}$) and octacosane ($C_{28}H_{58}$) grown from different solvents.

Crystal/ solvent	ΔH_d (kJ/mol)	ΔS_d (J/mol K)	X at 25°C	N₀ at 25°C	x_{seq} at 25°C
$C_{36}H_{74}$/D	128.9	372	6.91×10^{-4}	2.19×10^{18}	1.91×10^{-3}
$C_{36}H_{74}$/H	122.7	354	9.68×10^{-4}	3.98×10^{18}	3.48×10^{-3}
$C_{36}H_{74}$/P	124.5	360	9.63×10^{-4}	5.01×10^{18}	4.38×10^{-3}
$C_{36}H_{74}$/PE	128.5	373	9.15×10^{-4}	4.56×10^{18}	3.99×10^{-3}
$C_{28}H_{58}$/PE	99.3	301	2.08×10^{-2}	9.75×10^{18}	6.32×10^{-3}

Table 3 : Values necessary for the caluclation of x_{seq} and of the α factor (eq. (8)) for five crystal-solvent systems : ΔH_d and ΔS_d are the enthalpy and entropy of dissolution of the crystal while X and N_o are the solute mole fraction and the number of solute molecules per unit volume of solution. The solvents are decane (D), heptane (H), pentane (P) and petroleum ether (PE).

In table 3 are some of the values necessary for the calculation of α at 25°C[35-37]. The solutions are not exactly ideal but deviation from ideality is negligible as shown by comparison of ΔH_d and ΔS_d with ΔH_{fus} and ΔS_{fus}[35-37]. For $C_{28}H_{58}$ and $C_{36}H_{74}$, the ΔH_{fus} values are 100.00 and 129.20 kJ/mol respectively, while the ΔS_{fus} values are 301 and 370 J/mol K respectively. The mole fraction X is therefore calculated from $\ln X = (-\Delta H_d/RT) + \Delta S/R$, where R is the molar gas constant. The number of molecules per unit volume of crystal can be obtained from the crystal structures[38]; these values are about 1.45 and $1.14 \times 10^{21}/cm^3$ for $C_{28}H_{58}$+ $C_{36}H_{24}$ respectively. The energy terms can be calculated by means of a LENNARD-JONES 6-12 potential[38,39]. E_{cr} is about 203 and 261 kJ/mol for octacosane and hexatriacontane. For hexatriacontane, E^{sl}_{001} =251, E^{sl}_{110} = 72, E^{sl}_{010} = 70 and E^{sl}_{110} =13 kJ/mol while for octacosane the same slice energies are 192, 56, 54 and 10 kJ/mol. The α values corresponding to the different solvent-solute systems and to the different crystal faces are reported in table 4. They have been calculated with eq. (8) using ΔH_d (preferred over ΔH_{fus})

Crystal/Solvent	α_{001}	α_{110}	α_{010}	α_{100}
$C_{36}H_{74}$/D	57.0	16.3	15.0	2.8
$C_{36}H_{74}$/H	56.1	16.1	15.6	2.8
$C_{36}H_{74}$/P	55.2	15.8	15.4	2.8
$C_{36}H_{74}$/PE	55.3	15.8	15.4	2.8
$C_{28}H_{58}$/PE	40.8	11.9	11.5	2.0

Table 4 : The α factors for different crystal faces of hexatriacontane and octacosane in different solvents (calculations from eq.(8)).

The following comments can be made. Since ΔH_{fus} (or ΔH_d) is high, the $\Delta H_{fus}/k_BT$ ratio is always high compared to $\ln x_{seq}$ in eq. (8). The α values are therefore determined by the solute and the crystal, independent of the solvent. Another point is that the α values are high, especially for the (001) face. There are two reasons for such high values. First, entropy terms are usually neglected and ΔH_{fus} rather than the more proper $\Delta G_{fus} = \Delta H_{fus} - T\Delta S_{fus}$ is used. Since long chain normal alkanes have very high entropies of fusion (dissolution), the ΔG terms are very different from the

ΔH terms. As an example, for hexatriacontane, ΔH_{fus} = 129.2 kJ/mol whereas ΔG_{fus} = 18.9 kJ/mol at 25°C. Using ΔG_{fus} in eq. (8) for growth from a petroleum ether solution, we get α_{001} = 13.3, α_{110} = 3.8, α_{010} = 3.7 and α_{100} = 0.7. These values are four times lower than calculated with the enthalpy terms only. In addition, $\ln x_{seq}$ is no longer negligible with respect to $\Delta G_{fus}/k_BT$. For the same system $\Delta G_{fus}/k_BT \simeq 7.6$ and $-\ln x_{seq} \simeq 6.3$ and the solvent has a non negligible effect on α values. The second reason for high α values concerns mainly the (001) face for which $E^{sl}_{001} \simeq E_{cr}$. This is due to the fact that we consider that on (001) the adsorbed molecule has the same position as in the solid state. We shall see later (section 7.2) that it is more likely for the molecule to be parallel to the surface rather than upright. In some cases, it is therefore possible that incorrect growth mechanisms could be predicted on the basis of the α values. Sometimes α is used to classify the relative growth rates of the crystal faces. When α decreases, the surface roughness increases and accordingly the growth rate increases. From this point of view, the α values in table 4 are in the proper order, the growth rate of (001) being much lower than the growth rate of all other faces. But the same result would have been obtained by considering only the attachment energies E^{att}_{hkl} from which it is also possible to deduce the growth morphology of a crystal[40,41].

At the present state of the theory and due to the lack of sufficient experimental evidence, α should only be considered as a factor describing a tendency for the faces to be more or less rough and to grow more or less rapidly.

VII. THE ADSORPTION LAYER

VII.1. Theoretical Aspects

One of the most important parameters appearing in the growth rate equations is the number n_{s_o} of molecules adsorbed per unit area of the crystal face. An equation for n_{s_o} valid for the vapor phase, was first given by the BCF theory[42], and allows order of magnitude estimation of n_{s_o} :

$$n_{s_o} = n_o \exp - \frac{\Delta H_{ks}}{k_BT} \qquad (11)$$

where ΔH_{ks} is enthalpy of kink-surface equilibrium and n_o is the maximum number of adsorption sites available for adsorption. Eq. (11) can be used for solution growth only if ΔH_{ks} takes the solvent into account (eq. (3c)). Since ΔH_{ks} is always a fraction of ΔH_d for solution growth, the most simple case (eq. (4)) is given by

$$n_{s_o} = n_o \exp - \alpha \qquad (12)$$

A correct calculation of α allows calculation of $n_{s\circ}$, one of the most useful terms relevant to crystal growth. Since the growth rate of a face is proportional to $n_{s\circ}$ the lower the dissolution enthalpy (i.e the higher the solubility), the lower α, and the higher the growth rate. The validity of eq. (11) is also questionable because entropy terms are not included. A better approximation of $n_{s\circ}$ would be obtained by using the free energy such that :

$$n_{s\circ} = n_\circ \exp - \frac{\Delta G_{ks}}{k_B T} \tag{13}$$

This was suggested[43,44], but the main difficulty is to evaluate ΔS_{ks} which ranges between 0 and ΔS_d since the highest entropy change results from the transfer of a molecule from a kink to the solution.

Another expression for $n_{s\circ}$[26,45] originates from the equations describing equilibrium between the adsorption layer and solution :

$$\frac{n_{s\circ}}{\tau_{des}} = \frac{l\, N_\circ}{\tau_{desolv}} = \frac{D_v N_\circ}{\Lambda} \tag{14}$$

The quantities not yet defined are τ_{des} and τ_{desolv}, the relaxation times for desorption from the adsorption layer and for desolvation of a solute molecule respectively. The length l can be considered the mean free path of a solute molecule in the solution. Since the relaxation times can be written as a function of a frequency term and of an activation free energy ($\tau = \nu^{-1} \exp(\Delta G/k_B T)$, $n_{s\circ}$ can be written as :

$$n_{s\circ} = l N_\circ \exp \left(\frac{\Delta G_{des} - \Delta G_{desolv}}{k_B T}\right) \tag{15}$$

where it is assumed that the frequencies for desorption and desolvation are the same. In addition it is often assumed that l is equal to the length "a" of a molecule in the crystal. Finally, in eq. (14), D_v is the coefficient of volume diffusion and Λ the length that a molecule must travel before passing from the solution to the surface. The ratio D_v/Λ is therefore the drift velocity for entering the adsorption layer, Λ being defined by $\Lambda = a\, \tau_{desolv}/\tau_v$ where τ_v is the relaxation time for volume diffusion.

Returning to eq. (15), the main problem is once more calculation or estimation of ΔG terms. We can assure that eqs. (13) and

(15) are satisfactory to obtain the correct order of magnitude for $n_{s\circ}$, but that calculations of more accurate values are improbable.

From results obtained by growing long chain alkanes from organic solvents, an attempt was made to give a more precise thermodynamic formulation of the adsorption layer theory.[46,47] With this theory the concentration of the molecules in the adsorption layer corresponds to the Gibbs interface excess. However, this new formalism leads to rather cumbersome equations, and only the variation of $n_{s\circ}$ with temperature can be determined. In the most simple case :

$$\frac{\partial n_{s\circ}}{\partial(1/T)} = - \left(\frac{\Gamma_1^\circ a_2 \; H^\circ_{1des} + \Delta H_{ks}}{R}\right) \tag{16}$$

The product $\Gamma_1^\circ \, a_2$ is the number of solvent molecules which must be desorbed in order to allow the adsorption of one solute molecule. H°_{1des} is the enthalpy of desorption of the solvent and ΔH_{ks} the enthalpy for transfering a crystal molecule from a kink to the surface. Comparison of eqs. (13), (15) and (16) does not show a simple relationship between them[46] since we should have :

$$\frac{\partial(\Delta G_{ks}/T)}{\partial(1/T)} = \Gamma_1^\circ \, a_2 \Delta H^\circ_{1des} + \Delta H_{ks} \tag{17}$$

$$\frac{\partial \ln l}{\partial(1/T)} = - \frac{(\Gamma_1^\circ \, a_2 \; \Delta H^\circ_{1des})}{R} \tag{18}$$

From this we must bear in mind that $n_{s\circ}$ can be calculated within some important approximations : if the entropy terms are negligible, if lattice models are used, and if we can imagine the adsorption layer. However, these points are realized simultaneously only in the case of small molecules and when the crystal has a low entropy of dissolution.

VII.2. Validity of the models

The models used to describe the adsorption layer, especially the lattice models, can easily fail when growth occurs from non aqueous solutions because large molecules, with anisotropic dimensions are often involved. In the most favorable cases it is possible to clarify the interface model by calculating the interaction energies between the adsorbed molecule and the crystal face.

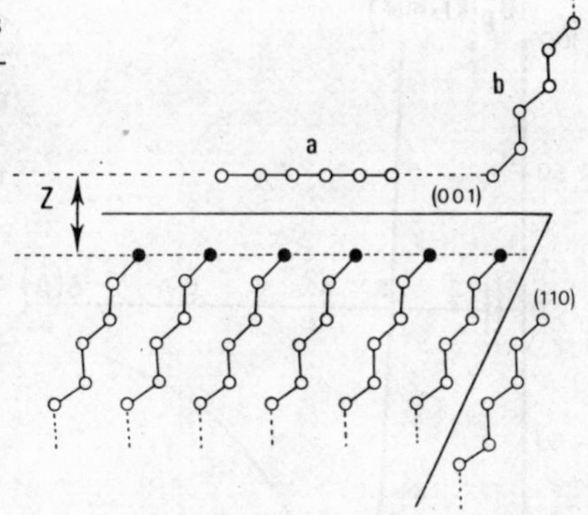

Fig. 3 : Schematic representation of the adsorption positions on a monoclinic long chain alkane crystal limited by (001) and (110) faces (after ref.[49])

As an example, we consider the growth of long chain normal alkanes from a solution of short chain normal alkanes. The crystals are flat, thin, exhibiting large (001) faces and small (110) faces. Their structure is either orthorhombic or monoclinic, but in both cases they are made up of layers parallel to (001) with the molecules inside running parallel to the c axis. As a consequence, the molecules are parallel to (110) and either normal to (001) in the orthorhombic modification or inclined with an angle of 30° relative to (001) in the monoclinic modification. The best adsorption position on (110) is the usual crystallographic position, the molecule being parallel to the face (fig. 3). The lattice model works rather well as shown for hexatriacontane[48] and octacosane[44] crystals. The adsorption energy on (110), or attachment energy E^{att}_{110} in eq. (10), is nearly proportional to the chain length, about 137 and 176 kJ/mol for octacosane and hexatriacontane respectively[38,44]. The lattice model fails, however, in the case of adsorption on (001). If the adsorbed molecule occupies its normal crystallographic position (fig. 3b), i.e. standing upright onto the surface, its adsorption energy is independent of chain length provided that the chain contains more than about seven carbon atoms. The adsorption energy would be about 10 kJ/mol, compared to the lattice energy of about 261 kJ/mol. In addition, it is unlikely that a molecule keeps this position while it diffuses towards the kinks. For this reason, it was suggested[49] that in the adsorption layer the molecule lies and moves parallel to the surface (fig.3a). The adsorption energy then becomes a function of the chain length and of the distance Z passing through the carbon planes (fig. 3). The variation of adsorption energy with Z is shown in fig. 4 for an octacosane molecule located at the best adsorption position on an infinite (001) plane of a normal alkane crystal. If the adsorption energies correspond to the potential wells, they are much higher than previously : 114 and 146 kJ/mol for octacosane and hexatriacontane respectively[49], instead of 10 kJ/mol. These values are perhaps the upper limit but they are more plausible than those arising from the lattice model.

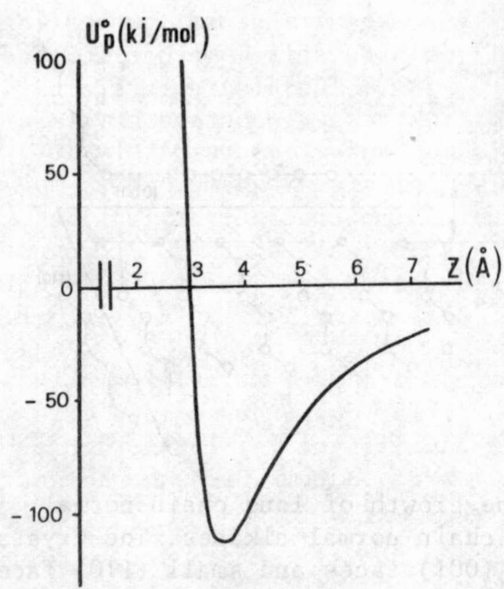

Fig. 4 : Potential energy curve of a $C_{28}H_{58}$ molecule adsorbed parallel to the (001) face of a normal alkane crystal, as a function of the distance to this face (after ref.[49])

There is no direct proof that the molecules are adsorbed parallel to the (001) faces, but this occurs in the case of the adsorption from solution of long chain hydrocarbons onto the (0001) face of graphite[50,51]. It was found that an ordered boundary layer forms at the graphite surface. Within the layer, normal alkanes have their chain axis oriented parallel to the surface[50]. From the shape of the adsorption isotherms it is concluded[51] that the long chain alkanes $C_{22}H_{46}$, $C_{28}H_{58}$ and $C_{32}H_{66}$ are strongly adsorbed and that the greater the chain length, the lower the bulk concentration at which the plateau of the isotherms is reached. Therefore adsorption is mainly caused by strong lateral interactions between molecules extended side by side, parallel to the graphite basal plane. Ordering increases with chain length. On the other hand, squalane (hexamethyltetracosane, $C_{30}H_{62}$) is not strongly adsorbed probably because the molecules are not adsorbed parallel to the surface. The methyl side groups cause such a configuration to be energetically unfavorable and prevent the adsorption layer to be stabilized by lateral interactions.

VIII. GROWTH KINETICS

VIII.1. Theoretical considerations

The development of different growth rate theories is beyond the scope of this paper and only some elementary notions are considered here. The only question we are concerned with is the following Are we able to identify the growth mechanism of a face and consequently the interfacial processes from measurements of the growth rate R as a function of the supersaturation σ ?

The crystals can grow either by a two dimensional nucleation mechanism (birth and spread model) or by a spiral mechanism. The former concerns perfect crystals[43,52,53]. The R(σ) curves are exponential, e.g. (110) faces of thick hexatriacontane crystals from petroleum ether.[48] Since this mechanism occurs rarely it is not useful to describe further. The spiral growth mechanism is much more prevalent since it concerns imperfect crystals. The theory of this growth mechanism[42] is now called the BCF theory. From this theory, several particular cases were derived[26,54] while a more general formalism includes all possibilities[45].

When a screw dislocation emerges on a face it originates a growth spiral, the steps of which spread over the surface with a lateral velocity v. The steps are equidistant (at least theoretically), their separation being $y_\circ$, so that if d is the height of the step :

$$R = \frac{vd}{y_\circ} \tag{19}$$

$$\text{with } y_\circ = \frac{c'\lambda a}{k_B T \ln(X/X_s)} \simeq \frac{c'\lambda a}{k_B T\sigma} \tag{20}$$

when the supersaturation ratio X/X_s is not too high. The factor c' depends on the spiral shape, λ is the specific edge free energy of the step (per molecule) and a the length of a molecule. A clear definition of the supersaturation in solution[55,60] is often difficult mainly because the supersaturation ratio should be expressed in terms of activities. Since they are often unknown, one possibility is to use the mole fractions X. In the case of ideal solutions, σ is as above :

$$\sigma = \frac{X_s - X_c}{X_c} = \Delta H_d \left(\frac{T_s - T_c}{R T_c^2}\right) \tag{21}$$

The step motion depends on the diffusion rate of the solute molecules through the solution, the exchange kinetics of the solute between the bulk and the adsorption layer, the diffusion rate on the surface, and the incorporation kinetics into the kinks. Assuming the mean free path x_s for surface diffusion is much lower than Λ, the distance for entering the adsorption layer, the general growth rate equation is[45] :

$$R = \frac{N_\circ \, \Omega \, D_v \, \sigma}{\Lambda + \delta + \Lambda\Lambda_s \frac{y_\circ}{x_s^2} + \Lambda\left[\frac{y_\circ}{2x_s} \coth\left(\frac{y_\circ}{2x_s}\right) - 1\right]} \qquad (22)$$

The four terms in the denominator have dimensions of length that can be respectively regarded as the impedance of the adsorption reaction, the impedance in the unstirred layer, the impedance for entering the steps, while the last one is the impedance for surface diffusion. The definition of Λ was given in eq. (14). Similarly, Λ_s is a phenomenological coefficient describing the exchange kinetics between the steps and the adsorbed layer ; $\Lambda_s = a\,\tau_k/2\tau_s$, τ_k and τ_s being the relaxation times for entering a kink and for surface diffusion respectively. The other quantities not yet defined are Ω the volume of a molecule in the crystal and δ the thickness of the boundary layer where the solution concentration is influenced by the growth of the crystal. Some relationships between the different parameters were given previously (eqs. (13,14,15). Other useful relationships involve the coefficient for surface diffusion D_s :

$$D_s = \frac{a^2}{\tau_s} = a^2 \, \nu_s \, \exp - \frac{\Delta G_s}{k_B T} \qquad (23)$$

$$\text{and } x_s = (D_s \tau_{des})^{1/2} = a \exp - \left(\frac{\Delta G_{des} - \Delta G_s}{2k_B T}\right) \qquad (24)$$

From the different equations we obtain :

$$\frac{D_s}{\Lambda_s} = \frac{2a}{\tau_k} \qquad (25)$$

which is the drift velocity of an adsorbed molecule towards the steps.

From eq. (22), the growth rate increases when the solubility increases. This point was discussed elsewhere [34] by consideration of the growth rates of the (110) faces of hexatriacontane crystals grown from pentane, heptane and decane. Normalized to pentane, the growth rate ordering is 1 : 0.66 : 0.33 while the solubility ordering is 1 : 0.80 : 0.50. R is proportional to σ which is also contained in y_o (eq. (20)) so that R may be either a linear or a parabolic function of σ :

$$R = b_1\sigma \quad \text{or} \quad R = b_p\sigma^2 \tag{26}$$

The significance of the kinetic coefficients b_1 and b_p changes depending on the different growth mechanisms; the growth mechanism is determined by the impedance of greatest influence.

The different possibilities arising from eq. (22) were examined in order to interpret the growth rates of the (001) faces of hexatriacontane crystals grown from heptane[63]. All but two mechanisms could be excluded, and it was not possible to decide whether volume diffusion or a mixed process involving volume and surface diffusion were taking place. Since we are dealing with interface reactions, only the growth mechanisms involving surface processes will be considered. This means that in addition to the condition $\Lambda \gg x_s$ we have also $\Lambda \gg \delta$ so that growth is not disturbed by the boundary layer. In this case, eq. (22) is simplified, the new equation including some previous results[42,54] :

$$R = \frac{N_o \Omega D_v \sigma}{\Lambda}\left[\frac{1}{\dfrac{\Lambda_s y_o}{x_s^2} + \dfrac{y_o}{2x_s}\coth\dfrac{y_o}{2x_s}}\right] \tag{27}$$

Eq. (27) can also be written in a manner suitable for further discussion :

$$R = \frac{N_o \Omega D_v \sigma}{\Lambda}\left[\frac{2x_s}{y_o}\tanh\frac{y_o}{2x_s}\right]\left[1 + 2\frac{\Lambda s}{x_s}\tanh\frac{y_o}{2x_s}\right]^{-1} \tag{28}$$

Using eq. (20) for y_o, we have two main possibilities depending on the value of the Λ_s/x_s ratio.

i) if $\Lambda_s \gg x_s$:

$$R = \frac{N_o \Omega D_v}{\Lambda}\,\frac{x_s^2}{\Lambda_s}\left(\frac{k_B T}{c'\lambda a}\right)\sigma^2 \tag{29}$$

which simplifies to :

$$R = \frac{2n_{s\circ}\Omega}{\tau_k}\left(\frac{k_BT}{c'\lambda}\right)\sigma^2 \tag{30}$$

The growth rate is a parabolic function of the supersaturation and depends on the kink integration kinetics.

ii) if $\Lambda_s \ll x_s$:

$$R = \frac{N_\circ\Omega D_v 2x_s}{\Lambda}\left(\frac{k_BT}{c'\lambda a}\right)\tanh\left(\frac{y_\circ}{2x_s}\right)\sigma^2 \tag{31}$$

which simplifies to :

$$R = \frac{2n_{so}D_s\Omega}{x_s}\left(\frac{k_BT}{c'\lambda a}\right)\tanh\left(\frac{y_\circ}{2x_s}\right)\sigma^2 \tag{32}$$

This is the original BCF equation[42] which predicts that the growth rate of the face depends on surface diffusion. At low supersaturation ($y_\circ \gg x_s$), $\tanh(y_\circ/2x_s)$ is unity so that R is parabolic with the supersaturation :

$$R = \frac{2n_{so}D_s\Omega}{x_s}\left(\frac{k_BT}{c'\lambda a}\right)\sigma^2 \tag{33}$$

At high supersaturation ($y_\circ \ll x_s$), $\tanh(y_\circ/2x_s)$ is approximately $y_\circ/2x_s$ so that :

$$R = \frac{n_{so}D_s\Omega}{x_s^2}\sigma \tag{34}$$

The growth rate is a linear function of the supersaturation. When σ increases, two growth laws may be observed successively. The transition from parabolic to linear laws occurs at the value $\sigma = c'\lambda a/2k_BTx_s$.

It should be mentioned that the original BCF theory[42] allows the possibilities of obtaining eq. (30) from eq. (32) by multiplication of the numerator by a factor β:

$$\beta = \left[1 + \frac{D_s \tau_k}{a\, x_s} \tanh \left(\frac{y_\circ}{2x_s} \right) \right]^{-1} \tag{35}$$

β accounts for the difficulty encountered by a molecule to enter kinks. If this difficulty is great the second term in brackets is much larger than one and $\beta << 1$. Insertion of this term in eq. (32) leads to eq. (30).

VIII.2. Application to Real Systems

As we have seen above regarding surface roughness and adsorption layers, it is difficult to calculate or even to evaluate the entropy terms corresponding to different activation free energies. This is also true of the entropy terms involved in the growth rate equations. Calculation of only the enthalpy terms is easier to perform. If growth rate measurements are made at different temperatures, it is possible to get an experimental value E_{exp} of the crystallization energy and compare it to the calculated value E_c:

$$E_c = - R \left(\frac{\partial (ln\, b)}{\partial (1/T)} \right) \tag{36}$$

where b is one of the kinetic coefficients defined in eq. (26). The growth mechanism should unambiguously be determined from the best agreement between E_{exp} and E_c.

For instance, we may consider two growth mechanisms resulting from surface reactions, R being in both cases a parabolic function of the supersaturation. When growth is determined by kink integration kinetics (eq. (30)), or by surface diffusion (eq. (33)), the crystallization energies are respectively :

$$E_c = RT + \Delta H_k + \Delta H_{ks} \tag{37}$$

$$E_c = RT + \frac{1}{2} (\Delta H_s + \Delta H_{des}) + \Delta H_{ks} \tag{38}$$

if eq. (13) is used for the calculation of n_{s_o}. These equations transform into :

$$E_c = RT + \Delta H_k + \Gamma_1^\circ a_2 \ \Delta H^\circ_{des} + \Delta H_{ks} \qquad (39)$$

$$E_c = RT + \frac{1}{2}(\Delta H_s + \Delta H_{des}) + \Gamma_1^\circ \ a_2 \Delta H_{1\,des}^\circ + \Delta H_{ks} \qquad (40)$$

if eq. (16) is used for n_{s_o}. In all cases it was assumed that the variation of the edge free energy λ with T is negligible.

There are unfortunately only a few examples where the crystallization energies are taken into account as a tool to recognize the growth mechanisms. There are even fewer limited to crystallization from non aqueous solutions. This was attempted for long chain normal alkane series grown from different solvents, especially from short chain alkanes. The similarity between solute and solvent molecules causes the solution to be nearly ideal. Moreover, the calculation of interaction energies between molecules can be performed with good accuracy due to the short-range van der Waals forces. But even for this case there are some difficulties arising from the E_c values. Comparing eqs (37) and (38) or eqs (39) and (40) we see that ΔH_k must be significantly different from $(\Delta H_s + \Delta H_{des})/2$ in order to unambiguously determine the most probable growth mechanism. In the case of (110) faces of octacosane crystals[44] grown from petroleum ether, a growth mechanism determined by kink integration kinetics could not be deleted on the basis of E_c values. By some additional considerations, it was shown that only a mechanism involving surface diffusion could explain the growth rate curves. On the other hand, it was not possible to decide[63] whether (001) faces of hexatriacontane grown from heptane involved a growth mechanism with only volume diffusion or volume and surface diffusion at the same time. However, in both cases it was possible to eliminate other mechanisms on the basis of E_c and E_{exp} values.

To point out the difficulties encountered when determining growth mechanisms and consequently the interfacial processes we can mention two numerical applications. When the crystallization of hexatriacontane crystals occurs from pentane, heptane, or decane, the growth rates of the (110) faces are all parabolic functions of the supersaturation[61,62]. The experimental crystallization energies are 122, 149 and 205 kJ/mol respectively. If growth were determined by the kink integration kinetics, eq. (39) leads to the following values of E_c : 130, 135 and 136 kJ/mol. The agreement with the experimental values is not good because the E_c values do not vary enough when the solvent is changed. On the other hand, if the rate determining step is surface diffusion, E_c can be fitted almost perfectly to E_{exp} provided that two important assumptions are made. The first one is that surface diffusion is the same in the three solvents. The enthalpy for surface diffusion should be about 100 kJ/mol. The second is that the adsorption layer should be dilute when growth occurs from decane whereas it should be more concentrated when growth occurs from heptane and even more so from pentane. This hypothesis can be justified since the solubility of hexatriacontane is much lower in decane than in pentane.

The $(001)_{64}$ faces of hexatriacontane crystals grown from petroleum ether have a more complex behavior since they exhibit, at high supersaturation, a superimposition of a parabolic and an exponential growth law. The choice of growth mechanism corresponding to the parabolic region is not too difficult to determine. The crystallization energy E_{exp} is rather low, only 60 kJ/mol. From the calculation of E_c by means of eqs. (39) and (40), the best fit is obtained when growth is controlled by kink integration kinetics (E_c = 97 kJ/mol) whereas the disagreement is larger if we suppose that it is controlled by surface diffusion (E_c = 142 kJ/mol). A further support of the kink mechanism concerns the retardation factor β. β must be very small in order to be inserted in eq. (32) from which the crystallization energy (eq. (39)) is obtained.

IX. CONCLUSION

The main interest of studying crystal growth from non aqueous solutions is that the use of a solvent series allows observation of solubility and solvent effects. Crystal growth theories do not differentiate between growth from aqueous or non aqueous solutions because the mechanisms are fundamentally identical. However in the latter case, some difficulties may arise due to the large size of the solvent and solute molecules. Sometimes the adsorption layer cannot be clearly defined, implying that lattice models can easily fail. In some favorable cases, it is possible to better understand the crystal-solution interface from calculation of interaction energies between molecules. Due to the lack of direct experimental data with regard to the interface, only order of magnitude estimates of parameters can be obtained.

Acknowledgement

The author wishes to thank Mrs M.C. TOSELLI and Miss D. THIVEND for technical assistance.

REFERENCES

1 . PARKER A.J., Quarterly Reviews 16 (1962) 163
2 . TCHOUBAR B., Bull. Soc. Chim. 9 (1964) 2069
3 . AGAMI C., Bull. Soc. Chim. 11 (1967)4031
4 . SURZUR J.M., in *Le Rôle du Solvant en Chimie Organique*, I.P.S.O.I., Université Aix-Marseille III, 1969.
5 . LECI C.L., GARTI N. and SARIG S., J. Crystal Growth 51 (1981) 85.
6 . Handbook of Chemistry and Physics, 56th edition, Weast R.C. ed., C.R.C. Press, Cleveland, (1975)
7 . MILLER J. and PARKER A.J., JACS 83 (1961) 117
8 . ALEXANDER R., KO E.C.F., PARKER A.J. and BROXTON T.J., JACS 90 (1968) 5049
9 . DUNNING W.J. and NOTLEY N.T., Z. Elektrochem. 61 (1957) 55
10. VOLMER M. and WEBER A., Z. Phys. Chem 119 (1926) 277
11. KAISCHEW R. and STRANSKI I.N., Z. Phys. Chem. 26 (1934) 317
12. BECKER R. and DÖRING W., Ann. Physik 24 (1935) 732
13. SAPPOK R., in *Ullmanns Encyklopädie der technischen Chemie*, Band 18, Verlag, Weinheim, 1979, p 501
14. SAPPOK R., J. Oil Chem. Assoc. 61 (1978) 299
15. HONIGMANN B. and HORN D., in *Particle Growth in Suspensions*, Smith A.L. ed., Academic Press, London, 1973, p 283
16. HORN D. and HONIGMANN B·, in *Polymorphie des Kupferphtalocyanins* XII Fatipec Congress, Verlag Chemie, Weinheim, 1974, p 181
17. Von SYDOW E., Acta Chem. Scand. 9 (1955) 1685
18. O'CONNOR R.T., in *Fatty Acids*, Part I, Markely K.S. ed., Interscience, New York, 1960, p 306
19. SATO K. and OKADA M., J. Crystal Growth 42 (1977) 259
20. GARTI N., SARIG S. and WELLNER E., Thermochimica Acta 37 (1980) 131
21. GARTI N., WELLNER E. and SARIG S., Kristall und Technik 15 (1980) 1303
22. GARTI N., WELLNER E. and SARIG S., ICCG-6 Abstract Booklet, vol. IV, Moscow, 1980, p 81
23. WATSON D.H., in *Proceedings Third International Conference Electron Microscopy*, London, 1954, p 497
24. JACKSON K.A., in *Liquid Metals and Solidification*, Amer. Soc. Metals, Cleveland, 1958, p 174
25. GILMER G.M. and BENNEMA P., J. Crystal Growth 13/14 (1972)148
26. BENNEMA P. and GILMER G.H., in *Crystal Growth an Introduction*, Hartman P. ed., North Holland, Amsterdam, 1973, p 263
27. BOURNE J.R. and DAVEY R.J., J. Crystal Growth 36 (1976) 278
28. BOURNE J.R. and DAVEY R.J., J. Crystal Growth 36 (1976) 287
29. BOURNE J.R. and DAVEY R.J., J. Crystal Growth 43 (1978) 224
30. BENNEMA P. and van der EERDEN J.P., J. Crystal Growth 42 (1977) 201
31. HUMAN H.J., van der EERDEN J.P., JETTEN L.A.M.J. and ODERKERKEN J.G.M., J. Crystal Growth 51 (1981) 589

32. HARTMAN P. and BENNEMA P., J. Crystal Growth 49 (1980) 145
33. WATANABE T., J. Crystal Growth 50 (1980) 729
34. SIMON B. and BOISTELLE R., J. Crystal Growth 52 (1981) 779
35. BOISTELLE R. and LUNDAGER MADSEN H.E., J. Chem. Eng. Data 23 (1978) 28
36. LUNDAGER MADSEN H.E. and BOISTELLE R., J. Chem. Soc. Faraday I, 75 (1979) 1254
37. LANDOLT-BÖRNSTEIN in *Zahlenwerte und Funktionem* 6 Aufl., II Band, 4 Teil, Springer, West Berlin, 1961
38. BOISTELLE R. in *Current Topics in Materials Science,* vol. 4, chap. 8, Kaldis E. ed., North Holland, Amsterdam, 1980, p 413
39. LUNDAGER MADSEN H.E. and BOISTELLE R., Acta Cryst. A32 (1976) 828
40. HARTMAN P., J. Crystal Growth 49 (1980) 157
41. HARTMAN P., J. Crystal Growth 49 (1980) 166
42. BURTON W.K., CABRERA N. and FRANK F.C., Phil. Trans. Roy. Soc. A 243(1951) 299
43. KAISCHEW R., Acta Phys. Hung. 8 (1957) 75
44. BOISTELLE R. and DOUSSOULIN A., J. Crystal Growth 33 (1976)335
45. GILMER G.H., GHEZ R. and CABRERA N., J. Crystal Growth 8 (1971) 79
46. LUNDAGER MADSEN H.E., J. Crystal Growth 39 (1977) 250
47. LUNDAGER MADSEN H.E., H. Crystal Growth 46 (1979) 495
48. SIMON B., GRASSI A. and BOISTELLE R., J. Crystal Growth 26 (1974) 77
49. BOISTELLE R. and LUNDAGER MADSEN H.E., J. Crystal Growth 43 (1978) 141
50. KERN H.E., V. RYBINSKI W. and FINDENEGG G.H., J. Colloid and Interface Science, 59 t.2 (1977) 301
51. KERN H.E. and FINDENEGG G.H., J. Colloid and Interface Science 75 t. 2 (1980) 346
52. VOLMER M., in *Kinetik der Phasenbildung*, Steinkopf, Leipzig, 1939
53. HILLIG W.B., Acta Met. 14 (1966) 1968
54. CHERNOV A.A., Soviet Physics Usp. 1(4) (1961) 126
55. MULLIN J.W. and SÖHNEL O., Chem. Eng. Sci. 32 (1977) 683
56. MULLIN J.W. and SÖHNEL O., Chem. Eng. Sci. 33 (1978) 1535
57. SÖHNEL O. and GARSIDE J., J. Crystal Growth 46 (1979) 238
58. van LEEUWEN C.,J. Crystal Growth 46 (1979) 91
59. van LEEUWEN C. and BLOMEN L.J.P.J., J. Crystal Growth 46 (1979) 96
60. CARDEW P.T., DAVEY R.J. and GARSIDE J., J. Crystal Growth 46 (1979) 534
61. PETINELLI J.C., in *Rôle du Solvant et des Impuretés sur les Cinétiques de Croissance et de Nucléation des n-Paraffines en Solution,* Thesis, Université Aix-Marseille III, 1977
62. BOISTELLE R. and PETINELLI J.C., J. Crystal Growth , in preparation
63. RUBBO M. and BOISTELLE R., J. Crystal Growth 51 (1981) 480
64. LUNDAGER MADSEN H.E. and BOISTELLE R., J. Crystal Growth 46 (1979) 681

ELECTROCRYSTALLIZATION

EUGENI BUDEVSKI

Central Laboratory of Electrochemical Power Sources
Bulgarian Academy of Sciences
Sofia 1040, Bulgaria

I. INTRODUCTION

The term electrocrystallization reflects the idea that mass transfer in the process of crystallization is accompanied by charge transfer. This term was coined by H. Fischer in the 40's and is composed of two words derived from two different fields of science. This makes it appropriate to interpret some basic concepts of the field of crystallization as used in the language of electrochemistry.

The linear crystallization rate $v[\text{cm sec}^{-1}]$ is manifested by the current, or more precisely, by the current density $i[\text{A cm}^{-2}]$

$$i = vzF/v_m$$

while the deposited amount of material m is defined by the electric charge transferred across unit area phase boundary

$$q = \int i dt = mzF/v_m$$

where v_m is the molar volume and z the number of Faradays F required for deposition of one mole. Note that m is given in this case as the volume of material deposited per unit area. The equilibrium of a crystal with its ambient phase is given by the equilibrium electrode potential E_{rev}. The supersaturation, i.e., the deviation $\Delta\mu$ of the chemical potential of the ambient phase from its

B. Mutaftschiev (ed.), Interfacial Aspects of Phase Transformations, 559–604.

equilibrium value, is measured by the so-called overvoltage η_c equal to the deviation of the electrode potential from its equilibrium value : $\eta_c = E_{rev} - E = \Delta\mu/zF$.

Both quantities, current and potential, can be controlled in time and amplitude to very high accuracy. Two additional parameters accessible for experimental measurement are the exchange rate of particles of the crystal with its ambient phase and the concentration of adions or adatoms. The adatom concentration is influenced by the potential and can therefore be detected as a capacitance in the impedance of the electrode while the exchange rate is manifested as a resistance in series with the adatom capacitance. The exchange rate is often given as an exchange current density $i_o\,[A\ cm^{-2}]$ defined at the equilibrium potential.

Because of the existence of an electric double layer at the phase boundary, the ion transfer across it encounters a potential dependent energy barrier $\Delta G^*_{dep} + \beta z e_o E/k_B T$ for the process of deposition and $\Delta G^*_{diss} - (1 - \beta)\ z e_o E/k_B T$ for the process of dissolution. The charge of the electron is e_o and β is the transfer coefficient ($0 < \beta < 1$). β determines to what extent the energy barriers are changed by the electrode potential for the cathodic (deposition) and the anodic (dissolution) processes respectively. The Gibbs free energy change is equal to the difference of height of the two barriers $\Delta G = \Delta G^*_{diss} - \Delta G^*_{dep} - z e_o E/k_B T$ and is related via E to the supersaturation.

II. KINETICS OF ATOM INCORPORATION

The incorporation of atoms into the crystal lattice, i.e., the process of growth of a crystal, depends on the crystallographic character of the growing surface and is largely affected by the surface structure. The surface of a solid metal electrode has a complex character. One can distinguish areas occupied by crystallites with different crystallographic orientations and intergrain boundary regions. On the other hand, crystal defects, inclusions, adsorbed molecules, oxide layers, etc. complicate the state of the surface. The simplest case is that of a single crystal surface.

At low temperatures in equilibrium, a crystal is surrounded by flat, atomically smooth, low index faces, or singular faces according to Frank's classification.[1] These faces are given by the Gibbs definition of equilibrium form, requiring a minimum of the surface free energy of the crystal at constant volume.

At higher temperatures, due to thermal fluctuations, singular faces can acquire a surface roughness characterized by the formation of adatoms and surface vacancies, di- or polyatomic surface clusters and vacancies, and steps with kinks. The surface roughness becomes appreciable only above a critical "roughening" temperature T_r. The theoretical value of T_r for a simple cubic crystal lattice,[2] $T_r = 0.6\psi_1/k$, where ψ_1 denotes the bond energy between two first neighbor atoms, is near or above the melting point. In reality, T_r is somewhat below the melting point. At room temperature singular faces of metals have a relatively low degree of roughening showing only adatoms and monatomic vacancies in the smooth regions and kink positions along the steps. Calculated from experimental ledge energies at room temperatures, every fourth to fifth atom along a step is in a kink position (see,e.g., Eq. (4)).

II.1. The Equilibrium Potential

While the separation work ψ_k from a kink site of the surface (i.e., the bond energy per atom of an infinite lattice) is the main parameter governing the saturation pressure of a large crystal, in the case of electrocrystallization the charge transfer across the electric field of the double layer and the ion-solvent interaction must also be considered. According to KAISCHEW,[3] the frequencies of deposition and dissolution of atoms on/from a site i of the crystal surface are given here by

$$\omega_{dep,i} = p_i c \exp(-\Delta G^*_{dep,i}/k_B T) \exp(-\beta z e_o E/k_B T) \qquad (1)$$

and

$$\omega_{diss,i} = q_i \exp(-\Delta G^*_{diss,i}/k_B T) \exp\left[(1-\beta) z e_o E/k_B T\right] \qquad (2)$$

where E is the electrode potential. The ΔG^*'s are the activation energies of deposition and dissolution of an atom on/from the site i at E = 0 and a standard concentration c. Activation energies for any particular site are related to the energies $\Phi_{i,s}$ for transfer of an atom from that particular site i to the solution

$$\Phi_{i,s} = -(\Delta G^*_{dep,i} - \Delta G^*_{diss,i})$$

In contrast to a homogeneous (liquid) surface, for deposition on a crystal face for any particular site a different value of the deposition and dissolution frequencies ω can be ascribed, determined by the energy barriers ΔG^*_{dep} and ΔG^*_{diss} and the constants p_i and q_i. If there are n_i sites per unit area of type i, an exchange current density $i_{o,i}$ per unit area of atoms deposited on these sites with the ions in the solution may be defined in analogy to a homogeneous surface :

$$i_{o,x} = \omega^{rev}_{dep,x} ze\, n^{o}_{x} = \omega^{rev}_{diss,y} ze_o n^{o}_{y} \qquad (3)$$

where n_x is the number of sites of type x per unit area which by addition of one atom are transformed to sites of type y (e.g., a free surface site "o" is transformed by addition of one atom to an adatom site "ad", or a vacancy "v" to a free surface site "o"). The equality $\omega^{rev}_{dep,x} n^{o}_{x} = \omega^{rev}_{diss,y} n^{o}_{y}$ is fulfilled only at the equilibrium potential when equilibrium concentrations n^{o}_{x} and n^{o}_{y} are established. In the case of kink sites (k), the equality $n^{o}_{k} = n^{o}_{k+1}$ always holds, because the addition of one atom to a kink site does not change the number of kink sites. Hence, the equilibrium potential is defined by the equality of dissolution and deposition frequencies only, $\omega^{rev}_{diss,k} = \omega^{rev}_{dep,k}$, irrespective of the number of the kink sites n_k. However, the intensity of the exchange of kink atoms per unit area with ions in the solution, i.e., the exchange current density of kink atoms is proportional to the number of kink sites n_k (see Eq. 3) and hence to the step density the step roughness being defined by the temperature.

As mentioned, at the equilibrium potential of an infinitely large crystal with its ambient phase, the deposition and dissolution frequencies of kink atoms must be equal. From Eqs. (1) and (2), recalling

$$-(\Delta G^{*}_{dep,k} - \Delta G^{*}_{diss,k}) = \Phi_{k,s},$$

it follows that :

$$ze_o E_r = \Phi_{k,s} + k_B T \ln(p_k/q_k) + k_B T \ln c \qquad (4)$$

The reversible potential E_r is thus related to the dissociation energy of a kink atom $\Phi_{k,s}$ and contains, besides the energy of the broken bonds Φ_k, additional terms such as energy of ionization, energy of hydration etc. in the considered case of electrocrystallization.

II.2. Concentration and Mean Displacement of Adatoms

The equilibrium adatom concentration C^{o}_{ad} [mol cm^{-2}] on the surface of an infinite crystal is given by

$$n_{ad} = C^o_{ad} N_A = n_o \exp(-Q/k_B T) \qquad (5)$$

where the pre-exponential term contains the entropy factor and is roughly equal to the surface atomic density n_o.[2]

$$W = \Phi_{k,s} - \Phi_{ad} \qquad (6)$$

is the energy of the transfer of an atom from a kink position to the adsorption site. $\Phi_{k,s}$ is, as mentioned, a constant characteristic for the crystal, while the desorption energy Φ_{ad} depends on the crystallographic character of the face and contains a part of the hydration energy of adatoms.[5,6] This latter fact makes the assessment of C^o_{ad} rather difficult.

Adatoms are in continuous exchange with ions in solution. The deposition and dissolution rates expressed in terms of current densities [A cm^{-2}] can be written as :

$$i_{c,ad} = \omega_{dep,ad} z e_o n_o \quad \text{and} \quad i_{a,ad} = \omega_{diss,ad} z e_o n_{ad} \qquad (7)$$

The density of adatoms n_{ad} is assumed to be much smaller than n_o and is neglected with respect to it in the first equation. At the potential E the rates of dissolution and deposition are equal, $i_{c,ad} = i_{a,ad}$. From equations (7), (1) and (2), one obtains :

$$C_{ad}(E) = \frac{n_o}{N_A} \frac{p_{ad}}{q_{ad}} c \exp(\Phi_{ad}/k_B T) \exp(-z e_o E/k_B T) \qquad (8)$$

where $\Phi_{ad} = -(\Delta G^*_{dep,ad} - \Delta G^*_{diss,ad})$ is the desorption energy of an adatom. At the equilibrium potential, $E = E_{rev}$ the adatom concentration $C_{ad}(E_{rev}) = C^o_{ad}$ is a constant independent of c or E_{rev} because of their relation through the Nernst equation (see Eq. (4)). The potential independence of C^o_{ad} also follows directly from Eq. (5).

Generally, when the exchange rate between adatoms and kink atoms is inhibited (e.g., large step distances) the local adatom concentration C_{ad} can change with E or $\eta_c = E_{rev} - E$

$$C_{ad}(\eta) = C^o_{ad} \exp(-z e_o \eta/k_B T) \qquad (9)$$

The mean residence time of adatoms on the face is given by $1/\tau_r = \omega_{diss,ad}$. Taking into account Eq. (7) for any arbitrary potential when a local equilibrium $i_{a,ad} = i_{c,ad}$ is established :

$$\tau_r = C_{ad} zF/i_{a,ad} = C_{ad} zF/i_{c,ad} \tag{10}$$

Hence the mean residence time depends on potential. At the equilibrium potential $\tau_r^o = C_{ad}^o\ zF/i_{o,ad}$, where $i_{o,ad}$ is the exchange current density, τ_r^o is a function of the bulk concentration or E_{rev} through $i_{o,ad}$.

During residence on the surface, adatoms migrate at random across it making an average displacement λ_s given by

$$\lambda_s^2 = D_s \tau_r = \frac{zFC_{ad}^o}{i_{o,ad}} D_s \exp\left[-(1-\beta) ze_o \eta_c / k_B T\right] \tag{11}$$

where D_s is the surface diffusion coefficient. The mean displacement distance λ_s is a function of the overpotential. At the equilibrium potential $\lambda_s^o = (zFC_{ad}^o D_s / i_{o,ad})^{1/2}$.

II.3. Propagation Rate of Steps

If the potential of an atomically smooth (stepless) singular face is changed, e.g., below the reversible potential, the enhanced deposition rate increases the adatom concentration above its equilibrium value C_{ad}^o until the opposite reaction $i_{a,ad}$, increasing with C_{ad}, balances with $i_{c,ad}$ (see Eq. (9)).

If steps are present, in close vicinity of the step edge, the kink atom - adatom exchange reaction is supposed to keep the adatom concentration at its equilibrium value C_{ad}^o. At a distance greater than λ_s, however, adatoms deposited on the face cannot reach the step edge during their residence time. The adatom concentration keeps the value $C_{ad}(\eta)$ unaffected by the presence of the step. This causes a diffusion flux $D_s(C_{ad} - C_{ad}^o)/\lambda_s$ [moles cm^{-1} sec^{-1}] of adatoms towards the step edge where they are incorporated into the crystal lattice. At room temperatures[4] where the average kink distance is of the order of several atoms, diffusion along the step edge can be neglected. The incorporation of atoms in the step propagates its edge with a rate $v_{sd} = zF/q_{mon})D_s \Delta C_{ad}/\lambda_s$ [cm sec^{-1}] where $q_{mon} = ze_o\eta_o$ is the electrical charge of a complete monolayer of area unity, or with Eqs. (9) and (11) :

$$v_{SD} = \frac{2i_{o,ad}\lambda_s}{q_{mon}} V(\eta) \tag{12}$$

where

$$V(\eta) = \exp\left[(1-\beta)ze_o\eta/k_BT\right] - \exp(-\beta ze_o\eta/k_BT) \tag{12a}$$

or for $ze_o\eta/k_BT << 1$

$$V(\eta) = ze_o\eta/k_BT. \tag{12b}$$

The factor 2 in Eq. (12) accounts for the two equivalent fluxes from both sides of the step.

Part of the atoms incorporated into the step edge at a cathodic overpotential are supplied by direct transfer from the solution. The rate of this direct transfer reaction is determined by the frequencies of deposition and dissolution (1) and (2) applied to the step sites "st", i.e. sites adjacent to the step edge. We can define cathodic, anodic and exchange current densities for these sites in analogy to the adsorption sites (see Eq. (8)). The partial propagation rate v_{DT} due to direct transfer is given by

$$v_{DT} = \frac{i_{o,st}\delta}{q_{mon}} V(\eta) \tag{13}$$

where $i_{o,st} = \omega_{dep,st}(E_{rev})ze_o/\delta a = \omega_{diss,st}(E_{rev})ze_o/\delta a$

is the exchange current density of the step sites atom; δ is the width of an atomic row added to the step edge and a the length per atom along the step ($\delta a = 1/n_o$).

If the exchange current densities $i_{o,ad}$ and $i_{o,st}$ are of the same order of magnitude, the ratio λ_s/δ predetermines a much higher contribution of the surface diffusion.

II.4. Exact Solution of the Surface Diffusion Problem

If two steps are closer than twice the mean displacement distance of adatoms, $x_o < \lambda_s$, the regions of depleted adatom concentration overlap and the surface diffusion rate decreases from that of Eq. (12).

The surface diffusion problem can be treated in greater detail by solving Fick's diffusion equations. This problem has been treated by a number of authors[7-12] coming to almost identical results. We will follow a treatment similar to DESPIC and BOCKRIS[8] but also see Ref. (10).

Consider an infinitesimal area dxdy of the surface at a distance x from the step. The atom fluxes to and from this surface part are defined as follows :

$$\frac{dC_{ad}(x,t)}{dt} = D_s \frac{d^2C_{ad}(x,t)}{dx^2} - \frac{i(x,t)}{zF} \tag{14}$$

$$i(x,t) = i_{o,ad} \frac{C_{ad}(x,t)}{C^o_{ad}} \exp(1-\beta)ze_o\eta/k_BT - \exp(-\beta ze_o\eta/k_BT) \tag{15}$$

$$i - C_{DL}d\eta/dt = \frac{1}{x_o}\int_o^{x_o} i(x,t)dx \tag{16}$$

The $C_{CL}d\eta/dt$ term accounts for double layer charging. The right hand part of Eq. (16) represents the mean Faradaic current to the crystal face.

The boundary conditions can be formulated in accordance with the incorporation kinetics of adatoms into the step edge.

a) With a high incorporation rate constant, the adatom concentration in close vicinity of the step edge changes insignificantly during growth and the boundary conditions are

$$C_{ad}(x,0) = C_{ad}(0,t) = C^o_{ad} \;;\; [dC_{ad}(x,t)/dx]_{x=x_o} = 0 \tag{17}$$

b) With a low incorporation rate constant, the incorporation process needs an increase of adatom concentration in close vicinity of the step edge and the boundary conditions are in this case :

$$C_{ad}(x,0) = C^o_{ad}\;,\quad [dC_{ad}(x,t)/dx]_{x=x_o} = 0 \quad \text{and}$$

$$D_s[dC_{ad}(x,t)/dx]_{x=0} = v_{inc}[C_{ad}(0,t) - C^o_{ad}]/C^o_{ad} \tag{18}$$

where $[v_{inc}$, mol cm^{-1} s$^{-1}]$, is the incorporation rate constant.

This equation system has been solved for the potentiostatic case.[7-10] Many authors have solved this system under different additional assumptions and approximations, e.g., refs. (13,14). It is out of the scope of this chapter to go into details of the problem and the reader is advised to consult the cited authors. A comprehensive review can be found in Vetter's "Electrochemical Kinetics".[10]

Three points may be worth emphasizing here :

a) The adatom concentration distribution $C_{ad}(x)$ at steady state has been solved by a number of authors[7,8,10] and can be represented in the form :

$$\frac{C_{ad}(x)}{C^{o}_{ad}} = \exp\left(\frac{ze_{o}\eta}{k_{B}T}\right) + \left[1 - \exp\left(-\frac{ze_{o}\eta}{k_{B}T}\right)\right]\frac{\mathrm{ch}(x_{o} - x)/\lambda_{s}}{\mathrm{ch}(x_{o}/\lambda_{s})} \tag{19}$$

Where the substitution $\lambda_{s} = \lambda^{o}_{s} \exp[-(1 - \beta)ze_{o}\eta/2k_{B}T]$ with

$\lambda^{o}_{s} = (zFD_{s}C^{o}_{ad}/i_{o})^{1/2}$ has the meaning of the surface diffusion penetration depth and is equal to the mean displacement distance of adatoms defined by Eq. (11). (Note that λ_{s} is a function of the overpotential). Relation (19) is represented in Figs. 1a and 1b for different η and λ^{o}_{s}/x_{o} values. At $x > 4\lambda^{o}_{s}$, for sufficiently large x_{o} values, $C_{ad}(x)$ reaches its saturated value defined by Eq. (11) as seen, e.g., for the case $x_{o}/\lambda^{o}_{s} = 10$ and $\eta = 10$ mV in Fig. 1b.

The local current densities i(x) are easily found by inserting Eq. (19) in Eq. (15) and are represented for different η and λ^{o}_{s}/x_{o} values in Figs. 1c and 1d.

b) The step propagation rate v[cm s^{-1}] can easily be obtained by differentiation of Eq. (19) bearing in mind Eq. (11) :

$$v_{SD} = \frac{2\lambda_{s} i_{o,ad}}{q_{mon}} V(\eta) \tanh(x_{o}/\eta_{s}) \tag{20}$$

For $x_{o} >> \lambda_{s}$ this equation coincides with Eq. (12).

c) The galvanostatic transient $\eta(t)_{i}$ can give information about the otherwise unknown adatom concentration. From the equivalent electric circuit simulating the deposition mechanism[15] described by Eqs. (14) to (16), it can be assumed that after the double layer charging is achieved ($t > 4C_{DL}R_D$), the initial slope i/C_{DL} should change to $i/(C_{DL} + C_A)$, where $R_D = RT/zFi_o$ is the resistance of the ion transfer across the double layer and $C_A = C^o_{ad}z^2F^2/RT$ is the adatom pseudocapacitance. Before coming to saturation the transient should follow a Warburg impedance-like $\sqrt{t}$ law due to the surface diffusion evolution. Under given circumstances depending on the ratio λ_s/x_o, this can hinder the evaluation of C_A. A large value of C_A, $C_A > C_{DL}$ (with $C_{DL} \simeq 20\ \mu F/cm^2$, this requires $C^o_{ad} > 5.10^{-12}mol\ cm^{-2}$), and a large value of x_o/λ_s are a prerequisite for reliable determination of C_A and C^o_{ad}.

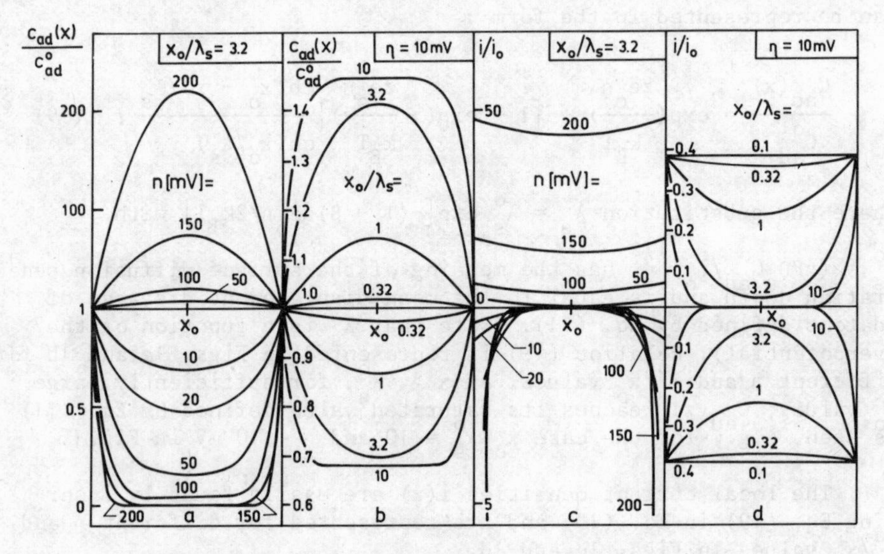

Fig. 1 : Adion concentration profiles $C_{ad}(x)/C^o_{ad}$ and current density distribution $i(x)/i_o$ between two growth steps at different x_o/λ_s and η values (DESPIC and BOCKRIS[8]).

II.5 Current Density on a Stepped Crystal Face.

Consider a stepped singular face with a regular step spacing $2x_o$. The step density $L_s[cm^{-1}]$ or its equivalent, the total step length per unit area$[cm\ cm^{-2}]$, is $L_s = 1/2x_o$. At a given overpotential all steps propagate with a rate v determined by Eq. (20) or Eq. (13) or the sum of both. The current density is

$$i = v\,L_s\,q_{mon} \quad (21)$$

According to the dominating mechanism we obtain :

a) For the surface diffusion mechanism with υ given by Eq.(20)

$$i = i_{o,ad} \; V(\eta) \; \frac{\lambda_s}{x_o} \; \tanh(x_o/\lambda_s) \qquad (22)$$

The first theoretical treatment of this mechanism was given by LORENZ[7] who adapted the theory of BURTON, CABRERA and FRANK[2] to the case of metal deposition. Later the problem was treated by VERMILYEA,[12] FLEISCHMANN and THIRSK.[11] DESPIC and BOCKRIS,[8] DAMJANOVIC and BOCKRIS[9] and SCHNITTLER[16] who brought the theory of surface diffusion to its present state.

Two limiting cases can be considered here :

i) $\lambda_s >> x_o$: Eq. (22) yields a Butler-Volmer relation $i = i_o \; V(\eta)$, i.e., charge transfer is the rate determining step and current density is unaffected by surface morphology (liquid-like behavior).

ii) $\lambda_s << x_o$: The tanh (x_o/λ_s) term in Eq. (22) is equal to unity and

$$i = 2 \; i_{o,ad} \; L_s \lambda_s \; V(\eta) \qquad (23)$$

Surface diffusion controls the process and the current density is proportional to the step density L_s.

b) For the direct transfer mechanism from Eqs. (21) and (13) we obtain :

$$i = i_{o,st} \; \delta \; L_s \; V(\eta) \qquad (24)$$

This equation is homologous to Eq. (23); both are linear with respect to the step density L_s.

The direct transfer mechanism was first suggested by VOLMER[17] and was later discussed in the works of LORENZ[7], FLEISCHMANN and THIRSK,[11] MOTT and WATTS-TOBIN[18] and others. A comprehensive treatment of the problem may be found in VETTER.[10]

As seen from Eqs. (23) and (24) in both cases, direct transfer or surface diffusion (with $\lambda_s < x_o$), the current density depends on the step density and therefore surface morphology. A high-indexed surface zone with a high density of steps or growth sites (kink positions), where $\lambda_s >> x_o$, would follow the same growth law as a liquid metal, i.e., the Butler-Volmer relation. Special growth behavior is expected for perfect, defect-free singular faces.

III. METAL DEPOSITION ON A PERFECT CRYSTAL FACE

A perfect crystal bound by singular faces exhibits no growth sites. A 2D nucleation process is required for the deposition of every new layer as first noted by GIBBS[19] and developed later by VOLMER and WEBER,[20] BRANDES[21] and KAISCHEW.[22] This is because one atom deposited on the crystal face has a smaller bond energy to the crystal than a kink atom and it stays only temporarily on the surface as an adatom. An important feature is the tendency of these surface atoms to cluster together, thereby increasing their stability.

Only after reaching a critical size can these clusters grow spontaneously under the existing overpotential and give birth to new lattice nets. Using the classical Gibbs equation

$$\Delta G(N) = - Nze_o\eta_c + \Phi(N) \tag{25}$$

connecting the free energy of formation of a cluster to the supersaturation η_c and its size N, with $\Phi(N)$ accounting for the excess surface free energy, the size as well as the free energy of formation of this critical cluster can be easily evaluated :

$$N^* = bs\varepsilon^2/(ze_o\eta_c)^2 \tag{26}$$

and

$$G^* = bs\varepsilon^2/ze_o\eta_c \tag{27}$$

where ε is the specific ledge-free energy, b is a geometrical factor relating the surface area of the nucleus to its perimeter and s is the area occupied by one atom on the surface of the nucleus. When the geometrical form of the nucleus is a circle or a regular polygon, the size can be given by the inscribed radius ρ

$$\rho^* = s\varepsilon/ze_o\eta_c = \varepsilon/q_{mon}\eta_c \tag{28}$$

where the constant ze_o/s can be replaced by its equality q_{mon} giving the amount of electricity per unit area [As cm^{-2}] needed for the deposition of one monoatomic layer.

The rate of 2D nucleation, i.e., the rate of formation of clusters big enough to ensure spontaneous growth, is proportional to the population of critical nuclei and hence is given by the expression

$$J = k_1 \exp(-\Delta G^*/k_BT) = k_1 \exp(-bs\varepsilon^2/ze_o k_BT\eta_c) \quad (29)$$

first noted by ERDEY-GRUZ and VOLMER.[23] Further development of the theory (KAISCHEW[22] and HIRTH[24]) revealed that the pre-exponential term k_1 is a product of the so called Zeldovich nonequilibrium factor $\Gamma = (\Delta G^*/3\pi k_BT N^{*2})^{1/2}$, of the probability ω^* for an adatom to join the critical nucleus and of the surface atomic density n_o. Γ and ω^* are weakly dependent on overpotential and are usually considered constants. The $J - \eta_c$ relationship according to Eq. (29) shows a sharp rise of the nucleation rate at a critical value, below which the rate is negligibly small (Fig. 2).

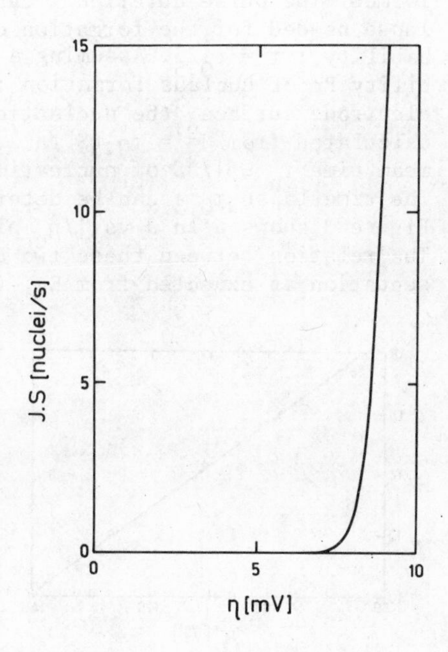

Fig. 2 : Overpotential dependence of the 2D nucleation rate J nuclei cm^{-2} (Eq. (29)). $k_1 = 10^{13} s^{-1} cm^{-2}$, $\varepsilon = 2 \times 10^{-13}$ J cm^{-1}, b = 4, T = 318 K with $ze_o/s = q_{mon} = 2 \times 10^{-4}$ A s cm^{-2} for a Ag(100) face, S = 2 × 10^{-4} cm^2 from experimental data.[25]

The existence of a supersaturation or an overpotential threshold in the growth of a crystalline phase is a feature characteristic of nucleation induced processes.

Based on this phenomenon a double pulse technique, originally developed for the study of the electrolytic phase formation by SCHELUDKO and TODOROVA[28] (see also section V.2.) has been used by BUDEVSKI et al.[25-27,29] for the investigation of 2D nucleation kinetics on screw-dislocation free faces. In order to prepare a dislocation-free face, an appropriately oriented single crystal is enclosed in a glass tube with a capillary end and is used as a cathode in an electrolytic cell. The seed crystal is grown electrolytically into the capillary end of the tube. Using an a.c. super imposed on the cathodic current, the front face can be grown as a perfect dislocation-free face, filling the whole cross section of the capillary (usually about 0.2 mm diameter). The preparation technique of dislocation-free faces of silver single crystals and their properties are described elsewhere.[25-27] The nucleation overpotential threshold in 6M $AgNO_3$ at 45°C has been found to be about 5-8 mV. Below this threshold no current is observed. If an overpotential pulse exceeding the threshold value is applied on a cell polarized at, e.g., 3-4 mV, a current following this pulse shows that one or more nuclei are formed. Under the polarizing overpotential the nuclei grow to the capillary wall, and the initial state is restored. The pulses can be adjusted in amplitude η_c and duration τ in such a way that only in half of the cases a nucleus is formed. The pulse duration τ can then be considered as the time lapse needed for the formation of a single nucleus with a 50 % probability : $\tau = \tau_{0.5}$. Assuming a Poisson distribution of the probability Pr of nucleus formation : $\ln(1 - Pr) = -JS$, S being the electrode surface, the nucleation rate J [nuclei s^{-1} cm^{-2}] can be calculated from $J^{-1} = \tau_{0.5} S \ln 2$. Note that $\tau_{0.5}$ differs from the mean time $\tau_n \equiv 1/JS$ of nucleation as pointed out by Toshev et al.[30] The time lapse $\tau_{0.5}$ can be determined at different overpotentials. Figure 3 shows a ln J vs $1/\eta_c$ plot of data obtained in this way. The relation between these two quantities is linear in this representation as expected from Eq. (29).

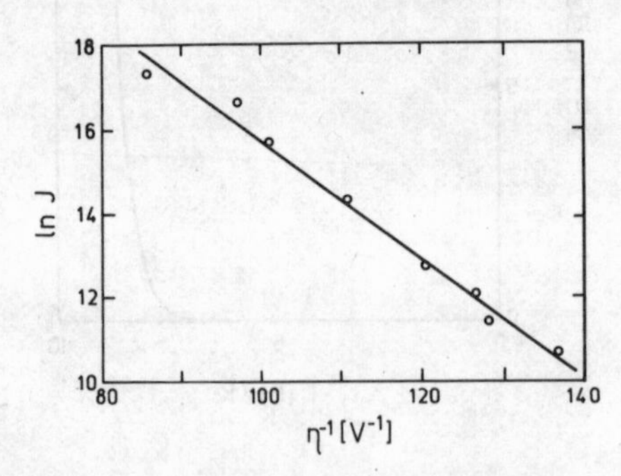

Fig. 3 : Logarithm of the 2D nucleation rate J, nuclei s^{-1} cm^{-2} as function of the reciprocal of the overpotential for (100) Ag in 6M $AgNO_3$ at 45°C.[25-27]

The contribution of the overpotential dependence of the Zeldovich factor Γ and the attachment frequency ω are obviously small enough to produce an appreciable effect on the general $\ln J - 1/\eta_c$ relation. From the slope, the specific edge energy $\varepsilon = 2\ 10^{-13}$ J cm^{-1} can be calculated, where a square form (b = 4) of the critical nucleus has been assumed. From Eqs. (26) and (27), ΔG_c and N_c have been found to vary in the range of 10×10^{-20} to 6.9×10^{-20} J and 80 to 36 atoms respectively. Also of interest is the frequency factor $k_1 = n_o \Gamma \omega_c$ in the rate equation (29) which can be calculated from the intercept of the $\ln J - 1/\eta_c$ curve.

The following values of the specific edge energy and the frequency factor k_1 are currently considered most probable for the cubic and octahedral faces of silver in 6M $AgNO_3$ at 45°C :

	(100)	(111)
$\varepsilon\,[J\ cm^{-1}]$	$(1.9 \pm 0.2) \times 10^{-13}$	$(2.2 \pm 0.2) \times 10^{-13}$
$k_1\,[s^{-1}\ cm^{-2}]$	$10^{12} - 10^{15}$	$10^{10} - 10^{11}$

The variation of ε and k_1 for both faces is within the limits of experimental error so that no speculations can be made from the variation at this stage of experimental knowledge. The variation of k_1 over an interval of several orders of magnitude indicates that not all adsorption sites n_o are equivalent for the nucleation process and also that their number or activity changes from experiment to experiment.

III.1. Kinetics of Step Propagation and Mechanism of Metal Deposition.

In experiments with dislocation-free faces described in the preceeding section, after the nucleation pulse a current indicating formation of a nucleus, is observed (e.g. Figure 9). Generally the current initially increases linearly with time goes through a maximum and then falls to zero within a few seconds. The integral of the current-time curve gives an amount of electricity equivalent to the deposition of a monatomic layer. The shape of the i-t curves follows the growth and decay of the peripheral edge of the spreading monolayer growing over and out of the face boundaries.[25-27] In the beginning, before the step has reached the face boundaries, the current should follow the linear law

$$i = q_{mon}\ 2bv^2t \qquad (30)$$

whith b the geometrical factor defined in the preceeding section. The propagation velocity v of the step can be evaluated from the slope $(di/dt)_{t=0}$, if the geometrical form of the spreading layer is known.

A precise technique, which gives information not only on the propagation rate but also on orientation and form of the peripheral step confining the spreading monolayer, was developed by BOSTANOV et al.[31,32] by growing nomatomic layers on dislocation-free faces with a rectangular elongated shape. For this purpose the crystal is grown, in the usual way, into a capillary of rectangular (100 × 400 μm) cross section to produce a dislocation-free face. As expected, it was found that the current-time curve following the formation of the nucleus has a plateau, reflecting the fact that for a certain period of time, the step propagates with a constant length defined by the shorter side of the rectangular face.

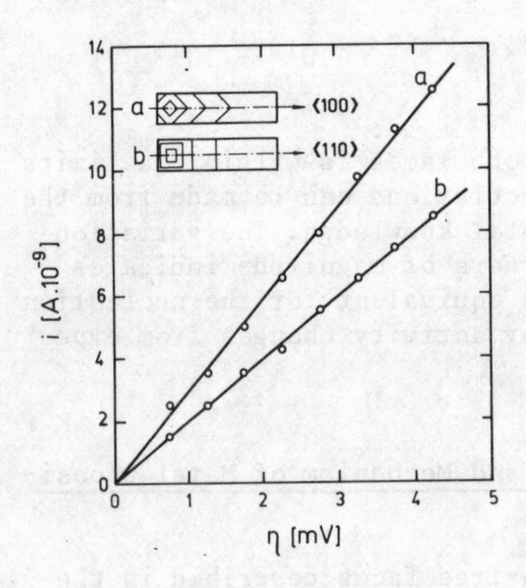

Fig. 4 : Current plateau values as function of the overpotential for two different orientations of the cross-section axis of a rectangular (0.2 × 0.4 mm) capillary with respect to the Ag single crystal; a-along the <100> and b-along the <110> direction in the plane of the (100) front face.[31,32]

Figure 4b shows a linear relation between the current plateau and the overpotential permitting determination of the rate constant of step propagation, $v/\eta = \kappa_y$ $[\mathrm{cm\ s^{-1}\ V^{-1}}]$. For this purpose, again, the form of the step and its orientation with respect to the rectangular boundaries of the face is needed. To solve the problem, two orientations of the rectangular cross section of the capillary with respect to that of the seed crystal have been used. In the case of the cubic face, the orientations were chosen at an angle of 45° along the [100] and the [110] directions. The slopes of the current-overvoltage curves of Fig. 4 would be independent of orientation if the spreading monolayer has a circular form. Experiment showed, however, that the slopes for both orientations differ by a factor of $1.42 = \sqrt{2}$, the orientation [100] giving the higher value. It is evident that the growing monatomic layer has the form of a square with the sides oriented along the [110] direction.

Knowing the forms and the orientations of the propagating steps, exact values of the propagation rate constants were found for monatomic steps on the (100) and (111) faces of silver crystals.

$$\kappa_v = (1.00 \pm 0.05) \text{ cm s}^{-1} \text{ V}^{-1}$$

Under certain conditions, e.g., if a higher overvoltage pulse of a longer duration was applied to the cell, an "activation" of the surface was observed.[32] On activated surfaces, the value of κ_v is roughly doubled, e.g., $\kappa_v(100) = 1.9$ cm $s^{-1}V^{-1}$.

Under the cited conditions (higher pulses of longer duration) in some cases, only one polyatomic step remained, advancing on the face at the lower growth overvoltage. With the Nomarski differential contrast technique used in these experiments, some of these thicker steps were observed directly. Figure 5 shows a 15-atomic step moving along a (100) face in two consecutive positions over an interval of two seconds. From the current and the propagation rate, which is easily determined from the photograph, the height of the step is obtained.

A rather unexpected fact is that the rate of propagation was independent of step height (up to heights of 50 to 100 Å, at which point on concentration polarization in the solution begins to interfere) giving a value of $\kappa_v = 1.9$ cm $s^{-1}V^{-1}$, very close to that found on activated surfaces with monatomic layers. This, together with the assumption that the edge of the polyatomic step is clustered down to atomic dimensions could be taken as evidence for a direct transfer mechanism of ion incorporation. However, the steepness of the step front cannot be estimated from the interference picture obtained with the differential contrast technique, and if the individual monatomic steps forming the polyatomic step front are sufficiently far apart, surface diffusion can not be excluded.

The steepness of the front has been estimated by BOSTANOV et al.[32] using the decay curve of the current at the time when the step begins to disappear on the edge of the crystal face (Fig. 6). Knowing the propagation velocity, the current gradient at the end of the current-time curve gives information on the mean step distance of the step train. A value of about 160 Å for the mean step distance was found for the case shown in Figure 5. However, this value is the highest possible by this estimation technique because a very steep but slightly curved (e.g., radius of curvature of 25 cm for the 0.1 mm long front) would produce the same gradient as that shown in Fig. 6 . Hence, we can assume that the mean step distance $2x_o$ is less than 160 Å. The step height independence of the propagation rate indicates that $\lambda_s < 80$ Å.

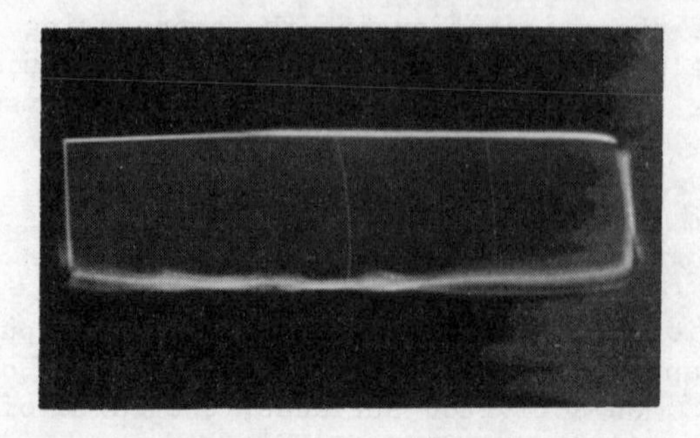

Fig. 5 : Two overlapping flashlight photographs of a 15 atomic layer thick step advancing on a dislocation-free Ag(100) face in a rectangular glass capillary (0.1 × 0.4 mm cross-section) taken in an interval of two seconds.[32]

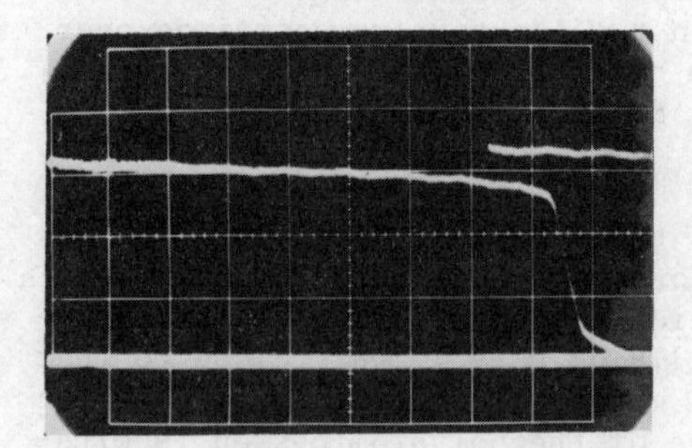

Fig.6 : Oscillogram of the current decay at the moment of collision of a thick step on the face boundary edge. Time axis 20 ms cm^{-1}, current axis 10^{-7} A cm^{-1}.[32]

To estimate the surface diffusion contribution, the exchange current density of adatoms on dislocation-free faces is needed. Exchange current densities were measured by VITANOV et al.[33,34] From the frequency dependence of the impedance on dislocation-free faces. A value of $i_{o,ad} = 0.06$ A cm^2 was found in 6M $AgNO_3$ at 45°C on (100) silver single crystal face. With this value, and $\lambda_s < 80$ Å, the propagation rate constant $\kappa_v = v_{SD}/\eta$ can be calculated from Eq. (20), linearizing V(η), and setting $\tanh(x_o/\lambda_s) = 1$: $\kappa_v < 1.8 \times 10^{-2}$ cm $s^{-1}V^{-1}$ which is about 60 times lower than the experimental value 1.04 cm $s^{-1}V^{-1}$. We conclude that surface diffusion plays a subordinate role in the electrolytic deposition of silver.[32,35] If direct transfer is dominant, from Eq. (13) with $\delta = 2.88$ Å and $\kappa_v = 1.04$ cm $s^{-1}V^{-1}$, we obtain $i_{o,st} = 200$ A cm^{-2}, an exchange current density for the step atoms of about four orders of magnitude higher than for the adatoms (see also section IV.1.).

An intact dislocation-free face as cathode represents an

ideally polarizable electrode in the potential interval between the reversible potential and the nucleation overpotential threshold. In addition to impedance measurements, galvanostatic transients are useful for determination of the double layer capacitance C_{DL} and the adsorption pseudocapacitance C_A. For this purpose the slopes $d\eta/dt$ at $t = 0$ and at $t \simeq 4\tau_{DL}$ can be used, where $\tau_{DL} = C_{DL}R_D$ is the time constant for the double layer charging and R_D the resistance of the ion transfer reaction.
The transient on a dislocation-free face should be linear after the deflection at $t \simeq \tau_{DL}$ up to a nucleation overpotential of 5-8 mV. Due to appreciable penetration of the electrolyte between the walls of the glass capillary and the crystal, the transient is distorted to a $\sqrt{t}$ dependence at a very early stage. Because the second linear part of the transient is distorted by this effect, VITANOV et al.[34] used $d\eta/dt$ at $t = 4\tau_{DL}$ to estimate the value of C_A. In agreement with impedance measurements,[36] a value of $C_{DL} = 30\ \mu F\ cm^{-2}$ for the cubic face of silver was found from the initial slopes of the transients. From the slope at $t = 4\tau_{DL}$ (where $i/(d\eta/dt)_{4\tau_{DL}} = C_A + C_{DL}$), $C_A = 10 - 20\ \mu F\ cm^2$, indicating an ad-atom concentration of $C^o_{ad} = (2 \text{ to } 5) \times 10^{-12}\ mol\ cm^{-2}$.[35] This very rough estimation gives an upper experimental limit of C^o_{ad} because neglect of the penetration effect produces a false increase in C_A.

III.2. Deposition Kinetics on Perfect Crystal Faces

The kinetics of metal deposition on crystal faces free from screw-dislocations is determined by the rate of nucleation J [nuclei $cm^{-2}s^{-1}$] of new lattice nets and their rate of propagation v [$cm\ s^{-1}$] over the face. Almost simultaneously NIELSEN[37], CHERNOV[38] and HILLIG[39] showed that depending on the values of these two parameters, two different mechanisms can be distinguished, with the extension of the surface playing a significant role : (i) layer by layer growth and (ii) multinuclear multilayer growth.

At low overpotentials, where the nucleation time $\tau_n = (JS)^{-1}$ is much larger than the time $\tau_p = S^{1/2}v^{-1}$ for propagation of one monatomic layer over the whole surface, each nucleus has sufficient time to spread over the surface before the next nucleus is formed. Under these conditions each layer is formed by a single nucleus. The current following the development and the decay of the peripheral edge of each monolayer is not stable and fluctuates with nucleation and spreading over the face. The mean current density is determined by the nucleation frequency, $1/\tau_n = JS$, and is independent of the rate of propagation v:

$$i = \frac{1}{\tau_n} \int_o^{\tau_n} i(t)dt = JSq_{mon} \qquad (31)$$

With J given by Eq. (29), this equation originated with ERDEY-GRUZ and VOLMER.[23]

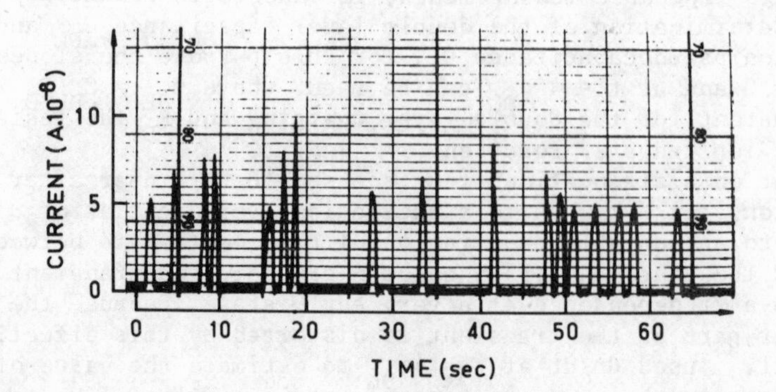

Fig. 7 : Fluctuations of the current on a dislocation-free Ag(100) face at 8.5 mV, indicating periodic nucleation and spreading of monatomic layers. The integrals under the i-t pulses are equal to $q_{mon} \simeq 2 \times 10^{-4}$ A cm^{-2}. Different heights and forms of the pulses indicate that there are no preferential sites of nucleation on the face.

Figure 7 shows the recorded fluctuations of the current from the deposition of silver on a (100) dislocation-free face at 8.5 mV. The time integral of the current in each fluctuation is equivalent to the deposition of one monatomic layer and the mean time interval between two acts of nucleation is inversely proportional to J.[40]

With growing overpotentials, the nucleation rate J increases much faster than the propagation rate (see Eqs. (13) or (20) and (29)) so that the mean nucleation time τ_n becomes much shorter than the propagation time τ_p. The deposition of each layer proceeds with formation of a large number of nuclei. The kinetics of deposition are determined in this polynuclear mechanism both by J and v.

The kinetics of formation of one monatomic layer on an intact face are given by[41,42]

$$i = q_{mon} bJ\upsilon^2 t^2 \exp(-bJ\upsilon^2 t^3/3). \tag{32}$$

The pre-exponential term gives the current flowing to the peripheries of the independently growing nuclei, progressively formed with time (see Eq. (30)). The exponential term takes into account the overlapping effect in terms of the Kolmogorov-Avrami formulation. In an early stage, before overlapping has occured, the current follows the quadratic equation given by the pre-exponential term.

Fig. 8 : Current-time transient in an i vs t^2 plot obtained on a dislocation-free Ag(100) face at 14 mV.[40]

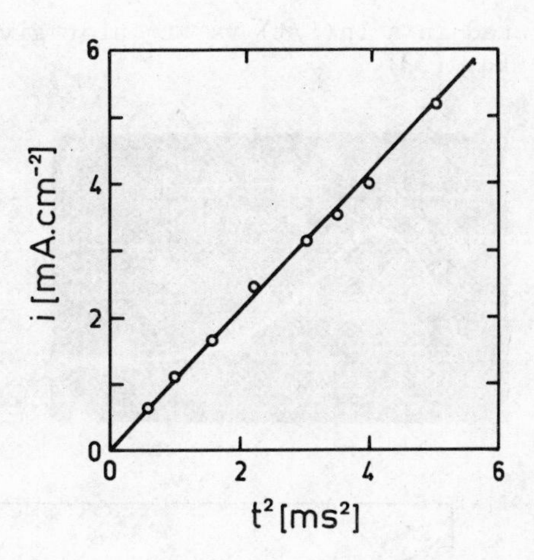

Figure 8 shows a current time transient in an i vs t^2 plot obtained on a (100) face of a silver single crystal.[43] From the slopes of a series of i-t^2 curves recorded at different overpotentials, the value of $J\upsilon^2$ can be obtained as a function of η. The propagation rate υ and its η dependence are known (see section III). A ln J vs. $1/\eta$ analysis can then be made as described in the preceding section.[43] A value of $\varepsilon = 2$ to 2.8×10^{-13}J cm^{-1} was found,[43] in good agreement with that obtained from the time lag of formation of a single nucleus.

In some cases of 2D growth where active centers play a role, nucleus formation can proceed instantaneously from the beginning of the pulse. In this case the current density is given according to FLEISCHMANN and THIRSK[42] by

$$i = q_{mon} 2bN_o \upsilon^2 t \exp(-bN_o \upsilon^2 t^2) \tag{33}$$

The pre-exponential term gives the current flowing to the peripheries of the N_o independently growing nuclei (see Eq. (30)). The exponential term takes into account the overlapping effect. On dislocation-free faces the instantaneous nucleation can be simulated, thus extending the possibilities of investigation of the 2D nucleation.

For this purpose a nucleation pulse (τ_{nucl} > 8 mV and duration τ_{nucl}) is applied to the dislocation-free face prepolarized at a growth overpotential η < 5 mV. The nuclei created during the nucleation pulse are grown at the lower overpotential of growth and the current transient is recorded. Figure 9 shows a transient recorded in this way. In Figure 10 the same transient is represented in a ln(i/t) vs t^2 plot giving a straight line as required by Eq. (33).

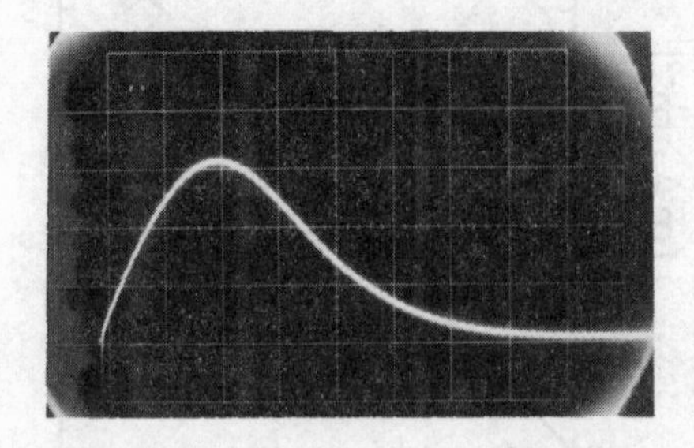

Fig. 9 : Current transient on a Ag(100) dislocation-free face at η = 4 mV following a nucleation pulse η_{nucl} = 17 mV, τ_{nucl} = 120 μs. Time scale 10 ms/div, current scale 0.4 μA/div.[43]

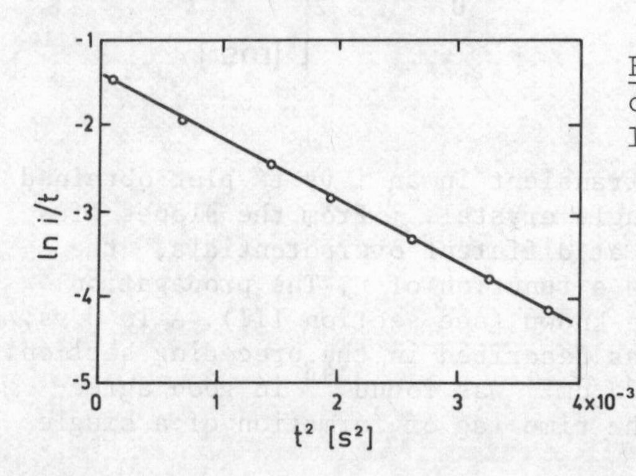

Fig. 10 : The current transient of Figure 9 represented in an ln(i/t) vs t^2 plot.[43]

From its slope, N_o and $J = N_o/\tau_{nucl}$ are obtained. A lnJ vs $1/\eta$ analysis of J-data obtained at different τ_{nucl} values renders again ε ranging between 2 and 2.8×10^{-13} J cm^{-1}.[43]

Equation (32) gives the kinetics of monatomic layer extension. As nucleation proceeds continuously, however, new nuclei are formed on top of the deposited first layer long before it has covered the whole area of the face. The same happens with the third, fourth and so on layers as soon as they are large enough to make the nucleation possible. The deposition process under these conditions is known as "multinuclear multilayer growth". This mechanism was extensively theoretically investigated by BOROVINSKII and ZINDERGOSEN,[44] ARMSTRONG and HARRISON[45] and RANGARAJAN[46] all based on the Kolmogorov-Avrami theory, and by BERTOCCI[47] and GILMER[48] using Monte Carlo simulation techniques. Calculations have shown that after the initial rise, the current-time transient passes through several oscillations to steady down after the deposition of several layers to the final value i (Fig. 11). The steady state current density is given by

$$i_\infty = q_{mon}\beta(bJv^2)^{1/3} \qquad (34)$$

where β is a constant varying around unity depending on the theory.

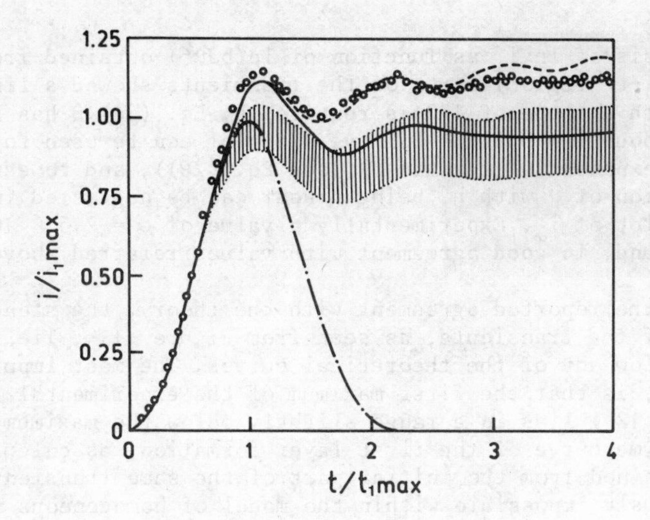

Fig. 11 : Theoretical and experimental current-time transients for multinuclear multilayer growth. Solid line;[45] dashed line;[44] circles Monte Carlo simulation;[48] dash-dot-dash line first monolayer formation. Hatched area represents the range of variation of experimental i-t curves obtained at different overpotentials on dislocation-free Ag(100) faces.[40,49] All transients normalized to the i_{max} and t_{max} at the first layer formation.

The transients of Figure 11 are normalized to the parameters at the maximum current of the first layer formation according to Eq.(32): $t_{1,max} = (2/bJv^2)^{1/3}$ and $i_{1,max} = q_{mon}(4bJv^2)^{1/3}\exp(-2/3)$. As seen from Figure 11, in the initial monolayer region all calculated transients follow the quadratic law from the pre-exponential term of Eq. (32). The theoretical curves eventually go slightly astray which seems to be connected with the different approximation used in calculating the contribution of the n+1-th layer on top of the n-th layer. The most reliable results seem to be obtained by the nucleation and disc growth Monte Carlo simulation (circles in Figure 11) calculated by GILMER.[48]

Values of Jv^2 obtained from the slopes of i vs t^2 plots (see Figure (8)) can be used for the calculation of $t_{1,max}$ and $i_{1,max}$.

These values are needed for normalization of the experimental potentiostatic i-t transients (Figure 12). Potentiostatic transients recorded at different overpotentials and normalized in the ascribed way give a family of curves lying in the hatched region of Figure 11 .[49]

An analysis of $\ln i_\infty$ as function of $\ln(bJv^2)$ obtained from the initial (i-t^2 region) parts of the transients showed a linear dependence with a slope of 1/3 as required by Eq. (34). β has been found to be about $\beta = 0.72$. A $\ln i_\infty$ vs $1/\eta$ plot can be used to evaluate the exponential term in Jv (see Eq. (29)), and to estimate ε. The variation of v with η, being linear can be neglected in the logarithmic plot of i_∞. Experimentally a value of $\varepsilon = 2.5 \times 10^{-13}$ J cm^{-1} was found, in good agreement with values referred above.[35,40]

Despite the reported agreement with the theory, the steady-state parts of the transients, as seen from Figure 11 , lie in a range well below any of the theoretical curves. The most important fact, however, is that the first maximum of the experimental transients (Fig. 12) lies in a range slightly below the maximum of the current-time curve of the first layer formation, as calculated from Jv^2, obtained from the initial part of the same transient. This is obviously impossible within the model of homogeneous nucleation. One possible explanation is to assume that nucleation occurs on nucleation centers. There is much evidence that active sites play a role even in this very simple case of metal deposition.

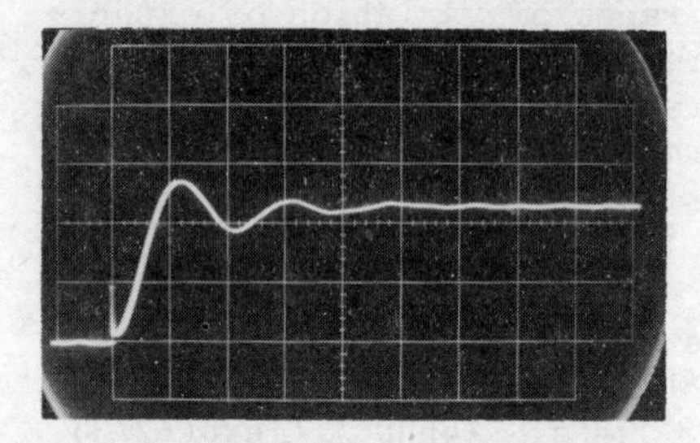

Fig. 12 : Experimental current transient for multinuclear multilayer growth. η = 14 mV. Vertical sensitivity : 2 μA/div. Time base : 5 ms/div.

The logarithmic scale analysis, used in the referred experiments, is sensitive to the energetics of the nucleation process. The good agreement between ε-values obtained in so many different ways shows that the energetics are only slightly affected by eventual changes of the state of the surface. Larger deviations were observed in the kinetic factor k_1, showing a strong influence of the state of the surface due mainly to active centers. This conclusion seems to be supported by the rather small differences of "fresh" and "aged" surfaces on the propagation rate, varying not more than by a factor of two.

IV. METAL DEPOSITION ON FACES WITH EMERGING SCREW-DISLOCATIONS

The most important factors for the process of growth in presence of screw-dislocations is the Burgers vector of the screw-dislocation and the rotation period T_r of the spiral of growth. At steady-state when the form of the spiral remains unchanged with time, the current density is given by

$$i = q_{mon} \nu T_r^{-1} \quad (35)$$

where ν is the height of the spiral step in monatomic layer units and is related to the value and direction of the Burgers vector. The reciprocal of the rotation period gives the frequency of generation of new layers. When more than one spiral is operating on the surface, the current is given by the sum of the contributions of the individual spirals. At steady-state T_r^{-1}, i.e., the normal growth rate in monolayer units, must be constant for all spirals. The contribution of each spiral is proportional to its area of operation. The average area of operation of each spiral, however, is inversely proportional to the number of the dislocations, so that at steady-state the current density becomes independent of the number of screw-dislocations intersecting the surface of the growing face.

Analytical evaluation of the form and period of rotation of a spiral produces some difficulties. Different approximations have been made for the isotropic case. BURTON, CABRERA and FRANK[2] found two approximations of the solution $r(\Theta)$ (r and Θ are the two-dimensional polar coordinates) describing the form of the stationary rotating spiral. The first approximation yields an Archimedian spiral with a radius of curvature at the center $\rho = \rho_c$. The step spacing is in this case uniform for the whole spiral $d = 4\pi\rho_c = 12.6\rho_c$. This result corresponds to a constant radial propagation rate, except for steps for which ρ is smaller than ρ_c. A second approximation derived by the same authors combines the solution for small r values, where the $v(\rho)$ relation is taken into account, with the former solution valid for large r values. This solution yields 19.9 for the factor connecting d and ρ_c, which was later corrected to 19.0 by CABRERA and LEVINE.[50]

The case of polygonized spirals has been discussed by CABRERA,[51] KAISCHEW et al.,[52] and by CHAPON and BONISSENT.[53] In all these treatments the step propagation rate was considered constant, independent of the step length, except for steps shorter than the side length l_c of the 2D nucleus at the given overpotential. Calculation of the period of rotation T_r and the step spacing $d = T_r v_\infty$ of the polygonized spiral is simplified in this case. For a quadratic spiral, the period T_r is 4 times the time lapse τ_c

needed for a new step at the center to reach the curb length l_c. Since the propagation rate was assumed constant and the form of the spiral orthogonal, the time lapse $\tau_c = l_c/v_\infty$, hence $d = 4l_c$ or $d = 8\rho_c$. The spiral ledge spacings are constant, independent of their distance from the center of the spiral. For a k-cornered spiral with $l_c = 2\rho_c \tan(\pi/k)$, the ledge spirals are given by $d = 4k\rho_c \sin^2(\pi/k)$.

The assumption $v = v_\infty$ for $l = l_c$ and $v = 0$ for $l > l_c$ is an oversimplification. It is obvious that the time lapse τ_c and hence the period of rotation should in reality be larger because the step adjacent to the new one propagates with a speed smaller than v_∞. The calculation is complicated because the propagation rate of all the adjacent steps are also l dependent so that the propagation rate of the spiral is given by a system of differential equations. A numerical calculation of this system was given by BUDEVSKI et al.[54] who found that, irrespective of the form of the spiral, the period of rotation of a polygonized spiral is

$$T_r = 19\rho_c/v_\infty \tag{36}$$

and hence from $d = T_r v_\infty$ with eq. (28)

$$d = 19\rho_c = 19\varepsilon/q_{mon}\eta_c \tag{37}$$

The situation becomes more complicated when two or more dislocations with their centers located at distances comparable to the site length l_c of the critical nucleus are operating simultaneously. It is beyond the scope of this lecture to go into details of the problem and the reader is advised to consult the review article of BURTON, CABRERA and FRANK.[2]

The step density L_s, important for the calculation of the current density according to Eq. (22) or (24), is given by (see Eq. (37))

$$L_s = d^{-1} = q_{mon}\eta_c/19\varepsilon \tag{38}$$

The spirals of growth, consisting of steps with a height of the order of the atomic radii, are unobservable with normal microscopic techniques. A pyramidal pattern is observed instead, with slopes of pyramids determined by the step height and the step distance :

$$\tan\alpha = \nu h_{mon}/d = \nu h_{mon} q_{mon} \eta_c / 19\varepsilon \; ; \qquad (39)$$

h_{mon} is the height of a monatomic layer on the respective face and ν the number of layers forming the spiral step, related to the Burgers vector of the dislocation. As seen from Eq. (39), the slope of a pyramid of growth is proportional to overvoltage.

In a number of electrocrystallization cases, pyramidal growth patterns as well as spirals of growth have been observed.[52,56,57,27] Generally, the relief of growth has a complex character. Simpler conditions are found only on a relaxed surface, where only few dislocations are operating. Figure 13a shows the quadratic pyramids of growth observed on cubic face.

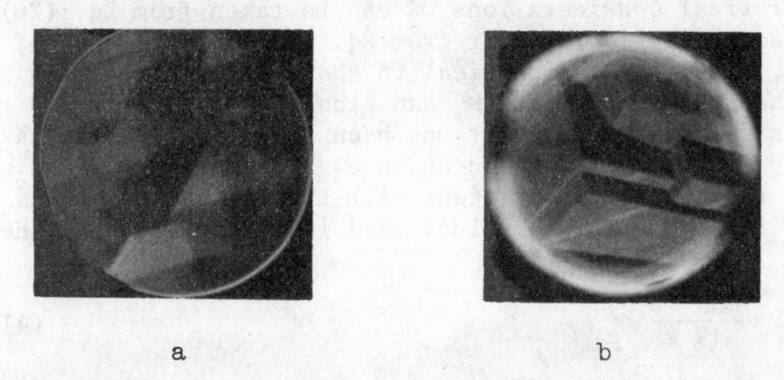

a b

Fig. 13 : Pyramids of growth on an Ag(100) face (a) and on an Ag(111) face (b). A triangular stripe is visible on each of the triangular pyramids on the (111) face (b) propagating down to the base of the pyramids, produced by a short change of the overpotential during growth at constant overpotential. The photograph is made one second after the pulse application and can be used for a direct estimation of the spiral step propagation rate.[58]

On an octahedral face the pyramids are triangular as shown in Figure 13b. The pyramids of growth are considered to be a macroscopic picture of the spirals of growth. The reasons for the latter statement are : (i) the slopes of the pyramids depend on overpotential (see Figure 13b); (ii) the pyramids appear always on the same sites of the face. For anodic dissolution, an etch pit is produced at each of these sites.

IV.1. Current Density and Morphology of Growth

On surfaces with a higher degree of perfection (e.g., Figure

13), the growth patterns are relatively stable. The pyramids have equal and uniform slopes. The step density defined by η is also uniform. The current density as function of the overpotential is then easily obtained by insertion of Eq. (38) into Eq. (22),(23) or (24). For small η_c values, the exponential functions $V(\eta)$ and λ_s can be linearized giving a quadratic dependence of i on η_c.

From an experimental point of view it is better to represent the current density in terms of the propagation rate v_∞ which under certain circumstances is experimentally directly accessible. From Eqs. (35) and (36) with Eq. (28) one obtains :

$$i = \frac{\nu q_{mon}^2}{19\ \varepsilon}\ v_\infty \eta_c \tag{40}$$

For theoretical considerations v_∞ can be taken from Eq. (20) for the surface diffusion case or from Eq. (13) for the case of direct transfer, with results identical to those of the preceding paragraph. The propagation rate v_∞ can also be obtained from direct experimental measurements. It has been shown that in the case of silver, v_∞ is linearly dependent on overpotential : $v_\infty = x_v \eta_c$. This is, incidentaly, consistent with theoretical equations (20) and (13) for the small η_c values used in the experiments. Hence

$$i = \frac{\nu q_{mon}^2 x_v}{19\ \varepsilon}\ \eta_c^2 \tag{41}$$

This quadratic overvoltage relation has been experimentally investigated for silver deposition on (100) faces of Ag crystals by VITANOV et al.[34] From the experimental $i - \eta_c^2$ plot, shown in Figure (14),[29] it was found that $\varepsilon = 2.4 \times 10^{13}$ J cm^{-1}.

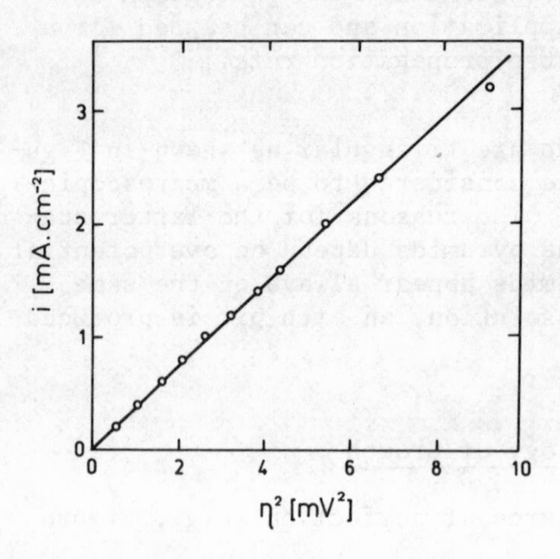

Fig. 14 : Steady-state current-overvoltage dependence in an i vs. η^2 plot on an Ag(100) face in 6M $AgNO_3$ at 45°C. Screw-dislocation density $1 \div 5 \times 10^{-4}$ disl. cm^{-2}.[29] Overvoltage is iR corrected.

The spiral steps have been assumed as monatomic (i.e., $\nu = 1$) and x_v has been taken from propagation rate measurements of single monatomic steps (see section III) or from the spiral step propagation rate directly as obtained from experiments illustrated in Figure (13).

The experiments described are a good quantitative verification of the theory of Burton, Cabrera and Frank.

Verification of the theory of spiral growth makes it possible to grow crystal faces at a given overvoltage and, after a regular surface profile is obtained, to stop the growth and calculate the step density from Eq. (38). In addition to the technique for preparation of dislocation-free, atomically smooth faces, the described technique for preparation of electrode surfaces of known profile is important for the investigation of electrochemical phenomena on solid surfaces.

An instructive example is the exchange rate measurements on faces prepared by growth at different overpotentials.

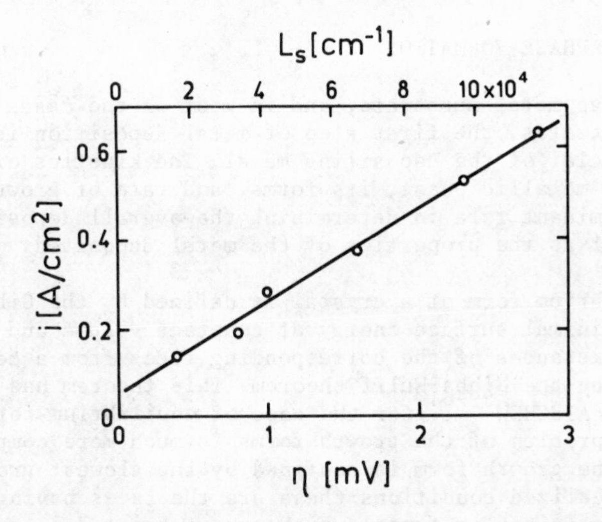

Fig. 15 : Dependence of the exchange current density on step density on an Ag(100) face intersected by screw-dislocations (ca. 5×10^{-4} disl. cm^{-2}). Before each series of impedance measurements for the evaluation of the exchange current density, the face was grown at the indicated overpotential (iR-corrected) until a steady-state profile was obtained. Step densities L_s calculated from the overpotential of growth according to Eq.(38).[33,34,29]

It was found by VITANOV et al.[34] that the exchange current density is strongly influenced by the surface profile and is proportional to step density, as shown in Figure (15). The step density was calculated from the overvoltage at which the faces were grown and Eq. (38). As seen from Figure 15, the overall exchange current density depends linearly on step density

$L_s : i_o = i_{o,ad} + i_{o,st}^{(T)} L_s$, giving for the extrapolated (for $L_s = 0$) value an exchange current density equal to that obtained on dislocation-free faces : $i_{o,ad} = 0.06$ A cm^{-2}.

From the slope of the $i_o - L_s$ dependence, the value of $i_{o,st}^{(L)} = 5.2 \times 10^{-6}$ A cm^{-1} has been found. This value can be used for the evaluation of the exchange current density of step atoms assuming that the observed catalytic effect of the step is extended to a distance not higher than one atomic diameter. With $d_{Ag} = 2.88$ Å, a value of $i_{o,st} = 180$ A cm^{-2} can be found, corresponding perfectly to the value obtained from the propagation rate measurements described in the section III.

V. ELECTROLYTIC PHASE FORMATION

On a foreign metal substrate, and in most of the cases of like-metal substrates, the first step of metal deposition is the formation of nuclei of the depositing metal. The kinetics of nucleation of the new metallic phase, its forms, and rate of growth, often have a dominant role in determining the overall deposition kinetics as well as the properties of the metal deposited.

The equilibrium form of a crystal is defined by the Gibbs-Curie condition for minimal surface energy at constant volume and is described by the distances of the corresponding faces from a central point as given by the Gibbs-Wulff theorem. This theorem has been generalized by KAISCHEW[59-61] for the case of equilibrium form on a substrate. The problem of the growth forms is much more complicated. Theoretically the growth form is confined by the slowest growing faces. Under idealized conditions these are the faces having a close packed structure (singular faces) which are restricted to grow over the 2D nucleation mechanism. Crystal defects, particularly screw-dislocations, can enhance the rate of growth of singular faces, while adsorption effects inhibit the growth of nonsingular faces. In electrolytic metal deposition, diffusion and electric field effects can additionally influence the forms of growth of crystals making prediction of the growth forms a difficult task at this stage of knowledge.

The formation of nuclei of a new phase can be considered as a sequence of bimolecular reactions in which every cluster of one class transforms into the next higher or lower by attachment or detachment of one atom :[62]

$$A_1 \underset{\longleftarrow}{\overset{\pm A_1}{\longrightarrow}} A_2 \underset{\longleftarrow}{\overset{\pm A_1}{\longrightarrow}} A_3 \ldots A_N \ldots \qquad (42)$$

At equilibrium, all rates of the attachment and detachment reactions are equal

$$\omega_a(N-1)\, Z^o(N-1) = \omega_d(N) Z^o(N) \qquad (43)$$

Where $\omega_a(N)$ and $\omega_d(N)$ $[s^{-1}]$ are the frequencies of the attachment/detachment of one atom to/from a cluster of N atoms. $Z^o(N)$ $[cm^{-2}]$ are the corresponding cluster concentrations at equilibrium. $Z(N = 0)$ corresponds to the number of adsorption sites Z_o of the substrate per unit area and $\omega_a(N = 0) \equiv \omega_a(0)$ to the impingement rate of atoms from the parent phase per adsorption site. If all $\omega_a(N)$ and $\omega_d(N)$ are known, the equilibrium cluster partition function $Z^o(N)$ can be calculated.

If an overpotential is applied, a net flux of clusters to higher classes is set. At steady-state, the flux

$$J = \omega_a(N-1)\, Z(N-1) - \omega_d(N)\, Z(N) \qquad N = 1,2,3,\ldots S$$

as well as the concentrations $Z(N)$ are constant. The BECKER and DÖRING[62] approach leads to an equation giving the nucleation rate as a function of the attachment/detachment frequencies (see also refs. 63 and 22) :

$$J = \omega_a(0) Z_o / \left[1 + \sum_{N=1}^{S-1} \prod_{i=1}^{N} \omega_d(i)/\omega_a(i)\right] \qquad (44)$$

Using the attachment/detachment frequencies as defined by KAISCHEW[3] (see Eqs. (1) and (2)) it is easy to show[64,65] that the products in the denominator contain the dissociation energies $\Phi_{i,s}$ and the overpotential η_c as an exponential term

$$\prod_{i=1}^{N} \frac{\omega_d(i)}{\omega_a(i-1)} = k_N \exp\left[(N\Phi_{k,s} - \sum_{i=1}^{N} \Phi_{i,s})/k_B T\right] \exp(-Nze_o\eta_c/k_B T) \qquad (45)$$

where k_N includes the product of the N preexponential constants from the ω_d/ω_a ratios, as well as an entropy term arising from the use of dissociation energies. The sum $\Sigma\Phi_{i,s}$ gives the dissociation energy of the cluster including its interaction with the substrate. The term $N\Phi_{k,s} - \Sigma\Phi_{i,s} = \Phi(N)$ represents the difference in the energy of N atoms of the bulk crystal and that of the N atoms in the cluster. This difference can be considered as the excess surface free energy of the cluster because it arises from the unsaturated bonds of the surface atoms.[66] A comparison with Eq. (25) shows then that the exponent of Eq. (45) is equal to the energy of formation of a cluster of size N :

$$\Delta G_N = -Nze_o\eta_c + (N\Phi_{k,s} - \Sigma\Phi_{i,s}) \tag{46}$$

With increasing N the sum $\Sigma\Phi_{i,s}$ tends to $N\Phi_{k,s}$ because more and more atoms are found in a kink position in the process of disintegration of the cluster. Therefore the ΔG_N value, being positive at small N values, rises initially, passes over a maximum and becomes negative only at larger N values. The size N^* of the cluster where ΔG_N reaches its maximum value ΔG^* is known as the critical cluster size because further growth is connected with a decrease of ΔG and proceeds spontaneously under the action of the overpotential η_c.

If the bond energies between the crystal atoms themselves and those between them and the substrate are known, the value of ΔG_N can be calculated as a function of N. In Figure 16, ΔG_N is calculated as by STOYANOV[65] as a function of N at different $\Delta\mu(= ze_o\eta_c)$ values; Φ_k and Φ_i are calculated from the respective bond energies between first neighbors ψ_1 for the cluster atoms and ψ_a for the cluster atom-substrate interaction (no solvation effects have been considered in this calculation). The procedure is complicated if a fixed crystal lattice arrangement of the atoms is imposed.[67] But this has little or no physical meaning because at room temperatures small atomic aggregates have an increased mobility, so that only close packed arrangements can be considered as in the referred calculation of Stoyanov. It is of interest to note that with changing overpotentials, N^*, being an integer, changes by discrete values only. Hence, a supersaturation interval of stability can be ascribed to each discrete value of N^*.

With increasing N, however, this atomistic method based primarily on lattice bond energies becomes quite intricate and can be replaced by classical thermodynamic methods where macroscopic quantities can reasonably be applied.

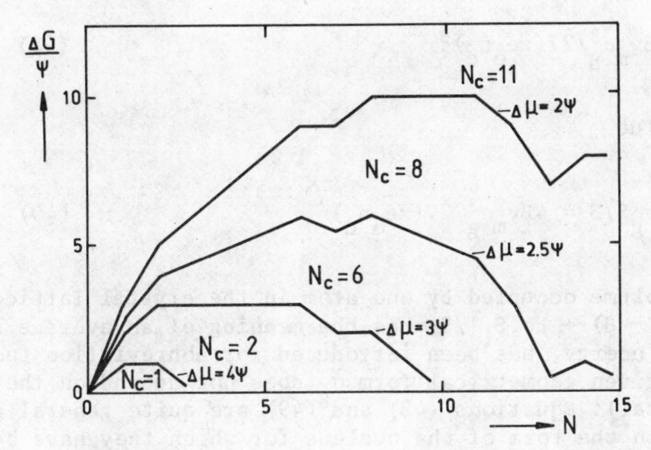

Fig. 16 : Gibbs free energy of formation of clusters as function of size N at different supersaturations $\Delta\mu$. Φ_k and Φ_i are calculated from ψ_1 and $\psi_a = \psi_1$ and the number of first neighbors of the cluster configuration of lowest ΔG value (five fold symmetry, close-packed structure) is given according to STOYANOV.[65]

As shown already, the term $N\Phi_{k,s} - \Sigma\Phi_{i,s}$ gives the excess free energy σS of the small crystal attributed to the unsaturated bonds of the surface atoms. If a crystalline cluster in contact with a substrate is considered, the different crystallographic faces and the interface boundary have different σ values, so that $\Phi(N) = \Sigma\sigma_i S_i$, or $\Phi(N) = S_j^*(\sigma_j - \beta) + \Sigma\sigma_i S_i$ in the case of heterogeneous nucleation. Here, S_j^* is the area of the crystal-substrate interface, σ_j the surface free energy of the crystal face j in contact with the substrate and β is the crystal-substrate adhesion energy.[59-61]

In this case ΔG_N can be given by

$$\Delta G_N = -Nze_o\eta_c + S_j^*(\sigma_j - \beta) + \Sigma\sigma_i S_i \tag{47}$$

The maximum value of ΔG_N can be obtained from $d\Delta G_N/dN = 0$. But differentiation of Eq. (47) with respect to N is possible only if a relation between S_i, S_j^* and N is found. Such a relation exists only if a specific geometrical form is considered. For any given geometrical form, the surface area S of the form is related to its volume by $S^3 = BV^2$, where B is a factor depending on the form, e.g., $G = 36\pi$ for a sphere and $B = 6^3$ for a cube. The maximum value of ΔG is found then at

$$N^* = 8Bv_m\sigma_g^3/27(ze_o\eta_c)^3 \tag{48}$$

with the value

$$G^* = \sigma_g S/3 = 4Bv_m^2\sigma_g^3/27(ze_o\eta_c)^2 \tag{49}$$

v_m is the volume occupied by one atom in the crystal lattice and $\sigma_g = [S_j^*(\sigma_j - \beta) + \Sigma\sigma_i S_i]/S$, has the meaning of an average specific surface energy, has been introduced for abbreviation (note that for a given geometrical form σ_g does not depend on the size of the crystal). Equations (48) and (49) are quite general and do not depend on the form of the nucleus for which they have been used. The only factors that have to be considered are the geometrical factor B and the mean specific surface energy σ_g, which vary from form to form changing the ΔG^* and N^* values accordingly.

The lowest value of ΔG^* of interest for the evaluation of the nucleation rate equation is that for which $\Phi(N)$ has a minimum, i.e., the ΔG^* value of the Gibbs-Wulff-Kaischew equilibrium form defined by $\sigma_i/h_i = (\sigma_j - \beta)/h_j$ = const, where h_i and h_j are the distances of the faces i and of the contact face j from the Wulff point respectively. The term $\sigma_g S/3$ in Eq. (49) can easily be obtained from this relation as function of σ_1/h_1 and the volume of the crystal $V^* = (1/3)\ \Sigma_{i,j} h_i S_i$. Then bearing in mind $V^* = v_m N^*$ and $S^3 = BV^{*2}$, one obtains, as shown by KAISCHEW,[59-61]

$$\sigma_i/h_i = (\sigma_j - \beta)/h_j = ze_o\eta_c/2v_m. \tag{50}$$

This is a general form of the well known Gibbs-Thomson equation relating the size of the critical nucleus and its equilibrium form to the overpotential.

According to KAISCHEW, the Gibbs free energy ΔG_c^* of nucleus formation in the heterogeneous case is related to ΔG_c for the nucleation in a homogeneous phase by

$$\Delta G^* = \Delta G_o^*\ V^*/V_o^* \tag{51}$$

where V_o^* and V^* are the volumes of the nucleus in the homogeneous phase and that in contact with the substrate. Equation (51) follows directly from Eqs. (48) and (49). It is self-evident that Eq. (51)

is valid only if clusters of the same geometrical form, e.g., the equilibrium form, are compared.

The linear dimensions h_i of the critical cluster are determined according to Eq. (50) by the overpotential. The presence of the substrate affects only the value of h_j, i.e., the distance of the Wulff point from the contact interface. The larger the value of the adhesion energy β is, the smaller is the thickness and hence the volume V_c^* of the critical cluster. If different orientations of the cluster with respect to the substrate are considered, this would mean that the orientation showing the highest crystal-substrate interaction β would have the smallest volume and according to Eq. (51) the lowest energy of formation. When the substrate is noncrystalline and does not exert orientational effects on the nucleus, the face with the most close-packed structure would correspond to this requirement. Or from the point of view of the energetics of the nucleation process on a nonstructured surface the most favorable orientation is that at which the nucleus is in contact with its most close-packed singular face with the substrate. As a matter of fact, no change of orientation of the nucleus can be expected going to higher overpotentials for energetical reasons only.

V.1. The Nucleation Rate Overpotential Relation.

The interpretation of the nucleation rate equation in terms of its overpotential dependence is rather illigible as represented in the form of Eq. (45). It has been shown in the preceeding section that the terms $\prod_{i=1} \omega_d(i)/\omega_a(i)$ contain the energy of formation of the cluster of class N and also that at a given overpotential ΔG_N has a maximum for the critically sized cluster. If all terms other than that of the critical cluster are neglected, Eq. (45) reduces to

$$J = k_1 \exp(-\Delta G^*/k_B T) \tag{52}$$

where k_1 includes constants which can be derived from Eq. (45) : $k_1 = k^* \omega_a^* Z_o$. Equation (52), known as the VOLMER-WEBER[20] equation, gives through ΔG^* the overvoltage dependence of the nucleation rate. In the derivation here, two sighificant approximations are made clear : (i) the contribution of all terms, other than that of the critical nucleus to the nucleation rate, is neglected and (ii) the existence of parallel reactions to the equilibrium form reaction has been disregarded.

For low overpotentials, where approximation (i) is rather poor, the size of the nucleus becomes larger and the sum in the denominator of Eq. (45) is composed of a large number of terms, it can be replaced by an integral. The integration, done by BECKER and DORING[62] for liquid droplets, and by KAISCHEW[22,68,69] for cubic crystals, leads to an expression of the form

$$J = \omega_a^* Z_o k^* \Gamma \exp(-\Delta G^*/k_B T) \qquad (53)$$

The comparison of this equation with Eq. (52) shows that the contribution of the omitted terms in the derivation of Eq. (52) can be accounted for by the introduction of the factor $\Gamma = (\Delta G_c/3\pi k_B T N_c^2)^{1/2}$ known as the Zeldovich nonequilibrium factor. The attachment frequency to the nucleus the critical size, ω_a^*, depends on the form of the nucleus and on the chosen mechanism of deposition (direct transfer, surface diffusion, etc.). Z_o is the number of sites per unit area where nucleation can start. For a homogeneous surface Z_o is given by the adsorption sites for adatoms and is of the order of $10^{15}cm^{-2}$. The factor k^* contains the vibrational frequencies of the cluster atoms in the positions i with respect to the frequency of vibration of the kink atoms[64] and can be assumed to be of the order unity. However, the inclusion of an entropy term, particularly when dissociation energies are used for the calculation of ΔG^*, makes the k^* value largely uncertain. It was emphasized that configurations other than equilibrium (Wulff-Kaischew) can contribute to the overall nucleation rate. Kaischew and Stoyanov have shown that at low overpotentials this contribution becomes appreciable.[70]

The uncertainty in Z_o, ω_a^*, k^*, and the contribution ratio of nonequilibrium configurations discredits to a large extent the attempt to quantitatively evaluate the nucleation rate. Nevertheless, better understanding of the factors determining the pre-exponential term in Eq. (52) has obviously been achieved. For more details, refer to the review articles of HIRTH and POUND[71] NEUGEBAUER[72] and TOSHEV.[73]

Experimentally, however, it seems that little progress has been made since the evaluation of the Volmer-Weber equation in 1926. The experimentalist is bound to use Eq. (52), disregarding the small dependence of k_1 from η_c (derived from the Zeldovich factor or the attachment frequency ω_a^*).

Introducing ΔG^* from Eqs. (51) and (49) into Eq. (52), the overpotential dependence of the nucleation rate can be obtained in the form

$$\log J = \log k_1 - k_2/\eta_c^2 \qquad (54)$$

where

$$k_2 = 4Bv_m^2 \sigma_g^3 (V^*/V_o^*)/27(ze_o)^2 k_B T \quad (55)$$

includes, through V^*/V_o^*, the cluster-substrate interaction. The J-η_c relation[54] shows a sharp rise in J above a critical value, below which the rate of nucleation is negligibly small. The shape of the curve is very similar to the ones discussed in section III and presented in Figure 2.

With increasing overpotentials, the number of atoms N^* forming the critical nucleus, reduces dramatically to only several atoms. Macroscopic quantities, such as volume, surface, surface energies, etc. lose their physical meaning in such cases, and the use of atomic theory becomes reasonable. The atomistic approach for calculation of nucleation rate dependence on supersaturation was first used by WALTON[74,75] and was developed later into a general nucleation theory by STOYANOV et al.[64,76]

For small clusters, STOYANOV[65,76] shows that all terms in the denominator of Eq. (44) can reasonably be neglected with respect to that of the critical nucleus where ΔG_N has its maximum. Further simplification results when only close packed structures or atomic configurations are considered, because of the high mobility of the atoms in small atomic aggregates. Equation (44) with Eqs. (45) and (46) gives

$$J = Z_o \omega_a^* k^* \exp\left[N^* ze_o \eta_c - (N^* \Phi_{k,s} - \sum^{N} \Phi_{j,s})/k_B T\right] \quad (56)$$

where the asterisk denotes the values ω_a, k and N for the critically sized nuclei. The attachment frequency of an atom to the critical cluster, proportional to the cathodic component of the exchange current, is given by : $\omega_{a,c} \simeq \exp(1 - \beta) ze_o \eta_c / k_B T$. Equation (56) can then be rewritten

$$\log J = A(N^*) + \left[N^* + (1 - \beta)\right] ze_o \eta_c / k_B T \quad (57)$$

where $A(N^*)$ includes all constants in the pre-exponential term of Eq. (56) as well as the N^* dependent term $-(N^* \Phi_{k,s} - \sum^{N} \Phi_{i,s})$. In Eq. (57), N^* is overpotential dependent, but clearly can take only discrete values. As seen from Figure 16, an overpotential interval of stability can be found for any discrete value of N^*. Therefore, while the second term in Eq. (57) changes linearly with η_c with a slope depending on N^*, the first term (giving the intercept of the log J - η_c relation) takes on only discrete values depending on N^*. Consequently, the log J - η_c dependence at high overpotentials is composed of a series of linear sections with slopes changing with the size of the cluster passing from one overpotential interval to the other.

V.2. Comparison with Experimental Data

Several techniques have been developed to investigate the rate of nucleation. Earlier methods were based on galvanostatic techniques. The first to use this technique were SAMARTZEV and EVSTROPIEV.[77] Later it was refined in experiments of SCHOTTKY[78] and GUTZOV,[79] but because of the high sensitivity of the nucleation on overpotential, the results are difficult to interpret. Potentiostatic techniques give better results. The simplest potentiostatic technique, used by KAISCHEW, SCHELUDKO and BLIZNAKOV[80] and by SCHELUDKO and BLIZNAKOV,[81] is to form an overpotential pulse on the electrode and to measure the time that a current begins to flow. The technique allows the estimation of the average time τ_1 needed for the formation of the first nucleus. Determination of τ_1 by J^{-1} involves difficulties connected with non-stationary processes, discussed in detail by TOSCHEW.[73]

The most straightforward test of the nucleation rate equation is to plot the number of nuclei per unit area as a function of time at different overpotentials. The double pulse technique developed by SCHELUDKO and TODOROVA[28] was successfully used in a wide range of experiments.[82-85] The electrode is conditioned at the equilibrium potential of the deposited metal and a short overpotential pulse is applied. The nuclei formed during this first nucleation pulse are grown with a second, lower overpotential pulse. The amplitude of the second growth pulse is at an overpotential lower than the critical value so that no nucleation can proceed. The duration of this pulse is chosen so that nuclei formed during the first nucleation pulse can grow to a size visible under a microscope. Figure 17 shows Hg droplets formed on a spherical Pt single crystal electrode. The nuclei were formed with a cathodic pulse of 87 mV and a pulse duration of 1.5 msec and grown at 20 mV for about 3 sec until they became visible. The nuclei density is determined simply by counting the number of droplets for a given surface area of the electrode. The technique, being based on visual counting, allows determination of nucleation rate on different crystallographic zones of single crystal electrodes. In other words, it allows the discrimination of differently active zones on the electrode surface.[84,85]

A modification of the double pulse method is to record the current transient following the first pulse. From the slope of the i vs. $t^{1/2}$ plot, the number of the growing nuclei can be determined.

This technique has been successfully used by GUNAWARDENA, HILLS and McKEE[86] and GUNAWARDENA, HILLS and MONTENEGRO[87,88] for calculating the number of nuclei following a galvanostatic or a potentiostatic pulse. From an experimental point of view this modification is more easily performed especially where large number of nuclei have to be counted. However, this method, based on additional assumptions and calculations, should always be carefully checked by the visual technique described above.

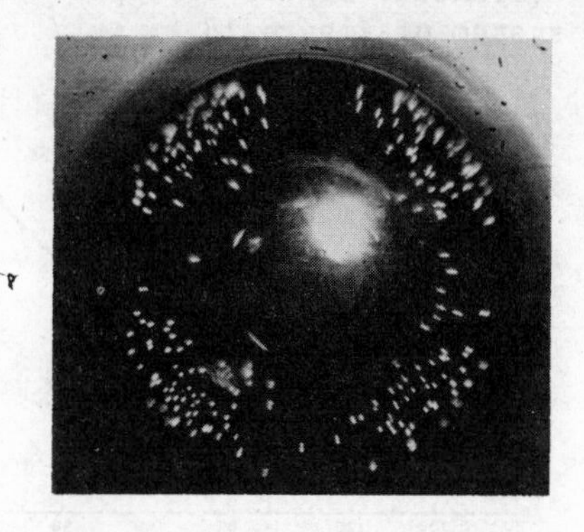

Fig. 17 : Hg droplets formed on a spherical Pt single crystal electrode. Nuclei formed at $\eta_{nucl.}$ = 87 mV, $\tau_{nucl.}$ = 15ms, η_{growth} = 20 mV, τ_{growth} = 3 sec.[86]

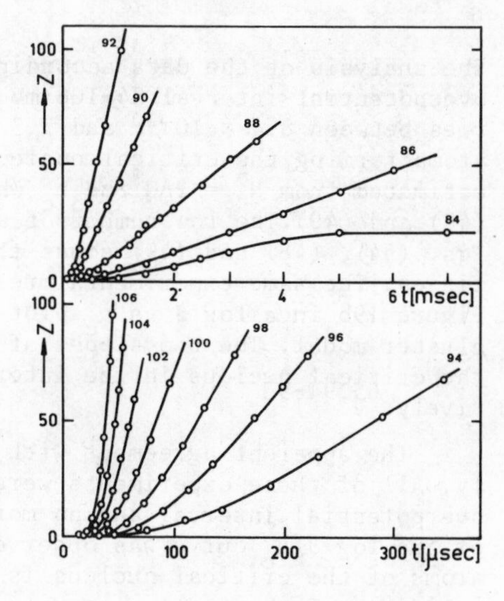

Fig. 18 : Experimental plots of the number of nuclei vs. time in the electrodeposition of mercury on Pt after TOSHEV and MARKOV.[89] The figures denote the corresponding nucleation overvoltage.

Figure 18 shows experimental plots of the number of nuclei vs. time (i.e., vs. the nucleation pulse duration) at different overpotentials in the electrodeposition of mercury on platinum.[89] The steady-state nucleation rate (dN/dt = const) is clearly set only after an induction period. This induction period has been treated in terms of nonstationary effects associated with the readjustment of the surface to the new overpotential conditions.

A detailed mathematical analysis based on the Zeldovich-Frenkel formulation of the nucleation kinetics was given by KASHCHIEW[90] and TOSHEV.[73]

From the linear part of the N/t curves (Figure (18)), the steady-state nucleation rate J for different overpotentials can be evaluated. Figure 19 represents a typical J vs. $1/\eta_c$ plot for the system of Figure 18.

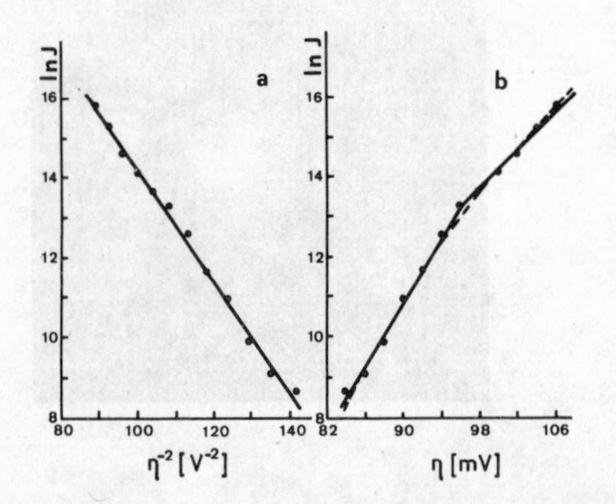

Fig. 19 : Experimental steady-state nucleation rate J values calculated from the data of Figure 18; (a) in an ln J vs $1/\eta^2$ plot and (b) in an ln J vs η plot.[91]

The analysis of the data according to Eq. (54) shows that in the overpotential interval 84-106 mV studied, the value of ΔG^* varies between 8.4×10^{-20} and 5.3×10^{-20} J, while the number of atoms forming the critical nucleus varies between 13 and 6 (N^* is estimated from $N^* = 2\Delta G^*/ze_o\eta_c$ which follows directly from Eqs. (48) and (49). The low number of atoms does not permit the use of Eqs. (54), (48) and (49) since they are based on macroscopic quantities. The same experimental results are represented again in Figure 19b in a log J vs η_c plot according to Eq. (57) of the small cluster model. The two slopes of the plot give 10 and 6 atoms for the critical nucleus in the intervals 84-96 and 96-106 mV respectively.[65,91-93]

The apparent agreement with theory should be treated cautiously. All of these experiments were done over a relatively small overpotential interval and no more than one cusp, i.e., two slopes, of the log J/η_c curve was observed. Determination of the number of atoms of the critical nucleus is not a good criterion for the validity of the small cluster model equation. Furthermore, it does not lead to any conclusion about the free energy of nucleation or forces of interaction among the cluster atoms or between them and the substrate, which would be the ultimate goal of a nucleation rate study.

Experimental Data	Ref.	Classical Model			Atomistic Model	
		η/V	ΔG_c erg	N_c	η V	N_c
Ag/Pt(100)	(88)	0.200 - 0.280	$(3.7 - 1.9) \times 10^{-13}$	2 - 1	0.200 - 0.240 0.240 - 0.280	1 0
Ag/Pt(111)	(88)	0.200 - 0.250	$(4.8 - 3.1) \times 10^{-13}$	3 - 2	0.200 - 0.220 0.220 - 0.250	3 1
Hg/Pt	(88)	0.090 - 0.102	$(10.1 - 7.9) \times 10^{-13}$	14 - 10	0.090 - 0.102	11
Hg/Pt	(95)	0.084 - 0.106	$(8.4 - 5.3) \times 10^{-13}$	13 - 6	0.084 - 0.096 0.096 - 0.106	10 6
Ag/Pt	(100)	0.170 - 0.210	$(8.0 - 5.2) \times 10^{-13}$	6 - 3	0.170 - 0.193 0.193 - 0.210	5 2
Cu/Pt	(101)	0.047 - 0.080	$(2.7 - 0.9) \times 10^{-13}$	7 - 1	0.047 - 0.051 0.053) 0.080	14 2

Table 1.
Gibbs free energies of 3D nucleation ΔG_c and number of atoms N_c forming the critical nucleus in the deposition of different metals on Pt as calculated from the classical and the atomistic model.[91] η indicates the overpotential interval used in the calculation.

Nevertheless, the comparison of theory to a large number of experiments by different authors leads to the important conclusion that log J vs $1/\eta_c^2$ plots give reasonable results with respect to the number of atoms and free energy of formation of the critical nucleus. Table I, collected by MILCHEV and STOYANOV[65,91] shows results treated according to both models. The number of atoms of the critical cluster in all cases is far too low to make a classical treatment acceptable, but coincides well in both treatments. A comparison of the atomistic with the classical treatment is given in Figure (19b). The dashed line represents the straight line of the log J vs $1/\eta_c^2$ plot of Figure (19a) obtained by the least square method and plotted in the coordinates of Figure (19b). The closeness of both curves, in most of the cases, suggests that both the classical equation, Eq. (54), and the atomistic equation, Eq. (57), coincide in the high overpotential region. The atomistic model better represents the experimental results with discrete lines, as required by Eq. (57). By lowering the overpotential as the η_c intervals of constant cluster size N_c become narrower, the experimental results are expected to be represented better by the classical equation, Eq. (54). How far the values of ΔG_c and the quantity $\sigma^3 V_g^* / V_c$ obtained from the classical plot at higher overpotentials can be trusted is still an open question. A detailed analysis of this problem was recently done by MILCHEV and VASSILEVA.[93]

V. CONCLUSIONS AND OUTLOOK

The basic principles of electrocrystallization seem to be well established. Significant progress has been made in the last two decades forward an understanding of the energetics of the nucleation and growth processes. The kinetics of growth has been well developed theoretically and confirmed experimentally in isolated cases. Further development of concepts of the state and behavior of the crystal-solution interface as well as in the understanding of the activation and inhibition of the nucleation and growth processes is needed.

Three dimensional nucleation theory is still at a very early stage of development. Significant progress has evidently been made by the introduction of the atomistic theory. We are obviously still far from an adequate understanding of the processes of nucleation, cluster orientation, grain growth interaction, dendrite formation and properties of bulk deposits.

ACKNOWLEDGEMENT.

The author is indebted for many helpful discussions to Dr. A. Milchev, Dr. W. Bostanov, Dr. G. Staykov, and Dr. Z. Stoinov as well as to Mrs. M. Kermekchieva and Mr. P. Yankulov for the technical help in the preparation of the manuscript.

REFERENCES

1 . F.C. FRANK, in *Growth and Perfection of Crystals*, Doremus et al. eds., J. Wiley, New York, 1958, p. 3.
2 . W. BURTON, N. CABRERA and F. FRANK, Phil. Trans. A 243(1958) 299
3 . R. KAISCHEW, Annuaire Univ. Sofia, Fac. Physicomathem. (Chimie) 42(1945/1946)109
4 . A.A. CHERNOV, Ann. Rev. Materials Science, 3(1973)397
5 . B. CONWAY and J. O'M. BOCKRIS, Proc. Royal Soc., 248(1958)394; Electrochim. Acta, 3(1960)340
6 . J.O'M BOCKRIS and A. DAMJANOVIC, in *Modern Aspects of Electrochemistry*, Bockris and Conway eds., Butterworts, London, 1964, p. 224
7 . W. LORENZ, Z. Naturforsch., 9a (1954) 716
8 . A. DESPIC and J. O'M. BOCKRIS, J. Chem. Phys., 32(1960)389
9 . A. DAMJANOVIC and J.O'M BOCKRIS, J. Electrochem. Soc. 110(1963) 1035
10. K. VETTER, in *Electrochemical Kinetics*, Acad. Press, N. Y., 1967, p. 283
11. M. FLEISCHMANN and R. THIRSK, Electrochim. Acta. 2(1960)22
12. D.A. VERMILYEA, J.Chem. Phys. 25(1956)1254
13. W. MEHL and J. O'M BOCKRIS, J. Chem. Phys. 27(1957)817
14. W. MEHL and J. O'M BOCKRIS, Can. J. Chem. 37(1959)190
15. E. BUDEVSKI and Z. STOINOV, Izv. Inst. Phys. Chem., Bulg. Acad. Sci. 4(1964)35
16. CH. SCHNITTLER, Thesis, Techn. Hoschschule Ilmenau, GDR (1969)
17. M. VOLMER, in *Actualites Scientifiques et Industrielles*, Vol. 85, Herrmann Verl., Paris, 1933, pp. 1-12
18. N. MOTT and R. WOTTS-TOBIN, Electrochim. Acta, 4(1961)79
19. J. W. GIBBS in *Collected Works*, Vol. 1, Yale Univ. Press, New Haven (1948)
20. M. VOLMER and A. WEBER, Z. Phys. Chem. 119(1926)277
21. H. BRANDES, Z. Phys. Chem., 126(1927)196
22. R. KAISCHEW, Acta Phys. Acad. Sci. Hung. 8(1957)75
23. T. ERDEY-GRUZ and M. VOLMER, Z. Phys. Chem. 157(1931)165
24. I. P. HIRTH, Acta Met. 7(1959)755
25. E. BUDEVSKI, V. BOSTANOV, T. VITANOV, Z. STOYNOV, A. KOTZEVA, R. KAISCHEW, Electrochim. Acta 11(1966)1697.
26. E. BUDEVSKI, V. BOSTANOV, T. VITANOV, Z. STOYNOV, A. KOTZEVA, R. KAISCHEW, Phys, Stat. Solidi, 13(1966)577
27. R. KAISCHEW and E. BUDEVSKI, Contemp. Phys. 8(1967)489
28. A. SCHELUDKO and M. TODOROVA, Bull. Acad. Sci. Bulg. (Phys.) 3(1952)61
29. E. BUDEVSKI, in *Progress in Surface and Membrane Science*, Vol. 11, Acad. Press, New York, 1976, p. 71
30. S. TOSHEV, A. MILCHEV and S. STOYANOV, J. Cryst. Growth, 13/14 (1972)123

31. V. BOSTANOV, R. ROUSSINOVA and E. BUDEVSKI, Chem. Ing. Techn., 45(1973)179 (-22)
32. V. BOSTANOV, G. STAYKOV, D. K. ROE, J. Electrochem. Soc. 122 (1975)1301 (-22)
33. T. VITANOV, E. SEVASTIANOV, V. BOSTANOV and E. BUDEVSKI, Elektrokhimiya 5(1969)451
34. T. VITANOV, A. POPOV and E. BUDEVSKI, J. Electrochem. Soc., 121(1974)207
35. V. BOSTANOV, J. Cryst. Growth 42(1977)194
36. T. VITANOV, E. S. SEVASTIANOV, V. BOSTANOV and E. BUDEVSKI, Elektrokhimia (russ.) 5(1969)451
37. A.E. NIELSEN, in *Kinetics of Precipitation*, Pergamon Press, N. Y., 1964, p. 40
38. A.A. CHERNOV and B. J. LYUBOV, Cryst. Growth, USSR, Moscow, Nauka, 5(1965)11
39. W. B. HILLIG, Acta Met., 14(1966)1868
40. E. BUDEVSKI, V. BOSTANOV and G. STAYKOV, Ann. Rev. Materials Science, Vol. 10(1980), in press.
41. A. BEWICK, M. FLEISCHMANN and H. R. THIRSK, Trans. Faraday Soc. 58(1962)2200
42. M. FLEISCHMANN and H. R. THIRSK, in *Advances in Electrochemistry and Electrochemical Engineering*, ed. P. Delahay, Vol. 3 J. Wiley, N. Y.,1963, p. 123
43. G. STAIKOV, V. BOSTANOV and E. BUDEVSKI, Electrochim. Acta, 22(1977)1245
44. L.A. BOROVINSKII and A.N. ZINDERGOSEN, Dokl. Acad. Nauk,USSR, 183(1968)1302
45. R.D. ARMSTRONG and I.A. HARRISON, J. Electrochem. Soc. 116 (1969)328
46. S.K. RANGARAJAN, J. Electroanal. Chem. 46(1973)125.
47. U. BERTOCCI, J. Electrochem. Soc. 119(1972)822
48. G. GILMER, J. Cryst. Growth, in press
49. W. BOSTANOV, W. OBRETENOV, G. ŠTAIKOV, D. ROE and E. BUDEVSKI in preparation for J. Cryst. Growth.
50. N. CABRERA and M. M. LEVINE, Phil. Mag. 1(1956)450
51. N. CABRERA, in *Structure and Properties of Solid Surfaces*, Chicago Univ. Pres, 1953, Chicago, p. 295
52. R. KAISCHEW, E. BUDEVSKI and J. MALINOVSKI, Z. Phys. Chem. 204 (1955)348
53. C. CHAPON and A. BONISSENT, J. Cryst. Growth 18(1973)103
54. E. BUDEVSKI, G. STAIKOV, V. BOSTANOV, J. Cryst. Growth, 29 (1975)316
55. BOSTANOV V., E. BUDEVSKI and G. STAIKOV, Faraday Symp. of the Chem. Soc., Nr 12(1977)p . 83
56. H. SEITER, H. FISCHER and L. ALBERT, Electrochimica Acta, 2 (1960)167
57. E. BUDEVSKI, T. VITANOV and W. BOSTANOV, Phys. Stat. Sol.,8 (1965)369.
58. W. BOSTANOV, R. ROUSSINOVA, E. BUDEVSKI, Comm. Dept. Chem., Bulg. Acad. Sci. 2(1969)885

59. R. KAISCHEW, Bull. Acad. Bulg. Sc. (Phys) 1(1950)100
60. R. KAISCHEW, Bull. Acad. Bulg. Sc. (Phys) 2(1951)191
61. R. KAISCHEW, Arbeitstagung Festkörperphysik, Dresden 1952,p.82
62. R. BECKER and W. DÖRING, Ann. Phys. 24(1935)719
63. M. VOLMER, in *Kinetik der Phasenbildung*, Theodor Steinkopf Verl., Leipzig-Dresden, 1939
64. A. MILCHEV, S. STOYANOV and R. KAISCHEW, Thin Solid Films, 22 (1974)255,267
65. S. STOYANOV, in *Nucleation Theory for High and Low Supersaturations*, Current Topics in Materials Science, Vol. 3, E. Kaldis ed., North-Holland Publ. Co. 1978, Chapter 4.
66. I. N. STRANSKI, Ann. Univ. Sofia, Fac. Phys.-Math., Livre 2 (Chemie), 30(1936/37)367
67. B. LEWIS, Thin Solid Films 1(1967)85.
68. R. KAISCHEW, Z. Elektrochem. 61(1957)35
69. R. KAISCHEW, in *Festkorperphysik und Physik der Leuchtstoffe*, Acad. Verl. Berlin, 1958, p. 133
70. R. KAISCHEW and S. STOYANOV, Comm. Dept. Chem., Bulg. Acad. Sc. 2(1969)127
71. J.P.HIRTH and G.M. POUND, in *Condensation and Evaporation*, Mac Millan, N. Y., 1963
72. C. A.NEUGEBAUER, *Handbook of Thin Film Technology*, McGraw-Hill, 1970
73. S.TOSHEV, in *Homogeneous Nucleation in Crystal Growth : an Introduction*, Hartman ed. p. 1, North-Holland Publ. Co., 1973
74. D. WALTON, J. Chem. Phys. 37(1962)2182; Phil. Mag. 7(1962)1671
75. D. WALTON, T. RHODIN and R.W. ROLLINS, J. Chem. Phys. 38(1963) 2698
76. S. STOYANOV, Thin Solid Films, 18(1973)91
77. A.G. SAMARTZEV and K.S. EVSTROPIEV, Izvestia Acad. Sc. USSR, 603(1934)
78. W. SCHOTTKY, Z. Phys. Chem. (N.F.) 31(1962)40
79. I. GUTZOW, Bull. Inst. Chim. Phys., Acad. Bulg. Sc., 4(1964)69
80. R. KAISCHEW, A. SCHELUDKO and G. BLIZNAKOV, Bull. Acad. Bulg. Sc. (Phys) 1(1950)137
81. A. SCHELUDKO and G. BLIZNAKOV, Bull. Acad. Bulg. Sc.(Phys)2 (1951)227
82. R. KAISCHEW and B. MUTAFTSCHIEV, Electrochim. Acta. 10(1965) 643; Z. Phys. Chem. 204(1955)334
83. R. KAISCHEW, S. TOSCHEW and I. MARKOV, Comm. Dept. Chem., Bulg. Acad. Sc., 2(1969)463
84. R. KAISCHEW and B. MUTAFTSCHIEV, Bull. Acad. Bulg. Sc., 4(1954) 105; ibid 5(1955)77
85. S. TOSCHEW and B. MUTAFTSCHIEV, Electrochim. Acta, 9(1964)1203
86. G.A. GUNAWARDENA, G.J. HILLS and S. McKEE,J. Electroanal. Chem.
87. G.A. GUNAWARDENA, G.J. HILLS and J. MONTENEGRO, Electrochim. Acta.
88. G.A. GUNAWARDENA, G.J. HILLS and J. MONTENEGRO, Disc. Faraday Soc. 12(Southampton, 1977)90

89. S. TOSHEV and J. MARKOV, Ber. Bunsenges. Phys. Chem. 73(1969) 184
90. D. KASHCHIEV, Surface Sc. 14(1969)209
91. 1. MILCHEV and S. STOYANOV, J. Electroanal. Chem. 72(1976)33
92. A. MILCHEV, E. Vassileva and V. KERTOV, J. Electroanal. Chem. 107(1980)323
93. A. MILCHEV and E. VASSILEVA, J. Electroanal. Chem. 107(1980) 337.

RECRYSTALLIZATION

H. GLEITER

University of Saarbrücken
6600 Saarbrücken, West Germany.

I. INTRODUCTION

We shall be concerned in this lecture with the discontinuous recrystallization mechanisms by which pure metals and alloys repair the structural damage caused by mechanical deformation. For comprehensive reviews of the existing body of experimental observations, the theories proposed, and discussions of the open issues, the reader is refferred to the existing literature.[1,2] The recrystallization process involves the nucleation and growth of new crystals in a material with a high density of defects. Hence, attention has to be focussed primarily on two problems : (i) the formation of the new crystals and (ii) the growth process of the new crystals.

II. NUCLEATION OF NEW CRYSTALS

Any theory of discontinuous recrystallization must account for the experimental observation that discontinuous recrystallization always involves the formation of new crystals with a low defect density growing by the migration of an incoherent interface (high angle grain boundary) into the surrounding deformed lattice. Five models have been advanced to explain the nucleation of the new crystals.

a) *Classical nucleation* theory has been applied to recrystallization[3]. The new crystals are formed by fluctuations in position of a group of atoms sufficiently large to release strain energy larger than the interfacial energy created at the matrix/nucleus interface. For this model to be realistic, large local elastic strains (≈ 20%) have to be postulated.

B. Mutaftschiev (ed.), Interfacial Aspects of Phase Transformations, 605–619.

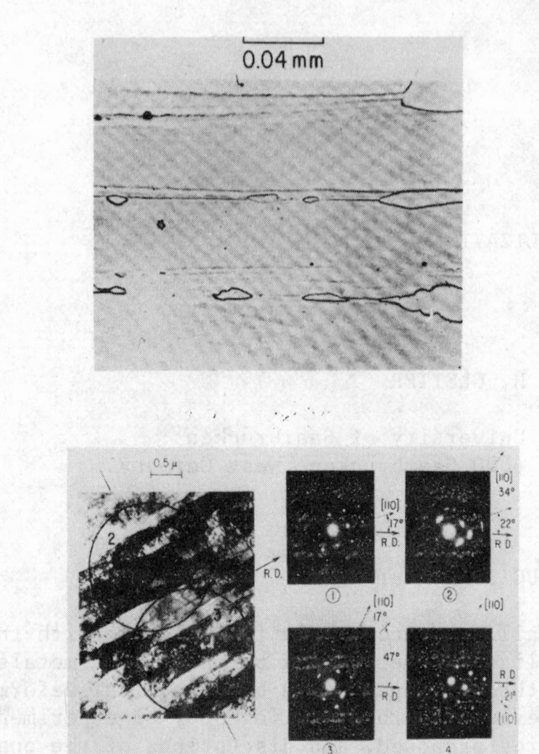

Fig. 1 : (a) Optical micrograph of a Fe 3 % Si crystal deformed by rolling and annealed for 25 min. at 600°C. In the deformation bands (horizontal lines) the recrystallization nuclei have formed first.[9] (b) Microstructure and diffraction information across a deformation band (Fe-3% Si annealed for 1280 min at 400°C). The subgrains are highly elongated in the rolling direction. The diffraction patterns 1 and 4 were taken on either side of the band. The total misorientation across the band is 38° in a distance of 4 μ.[9]

Homogeneous classical nucleation was shown to be impossible[4] due to the small driving force and the high energy of the matrix/nucleus interface. Heterogeneous nucleation is in principle possible on preexisting boundaries if the local strain energy is sufficiently high[5].

b) *The growth of polygonized subgrains*[6,8] either from sharply curved regions in a deformed crystal which polygonize most rapidly, or from a subgrain growing in a recovered cell structure (Figs. 1a,b).

In both cases the effective nucleus must differ substantially in orientation (10° to 15°) from the surrounding crystal in order to have a high mobility interface.

The nucleus may acquire the critical misorientation of 10° to 15°, either by subgrain coalescence[9] or by slow subgrain growth.[10,11] Evidence in favor of both processes has been presented. Apparently, the structure of the deformed material, the mobility and structure of the subboundary, the stacking fault energy, and the impurity content are some of the factors that control which mechanism is more likely.

c) The process of *strain-induced boundary migration*.[12,14] In this process the subgrain or subgrains of one grain adjacent to a pre-existing high-angle boundary grow by migration of that boundary into the neighboring grain (Figs. 2a,b).

d) The *martensitic mechanism*, the so-called "reverse Rowland" transformation.[15,17]

e) *Catalysed nucleation*.[18] Subgrains, capable of growth, do not grow until the potential nucleus is touched by a migrating boundary. This catalysed nucleation is assumed to be due to the release of internal stresses.

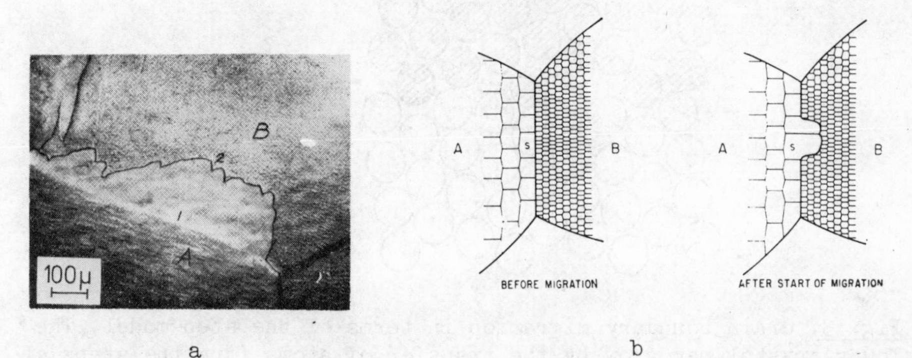

Fig. 2. (a) Strain induced boundary migration[64]; (b) Model for the strain induced boundary migration.[64]

The existing large body of experimental evidence (for a review see Ref. 19) seems to suggest that all nuclei exist within the deformed matrix. These nuclei require a high (10° to 15°) local misorientation. This is present within the material, for example at pre-existing grain boundaries, deformation bands, shear bands or generally within the deformed material if the deformation was severe. Probably, no single one of the above models is correct for all metals and all degrees of deformation. The model c) seems to be dominant at small strains. At higher strains, the formation of new grains by one of the other mechanisms seems to be the faster process.

III. CRYSTAL GROWTH

After new crystals have been nucleated in the deformed material, recrystallization proceeds by the growth of these newly formed crystals. The understanding of this growth process is primarily concerned with the atomistic mechanisms by which the nucleus/matrix interface migrates into the surrounding strained material. In principle, the growth process involves the migration of a high angle grain boundary. Therefore, attention will be focussed on this problem. Two models for the description of the atomistic processes involved in migration of such a boundary have been proposed.

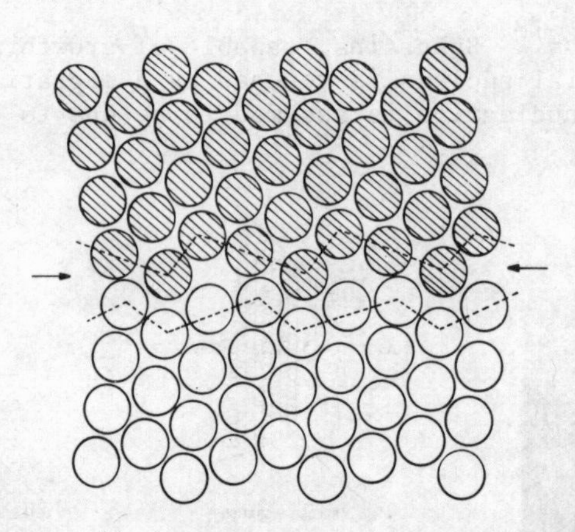

Fig. 3. Grain boundary migration in terms of the step model. The lower crystal may grow by the transfer of atoms form the steps at the surface of the upper crystal to the steps on the surface of the lower crystal or vice-versa.[20] The two "stepped surfaces" are indicated by broken lines.

In the step model of boundary migration,[20] grain boundary migration is envisioned as the transfer of atoms from atomic steps on the surface of the shrinking crystal to steps on the surface of the growing crystal (Fig. 3). The dislocation model of boundary migration[21] proposes that the migration of a grain boundary occurs by the climb of boundary dislocations (Fig. 4) and, thus, involves the transfer of atoms from one crystal to the adjacent crystal across the boundary.[21]

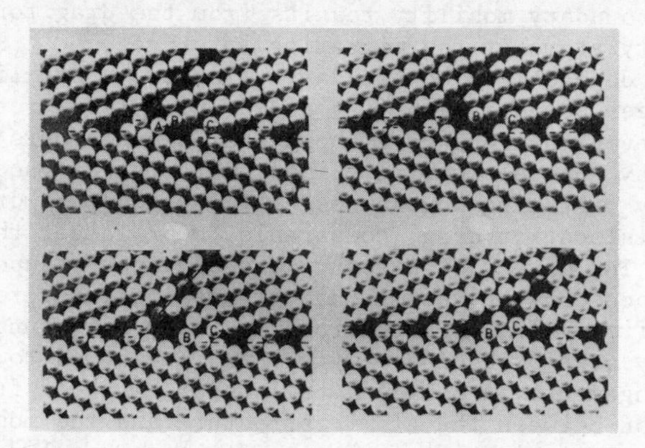

Fig. 4 : Ball model suggesting the kind of atom movements that may accompany grain boundary migration according to the dislocation model. The atom "A" is removed from the upper crystal by diffusion and the atom "B" by shear and a shuffle is transferred to the lower grain. The extra half plane of the dislocation is indicated by a line and the darker balls. The dashed line indicates the boundary position.[21]

Physically, both mechanisms are interrelated. In fact, they seem to be two descriptions in different terms of the same physical process. The apparent difficulty of the dislocation model for migration is the necessity for shear to accompany migration since the Burgers vector of the climbing dislocation in general has a component parallel to the boundary that causes boundary shear. Since this is not observed in most cases, the model requires the additional assumption that for the appropriate dislocation configuration the shears caused by migration sum to zero. If the boundary conformation[22] or the migration mechanism[23] remain constant, the migration rate of a boundary is linearly related to the difference in the Gibbs free energy ("driving force") between the strained matrix and the growing crystal. The proportionality factor is called the boundary mobility (m). The boundary mobility has been shown (in the case of dilute Cu-Cd-alloys) to be the same for recrystallization as for other driving forces[24]. Experimental studies of grain boundaries in different metals[25,28] (e.g. Au, Al, Cu, Ni, Pb, Cd, Bi, W, Z,) show that decreasing temperature and vacancy supersaturation as well as increasing pressure[27] and solute concentration result in lower boundary mobilities and, hence, smaller recrystallization rates. Special (low energy) boundaries are observed to have a higher mobility than random (high energy) boundaries provided the other conditions (temperature, solute concentration etc.) are the same. These findings have been rationalized in terms of the following processes.[25,26] The effect of solute atoms on the boundary mobility results from the drag forces exerted by the impurity atoms at the migrating interface.[28,30] Since the impurity drag depends on the boundary velocity, a non-linear relationship between boundary mobility and driving force results. However, recent investigations have shown that even in pure materials, drag forces may control mobility. The generation of vacancies and dislocation by migrating boundaries[31,33] may generate drag forces (vacancy or dislocation drag) comparable to or larger than impurity drag effects. The effect of pressure and of vacancy supersaturation ahead of the boundary on the migration rate was interpreted in terms of the free volume fluctuations in the grain boundary core. The elementary step of boundary migration is assumed to involve atomic rearrangements in boundary regions of enhanced free volume. The correlation between the atomic structure and the mobility of a boundary results primarily from two effects. In solid solutions, solute segregation (and, hence, solute drag) depends strongly on the boundary structure in the sense that high energy boundaries experience the maximum solute drag force. The second effect is due to the intrinsic boundary structure. Theoretical and experimental observations suggest high energy boundaries to have a larger free volume than low energy structures and, hence, the atomic rearrangements involved in the migration process may be easier in these interfaces.

IV. STRUCTURE OF MOVING BOUNDARIES

In the discussion so far (and in most papers on recrystallization) it was assumed that the structures of grain boundaries are similar either at rest or moving. However, there is now a body of evidence (enhanced dislocation absorption[34], enhanced boundary diffusion,[35,36] precipitate growth[37,38]) which do not agree with this assumption. Grain boundaries may have structures, either when moving or when static but at high temperatures, which are different from their static and low temperature counterparts. This view is also supported by recent observations of the discontinuous precipitation reaction at high and low energy boundaries[39]. However, for the present, it is difficult to derive a specific model for the structural changes that may occur when a boundary starts to migrate. In addition, there are also other factors in which a moving interface may differ from a static one - e.g. in defect concentration, dislocation configuration, or solute distribution.

V. REDUCTION OF DEFECT DENSITY DURING RECRYSTALLIZATION

Any theory of recrystallization must account for the following two facts. (i) The defect density in the recrystallized state is several orders of magnitude (typically 10^6-10^7 cm^{-2}) lower than in the deformed state. (ii) The density of certain defects (e.g. dislocations, coherent twin boundaries) in the recrystallized material is higher than the thermodynamically expected equilibrium density.

Reduction of the defect density during recrystallization may be understood in terms of the structural (non-equilibrium) defects in grain boundaries discussed in the chapter "On the Structure of Grain Boundaries in Metals". According to these results, two mechanisms of absorption of vacancies by a migrating grain boundary may be distinguished. In boundaries deviating little from a low energy structure, vacancy absorption occurs by the climb of (equilibrium) structural boundary dislocations. In high energy boundaries, vacancy absorption may occur by the delocalization of the vacancy, involving atomic shuffling processes in the sense that the free volume of a lattice vacancy entering the boundary is partly spread in the boundary and partly removed by a relaxational motion of the two crystals forming the boundary towards one another (cf. Figs. 19 and 20 of lecture on the structure of grain boundaries). The modelling of the absorption of lattice dislocations by grain boundaries is still not fully resolved. Three models have been advanced (cf. chapter "On the Structure of Grain Boundaries in Metals"). The *dissociation model* describes the incorporation of lattice dislocations into grain boundaries in terms of a dissociation reaction into (intrinsic) boundary dislocations.[40] In the *delocalization model*,[41] the incorporation process is proposed to depend on the boundary structure. In the case of high energy boundaries (with closely spaced intrinsic dislocations), the incorporation of a lattice dislocation in a boundary is envisaged to occur by widening of the dislocation core in the boundary plane. A dissociation description loses its physical significance due to the high dislocation density. In the *strain sharing model*,[43] the lattice dislocation entering the boundary repels the neighboring structural dislocations so that its strain field is effectively spread out in the boundary plane.

The non-equilibrium density of dislocations and twin boundaries in recrystallized materials has been explained in terms of growth accidents or boundary migration processes[43] during recrystallization (for a review cf. ref. 44). Recent investigations of the defect structure in crystals swept by a migrating boundary led to the conclusion that the crystal behind a migrating boundary contains dislocations[33] (Fig. 5) twin boundaries[44] (Fig. 6) and also vacancies[31,32] which are generated during the transfer process of atoms across a migrating boundary. The generation of these defects seems to have the following reason.[44]

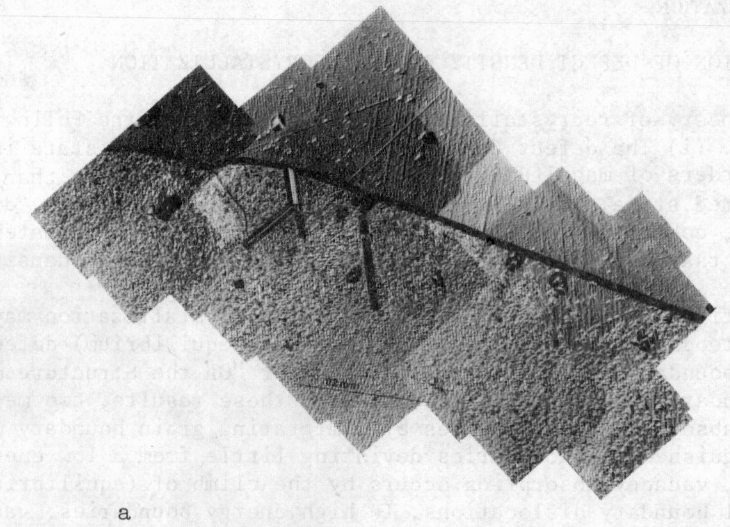

a

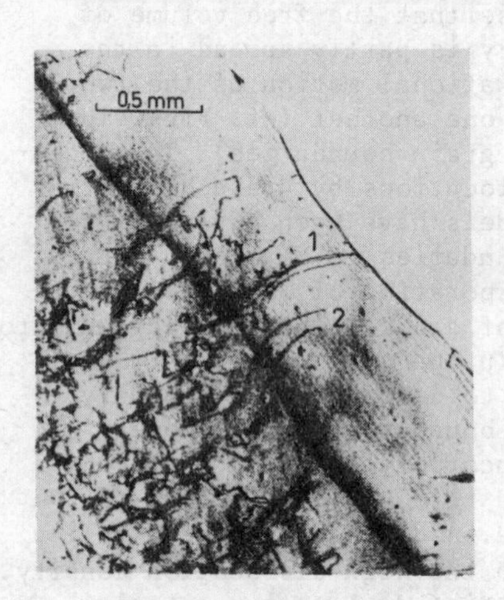

b

Fig. 5 : Micrograph (a) and synchroton X-ray topography (b) of dislocations generated during grain boundary migration. Fig. 5a shows the dislocations (revealed by etch pits) generated by a grain boundary moving through a pratically dislocation free single crystal of Si.[33] The boundary migrated in the upper direction. Fig. 5b shows the dislocations in a grain during recrystallisation of Al.[43]

If all atoms transferred across a recrystallization front (migrating boundary) would be perfectly incorporated into the lattice of the growing crystal, a defect free crystal would result. Howether, with increasing boundary migration velocity, an increasing fraction of atoms is not incorporated into perfect lattice sites (growth accidents). Depending on the position of the displaced atoms, a point defect, a dislocation, or a two-dimensional defect may result (cf. ref. 44). In addition to growth-accident formation, twin boundaries may also be formed by a dissociation reaction of a pre-existing boundary into a twin boundary plus a new high-angle interface[45] (Fig. 6). Recent observations by synchroton X-ray topography[43,46] suggest that stresses and spiral growth processes coupled with the solid state growth of a crystal result in the emission of dislocations from a migrating interface. Since these mechanisms are also likely to operate during recrystallization, the non-equilibrium density of dislocations observed in as recrystallized materials may be in part due to a process of this type.

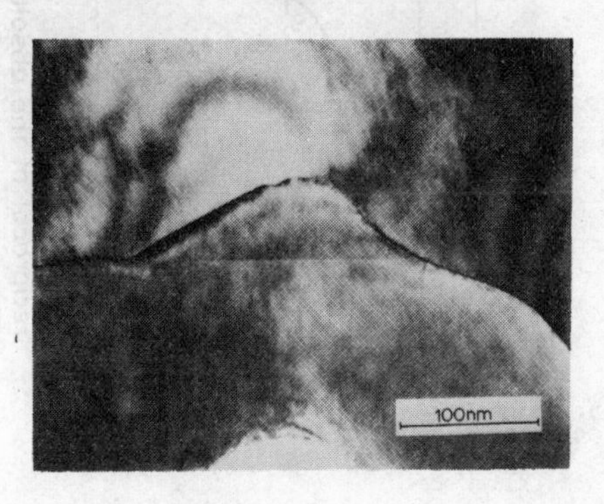

Fig. 6 : Electron micrograph of a grain boundary (53°<110>) that "decomposed" into a coherent twin boundary (straight horizontal line) and a new boundary (bulge)[45].

VI. RECRYSTALLIZATION OF MULTI-PHASE MATERIALS

Some of the effects involved in the recrystallization of materials with more than one phase will be outlined by considering the following three modes of recrystallization of multiphase materials.

(i) The second phase forms during recrystallization (of supersaturated solid solutions).

(ii) The second phase is present in the form of dispersed particles before the deformation occurs.

(iii) The second phase is present in the form of a duplex structure.

For a comprehensive review of this field (also covering other recrystallization modes), the reader is referred to refs. 47 - 50.

VI. 1. Recrystallization of Supersaturated Solid Solutions

The recrystallization kinetics of supersautrated solid solutions are controlled by the mutual influence of precipitation and recrystallization process as shown in Figs. (7a,b)[48]. On the basis of this diagram, three regimes of the recrystallization behavior may be distinguished.

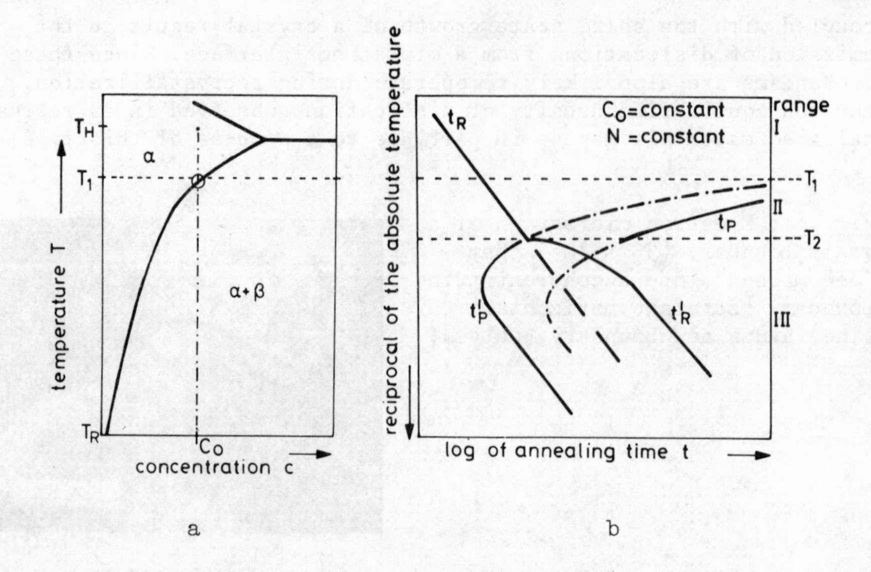

Fig. 7 : Recrystallization kinetics of supersaturated solid solutions. (a) Schematic phase diagram. An alloy of concentration c_o is homogenized at the temperature T_H subsequently quenched to T_R and plastically deformed in the supersaturated state. (b) Temperature dependence of the start of precipitation and recrystallization. t_v start of precipitaiton in the undeformed alloy, t'_p at dislocations, t_R start of recrystallization in a solid solution t'_p same process influenced by precipitated particles [48].

$T > T_1$: recrystallization is affected by segregation only and no precipitation occurs. $T_1 > T > T_2$: same as in the first regime. However, precipitation occurs after completion of recrystallization. $T < T_2$: precipitates influence the rearrangement of defects in the deformed matrix and the kinetics of the recrystallization process. During recrystallization, the volume fraction of precipitates grows. This may result in a slow-down of recrystallization. If the retarding force excerted by the precipitates and solute atoms on the migrating boundary exceeds the free energy difference between the deformed and recrystallized material, the discontinuous recrystallization process comes to an end. The various types of recrystallization diagram (Fig. 7) have been experimentally confirmed for different supersaturated deformed solid solutions [48,49].

VI. 2. Influences of Dispersed Particles on Recrystallization.

Experimental observations on the acceleration and retardation of recrystallization by second phase (non-deformable) particles suggest the following four parameters are crucial for recrystallization kinetics of those materials.

Interparticle spacing. Closely spaced particles may retard or inhibit recrystallization whereas widely spaced particles accelerate the process.[50]

Particle size. The dependence on particle size is less well established. For alloys with large interparticle spacing showing accelerated recrystallization, the recrystallization time (and temperature) is reduced as the particle size increases.[50,52,53] In general, the particle size and spacing varies during recrystallization due to coarsening processes which may result either from particle drag by the migrating boundary (Fig. 8) (for a review we refer to ref. 59) or from dissolution and re-precipitation of particles.[36,60]

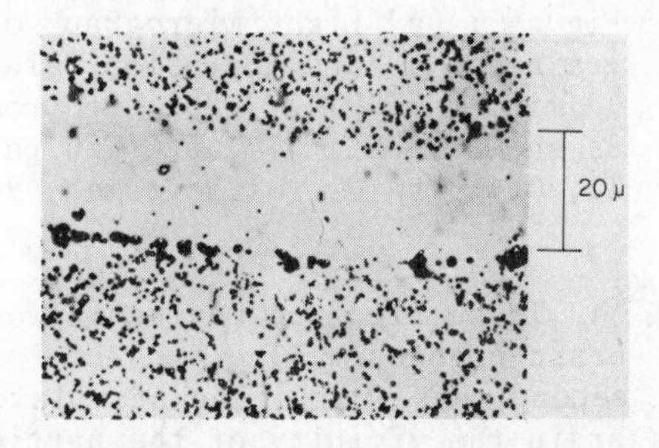

Fig. 8 : Micrograph of a particle free band behind a boundary that migrated toward the lower edge of the figure. The particles in the band were dragged along by the moving interface.[59]

Particle strength. Deformable particles and gas bubbles (pores) generally retard recrystallization[54,55] whereas the inhomogeneous strain in the vicinity of non-deformable (or fractured) hard particles promotes recrystallization,[56] except at very high strains where the refinement of dispersion becomes so large that retardation effects similar to deformable particles dominate[51]

The effect of particles on the *deformed state* may be summarized as follows. At small strains ($\varepsilon < 0.3$) widely spaced particles (diameter > 0.2 μ) result in large local lattice rotations which join up to form deformation zones as the particle distance decreases. For small (diameter < 0.1 μ) particles, the dislocation density is higher, the cell structure more diffuse, and the cell size is smaller than for a particle free material under comparable conditions. Stored energy measurements for both types of materials[57] indicate that except at lowest strains, the major influence of the particles on the deformation structure is on the dislocation distribution rather than on the total dislocation density.[47]

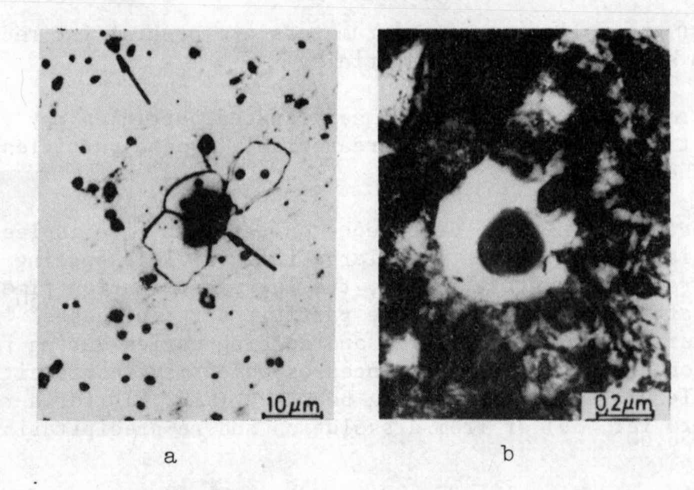

Fig. 9 : (a) Light micrograph of the formation of recrystallized grains at a hard inclusion (slag) (Fe 0.23% C, 50% deformed recryst. 42 min at 550°C).[51] (b) Electron micrograph of the same process, as shown in Fig. 9a at a TiC particle in a HSLA-steel with 0.26 wt% Ti, 0.14 wt% C, 90% deformed, recryst. 650 h at 550°C.[51]

These findings may be rationalized in terms of the following basic processes. The presence of large lattice rotations at (hard) second phase particles stimulates the nucleation of new grains at or in the vicinity of the particles as observed in several alloys (Fig. 9). Although accelerated nucleation kinetics are expected in the presence of large hard particles, the overall recrystallization kinetics depend on the particle spacing. At large particle spacings,the growing nuclei interfer little with the particles and, hence, an overall accelerated recrystallization is observed. However, if the average interparticle spacing is very small, the particles are likely to interfere with the growing crystal so that the crystal size is limited to the average particle spacing.[47]

In alloys containing small particles, the nucleation sites must be presumed to be similar to those in single phase material. The growth kinetics of a nucleus in such a material are then controlled by two opposing effects. The stored energy increases due to the presence of the particles and the retarding force (Zener drag[58]) due to the interaction of the moving grain boundaries with the particles. If the first effect is outweighted by the second one, retardation results, otherwise the recrystallization kinetics are accelerated. In addition to Zener drag, dissolution and reprecipitation of particles by a migrating grain boundary may be of importance for the recrystallization process[60] (Fig. 10). The observed kinetics of the dissolution process led to diffusion rates in migrating boundaries that were several orders of magnitude larger than in stationary interfaces of the same type.[36] So far the physics of this extremily rapid diffusion process is poorly understood.

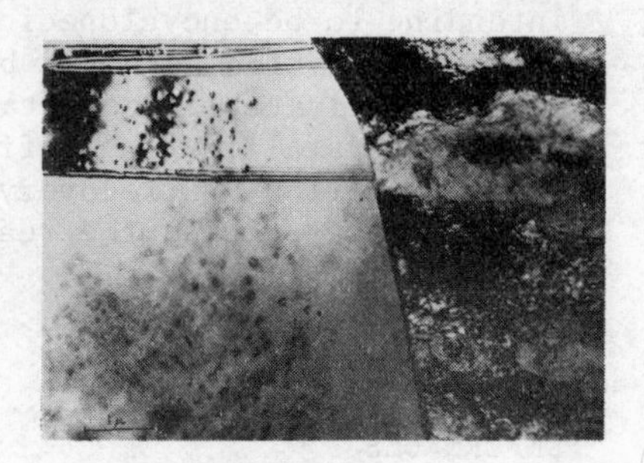

Fig. 10 : Dissolution of Ni_3Al particles by a recrystallization front. The deformed matrix ahead of the front contained coherent Ni_3Al particles that were dissolved by the recrystallization front and precipitated in the recrystallized material.[60]

VI. 3. Recrystallization of Deformed Coarse Duplex Structures

Duplex structures are of interest since they provide a favorable combination of relatively high strength and fracture toughness. They differ from dispersion hardened alloys in the sense that the volume fraction of the constituent phases is comparable and the topological arrangement may be described as a polycrystalline mixture of crystals of all phases. The recrystallization of deformed duplex structures has been investigated considerably in the case of α/β -brass as a model system.[61,63] Concerning the recrystallization modes of this duplex system, two cases may be distinguished. The most simple case results if the structural equilibration and the recrystallization temperatures coincide. If they do not, recrystallization and precipitation occur simultaneously, as the volume fraction and composition of the constituent phases vary in order to equilibrate according to the temperature variation. In the simple case of identical recrystallization and equilibration temperature, slightly ($\varepsilon < 40\%$) deformed α/β-brass alloys exhibit continuous recrystallization of the β-phase followed by a discontinuous recrystallization process of the α-phase. The long range ordering reaction in the β-phase (below 454°C) reduces the incubation time for recrystallization in the α-phase. The α-phase of highly deformed ($\varepsilon > 70\%$) alloys recrystallize above 500°C by a combined discontinuous transformation and recrystallization of the deformed β^*-phase (β^* is the phase into which β transforms upon heavy deformation; β^* has a distorted f.c.c. lattice) into dislocation-free β. Below 500°C recrystallization involves simultaneous precipitation and recrystallization starting from the regions of the former phase boundaries. This generates a fine-grained two-phase microstructure. If the equilibration and recrystallization temperatures are not identical, the thermally induced interface boundary migration modifies to the recrystallization processes described above.

VII. FINAL REMARK

The purpose of this lecture was to highlight some aspects of our present understanding of recrystallization phenomena without intending to be encyclopedic. Thus, a number of crucial aspects of recrystallization have not been considered, e.g. the structure of the deformed state, recrystallization textures, secondary and tertiary recrystallization, effects of trace impurities, dynamic recrystallization and recovery. Concerning these problems and the relevant literature, the reader is referred to the review articles mentioned.[1,2]

REFERENCES

1 . CAHN R.W., in *Physical Metallurgy*, North Holland, Publ. Comp., Amsterdam, 1974, p. 1129
2 . HAESSNER F., in *Recrystallization of Metallic Material*, Dr. Riederer Verlag, 1978, Stuttgart
3 . BURKE J.E. and TURNBULL D., Progress Metal Physics 3 (1952) 220
4 . CAHN R.W., in *Recrystallization Grain Growth and Textures*, H. Margolin ed., ASM Metals Park, Ohio, 1966, p. 99
5 . DOHERTY R.D., in ref. 1 (1978) p. 41
6 . CAHN R.W., Proc. Roy. Soc. 60A (1950) 323
7 . BECK P.A., J. Appl. Phys. 20 (1949) 637
8 . COTTRELL A.H., Progress Metal Physics 4 (1953) 255
9 . HU, H., in *Recovery and Recrystallization of Metals*, Interscience Publ., New York, 1963, p. 311
10. WALTER J.L. and KOCH E.F., Acta Met 11 (1963) 923
11. SMITH C.J.E. and DILLAMORE I.L., Metal Science 4 (1970) 161
12. BECK P.A. and SPERRY P.R., J. Appl. Phys. 21 (1950) 150
13. BAILEY J.E. and HIRSCH P.B., Proc. Roy. Soc. 267A (1962) 11
14. BAILEY J.E., Phil. Mag. 5 (1960) 833
15. VERBRAAK C.A. and BURGERS W.G., Acta Met. 5 (1957) 765
16. VERBRAAK C.A., Acta Met. 6 (1958) 580
17. SLAKENHORST J.W.H.G., Acta Met. 23 (1975) 301
18. RAY R.K., HUTCHINSON W.B. and DUGGAN B.J., Acta Met. 23 (1975) 831
19. DOHERTY R.D., in ref. 1, (1978) p. 23
20. GLEITER H., Acta Met. 17 (1969) 565
21. SMITH D.A., in *Grain Boundary Structure and Kinetics*, ASM Metals Park, Ohio, 1979, p. 337
22. BJORKLUND S. and HILLERT M., Metal Science 9 (1975) 127
23. AUST K.T. and RUTTER J.W., in *Ultra High Purity Metals*, ASM Metals Park, Ohio, 1962, p. 115
24. RATH B.B. and Hu. H., Trans Met. Soc. AIME 245 (1969) 1243
25. HAESSNER F. and HOFFMANN S., in *Recrystallization of Metallic Materials*, Dr. Riederer Verlag, Stuttgart, 1978, p. 63

26. GLEITER H. and CHALMERS B., Progr. Mat. Sci. 16 (1972) 127
27. HAHN H. and GLEITER H., Scripta Met. 13 (1979) 3
28. CAHN J.W., Acta Met. 10 (1962) 789
29. HILLERT M., Monograph and Report Series no. 33, Inst. of Metals, (1969) p. 231
30. HILLERT M., Metals Sci. 13 (1979) 118
31. GLEITER H., Acta Met. 27 (1979) 1749
32. GOTTSCHALK CH,, SMIDODA K. and GLEITER H., Acta Met. 28 (1980) 1653
33. GLEITER H., MAHAJAN S. and BACHMANN K.J., Acta Met. 28 (1980) 1603
34. PUMPHREY P. and GLEITER H., Phil. Mag. 32 (1975) 881
35. HILLERT M. and PURDY G.R., Acta Met. 26 (1978) 333
36. SMIDODA K., GOTTSCHALK CH. and GLEITER H., Acta Met. 26 (1978) 1833
37. FOURNELLE R.A. and CLARKE J.B., Met. Trans. AIME 3 (1972) 2757
38. SIMPSON C.J., AUST K.T. and WINEGARD W.C., Met. Trans. AIME 1 (1972) 1482
39. WIRTH R. and GLEITER H., Acta Met., in press
40. POND R.C. and SMITH D.A., Phil. Mag. 36 (1977) 353
41. PUMPHREY P. and GLEITER H., Phil. Mag. 30, (1974) 593
42. HORTON C.A.P., SILCOCK J.M. and KEGG G.R., Phys. Stat. Sol. 26(a) (1974) 215
43. GASTALDI J. and JOURDAN C., J. Crystal Growth 35 (1976) 17
44. GLEITER H., Progr. Mat. Sci. 24 (1981) 125
45. GOODHEW P.J.; Metal Science 13 (1979) 108
46. JOURDAN C. and GASTALDI J. , Scripta Met. 13 (1979) 55
47. HUMPHREYS F.J., Metals Science 13 (1979) 136
48. HORNBOGEN E. and KÖSTER U., in ref. 2 (1978) p. 159
49. HORNBOGEN E. and KREYE H., in *Textures in Research and Practice* Springer Verlag, Berlin, 1969, p. 274
50. COTTERILL P. and MOULD P.R., in *Recrystallization and Grain Growth in Metals*, Surrey University Press, London, 1976
51. KAMMA C. and HORNBOGEN E., J. Mat. Sci. 11 (1976) 2340
52. HUMPHREYS F.J., Acta Met. 25 (1977) 1323
53. HANSEN N. and BAY B., Mat. Sci. 7 (1972) 1351
54. PHILLIPS V.A., Trans AIME 236 (1966) 1302
55. PHILLIPS V.A., Trans AIME 239 (1967) 1955
56. KLEIN H.P., Z. Metallk. 61, (1970) 564
57. CHIN L.I.J. and GRANT N.J., Powder Metall. 10 (1967) 344
58. ZENER C., cited in SMITH C.S., Trans AIME 175 (1948) 15
59. GLEITER H. and CHALMERS B., ref. 26 (1972) p. 169
60. KREYE H. and HORNBOGEN E., Phys. Stat. Sol. 97(a)(1970) 189
61. HONEYCOMBE R.W.K. and BOAS W., Austr. J. Sci. Res. A1 (1948)70
62. CLAREBROUGH L.M., Austr. J. Sci. Res. A3 (1950) 72
63. MÄDER K. and HORNBOGEN E., Scripta Met. 8 (1974) 979
64. BECK P.A. and SPERRY P.R., J. Appl. Phys. 21 (1950) 150

IMPURITY EFFECTS IN CRYSTAL GROWTH FROM SOLUTION

R. BOISTELLE

Centre de Recherche sur les Mécanismes de la Croissance Cristalline
CNRS - Campus de Luminy, Case 913
13288 MARSEILLE - Cedex 9 - France.

ABSTRACT

Impurities can affect all stages of the crystallization process. Since they simultaneously influence kinetic and thermodynamic factors, they induce, at least theoretically, conflicting effects on nucleation and growth mechanisms. Kinetics are mostly slowed down by the presence of the impurities and only a few examples are known where they enhance growth rates. Adsorption of impurities occurs in kinks, steps or on the surfaces between the steps. The impurity molecules are either separated or linked to each other in the two dimensional adsorption layers which may include solute and solvent molecules. When adsorption differs on different crystal faces, the relative growth rates of the faces can change, sometimes resulting in habit modifications. Adsorption models and parameters relevant to habit changes are reviewed and a few examples are noted.

I. INTRODUCTION

Impurity and additive are terms which often have the same meaning in the vocabulary of crystal growth from solution. Even if the latter is generally more suitable, it appears that the former is more commonly used. Impurities play a role at all stages of the crystallization process. Their most visible effects concern the nucleation and growth rates of the crystals, often resulting in important habit changes. Impurities can affect solubility and consequently supersaturation when they interact with the solute, the solvent and the crystal. Complexes in the solution or mixed adsorption compounds at the crystal-solution interface can also form. Due to the adsorption, some conflicting effects occur between

B. Mutaftschiev (ed.), Interfacial Aspects of Phase Transformations, 621–638.

the kinetic and thermodynamic factors involved in crystallization. The mechanisms of the impurity action were often investigated but in recent years there have been fewer studies devoted to this field of basic research. On the other hand, impurities are more and more used in the chemical industry to control nucleation and growth. The use of impurities allows improvement of the quality of the products by changing properties such as size, size repartition, shape and apparent density of the crystals. Impurities are also employed to prevent caking and to facilitate storage and package. The different aspects of the impurity effects have often been reviewed in journals [1-14] or in specialized books. [15-18] In the remaining chapter we will summarize the essential features of crystallization in presence of impurities.

II. THERMODYNAMIC AND KINETIC EFFECTS

Some impurities are practically immovable from the crystal surface whereas others can move easily. The different exchange rates with the bulk of the solution results in different effects on the crystallization process. If the exchange rate of the impurity molecules is much higher than the exchange rate of the growth units, then adsorption mainly affects the surface and edge free energies of the crystals. Conversely, if the exchange rate of the impurity is low with respect to that of the solute, then adsorption has an important influence on the kinetic factors involved in nucleation and growth.[8]

In order to discuss these impurity effects, we will examine the simplest nucleation and growth rate equations (full developement of these equations is beyond the scope of this paper). The nucleation frequency J may be written [19-21] as :

$$J = K \exp - \frac{c' \, \Omega^2 \, \gamma^3}{(k_B T)^3 \, (\ln X/X_s)^2} \qquad (1)$$

where c' is a shape factor of the crystal and Ω the volume of a molecule in the crystal. The parameters k_B and T are the Boltzmann constant and the temperature. If activities are not available, the supersaturation ratio X/X_s can be expressed in terms of mole fractions.

We see from eq. (1) that an impurity can affect the nucleation rate either through the kinetic factor K or through the thermodynamic parameter γ (the surface free energy of a crystal face). Within a very simplified model, where the face is made up of only one type of adsorption site on which the adsorption follows a Langmuir isotherm, the variation of γ is[8,22-24] :

$$\Delta\gamma = \frac{k_B T}{s} \ln(1 - \theta) \tag{2}$$

where s is the area of the adsorption site and θ the degree of coverage of the surface by the impurity.

A crystal face grows either by two dimensional nucleation or by a spiral mechanism. In the former case we can write[18,25-27]:

$$R_m = d S J_2 \tag{3}$$

$$R_p = d v^{2/3} J_2^{1/3} \tag{4}$$

where R_m and R_p are the growth rates resulting from a mononucleation or a polynucleation mechanism. In the former case, it is assumed that every two dimensional nucleus of height d has enough time to spread over the face of area S before another one is formed. In the latter case, several nuclei form at the same time and the growth rate becomes a function of the lateral velocity v of the steps of the nuclei. The two dimensional nucleation frequency is :

$$J_2 = K_2 \exp - \frac{c' \lambda^2}{(k_B T)^2 \ln X/X_s} \tag{5}$$

where c' and K_2 are also shape and kinetic factors, while λ is the edge free energy of the nuclei.

Finally, if growth takes place by means of a spiral mechanism[28]:

$$R = \frac{vd}{y_\circ} \tag{6}$$

where $y_\circ$ is the equidistance between the steps of the growth spiral.

$$y_\circ = c' r_2^* = \frac{c' \lambda a}{k_B T \ln X/X_s} \simeq \frac{c' \lambda a}{k_B T \sigma} \tag{7}$$

the approximation being true only if the supersaturation ratio is low ($\sigma = (X - X_s)/X_s$). The terms which have not yet been defined are "a" the length of a molecule in the crystal and r_2^* the critical radius of a circular two dimensional nucleus. The factor c' depends on the spiral shape, being c' = 19 if the growth spiral is circular.

Returning now to the impurity effects on the growth rates, we see that adsorption influences either the kinetic parameters (K_2,v) or the thermodynamic parameters λ on which the size of the critical nuclei and the equidistance between the steps of the growth spiral depend.

II.1. Impurity Effects on Nucleation Rates

As an example, consider the crystallization of sodium perborate ($NaBO_2 \cdot H_2O_2 \cdot 3H_2O$) from an aqueous solution containing as a surfactant the sodium salt of the sulphated butyl ester of oleic acid.[30] Two kinds of nucleation experiments were performed, in both cases the mass of the precipitated sodium perborate being recorded as a function of time. The first was made at constant impurity concentration (30 ppm) with different supersaturations (Fig. 1).

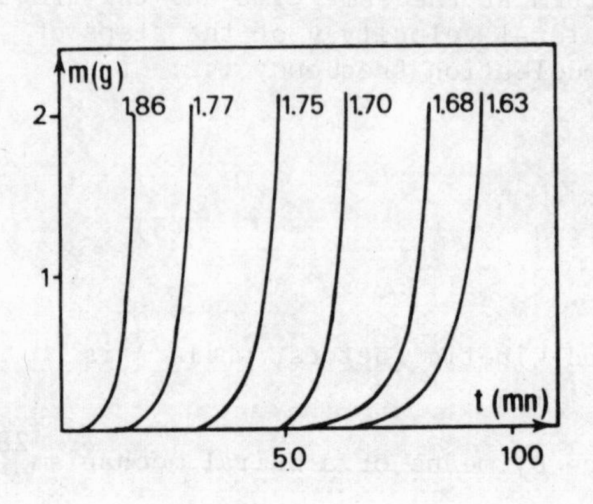

Fig. 1 : Mass of sodium perborate precipitated as a function of time, at different supersaturation ratios, from aqueous solution containing 30 ppm of the sodium salt of the sulphated butyl ester of oleic acid (SEO) (after ref. 30)

It clearly appears that the higher the supersaturation ratio, the lower the induction time, i.e. the easier the nucleation. All curves are similar, suggesting that the nucleation mechanism is not altered by the presence of the impurity. Only the nucleation rate is slowed down with respct to the pure solution.
The second kind of experiment was made at constant supersaturation with different impurity concentrations (Fig. 2). From here it is seen that nucleation becomes very slow, requiring larger induction times when the impurity concentration increases.

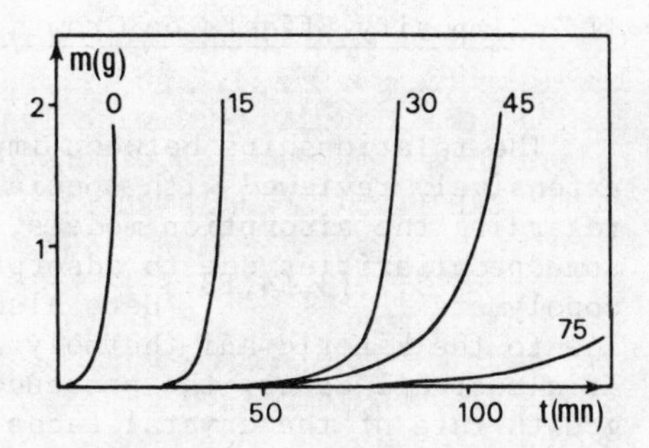

Fig. 2 : Mass of sodium perborate precipitated as a function of time, at a constant supersaturation ratio (1.68), from aqueous solutions containing different concentrations (ppm) of SEO (after ref. 30).

The impurity has three main effects : it decreases the nucleation rate, increases the induction period (time before nucleation occurs) and disturbs the nucleation process which becomes more or less irregular. The two former points are generally valid whereas the latter one seems to have some connection with the nature and size of the impurity. Since experimentally the nucleation rate decreases in the presence of the impurity, one should conclude that the kinetic factor K decreases more rapidly than the thermodynamic parameter γ (eqs. (1,2)). However, eqs. (1) and (2) give, at least theoretically, the possibility to get a larger nucleation rate due to the impurity. As a matter of fact, the size (e.g. expressed by the radius of the critical nuclei) depends on the surface free energy and the supersaturation according to the Gibbs-Thomson equation :

$$r^* = \frac{2 \gamma \Omega}{k_B T \, ln \, X/X_s} \quad (8)$$

When adsorption occurs it decreases γ (eq. (2)) and consequently r^*. Thus, the size of the critical nucleus in an impure solution is smaller than the size of the critical nucleus in a pure solution at the same supersaturation. The decrease of both nucleus size and surface free energy causes a lowering of the activation free energy of the process (eq. (1)). Nevertheless, this has never been observed and two points can account[8] for the fact that active impurities always inhibit the nucleation. First, adsorption mostly takes place at the growth sites of the clusters. Therefore, only a few impurity molecules are necessary to inhibit their growth, while the effect on the surface free energy γ is still negligible. Second, the impurities, especially long chain compounds and surfactants, are able to form compact adsorption layers which can drastically decrease the exchange kinetics of the solute molecules between the bulk and the cluster surface, without noticeable variation of γ.[31]

II.2. Impurity Effects on Growth Rates

The relationships between impuriies and growth rates have been extensively reviewed with special attention to theoretical problems, regarding the adsorption models, the nature of adsorption sites or some peculiarities due to adsorption of long chain compounds and copolymers[5,7,13,14,18]. Here also we can expect conflicting effects due to the kinetic and thermodynamic parameters in eqs. (3 - 7). In almost all cases, the presence of impurities decreases the growth rate of the crystal faces. There are a few exceptions to the rule such as the growth from aqueous solution of the (100) and (111) faces of lead nitrate ($Pb(NO_3)_2$) in the presence of methylene blue[32]. At constant supersaturation $\sigma = 0.08$, the growth rate increases with impurity concentration, passes through a maximum when the impurity concentration is in the range of 5g/l, then it decreases rapidly and is nearly zero at an impurity concentration about 50g/l. Another example concerns the growth of sodium triphosphate hexahydrate $Na_5P_3O_{10}\cdot 6H_2O$ from aqueous solutions containing dodecylbenzene sulphonate (DOBS)[33]. At $\sigma = 0.80$ for instance, the growth rate of the (0$\bar{1}$0) and (001) faces decreases continuously when the DOBS concentration increases. At a DOBS concentration of 20mg/l, (0$\bar{1}$0) growth is completely blocked. On the other hand, (100) grows faster than in a pure solution up to a DOBS concentration of about 10mg/l and decreases again for higher concentrations. In both these examples, adsorption influences first the edge free energies of the growth layers whereas the effect on the kinetic parameters becomes prevalent at high impurity concentration.

The general rule is, however, that impurities decrease the growth rates even at very low concentration. Adsorption seems not to change the growth mechanisms of the crystal faces. The growth rate curves, $R(\sigma)$, keep the same shape regardless of the growth mechanism, being only shifted towards higher supersaturation values when the impurity concentration increases. Examples can be found in the case of spiral growth (sucrose grown in the presence raffinose[13,34]) or in the case of growth by two dimensional nucleation (hexatriacontane grown in the presence of dioctadecylamine)[35]. The above conclusions are valid when the impurity does not induce steric effects. It concerns only small impurity molecules easily removed from the crystal surface. In this case, growth is regular : to each pair of supersaturation and impurity concentration corresponds a given growth rate.

This is not always true when large impurity molecules are involved such as long chain compounds, surfactants and copolymers. There are important steric effects since the molecules are anchored at several points of the crystal surface. Growth becomes irregular, sometimes irreversible. The crystal surfaces are strongly disturbed eventually exhibiting large growth hillocks.

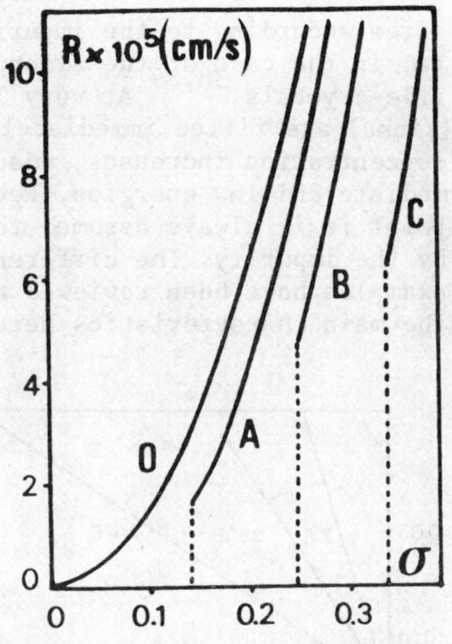

Fig. 3 : Growth rate of the (110) face of hexatriacontane grown from heptane in presence of a copolymer of ethylene vinyl acetate. The concentrations are 0, 50, 100 and 150 ppm for the curves 0,A, B and C respectively (after ref. 36)

Such phenomena were observed on hexatriacontane crystals grown from heptane in the presence of a copolymer of ethylene vinyl acetate[36] and on sodium perborate crystals grown from aqueous solutions in presence of a surfactant.[37] The R(σ) curves can be explained in the following way. As the impurity concentration increases, they are shifted towards higher values of the supersaturation (Fig. 3). They can be more or less parallel but in all cases critical supersaturations must be overcome for growth to occur. In some cases, the growth rate is either zero (below the critical supersaturation) or the same as that observed in a pure solution (above the critical supersaturation). This means that the curves A, B, and C in fig. 3, if shifted laterally can coincide with curve 0. The interpretation[14] is that both the adsorption kinetics and the exchange kinetics with the bulk of the solution are low, resulting in a progressive poisoning of the faces. Growth can start only from the clean areas between the impurities. This is possible only at high supersaturations where the critical size of the two dimensional nuclei is small enough.

III. ADSORPTION SITES AND ADSORPTION LAYERS

According to the nature of the solvent and of the crystal, impurity molecules adsorb separately onto the crystal surfaces or in the kinks and steps. They can also condense to form two dimensional adsorption compounds made up either by impurity molecules or by associations between impurity, solvent and solute. To distinguish between the different cases is often difficult because the crystal-solution interface cannot be investigated directly. Actually it is likely that adsorption occurs successively in different sites according to the impurity concentration. This was demonstrated in the case of the adsorption of cadmium ions on sodium chloride crystals.[38,39] At very low concentration, the high energy sites (kinks) are filled immediately (Fig. 4). Then, when the cadmium concentration increases, adsorption takes place in sites of intermediate and low energies. Nevertheless, from a theoretical standpoint it is always assumed that only one type of sites is occupied by the impurity. The different adsorption possibilities and several examples have been reviewed recently[7-9] and we will describe only the main characteristics hereafter.

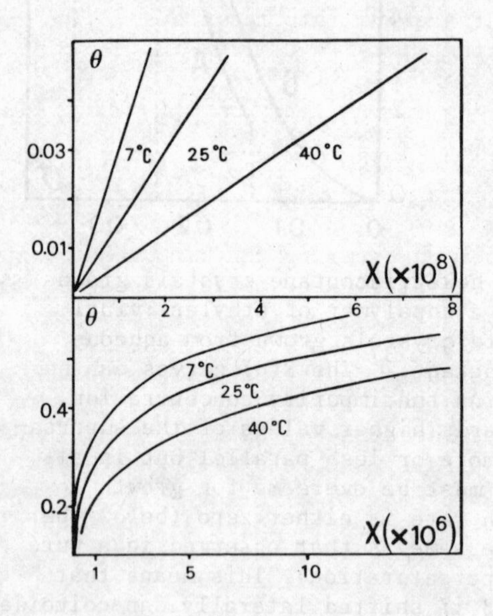

Fig. 4 : Adsorption isotherms of cadmium ions on the {111} faces of sodium chloride crystals. Degree of coverage as a function of the impurity mol fraction (after ref. 38)

III.1. Adsorption in Kinks

Since growth proceeds only through kinks, adsorption of impurities in kink sites decreases the number of kinks available for growth. Due to partial poisoning of the kinks, the distance between those free of impurity increases. The effect is more pronounced when the adsorption energy is high, i.e. when the exchange kinetics with the bulk of the solution or with the surface is low. In order to account for the difficulty encountered by a molecule to find and enter a kink, several retardation factors were introduced in the growth rate theories.[28,40-42] For instance, if growth is controlled by surface diffusion, the BCF theory[28,42] includes two factors which can play a role for growth in pure or in impure solution. The growth rate is :

$$R = \frac{2c_{\circ}\beta n_{s\circ}\Omega D_{s}\sigma}{x_{s}y_{\circ}} \tanh \frac{y_{\circ}}{2x_{s}} \tag{9}$$

where $n_{s\circ}$ is the number of solute molecules adsorbed per unit area, Ω the volume of a molecule in the crystal, while D_s and x_s are the coefficient and the mean free path of surface diffusion respectively. The retardation factor $c_{\circ}$ which is the probability of finding a step and a kink, is rather complicated.[28] Some values have been tabulated as a function of the distance between kinks and of the diffusion lengths onto the surface and into the steps.[18] The retardation factor β for entering the kinks is simpler :

$$\beta = \left[1 + \frac{D_s \tau_k}{a\, x_s} \tanh\left(\frac{y_{\circ}}{2x_s}\right)\right]^{-1} \tag{10}$$

where τ_k is the relaxation time before entering the kink. When β is very small, i.e. when 1 is negligible in eq. (10), τ_k becomes the rate determining step in eq. (9) (assuming that $c_{\circ} \cong 1$).

Adsorption in kink sites was assumed in the case of ammonium sulphate crystals, $(NH_4)_2SO_4$, grown from aqueous solutions poisoned by chromium ions.[43] The growth rate decreases drastically and there is polygonization of the growth spirals which are elliptical when growth occurs in pure solution. Kinks are also poisoned in the case of sodium chlorate crystals grown in the presence of a sodium sulphate impurity.[44-46] It is possible to achieve complete growth inhibition by poisoning only 5 % of the surface as shown by the adsorption isotherms of trinitroli (methylene phosphonic acid) adsorbed on barium sulphate crystals.[47] If growth is carried out by two processes, through free kinks on one hand and through poisoned kinks on the other hand, the total growth rate of the face is[44] :

$$R_i = R(1 - \theta) + R_{\infty}\theta \tag{11}$$

R_{∞} being the lowest possible rate when $\theta = 1$. The degree of coverage of the face depends on the impurity concentration C_i. If the adsorption isotherm is a Langmuir type :

$$\theta = \frac{BC_i}{1 + BC_i} \tag{12}$$

(where B is a constant), eq. (11) can be rewritten as :

$$\frac{1}{\eta_i} = \frac{B}{1 - \eta_\infty} \frac{1}{C_i} + \frac{1}{1 - \eta_\infty} \tag{13}$$

with $\eta_i = R_i/R$ and $\eta_\infty = R_\infty/R$.

A plot of $1/(1 - \eta_i)$ vs $1/C_i$ gives straight lines from which it is possible to determine B and η_∞.

III.2. Adsorption in Steps

Adsorption in steps has been assumed when ADP crystals are grown from aqueous solutions in presence of chromium ions.[48,49] With 15 ppm of impurity, the lateral velocity of the steps is reduced three fold. Individual steps are promoted instead of step bunches as in a pure solution. The interpretation is that bunching is only possible if there is a cooperation between dislocation groups. This happens if the distance between two dislocations is smaller than twice the critical radius r_2^* of the two dimensional nucleus stable at the given supersaturation (Eq. (7)). Since adsorption decreases both the edge free energy of the steps and r_2^*, bunching is prevented. Similar changes of the step shape were also observed on sucrose crystals grown from aqueous solutions in the presence of raffinose.[50]

The reduction of the step spreading rate which results from impurity adsorption is interpreted[50-52] on the basis that a step section can move only if the distance d separating two impurity molecules adsorbed along the step is $d < 2\,r_2^*$. If this condition is fulfilled, the step can bow out and pass between the anchored impurities. Its curvature increases and its velocity decreases. The average step velocity is therefore proportional to the fraction of unpinned step length. If v_i and v are the step velocities in presence of impurities and in a pure solution respectively, we have, for a given impurity distribution[50,51] :

$$v_i = v\,(d - pd + p)\,p^d \tag{14}$$

where $p = 1 - \theta$ is the probability of finding a site along the step whithout impurity, when the degree of coverage of the steps by adsorbed impurities is θ. The probability of finding a gap of length m (in molecular spacings) is $\theta^2 p^m$. The step velocity is proportional to the fraction of step length which can move, i.e. the sum of all lengths greater than d times the probability of each.

Another expression also proposed for the reduction of the step velocity, taking into account the radius of curvature of the step[52] is :

$$v_i = v\,(1 - 2\,r_2^*/z)^{1/2} \qquad (15)$$

which is an approximation[28] of the equation :

$$v_i = v\,(1 - 2\,r_2^*/r) \qquad (16)$$

An analogy is made between the radius of curvature r of the step and the average impurity separation z which is assumed to be uniform. If r and z are larger than $2r_2^*$, the step can filter through the impurities. Equations (15) and (16) are identical[5] for $z = 2r_2^*$ and $z = \infty$. If the impurities are arranged in a two dimensional lattice of[52] spacing z and density d, we have $z = d^{-1/2}$ and eq. (15) becomes :

$$v_i = v\,(1 - 2\,r_2^*\,d^{1/2})^{1/2} \qquad (17)$$

At any impurity concentration there is a critical supersaturation depending on r_2^*, below which growth is completely inhibited.

III.3. Adsorption on Surfaces

Adsorption on surfaces between steps was assumed when ADP crystals grow in the presence of aluminium and iron ions.[48,49] The edge free energies are assumed to be unchanged, this view being supported by the observation that the activity of the growth centers does not change significantly. Actually, it is difficult to know whether adsorption takes place into the steps or onto the surfaces. If the impurities are strongly adsorbed on the surface, they are almost immobile and the steps which move along are rapidly poisoned and stopped by a pair of impurities. The model proposed above[52] is also valid in this case. If adsorption takes place on the surface, the number of sites available for the solute molecules is reduced in such a way that there is a significant decrease in the flux of growth units towards the steps and the kinks. An approximate expression for the growth rate of a face covered by surface adsorption has been given elsewhere and applied to experimental results.[7,48] The number of growth units is assumed to be directly proportional to the fraction of free surface so that the step velocity is :

$$v_i = v\,(1 - \theta) \qquad (18)$$

III.4. Adsorption Layers

A special kind of adsorption on the surface concerns the two dimensional adsorption layers which form when lateral interaction energies between molecules are strong. An important property of these layers is that they are stable even when the concentration in the bulk of the solution is very low[3,8], lower than the concentration necessary to stabilize the corresponding three dimensional phase. In most cases, the composition of these layers is unknown. Several examples have been discussed on the basis of crystallographic aspects[2]. In a very favorable case, owing to adsorption isotherms[38] and epitaxy studies[39] it was shown that cadmium does not adsorb alone on the sodium chloride faces, but that it is associated with sodium and water in the shape of a mixed salt $Na_2CdCl_4 \cdot 3H_2O$. This conclusion was drawn from the fact that the number of cadmiums ions adsorbed on different crystal faces of sodium chloride are the same as those contained in the contact planes of the mixed salt in epitaxy on these faces.

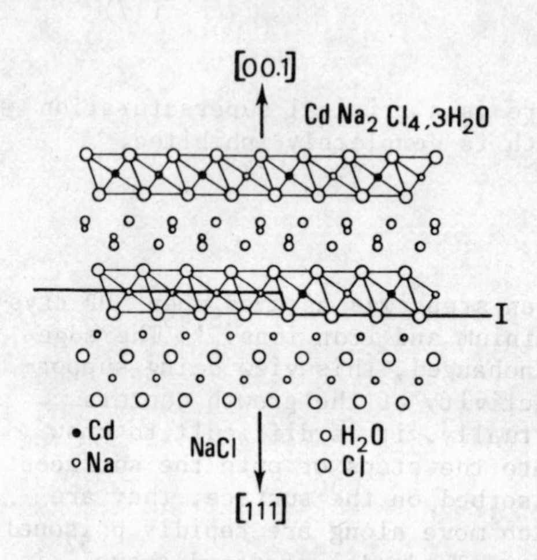

Fig. 5 : Two dimensional adsorption layer of $Na_2CdCl_4 \cdot 3H_2O$ on the (111) face of sodium chloride (after ref. 39)

Figure 5 shows a two dimensional adsorption layer on the (111) face of NaCl when $CdCl_2$ is added to the solution. Close to the NaCl surface, cadmium ions can take the place of sodium ions within the chlorine octahedra above which the adsorption layer extends by sodium and water. The whole layer has the composition, the structure, and the thickness of a slice $d_{00.3}$ of the mixed salt $Na_2CdCl_4 \cdot 3H_2O$. The fit between adsorption layer and crystal surface is perfect even if there are steps as depicted on fig. 5. Similar layers were also assumed to exist on other NaCl faces : $Na_2HgCl_4 \cdot H_2O$[53,54] on (100) and $Na_4Fe(CN)_6 \cdot 3H_2O$[53] on (100) when $HgCl_2$ and $Fe(CN)_6^{4-}$ ions are added to the solutions. Mixed salts also occur frequently when urea crystallizes from aqueous solutions in the presence of inorganic salts.[55]

IV. NATURE OF THE IMPURITIES AND HABIT CHANGES

The overall shape of a crystal is made only of faces with the slowest growth rates. When the impurities have different growth inhibiting efficiency on different crystal faces, theses faces do not maintain the same relative growth rates with respect to each other. Accordingly, habit changes become possible. Since growth kinetics are involved, the interpretations should be based on dynamic considerations. But due to lack of information about the parameters concerning the crystal-solution interface, they are instead based on crystallochemical, i.e. static considerations. Important difficulty still arises from the fact that the impurity nature is not always exactly known. The active impurity is not always what has been added to the solution. Several variables of the solvent-solute-impurity system can act simultaneously. The different experimental possibilities to get habit changes have been reviewed previously.[9] We give only a summary in the following sections.

IV.1. Nature of Impurities

The effects of supersaturation are often superimposed on solvent and solute effects but they are not well explained because adsorption energies and exchange kinetics at the interfaces are unknown. In some cases, only one new crystal form appears on the crystal habit at some critical value of the supersaturation. In other cases many new faces appear.[57] Supersaturation effects are often related to the desolvation kinetics of the solute or of the crystal[58] which is related to the dielectric constant of the solvent. Solvent effects can also lead to habit changes if they modify the growth mechanism of the faces. The solute itself is sometimes an impurity when, as for electrolytes it is possible to change the stoichiometry of the solution. Calcite crystals are elongated or tabular according to whether there is an excess of calcium or carbonate ions in the solution.[59] In some cases, it happens that there is an association of solute molecules resulting in different solution structures.[60]

Obviously, solvent, solute and supersaturation cannot really be considered as impurities even if they sometimes originate habit changes. In the same way, the crystallization temperature sometimes plays a role but in most cases all crystal faces are influenced identically. Nevertheless, anomalous temperature dependences of growth rates have been observed on many crystalline species.[61-64] The effects of all the factors quoted above are mainly combined with those of real impurities, i.e. of additives. But the interpretation of impurity effects poses the problem of the exact nature of the chemical species which adsorbs and eventually induces a habit change. This problem is sometimes simplified if the impurity efficiency changes with the crystallization conditions. As an example, we can consider the influence of glycine, NH_2-CH_2-COOH, on the growth habit of sodium chloride.[53,65]

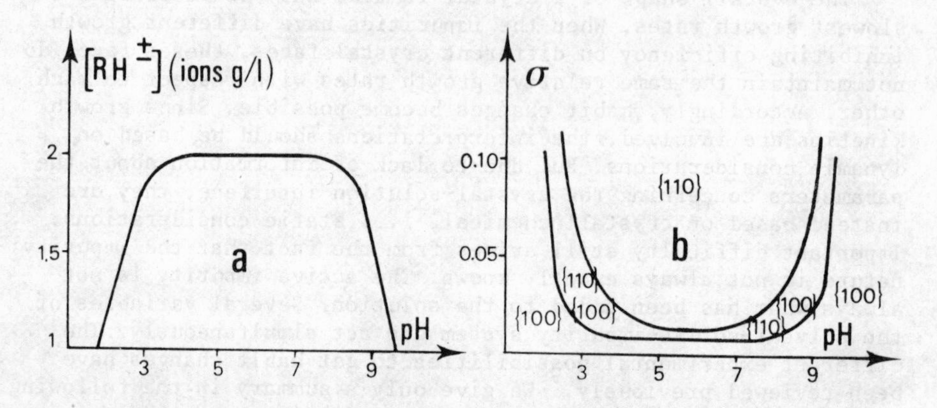

Fig. 6 : Concentration $RH^{\pm}$ in zwitterions $H_3^+N-CH_2-COO^-$ in an aqueous solution containing 150g/l of glycine (a) and supersaturation (σ) necessary to get a habit change of sodium chloride crystals (b) as a function of pH (after ref. 53,65)

When glycine is added to an aqueous solution, it may exist as zwitterions ($H_3^+N-CH_2-COO^-$) according to the pH of the solution (Fig. 6a). Around the isoelectrical point at pH = 5.7, the concentration of zwitterions is maximum. This pH range also corresponds to the highest efficiency of the impurity (Fig. 6b). The critical supersaturation necessary to get the {110} form on sodium chloride crystals is very low in the same pH range, whereas it increases drastically outside of it. The interpretation is based on the existence of chains of zwitterions[53] adsorbed in the steps of the (110) faces along the <001> directions. This view is supported by the observation of many striations along these directions. The impurity nature can also change by formation of complexes like those quoted in section III.4. Sometimes there is a twofold step in the process. Thus, if $HgCl_2$ is added to a sodium chloride solution, 98 % of it transforms into $HgCl_4^{2-}$ complexes which adsorb on the (110) faces of the NaCl crystals and form the two dimensional adsorption layer $Na_2HgCl_4 \cdot H_2O$. It is this chemical species which is responsible for the habit change of sodium chloride[54] and not $HgCl_2$ as supposed previously.[53] This point is supported by the fact that, for a given impurity concentration, the efficiency is maximum when the concentration ratio $NaCl/HgCl_2$ is about 2.

IV.2 Habit Changes

The habit changes of crystals are due to kinetic phenomena, the microscopic parameters of which are difficult or impossible to determine. For an impurity to be active, it is sufficient that it adsorbs temporarily on the crystal faces. During this adsorption, the growth rates are slowed down and eventually habit changes occur. The impurity action is explained by means of the PBC (periodic bond chain) theory.[66-68] Many examples have been treated in such a way.[2,6] The main points of the interpretations are the following. On a growing crystal normally there exist only F faces, i.e. flat faces, which advance layer by layer either by two dimensional nucleation or by a spiral mechanism. Only on these faces are two dimensional nuclei limited by at least two ledges with positive ledge free energies.[22-24] The S and K faces are not flat. They develop either by independent completion of rows or by continuous growth. They have one or no stable edge. In the PBC theory, the F, S and K faces contain respectively two, one, or zero PBCs. Since the impurities are able to stabilize the S and K faces and to induce their appearance on the growth shape, it may be concluded[2] that these faces have formally been transformed into F type. This could be achieved by formation of mixed PBCs between the molecules of the impurity and those of the crystal face.[69] The efficiency of the impurities generally increases with increasing supersaturation. For each impurity concentration there is a critical value of the supersaturation at which a given S or K face appears.

V. CONCLUSION

During the three last decades, the effects of impurities on the crystallization processes for several hundred crystal-solution systems have been investigated. Nevertheless, the exact mechanism by which they influence nucleation and growth has not been exactly demonstrated because no direct experiments can be carried out at the crystal-solution interface. There is therefore a lack of direct information about the kinetic and thermodynamic parameters which play a role at different crystallization stages. During the last decade, there have been no important advances either in the adsorption models or in the theories that include impurity effects. When elaborated for small molecules, the models give a particularly simple picture of adsorption (which is assumed to take place in only one kind of site). Impurities are often chosen among long chain compounds, surfactants and copolymers, especially in the case of industrial crystallization. Due to the great flexibility of these molecules and to the numerous conformations they can assume, adsorption occurs simultaneously at many different points. The adsorption models do not account for this type of impurity effect. However, some progress has been made in the understanding of the phenomena. Complete dynamic interpretations are not yet possible, but based on the idea that adsorption is temporary, static interpretations allow some guidelines regarding the choice of impurities to be made. The best ones are those which exhibit similarities with the crystal structures or which are able to form mixed compounds with the solute and/or with the solvent at the crystal-solution interface.

ACKNOWLEDGEMENT

The author whishes to thank Mrs M.C. Toselli and Miss J. Fayol for assistance with the manuscript.

REFERENCES

1 . SHEFTAL N.N., in *Growth of Crystals*, Shubnikov A.V. and Sheftal N.N. ed., Consultants Bureau, New York, 1 1958, p 5 and 3 1962 p 3
2 . HARTMAN P., in *Adsorption et Croissance Cristalline*, CNRS, Paris, 1965, p 477
3 . KERN R., Bull. Soc. Fr. Min. Crist. 91(1968)247
4 . KERN R., in *Growth of Crystals*, N.N. Sheftal ed., Consultants Bureau, New York, 8, 1969, p 3
5 . PARKER R.L., Solid State Physics 25 (1970) 151
6 . HARTMAN P., in *Crystal Growth an Introduction*, Hartman P. ed., North Holland, Amsterdam, 1973, p 367
7 . DAVEY R.J., J. Cryst. Growth 34 (1976) 109
8 . MUTAFTSCHIEV B., in *Critical Reviews in Solid State Sciences*, vol. 6, issue 2, CRC Press, Cleveland, 1976, p 157
9 . BOISTELLE R., in *Industrial Crystallization*, Mullin J.W. ed., Plenum Press, New York, 1976, p 203
10. KHAMSKII E.V., in *Industrial Crystallization*, Mullin J.W. ed., Plenum Press, New York, 1976, p 215
11. MULLIN J.W., in *Industrial Crystallization 78*, de Jong E.J. and Jancic S.J. ed., North Holland, Amsterdam, 1979, p 92
12. KAMSKII E.V., in *Industrial Crystallization 78*, de Jong E.J. and Jancic S.J. ed., North Holland, Amsterdam, 1979, p 105
13. DAVEY R.J., in *Industrial Crystallization 78*, de Jong E.J. and Jancic S.J. ed., North Holland, Amsterdam, 1979, p 169
14. SIMON B. and BOISTELLE R., J. Cryst. Growth 52 (1981)779
15. BUCKLEY H.E., in *Crystal Growth*, Wiley, New York, 1951
16. DOREMUS R.H., ROBERTS B.W. and TURNBULL D., in *Growth and Perfection of Crystals*, Wiley, New York, 1958

17. MULLIN J.W., in *Crystallization*, 2nd ed., Butterworths, London, 1972
18. OHARA M. and REID R.C., in *Modeling Crystal Growth Rates From Solution*, Prentice Hall, Englewood Cliffs, New Jersey, 1973
19. VOLMER M. and WEBER A., Z. Phys. Chem. 119 (1926) 277
20. KAISCHEW R. and STRANSKI I.N., Z. Phys. Chem. 26 (1934) 317
21. BECKER R. and DORING W., Ann. Phys. 24 (1935) 732
22. STRANSKI I.N., Bull. Soc. Fr. Min. Crist. 79(1956)359
23. KNACKE O. and STRANSKI I.N., Z. Elecktrochem. 60(1956)816
24. LACMANN R. and STRANSKI I.N., in *Growth and Perfection of Crystals*, Doremus R.H., Roberts B.W. and Turnbull D. ed., Wiley New York, 1958, p 427
25. VOLMER M., in *Kinetik der Phasenbildung*, Steinkopf, Leipzig, 1939
26. KAISCHEW R., Acta Phys. Hung. 8 (1957) 75
27. HILLIG W.B., Acta Met. 14 (1966) 1968
28. BURTON W.K., CABRERA N. and FRANK F.C., Phil. Trans. Roy. Soc. A 243 (1951) 299
29. CABRERA N. and LEVINE N.N., Phil. Mag. 1 (1956) 450
30. DUGUA J. and SIMON B., J. Cryst. Growth 44 (1978) 265
31. CASES J.M. and MUTAFTSCHIEV B., Surface Science 9 (1968) 57
32. BLIZNAKOV G. and KIRKOVA E., Z. Phys. Chem. 206 (1957) 271
33. TROOST S., J. Cryst. Growth 3/4 (1968) 340
34. SMYTHE B.M., Aust. J. Chem. 20 (1967) 1097
35. SIMON B., GRASSI A. and BOISTELLE R., J. Cryst. Growth 26 (1974) 90
36. PETINELLI J.C., Revue de l'Institut Français des Petroles, Vol. 34 n° 5,(1979), p 791
37. DUGUA J. and SIMON B., J. Cryst. Growth 44 (1978) 280
38. BOISTELLE R., MATHIEU M. and SIMON B., Surface Science 42(1974) 373
39. BOISTELLE R. and SIMON B., J. Cryst. Growth 26 (1974) 140
40. CHERNOV A.A., Soviet Physics Usp. Vol 4 n° 1, (1961) 116
41. CHERNOV A.A., in *Adsorption et Croissance Cristalline*, CNRS, Paris, 1965, p 265
42. BENNEMA P. and GILMER G.H., in *Crystal Growth an Introduction*, Hartman P. ed., North Holland, Amsterdam, 1973, p 263
43. LARSON M.A. and MULLIN J.W., J. Cryst. Growth 20 (1973) 183
44. BLIZNAKOV G., in *Adsorption et Croissance Crystalline* CNRS, Paris, 1965, p 291
45. BLIZNAKOV G. and KIRKOVA E. , Z. Phys. Chem. 206 (1957) 271
46. BLIZNAKOV G. and KIRKOVA E., Kristall und Technik 4/3 (1969) 331
47. LEUNG W.H. and NANCOLLAS G.H., J. Cryst. Growth 44 (1978) 163
48. DAVEY R.J. and MULLIN J.W., J. Cryst. Growth 26 (1974) 45
49. MULLIN J.W. in *Industrial Crystallization*, Mullin J.W. ed. , Plenum Press, New York, 1976, p 245
50. DUNNING W.J., JACKSON R.W. and MEAD D.G., in *Adsorption et Croissance Cristalline*, CNRS, Paris, 1965, p 303
51. ALBON N. and DUNNING W.J., Acta Cryst. 15 (1962) 474

52. CABRERA N. and VERMILYEA D.A., in *Growth and Perfection of Crystals*, Doremus R.H., Roberts B.W. and Turnbull D. ed., Wiley New York, 1958, p 393
53. BIENFAIT M., BOISTELLE R. and KERN R., in *Adsorption et Croissance Cristalline*, CNRS, Paris, 1965, p 577
54. REDOUTE M., BOISTELLE R. and KERN R., C.R. Acad. Sc Paris 262 (1966) 1081
55. GAEDEKE R., WOLF F. and BERNHARDT G., Kristall und Technik 15/5 (1980)557
56. KERN R., Bull. Soc. Fr. Miner. Crist. 76(1953)325
57. ESPIG H. and NEELS H., Kristall und Technik 2/3 (1967) 401
58. BLIZNAKOV G., KIRKOVA E. and NIKOLAEVA R., Kristall und Technik 6/1(1971)33
59. KIROV G.K., VESSELINOV I. and CHERNOVA Z., Kristall und Technik 7/5 (1972) 497
60. TROOST S., J. Cryst. Growth 13/14 (1972)449
61. SIPYAGIN V.V., Kristallografiya 12 (1967) 678
62. SIPYAGIN V.V. and CHERNOV A.A., Kristallografiya 17 (1972)1009
63. CHERNOV A.A., Annual Review of Materials Science 3 (1973)397
64. CHERNOV A.A. and SIPYAGIN V.V., in *Current Topics in Materials Science 5*, Kaldis E. ed., North Holland, Amsterdam, 1980,p 279
65. REDOUTE M., BOISTELLE R. and KERN R., C.R. Acad. Sc. Paris 260 (1965) 2167
66. HARTMAN P. and PERDOK W.G., Acta Cryst. 8 (1955) 49 and 521
67. HARTMAN P., Z. Krist. 119 (1963) 65
68. HARTMAN P., Fortschr. Mineral. 57 (1979) 127
69. HARTMAN P. and KERN R., C.R. Acad. Sc. Paris 258 (1964) 4591

DISSOLUTION PHENOMENA

Blaise SIMON

Laboratoire de Dynamique et Thermophysique des Fluides
Université de Provence, Centre de St Jérome
13397 Marseille Cedex 4 France

ABSTRACT

Dissolution morphologies, in particular the apparition of curved surfaces, are discussed, both for the dissolution of spheres and of cavities. The theory of LACMANN, HEIMANN and FRANKE, based on hypotheses of the dissolution mechanisms on a molecular level is discussed, together with its experimental confirmations.

Crystal imperfections have a greater influence on dissolution than on growth kinetics. The preferential negative nucleation at edge dislocations is probably at the origin of etch pits.

Even for crystals with high solubilities, the dissolution kinetics are probably not perfectly bulk diffusion limited. The influence of impurities is described with emphasis on the importance of adsorption kinetics.

INTRODUCTION

Dissolution phenomena are commonly encountered in everyday life. The wide applicability of these phenomena can be illustrated by two examples. The decalcification of teeth and bones is a particular case of liquid phase corrosion, and research of appropriate dissolution inhibitors is an important branch of medical science. In a much different area, such as metallurgy or solid-state physics, the observation of etch pits formed at early stages of dissolution gives important information about dislocations, symmetry of faces, and grain boundaries.

B. Mutaftschiev (ed.), Interfacial Aspects of Phase Transformations, 639–650.

Dissolution is often considered fundamentally different from growth. Crystals grow with flat faces, and dissolve with rounded ones. Dissolution is more rapid than growth, and activation energies are lower. These differences are real, but it should be remembered that growth and dissolution may be compared only in very special conditions, which are usually not met.

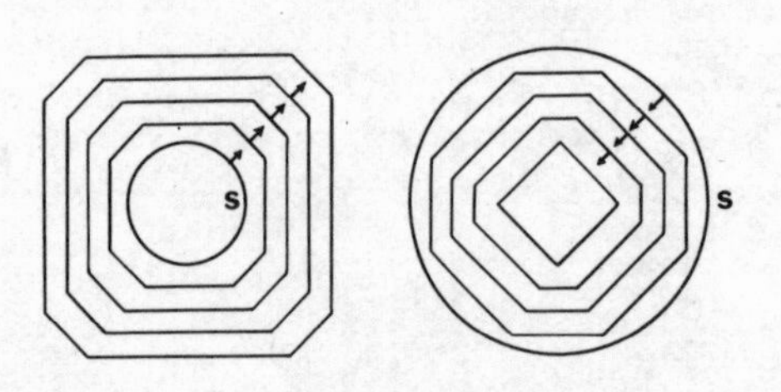

Fig. 1 : a) The growth of a sphere gives an identical final shape as the dissolution of a cavity. b) The dissolution of a sphere leads to the same shape as the inward growth of a cavity.

Fig. 1a shows that the growth of a crystal sphere gives a polyhedron bounded by faces with the lowest growth rates. The dissolution of a cavity similarly gives a polyhedron bounded by the faces with the lowest dissolution rates. Experimentally it is found[1] that these polyhedra are identical. This indicates that fastest growing faces also have the fastest dissolution rates ; but whether these rates are equal is not known. On the other hand, the dissolution of a sphere (Fig 1b) yields a polyhedron bounded by the faces having the highest dissolution rate. The same morphology is obtained by the inward growth of a cavity where a supersaturated solution is injected. These phenomena will be described in more detail later, but the important point now is that the growth of a polyhedron must be compared, regarding both morphology and kinetics, only to the dissolution of a cavity. The interest of dealing with cavities[1], is that they give the only possibility to determine growth forms of crystals which for practical reasons cannot be grown as polyhedra.

II. MORPHOLOGY

The dissolution morphology of crystals, especially the apparition of curved surfaces, is closely related to the dissolution mechanisms at a molecular scale. We will describe two methods of predicting dissolution morphologies.

II.1. FRANK Method

The FRANK method[2] is based on a theorem stating that if a crystal face has a rate (of growth or dissolution) depending only on its orientation, the trajectories of surface elements with a given orientation are straight lines. Nothing is assumed about the nature of the face, in particular as to its relation with the crystal structure. From the experimental dissolution rates (DR) of a number of faces, a polar diagram of the reciprocal DR (the reluctance diagram) is constructed. The construction of the dissolution form is based on a second theorem : the trajectory of a surface element is parallel to the normal drawn from the reluctance diagram for this orientation. Such a construction was made by FRANK and IVES[3] for the dissolution of cavities in germanium from the dissolution rates measured by BATTERMAN.[4] The initial form progressively changes from a sphere to a rounded form with flat (111) facets. The important result, in agreement with the experimental morphology, is that the rounded surfaces do not disappear during dissolution. This can be understood from the polar diagram of the DR : the persistence of curved surfaces is due only to the fact that the dissolution rates of different faces are about the same (the polar diagram of the DR has not many sharp cusps)

II.2. Method of LACMANN, HEIMANN and FRANKE

Unlike the previous method, where nothing is assumed about the molecular mechanisms of dissolution, this method[5-7] is based on the hypothesis that dissolution mechanisms are related to the crystal structure. It has been used for the prediction of dissolution forms

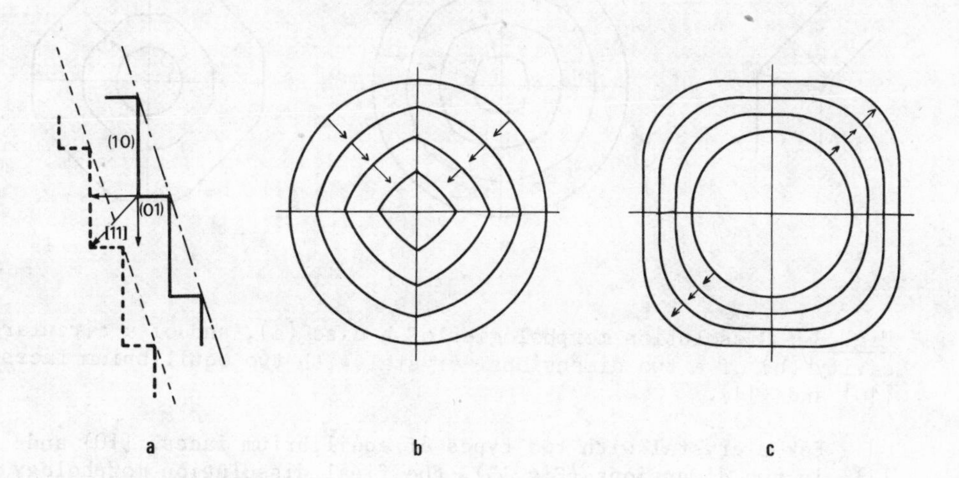

Fig. 2 : Dissolution morphologies of a non equilibrium (stepped) face (a), of a circle (b), and of a circular cavity (c) of a two dimensional crystal with only one type of equilibrium face. Curved surfaces have a constant radius of curvature.

of spheres, and has received convincing experimental confirmation by inspection of surfaces with the scanning electron microscope. Two different models have been proposed.

In the first model, it is supposed that the sphere is made up of facets of the equilibrium form, which are stable during dissolution (Fig. 2a). This is possible if corner and edge molecules are bound to the crystal with energies higher than that in a kink position. This model predicts that a polyhedral crystal with the equilibrium form (not a sphere) will keep its flat singular faces during dissolution. Regarding non singular orientations, Fig 2a shows in two dimensions how, according to this model, every point of a macroscopically straight edge translates in the $|11|$ direction with a velocity independent of the edge orientation. If the starting shape is a circle (Fig. 2b) the resulting shape is limited by curved {11} faces built up by {10} facets. In three dimensions, a sphere made up of {111} equilibrium facets will give curved {100} surfaces. This is effectively the case for silicon.[8] In general the final form is obtained by truncating the corners. The dissolution of a circular cavity is indicated in Fig 2c. Equilibrium faces appear during dissolution, but curved surfaces with a constant area persist. This has been demonstrated for germanium.

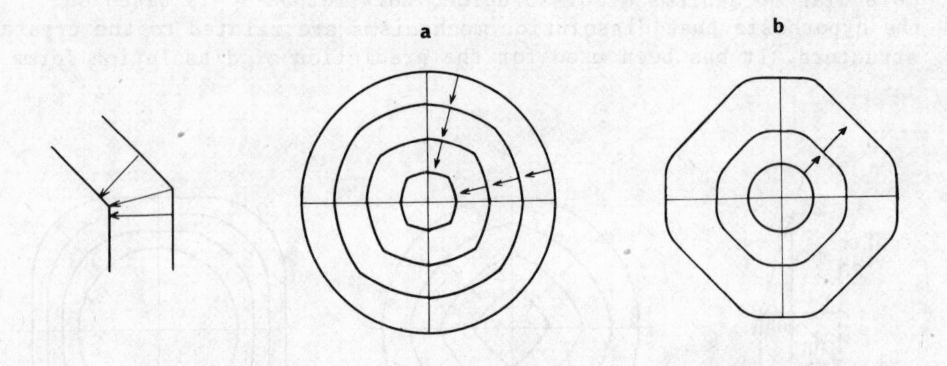

Fig. 3 : Dissolution morphologies of a disc (a), and of a circular cavity (b) of a two dimensional crystal with two equilibrium faces {10} and {11}.

For a crystal with two types of equilibrium faces, {10} and {11} in two dimensions (Fig. 3), the final dissolution morphology of a disc consists of only one non crystallographic form, with an orientation depending on the relative DR of the equilibrium facets. An example is MgO[9], with possible equilibrium faces {100} {111} and {110} depending on the solvent. The dissolution forms contain rounded faces {hhl} , depending also on the solvent. The dissolution of a cavity gives a shape limited by the two types of equilibrium faces with relative sizes depending on their velocities, together with curved surfaces.

For this model, if the faces have not too different DR's, as assumed till now, curved surfaces do not disappear from the dissolution form of a cavity. Such curved surfaces will of course disappear during the growth of a sphere, so that the morphological identity assumed in the introduction between growth of spheres and dissolution of cavities is not observed. The disappearance of curved surfaces from the dissolution forms (spheres and cavities) is expected only when equilibrium faces have very different DR's (Fig. 4). This is the case for NaCl and alum[1]. With these conditions, the growth morphology of a sphere is identical to that of a dissolved cavity.

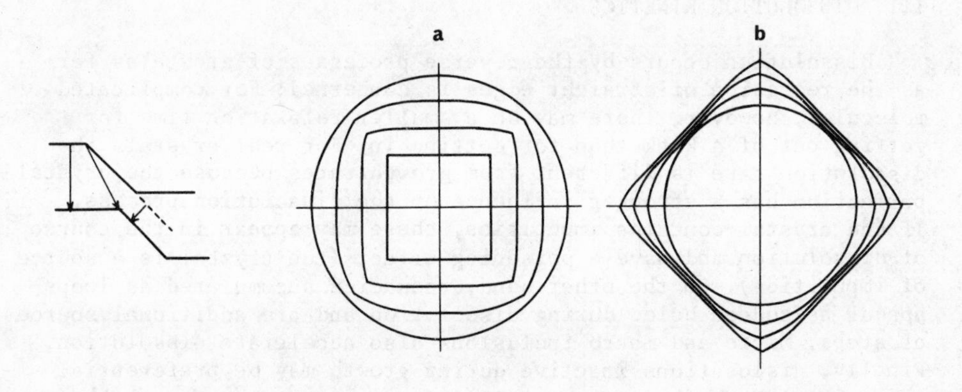

Fig. 4 : Dissolution morphologies of a sphere (a) and of a cavity (b) of a crystal where the DR of two equilibrium faces are very different.

In the second model of LACMANN, HEIMANN and FRANKE, the equilibrium facets present on the initial sphere are not stable, because of the preferential detachment of molecules from corners and edges. So, the curved surfaces are made up of vicinal faces having DR's proportional to their kink densities, which are changing with time. In this model a polyhedral crystal will dissolve with curved corners and edges. Fig. 5 shows that flat non equilibrium crystallographic faces, those with highest kink density, appear in the dissolution of a sphere. The dissolution of a cavity produces a polyhedron bounded by the equilibrium faces with the smallest kink density, without curved surfaces.

It is thus clear that a close study of dissolution forms of

spheres and cavities gives essential information on the dissolution mechanisms of the faces.

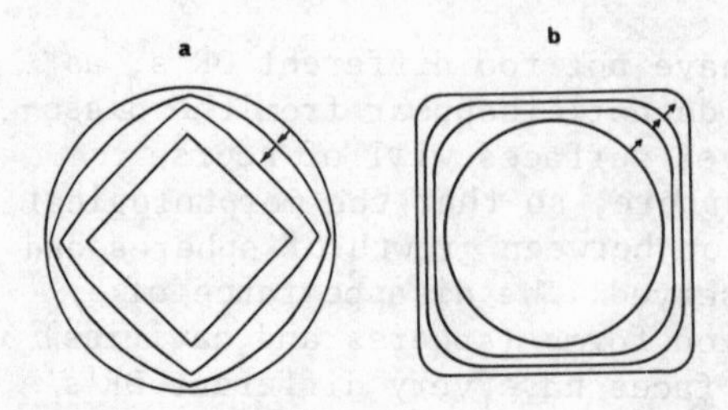

Fig. 5 : Dissolution morphologies of a sphere (a) and of a cavity (b) when the equilibrium faces are not stable.

III. DISSOLUTION KINETICS

Dissolution occurs by the reverse processes of growth as far as the recession of straight edges is concerned. For complicated molecules, however, there may be a smaller relaxation time for getting out of a kink than for getting in. For real crystals the dissolution rate is different from growth rates because the crystal perfection has a stronger influence on the dissolution process. If the crystal contains imputities, these may appear in the course of dissolution and have a poisoning effect (the crystal is a source of impurities). On the other hand, vacancies accumulated as loops appear as sudden holes during dissolution and are additional sources of steps. Micro and macro inclusions also accelerate dissolution. Finally, dislocations inactive during growth may be preferential sites for dissolution. These differences between growth and dissolution rates will probably not be apparent in the dissolution and in the initial stages of growth of spheres because the surface is composed of a great number of facets each dissolving independently.

III.1. Influence of Dislocations

It is well established that dislocations are preferential sites for the apparition of etch pits; edge and screw dislocations have a nearly equal efficiency in this respect[10]. They are considered also as preferential sites for negative nucleation, but this is true probably only for edge dislocations. Indeed the formation of etch pits at screw dislocations is an accelerated dissolution rate of edges already present rather than a true nucleation.

The Gibbs free energy of formation of a negative circular hole (nucleus) on a perfect surface is given by

$$\Delta G_p = - \pi r^2 \ d \ \Omega^{-1} \ \Delta\mu + 2\pi r\gamma$$

where d and r are the height and radius of the hole respectively, Ω is the volume of a molecule, γ the edge free energy, $\Delta\mu$ the difference in chemical potential of a molecule in the solution and of one in the perfect crystal.

For a nucleus forming at the emergence of an edge dislocation, the additional energy of the dislocation must be taken into account. CABRERA et al.[11] consider only the elastic energy, corresponding to the deformation of the crystal far from the dislocation, and neglect the core energy. This approximation is valid for large nuclei. With the addition of the elastic energy,

$$E_d = - d\,\mu\, b^2 [4\pi(1-\nu)]^{-1} \ln (r/r_o)$$

(b is the Burgers vector, μ the shear modulus, ν the Poisson ratio, r_o the radius of the core of the dislocation), to ΔG_p, a function ΔG_d is obtained (Fig. 6a).

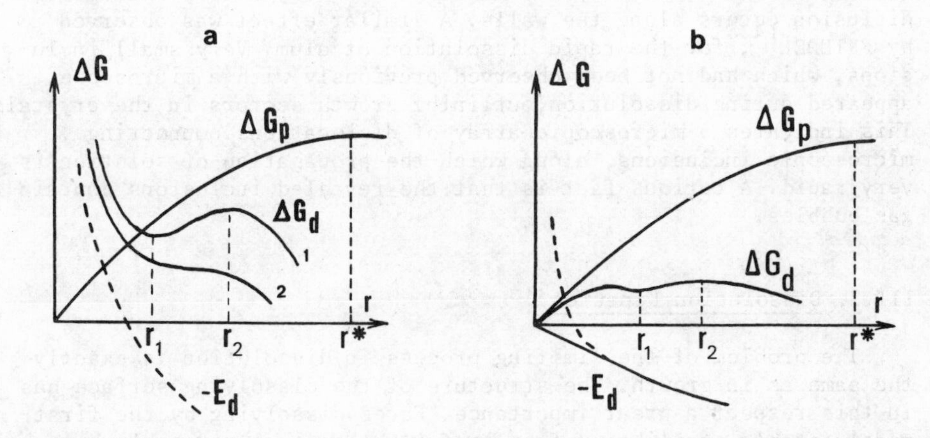

Fig.6 : a) Gibbs free energy of formation of a two dimensional hollow nucleus. Curve 1 : low undersaturation. Curve 2 : higher undersaturation. b) Core energy is taken into consideration.

Holes with size r_1 are stable. The critical size r_2 above which the development is irreversible is smaller than the critical size r^* on a perfect surface, as is the activation energy. It can be stated that the negative nucleation rate is higher on dislocations than on a perfect surface. For higher undersaturation (curve 2 of Fig. 6a) the dislocation opens up spontaneously (hollow core).

In the case of small negative nuclei, the core energy of the dislocation must be considered. SCHAARWÄCHTER[12] has shown that the elastic energy previously described should be used for $r > r_o$ only, and that for $r < r_o$ the energy is given by

$$E_d = 2\, d\, \mu\, b^2 r / r_o .$$

The resulting ΔG_d curve is given in Fig. 6b. It shows a still smaller free energy of activation for negative nucleation, but a spontaneous opening is not expected. The enhanced nucleation rate along a dislocation line explains the formation of etch pits. When the dislocation line is not perpendicular to the surface, the pit is not symmetric. SCHAARWÄCHTER[12] has discussed the form of the pit as a function of the negative nucleation rate and of the lateral propagation rate of the steps.

Another phenomenon related to preferential dissolution rates along dislocations is the formation of negative whiskers[13]. They represent chanels, so long and narrow that it is difficult to imagine how the undersaturation can be kept the same on the bottom of the hole as on the crystal surface. Probably a very strong surface diffusion occurs along the walls. A similar effect was observed by FRIEDEL[14], for the rapid dissolution of alum. Very small inclusions, which had not been observed previously with a microscope, appeared during dissolution, outlining growth sectors in the crystals. This indicates a microscopic array of dislocations connecting microscopic inclusions, along which the propagation of solution is very rapid. A curious fact is that the revealed inclusions contain gas bubbles.

III.2. Dissolution Kinetics

The problem of the limiting process in dissolution is exactly the same as in growth. The structure of the dissolving surface has in this respect a great importance. Faces dissolving by the first model (stable equilibrium faces) of LACMANN et al. probably have surface limited kinetics. On the other hand, the dissolution of non equilibrium vicinal faces is presumably limited by bulk diffusion. Both cases are found experimentally.

Surface limited dissolution is often found for crystals with low solubilities.[15-17] This gives dissolution rates as parabolic functions of the undersaturation. There are even cases where a two dimensional negative nucleation has been detected.[18] It is clear that for low solubilities, it is not possible to have large values of $\Delta\mu$, so that the rate of dissolution can be measured only in the non linear part of the $R(\sigma)$ curve (R is the rate of dissolution, σ the undersaturation).

Dissolution nearly limited by bulk diffusion occurs rather for substances with high solubilities.[19,20] The dissolution rates are linear functions of the undersaturation (this alone is not proof of a bulk diffusion limitation), and the heat of activation is about the same as the activation energy for bulk diffusion (a better indication). In first approximation (sufficient for the prediction of dissolution rates with reasonable accuracy) a volume limitation for dissolution may be accepted. But in fact, there are arguments against this hypothesis : The most convincing is that different faces of a crystal have different dissolution rates.[19]

Here we will show that even for crystal surfaces with a very high kink density, the dissolution kinetics may not be completely limited by volume diffusion.[20] Compressed discs of NaCl powder with randomly oriented crystals were dissolved in water following a technique similar to the rotating disc method. After a partial dissolution, the overall surface is flat, and an inspection with the scanning electron microscope indicates that each crystallite has a surface parallel to that of the sample ; no facets are detected, and thus it can be assumed that the average kink density is very high. The dissolution rate is a linear function of the concentration difference $c - c_s$ (c and c_s are the actual concentration and the saturation concentration respectively) with an activation energy of 5 kcal/mol. Assuming that the dissolution is perfectly limited by volume diffusion, the dissolution rate R can be calculated *a priori* from the diffusion equation

$$R = D^{2/3} \nu^{-1/6} \rho_c^{-1} \rho \ln \left[(1 - W_s)/(1 - W)\right]$$

where ν is the kinematic viscosity, ρ_c and ρ are the densities of the crystal and the solution respectively, W_s and W are the mass fractions of salt in the solution and at equilibrium respectively.[21] This formula takes into account the convective flux of solute, and only differs from the rotating disc formula by a numerical factor. In this hypothesis, a plot of R versus $\rho \ln \left[(1 - W_s)/(1 - W)\right]$ should give a straight line ; in practice there is a systematic deviation from linearity. Moreover, the calculated theoretical heat of activation amounts to 3 Kcal/mol which is lower than the experimental value. These two observations indicate that the dissolution is not perfectly volume diffusion controlled.

III.3. Dissolution in the Presence of Imputities

Dissolution, as well as growth rates, are changed by the addition of impurities, and changes in morphology may result.

One example is the dissolution of cavities in NaCl described by FRIEDEL[1]. In pure water, cubic cavities form. The addition of urea gives octahedral cavities. Similar experiments were made with germanium[8] dissolved in a solution composed of HNO_3 - HF - CH_3COOH in various proportions. Growth forms, deduced from the dissolution of hemispherical cavities, are made up of {100} and {111} faces with relative sizes depending on the solvent composition. For each composition, the dissolution form of spheres was found to correspond exactly to that predicted from the experimental growth form.

These experiments show that impurities act on the same faces and have a similar impeding effect for both growth and dissolution.

The influence of the impurity is of course due to adsorption and two mechanisms have been proposed. CABRERA et al.[22] assumed that impurities are strongly adsorbed on the surface and immobile. It is clear that such isolated molecules are obstacles for the propagation of steps during growth. But it is difficult to see why these impurities should slow the recession of steps during dissolution, unless their concentration is very high, with adsorption bonds parallel to the crystal surface.

The most probable mechanism is a poisoning of kinks. This has been described in detail by GILMAN et al.[10], for the inhibition of the dissolution of LiF by Fe^{3+}. This ion is adsorbed on kinks, taking the place of Li^+ ions with the same ionic radius. Three Fe-F bonds are formed with the crystal. It is possible then that three F^- ions from the solution adsorb at the same site, producing a stable $FeF_6{}^{3-}$ ion. The detachment of this stable complex ion from the kink would require breaking of three F-Li bonds with the crystal. The specificity of the impurity (ionic radius, stability of the complex) could be explained by such a model.

A particular effect of impurity, discussed by FRANK[2] is the kinetic aspect of adsorption, together with the kinetics of ledge motion. For high dissolution rates, fresh new surfaces are formed rapidly and remain clean if adsorption kinetics are not very rapid, so that the role of the impurity is negligible. On the other hand, for slow edge motion rates, the adsorption has sufficient time to take place and the influence of the impurity is greater. It can even be expected that the effect will be increasingly important as dissolution proceeds. There are many experimental facts confirming such a time-dependent adsorption.

a) It is difficult to prepare a saturated solution by the dissolution of a salt in presence of impurities[23] because as the undersaturation decreases, the influence of the impurity is stronger. The only way to dissolve more salt with a reasonable rate is to add new crystals to the solution or to create new surfaces by crushing.

b) The time dependent influence of impurities was studied by IVES and HIRTH,[24] who observed that for etch pits, the trajectories of surface elements with a given orientation are not straight lines, indicating that in presence of impurities, the dissolution rates do not depend on the orientation only. The pits have curved profiles, due to the less rapid propagation of steps far from the center. In fact, the relative variation of length of ledges dl/l is greater at the center of the pit, and it is here that the influence of the impurity is the least. On the other hand, far from the pit center, dl/l is smaller so that the impeding influence of the impurity is more important. The presence of impurities is a necessary conditions for the formation of "good" etch pits.

c) The poisoning effect of Cd^{++} ions on the dissolution rate of sodium chloride is greater at lower dissolution rates. This was shown by direct measurements of dissolution kinetics,[20,25] but the concentrations of cadmium were quite different in these two experiments. In the experiments of MUTAFTSCHIEV et al.[25] the crystals were left for long times in the saturated solution with the impurity, so that adsorption had sufficient time to take place (probably on equilibrium faces). Then pure water was added to the system. With these conditions, extremely small concentrations of Cd^{++} (about 10^{-3}%) are sufficient to produce a strong effect on dissolution.

On the other hand, in the experiments of SIMON[20] the sample, where the exposed surface probably had a very high kink density, was suddenly dropped into the undersaturated solution with the impurity. The dissolution was so rapid that the influence of the impurity was extremely small. It was necessary to use Cd^{++} concentrations one thousand times higher than in ref.25 in order to have a noticeable effect.

REFERENCES

1 . FRIEDEL J., in *Leçons de Cristallographie*, Blanchard., Paris, 1964

2 . FRANK F.C., in *Growth and Perfection of Crystals*, Wiley, N. Y., 1962, p. 411

3 . FRANK F.C. and IVES M.B., J. Appl. Phys. 33(1960) 1996

4 . BATTERMAN B.W., J. Appl. Phys. 18(1957) 1236

5 . LACMANN R., FRANKE W. and HEIMANN R., J. Cryst. Growth 26(1974) 107

6 . LACMANN R., HEIMANN R. and FRANKE W., J. Cryst. Growth 26(1974) 117

7 . For a general review of crystal dissolution see HEIMANN R., *Auflösung von Kristallen*, Springer, Wien, 1975

8 . FRANKE W.R., HEIMANN R. and LACMANN R., J. Crystal Growth 28 (1975) 145
9 . HEIMANN R., FRANKE W.R. and LACMANN R., J. Cryst. Growth 28 (1975) 151
10. GILMAN J.J., JOHNSTON W.G. and SEARS G.W., J. Appl. Phys. 29 (1958) 747
11. CABRERA N., LEVINE M.M. and PLASKETT J.S., Phys. Rev. 96 (1954) 1153
12. SCHAARWACHTER W., Phys. Stat. Sol. 12 (1965) 375
13. SEARS G.W., J. Chem. Phys. 32 (1960) 1317
14. FRIEDEL G., Bull. Soc. Fr. Min. Crist. 48 (1925) 6
15. BARTON A.F.M. and WILDE N.M., Trans. Farad. Soc. 67 (1971) 3590
16. CAMPBELL J.R. and NANCOLLAS G.M., J. Phys. Chem. 73 (1969) 1735
17. BOVINGTON C.M. and JONES A.L., Trans. Farad. Soc. 66 (1970) 764
18. CHRISTOFFERSEN J., J. Cryst. Growth 49 (1980) 29
19. HAUSSUHL S., and MULLER W., Kristall u. Technik 7 (1972) 533
20. SIMON B., J. Cryst. Growth 52 (1981) 789
21. BIRD R.B., STEWART W.E. and LIGHTFOOT E.N., in *Transport phenomena*, Wiley, N. Y., 1960
22. CABRERA N. and VERMILYEA D.A., in *Growth and Perfection of Crystals*, Wiley, N. Y., 1962
23. STEINIKE U., Z. Anorg. Allg. Chem. 317 (1962) 186
24. IVES M.B. and HIRTH J.P., J. Chem. Phys. 33 (1960) 517
25. MUTAFTSCHIEV B., CHAJES H. and GINDT R., in *Adsorption et Croissance Cristalline*, CNRS, Paris, 1965, p. 421.

CRYSTALLOGRAPHIC APPLICATIONS OF HOLOGRAPHIC INTERFEROMETRY

F. BEDARIDA

Istituto di Mineralogia dell'Università
Palazzo delle Scienze, Corso Europa
16100 Genova (Italy)

ABSTRACT

Holographic interferometry has proved to be a very powerful tool in Crystallography.
Two methods, double exposure interferometry and depth contouring, allow reconstruction of rather precise topography of crystal faces by reflection interferometric patterns.
By holographic interferometry in transmission applied to crystal growth from solution, it is possible to visualize patterns not visible with normal optics (for instance convective plumes and boundary layers) and from interference fringes, mean concentration gradients and diffusion coefficients of solution growth systems can be measured.
A fourth method which will be introduced, is called multidirectional holographic interferometry and at present is a matter of active research and will permit point by point mapping of concentration values in the volume of solution where the crystal grows.

I. INTRODUCTION.

The coherence of a wave describes the accuracy with which it can be represented by a pure infinite sine wave. This was presumed in some books for common light, which however, is very far from accurate. Ordinary light in fact is a pure sine wave if it is observed in a limited space or for a limited period of time and may then be called a partially coherent wave.
Coherence is both a function of space and time. We may therefore define two different criteria of coherence.
The first criterion expresses the expected correlation between a

B. Mutaftschiev (ed.), Interfacial Aspects of Phase Transformations, 651–667.

wave now and a certain time later. The other between here and a certain distance away, to use the definition given by Lipson and Lipson.[1] The coherence time for ordinary light is $\simeq 10^{-8}$ sec while for laser light is $\simeq 10^{-2}, 10^{-3}$. This means that the coherence length for ordinary light is of the order of a cm and for laser light is of the order of a km.

Coherence in time guarantees a perfect monochromaticity of the laser light. Its importance will be clear later when the concentration gradients of a solution will be measured as a function of its refractive index. Coherence in space guarantees a strict constancy of phase difference when two beams interfere regardless of their path. As a consequence of these two coherence properties, laser light was chosen for interferometric experiments in crystallographic applications.
Three methods using holographic interferometry shall be described in detail. Two are applied to study microtopography of crystal surfaces and make use of laser light in reflection, one is applied to crystal growth from solution and makes use of laser light in transmission. A fourth transmission method, at the moment matter of active research, will be introduced.

II. MICROTOPOGRAPHY of CRYSTAL SURFACES by HOLOGRAPHIC TECHNIQUES

Microtopography of crystal surfaces[2,3] has been studied with light interferometry for a long time. One of the highest levels in this research was the work of the late S. Tolansky at the London University.[4]

Using laser light, new types of interferometry may be realized that actually have some advantages over the ones using common monochromatic light.

Two different methods will be described here. The first one is a double exposure method, the second makes use of depth contouring.

II.1. Double Exposure Method

A modified Mach-Zender type interferometer can be used in this method. Fringes are obtained on a holographic plate and by the interfringe distance and some geometrical parameters of the system it is possible to measure dihedral angles between crystal faces and facets. The system works even when the surface of the crystal studied is not highly reflecting.

The laser beam passes through the beam splitter A (fig. 1) : beam AB (reference beam), reflected by mirror B and spread by objective L_1, illuminates the plate.

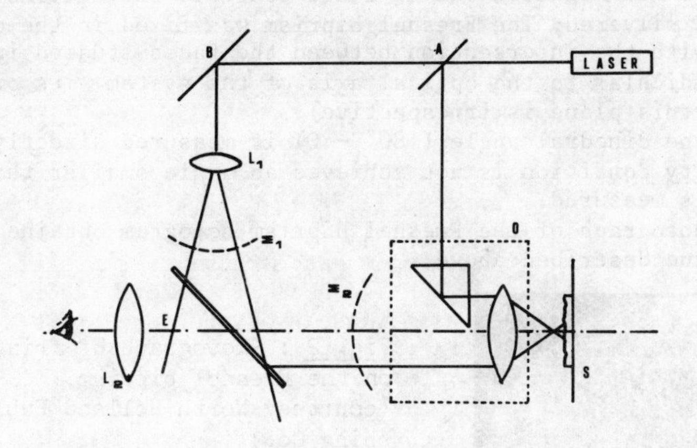

Fig. 1 : Schema of the optical apparatus.
(courtesy North-Holland Publishing Co.)

The beam reflected by A goes through objective 0 (opak illuminator) and hits the crystal surface S. The real image enlarged by objective 0 is recorded on the photographic plate (at 45° on the optical axis SE). Given its position, the plate records the wave front diffracted by the crystal surface at different distances from the object. The inclination of the plate has no effect in this experiment. On the photographic plate the two interfering wave fronts Σ_1 and Σ_2 produce a hologram.

Before removing the photographic plate it is possible to superimpose on it a new interferogram between Σ_1 and Σ'_2, where Σ'_2 is the wave front reflected by the surface S of the crystal shifted by Δx (some tens of microns) along the normal to the optical axis.

After the photographic plate is developed, it is put again in the same position in the optical system and illuminated only with Σ_1. In this way one can see the interference fringes between Σ_2 and Σ'_2 with lens L_2 (several centimeters focal length). The fringes may be photographed with a conventional camera adapted to macrophotography.
The interfringe distances of the pattern and the magnification of the system may be measured from the photographs, while Δx is measured on the micrometric screw.
The measurement of the magnification is done directly by taking a photograph of a hologram of a rule with known interval (in this experiment 1/100 mm).

As stated previously, this method is suitable for measuring dihedral angles between faces or facets of a crystal.

In order to check the method, the angle between the faces of a Fresnel biprism was measured and to avoid spurious reflections the faces were silvered. The Fresnel biprism was fixed in the optical system with the intersection between the faces studied in a plane perpendicular to the optical axis of the system (its orientation within this plane is irrespective).
In this case the dihedral angle (180° - Θ) is measured directly. If the normality condition is not achieved an angle smaller than the real one is measured.
Fig. 2 is a photograph of the Fresnel biprism hologram obtained by the technique described above.

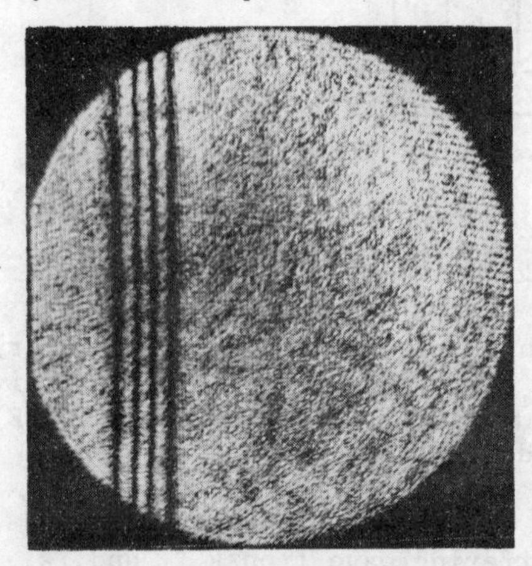

Fig. 2 : Photograph of fringes on the Fresnel biprism. (courtesy North-Holland Publishing Co.)

The interference fringes are parallel to the line of intersection between the faces studied. From the interfringe distance, it is possible to obtain the value of the dihedral angle of the Fresnel biprism using the simple formula

$$\Theta = \lambda / 4 \Delta 1$$

where λ is the wavelength of the light and $\Delta 1$ the interfringe distance. The measured angle (178° 50') has been compared with the value obtained by the usual goniometric methods; the agreement is within 2 %.

The angles between the faces of a diamond trigon have been measured. The diamond is fixed on a glass slide, then the image of the trigon is focused on the photographic plate. In general, none of the three edges of the negative pyramid is normal to the optical axis of the system. Fig. 3 is a typical photograph obtained by a doubly exposed hologram.
The photograph gives the following information :

i) the 3 angles between the projections of the edges on a plane perpendicular to the optical axis (this is the plane of the photograph)

ii) the 3 intersection angles of the dihedrons, each with a plane parallel to the optical axis and normal to the projection of the edge of the dihedron on the photographic plate (by the interfringe distance measurement and as was the case of the dihedral angle of the Fresnel biprism randomly oriented).

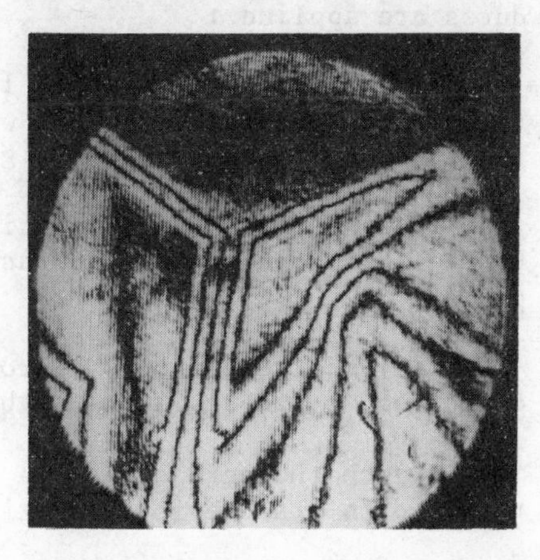

Fig. 3 : Photograph of trigon fringes.
(courtesy North-Holland Publishing Co.)

It should be pointed out that one of these 3 angles is the real dihedral angle if its edge is normal to the optical axis. However, the geometry of the trigon is completely characterized by the six angles since one can obtain from them the three dihedral angles formed by the faces of the trigon.
In this case the values of the dihedrons are 176°12', 175°58', 176°05'. This means that the trigon is very flat.

The holographic technique just described offers some advantages if compared with conventional interferometry.
There is no need for a flat reference surface (which even if rectified, is never perfectly flat). For this reason a higher precision is achieved.
Normal interferometry requires highly reflecting surfaces (while the reference surface must be semireflecting and its silvering is very critical). In the present method, fringes are obtained even if the surface studied has a very poor reflectivity and thus there is no critical need for silvering.
Therefore the range of application of the present method is larger than for conventional interferometry.

II. 2. Depth Contouring Method

The principle of this method[3] is the following. A hologram of an object is formed with the wavelength λ_1. Then the hologram as well as the object is illuminated with a different wavelength λ_2 in such a way that the resulting image of the hologram is nearly coincident with the image of the object.

The interference between the wavefront from the hologram and the wavefront from the object is such that fringes appear on the image of the object. The fringes are true contours of constant level[5-12] if a proper optical arrangement and the following procedures are applied :

i) For the λ_1 wavelength light, the object is illuminated through the beam splitter BS_2 with a plane wave normal to the z axis of the optical system (fig. 4a)

ii) the image (not drawn in fig. 4a) of the object is settled by the telescope system T.S. as near as possible to the holographic plate,

iii) the reference wavefront is plane (the angle between its direction of propagation and the z axis is Θ_r, (fig. 4a),

iv) the holographic plate exposed and developed is carefully replaced in its previous position.

For the λ_2 wavelength light,

v) the holographic plate is illuminated with a plane reconstruction wavefront, as is the surface of the object,

vi) because of the grating effect of the hologram due to the different wavelength used, the reading out of the hologram by the wavelength λ_2 causes both angular and axial displacements of the holographic image from the image of the object given by the optical system.

vii) to compensate the angular displacement, the mirror M_3 is rotated so that the angle between the direction of propagation of the reconstruction beam and the z axis (fig. 4b) takes the value Θ'_r satisfying[8] the equation

$$\sin \Theta'_r = (\lambda_2/\lambda_1) \sin \Theta_2 \qquad (1)$$

In this way the angular displacement is eliminated and the axial displacement produces the interference between the two wavefronts from the holographic image and from the image of the object. The axial distance between two consecutive contour fringes is given by[8]

$$\delta z = \lambda_1\lambda_2/2|\lambda_2 - \lambda_1| \qquad (2)$$

The spacing δl between two successive fringes is measurable, so we can obtain the equation

$$\tan \phi = \delta z/\delta l \qquad (3)$$

For a small angle (as in the present case)

$$\phi = \delta z/\delta l \qquad (4)$$

where ϕ is the angle between the unit vectors of the plane normal to the z axis and of the surface under consideration. The last one is contained in a plane normal to the fringe family.

The method does not directly give the sign of the angle, which may easily be checked by micrometric movements of the object and considering the interfringe variation.

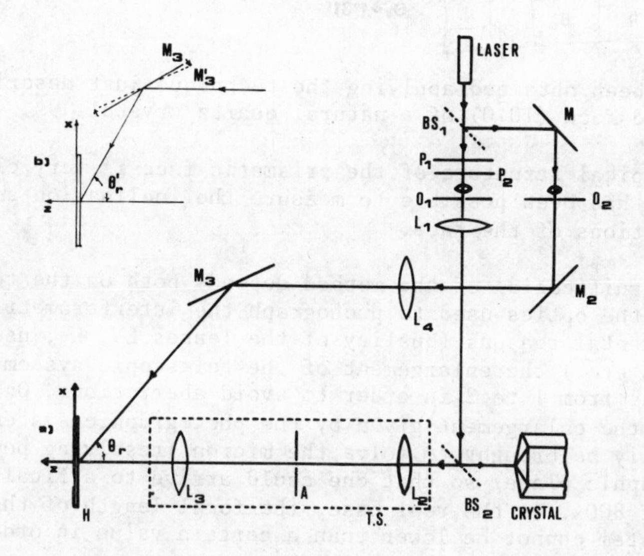

Fig. 4 : a) Schema of the apparatus for two wavelength holographic contour mapping : BS = beam splitters, P = polarizers, O = objectives, L = lenses, A = aperture, T.S. telescope system, M = mirrors, H = holographic plate, Θ_r = angle between the reference beam and the z axis. b) Position of the mirror M in the reconstruction stage of the hologram : Θ_r = angle between the reconstruction beam and the z axis.
(courtesy North-Holland Publishing Co.)

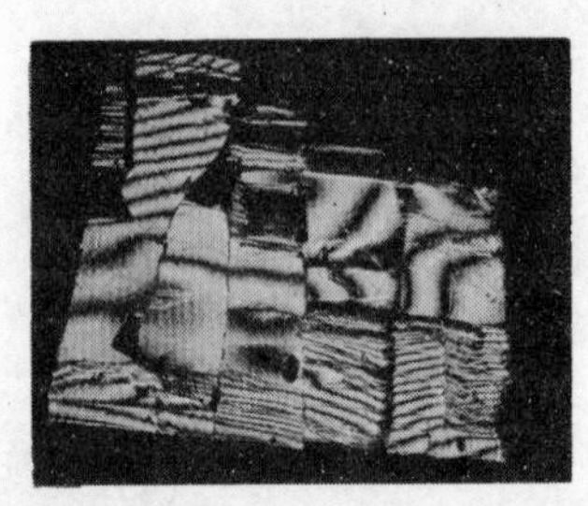

Fig. 5 : a) Contour hologram of the silvered face $(10\bar{1}0)$ of a natural quartz crystal, magnification 20x; b) Map of the face $(10\bar{1}0)$ of the quartz crystal with some measured slopes. (courtesy North-Holland Publishing Co.)

a

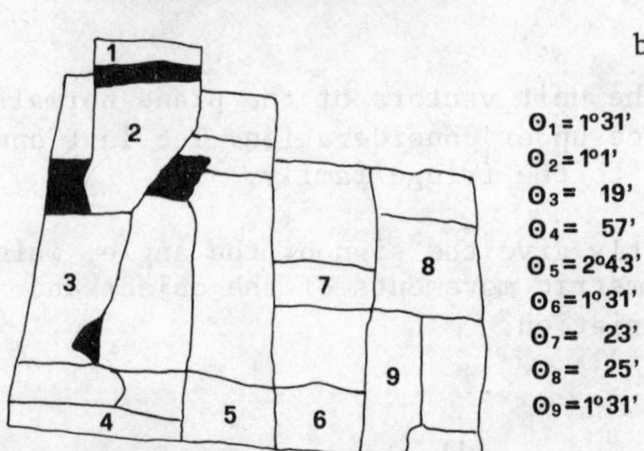

b

Fig. 5 has been obtained applying the technique just described to the silvered face $(10\bar{1}0)$ of a natural quartz crystal.

The typical structure of the prismatic face is very clearly seen and it has been possible to measure the inclination angles of some portions of the face.

The magnification of the method depends both on the telescopic system and the optics used to photograph the interferometric image. For experimental reasons (quality of the lenses L_2, L_3, use of the diaphragm A etc.) the enlargement of the telescopic system must have a value from 1 to 2 in order to avoid aberrations. On the other side the enlargement given by the photographic system may theoretically be brought to solve the micron (resolving power of the holographic plate) so that one could arrive to a final magnification of 800x. In the real case, the focal length of the photographic system cannot be lower than a certain value in order to avoid the light of the reconstruction beam.

The range of use of this method[8] has been calculated to be dependent on the following condition

$$|Z|(\tan \phi)^2 < \lambda^3/2(\lambda_2 - \lambda_1)^2$$

where Z is the depth of the image of the object and ϕ is the slope of the surface as defined above.
In the inclination range $0 < \phi < 10°$ and in the best experimental conditions it is possible to measure in the slope variations down to 1/10 of a degree.
The method is better suited for very small angles since the system works best for nearly paraxial rays.
The error in the evaluation of ϕ is due to the measurement of the interfringe distance δl and is between 5 and 10 %.
It should be noticed that to apply the technique just described, rather sophisticated tools are required as well as good familiarity with holographic interferometry.

The common feature of the two methods described above is that the interferometric image is found immediately, either by using one wavelength but slightly shifting the position of the crystal, or by using light with two different wavelengths but keeping the crystal position fixed.
With the first method it is possible to measure local slopes; the second method results in a depth map on the whole surface.
The two methods are then complementary and may be performed directly on even a poorly reflecting sample, without any silvering and matching flat as in multiple beam interferometry. Silvering of the sample is needed only if the crystal is a transparent one, in order to avoid spurious reflections.

III. HOLOGRAPHY IN CRYSTAL GROWTH FROM SOLUTION

Holographic interferometry has proved to be a very powerful tool to study crystal growth from solution.

A particular optical schema has been realized[13,14] to check what happens in a water solution where a crystal of $NaClO_3$ grows. According to this procedure it has been possible to study two cases growth by molecular diffusion and growth by convective movements plus diffusion. This has been achieved with a horizontal and a vertical configuration of the thin vessel with the solution.

In a horizontal setting of the system, given in the schema of fig.6, 2 types of growth were inferred from the shape of the interference fringes that are equiconcentration lines in the solution.

a) regular layer growth (the crystal keeps flat faces); the fringes, parallel to the faces, cross the faces near the edges (fig. 7).

b) dendritic growth (the crystal has concave faces); the fringes outline the crystal without intersecting its edges (fig. 8).

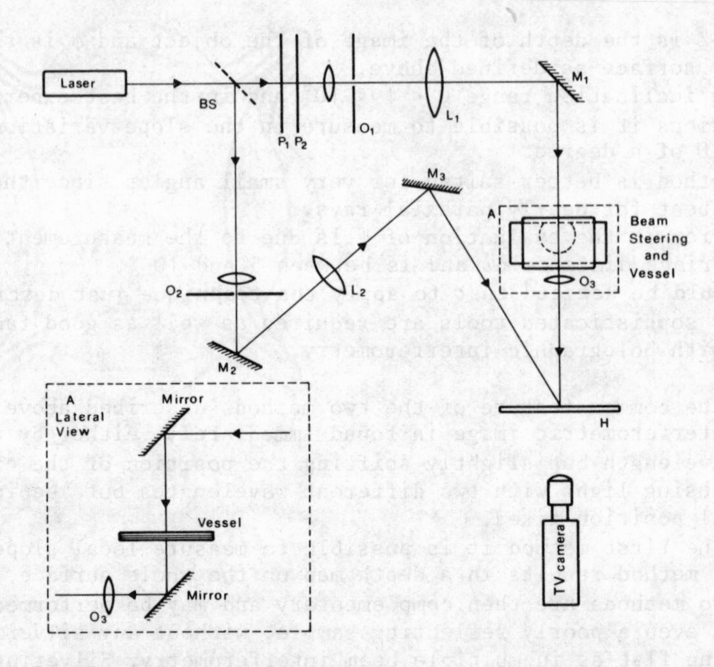

Fig. 6 : Schema of the apparatus used with horizontal placement of the vessel : BS = beam splitter, P = polarizers, O = objectives, L = lense, M = mirrors, H = holographic plate, A = beam steering and vessel ; a vertical view of A is given in the bottom left.

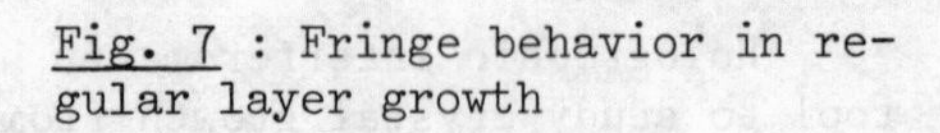

Fig. 7 : Fringe behavior in regular layer growth

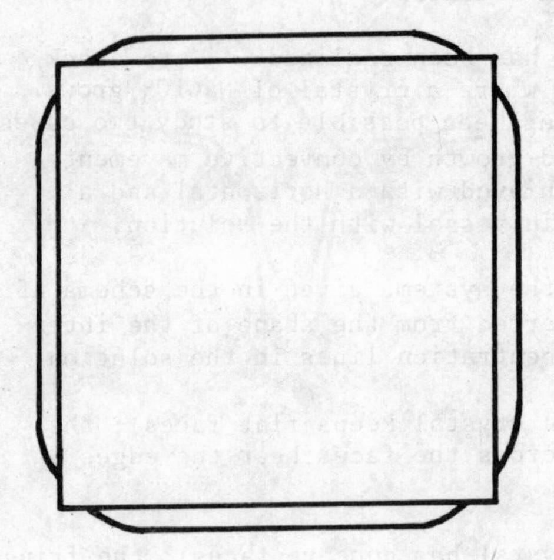

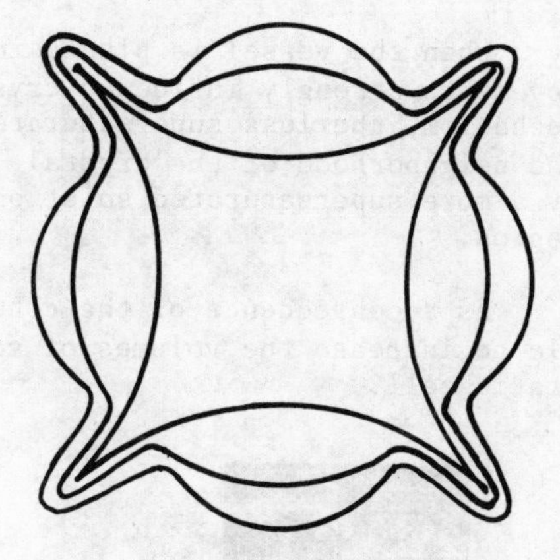

Fig. 8 : Fringe behavior in dendritic growth.

BOSCHER[15] obtained a qualitative agreement between the equiconcentration contours deduced from Fick's second law for two dimensions and the contours obtained experimentally.[16]

The growth velocity measured at the center of a face and the one calculated from Fick's first law for one dimension do not fit. This is because many factors are not taken into account. First of all, convection at different regimes and with different geometries strongly modifies the mechanism of growth. Also, no difference is usually considered as to the type of the face (F,S,K) and of the possible presence of defects.

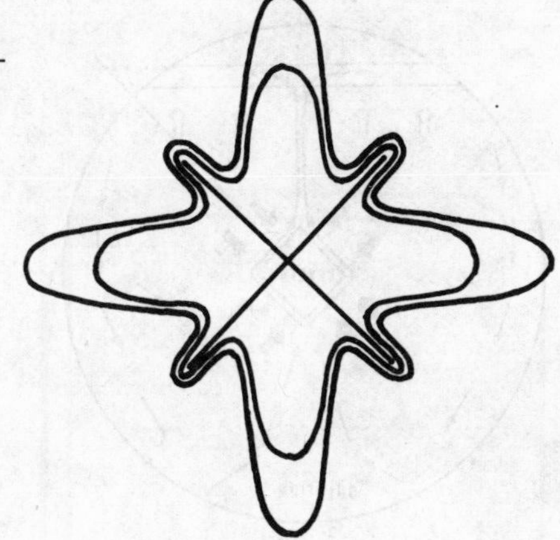

Fig. 9 : Fringe behavior in skeletal growth.

It is worth pointing out that when the growth goes on as skeletal, the fringes of equiconcentration have the shape of Fig.9.

The transition from the stage of Fig. 9 is not smooth; there is rather an intermission with high turbulence in the solution.

When the vessel is placed in a vertical position, convective movements strongly influence crystal growth. By this convective mechanism, the less supersaturated solution is thrown out from the neighborhood of the crystal, goes up in a plume and is replaced by a more supersaturated solution coming from the surrounding region.

As a consequence of the coherence of laser light, it is possible to increase the volumes of solution up to 30 cm^3 in a thermostatic cell.

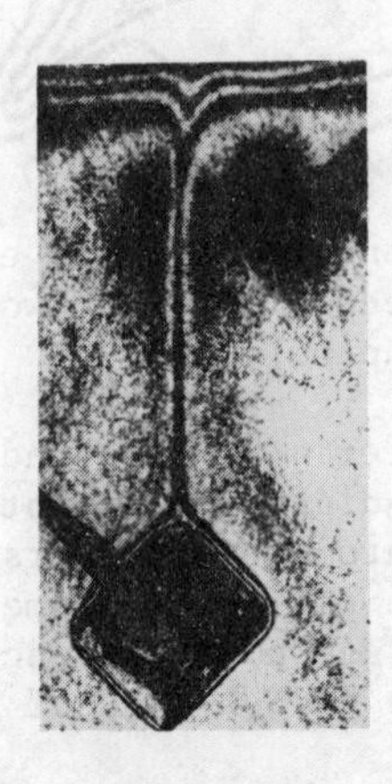

Fig. 10 : Photograph of fringes produced by convection

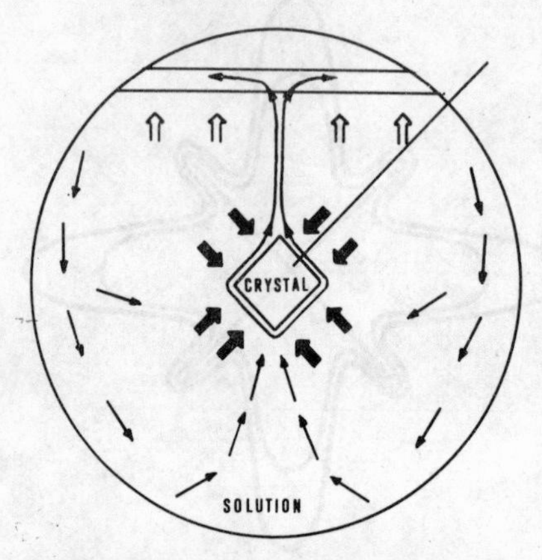

Fig. 11 : Scheme of the flows in the liquid.

From the fringes shown in fig. 10 it is possible ot obtain a schema (fig. 11) of the movements that take place (in a particular regime) in the solution during the crystal growth process.
By natural convection, the exhausted solution goes up in a plume to the surface where it forms a thick layer.
The arrows show a double effect : the motion of the solution that replaces the lighter one gone away from the crystal by the convec-

tive plume and the diffusion of the solute, caused by the gradient of concentration around the growing crystal.

The evolution of crystallization heat causes a thermal gradient which, measured with thermocouples, has a value of the order of one hundredth of a degree. This confirms, at least in this case, that convective movements are mainly due to the concentration gradient and negligibly affected by the thermal gradient.

The spatial configuration of the solution concentration is probably very complex inside the plumes and around the crystal and different at different regimes, but in the cases observed for $NaClO_3$ may be not too far from the one given in the scheme of Fig. 11. A real knowledge of this concentration distribution would help very much to understand the correlation between the growth rate of the faces of a crystal and the concentration gradient in the solution.

With the type of holographic interferometry just described, it is impossible to achieve this result, but modifying the optical schema and using mathematics these difficulties may be bypassed, as will be sketched below. The intensity transmitted by a real time interferogram may be described by the equation

$$I(x,y) = k\left[1 + \cos\{\phi_1(x,y) - \phi_o(x,y)\}\right]$$

where (x,y) is a point of the image plane, k is a constant and $\phi_1(x,y) - \phi_o(x,y) = \Phi(x,y)$ is the phase difference due to a variation of the refractive index caused by the change in the state of the object.

Observing the holographic interferogram along the z direction normal to the image plane, the function $\Phi(x,y)$ is defined by the equation

$$\Phi(x,y) = \frac{2\pi}{\lambda}\int_r f(x,y,z)\,dz, \qquad (5)$$

where $f(x,y,z) = n(x,y,z) - n_o$ is the change of the index of refraction, λ is the wavelength of the laser used, n_o is the refraction index at room temp. The integration is made along the entire object field.

If the solution is crossed by a plane wave, equation (5) may be inverted and one obtains the three dimensional field of the refractive index in two cases : when the field is constant along the z direction or if it has a radial symmetry.

If the field is not constant, with this technique a mean value may be measured.
A test has been performed for a field with radial symmetry. A vessel with parallel faces contains a particular gel (in this case tetramethoxysilan).
A fine constantan wire is tightened vertically in the middle of the vessel. If current is passed through the wire, the heat diffuses isotropically in the gel increasing its temperature. Density decreases as temperature increases and therefore the refractive index varies.

Fig. 12 : Hologram showing the variation of the index of refraction in a heated gel.

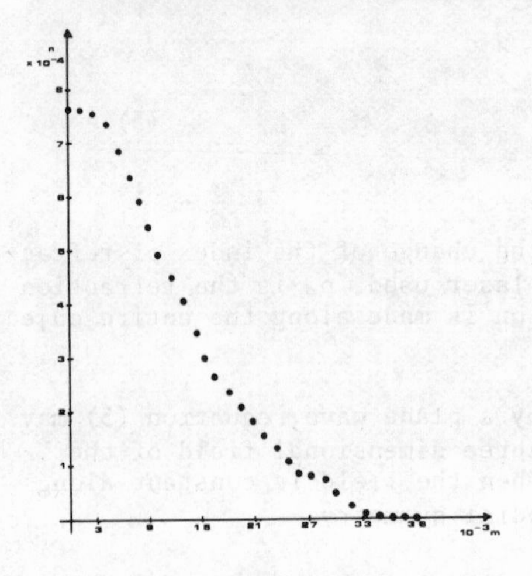

Fig. 13 : Plot of the radial variation of the refractive index.

The variation $n(x,y,z)-n_o$ has been obtained by a holographic interferogram (fig. 12).
The plot of the light intensity of this interferogram transformed with some mathematical techniques into a radial variation plot of the refractive index is given in fig. 13.

IV. MULTIDIRECTION HOLOGRAPHIC INTERFEROMETRY

This technique uses a divergent pencil of rays and one can register at the same time the phase variations of a set of plane waves that cross the solution along different directions. These may be obtained, for instance, through a diffuser or a phase grating or with other optical tools.
The point is that the hologram may store all the information contained in a general wavefront, no matter how complicated it is, as it happens for the wavefront from a diffuser or for the wavefront diffracted by a phase grating.

In order to select and register on a photograph or a video tape the waves that propagate in different directions, an optical system with a spacial filter may be needed.

In the case of this experiment it is possible to write down a system of equations of the type

$$\Phi_k = \frac{2\pi}{\lambda} \int [n(x,y,z) - n_o]\, dz_k \qquad (6)$$

where z_k is the k-th direction of observation and Φ_k is the phase variation relative to the direction z_k, and the integrals are calculated on straight lines with the approximation of no refraction. The solution of the set of equations (6) can give the distribution of the refractive index in every type of field, either symmetric or asymmetric.

SWEENEY and VEST[17,18] and MATULKA and COLLINS[19] have demonstrated the possibility of reconstructing asymmetric fields in fluids as a function of a discrete series of data collected on optical directions contained in an angle of view of 60°. The condition for such a reconstruction is that it must be possible to disregard refraction of the light, since it is negligible on the propagation directions of the waves inside the fluids. In the experiments of crystal growth from solution an assumption of the refractionless condition may also be reasonably done for the convective solution plumes.

Experiments have been made using a diffuser or a phase grating and registering double exposure holographic interferograms.

The phase grating has the advantage that the angle of spreading is larger and the direction of observation of each diffraction order (itself a hologram) is not critical.

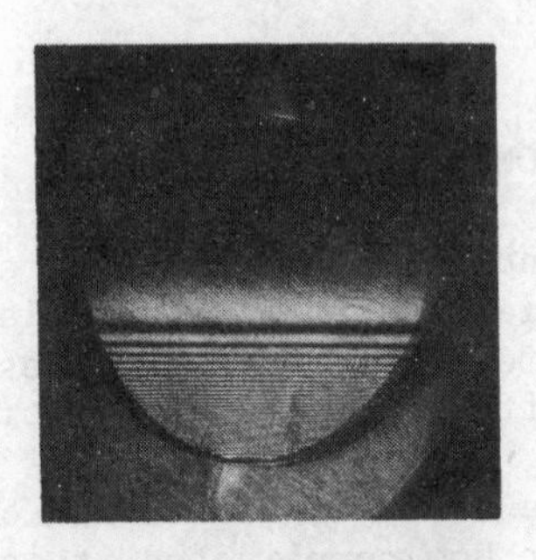

a

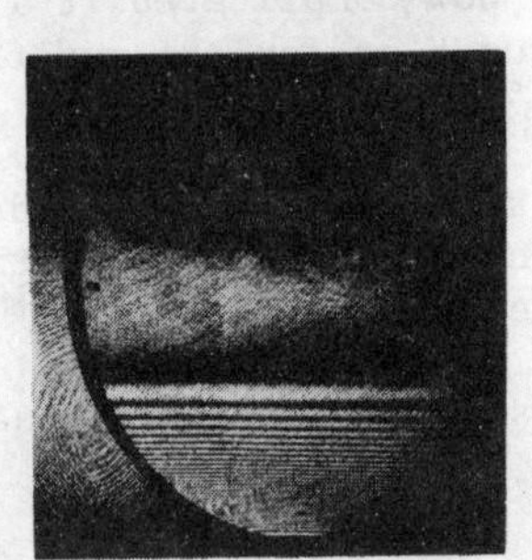

b

Fig. 14 : a) and b) Two different views obtained by different diffraction orders of a phase grating. With eight of such different images it is possible to reconstruct point by point the concentration field. The vessel used is 3mm thick.

Two of the first examples obtained with a phase grating are given in fig. 14. The angles of view in a and b are different, hence the patterns are different.
The experiment may be fully described with mathematical methods which are matter of research at this moment. The goal to be achieved is a three dimensional mapping of the concentration[20] point by point in the solution during the crystal growth.

ACKNOWLEDGEMENT

I wish to thank warmly my colleagues P. Boccacci C. Pontiggia and L. Zefiro, since this lecture is the result of the cooperative work of our group of research. I gratefully aknowledge the kind permission for the use of VDC 501 apparatus in the Bioengineer Laboratory of the Electronics Institute (Genova University). I am indebted to Madam Lefauchaux and Madam Robert of the University of Paris VI for the suggestion of the use of a gel in some part of the experiment.

REFERENCES

1 . LIPSON S.G. and LIPSON H., in *Optical Physics*, Cambridge Press, 1969
2 . BEDARIDA F., ZEFIRO L. and PONTIGGIA C., J. Cryst. Growth, 24/25 (1974) 327
3 . BEDARIDA F., ZEFIRO L. and PONTIGGIA C., J. Cryst. Growth, 34 (1976) 79-84
4 . TOLANSKY S., in *An Introduction to Interferometry*, Longmans, 1960
5 . MEIER R.W., J. Opt. Soc. Am., 55 (1965) 987
6 . CHAMPAGNE E.B., J. Opt. Soc. Am., 57 (1967) 51
7 . HAINES K.A. and HILDEBRAND B.P., Phys. Letters, 19 (1965)10
8 . SELENKA J.S. and VARNER J.R., Appl. Op., 7 (1968) 2107
9 . VARNER J.R., Developments in holography, S.P.I.E., Seminar Proc. 25 (1971) 239
10. ROBERTSON E.R. and ELLIOT S.B., in *Applications de l'Holographie*, Besancon, 1970
11. HEFLINGER L.O. and WUERKER R.F., Appl. Phys. Letters 15 (1969) 28
12. FRIESEM A.A., LEVY U. and ZILBERBERG Y. Conf. on engineering uses of coherent optics, Univ. Strathclyde, Glasgow, 1975
13. BEDARIDA F. and CIMMINO F., Rend. Soc. It. Min. Petr., XXX (1975) 483-493
14. BEDARIDA F. and DELLA GIUSTA A., Per. Min., 37 (1968)503-509
15. BOSCHER J., Annales de l'Association internationale pour le calcul analogique, 4 (1965) 117
16. GOLDZTAUB S., ITTI R. and MUSSARD F., J. Cryst. Growth, 6 (1970) 130-134
17. SWEENEY D.W. and VEST C.M., Appl. Optics, 12 (1973) 2649-2664
18. SWEENEY D.W. and VEST C.M., Int. J. Heat Mass Transfer (1974) 1443-1454
19. MATULKA R.D. and COLLINS D.J., J. Appl. Phys. 12 (1971) 1109-1119
20. BEDARIDA F., BOCCACCI P., PONTIGGIA C. and ZEFIRO L., A review on the applications of holographic interferometry to hydrodinamic phenomena in crystal growth. In press on "Physico Chemical Hydrodinamics- The international Journal".

STUDIES OF SURFACE MORPHOLOGY ON AN ATOMIC SCALE

H. BETHGE

Akademie der Wissenschaften der DDR,
Institut für Festkörperphysik
und Elektronenmikroskopie
GDR-402 Halle/Saale.

ABSTRACT

Electron microscopical methods for imaging of step structures on surfaces of macroscopic crystals are described. The decoration technique is given special emphasis because it has been used in the most detailed experiments. Prerequisites for the decoration effect are described and special conditions under which the decoration technique can be applied to different crystals are defined.

For a so-called nearly perfect surface only step structures on an atomic scale are determining. Generation of such step structures, dependent on the pre-treatment or on the formation of the surface, is discussed for the investigated surfaces of NaCl crystals.

I. INTRODUCTION AND SURVEY OF METHODS FOR IMAGING OF STEP STRUCTURES ON NEARLY PERFECT SURFACES

There are many methods of investigating surface structure and much of the work done in surface physics during the past ten years was the development of new characterization methods. Besides the problem of the lattice structure of the clean surface, understanding of the sorption processes, including the interaction at the interface, was of special interest. Recently, the growth of thin films has been studied increasingly by means of different methods of surface physics.

Especially for the initial state of adsorption and growth of thin films it is necessary to study the morphological structure of the surface in addition to its real lattice structure. On a real

B. Mutaftschiev (ed.), Interfacial Aspects of Phase Transformations, 669–696.

surface, the atomically smooth low-index regions are relatively small and always limited by steps. Expansion of the smooth regions is realized by specific preparation or by recovery due to special treatment of the surface. We should distinguish between elementary steps and higher steps. As seen schematically in Fig. 1, elementary steps are either the height of one lattice plane distance or the height of the lattice unit.

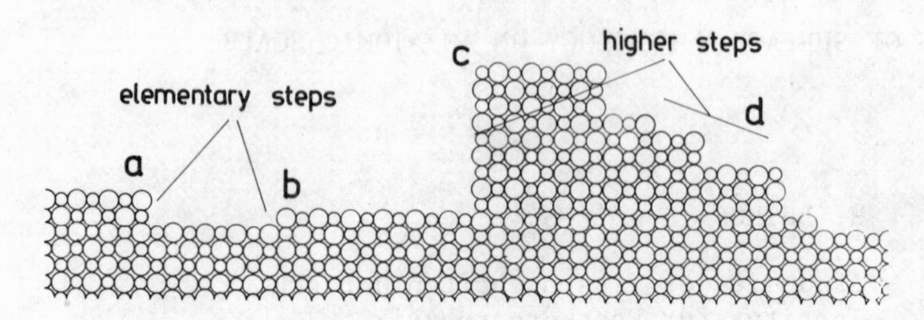

Fig. 1 : Schematic drawing of the cross-section of a surface containing steps.

Which of the two cases is involved in a given surface structure depends on the process of formation of the steps, and if this process is determined by dislocations, then the Burgers vector is decisive. Higher steps, e.g., cleavage steps, show a terraced fine structure as a rule. It should be mentioned, however, that no method exists to investigate a microstructure according to Fig. 1d. A surface which only shows elementary steps will be called a nearly perfect surface. Only such surfaces and step structures will be dealt with in the following.

The decoration technique is a special electron microscopic method which is easy to apply for studying step structures on a nearly perfect surface. Our own studies[1] were performed with the aim of "developing" step structures in such a way that the same material is evaporated on a surface under conditions such that preferred growth takes place at the steps. An example of this "homogeneous" decoration is given in Fig. 2. NaCl was evaporated onto the cleavage face of an NaCl crystal at a temperature of 200°C. A small-angle grain boundary ran horizontally and the cleavage direction was from bottom to top. One sees that the dislocations arranged within the boundary act as points of preferred growth and that, due to this preferred growth, a wall-like reinforcement appears at the steps proceeding from the screw dislocations. The predominant Burgers vector for dislocations in an NaCl crystal is a/2 <110>, and the height of the steps should therefore be only one atomic layer (2.81 Å). Critical for the application is the control of various parameters, among which the evaporation rate is of special importance.

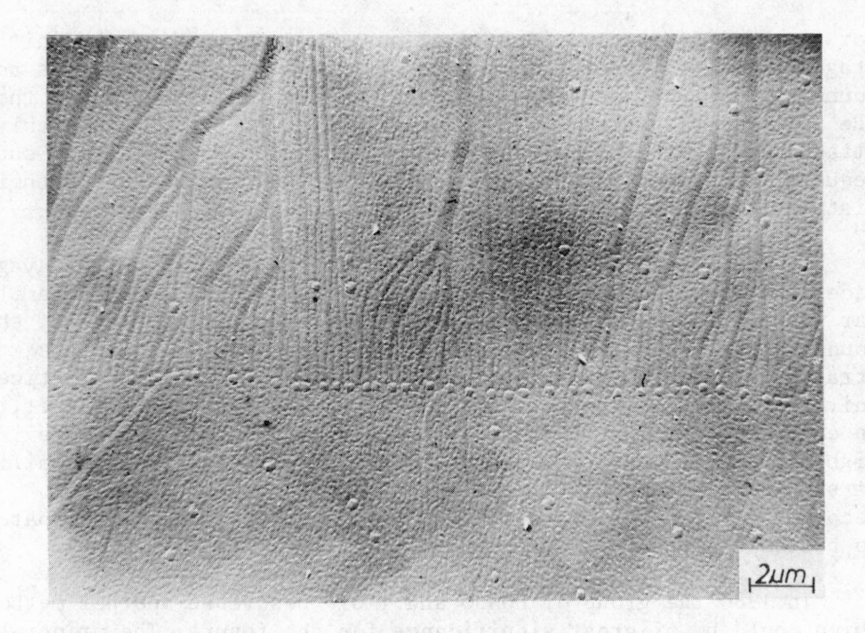

Fig. 2 : Development of monatomic steps on an NaCl (100) cleavage face due to NaCl evaporation.

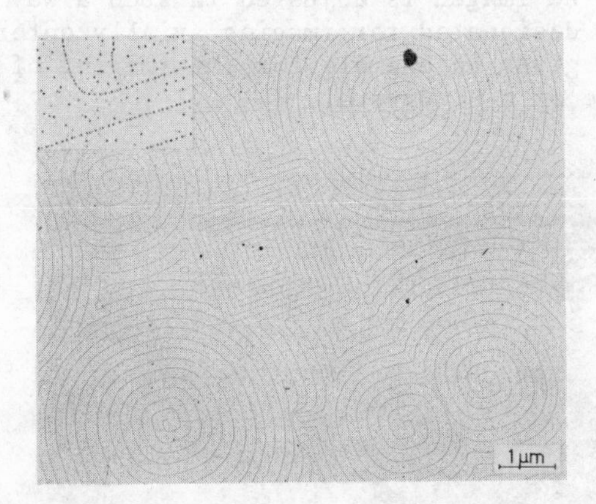

Fig. 3 : Dislocation-induced step systems on NaCl (100) cleveage face after evaporation (90 min, 400°C). Au decoration.

Simultaneously with our studies, a decoration effect by means of gold evaporation was described by BASSET.[2] Compared to homogeneous decoration, decoration with heavy metals is much easier to carry out and gives better contrast and higher lateral resolving power for electron microscopy investigations. Bassett studied the initial stage of the growth of thin fcc metal layers on alkali halides and found that three-dimensional nucleation initially takes place, then the small metal nuclei, with diameters of 10 Å, become preferably attached along the steps on a surface. The gold particles are subsequently embedded in a carbon replica and by observation of their distribution, the step structure of the surface is revealed.

In Fig. 3, Au decoration is illustrated. A NaCl(100) cleavage face is evaporated in a vacuum, thereby forming well defined lamellar systems.[3] It should be emphasized that the curved steps of the round spiral systems are one atom layer high (Fig. 1a) and the straight steps are twice as high, i.e., the height of the lattice unit (Fig. 1b). Where these two different lamellar systems meet, in each case two curved steps join one straight step. The more highly magnified part shows the decoration effect in more detail. It should be noted that the size and density of the Au nuclei decorating the steps are different for monatomic steps as compared to biatomic steps.

In 1980 the group of HONJO and YAGI[4] described another method which could be of great significance for the future. The principle idea is to improve the well-known method of electronic imaging of surfaces by electrons reflected at a flat angle[5] by using the electrons diffracted under a defined Bragg angle for imaging. The surface of the crystal to be imaged is adjusted in such a way that the diffraction reflex designated for imaging axially enters the objective lens. Fig. 4 gives an example for the imaging of atomic steps on the (111) face of a Si crystal.

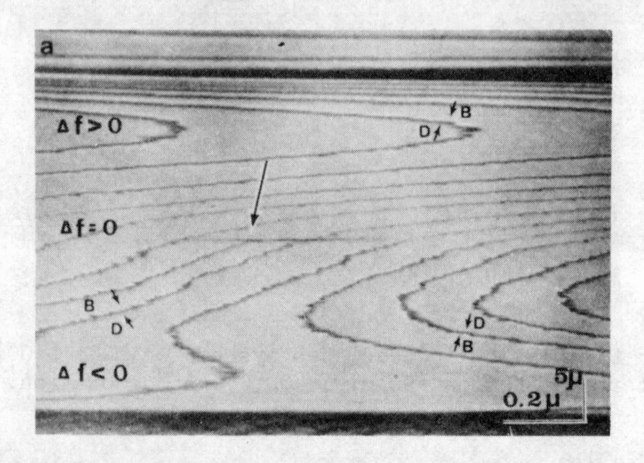

Fig. 4 : A reflection electron micrograph of (111) silicon surface taken with a (444) Bragg reflection. The arrow marks a screw dislocation emerging to the surface. (Adopted from Ref. (4)).

Because imaging of the step structure is possible only with clean surfaces, it is necessary for the specimen to be placed in UHV. The so-called Fresnel diffraction effect as an explanation for the remarkable contrast of the steps is emphasized. A more detailed treatment based on dynamical theory is still lacking. It is also an open question why sorption layers on a crystal hinder imaging of steps or at least strongly influence the contrast. Possibly, the configuration of the atoms in the step is influenced by the sorption layers, and thereby the elastic distortions of the atomic configuration are altered. A further notable result is that during a thermal treatment, the phase transition process from the (1 × 1) to the (7 × 7) structure could be observed, with reconstruction always beginning at the steps.

In a recently published work,[6] the Honjo-Yagi group described the imaging of steps on (100) MgO surfaces using reflection electron microscopy. The observed shape of the step structures is quite different from well-known cleavage structures (as discussed in section II) but the special cleavage process by electron beam flashing possibly influences the formation of steps.

The decoration technique and reflection electron microscopy permit investigation of the surfaces of bulk crystals. In special cases, imaging of step structures on the surface is possible with the conventional transmission electron microscope, in order to study thin films or two-dimensional extended small particles.

Fig. 5 : Transmission electron micrograph of an Au(111) film epitaxially grown on a (111) Ag substrate.

Imaging of monatomic steps was first demonstrated by CHERNS[7] with a dark-dield technique using forbidden reflections and by LEHMPFUHL[8] with a weak-beam technique. In our laboratory, the extinction contrast of strongly excited systematic (111) and (200) reflections was used in bright- and dark-field imaging of monatomic steps for very thin Au(111) films with a mean thicknesses of 3 - 30 monolayers.[9] Fig. 5 shows a Au(111) film of a mean thickness of 10 Å, grown by vapor deposition on the (111) face of Ag at 150°C. Monolayers of gold start either from high substrate steps of form growth spirals at screw dislocations. In the bright-field image an increase of contrast with increasing thickness is seen at the terraces; this increase agrees with the calculated thickness dependence of the extinction contrast for the present dynamical many-beam case. The additional enhancement and reversal of contrast at the steps is possibly connected with a Fresnel diffraction effect.

The decoration technique is discussed more in detail in the following. Using this method, some selected investigations are described which are of general interest for the interpretation of processes on surfaces.

II. APPLICATION OF THE DECORATION TECHNIQUE TO IMAGING OF STEPS ON NaCl SURFACES

Most alkali halides can be decorated without difficulty. If conventional oil-diffusion vacuum pumps are used, heating the crystal to about 150°C is recommended to avoid contamination of the surfaces. The results described in this section were obtained under these simple conditions. For other crystals or for special requirements, the decoration is much more complicated, as discussed in section IV.

Alkali halides are advantageous for obtaining crystallographically defined surfaces because cleavage faces are easy to produce. But for NaCl and some other alkali halides, a cleavage face produced in free atmosphere after decoration is more or less reconstructed and the step structures generated have nothing to do with the cleavage process. This is demonstrated in Fig. 6 and 7. In Fig. 6 the micrograph shows the typical structures after cleavage in free atmosphere. Water vapor has dissolved the surface, and the subsequent heating in vacuum has recrystallized it. Thus, the decorated steps are growth steps.[10]

For detecting real cleavage structures of an NaCl crystal, it is advisable to perform the cleavage in vacuum, otherwise care must be taken that after cleavage no water condenses on the as-cleaved face. The typical patterns of cleavage structures on (100) cleavage faces are shown in Fig. 7. The arrows indicate the respective direction of cleavage. The lightining-shape cleavage steps seen in Fig.7b are called elementary cleavage structures.[11]

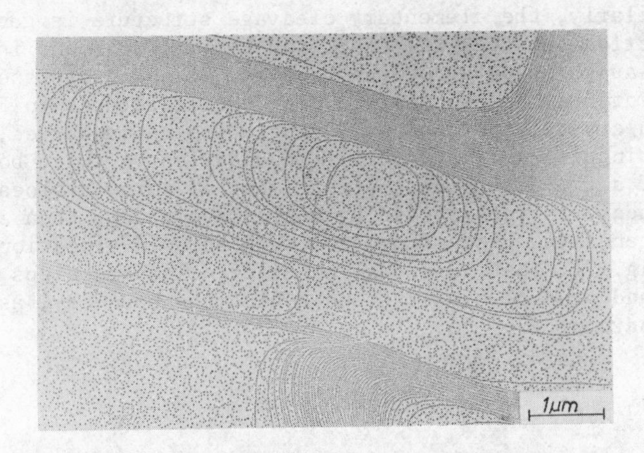

Fig. 6 : Step structures formed by recrystallization after dissolution of the surface of an NaCl crystal in free atmosphere.

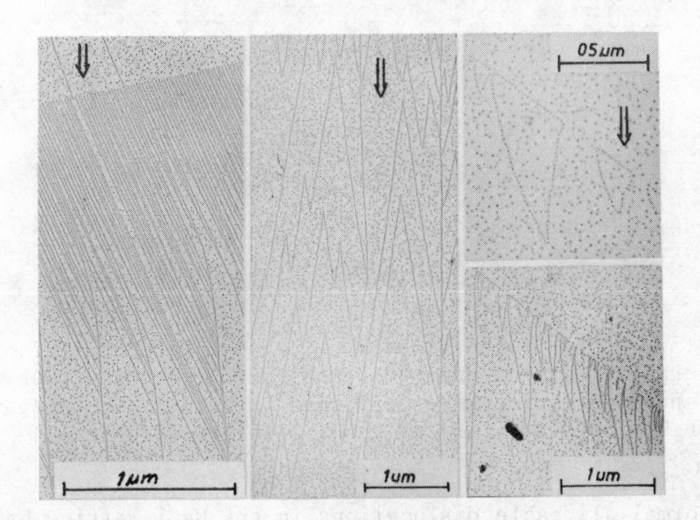

Fig. 7 : Typical patterns of cleavage structures on an NaCl (100) cleavage face.

The steps proceeding from a tip are monatomic. The density of these steps, i.e., the number of steps per face unit, as well as the enclosed angle, depend on the velocity of the cleavage crack. To produce a cleavage face having as small as possible density of cleavage steps, a crack velocity as high as possible is necessary.

The structures in Fig. 7c can be explained by special dislocation processes which take place in the crack tip during crack propagation. Similarly, the elementary cleavage structure is generated by interactions between the crack and the dislocations in the crystal.[12] Cleavage steps according to Figs. 7c and d are special cases of the more common process yielding the structures in Fig. 7b. Normally, each crystal contains small-angle grain boundaries, among which one distinguishes twist and tilt boundaries. A twist boundary consists of a two-dimensional arrangement of screw dislocations. If such a boundary is intersected by a cleavage crack, then a cleavage step is generated proceeding from each screw dislocation. This is shown in Fig. 7a. The individual monatomic cleavage steps often run together and produce higher steps behind the boundary, as seen in the lower part of Fig. 7a.

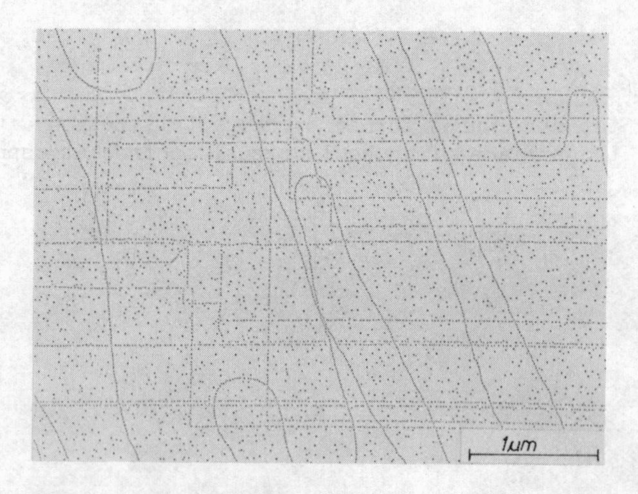

Fig. 8 : Glide steps of individual moving dislocations ; after a special thermal treatment numerous cross slip processes took place. Annealing and deformation were performed in vacuum. The steps running diagonally belong to the original cleavage structure.

The normal slideable dislocations in the NaCl lattice have Burgers vectors of a/2 <110> . Such dislocations are often activated by elastic strains immediately after cleavage. If a moving dislocation emerges to the surface and the pertinent Burgers vector has a component normal to the surface, then the dislocations produce a glide step. According to the slip planes possible in the NaCl lattice, the glide steps on the surface always run in <100> directions. Such glide steps thus permit the description of the crystallographic orientation of surface regions in the electron-microscopic dimension. Fig. 8 represents an image showing a number of glide steps. The more or less curved steps running almost diagonally belong to the original cleavage structure which, as described above, was drastically changed by a weak H_2O dissolving effect after cleavage in free atmosphere. The slip structures were produced during deformation in vacuum. After special thermal pretreatment the glide steps show typical processes of cross slip. It should be mentioned that the decoration technique is an excellent method for studying the kinetics of individual dislocations. By means of special techniques, it is even possible to detect the kinetics of slip processes.[13]

Cleavage structures, mostly connected with step structures which are due to deformation processes, occur in a great variety.[14] More defined surfaces, e.g., with regard to a relatively well defined density of steps, are obtained if an NaCl crystal is evaporated in vacuum. At temperatures above about 300°C in a high vacuum, the surface of an NaCl crystal is decomposed by evaporation. Under clean conditions, i.e., cleavage in vacuum, the evaporation takes place first at the steps which thereby changes their shape and position. Besides decomposition at the steps on the free face, "Lochkeime" originate, i.e. diskshaped holes of monatomic depth.

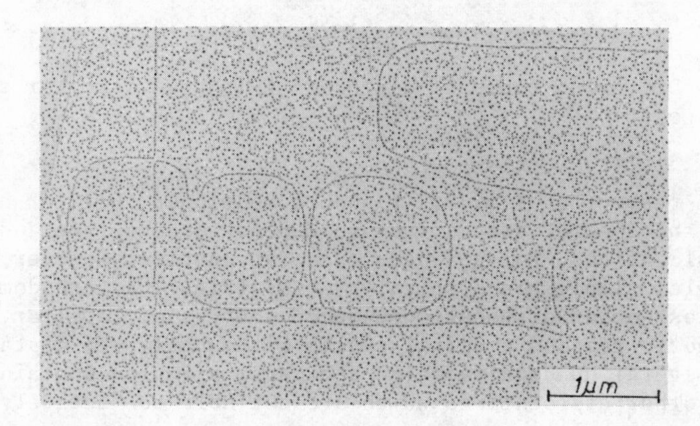

Fig. 9 : Change of the shape of cleavage steps and generation of "Lochkeime" as first stage of evaporation (cleavage in vacuum, thermal treatment : 320°C, 3 hours).

Fig. 9 shows corresponding evaporation structures formed in the first monolayer of a cleavage. With increasing evaporation, starting from the emergence points of dislocations, the lamellar systems generated by the dislocations eventually determine the step structure of the surface, as seen in Fig. 3. The individual types of lamellar systems determined by the dislocations are represented in Fig. 10. Without going into detail (see Ref. (15)) it may be mentioned that the observed systems can be well interpreted by the dislocations existing in an NaCl crystal and by the splitting of dislocations just below the surface (types b, d, and e respectively). The circular lamellar system belongs to a dislocation with a Burgers vector parallel to the surface, i.e., it belongs to an edge dislocation.[15]

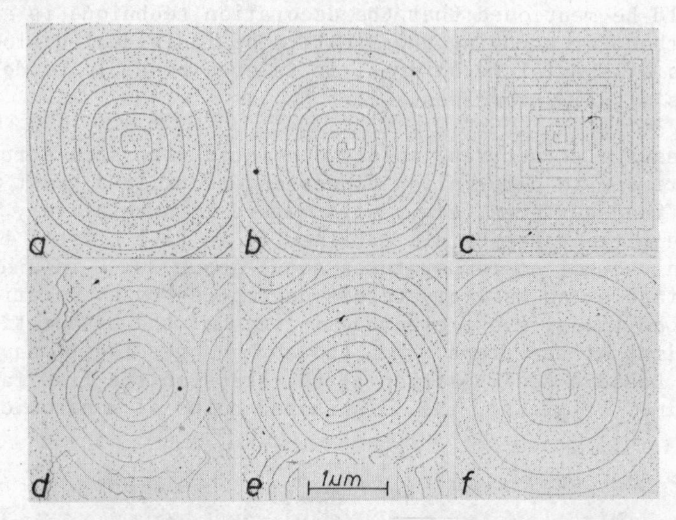

Fig. 10 : Different types of dislocation-induced lamellar systems after evaporation.

With respect to Fig. 3, it should mentioned that for a given surface, the special step structure depends on the type and density of dislocations. For a longer evaporation time, however, as a rule single, simple round spiral systems (Fig. 10a) are dominating and suppress other dislocation-induced sources of lamellar systems. For energetic reasons a spiral system with small step distance can produce several hundred lamellae and cover a region containing numerous other dislocations which are not able to "actively" produce their own lamellar systems.

Previously it was assumed that the steps of round spirals or circular steps of the cleavage structure and the glide steps of single moving dislocations are of monoatomic height. This conclusion is justified because in all the cases mentioned, the Burgers vector of the dislocations determines the step height. The glide steps on the surface can be used as an indicator for obtaining information on the height of steps when a glide step intersects steps of unknown height. By slight dissolution with water vapor or by slight evaporation at the crossing points, the steps are rounded and the rounding patterns thus produced can be used to determine the step height.[16] In Fig.11a three steps, A, B, and C, cross step G running horizontally.

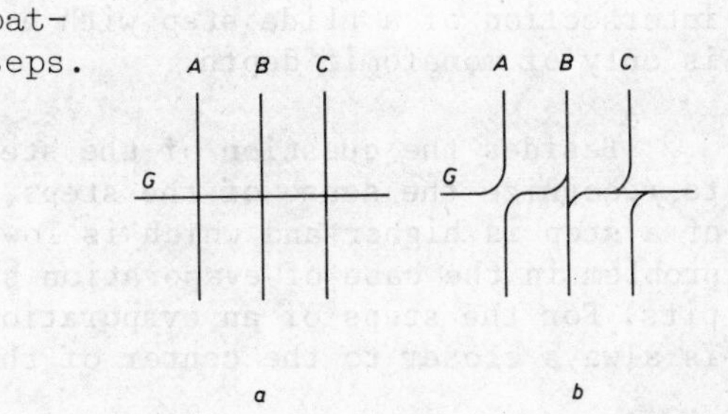

Fig. 11 : Schematic demonstration of the formation of rounding patterns at crossing points of steps.

Fig.11b shows one possible image of decoration after a treatment producing the rounding structures. By the interaction of step A with step G, it follows that both steps must be of the same height. One sees that step B is higher than step G, and as shown on the right side of Fig. 11b, step G must be relatively higher than step C. But if step G is a glide step and if it also shows a cross slip, then it must be a monatomic glide step; therefore it follows that step A must also be monatomic. To give more than a relative statement for the cases B and C is more complicated. For more detail, refer to previous work in which we have "reconstructed" the step structure of a given surface.[17]

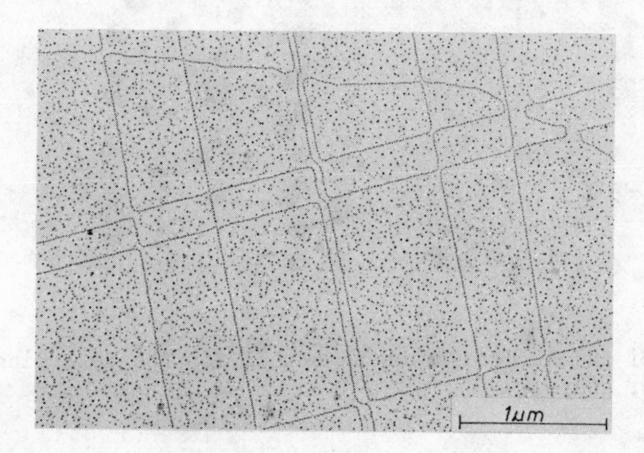

Fig. 12 : Rounding patterns according to Fig. 11 between glide and cleavage steps, produced by slight evaporation.

The simple example shown in Fig. 11 is given in Fig. 12 for the interaction of glide steps with cleavage steps. The rounding structures were produced by slight evaporation. Fig. 9 should also be mentioned. In that micrograph, the rounding pattern of the intersection of a glide step with a "Lochkeim" reveals that this is only of monatomic depth.

Besides the question of the step height, it is often necessary to recognize the sense of the steps, i.e. to determine which side of a step is higher and which is lower. There is, of course, no problem in the case of evaporation structures because they form pits. For the steps of an evaporation spiral, e.g., the lower level is always closer to the center of the spiral.

Starting from one step with known sense, the senses of the neighboring steps can be determined by means of the dissolving growth structures discussed above. The interpretation is not always easy; moreover, the rounding structures must first be generated which changes the original structure. The method of additional decoration with a low-melting metal has proven successful in avoiding these difficulties. Before decoration with a heavy metal, decoration with a low-melting point metal is first carried out which gives much bigger and more distant particles; the center of these particles is always situated beside the step, as demonstrated in Fig. 13.

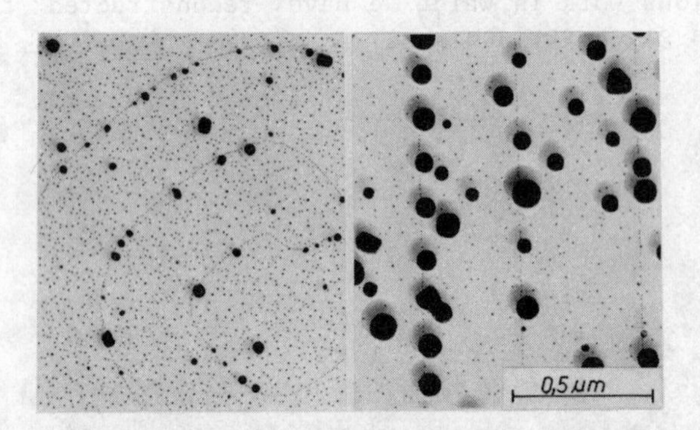

Fig. 13 : Bi and Au decoration as a tool for the determination of the sense of steps.

The gold decoration shows the exact position of the step, and quite obviously the décoration of the step by Bi particles is assymetrical. In the left-handed picture exhibiting decorated evaporation structures, it is clearly observed that the Bi particles are always located on the bottom part of the steps. This result, valid for low melting point metals, does not concern the position of the smaller particles of the heavy metals. According to METOIS[18] who first used the deposition of two different metals to determine the position of two different metals to determine the position of the decorating particles with respect to the step level, found that gold on KCL occupies the upper level of the step.

III. INVESTIGATIONS ON THE KINETICS OF STEPS

Fundamental work by VOLMER[19] in the early twenties proved that surface diffusion plays a decisive role in the growth of crystals from the vapor phase. Later, in 1928, KOSSEL[20] and STRANSKI[21] gave atomic explanations.According to Fig. 14 an atom or molecule coming from the vapor is adsorbed for a certain time on the surface.

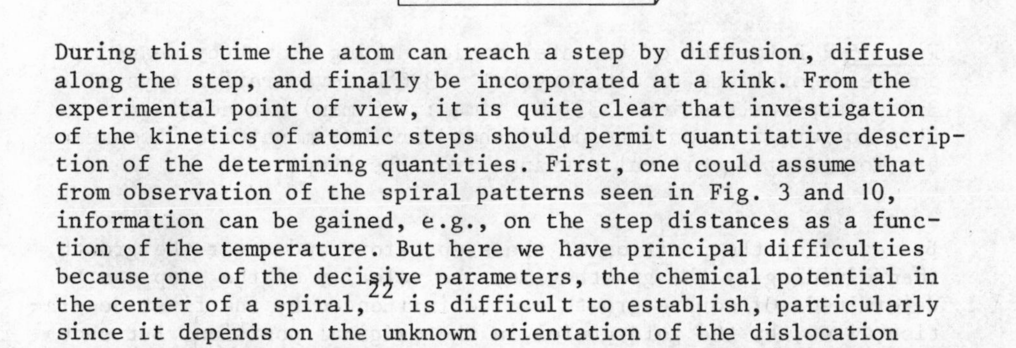

Fig. 14 : Schematic demonstration of a monomolecular step during vapor growth.

During this time the atom can reach a step by diffusion, diffuse along the step, and finally be incorporated at a kink. From the experimental point of view, it is quite clear that investigation of the kinetics of atomic steps should permit quantitative description of the determining quantities. First, one could assume that from observation of the spiral patterns seen in Fig. 3 and 10, information can be gained, e.g., on the step distances as a function of the temperature. But here we have principal difficulties because one of the decisive parameters, the chemical potential in the center of a spiral,[22] is difficult to establish, particularly since it depends on the unknown orientation of the dislocation emerging from the surface.

The investigation of the motion of steps which are independent of the spiral mechanism is much more transparent. Two techniques have been developed for observing step motion and for measuring the step velocity.

Our former method is the so-called "double decoration", an example of which is demonstrated in Fig. 15. The technique for studying growth is as follows : A crystal is cleaved in vacuum and one of its surfaces is subsequently evaporated to get defined step structures. On this face, a first decoration is then performed (the broken line in Fig. 15). Then the two crystals are opposed with a separation of only 1 mm. The crystals are heated to different temperatures, e.g., the previously evaporated and predecorated crystal to 330°C, and the opposite one to 350°C. Molecules are evaporated from the warmer crystal to the colder crystal. The rather well defined supersaturation conditions produce a well-controlled growth on the colder crystal. After a given growth time the surface of the grown crystal is decorated again (full line in Fig. 15).

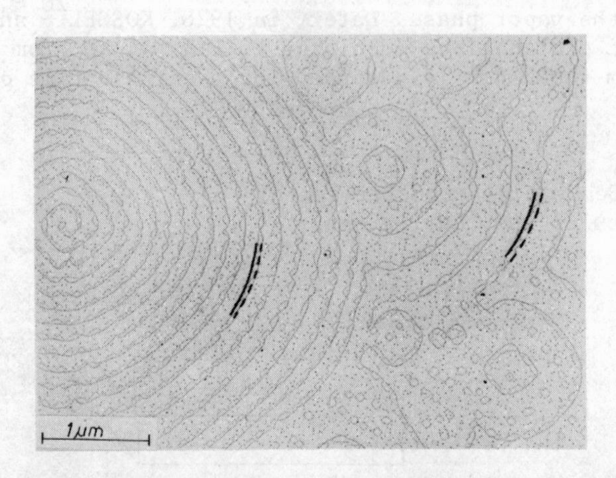

Fig. 15 : Detection of the step motion during growth by means of double decoration (growing crystal : 330°C, evaporation crystal : 350°C, supersaturation : 337 %, time : 7.5 min). In addition to step motion, it should be noted that circular monatomic "Lochkeime" are formed at a critical distance from the steps.

By measuring the position of the step before and after the growth, then dividing by the growth time, the velocity of the step can be determined. Of course growth can be influenced by the first decoration, i.e., by the gold nuclei in the original position. But such an influence can be eliminated if two experiments with different growth times are performed and only the differences are used for determining the step velocity.

From Fig. 15 one sees that the step velocity depends on the distance from the neighboring steps. An evalution of a number of micrographs is shown in Fig. 16 for different supersaturations. The step velocity increases with increasing step distance and finally reaches a constant value for a critical step distance y_o^*.[23] A simple explanation for this is given in Fig. 17. The hatched regions of width x_e are called trapping ranges. A molecule from the vapor impinging on the surface within region x_e can reach the step by diffusion and is trapped. For a molecule impinging on the area between the two x_e regions, the diffusion range is too small for the molecule to reach the step.

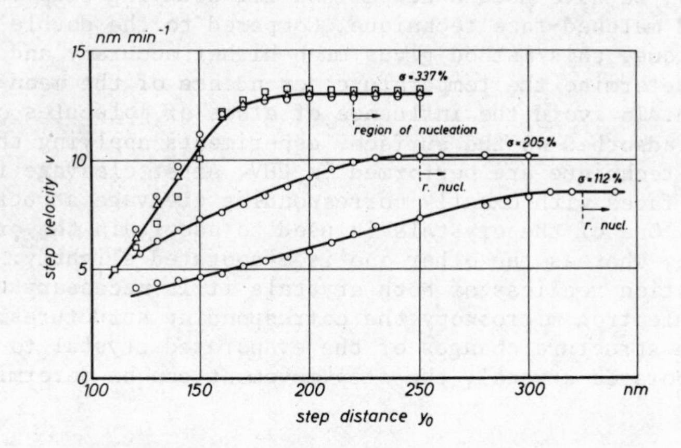

Fig. 16 : Dependence of step velocity on step distance at differing supersaturations.

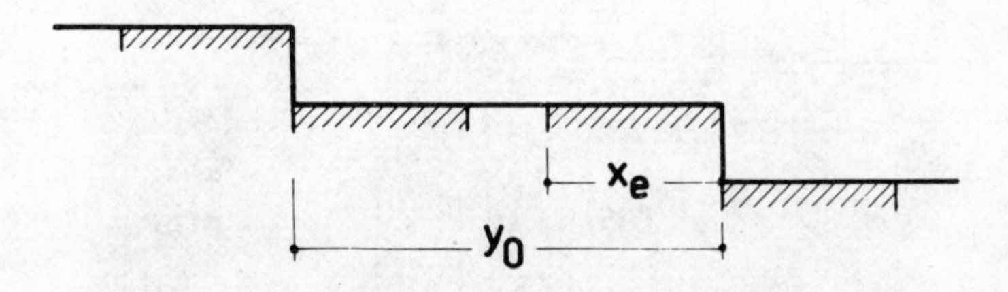

Fig. 17 : Scheme demonstrating the dependence of step velocity on step distances.

Therefore, it can be seen that not all molecules that impinge on the surface can reach a step. This means that in the case illustrated, the step distance y_o is within the constant region of the measured curves. It is also clear that with diminishing y_o and overlap of the trapping ranges x_e, the step velocity must decrease. Moreover, with regard to the measured curves, the step distance now lies within the increasing part of the curves. From this simple consideration, it follows that the critical step distance y_o^* at which the curves reach the saturation value correspond to approximately twice the width of the trapping range x_e. Of course, the trapping range is not identical to the conventional quantity of the mean diffusion length $\bar{x}_s$. From the well-known BCF theory[24] in which the step velocity is dealt with from a theroetical point of view, it follows that $\bar{x}_s$ is about half of x_e. We note that the smallest diffusion lengths were measured for the high supersaturation of 337 %. The trapping range length is then 900 Å and $\bar{x}_s$ about 430 Å.

Recently, we have used a new method for studying step velocity, the so-called matched-face technique. Compared to the double decoration technique, this method gives much higher accuracy and it is possible to determine the temperature dependence of the mean diffusion length. To avoid the influence of atoms or molecules of residual gas adsorbed on the surface, esperiments applying the matched face technique are performed in UHV. After cleavage in UHV, two cleavage faces with exactly corresponding cleavage structures are employed. One of the crystals is used to ascertain the original structure, whereas the other one is evaporated slightly. After making decoration replicas of both crystals it is necessary to find out by electron microscopy the corresponding structures. By comparing the structure changes of the evaporated crystal to those of the unevaporated crystal, the step movement can be determined exactly.[25,26]

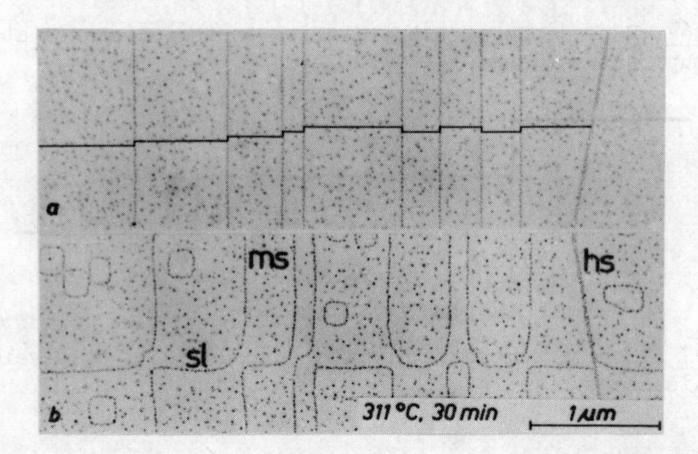

Fig. 18 : Corresponding structures for demonstrating the matched-face technique : (a) Original cleavage structure, the drawn line marks the surface profile; (b) corresponding surface after annealing (ms, monatomic step; hs, higher step; sl, slip line of a single dislocation).

Fig. 18 shows an example of step shifting after evaporation at a relatively low temperature. The higher step (hs) serves as fixed point for the correct correlation between the two micrographs. On the evaporated crystal, after cleavage, a dislocation runs along the face producing a slip line. As explained in section II, rounding at the points of intersection shows that the moved steps are monatomic.

For separate steps, we measured the maximum step velocity, v_{max}. The velocity of isolated step pairs was also determined. Following the principal theoretical consideration of the BCF theory, the mean diffusion length could be determined from experimental measurements. If the values of the determining quantities are known one can calculate the movement and interaction of parallel steps within trains and the kinetics to be expected for a given step system can be computer-simulated.[26] Micrograph (a) of Fig.19, shows the initial structure and micrograph (b) the final structure after 240 minutes of evaporation at 300°C. The computer-plotted profile images demonstrate good agreement between experimental and numerical results.

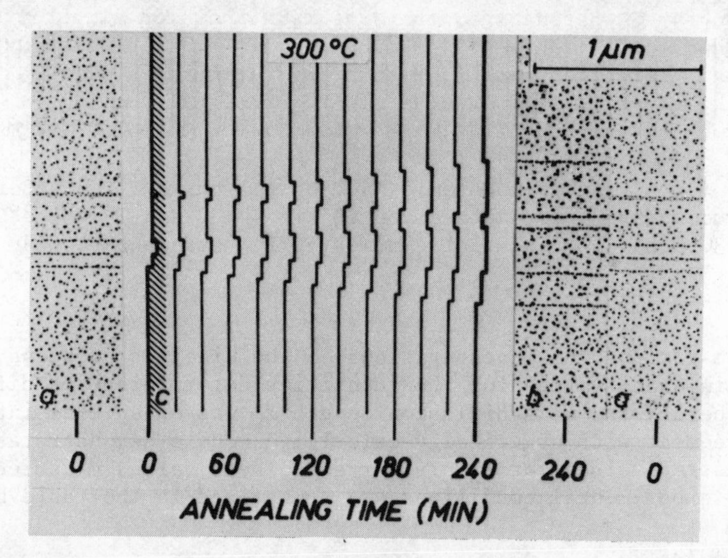

Fig. 19 : Comparison between observed movements of monatomic steps with mutual interactions and numerical simulations; (a) original position; (b) position after annealing at 300°C and 240 min.

The validity of the extended BCF-theory could also be verified by other experiments carried out at temperatures where evaporation in vacuum is completely determined by the propagation of steps associated with dislocations emerging to the surface. Let us consider, for instance, a monatomic spiral step developed around the emergence point of an a/2 <110> dislocation. In order to reveal surface diffusion-controlled interactions that occur between adjacent spiral turns, after a long-time evaporation at a certain temperature T_1 (steady-state spiral), the distance of successive spiral turns has been changed by a sudden jump in temperature from T_1 to a substantially lower value T_2. At this temperature the evaporation was continued for a certain time t_2. Between the inner spiral turns generated after the temperature decrease, (i.e., at

T_2) and the outer spiral turns created before (i.e.,at T_1) surface diffusion-controlled interactions take place which result in periodic step bunching as shown in Fig. 20a. The experimentally observed step configuration could be predicted numerically with a high degree of correspondence (see Fig. 20b) by use of a simplified step model based on the extended BCF theory.[24,27]

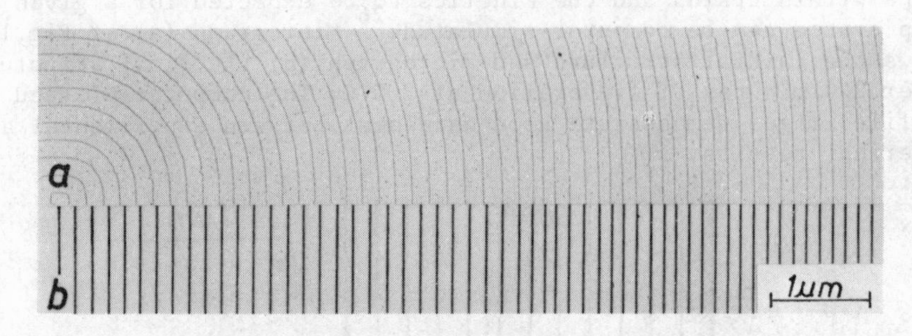

Fig. 20 : Bunching of monatomic NaCl spiral turns : (a) observed after successive free evaporation at T_1= 444°C and T_2 = 358°C (t_2 = 330 min); (b) caluclated numerically (straight step approximation).

The goal of the investigations on the kinetics of steps was to obtain exact values for the quantities determining the diffusion, whereby the mean diffusion length $\bar{x}_s$ was of special interest. The agreement with numerically calculated models demonstrates that the measured values for $\bar{x}_s$ are correct. These values measured in various experimental conditions are summarized in the table below.

Experimental conditions and techniques	Supersaturation or surface temperature	Trapping range width x_e	Mean 2D diffusion length $\bar{x}_s$
Growth (330°C) (double-decoration technique)	337 % 112 %	900 Å 1500 Å	430 Å 720 Å
Evaporation (330°C) (double-decoration technique)		5200 Å	2480 Å
Evaporation (matched-face technique)	314°C 278°C 263°C		2500 ± 400 Å 3600 ± 700 Å 4500 ± 1000Å

With regard to the usual theoretical considerations on the diffusion length, the experimental values are higher by one to two orders of magnitude. One reason for these discrepancies might be that diffusion kinetics based on parameters such as Debye frequency and elementary jump distance is oversimplified. Another open question is whether the diffusing species are simple molecules. The experiments reveal only mass transfer and it should be investigated whether complexes of molecules have an important role in the diffusion. For example, two molecules with a strong bond to each other, might have, as a diffusing particle, an interaction energy with the surface lower than that of individual molecules. Contrary to such an assumption is the fact that in the vapor flux of an evaporating NaCl crystal, primarily individual molecules were measured, but the question remains whether diffusing particles are identical with molecules in the vapor phase. Summarizing this short discussion, one can state that from the molecular-kinetic point of view it is necessary to better understand the individual stages of vapor growth and evaporation, including the diffusion .

A special kind of diffusion is edge self-diffusion, for example, the diffusion of NaCl molecules along an atomic step as outlined in Fig. 14. This effect can be studied by observing the change of the shapes on special step structures, particularly the steps of the elementary cleavage structure (Fig. 7b). Here again, the matched-face technique[25] was applied. Fig. 21 shows the original cleavage structure; for the given values for the crystal heating, no thermal influence can be proven. The corresponding structure, on the annealed counter surface, however, shows that all tips present in the lightning-shaped path of the steps exhibit distinct club-shaped rounding.

As demonstrated in the schematic drawing of Fig. 21c, the change of shape of the step structure in the region of a tip takes place only by diffusion at the step. The annealing temperature (275°C) is too low for evaporation; this means that thermal energy permits only detachment of molecules from kink positions and their diffusion at the step, but not detachment from the step onto the face. The driving force for diffusion along the ledges can be derived from the gradient of the chemical potential which is determined by locally differing step curvatures.

Given isotrope specific ledge free energy and using a step diffusion coefficient, a differential equation can be formulated[28] to describe the alteration of the tip structure shown in Fig. 21b. Because this partial differential equation is not analytically solvable, it was numerically integrated. Tip profiles determined experimentally and numerically are represented in Fig. 22. It should be pointed out that the apex angle at the tip influences the form of the club-shaped steady-state profiles. For tips with nearly parallel steps, the club-shaped tips are first tied off.

In a later stage isolated circle structures originate.

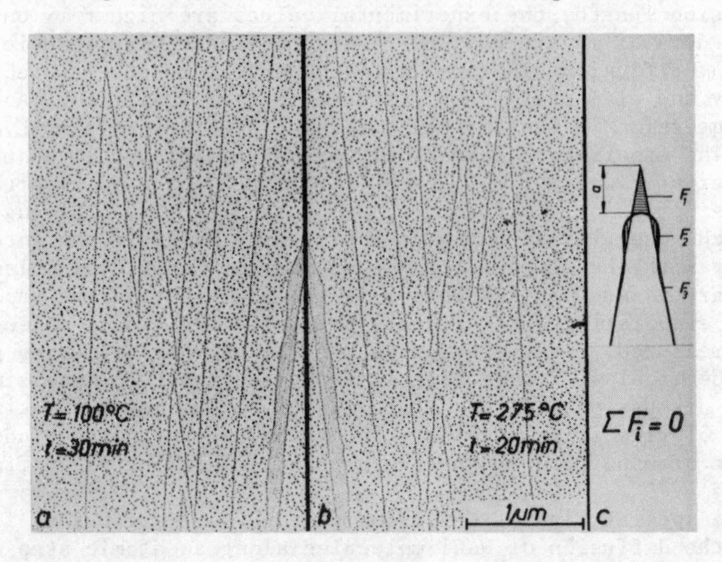

Fig. 21 : Matched faces of cleavage structures after different thermal treatment. (a) elementary cleavage tips, T = 100°C, t = 30 min; (b) thermally activated matched tips, T = 275°C, t = 20 min; (c) geometrically fitting tip profiles.

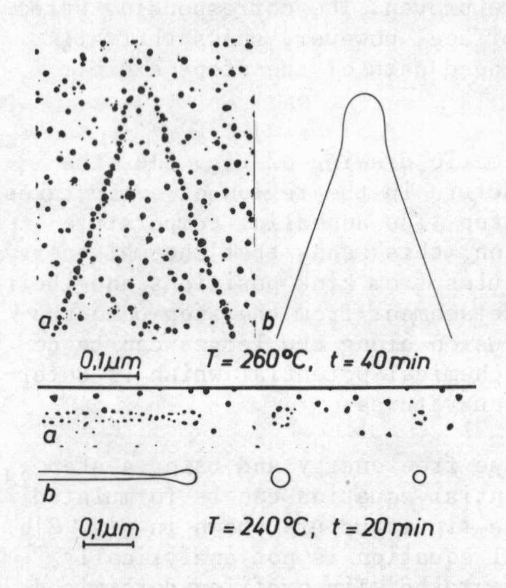

Fig. 22 : Experimentally observed and numerically simulated steady-state profiles in dependence on the enclosed angle in the tip.

Good agreement between experimental observation and theoretical treatment confirms the concept of morphological change by isotropic edge diffusion. The calculated activation energy of the edge diffusion along the monatomic NaCl surface steps is $E_r = 1.05$ eV.

The edge diffusion coefficient has also been determined.

IV. CONDITIONS FOR THE DECORATION EFFECT

For the decoration effect to be applied to the imaging of step structures as described above, two conditions should be fulfilled.

1. Formation of a metal nucleus at a step must be energetically preferred compared to nucleation on a smooth surface.

2. For the formation of a nucleus, diffusion of the metal atoms is necessary; this means that the surface conditions must not hinder diffusion. Generally, the surface between the steps should be atomically smooth and sorption layers which hinder diffusion should not be present.

The importance of these two conditions can be illustrated by some experiences. Most alkali halides are especially suitable for decoration even after a strong influence such as that of water vapor on easily dissolvable crystals, described in section II. Thermal treatment in vacuum not only perfectly "reconstructs" the surface region previously dissolved, but also leads to a desorption of the sorption layers. This is valid for NaCl, KCl, NaBr, NaI and KI but quite a different behavior is noted with LiF. Contrary to most of the above mentioned alkali halides, LiF is not dissolved by water vapor. In addition, it is impossible to decorate a LiF crystal cleaved in free atmosphere and subsequently vacuum annealed. In a "normal" vacuum (about 10^{-5} torr), decoration is successful only if it is carried out within a few seconds after production of the cleavage faces. Cleavage faces produced under UHV conditions can be decorated without any problems.One explanation may be that on a LiF surface, strongly bound sorption layers apparently form very quickly thereby influencing the decoration effect.

In addition to the two principal conditions mentioned, for electron microscopic investigation it is necessary that decoration particles give good contrast. Furthermore,for high resolution, e.g., wiht closely neighboring steps, decoration particles should be as small as possible. Hence it follows that heavy metals are preferred and that nucleation is three dimensional. The latter is the case, for example in gold decoration at steps on alkali halides. With regard to the interface conditions, gold on alkali halides are systems of weak interactions, which means three-dimensional nucleation and good imaging conditions.

Contrast and resolving power of the decoration effect were also treated quantitatively.[29] To recognize a step visually, certain relations of nucleation on a step with respect to the nucleation on the face must be known. The distance between nuclei along the steps must be smaller than the mean distance between nuclei on the free regions. The decoration contrast K has been shown to be :[29]

$$K = \frac{d_L n_L - d_F \sqrt{n_F}}{d_L n_L + d_F \sqrt{n_F}}$$

where n_L (cm^{-1}) is the line density of nuclei along the steps, n_F (cm^{-2}) is the surface density of nuclei in step-free regions, and d_L and d_F, the diameters of nuclei along steps and in the step-free regions respectively. The lateral resolution δ is the smallest distance at which two steps are imaged separately. If s_L is the mean distance between nuclei at a step, we have then $\delta = s_L = 1/n_L$; i.e., the lateral resolution is equal to the reciprocal value of the density of nuclei at the step.

Summarizing these considerations, one can state that for a good decoration effect with high contrast and high resolution for the step imaging, heavy metals (with a high scattering factor for the imaging electrons) are especially suitable. Conditions for evaporation should be chosen such that extremely small three-dimensional nuclei form on the surface to be decorated. Often the determining parameters are not easy to achieve. In platinum decoration of NaCl, for example, one observes that Pt nuclei tend to grow two-dimensionally. To avoid this, a higher substrate temperature is useful, but as a result the density n_L decreases. According to our experience with high resolution decoration of alkali halides and some other materials, simultaneous evaporation of Au and Pt under UHV conditions at room temperature gives the best results.

An illustration of high density step imaging is given in Fig. 23. The mean distance of the steps is 55 Å, in agreement with the dislocation density at the boundary, which can be calculated from the misorientation of two subgrains.

As previously mentioned, a general condition for decoration is that the surface to be decorated should be atomically smooth in order to permit diffusion of the decoration atoms. This condition limits some applications, particularly the decoration of metal surfaces. Until now, the decoration technique has been used only for investigation of Ag surfaces. Following earlier work by KAISCHEW et al.[30] well defined Ag surfaces have been obtained by electrolytic overgrowth of a monocrystalline Ag hemispherical cap.[31] In Fig. 24 such an Ag crystal with its growth faces is represented together with an electron micrograph showing the decoration of monatomic steps of a growth spiral.

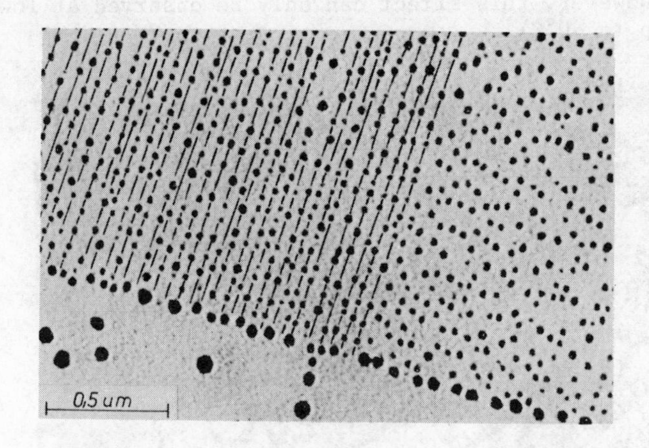

Fig. 23 : Highly resolved step structure produced by cleavage of a twist boundary. Cleavage in vacuum and decorated with Au at 100°C.

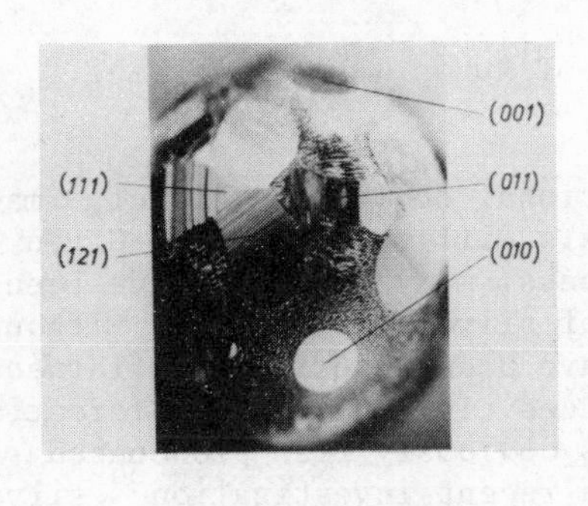

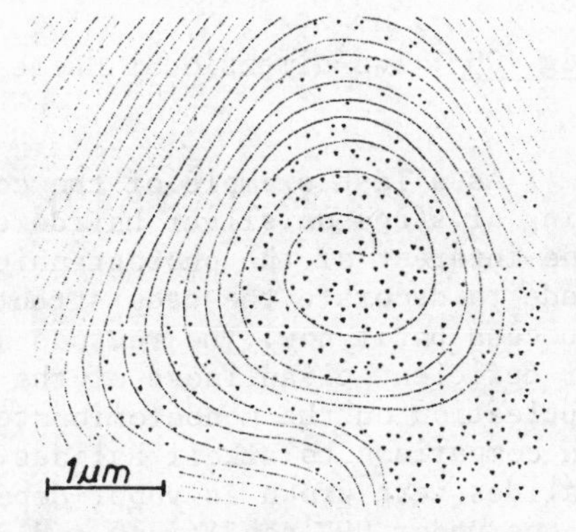

Fig. 24 : Electrolytically grown faces on a spheric Ag single crystal and the micrograph of a (111) Ag face decorated with Au.

By means of decoration imaging, some problems of the kinetics of steps were studied.[32] The system Au on Ag is a system of stronger interaction. Under clean conditions at a relatively low temperature, two-dimensional overgrowth takes place, as seen in the micrograph of Fig. 25. In order to obtain clean surfaces after electrolytic growth, the crystal was evaporated, i.e., the decorated steps are evaporation steps. It is remarkable that the extended flat nuclei at the steps are clearly formed on the bottom level. As an explanation for this fact, it was calculated that for a diffusing

Au atom, an additional activation energy of 0.15 eV is necessary for the jump from the plane into a position on the bottom face of the step.[33] However, this effect can only be observed at low temperatures (up to 60°C).

Fig. 25 : Two-dimensional nucleation of Au on a (111) Ag face.

As a last example of the conventional decoration effect, imaging of steps on silver halide crystals will be discussed. Given the interest of the photographic process, many attempts have been made to decorate the step structure of silver halides, but without success until now. The reasons may have been insufficient flatness or deficient cleanliness of the surfaces. In addition, preferential nucleation on the monoatomic steps is obviously less pronounced in comparison to alkali halides. In a recent investigation[34] silver halides were grown as vapor-deposited films onto NaCl(100) cleavage faces under UHV conditions and decoration was carried out at temperatures between 25° and 50°C. Fig. 26 shows that on the (100) surface of the AgBr crystals, numerous growth steps are present originating from emergence points of screw dislocations. Note that the emergence points are always distinguished by the formation of a large cluster of the decoration metal; this may be important for the formation of the latent image in the photographic process.

The decoration technique also implies that the carbon film can be detached from the crystal without great difficulty and that decoration particles are not extracted from the carbon film during its detachment. The latter is often the case with numerous decoration metals and also with gold. Least delicate is the Pt decoration. For the above mentioned reasons, the usual decoration technique cannot be applied to a number of crystals. But if the conditions described are fulfilled, i.e., heterogeneous nucleation and preferred deposition at the steps are attained, a decoration technique using the scanning electron micorscope can be employed.[35] The material chosen for the decoration must have a secondary emission coefficient essentially greater than that of the crystal whose surface is to be investigated. Accordingly, ionic crystals are a preferred decoration material.

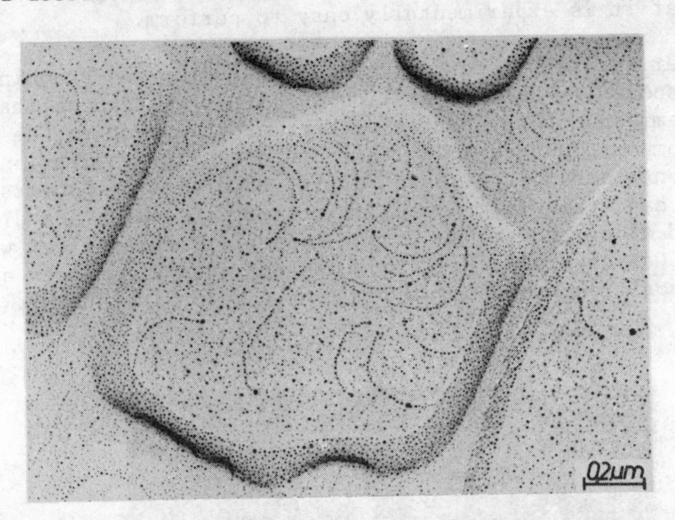

Fig. 26 : Growth of a AgBr film vapor-deposited onto NaCl substrate at 300°C. Au decoration.

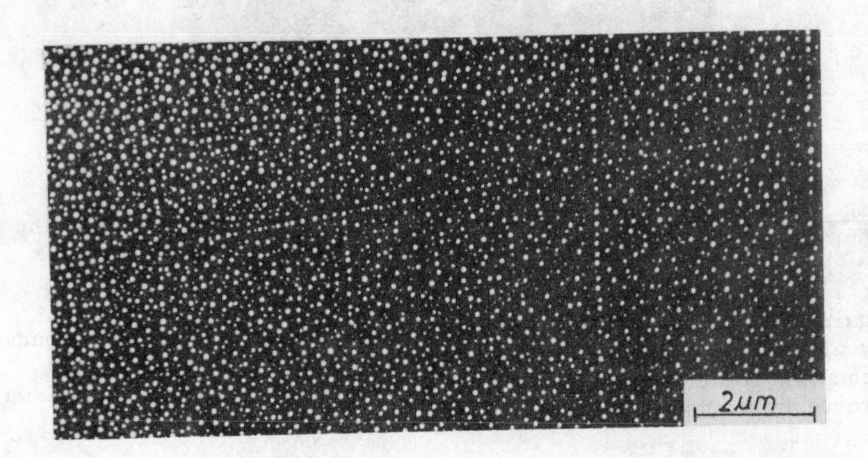

Fig. 27 : Cleavage steps on a surface of an Si crystal decorated by deposition of CsI directly imaged in the SEM.

As an example of this special technique, Fig. 27 shows glide and cleavage steps on the cleavage face of a Si crystal. As a decoration material, CsI was evaporated at 150°C. The high yield of secondary electrons permits easy recognition of the CsI nuclei. For each surface the appropriate decoration material must be determined. Concerning the resolving power, lateral resolution does not produce results as good as those achieved in the conventional decoration technique. But undoubtedly, the chief advantage of the method is that it is experimentally easy to perform.

Another decoration method carried out using the scanning electron microscope is "self-decoration".[36] It is well-known that some compound semiconductors undergoing thermal treatment in the surface region decompose in such a way that preferential evaporation of one component takes place. As a result, remaining components accumulate and can again form nuclei on the surface which are preferably trapped along the steps. In our laboratory this effect was studied on GeS crystals. Fig. 28 shows the step structures arising from decomposition and decorated by small germanium nuclei.

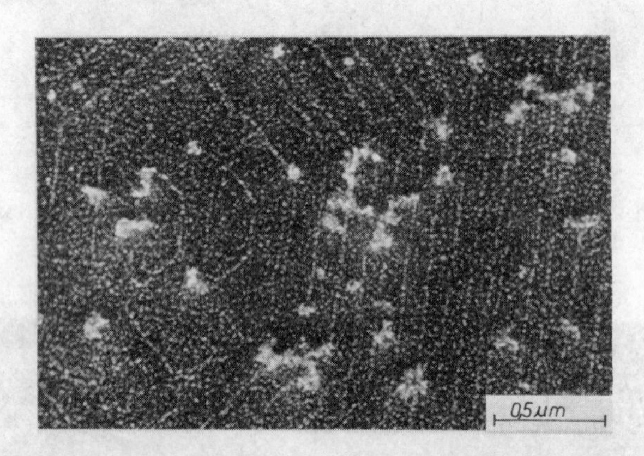

Fig. 28 : Evaporation pit on a GeS surface visible in the SEM by thermic self-decoration due to heat treatment.

Again, due to the different emission coefficients for the secondary electrons, this decoration effect is easy to observe in the scanning electron microscope. The presence of germanium nuclei was proven by means of x-ray microprobe and Auger spectroscopy.

V. CONCLUSION

Most molecular processes on surfaces are characterized by the

fact that for interactions, the surface cannot simply be regarded as a perfect and homogeneous lattice plane. A real surface has preferred sites in which interaction is drastically different from that in a perfect region. Among such distinct sites are elementary steps, always present on a real surface. By means of the decoration technique and using the electron microscope, the step structure of macroscopic crystals can be studied. In previous work, applying the decoration effect to detect step structure has been referred to as the first kind of decoration. Subsequently, decoration of the second and third kind[37] were defined.
Decoration of the second kind means that density, shape and orientation of the decoration nuclei on step-free surfaces allow distinction of different surface conditions, e.g., with respect to adsorption layers, between regions of different lattice structure or atomic roughness of a surface. Study of the third kind of decoration should eventually enable detection of discrete atomic defects in the surface.
It was not possible to discuss decoration of the second and the third kinds here, but new results[38] show that it is obviously interesting to study these phenomena in more detail. However, it will be necessary to combine investigations on nucleation aspects of the decoration effect with other actual methods of surface physics.

REFERENCES

1 . BETHGE H., Verh. 4. Int. Konf. Elektr. Mikroskopie, Springer Berlin, 1958, pp.409-414
2 . BASSETT G.A., Phil. Mag. 3(1958)1042-1045
3 . BETHGE H. and KELLER K.W., Z. Naturf. 15a(1960)271-272
4 . OSAKABE N., YAGI K. and HONJO G., Japan.J. Appl. Phys. 19(1980) 309-312
5 . BORRIES B. von, Z. Physik 116(1940)370-378
6 . OSAKABE N., TANSHIRO Y., YAGI K. and HONJO G., Surf. Sci. 102 (1981)424-442
7 . CHERNS D., Phil. Mag. 30(1974)549-556
8 . KAMBE K. and LEHMPFUHL S., Optik 42(1975)187-194
9 . KLAUA M. and MEINEL K., Ber. X Arbeitstagung Elektronenmikroskopie Leipzig, 1981, in press
10. BETHGE H. and KROHN M., *Adsorption et Croissance Cristalline*, C.N.R.S. Paris, 1965, pp. 389-406
11. BETHGE H., KÄSTNER G. and KROHN M., Z. Naturforsch. 16a(1961) 321-323
12. BETHGE H., Proc. Conf. on the Physical Basis of Yield and Fracture, Oxford, 1966, pp. 17-23
13. BETHGE H., Proc. 4th Europ. Reg. Conf. Electr. Micr. Rome, 1968, Vol 1, pp. 273-274

14. KROHN M. and BETHGE H., *Crystal Growth and Materials*, E. Kaldis and H.J. Scheel eds., Amsterdam, 1976, pp. 142-164
15. KELLER K.W., phys. stat. sol. 36(1969)557-565
16. BETHGE H. and KELLER K.W., Optik 23(1965)462-471
17. BETHGE H., KELLER K.W. and KROHN M., Izvest. Chim. Bulg. Akad. Nauk. 11(1978)489-516
18. METOIS J.J., HERAUD J.C. and KERN R., Surf. Sci. 78(1978)191-208
19. VOLMER M. and ESTERMANN J., Z. Physik 89(1921)13-17
20. KOSSEL W., Leipziger Vorträge : *Quantentheorie und Chemie*, Leipzig, 1928,pp. 1-46
21. STRANSKI I.N., Z. Phys. Chem. 136(1928)259-278
22. KELLER K.W., Proc. Int. Conf. Crystal Growth, Boston, 1967,pp. 629-633
23. BETHGE H., KELLER K.W. and ZIEGLER E., J. Cryst. Growth. 3 (1968)184-197
24. BURTON W.K., CABRERA N. and FRANK F.C., Proc.Roy.Soc.(London) A243 (1951)358
25. HOCHE H. and BETHGE H., J. Cryst. Growth. 33(1976)246-254
26. HOCHE H. and BETHGE H., J. Cryst. Growth 42(1977)110-120
27. BETHGE H., HOCHE H., KATZER D., KELLER K.W., BENNEMA P. and van der HOCK B., J. Cryst. Growth. 48(1980)9-18
28. HOCHE H., J. Cryst. Growth. 33(1976)255-265
29. KROHN M., GERTH G. and STENZEL H., phys. stat. sol. (a)55(1979) 375-383
30. KAISCHEW R., BUDEWSKI E. and MALINOWSKI J., C.R. Acad. Bulg. Sci. 2(1949)29-32
31. BETHGE H. and KLAUA M., Ann. Phys. (Lpz.) 17(1966)177-187
32. KLAUA M. and BETHGE H., J. Cryst. Growth. 3(1968)188-190
33. KLAUA M., Rost. Kristallov 11(1975)65-70
34. HAEFKE H., KROHN M. and PANOV A., J. Cryst. Growth 49(1980)7-12
35. BETHGE H., Proc. 9th Int. Conf. on Electron Microscopy, Vol. 1, Toronto, 1978, pp. 436-437
36. BETHGE H. and VETTER J., J. Cryst. Growth,(1981)in press
37. BETHGE H., Proc. Sept. Congr. Int. Microscopie Electronique, Vol. 2, Grenoble, 1970, pp. 365-366
38. BETHGE H., 4th EPS Gen. Conf., London, 1979, pp. 25-42.

INDEX

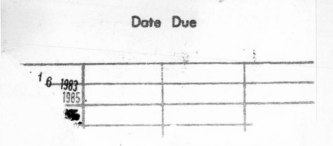